PHYLOGENY *and* EVOLUTION
of the MOLLUSCA

PHYLOGENY *and* EVOLUTION *of* *the* MOLLUSCA

Edited by

Winston F. Ponder
David R. Lindberg

UNIVERSITY OF CALIFORNIA PRESS
Berkeley   Los Angeles   London

University of California Press, one of the most distinguished university presses in the United States, enriches lives around the world by advancing scholarship in the humanities, social sciences, and natural sciences. Its activities are supported by the UC Press Foundation and by philanthropic contributions from individuals and institutions. For more information, visit www.ucpress.edu.

University of California Press
Berkeley and Los Angeles, California

University of California Press, Ltd.
London, England

Library of Congress Cataloging-in-Publication Data

Phylogeny and evolution of the mollusca / edited by Winston F. Ponder, David R. Lindberg.
    p. cm.
  Includes bibliographical references and index.
  ISBN 978-0-520-25092-5 (cloth : alk. paper)
  1. Mollusks—Phylogeny. 2. Mollusks—Evolution. I. Ponder, W. F.
II. Lindberg, David R., 1948–
  QL406.7.P49 2008
  594.13'8--dc22   2007027446

Cover photograph: A bivalve of the genus *Limaria* (Bivalvia, Limidae) from Panglao, Bohol, the Philippines. Photo by Sheila Tagaro, Panglao 2004 Marine Biodiversity Project.

IN MEMORY OF ELLIS L. YOCHELSON
AND ALASTAIR GRAHAM

CONTENTS

LIST OF CONTRIBUTORS

STEPHANIE W. AKTIPIS
Harvard University
Cambridge, Massachusetts, United States

JEFFREY L. BOORE
DOE Joint Genome Institute and Lawrence
    Berkeley National Laboratory
Walnut Creek, California, United States

DONALD J. COLGAN
Australian Museum
Sydney, New South Wales, Australia

BERNARD M. DEGNAN
University of Queensland
Brisbane, Queensland, Australia

JIŘÍ FRÝDA
Czech Geological Survey
Prague, Czech Republic

DANIEL L. GEIGER
Santa Barbara Museum of Natural History
Santa Barbara, California, United States

GONZALO GIRIBET
Harvard University
Cambridge, Massachusetts, United States

KENNETH M. HALANYCH
Auburn University
Auburn, Alabama, United States

GERHARD HASZPRUNAR
Zoologische Staatssammlung München
München, Germany

JOHN M. HEALY
Queensland Museum
Brisbane, Queensland, Australia

ANNETTE KLUSSMANN-KOLB
J. W. Goethe-University
Frankfurt am Main, Germany

DEMIAN KOOP
University of Queensland
Brisbane, Queensland, Australia

DAVID R. LINDBERG
University of California
Berkeley, California, United States

ROYAL H. MAPES
Ohio University
Athens, Ohio, United States

MONICA MEDINA
University of California
Merced, California, United States

PETER MORDAN
The Natural History Museum
London, United Kingdom

SHARON MOSHEL-LYNCH
University of California
Berkeley, California, United States

MICHELE K. NISHIGUCHI
New Mexico State University
Las Cruces, New Mexico, United States

ALEXANDER NÜTZEL
Bayerische Staatssammlung für Paläontologie
und Geologie
Munich, Germany

AKIKO OKUSU
Simmons College
Boston, Massachusetts, United States

PAVEL YU. PARKHAEV
Paleontological Institute
Moscow, Russia

WINSTON F. PONDER
Australian Museum
Sydney, New South Wales, Australia

PATRICK D. REYNOLDS
Hamilton College
Clinton, New York, United States

TAKENORI SASAKI
The University of Tokyo
Tokyo, Japan

CHRISTOFFER SCHANDER
University of Bergen
Bergen, Norway

ENRICO SCHWABE
Zoologische Staatssammlung München
München, Germany

W. BRIAN SIMISON
Center for Comparative Genomics, California
   Academy of Sciences,
San Francisco, CA, USA

LUIZ R. L. SIMONE
Museu de Zoologia da Universidade de São Paulo
São Paulo, Brazil

GERHARD STEINER
University of Vienna
Vienna, Austria

ELLEN E. STRONG
National Museum of Natural History
Washington DC, United States

CHRISTIANE TODT
University of Bergen
Bergen, Norway

VERENA VONNEMANN
Ruhr-Universität Bochum
Bochum, Germany

CHRISTOPHER WADE
University of Nottingham
Nottingham, United Kingdom

HEIKE WÄGELE
Rheinische Friedrich-Wilhelms-Universität
Bonn, Germany

PETER J. WAGNER
Field Museum of Natural History
Chicago, Illinois, United States

ANDREAS WANNINGER
University of Copenhagen
Copenhagen, Denmark

ACKNOWLEDGMENTS

We take this opportunity to thank Dr. Fred Wells for his considerable efforts in organizing the World Congress of Malacology in Perth (2004) and for inviting us to put together the symposium on molluscan phylogeny. We also thank the University of California Museum of Paleontology for financially supporting the symposium.

We thank the authors and the many reviewers for their considerable efforts, without which works such as this would not be possible, and the authors and our editors, Chuck Crumly and Francisco Reinking at the University of California Press, and Ted Young, Jr., of Publication Services Inc., for their patience while we pulled this volume together. We hope that it was worth the wait.

We also thank our long-suffering wives, Julie and Dixie, for their endurance and patience during the course of this work.

Last, but by no measure least, it is our pleasure and honor to dedicate this volume to two departed colleagues: Ellis L. Yochelson (U. S. Geological Survey and National Museum of Natural History) and Alastair Graham (University of Reading). Although the molluscs they studied were separated by hundreds of millions of years, they both dedicated their lives to understanding the form, function, and evolutionary histories of molluscs, and through their publications and mentoring provided generations of students with their insights, ideas, and friendship.

WINSTON PONDER
DAVID LINDBERG

1

Molluscan Evolution and Phylogeny

AN INTRODUCTION

Winston F. Ponder and David R. Lindberg

The Mollusca are the second largest phylum of animals, with about 200,000 living species. They have a remarkable fossil record reaching back to the earliest Cambrian, 543 million years ago (Mya), and putative molluscs are found in even earlier Precambrian rocks. They are also one of the most diverse groups of animals, with seven or eight classes of living molluscs (see below) recognized and at least two extinct class-rank taxa. The molluscan body plan is enormously varied, ranging from minute wormlike interstitial animals to giant squids and from microscopic snails to giant clams. It includes many taxa of great economic significance (e.g., oysters, scallops, squids) and others that carry diseases infecting humans. Molluscs comprise some of the most diverse creatures in the sea and also have a significant presence in most terrestrial and freshwater ecosystems. They are also among the most vulnerable of organisms, with more non-marine molluscs becoming extinct in historical times than all tetrapod vertebrates combined (e.g., Lydeard *et al.* 2004).

Molluscs were studied in some of the earliest zoological investigations, and they remain a popular group for scientific investigation as well as having a large amateur interest. Studies on molluscan evolution are facilitated by the excellent fossil record and the diversity of structure seen in the group and, like those of other biota, have been greatly assisted by the advent of molecular biology. Information on the study, collection, and preservation of molluscs can be found in Sturm *et al.* (2006).

This work was inspired by another synoptic edited volume (Taylor 1996) on molluscan evolution published 11 years earlier. A more recent volume on molluscan molecular systematics and phylogeography (Lydeard and Lindberg 2003) covers some common ground, but in this fast-moving field, updating is necessary. Other recent contributions that have dealt with synoptic overviews of molluscan evolution and diversity include Beesley *et al.* (1998) and Lindberg *et al.* (2004).

In the current volume we have attempted to cover each of the major groups of molluscs in a systematic way with a similar format. Each chapter provides some general information on its group, but authors were asked, in particular, to incorporate overviews of the evolution of their groups, looking at the contribution of data from different sources: the morphological (including ultrastructure), molecular, developmental, and fossil records. Most of the chapters dealing with a particular taxon also include a section on adaptive radiations (see below).

This book had its genesis in the symposium on molluscan phylogeny held at the World Congress of Malacology in Perth, Western Australia 2003 (Wells 2004). The authors presenting at that symposium were asked to contribute and, where possible, to collaborate as much as possible to ensure that a wide range of expertise was available for each of the overviews. The volume covers all extant classes, with the gastropods (which are by far the largest class) getting special treatment with seven chapters. In addition to the taxon groups, there are overviews of the Cambrian radiation, of relationships of the molluscs and the higher groupings within the phylum, and of two currently important areas of research: genomics and development.

A feature of many of the chapters is that they are the products of collaboration between authors from different scientific disciplines, such as molecular biologists, morphologists, and paleontologists. Although many of these chapters provide significant insights and breakthroughs in our understanding of molluscan evolution, they also highlight some of the current significant disagreements and gaps in knowledge that will act as a focus to stimulate future research. As in the production of these chapters, future research to resolve these disagreements and to fill gaps in our knowledge will also likely be collaborative and multidisciplinary. In malacology, as with other zoological disciplines, the various data sets (paleontological, morphological, ultrastructural, molecular, isotopic, developmental, behavioral, mineralogical, physiological, biochemical, behavioral, etc.) that are available, and that should be consulted and incorporated into our research programs, are too diverse for an individual researcher to master.

This book represents the work of 36 authors from 11 countries and reflects the globalization of molluscan evolutionary studies that has taken place during the last 50 years, a phenomenon that has been immensely beneficial for the field in providing greater training and collaboration opportunities, broader research perspectives, and so on. After a long history of domination of the field by a handful of workers in even fewer, mainly Northern Hemisphere countries, globalization of molluscan studies began in 1962, a year that saw the formation of the international society Unitas Malacologica and the advent of the journal *Malacologia*. One of the rationales for the founding of *Malacologia* was "To promote cooperation and to stimulate research in malacology through the medium of an international journal" (Berry *et al.* 1962), thereby distinguishing itself from the existing national and regional journals that were available at that time. A third major event was the creation of the World Congress of Malacology series (Washington DC 1998; Perth 2004; Antwerp 2007). These changes have not been without critics or problems, and the future of malacological journals, societies, and meetings is uncertain. We are convinced that in addition to more collaborative research, our field must be prepared to move forward through improving the dissemination of information—for example, by the consolidation of the plethora of molluscan specialty journals, holding world congresses in different major regions every three years, and the more visible presentation of molluscan work at general scientific meetings.

MOLLUSCAN PHYLOGENY AND SYSTEMATICS

Phylogenetics assumes that any group of organisms is related by descent from a common ancestor. We cannot *know* the phylogeny of any group; we can only hypothesize it based on the best available evidence. A central task of most of the chapters of this volume is to review that evidence. The more accurate the phylogeny, the more likely it is to be able to solve both scientific and practical problems. Phylogenetic information can be used to make predictions about the origin of novel characters or states, the evolution of diversity or the morphology, genetic makeup, physiology, or proteins of unstudied taxa.

The discovery of phylogeny is the goal of systematics. In the following paragraphs we briefly explore some of the ramifications of systematics in a general way.

Molluscan systematics has had a long evolution. Molluscs, like plants and butterflies, were some of the earliest natural history curiosities to be coveted by collectors. Pre-Linnean naming of the specimens in natural history volumes gave rise to some of the earliest animal classifications (see Dance 1986 for a detailed account of early malacological history).

Building on the landmark synoptic treatments of the 1900s (e g , Thiele 1929–1931, Wenz 1938–1944, Hyman 1967, Grassé 1960–1989, Moore 1960–1996), our understanding of molluscan phylogeny has undergone a remarkable transformation in the last few decades. The traditional strengths in morphology and paleontology had already given molluscan workers a leading edge. In the late twentieth century, advancing research technologies such as scanning (SEM) and transmission (TEM) electron microscopes, geometric morphometrics, and molecular techniques built on earlier tools such as histology and detailed anatomy. The combination of these new tools, together with new approaches (notably computer-assisted phenetics and cladistics, which allowed the testing of phylogenetic hypotheses using parsimony and other criteria) generated a renewed interest in phylogenetics. This began to be reflected in molluscan systematics in the 1980s and was well entrenched by the 1990s. Interestingly, it was a malacologist, Adolf Naef (1883–1949), who had an important role in the establishment of phylogenetic thinking and thereby contributed to significantly changing systematic malacology during the last century (see Naef 1919, 1921: 11–39).

The advent of molecular methodology applied to phylogenetics in the 1980s created a new revolution, and with these methods becoming increasingly inexpensive and sophisticated, they are now routine tools. Cladistic and other repeatable methodologies have also given rigor and respectability to phylogenetics that was previously lacking. The next decade will see many new advances as well as challenges.

However, as the use of these increasingly sophisticated tools becomes common, it also requires that researchers have access to facilities such as electron microscopes as well as molecular, isotopic facilities, thereby reducing the potential for "proletarian" participation in some aspects of molluscan systematics. While these requirements and expectations for modern biological phylogenetic systematics have substantially surpassed the resources available to most amateurs and para-professionals and even the resources of many smaller institutions, these communities continue to produce many important taxonomic contributions and other work.

Molluscan systematics encompasses three distinct activities: taxonomy, classification (which may or may not be intended to reflect phylogeny), and nomenclature. Although these three components are rigorously and distinctly practiced by systematists, they are often amalgamated under the term "taxonomy." Although the breadth of meaning of the term "taxonomy" is clearly understood among most practitioners, it can obscure the methodology and practices of modern systematics to others. On the other hand, not all systematists work across the full breadth of systematics. For example, they can be engaged in the study of molecular phylogenies without applying the results of their studies to the nomenclature of the group. Similarly, the resolution of nomenclatural issues can be carried out without a phylogenetic study of the species or the generation of a new classification, but usually not without extensive library resources.

Today the practice of molluscan systematics is very different from what it was for the previous two centuries. Today's classifications are treated as hypotheses, subject to recurrent testing on a scale never imagined or experienced by the monographers of the nineteenth and twentieth centuries. In the past, type material and allocated specimens might be examined once every 50 to 100 years, thereby testing the previous systematic monograph or classification of the taxon. Today's phylogenetic classifications are based on relatedness and sometimes tested yearly. The rate of addition of new character sets, especially molecular ones, as well as new taxa, continues to increase every year. Today's classifications are also valued for their predictive value with disciplines as disparate as ecology, physiology, biogeography, medicine, and conservation biology benefiting when evolutionary relationships are reflected in them.

In addition, the research breadth of today's systematic malacogists has broadened and deepened. Not only is the extraordinary knowledge of a specific taxon still required; additional knowledge of anatomy, developmental biology, molecular biology, paleontology, and other disciplines (all at multiple scales) is vital, as well as a working understanding of the methodological literature of character analysis and phylogenetic hypothesis testing. Today's systematists are (or should be) capable of solving primer problems, understanding the implications of various gap coding schemes, and appreciating the *in situ* staining patterns of *Hox* genes in addition to understanding anatomy and histology and providing identifications, authors, dates, and associated geographical and bathymetric distributions for a taxon.

Lastly, the molluscan research presented in this volume is predicated on classifications that reflect evolutionary history (phylogenies). While the majority of these chapters deal primarily with taxonomy and classification, all three components of molluscan systematics and their interactions are recognizable in most. Also well illustrated is the remarkable breadth and depth of many of the research programs in molluscan evolution and systematics.

EVOLUTIONARY DEVELOPMENT

Molluscs, especially gastropods, were important taxa in early developmental studies, and biologists in the late nineteenth century pioneered descriptive and comparative studies of cell lineages (reviewed by Guralnick 2002). However, while the general patterns they found were remarkably invariant for many taxa, more detailed investigations showed substantial variation at finer scales, ultimately contributing to the abandonment of this research enterprise (Guralnick 2002). Later ontogenetic stages of development received considerable attention in the 1930s, especially the workings of gastropod torsion (e.g., Garstang 1929; Crofts 1937). This focus remains prevalent, and it is perhaps not unexpected that only torsion and shell ontogeny were featured topics in contributions to the Ponder (1988) volume. However, by the 1993 London meeting (Taylor 1996), cell lineage patterns and their role in reconstructing gastropod relationships were once again being addressed (van den Biggelaar 1996).

The entire area of molecular evolutionary development has grown rapidly since the benchmark symposia of the 1993 London meeting. However, molecular evolutionary development research programs in molluscs have been very limited compared to other groups such as the Nematoda (*Caenorhabditis*), Arthropoda (*Drosophila*), Urochordata (*Clionia*), and Vertebrata (e.g., *Xenopus, Danio*). And it is not just the Mollusca that remain underutilized and understudied, but also all crown group members of the Lophotrochozoa, undoubtedly because of the lack of model organisms on this branch of the Tree of Life. The long-awaited availability of complete molluscan genomes (*Lottia, Aplysia, Octopus, Biomphalaria*) will undoubtedly improve the current situation and facilitate more comparative work. However, the disbanding of the molluscan research group at Utrecht (The Netherlands) following the retirement of Prof. Jo van den Biggelaar, and the untimely death of Prof. André Adoutte (France), represented serious setbacks for molluscan molecular evolutionary developmental research early in this decade. But with increasing molluscan research activity by B. Degnan, D. Jacobs, D. Lambert, M. Martindale, A. Wanninger, and other research groups around the world, the promise for future work in this important area of molluscan evolution may be brighter, and in recognition of this potential we have included a separate chapter on molluscan evolutionary development (Wanninger *et al.*, Chapter 16).

MOLECULAR PHYLOGENETICS

In the Taylor (1996) volume 17% of the papers discussing the phylogenetic relationships of living taxa used molecular characters; a little over 10 years later, 76% of the chapters in this volume use or discuss molecular results as part of their phylogenetic analyses. Molecular data sets have gone from single genes for a handful of taxa to multiple genes across many representatives of a group. Combined analyses using both morphological and molecular data are still uncommon, and where they have been done, congruence between molecular and morphological data is uncommon (but see Giribet and Wheeler 2002 for an exception) and the topology of the combined analysis tree(s) is often surprising relative to the trees for the separate data sets (*e.g.*, Aktipis *et al.*, Chapter 9).

It is now evident that molecular data has not been a silver bullet for previous problems associated with reconstructing molluscan history from morphology or stratigraphy. It appears just as susceptible to homoplasy as morphological data, and rates of change appear variable both within and between genes as well as taxa. However, as molecular data becomes even more prevalent in molluscan evolutionary studies, there is little doubt that some of the results will surprise and perplex us. These unanticipated outcomes should not be arbitrarily dismissed as flawed studies. Each should be carefully examined and considered in the context of what we currently know about the limitations and behavior of molecular data as well as in the context of our limited, but increasing, knowledge of molluscan and spiralian genomics.

The recent multigene analysis of the Mollusca by Giribet *et al.* (2006) is a case in point. They utilized five gene fragments from both nuclear (H3) and mitochondrial (18S rRNA,

28S rRNA, COI, 16S rRNA) genomes for seven outgroup taxa and 78 Mollusca. While monophyly of the Mollusca was supported, only the Scaphopoda, Cephalopoda, Caudofoveata, and Solenogastres were found to be monophyletic within the Mollusca. The Gastropoda were diphyletic, with the patellogastropods placed as the sister of Solenogastres and nested within a clade of Cephalopoda and Scaphopoda. The remaining gastropod taxa (Orthogastropoda) were nested between a clade of bivalves and a clade composed of bivalves plus Polyplacophora and Monoplacophora. While a sister relationship with the Solenogastres is surprising, the placement of the Patellogastropoda within a clade including the Cephalopoda and Scaphopoda has been suggested by others (Lindberg and Ponder 1996; Waller 1998; Haszprunar 2000; Steiner and Dreyer 2003).

While it is tempting to dismiss Giribet *et al.* (2006) as just another strange result of a molluscan molecular analysis, there are patterns here that defy such a cavalier response. Yes, the placement of the monoplacophorans among the polyplacophorans is suspect and based on only a single sequence (28S rRNA) (e.g., Haszprunar *et al.*, Chapter 2), but arguments for long-branch attraction or uninformative genes are not supported by some of the broader patterns. As noted above, the molecular data recover a monophyletic Scaphopoda (jackknife support value = 100). Moreover, the two major groups within the Scaphopoda (Dentaliida and Gadilida; Reynolds and Steiner, Chapter 7) are also present. The cephalopod clade is also well supported (jackknife support value = 100) and groupings within the clade uncontroversial (Nishiguchi and Mapes, Chapter 8). The presence of a gastropod group (Patellogastropoda) (jackknife support value = 100) within a clade including the Cephalopoda and Scaphopoda is not unexpected (see also Waller 1998; Haszprunar 2000). The diphyletic Gastropoda and Bivalvia also share an intriguing pattern. The Gastropoda precisely separate into Eo- and Orthogastropoda. The bivalves also separate into relatively older clades with early Paleozoic originations (e.g., Protobranchia and Pteriomorphia) and taxa with primarily late Paleozoic or Mesozoic originations (Heteroconchia) (see Giribet, Chapter 6); the former and older clade is also the sister taxon of the Polyplacophora (+ Monoplacophora), named by Giribet *et al.* (2006) as Serialia.

Thus, analysis of this multigene data set recovers numerous uncontroversial sets of relationships (Mollusca; Scaphopoda; Cephalopoda; Caudofoveata; Solenogastres; Scaphopoda-Cephalopoda-Eogastropoda) as well as producing some apparently bizarre results (diphyletic Gastropoda and Bivalvia). These results may also be indicative of something other than long-branch attraction, because the deepest divergences produce less disconcerting results than some later ones. Dismissal of this entire analysis as some aberrant outcome may miss something important; instead, alternative hypotheses need to be explored, including the possibility that there were substantial genomic changes within major gastropod and bivalve lineages.

Genomic reorganizations have occurred in molluscan mitochondrial genomes on several occasions (Bivalvia, Patellogastropoda, Heterobranchia) and have taken various forms (gene order, gene deletion, tandem repeats, secondary structure) (see Simison and Boore, Chapter 17, and references therein). While mitochondrial genomes are substantially less complex than nuclear genomes, the types of changes seen in these smaller genomes undoubtedly occur in the nuclear genome as well and at many different rates. For example, there are virtually no substantial changes in mitochondrial gene order between the Polyplacophora and Caenogastropoda (Simison *et al.* 2006; Simison and Boore, Chapter 17). However, within the Gastropoda, the Patellogastropoda (Simison *et al.* 2006) and, especially, the Heterobranchia (Kurabayashi and Ueshima 2000; Grande *et al.* 2004) have undergone substantial gene rearrangements, and bivalves also show major gene order changes (Serb and Lydeard 2003; Simison and Lindberg, unpublished data).

Thus, irregular and conflicting outcomes of molecular analyses such as that of Giribet *et al.* (2006) may be resolved as our understanding of molluscan genomics increases, methodological approaches are refined, and new ones are developed. In the meantime, such unconventional results need to be kept on the table for discussion, exploration, and testing.

PALEOBIOLOGY

Since the Taylor volume (1996) there have also been significant discoveries and advances in molluscan paleontology. The Precambrian, putative stem mollusc *Kimberella* was proposed (Fedonkin and Waggoner 1997), and there has been substantial research into the molluscan affinities of the halkieriids (Vinther and Nielsen 2005, and references therein) and reexamination of the several Burgess Shale taxa (Caron *et al.* 2006). With few exceptions, new Lagerstätte have failed to provide any new insights into molluscan evolutionary relationships. One notable exception is the Herefordshire Lagerstätte (Silurian, England), where new preparation techniques, combined with imaging and computer reconstruction, have provided striking three-dimensional reconstructions of a putative spiny mollusc as well as soft-part anatomy of a platyceratid gastropod (Sutton *et al.* 2001, 2006).

Despite the rarity of molluscan Lagerstätte, new material from conventional localities as well as the re-examination of the existing material has provided new morphological data, stratigraphic occurrences, and distributions for many molluscan groups over the last 10 years. Where appropriate, the paleontological record is reviewed and discussed in the taxon chapters, and in two cases—the Cambrian radiation of molluscs and Paleozoic gastropod evolution— these topics are treated independently.

Methodological advances to examine fossil structures include serial sectioning, computer-assisted three-dimensional reconstruction, and energy-dispersive X-ray spectrometry. Stable-isotope studies are being used to document the paleoecological and paleoclimatological parameters of past molluscan communities. There has also been substantial progress made in the area of high-precision geochronology, with uncertainties of a million years or less as far back as the Cambrian (Kerr 2003). Finer resolution of geologic time will provide a better understanding of molluscan evolution, including extinctions and recoveries, responses to climate change, adaptive radiations, and other events.

ADAPTIVE RADIATIONS

We asked the authors of the taxon chapters to include a section on adaptive radiations so that putative radiations could be discussed in the light of current phylogenies. We recognize that identifying an "adaptive radiation" can be controversial (e.g., Losos and Miles 2002). Adaptive radiations are typically defined as rapid speciation events in which a single taxon (or a few taxa) gives rise to numerous species that occupy various habitats or niches. Famous examples of adaptive radiations include Darwin's finches on the Galápagos Islands (Grant 1981) and the cichlid fish of the Great Rift Lakes of Africa (Sturmbauer 1998). Molluscan examples include the partulid land snail radiations on tropical Pacific Islands (Murray *et al.* 1993), the Pomatiopsidae in the Mekong River drainage (Davis 1979), and freshwater cerithioideans in Thailand (Glaubrecht and Kohler 2004).

On a phylogenetic tree, adaptive radiations are typically fan-shaped and can be unresolved polytomies because of the rapidity of the speciation event or well resolved because of character displacement and strong selection for specific characteristics in the different habitats (Gittenberger and Gittenberger 2005; Grant and Grant 2006). However, not all putative adaptive radiations fulfill the foregoing criteria, and multiple micro- and macroevolutionary processes can produce similar patterns. Time especially plays havoc with the general model; for example, extinction can erase or enhance taxon patterns, leaving highly diverse clades relative to depauperate sister taxa with high extinction rates, or produce living distributions that are suggestive of earlier adaptive radiations

into multiple habitats. Alternatively, this latter pattern could simply be the result of continuous, but low, speciation rates in marginal habitats; some of the deep-sea vent taxa may fit this category. In some cases other macroevolutionary processes are argued to be responsible. One of the most famous of these is the putative adaptive radiations of the land snails *Poecilozonites* and *Ceron* on Bermuda and in the Bahamas advocated by the late Steven J. Gould (e.g., Gould 1969), which he and coworkers later reinterpreted (Gould 2002; Gould and Woodruff 1986) in terms of nonadaptive evolution and contingency (see Tillier 2003 for discussion). Here our authors have often interpreted adaptive radiations more broadly, allowing them to encompass speciation and perhaps products of former radiation over longer periods of geological time.

FUTURE STUDIES

The chapters in this volume are overviews of work in progress, in part reflected by some divergent opinions expressed in different chapters in this volume: clearly; much remains to be done. The authors of each chapter provide a section outlining gaps in knowledge and recommendations for future studies to address these.

There is a particularly exciting time just ahead with genomic studies. The *Lottia* and *Aplysia* genomes will provide a new comparative molecular framework previously available only in model organisms such as *Drosophila, Xenopus,* and *Homo.* Two other taxa are currently being sequenced (*Octopus, Biomphalaria*), and as production costs continue to decline, other molluscan genomes will undoubtedly become available. However, to maximize the comparative power of these data and to more fully understand molluscan genomics, a better sampling of taxa is needed, especially from the aplacophorans, polyplacophorans, bivalves, and scaphopods.

A commonly held view is that malacology worldwide currently suffers from the continuing erosion of practitioners in museums and universities, is exasperated by the generally decreased funding opportunities for the "lesser" phyla, and a declining public interest in natural history. However, despite the setbacks, the malacological community, relatively small and fragmented as it is, has within its ranks some extremely innovative and productive scientists, both past and present. The Mollusca present wonderful opportunities to address a host of research questions and their fascinating diversity, unparalleled conservation issues, and economic importance.

In the United States the U.S. National Science Foundation is attempting to address some of these issues and concerns with special programs such as Partnerships for Enhancing Expertise in Taxonomy (PEET), the Assembling the Tree of Life initiative (AToL), and the Research Coordination Networks in Biological Sciences (RCN), which seek to enhance interactions between researchers to create new directions and advance the field. Although some similar programs have been established in other countries—for example, the Australian Research Council Linkage International encourages networks and collaborations between researchers, and the Australian Biological Resources Study (ABRS) funds systematics programs—these are disappointingly few.

To combat declining public interest and understanding of natural history we must also increase visibility and familiarity of molluscs through outreach and public education. Molluscs play important roles in many cultures and societies as objects of beauty, tradition, function, and food. There are a multitude of stories and exemplars of the importance of molluscs in everyday life, and these must be identified, developed, and presented to multiple audiences, including primary and secondary education venues as well as the general public.

There is no doubt that there will be new directions and challenges for molluscan phylogenetic studies in the future. Given the increasing sophistication and cost of research, there is need for more improved communication, collaborations, and coordination of objectives and programs. Communication of the results of work to a wide range of audiences is an essential ingredient now and will continue to be in the future. We hope that this volume is a small step in that direction.

TERMINOLOGY AND CONCEPTS

A brief, schematic guide to some of the terminology, methods and concepts used in the book follows. This compact summary is meant to assist readers who are not necessarily familiar with all aspects of the field. We recommend that the reader use resources available through the World Wide Web and literature for more comprehensive treatments.

THE MOLLUSCAN CLASSES

Seven or eight higher taxa ("classes") of living molluscs are recognized. These are listed in this section along with the major taxa recognized within each of them. See the appropriate chapters in this volume for more information.

Aplacophora: Considered to be paraphyletic by some authors and monophyletic by others, the spicule worms are a small group of taxa that comprise two distinct lineages. While these two groups are always treated as separate clades, they are often treated as separate class-level taxa:
   Solenogastres (= Neomeniomorpha)
   Caudofoveata (= Chaetodermomorpha)
Polyplacophora: Living taxa in this class (chitons) have eight shells surrounded by a girdle. There are three major groups:
   Multiplacophora (with 17 shell plates, extinct)
   Paleoloricata (extinct)
   Neoloricata (all living chitons)
Monoplacophora (= Tryblidia): These are deep-sea, limpet-like molluscs with a single shell and serial repetition of some organs.

The first two of the preceding three groups, together with some fossil taxa, have been informally termed the "placophorans"— a grade of molluscs with distinct linear organization of an anterior mouth and posterior anus. This term has also been used for the Polyplacophora alone. Extinct taxa that may be assignable to the "placophoran" grade are listed below—see "Early Fossil Taxa that may be Molluscs."

Bivalvia: These have a bivalved shell, exemplified by scallops, oysters, cockles, shipworms, and clams. Four to six major groups are usually recognized:

"Protobranchia" (now treated as two groups; see Chapter 6)

Pteriomorpha (mussels, scallops, oysters, arks)

Palaeoheterodonta (freshwater mussels, trigoniids)

Heterodonta (remaining bivalves, including Anomalodesmata)

Scaphopoda: These have a single, tusk-shaped shell open at both ends and are known as tusk shells. Two major groups are recognized:

Dentaliida

Gadilida

Cephalopoda: These have a single, chambered shell or have lost their shell; the animal has a head surrounded by arms, exemplified by *Nautilus,* squids, cuttlefish, and octopuses.

Nautilioidea (the nautiloids, including the pearly nautilus)

Ammonoidea (the ammonites, extinct)

Coleoidea (squids, octopus, cuttlefish)

Gastropoda: These have a single, often coiled, shell or have lost their shell; the animal is torted in development; the larva and some adults have an operculum. Examples include snails, limpets, slugs, and whelks. The following groups are recognized:

Eogastropoda

Patellogastropoda (the true limpets)

Orthogastropoda

Vetigastropoda (abalone, keyhole limpets, topshells)

Neritimorpha (nerites and relatives)

Caenogastropoda (the largest group of gastropods, including whelks, periwinkles, cowries, cones, volutes, creepers, and others)

Heterobranchia (some basal groups plus Euthyneura, which comprises the opisthobranchs [seaslugs and their relatives] and pulmonates [land slugs and many of the land snails])

Extinct major taxa attributed to molluscs include:

ROSTROCONCHIA: Rostroconchs began life with a single conch (or valve), but the single shell transformed ontogenetically into a non-hinged, bivalved shell that gaped at its margins. Based on muscle scar patterns, the animals appear to have had anterior feeding tentacles that were likely used for deposit or suspension feeding, and both motile and nonmotile forms have been recognized. Rostroconchs first occur in the Lower Cambrian and undergo an extensive Late Cambrian and Early Ordovician radiation. The last known rostroconchs occur in the Permian. Rostroconchs are thought to share common ancestry with the Bivalvia.

HYOLITHA: Hyoliths had cone-shape shells with a (presumably ventrally) flattened side. The aperture was closed by an operculum, and a single pair of anterior appendages (helens) was present in some taxa. The operculum was attached to the shell by pairs of symmetrical muscles. The shell was constructed of crossed-lamellar calcium carbonate crystals, and the apical tip bore a larval shell. A looped alimentary system connected the anterior ventral mouth to the anterior dorsal anus. Hyoliths were sessile, epifaunal deposit feeders, scraping sediments from the seafloor. They first appear in the Early Cambrian and became extinct at the end of the Paleozoic. Their status as molluscs is not universally accepted.

EARLY FOSSIL TAXA THAT MAY BE MOLLUSCS

Several Precambrian or Cambrian fossils may be molluscs but remain controversial (including within this volume). They are:

KIMBERELLA: It has been suggested that this Precambrian fossil, originally thought to be a jellyfish, is a stem-group mollusc (Fedonkin and Waggoner 1997). Specimens are bilaterally symmetrical and range between 3 and 100 mm in length. There is an antero-posterior ridge along the dorsal surface that is raised in the center. The margin of the fossil has small lobes that have been interpreted as putative gills.

ODONTOGRIPHUS AND WIWAXIA are two Middle Cambrian Burgess Shale taxa whose affinities have been argued to be either stem-group lophotrochozoan, molluscan, or annelid (Butterfield 2006; Caron *et al.* 2007). Both taxa possess a putative molluscan radula. Like *Kimberella, Odontogriphus* is thought to have had a cuticular dorsal exoskeleton with gills along the body margin (Caron *et al.* 2006). In contrast, *Wiwaxia* is covered by noncalcified sclerites that morphologically vary with their placement on the dorsum, as in the Halkieriidae.

HALKIERIIDAE: The worldwide presence of small, hollow, calcareous sclerites in numerous Precambrian and Cambrian sediments (collectively referred to as "small shellies") were an enigmatic component of the early fossil record. However, in the early 1990s an articulated fossil was found in the Lower Cambrian of Greenland that was covered with "small shellies." It was immediately apparent that what had been thought to be the remains of individual organisms were actually parts of a single larger animal. Recent work in the Cambrian of Europe, Asia, and Australia has greatly expanded our knowledge of the Halkieriidae. Vinther and Nielsen (2004) have highlighted the probable molluscan affinities of the halkieriids. The class Biplacophora has recently been introduced for this group.

SOME COMMON MORPHOLOGICAL TERMS USED IN MOLLUSCS

Although specialized terminology is mostly explained in the individual chapters, some terms are general to molluscs. The body of a mollusc consists of a *visceral* area covered with a sheet of tissue, the *mantle* (or *pallium*), which typically secretes the shell(s) or spicules. A locomotory muscular *foot* lies beneath the viscera and is the surface on which the animal moves or by which it adheres to the substrate.

A posterior cavity, the *mantle* (or *pallial*) *cavity*, contains the gills (which are typically *ctenidia*, the plesiomorphic molluscan gill) and the chemosensory *osphradium*. In some (e.g., chitons, monoplacophorans) there is no mantle cavity, and the gills lie in a *mantle* (or *pallial*) groove lying between the foot and the mantle. In gastropods the mantle cavity lies anteriorly as a result of larval *torsion* in which the head and foot are rotated 90–180 degrees relative to the viscera. The kidneys and genital duct typically open to the mantle cavity or mantle groove.

The *mouth* opens anteriorly, where a head may also be present. Inside the mouth is a cavity, the *buccal cavity*, typically containing a muscular *odontophore*, on which lies a ribbon-like *radula* comprising many small teeth. The buccal cavity opens to an esophagus, which in turn opens to a stomach and then the intestine and rectum. The typically paired, large *digestive glands* (also called midgut glands or liver) open to the stomach. The buccal cavity, odontophore, and radula are absent from bivalves.

The *heart* is surrounded by the saclike *pericardium* and usually consists of a *ventricle* and a pair of *auricles*. The kidneys (renal organs, nephridia) open to the mantle cavity and are typically also connected to the pericardium by *renopericardial canals*.

The *gonad* opens directly, or via one or both kidneys, to the mantle cavity. In more advanced taxa in which internal fertilization occurs, glandular structures are developed that coat the egg with protective layers after fertilization. In others, the eggs and sperm are ejected into the water column.

The nervous system typically consists of several *ganglia* (cerebral, pleural, pedal, visceral, etc.)— swellings where the nerve cell bodies lie—and nerve cords. Eyes have evolved multiple times in molluscs and can be cerebral (on the head), pallial (on the edge of the mantle/pallium), or in minute pores in the shell (as in chitons).

PHYLOGENY AND CLADISTICS

This work focuses on *phylogeny*, the hypothesized evolutionary relationships of organisms, the study of which is *phylogenetics* or phylogenetic systematics. Whereas phylogenies can be envisaged and created "by hand" using one's imagination, they are more often obtained by today's practitioners by using computer-assisted methods outlined in the following paragraphs.

Cladistic methodology (e.g., Kitching *et al.* 1998), can be used to create trees objectively "by hand" by mapping characters on them, but computer-assisted *maximum parsimony* methods are usually used today. Parsimony is the concept of minimizing the steps necessary to generate a tree, invoking the idea of preference for the least complex explanation. Cladistic methodology distinguishes between synapomorphy, homoplasy, and symplesiomorphy (see "Some Cladistic Terminology" below.).

Maximum likelihood (e.g., Philippe *et al.* 2005) is an approach used with molecular data. The optimal tree is the one that maximizes the statistical likelihood that the specified evolutionary model produced the observed character state data; the models specify the probabilities of character state changes during evolution.

Bayesian inference of phylogeny (e.g., Huelsenbeck *et al.* 2001) is based upon the posterior probability distribution of trees—the probability of a tree conditioned on the observations (using Bayes's theorem).

Tests of the results of cladistic methodology include the following:

Bootstrap (e.g., Newton 1996): Bootstrapping is used to assess the stability of taxon groupings in a phylogenetic tree. Bootstrap support values are calculated by randomly resampling characters, *with replacement,* and reanalyzing these bootstrapped samples. If a grouping is well supported in the original matrix, it will also be present in the bootstrapped samples, thereby giving the group more support in the consensus tree.

Jackknife (e.g., Goloboff *et al.* 2003): Like bootstrapping, jackknifing is a statistical method for evaluating support for groupings in a phylogenetic tree. Jackknifed statistics are derived by creating and reanalyzing new data matrices that have been constructed from the original data matrix *without replacement.*

Phenetic or *numerical taxonomy* (e.g., Sneath and Sokal 1973) methods were popular before cladistics became popular and computerized, and they are still used in molecular analyses. These analyses examine the overall similarity of characters without regard to cladistic principles.

Total evidence is the approach used when phylogeny is reconstructed by analyzing combined data of different kinds, such as morphological and molecular data. It is somewhat controversial, because gene phylogenies may be incongruent with organismal phylogenies.

All these methods analyze *characters*—features that might change through time and can include morphological (i.e., anatomical, shell, or similar features as well as cell ultrastructural details), developmental (e.g., cell lineage data or structural changes during organogenesis), or molecular characters (e.g., DNA sequence data, gene order).

SOME CLADISTIC TERMINOLOGY

Some of the phylogenetic (including cladistic) terminology that is commonly used in this volume may not be known to some readers and will be briefly explained here.

Some general terms include *taxon* (plural *taxa*), which relates to any taxonomic unit or grouping of organisms, and it usually has a name. Traditionally a taxon has a hierarchal *rank* (e.g., phylum, class, family, genus, species); however, the process of classification does not invoke ranking, and classifications may be presented unranked.

Some cladistic terms include *apomorphy,* a character that is derived or "advanced" relative to the ancestral condition; *synapomorphies* are derived characters that are shared by more than one taxon; *autapomorphies* are derived characters found in only one taxon; *plesiomorphies* are ancestral ("primitive") characters. It is requisite that individuals be compared at similar life stages, sexes, and so forth; in other words, characters are not scored for the larva of one species and the adult of another.

These comparable stages are referred to as *semaphoronts*.

The aim of phylogenetic studies is to identify *monophyletic* groups and establish their relationships. A *monophyletic group* (also called a *clade*) is one that includes the ancestor of that group and all of its descendants. By way of contrast, a *paraphyletic group* is one that includes the most recent common ancestor and some but not all descendants, and a *polyphyletic group* includes taxa that do not share a common ancestor.

Phylogenies are usually represented by *phylogenetic trees*. A phylogenetic tree is a diagram that graphically represents a hypothesis of evolutionary relationships among three or more taxa. Such a diagram is typically a *cladogram* (or *phylogram*) generated using cladistic methodology. A *consensus tree* is a tree that is congruent with all of the trees resulting from an analysis—typically the *most parsimonious trees*—those with the least number of steps (a measure of evolutionary divergence) needed to generate them. A *strict consensus tree* shows only those clades common to all trees. A *gene tree* is the phylogeny of a gene, but it may not necessarily reflect the phylogeny of the organisms with that gene. A *node* is a branching point on a tree. A *polytomy* is a branch point (node) in a tree with more than two descendant branches.

Sister groups (or *sister taxa*) are more closely related to the clade of interest than any other group in the tree (i.e., a sister group is the closest outgroup). The descendants of an ancestor are called *daughters,* and the siblings are called *sisters*.

An *outgroup* is a taxon (or group of taxa) used to *root* a cladogram. The root is placed between the outgroup(s) and the ingroup.

Homology of a character is similarity due to a common evolutionary origin, whereas *homoplasy* is similarity that is the result of independent evolutionary change. The latter can be the result of *parallelism* (*convergence*) when characters that look the same have evolved from different starting points, or *reversal* (reversion to resemble a primitive character state from a derived state).

Other terms that are sometimes used include *Bauplan,* a German word meaning "body plan" or "blueprint," used to refer to the common features of a taxon; *crown group,* meaning a (usually) living monophyletic group or clade; and *stem group,* applied to fossils that are more closely related to a particular crown group than to any other, but more basal than the crown group's most basal member. A *lineage* is a historical sequence from ancestor to descendants.

MOLECULAR BIOLOGY: A BRIEF INTRODUCTION TO SOME TERMS

We do not attempt to provide a detailed account of terms used in molecular work, but some of the commonly used ones are briefly explained in the following paragraphs.

DNA (deoxyribonucleic acid) is the key molecule involved in coding genetic information. DNA is composed of structural units called *nucleotides.* A DNA nucleotide consists of a molecule of a sugar and a phosphate group (which bind successive nucleotides together in a *strand*) and one of four *bases* (adenine, A; guanine, G; thymine, T; cytosine, C). A and G are similar in structure and called *purines;* T and C are also similar in structure to each other and called *pyrimidines.* The *sequence* of bases in a fragment of DNA, usually a *gene* or part of a gene, carries information, spelled out using just these four bases. DNA usually takes the form of a double strand in which the bases face each other, A matches up with T, and G matches up with C; this matching enables DNA to be copied (*replicated*). Double-stranded DNA is therefore usually measured in *base pairs* (bp) rather than nucleotides (nt).

Nuclear DNA (nDNA) is the DNA contained within the chromosomes in the cell nucleus; it encodes the largest part of the genome. *Mitochondrial DNA* (mtDNA) is found in the *mitochondria* (the organelles that generate energy for the cell), not in the nucleus. It is not inherited from both parents in the same fashion as nuclear DNA but is passed on by the female in the egg cell.

RNA (ribonucleic acid) is transcribed from DNA by enzymes (RNA polymerases) and is the template for the translation of genes into proteins. Three types are recognized: messenger RNA (mRNA), transfer RNA (tRNA), and ribosomal RNA (rRNA). RNA is similar to DNA in structure but contains a different sugar, often occurs as a single strand rather than double, and contains the base uracil (U) instead of the thymine (T) found in DNA.

The *genome* is the entire genetic material (coding and noncoding) that is present in each cell. *Coding regions* of genes are translated into proteins (via RNA and enzymes). A sequence of three successive nucleotides is a *codon* and represents one of 20 different *amino acids*. When the amino acids are assembled in the sequence specified by the DNA to form a protein molecule, their chemical properties make the molecule fold up into the correct shape to serve as a piece in the cell's structure and machinery. The part of the DNA that specifies the amino acid sequence is called an *exon*, while an *intron* is the part that separates coding regions. *Noncoding* areas are parts of DNA that do not code for genes; these are sometimes called "junk" DNA. When a protein is made from a gene, the gene is said to be *expressed*.

A *primer* is a short sequence of nucleotides (an *oligonucleotide*) used to start the replication of DNA in DNA sequencing and the *polymerase chain reaction (PCR)*. Using the PCR, a single copy of the DNA of interest, even if mixed with other DNA, can be amplified to obtain billions of replicates.

A *locus* is a point in the genome that is identified by a *genetic marker* and can be mapped by some means. It does not necessarily correspond to a gene.

DEVELOPMENT

The *ontogeny* or *morphogenesis* of an organism (or a particular organ or body system) describes its developmental history from its origin—a zygote (fertilized egg), for an organism, or precursor cell(s), for an organ or system—to its mature form. This is accomplished by cell division and the *differentiation* of the cells into specific types. Cell division is referred to as *cleavage*. The first

cleavage from the zygote produces two undifferentiated cells or *blastomeres*, the second cleavage four cells, and so on.

There are two general types of cleavage patterns in the Metazoa (multicellular organisms): *radial cleavage* and *spiral cleavage*. Molluscs have spiral cleavage, although it may be modified in especially yolky eggs, such as in the Cephalopoda. Equal or different amounts of cytoplasm may be provided to the blastomeres during the first two cell divisions, producing either *equal* or *unequal* cleavage patterns, respectively.

In molluscs (and other Spiralia) the cells produced by the third cell cleavage are substantially reduced in size. These cells are termed *micromeres*, whereas the four cells from the first two divisions are called *macromeres*. In addition, the micromeres are rotated relative to the macromeres and are located in the furrows between the larger cells. Subsequent divisions of the micromeres are also offset, and the term *spiral cleavage* refers to this division pattern.

Micromeres are also the first cells to become differentiated and have specific *cell fates*. In the molluscs, specific micromeres will give rise to certain tissue types, larval structures, and adult systems and organs. This early cell differentiation and specification is known as *mosaic development*.

Molecular techniques have greatly aided molluscan developmental studies in recent years. *Gene expression patterns* provide important data concerning the presence and activity of gene products during specific developmental stages. These patterns result from *gene cascades*, in which genes interact to activate or suppress other genes. These *developmental regulatory genes* are also referred to as *homeobox* genes and are broadly conserved across the Metazoa. Conserved homeobox genes include *Hox*, *ParaHox*, and *Nkx* genes.

Heterochrony is a developmental change over evolutionary time in the rate or timing of developmental events. Rice (1997, 2002) distinguished six forms of heterochronic change in a descendant relative to its ancestor: (1) *predisplacement*, in which a character begins developing earlier;

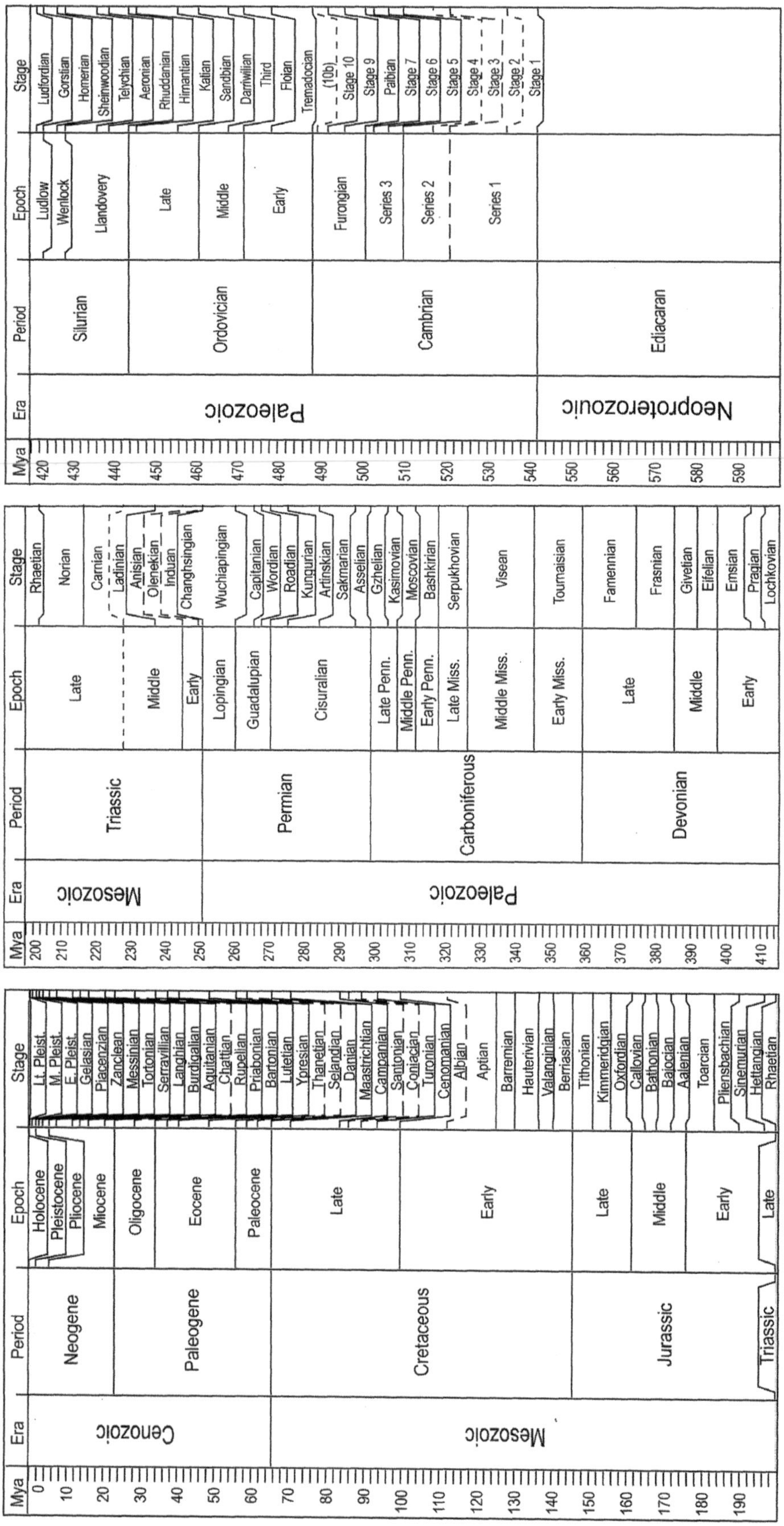

FIGURE 1.1. Diagram of geological stages showing the standard chronostratigraphy. NOTE: Based on the Geologic Time Scale 2004 (Gradstein et al. 2005) and created by TS-Creator© (Gradstein and Ogg 2006).

(2) *post-displacement,* in which a character begins developing later; (3) *neoteny,* in which a character develops at a lower rate; (4) *acceleration,* in which a character develops at a higher rate; (5) *hypermorphosis,* in which development of a character continues beyond the point at which it stopped in the ancestor; and (6) *progenesis,* in which the development of a character stops at an earlier stage.

FOSSILS AND THE FOSSIL RECORD

The time scale, showing the major *eras, periods, epochs,* and *stages,* is provided in Figure 1.1. The Cenozoic Era can also be divided into two "suberas" not shown in Figure 1.1: the *Quaternary* (since the beginning of the Pleistocene Epoch) and *Tertiary* (from the beginning of the Paleocene Epoch to the beginning of the Pleistocene). Also, the *eons* are not shown in the figure. The Phanerozoic Eon extends back from the present to 542 million years ago (Mya) (the beginning of the Cambrian), and the period before (the Precambrian) constitutes the Proterozoic Eon.

Historical time is referred to as *BCE* (before the Common Era). The Common Era (CE) begins with the year 1 on the Gregorian calendar. The notations CE and BCE are alternative notations for the commonly used AD and BC, respectively.

Fossils can be preserved in various ways. *Lagerstätten* are rare fossil deposits in which the preservation of soft body parts has occurred. *Steinkerns* are internal molds of fossil shells; usually the shell itself has been lost.

REFERENCES

Beesley, P. L., Ross, G. J. B., and Wells, A., eds. 1998. *Mollusca: The Southern Synthesis.* Fauna of Australia 5. Melbourne: CSIRO Publishing.

Berry, E. G., Burch, J. B., Carriker, M. R., Gismann, A., Robertson, R., Smith A. G., Sohl, N. F., and Taylor, D. W. 1962. Preface. *Malacologia* 1: 1.

Butterfield, N. J. 2006. Hooking some stem-group "worms": fossil lophotrochozoans in the Burgess Shale. *BioEssays* 28: 1161–1166.

Caron, J.-B., Scheltema, A., Schander, C., and Rudkin, D. 2006. A soft-bodied mollusc with radula from the Middle Cambrian Burgess Shale. *Nature* 442: 159–163.

——. 2007. Reply to Butterfield on stem-group "worms": fossil lophotrochozoans in the Burgess Shale. *BioEssays* 29: 201–202.

Crofts, D. R. 1937. The development of *Haliotis tuberculata,* with special reference to the organogenesis during torsion. *Philosophical Transactions of the Royal Society of London* B 228: 219–268.

Dance, S. P. 1986. *A History of Shell Collecting.* Leiden: E. J. Brill/Dr. W. Backhuys.

Davis, G. M. 1979. The origin and evolution of the gastropod family Pomatiopsidae, with emphasis on the Mekong River Triculinae. *Monograph of the Academy of Natural Sciences of Philadelphia* 20: 1–120.

Fedonkin, M. A., and Waggoner, B. M. 1997. The late Precambrian fossil *Kimberella* is a mollusc-like bilaterian organism. *Nature* 388: 868.

Garstang, W. 1929. The origin and evolution of larval forms. *British Association for the Advancement of Science, Report* 1928: 77–98.

Giribet, G., Okusu, A., Lindgren, A. R., Huff, S. W., Schrödl, M., and Nishiguchi, M. L. 2006. Evidence for a clade composed of molluscs with serially repeated structures: monoplacophorans are related to chitons. *Proceedings of the National Academy of Sciences of the USA* 103: 7723–7728.

Giribet, G., and Wheeler, W. C. 2002. On bivalve phylogeny: a high-level analysis of the Bivalvia (Mollusca) based on combined morphology and DNA sequence data. *Invertebrate Biology* 121: 271–324.

Gittenberger, A., and Gittenberger, E. 2005. A hitherto unnoticed adaptive radiation: epitoniid species (Gastropoda : Epitoniidae) associated with corals (Scleractinia). *Contributions to Zoology* 74: 125–203.

Glaubrecht, M., and Kohler F. 2004. Radiating in a river: systematics, molecular genetics and morphological differentiation of viviparous freshwater gastropods endemic to the Kaek River, central Thailand (Cerithioidea, Pachychilidae). *Biological Journal of the Linnean Society* 82: 275–311.

Goloboff, P. A., Farris, J. S., Kallersjo, M., Oxelman, B., Ramirez, M. J., and Szumik, C. A. 2003. Improvements to resampling measures of group support. *Cladistics* 19: 324–332.

Gould, S. J. 1969. An evolutionary microcosm: Pleistocene and recent history of the land snail *P. (Poecilozonites)* in Bermuda. *Bulletin of the Museum of Comparative Zoology* 138: 407–532.

——. 2002. *The Structure of Evolutionary Theory.* Cambridge, MA: The Belknap Press of Harvard University Press.

Gould, S. J., and Woodruff, D. S. 1986. Evolution and systematics of *Cerion* (Mollusca: Pulmonata) on New Providence Island: a radical revision. *Bulletin of the American Museum of Natural History* 182: 389–490.

Gradstein, F. M., and Ogg, J. G. 2006. TS-Creator ©-Chronostratigraphic data base and visualisation: Cenozoic-Mesozoic-Paleozoic integrated stratigraphy and user-generated time scale graphics and charts. *Geoarabia* 11: 181–184.

Gradstein, F. M., Ogg, J. G., and Smith, A. G., eds. 2005. *A Geologic Time Scale 2004*. Cambridge, UK: Cambridge University Press.

Grande, C., Templado, J., Cervera, J. L., and Zardoya, R. 2004. Molecular phylogeny of Euthyneura (Mollusca: Gastropoda). *Molecular Biology and Evolution* 21: 303–313.

Grant, P. R. 1981. Speciation and the adaptive radiation of Darwin finches. *American Scientist* 69: 653–663.

Grant, P. R., and Grant, B. R. 2006. Evolution of character displacement in Darwin's finches. *Science* 313: 224–226.

Grassé, P., ed. 1960–1989. *Traité de Zoologie*. Tome 5, Vol. 2: Mollusques Lamellibranches (1960); Vol. 3: Mollusques gastéropodes et scaphopodes (1968, 1960, 1976); Tome 5, Vol. 4: Céphalopodes (1989). Paris: Masson et Cie.

Guralnick, R. P. 2002. A recapitulation of the rise and fall of the cell lineage research program: The evolutionary-developmental relationship of cleavage to homology, body plans and life history. *Journal of the History of Biology* 35: 537–567.

Haszprunar, G. 2000. Is the Aplacophora monophyletic? A cladistic point of view. *American Malacological Bulletin* 15: 115–130.

Huelsenbeck, J. P., F. Ronquist, R. Nielsen, and J. P. Bollback. 2001. Bayesian inference of phylogeny and its impact on evolutionary biology. *Science* 294: 2310–2314.

Hyman, L. H. 1967. *The Invertebrates*. Vol. VI, *Mollusca 1*. New York, McGraw-Hill.

Kerr, R. A. 2003. A call for telling better time over eons. *Science* 302: 375.

Kitching, I. J., Forey, P. L., Humphries, C. J., and Williams, D. M. 1998. *Cladistics*, 2nd edition. Oxford: Oxford University Press.

Kurabayashi, A., and Ueshima, R. 2000. Complete sequence of the mitochondrial DNA of the primitive opisthobranch gastropod *Pupa strigosa*: systematic implication of the genome organization. *Molecular Biology and Evolution* 17: 266–277.

Lindberg, D. R., and Ponder, W. F. 1996. An evolutionary tree for the Mollusca: branches or roots? In *Origin and Evolutionary Radiation of the Mollusca*. Edited by J. Taylor. Oxford: Oxford University Press, pp. 67–75.

Lindberg, D. R., Ponder, W. F., and Haszprunar, G. 2004. The Mollusca: relationships and patterns from their first half-billion years. In *Assembling the Tree of Life*. Edited by J. Cracraft and M. J. Donoghue. Oxford and New York: Oxford University Press, pp. 252–278.

Losos, J. B. and Miles, D. B. 2002. Testing the hypothesis that a clade has adaptively radiated: iguanid lizard clades as a case study. *American Naturalist* 160: 147–157.

Lydeard, C., Cowie, R. H., Ponder, W. F., Bogan, A. E., Bouchet, P., Clark, S. A., Cummings, K. S., Frest, T. J., Gargominy, O., Herbert, D. G., Hershler, R., Perez, K. E., Roth, B., Seddon, M., Strong, E. E., and Thompson, F. G. 2004. The global decline of nonmarine mollusks. *Bioscience* 54: 321–330.

Lydeard, C. and Lindberg, D. R. (eds.) 2003. *Molecular Systematics and Phylogeography of Mollusks*. Washington, DC: Smithsonian Institution Press.

Moore, R. C., ed. 1960–1996. *Treatise on Invertebrate Paleontology*. Part I: *Mollusca 1* (1960); Part K: *Mollusca 3* (1964); Part L: *Mollusca 4* (1957, revised 1996); Part N: *Mollusca 6* (Bivalvia), Vols. 1 and 2 (1969); Vol. 3 (1971). Lawrence, KS: Geological Society of America and University of Kansas Press.

Murray, J., Clark, B., and Johnson, M. S. 1993. Adaptive radiation and community structure of *Partula* on Moorea. *Proceedings of the Royal Society of London Series B: Biological Sciences* 254: 205–211.

Naef, A. 1919. *Idealistische Morphologie und Phylogenetik; zur Methodik der systematischen Morphologie*. Jena: G. Fischer.

———. 1921. *Die Cephalopoden (Systematik)*. Fauna e Flora del Golfo di Napoli, Monograph 35. Naples: Pubblicata dalla Stazione Zoologica di Napoli.

Newton, M. A. 1996. Bootstrapping phylogenies: large deviations and dispersion effects. *Biometrika* 83: 315–328.

Philippe, H., Delsuc, F., Brinkmann, H., and Lartillot, N. 2005. Phylogenomics. *Annual Review of Ecology Evolution and Systematics* 36: 541–562.

Ponder, W. F., ed. 1988. *Prosobranch Phylogeny*. Malacological Review Suppl. 4. Ann Arbor, MI: Society for Experimental and Descriptive Malacology.

Rice, S. H. 1997. The analysis of ontogenetic trajectories: when a change in size or shape is not heterochrony. *Proceedings of the National Academy of Sciences of the United States of America* 94: 907–912.

———. 2002. The role of heterochrony in primate brain evolution. In *Human Evolution through Developmental Change*. Edited by N. Minugh-Purvis and K. J. McNamara. Baltimore: Johns Hopkins University Press, pp. 155–170.

Serb, J. M., and Lydeard, C. 2003. Complete mtDNA sequence of the north American freshwater mussel,

Lampsilis ornata (Unionidae): an examination of the evolution and phylogenetic utility of mitochondrial genome organization in Bivalvia (Mollusca). *Molecular Biology and Evolution* 20: 1854–1866.

Simison, W. B., D. R. Lindberg, and J. L. Boore. 2006. Rolling circle amplification of metazoan mitochondrial genomes. *Molecular Phylogeny and Evolution* 39: 562–567.

Sneath, P. H. A., and Sokal, R. R. 1973. *Numerical Taxonomy: The Principles and Practice of Numerical Classification*. San Francisco: W. H. Freeman.

Steiner, G., and Dreyer, H. 2003. Molecular phylogeny of Scaphopoda Mollusca inferred from 18S rDNA sequences: support for a Scaphopoda–Cephalopoda clade. *Zoologica Scripta* 324: 343–356.

Sturm, C. F., Pearce, T. A., and Valdés, A. 2006. *The Mollusks. A Guide to Their Study, Collection, and Preservation*. Boca Raton, FL: Universal Publishers.

Sturmbauer, C. 1998. Explosive speciation in cichlid fishes of the African Great Lakes: a dynamic model of adaptive radiation. *Journal of Fish Biology* 53: 18–36.

Sutton, M. D., Briggs, D. E. G., Siveter, David J., and Siveter, Derek J. 2001. An exceptionally preserved vermiform mollusc from the Silurian of England. *Nature* 410: 461–463.

———. 2006. Fossilized soft tissues in a Silurian platyceratid gastropod. *Proceedings of the Royal Society of London Series B: Biological Sciences* 273: 1039–1044.

Taylor, J. D., ed. 1996. *Origin and Evolutionary Radiation of the Mollusca*. Oxford: Oxford University Press.

Thiele, J., 1929–1931. *Handbuch der Systematischen Weichtierkunde*. Vol. 1. Jena, Germany: Gustav Fischer Verlag.

Tillier, S. 2003. From an evolutionary microcosm to general theory: land snails, Steve Gould and evolution. *Comptes Rendus Palevol* 2: 435–453.

van den Biggelaar, J. A. M. 1996. The significance of the cleavage pattern for the reconstruction of gastropod phylogeny. In *Origin and Evolutionary Radiation of the Mollusca*. Edited by J. Taylor. Oxford: Oxford University Press.

Vinther, J., and Nielsen, C. 2005. The Early Cambrian *Halkieria* is a mollusc. *Zoologica Scripta* 34: 81–89.

Waller, T. R. 1998. Origin of the molluscan class Bivalvia and a phylogeny of major groups. In *Bivalves: An Eon of Evolution—Palaeobiological Studies Honoring Norman D. Newell*. Edited by P. A. Johnston and J. W. Haggart. Calgary: University of Calgary Press.

Wells, F. E. 2004. Molluscan megadiversity: sea, land and freshwater. In *World Congress of Malacology, Perth, Western Australia. 11–16 July 2004*. Edited by F. E. Wells. Perth: Western Australian Museum.

Wenz, W. 1938–1944. *Gastropoda*. Edited by O. H. Schindewolf. Handbuch der Paläzooloigie Band 6, Teil 1-7. Berlin: Bonntraeger.

2

Relationships of Higher Molluscan Taxa

Gerhard Haszprunar, Christoffer Schander,
and Kenneth M. Halanych

The Mollusca is one of the best known and best defined metazoan phyla, comprising about 130,000 named extant, and about 70,000 described fossil, species, with probable actual extant diversity around 200,000. Aside from their high importance for humans as food, art, jewels, pests, and disease vectors, molluscs play an important role as model organisms in science, particularly in neurobiology and evolutionary biology. Moreover, as is virtually unknown to the broad public (and to many conservationists!), no other group of animals has so many species under threat of extinction by mankind.

The molluscan "bauplan" of textbooks usually lists or summarizes features shared by the majority of taxa. This idealistic bauplan differs substantially from all concepts of a hypothetical ancestral mollusc (HAM), which is a probabilistic reconstruction of a historical reality (Table 2.1; Lindberg and Ghiselin 2003). Of course, the HAM depends on the preferred internal phylogeny of Mollusca and thus differs among authors. In any case it is only the HAM (and certainly not the bauplan) that is suitable for consideration of any hypothesis on molluscan origins and early evolution, as discussed

subsequently in this chapter. Molluscs are characterized by a very distinct body plan, yet there is only a single, cytological character (rhogocytes, discussed below) that is diagnostic for all molluscan species.

Molluscs have a comprehensive fossil record, with Early Cambrian or even Late Proterozoic shelled or scaled fossils thought to have molluscan affinities. These shelled Cambrian fossils include monoplacophorans, gastropods, and bivalves (see also Parkhaev, Chapter 3). These fossils show unequivocally that the origin and early radiation of the Mollusca lie prior to the so-called "Cambrian explosion" (e.g., Runnegar 1996).

SYSTEMATIC POSITION OF THE MOLLUSCA

For more than a century authors controversially discussed whether Mollusca were more closely related to (unsegmented, acoelomate) turbellarian flatworms or to (segmented, coelomate) polychaetes. There is a growing consensus that the Mollusca reside in the clade Trochozoa (*sensu* Peterson and Eernisse 2001), which, in addition to the Mollusca, also includes Nemertea,

TABLE 2.1
Comparison of the Current Concept of HAM (Hypothetical Ancestral Mollusc) with the Typical Molluscan Bauplan as Described in Textbooks such as Brusca and Brusca (2003) or Ruppert et al. (2004).

FEATURES OF THE HYPOTHETICAL ANCESTRAL MOLLUSC (HAM)	TYPICAL FEATURES OF THE MOLLUSCAN BAUPLAN
Small body wormlike without distinct head or visceral hump	Large body divided in head, foot and visceral hump
Dorsal body covered by chitinous cuticle with aragonitic sclerites	Dorsal body covered with shell (plates), scales, or notum
Narrow foot sole, locomotion by cilia	Broad, sucking foot sole, locomotion by muscular waves
Hemocoel with rhogocytes, amoebocytes and colloblasts, no vessels	Free hemocoel (primary body cavity), no vessels
Mantle cavity small, terminal	Mantle cavity circumpedal, large or anterior
Ctenidia (if present) for ventilation only	Ctenidia (gills) for respiration
Coelomate gonopericardial (urinogenital) system with ultrafiltration from heart atria into the pericardium; simple nephroducts	Coelomate gonopericardial (urinogenital) system with ultrafiltration from heart atria into the pericardium; distinct nephroduct (kidney)
Carnivorous with distichous radular apparatus	Herbivorous, detritovorous, with a rasp tongue (radula) supported by cartilages
Simple, glandular, tube-like midgut	Esophagus, stomach with midgut gland(s), looped intestine
Tetraneuran, cord-like nervous system	From tetraneuran, ganglionate nervous system to true brains
With chemoreceptive osphradia, and possibly preoral sensory system	With eyes, statocysts, osphradia and head tentacles
Fertilization in mantle cavity (entaquatic)	Free spawners, brooders or with internal fertilization
Lecithotrophic trochophore larva	Planktotrophic trochophore or veliger

NOTE: HAM features based on Haszprunar 1992, 2000; Salvini-Plawen and Steiner 1996. Typical molluscan bauplan as described in Brusca and Brusca 2003; Ruppert et al. 2004.

Entoprocta, Sipuncula, and Annelida, groups that originally possessed a trochophore larva in their biphasic life cycle. Molecular data place this clade within a larger taxon, Lophotrochozoa, which also includes the Platyhelminthes (except the Acoelomorpha, which are considered by many authors to be the sister to the remaining bilaterian groups), the Gnathifera, Gastrotricha, Brachiopoda, Phoronida, and Bryozoa (= Ectoprocta) (for review and references see e.g., Jenner 2004; Halanych 2004). Unfortunately, relationships among the Trochozoa taxa remain largely unresolved; however, three clades (Annelida, Entoprocta, and Sipuncula, discussed below) have been suggested as the sister taxon of the Mollusca—each with varying degrees of support from different data sets. Figure 2.1 depicts competing hypotheses concerning the placement of molluscs in Lophotrochozoa.

All available morphological data, and, in particular, recent studies on monoplacophoran microanatomy (Haszprunar and Schaefer 1997) and polyplacophoran myogenesis (Wanninger and Haszprunar 2002a) and neurogenesis

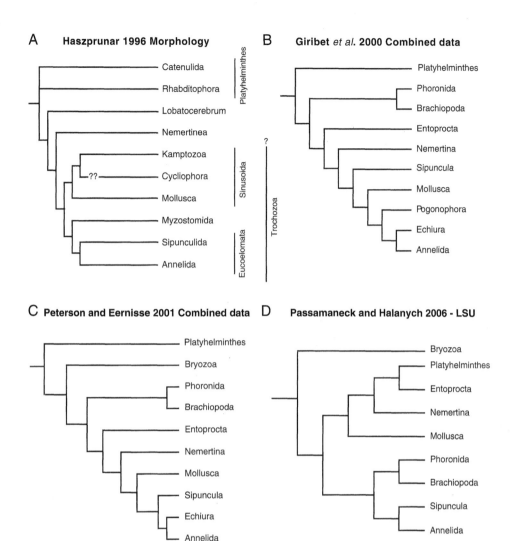

FIGURE 2.1. Competing alternative hypotheses of the position of Mollusca. (A) Haszprunar 1996, based on analyses of morphological data; (B) Giribet *et al.* 2000, and (C) Peterson and Eernisse 2001, combined morphology and ribosomal small subunit (SSU) data; (D) Passamaneck and Halanych 2006, ribosomal large subunit (LSU) data.

(Voronezhskaya *et al.* 2002; Friedrich *et al.* 2002) show Mollusca as primary nonsegmented animals, although various clades with distinct serial repetition do occur (Haszprunar and Schaefer 1997; Giribet *et al.* 2006; Nielsen *et al.*, 2007). Recent fine-structural and immunocytochemical studies provided evidence of a sister group relationship of Mollusca with Entoprocta (Haszprunar and Wanninger 2007; Wanninger *et al.*, in press). These two taxa share a chitinous cuticle and a sinusial circulatory system. In particular, the entoproct gliding larva has a ciliated foot sole with a molluscan-like pedal gland and anterior cirri (as in the aplacophoran Solenogastres[1]); serial,

ventrally intercrossing, dorsoventral muscle fibers (previously considered to be diagnostic for Mollusca; Haszprunar and Wanninger 2004); a tetraneural nervous system; unicellular notal glands; and a preoral, ciliary sensory system (as in Solenogastres) (see also Wanninger *et al.*, Chapter 16).

Analyses using molecular data are equivocal regarding the identity of the molluscan sister taxon. To date, molecular data bearing on mollusc origins has come largely from three sources: small (18S) nuclear ribosomal subunit (SSU), large (28S) nuclear ribosomal subunit (LSU), and

---

1. An alternative name is Neomeniomorpha.

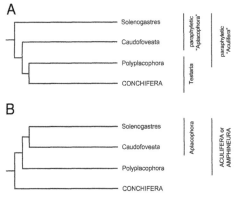

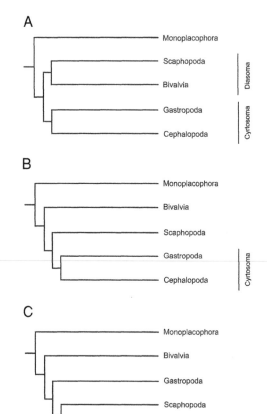

FIGURE 2.2. Hypotheses on interrelationships of basal molluscan groups based on morphology. (A) "Testaria"—hypothesis based on the cladistic analyses of Salvini-Plawen and Steiner (1996) and of Haszprunar (2000). (B) "Aculifera"—hypothesis based on the assumptions of Ivanov (1996), Scheltema (1993, 1996), and Scheltema and Schander (2006).

FIGURE 2.3. Hypotheses on the position of the Scaphopoda among the Conchifera. (A) "Diasoma/Cyrtosoma"—hypothesis based on Runnegar (1996), Runnegar and Pojeta (1985, 1992), and Salvini-Plawen and Steiner (1996). (B) "Cyrtosoma"—hypothesis based on Haszprunar (2000, Wanninger and Haszprunar 2001, 2002b). (C) Scaphopoda—Cephalopoda clade hypothesis after Waller (1998) and Steiner and Dreyer (2003).

mitochondrial data (mtDNA) in the form of gene arrangements or concatenated gene sequences.

The hypothesis that sipunculans may be sister to molluscs based on morphology (Scheltema 1993) is inconsistent with molecular (and morphological; see Gerould 1907) data. Sipunculids appear to be closely related to, and most likely within, annelids, based on mitochondrial (Boore and Staton 2002; Jennings and Halanych 2005; Bleidorn et al. 2006) and LSU data (Passamaneck and Halanych 2006).

Likewise, the morphological evidence for an entoproct/mollusc relationship is at odds with the molecular data. Cycliophorans (*Symbion*) appear to be the sister taxon for entoprocts based on LSU data (Passamaneck and Halanych 2006; see also Sørensen et al. 2000). Preliminary SSU data placed entoprocts close to rotifers (Winnepenninckx et al. 1998). This result was supported by Peterson and Eernisse (2001) but refuted by Zrzavy et al. (1998) and Giribet et al. (2000). Molecular analyses (reviewed in Halanych 2004; see also Boore et al. 2004) typically seem to place the mollusc, annelid, and brachiopod/phoronid clade(s) close together, in line with ideas of halkieriid fossils representing a stem ancestor for the group (Conway Morris and Peel 1995) (See also Simison and Boore, Chapter 17).

A recent phylogenetic analysis (Philippe et al. 2005) employed 71 genes and found strong support for grouping of two of the three lophotrochozoan taxa included: the annelids and molluscs, to the exclusion of platyhelminths (acoelomorphs were not included in this analysis). Use of this type of multigene analysis will undoubtedly increase in the future and be the most likely way to resolve the problematic question of molluscan origins.

Because there is some conflict between the molecular and morphological evidence to date, we present both views. Herein, G. H. reviews the morphological evidence, while C. S. and K. M. H. present the molecular view. A version of the evolutionary history of the Mollusca based on morphological inference is outlined and

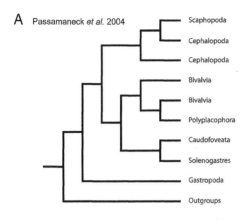

A Passamaneck *et al.* 2004

- Scaphopoda
- Cephalopoda
- Cephalopoda
- Bivalvia
- Bivalvia
- Polyplacophora
- Caudofoveata
- Solenogastres
- Gastropoda
- Outgroups

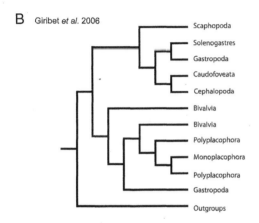

B Giribet *et al.* 2006

- Scaphopoda
- Solenogastres
- Gastropoda
- Caudofoveata
- Cephalopoda
- Bivalvia
- Bivalvia
- Polyplacophora
- Monoplacophora
- Polyplacophora
- Gastropoda
- Outgroups

FIGURE 2.4. (A) Summary of the LSU maximum likelihood tree from Passamaneck *et al.* (2004) on molluscan interrelationships. (B) Summary of tree produced by direct optimization in Giribet *et al.* (2006). See those papers for details.

illustrated in Figures 2.2 and 2.3, and an example of our molecular understanding is illustrated in Figure 2.4. However, we stress that each of these conclusions is preliminary.

MORPHOLOGICAL INFERENCE

The following account (Figure 2.2A) is mainly based on cladistic analyses of morphological data by Salvini-Plawen and Steiner (1996) and Haszprunar (2000).

HYPOTHETICAL ANCESTOR (MOLLUSCAN STEM SPECIES OR HAM)

As previously mentioned, it is important to note the major differences between what is

still called in many textbooks the molluscan bauplan or "archetype" (an idealistic concept without reality) and the hypothetical molluscan ancestor, which, though reconstructed, is an approximation of an actual ancestor (Table 2.1; see also Lindberg and Ghiselin 2003). The fossil record establishes that the molluscan ancestor was present in the Precambrian (Parkhaev, Chapter 3). Various scenarios for the molluscan stem species based on assumed ancestory have been advanced. The majority of authors considered either an advanced turbellarian flatworm or a somewhat reduced annelid; see reviews by Ghiselin (1988) and Haszprunar (1996). In the scenario outlined here, following Salvini-Plawen (1985, 1991, 2003), Haszprunar (1992, 2000), and Salvini-Plawen and Steiner (1996), it was probably a small (1–3 mm; Chaffee and Lindberg 1987), wormlike animal, dorsally protected by a chitinous cuticle with embedded aragonitic spicules or scales. A ciliated gliding sole with a large gland at its anterior end served as a locomotory organ. Underlying the sole epidermis was a muscular grid that consisted of outer circular, medial diagonal, and inner longitudinal muscles. In addition, the body was stabilized by dorsoventral fibers that intercrossed ventrally. The small mantle cavity at the rear end of the body had a respiratory function but possibly lacked gills. The buccal area of the gut possessed a chitinous radula, which was probably a series of bifid tongs rather than a rasping tongue. The midgut was a simple, thick, glandular tube with the posterior anus opening into the mantle cavity. The nervous system consisted of four cords: Two (pedal) supplied the pedal sole, two lateral (visceral) the inner organs. Both pairs emerged from the anteriorly placed cerebral and pedal centers, which formed a ring around the foregut as typical for lophotrochozoans. Sensory equipment included a preoral ("atrial") and a posterior ("osphradial") chemoreceptive system, whereas eyes and statocysts were both lacking. The circulatory system consisted of a dorsal heart with paired auricles and possibly fused ventricle (divided only in extant monoplacophorans). True vessels were lacking; instead, the blood, with hemocyanin as the respiratory molecule, flowed in sinuses and lacunae. The free hemocoel had amoebocytes (to attack invasive infections) and rhogocytes (pore-cells), the latter with a central role in the physiology of heavy metalaccumulation. Ultrafiltration took place in the pericardial wall of the auricles; the filtrate was released by a pair of pericardial (renal) ducts into the mantle cavity. The latter probably also released the gametes from the dorsally placed gonads. Development was by means of spiral cleavage and a lecithotrophic trochophore larva.*

The recent reinterpretation of the Middle Cambrian fossil *Odontogriphus omalus* from the Burgess Shale as an early mollusc (Conway Morris 1976; Caron *et al.* 2006, 2007) and of the Ediacaran *Kimberella* as the putative earliest member of the Mollusca (Fedonkin and Waggoner 1997) leads to a HAM that differs slightly from the hypothesis just presented. These fossils indicate that the first molluscs were not small but comparatively large animals. The radula of *Kimberella* has also been interpreted as being used for grazing mats of Cyanobacteria. Also, both *Odontogriphus* and *Kimberella* lack sclerites, but these are well developed in the similar *Wiwaxia* and in the Lower Cambrian *Halkieria evangelista*, which has also been interpreted as a mollusc (Vinther and Nielsen 2005; Scheltema and Schander 2006; however, see Butterfield 2006 and Caron *et al.* 2007 for alternative interpretations, and see Todt *et al.*, Chapter 4, and Parkhaev, Chapter 3).

RELATIONSHIPS WITHIN MOLLUSCA

The following scenario is based on the cladistic studies of Salvini-Plawen and Steiner (1996) and Haszprunar (2000). The main alternative hypothesis (Figure 2.2B), expressed by Scheltema (1988, 1993, 1996), considers the Aculifera (Polyplacophora and Aplacophora, the latter comprised of Solenogastres and Caudofoveata) as

*See Wanninger *et al.*, Chapter 16, for a detailed description of a larval molluscan HAM.

monophyletic and the Aplacophora as progenetic derivation from a polyplacophoran-like ancestor.

SOLENOGASTRES (= NEOMENIOMORPHA) AND CAUDOFOVEATA (= CHAETODERMOMORPHA)

Contrary to former ideas (e.g., Salvini-Plawen 1985, 1991, 2003), the Solenogastres are considered as the earliest offshoot of the Mollusca in the scenario adopted here. They largely agree with the bauplan of the molluscan ancestor outlined previously and alone have retained the plesiomorphic condition of the midgut. The only notable deviations (apomorphies) concern the genital system, which enables internal fertilization in these hermaphroditic animals. In addition, the foot sole is reduced to a narrow rim.

There are about 250 marine species (1 to 300 mm in length) of solenogasters, which feed nearly exclusively on Cnidaria. The cnidocysts pass intact through the gut. Larger species show respiratory folds, but never the typical molluscan gills with their ciliary bands for ventilation (ctenidia), in their rudimentary mantle cavity (see Todt *et al.*, Chapter 4, for further information).

After the Solenogastres diverged from the early molluscan lineage, the originally uniform midgut differentiated into the typical molluscan organization: stomach (sorting center), midgut gland (digestion), and intestine (transport). In addition, the evolution of true ctenidia provided ventilation of the mantle cavity and also respiration in larger individuals.

Because of their assumed plesiomorphic worm habit and the lack of a shell or shell plates, caudofoveates and solenogasters are often united as "aplacophorans," "worm-molluscs," or "spicule worms." However, morphology-based cladistic analyses (Salvini-Plawen and Steiner 1996; Haszprunar 2000) have shown that Caudofoveata is an independent offshoot with about 150 marine species between 2 and 80 mm in body length. The cephalic region of Caudofoveata is modified as a burrowing organ (head shield) for digging in soft sediments, and the foot sole is entirely lost. These gonochoristic animals shed their gametes via the pericardium and renal ducts (see Todt *et al.*, Chapter 4, for further information).

It is assumed that the plesiomorphic caudofoveate radula was composed of serial, tong-like structures that became specialized as forcep-like structures (Salvini-Plawen 1981); most extant species feed on foraminiferans. The bell-like mantle cavity bears a pair of ctenidia, which are expanded into the free water above the sediment when the animal is buried.

POLYPLACOPHORA (= LORICATA)

In this scenario, after both aplacophoran taxa split off the early molluscan lineage, several "typical" molluscan characters evolved. In this new lineage, a series of dorsal shell plates were formed, possibly by fusion of the spicule-forming cells. The former body cover (cuticle with scales) became reduced to the girdle surrounding the shell plates. A change from former carnivory to herbivory or detritovory is mirrored by a true rasping tongue with hollow, cartilaginous radular bolsters combined with a subradular sense organ for testing the food, while the intestine became a long tube with several loops. The foot sole was broad and formed a muscular sucker, enabling these animals to move on hard substrates. A groove, surrounding the foot and overlain by the girdle, is assumed to be the equivalent of the posterior mantle cavity.

Most of the *ca.* 1,000 species (3 to 500 mm body size) of the Polyplacophora (chitons) inhabit hard substrates of shallow marine or intertidal habitats, with a few also found in deep water. The most characteristic autapomorphies of the class are the eight dorsally placed shell plates surrounded by the girdle. Many are capable of rolling into a ball-like shape when dislodged to protect the foot. The circumpedal mantle cavity has 10 to 88 pairs of ctenidia. The pericardial ducts are differentiated as true renal organs, and the separated gonoducts form complex egg-layers with narrow pores to hinder polyspermy by the very elongated spermatozoa (Hodgson *et al.* 1988). Special features of the Polyplacophora are the aesthetes, complex organs in the shell plates, which play a role in storage of nutrients (e.g., Reindl *et al.* 1997). They often also bear photoreceptive structures that may be differentiated to true shell eyes (see Todt *et al.*, Chapter 4, for further information).

Scheltema (1996) and Ivanov (1996) have argued that the aplacophoran taxa with the Polyplacophora form a monophyletic group: Aculifera or Amphineura. However, morphological cladistic analyses (Salvini-Plawen and Steiner 1996; Haszprunar 2000) suggest that the aculiferans are a paraphyletic assemblage, whereas Polyplacophora and the conchiferan classes form a monophyletic group, Testaria. Most recently, Giribet *et al.* (2006) provided molecular evidence for a clade "Seriata" comprising Polyplacophora and Monoplacophora. Since this result, which violates the Conchifera concept, is based on only a short piece of 28S rDNA sequence, it needs confirmation before more serious consideration.

CONCHIFERA

Based on morphology, all remaining molluscan classes can be monophyletically united as Conchifera, with a (primary) single shell and paired statocysts as important synapomorphies (Haszprunar 2000).

The ultrastructures of locomotory cilia in various groups of molluscs have been studied by Lundin and Schander (1999, 2001a, b, c, unpublished). The cilia of the Solenogastres, Caudofoveata, and Polyplacophora are of the usual metazoan type with two ciliary rootlets oriented at almost 90° to each other and without an accessory centriole. In contrast, a single ciliary root occurs in monoplacophorans, gastropods, bivalves, cephalopods, and scaphopods (Lundin and Schander 2001a, b), indicating a more derived position and an additional synapomorphy of Conchifera.

MONOPLACOPHORA (= TRYBLIDIA)

Aside from the homogeneous shell, the 25 described species (1–40 mm body length) all live in deep waters and resemble chitons in several characters. This is true particularly for the eight pairs of shell muscles, the broad, sucker-like foot, the circumpedal mantle cavity with several pairs of ctenidia, and the buccal apparatus. After a long and controversial discussion (see Haszprunar, Chapter 5, for details) the multiplication of several organs (two pairs of auricles, three to six pairs of ctenidia, three to seven pairs of excretory organs) and the extension of the esophageal glands are considered as autapomorphies of Monoplacophora. Extant Monoplacophora live on deep, cold waters and feed on detritus, and certain species have symbiotic bacteria in the epidermis of the mantle (Haszprunar *et al.* 1995).

REMAINING GROUPS

It is difficult to find a clear synapomorphy of the remaining classes. Accordingly, the position of Bivalvia is uncertain, although they already show the trend to reduce (in parallel to the remaining groups—Scaphopoda, Gastropoda, and Cephalopoda) the number of shell muscles. Protobranch bivalves (and several "archaeogastropods" as well) have a cord-like nervous system; thus, a taxon "Ganglioneura" (Lauterbach 1984) is not justified.

BIVALVIA (= LAMELLIBRANCHIA, PELECYPODA, ACEPHALA)

Among the extant Mollusca, the two shell valves, each with a complex hinge and (originally two) shell adductor muscles, are diagnostic of bivalves, which may have originated from a monoplacophoran ancestor *via* a rostroconch-like stage (Runnegar and Pojeta 1985). Extant forms show six (plesiomorphic) to three (apomorphic) pedal retractor muscles (homologous with the shell muscles of monoplacophorans and chitons), but some Cambrian and Ordovician fossil bivalves had the original eight pairs (Runnegar and Pojeta 1992). The approximately 20,000 species (1–1,000 mm in size) have entirely lost the buccal apparatus, including the radula and salivary glands. Aside from these modifications, protobranch bivalves, in particular, are remarkably primitive. The morphological changes in higher bivalves largely reflect modifications relating to their suspension feeding habits (see also Giribet, Chapter 6).

Until recently the Bivalvia were united with the Scaphopoda as Diasoma, and the extinct Rostroconchia were considered as a common stem group (e.g., Runnegar 1996). Proposed synapomorphies of the living taxa included a quite similar nervous system, the burrowing foot type, and a seemingly early twofold shell formation in scaphopods. However, recent investigations of gene expression (Wanninger and Haszprunar 2001) have shown that Scaphopoda have a nondivided shell-anlage and are more closely related to Cephalopoda and Gastropoda rather than Bivalvia (see also Lindberg and Ponder 1996; Waller 1998; Haszprunar 2000; Steiner and Dreyer 2003; and discussion of molecular sequence data later in this chapter).

The main morphological synapomorphies of the remaining classes (Scaphopoda, Gastropoda, and Cephalopoda) include the presence of distinct head retractors and a muscular hydrostat system, which enable the animal to expand appendages or parts of the body by contractions of muscles (i.e., not only by compression of body fluid). All three classes, particularly the gastropods, also trend towards asymmetry in the genital system (Haszprunar 1988; Ponder and Lindberg 1997). Two further (coupled) morphological characters supporting this grouping include an elongated dorsal axis and an inverse U-shaped gut (ano-pedal flexure).

SCAPHOPODA (= SOLENOCONCHA)

There are about 600 species (3–150 mm body size) of scaphopods, all of which are marine and live in soft sediment. The shell forms a conical tube (resembling an elephant´s tusk) with two open ends. The foot is a powerful digging organ, and the mantle cavity lacks ctenidia, with ventilation achieved by ciliary strips. Up to 80 string-like tentacles ("captaculae") are diagnostic, and they usually act in catching foraminiferans that are crushed by the radula. Only the pericardial cavity remains of the heart, and the right excretory organ serves as the route for the gametes of these gonochoristic animals (see Reynolds and Steiner, Chapter 7, for further information).

CYRTOSOMA

Morphological data suggest a sister group relationship of Gastropoda and Cephalopoda, which together constitute the Cyrtosoma (Salvini-Plawen 1980; Haszprunar 2000; Figure 2.3). This grouping is diagnosed by the free head with cerebrally innervated eyes, whereas photoreceptors in Polyplacophora or Bivalvia are laterally or viscerally supplied. In contrast to the plesiomorphic condition of a circumpedal mantle cavity in Polyplacophora, Monoplacophora, Bivalvia, and Scaphopoda, the cyrtosoman mantle cavity is reduced, and the visceral commissure runs between (rather than around) the shell muscles, enabling concentration of the ganglia to form a brain.

GASTROPODA

Without doubt the Gastropoda, with more than 100,000 extant species of snails, slugs, whelks, and limpets (i.e., about 80% of all molluscs), is the most successful taxon of the Mollusca. Their enormous variability concerns all aspects of their anatomy, physiology, ecology, and reproduction, so that an accurate generalized statement about gastropods is difficult. Basic synapomorphies of this class are correlated with larval torsion, a 180° twist of the visceral part against the cerebropedal part of the body. This results in streptoneury, a twisting of the visceral loop into a figure 8), asymmetry of nephridia and (post-torsional right-side) gonad, rotation of the anterior esophagus, and (perhaps) the presence of a spirally structured, larval operculum.

Progress in our understanding of gastropod phylogeny during the last 20 years is outlined in the chapters dealing with aspects of gastropod phylogeny (Chapters 9 through 13) in this volume.

CEPHALOPODA

The cephalopods differ from all other molluscan classes by their primarily large size (according to the fossil record, 5 cm to 6 m body length; only certain tropical octopuses are smaller). They are a popular and economically important group, with about 1,000 extant species,

all being marine. The homology of the arms to cephalic (e.g., Salvini-Plawen 1985) versus pedal (e.g., Boletzky 1988) appendages of other conchiferans and the nature of the funnale are still somewhat doubtful. Innervation is equivocal (Budelmann and Young 1987), though mainly from the cerebral ganglia. The expression pattern of *Hox* genes show a cerebral origin, since none of the respective *Hox* genes is expressed in the gastropod foot (Lee *et al.* 2003), however, a shift of gene expression cannot be excluded.

Whereas the nautiloid external shell with septa is a plesiomorphic feature for the group, its enrolled shape is one of several autapomorphic features. In particular, the duplication of several organs (auricles, ctenidia, excretory organs, osphradia), which, although somewhat similar to the monoplacophoran condition, are (by parsimony) considered apomorphic because the closest conchiferan clades (Coleoidea, Gastropoda, Scaphopoda, Bivalvia) show single pairs of those structures.

EVIDENCE FROM FOSSILS

The fossil record of Mollusca is substantial and provides deep insight into the evolutionary history of various classes. However, because of the Precambrian origin of the phylum (Fedonkin and Waggoner 1997; Caron *et al.* 2006; Parkhaev, Chapter 3; Todt *et al.*, Chapter 4) and most major subgroups, "missing links" are not to be expected. In addition, simple shells often mask highly complicated and diverse anatomy (e.g., Monoplacophora and gastropod limpets). Aspects of the significance of the fossil record in understanding the early evolution of molluscs is outlined by Parkhaev (Chapter 3).

MOLECULAR INFERENCE

As noted in the preceding section, there has been surprisingly little effort to use molecular data to elucidate mollusc origins or the relationships of major molluscan taxa. However, there has been some progress since the review by Medina and Collins (2003). Initial analyses (e.g., Winnepenninckx et al. 1996) focused primarily on the SSU gene, which is, unfortunately, limited in phylogenetic signal for this region of the tree. Steiner and Dreyer (2003) used a more robust sampling of SSU data and found support for a clade including Scaphopoda plus Cephalopoda, a group supported by some morphological characters (see previous section). The Steiner and Dreyer analysis was primarily focused on relationships within Scaphopoda and did not include solenogasters nor caudofoveates, and they rooted the tree with chitons. This work was important for being the first strong molecular evidence to show that the Diastoma (Bivalvia and Scaphopoda) hypothesis is dubious, a view shared by more recent morphological work (see previous section).

Subsequent work on scaphopods by Dreyer and Steiner (2004) focused on information in the mitochondrial genome. Unfortunately, as indicated by this and another recent paper (Boore et al. 2004), mitochondrial data appears to be very limited in its ability to resolve relationships between mollusc lineages. In fact, in these analyses, the mitochondrial data fails to recover mollusc monophyly, as was also observed in early SSU studies (e.g., Ghiselin 1988). As in the SSU studies, it is possible that as more mitochondrial genomes are sampled, it will become possible to reconstruct a reliable evolutionary history. However, it should be noted that mollusc, and especially gastropod, mitochondria show an elevated evolutionary rate compared to other lophotrochozoans. This is particularly true for mitochondrial gene order (e.g., see Dreyer and Steiner 2004; Jennings and Halanych 2005; Simison and Boore, Chapter 17).

Passamaneck et al. (2004) provided a combined SSU and LSU study of all extant recognized classes except Monoplacophora. The SSU, LSU, and combined datasets all place Scaphopoda closer (or sister to) Cephalopoda, contra the Diastoma hypothesis and in agreement with Steiner and Dreyer (2003). This finding is significantly supported by the LSU

data set. Interestingly, cephalopod ribosomal genes have undergone a substantial increase in the rate of nucleotide substitution compared to other molluscs, and they display bias in nucleotide frequency. If uncorrected, both of these features can confound phylogenetic reconstruction. In the Passamaneck et al. analysis most cephalopod taxa were removed to control for these problems.

Although the ribosomal data tended to have weak support for deep internal branches, two additional results deserve mention. First, the LSU data places Polyplacophora within bivalves. Even though this placement is very unlikely to be correct, the results are consistent enough to call for a closer inspection of the relationship between these two taxa. Second, LSU data suggest that the two "aplacophoran" taxa may form a monophyletic group that may be more derived than expected. The derived nature of these taxa was also suggested (but not supported) by Giribet et al. (2006) (see subsequent discussion). For both studies, aplacophoran taxon sampling was very limited. Clearly, the placement of these groups has important implications for our interpretation of molluscan morphological evolution. As such, the placement of "aplacophoran" taxa deserves more attention.

Figure 2.4A summarizes the maximum likelihood tree for the LSU data from Passamaneck et al. (2004). Support for internal nodes is weak, but LSU data seems to provide more robust signal than SSU. In the combined tree, the LSU signal seems to be diluted by SSU data (see also Passamaneck and Halanych 2006). Interestingly, cephalopods and bivalves are not monophyletic. This situation with molecular data has been well documented for bivalves (Winnepenninckx et al. 1996; Giribet and Wheeler 2002); in contrast this cephalopod result seems novel to this analysis. However, Passamaneck et al. (2004) found that cephalopods had a nucleotide bias causing long-branch length problems for cephalopod LSU data.

Most recently Giribet et al. (2006) applied the first multigene analysis (SSU,

LSU, H3, 16S, COI) with representatives of all molluscan classes (Figure 2.4B), including a partial LSU sequence from a monoplacophoran. Their results did not support the scaphopod-cephalopod clade previously mentioned, and they call for a clade "Serialia" consisting of Polyplacophora and Monoplacophora alone. However, the conclusion suffers from the fact that only a partial LSU sequence was available for the monoplacophoran representative (*Laevipilina antarctica*) nor did it consider whether different data partitions were supportive of different topologies (e.g., SSU versus LSU; see Passamaneck *et al.* 2004). The study did not clear up any further interrelationship of molluscan classes, and neither Gastropoda nor Bivalvia was shown as monophyletic.

GAPS IN KNOWLEDGE

Despite recent advances, we still have a limited understanding of the relationships among major molluscan lineages. However, in the next few years our understanding of molluscan origins and relationships will change considerably as new data accumulates. We will have to sample molluscan lineages more densely to overcome problems of limited phylogenetic signal from some genes. In particular, expressed sequencing tag (EST) studies and genome projects will provide much needed new information. Nevertheless, morphology—particularly new studies on mollusc ontogeny combined with evo-devo studies—will continue to play an important role concerning both phylogenetic inference and in providing an evolutionary framework. If we really want to understand molluscan phylogeny and evolution, the genotype must meet the phenotype.

ACKNOWLEDGMENTS

The authors would like to thank David R. Lindberg and Winston F. Ponder for the invitation to the symposium at the World Congress of Malacology in Perth 2004 and for their editorial work. This work was supported by the National Science Foundation grant (DEB-0075618) to KMH. This work is AU Marine Biology Program contribution #16.

REFERENCES

Bleidorn, C., Podsiadlowski, L., Bartolomaeus, T. 2006. The complete mitochondrial genome of the orbiniid polychaete *Orbinia latreillii* (Annelida, Orbiniidae)—a novel gene order for Annelida and implications for annelid phylogeny. *Gene* 370: 96–103.

Boletzky, S. v. 1988. Cephalopod development and evolutionary concepts. In *The Mollusca*. Vol. 12: *Paleontology and Neontology of Cephalopods*. Edited by M. R. Clarke and E. R. Trueman. London: Academic Press, pp. 185–202.

Boore, J. L., Medina, M., and Rosenberg, L. A. 2004. Complete sequences of the highly rearranged molluscan mitochondrial genomes of the scaphopod *Graptacme eborea* and the bivalve *Mytilus edulis*. *Molecular Biology and Evolution* 21: 1492–1503.

Boore, J. L., and Staton, J. L. 2002. The mitochondrial genome of the sipunculid *Phascolopsis gouldii* supports its association with Annelida rather than Mollusca. *Molecular Biology and Evolution* 19: 127–137.

Brusca, R. C., and Brusca, G. J. 2003. *Invertebrates*. 2nd ed. Sunderland, MA: Sinauer Associates.

Budelmann, B. U., and Young, J. Z. 1987. Brain pathways of the brachial nerves of *Sepia* and *Loligo*. *Philosophical Transaction of the Royal Society of London Series B: Biological Sciences* 315: 345–352.

Butterfield, N. J. 2006. Hooking some stem-group "worms": fossil lophotrochozoans in the Burgess Shale. *BioEssays* 28: 1161–1166.

Caron, J. B., Scheltema, A., Schander, C., and Rudkin, D. 2006. A soft-bodied mollusc with radula from the Middle Cambrian Burgess Shale. *Nature* 442: 159–163.

———. 2007. Reply to Butterfield on stem-group "worms": fossil lophotrochozoans in the Burgess Shale. *BioEssays* 29:201–202.

Chaffee, C., and Lindberg, D. R. 1986. Larval biology of early Cambrian molluscs: the implications of small body size. *Bulletin of Marine Sciences* 39: 536–549.

Conway Morris, S. 1976. A new Cambrian lophophorate from the Burgess Shale of British Columbia. *Palaeontology* 19: 199–222.

Conway Morris, S., and Peel, J. S. 1995. Articulated halkieriids from the Lower Cambrian of North Greenland and their role in early protostome evolution. *Philosophical Transactions of the Royal Society of London Series B: Biological Sciences* 347: 305–358.

Dreyer, H., and Steiner, G. 2004. The complete sequence and gene organization of the mitochondrial genome of the gadilid scaphopod *Siphonondentalium lobatum* (Mollusca). *Molecular Phylogenetics and Evolution* 31: 605–617.

Fedonkin, M. A., and Waggoner, B. M. 1997. The late Precambrian fossil *Kimberella* is a mollusc-like bilaterian organism. *Nature* 388: 868–871.

Friedrich, S., Wanninger, A., Brückner, M., and Haszprunar, G. 2002. Neurogenesis in the mossy chiton, *Mopalia muscosa* (Gould) (Polyplacophora): evidence versus molluscan metamerism. *Journal of Morphology* 253: 109–117.

Gerould, J. 1907. The development of *Phascolosoma*. Studies on the embryology of the Sipunculidae. *Zoologische Jahrbücher, Anatomie* 23: 77–162, pls. 4–11.

Ghiselin, M. T. 1988. The origin of molluscs in the light of molecular evidence. In *Oxford Surveys in Evolutionary Biology*. Vol. 5. Edited by P. H. Harvey and L. Partridge. Oxford University Press, pp. 66–95.

Giribet, G., Distel, D. L., Polz, M., Sterrer, W., and Wheeler, W. 2000. Triploblastic relationships with emphasis on the acoelomates and the position of Gnathostomulida, Cycliophora, Plathelminthes, and Chaetognatha: A combined approach of 18S rDNA sequences and morphology. *Systematic Biology* 49: 539–562.

Giribet, G., Okusu, A., Lindgren, A. R., Huff, S. W., Schrödl, M., and Nishiguchi, M. L. 2006. Evidence for a clade composed of molluscs with serially repeated structures: monoplacophorans are related to chitons. *Proceedings of the National Academy of Sciences of the United States of America* 103: 7723–7728.

Giribet, G., and Wheeler, W. 2002. On bivalve phylogeny: a high-level analysis of the Bivalvia (Mollusca) based on combined morphology and DNA sequence data. *Invertebrate Biology* 121: 271–324.

Halanych, K. M. 2004. The new view of animal phylogeny. *Annual Review of Ecology and Systematics* 35: 229–256.

Haszprunar, G. 1988. On the orgin and evolution of major gastropod groups, with special reference to the Streptoneura (Mollusca). *Journal of Molluscan Studies* 54: 367–441.

———. 1992. The first molluscs—small animals. *Bollettino di Zoologia* 59: 1–16.

———. 1996. The Mollusca: Coelomate turbellarians or mesenchymate annelids? In: *Origin and Evolutionary Radiation of the Mollusca*. Edited by J. D. Taylor. Oxford: Oxford University Press, pp. 1–28.

———. 2000. Is the Aplacophora monophyletic? A cladistic point of view. *American Malacological Bulletin* 15: 115–130.

Haszprunar, G., and Schaefer, K. 1997. Anatomy and phylogenetic significance of *Micropilina arntzi* (Mollusca, Monoplacophora, Micropilinidae fam. nov.). *Acta Zoologica (Stockholm)* 77: 315–334.

Haszprunar, G., Schaefer, K., Warén, A., and Hain, S. 1995. Bacterial symbionts in the epidermis of an Antarctic neopilinid limpet (Mollusca, Monoplacophora). *Philosophical Transactions of the Royal Society of London Series B: Biological Sciences* 347: 181–185.

Haszprunar, G., and Wanninger, A. 2007. On the fine structure of the creeping larva of *Loxosomella murmanica*: Additional evidence for a clade of Kamptozoa (Entoprocta) and Mollusca. *Acta Zoologica (Stockholm)* (in press).

Hodgson, A. N., Baxter, J. M., Sturrock, M. G., and Bernard, R. T. F. 1988. Comparative spermatology of 11 species of Polyplacophora (Mollusca) from the suborders Lepidopleurina, Chitonina and Acanthochitonina. *Proceedings of the Royal Society of London Series B: Biological Sciences* 235(1279): 161–177.

Ivanov, D. L. 1996. Origin of Aculifera and problems of monophyly of higher taxa in molluscs. In *Origin and Evolutionary Radiation of the Mollusca*. Edited by J. D. Taylor. Oxford: Oxford University Press, pp. 59–65.

Jenner, R. A. 2004. Towards a phylogeny of the Metazoa: evaluating alternative phylogenetic position of Platyhelminthes, Nemertea, and Gnathostomulida, with a critical reappraisal of cladistic characters. *Contributions in Zoology* 73: 3–163.

Jennings, R. M., and Halanych, K. M. 2005. Mitochondrial genomes of *Clymenella torquata* (Maldanidae) and *Riftia pachyptila* (Siboglinidae): evidence for conserved gene order in Annelida. *Molecular Biology and Evolution* 22: 210–222.

Lauterbach, K.-E. 1984. Das phylogenetische System der Mollusca. *Mitteilungen der Deutschen Malakozoologischen Gesellschaft* 37: 66–81.

Lee, P. N., Callaerts, P., de Couet, H. G., and Martindale, M. Q. 2003. Cephalopod *Hox* genes and the origin of morphological novelties. *Nature* 425: 1061–1065.

Lindberg, D. R., and Ghiselin, M. T. 2003. Fact, theory, and tradition in the study of molluscan origins. *Proceedings of the California Academy of Sciences* 54:663–686.

Lindberg, D. R., and W. F. Ponder. 1996. An evolutionary tree for the Mollusca: Branches or roots? In *Origin and Evolutionary Radiation of the Mollusca*. Edited by J. Taylor. Oxford: Oxford University Press, pp. 67–75.

Lundin, K., and Schander, C. 1999. Ultrastructure of gill cilia and ciliary rootlets of *Chaetoderma nitidulum* Lovén, 1845 (Mollusca, Chaetodermomorpha). *Acta Zoologica* 80: 185–191.

———. 2001a. Ciliary ultrastructure of polyplacophorans (Mollusca, Amphineura, Polyplacophora). *Journal of Submicroscopic Cytology* 33: 93–98.

———. 2001b. Ciliary ultrastructure of neomeniomorphs (Mollusca, Neomeniomorpha = Solenogastres). *Invertebrate Biology* 120: 342–349.

———. 2001c. Ciliary ultrastructure of protobranchs (Mollusca, Bivalvia). *Invertebrate Biology* 120: 350–357.

Medina, M., and Collins, A. G. 2003. The role of molecules in understanding molluscan evolution. In *Molecular Systematics and Phylogeography of Mollusks*. Edited by C. Lydeard and D. R. Lindberg.Washington, DC: Smithsonian Institution, pp. 14–44.

Nielsen, C. L., Haszprunar, G., Ruthensteiner, B., and Wanninger, A. 2007. Early development of the aplacophoran mollusk *Chaetoderma*. *Acta Zoologica (Stockholm)* 88: 231–247.

Passamaneck, Y. J., and Halanych, K. M. 2006. Lophotrochozoan phylogeny assessed with LSU and SSU data: Evidence of lophophorate polyphyly. *Molecular Phylogenetics and Evolution* 40: 20–28.

Passamaneck, Y. J., Schander, C., and Halanych, K. M. 2004. Investigation of molluscan phylogeny using large-subunit and small-subunit nuclear rRNA sequences. *Molecular Phylogenetics and Evolution* 32: 25–38.

Peterson, K. J., and Eernisse, D. J. 2001. Animal phylogeny and the ancestry of bilaterians: inferences from morphology and 18S rDNA gene sequences. *Evolution and Development* 3: 170–205.

Philippe, H., Lartillot, N., and Brinkmann, H. 2005. Multigene analyses of bilaterian animals corroborate the monophyly of Ecdysozoa, Lophotrochozoa, and Protostomia. *Molecular Biology and Evolution* 22: 1246–1253.

Ponder, W. F., and Lindberg, D. R. 1997. Towards a phylogeny of gastropod molluscs—a preliminary analysis using morphological characters. *Zoological Journal of the Linnean Society* 119: 83–265.

Reindl, S., Salvenmoser, W., and Haszprunar, G. 1997. Fine structure and immunocytochemical studies on the eyeless aesthetes of *Leptochiton algesirensis*, with comparison to *Leptochiton cancellatus* (Mollusca, Polyplacophora). *Journal of Submicroscopic Cytology and Pathology* 29: 135–151.

Runnegar, B. 1996. Early evolution of the Mollusca: the fossil record. In *Origin and Evolutionary Radiation of the Mollusca*. Edited by J. D. Taylor. Oxford: Oxford University Press, pp. 77–87.

Runnegar, B., and Pojeta, J., Jr. 1985. Origin and diversification of the Mollusca. In *The Mollusca*. Vol. 10: *Evolution*. Edited by E. R. Trueman and M. R. Clarke. London, Academic Press, pp. 1–57.

———. 1992. The earliest bivalves and their Ordovician descendants. *American Malacological Bulletin* 9: 117–122.

Ruppert, E. E., Fox, R. S., and Barnes, R. D. 2004. *Invertebrate Zoology, A Functional Evolutionary Approach*. Belmont, CA: Brooks/Cole–Thomson Learning.

Salvini-Plawen, L. v. 1980. A reconsideration of systematics in the Mollusca (phylogeny and higher classification). *Malacologia* 19: 249–278.

———. 1981. On the origin and evolution of the Mollusca. In *Origine dei grandi phyla dei Metazoi. Atti dei Convegni Lincei (Roma)* 49: 235–293.

———. 1985. Early evolution and the primitive groups. In *The Mollusca*. Vol. 10, *Evolution*. Edited by E. R. Trueman and M. R. Clarke. London: Academic Press, pp. 59–150.

———. 1991. Origin, phylogeny and classification of the phylum Mollusca. *Iberus* 9: 1–33.

———. 2003. On the phylogenetic significance of the aplacophoran Mollusca. *Iberus* 21(1): 67–97.

Salvini-Plawen, L. v., and Steiner, G. 1996. Synapomorphies and plesiomorphies in higher classification of Mollusca. In *Origin and Evolutionary Radiation of the Mollusca*. Edited by J. D. Taylor. Oxford: Oxford University Press, pp. 29–51.

Scheltema, A. H. 1988. Ancestors and descendants: Relationships of the Aplacophora and Polyplacophora. *American Malacological Bulletin* 6: 57–61.

———. 1993. Aplacophora as progenetic aculiferans and the coelomate origin of mollusks as the sister taxon of Sipuncula. *The Biological Bulletin* 184: 57–78.

———. 1996. Phylogenetic position of Sipuncula, Mollusca and the progenetic Aplacophora. In *Origin and Evolutionary Radiation of the Mollusca*. Edited by J. D. Taylor. Oxford: Oxford University Press, pp. 53–58.

Scheltema, A. H., and Schander, C. 2006. Exoskeletons: tracing molluscan evolution. *Venus* 65: 19–26.

Sørensen, M. V., Funch, P., Willerslev, E., Hansen, A. J., and Olesen, J. 2000. On the phylogeny of the Metazoa in the light of Cycliophora and Micrognathozoa. *Zoologischer Anzeiger* 239: 297–318.

Steiner, G., and Dreyer, H. 2003. Molecular phylogeny of Scaphopoda (Mollusca) inferred from 18S rDNA sequences: support for a Scaphopoda-Cephalopoda clade. *Zoologica Scripta* 32: 343–356.

Vinther, J., and Nielsen, C. 2005. The early Cambrian *Halkieria* is a mollusc. *Zoologica Scripta* 34:81–89.

Voronezhskaya, E. E., Tyurin, S. A., and Nezlin, L. P. 2002. Neuronal development in larval chiton *Ischnochiton hakodadensis* (Mollusca, Polyplacophora). *Journal of Morphology* 444: 25–38.

Waller, T. R. 1998. Origin of the molluscan class Bivalvia and a phylogeny of major groups. In *Bivalves: An Eon of Evolution*. Edited by P. A. Johnston and J. W. Haggard. Calgary: University of Calgary Press, pp. 1–45.

Wanninger, A., Fuchs, J., and Haszprunar, G. In press. The anatomy of the serotonergic nervous system of an entoproct creeping-type larva supports a mollusc-entoproct clade. *Invertebrate Biology*.

Wanninger, A., and Haszprunar, G. 2001. The role of engrailed in larval development and shell formation of the tusk-shell, *Antalis entalis* (Mollusca, Scaphopoda). *Evolution and Development* 3: 312–321.

———. 2002a. Chiton myogenesis: Perspectives for the development and evolution of larval and adult muscle systems in molluscs. *Journal of Morphology* 252: 103–113.

———. 2002b. Muscle development in *Antalis entalis* (Mollusca, Scaphopoda) and its significance for scaphopod relationships. *Journal of Morphology* 254: 53–64.

Winnepenninckx, B., Backeljau, T., and De Wachter, R. 1996. Investigation of molluscan phylogeny on the basis of 18S rRNA sequences. *Molecular Biology and Evolution* 13: 1306–1317.

Winnepenninckx, B. M. H., Backeljau, T., and Kristensen, R. M. 1998. Relations of the new phylum Cycliophora. *Nature* 393: 636–638.

Zrzavy, J., Milhulka, S., Kepka, P., Bezdek, A., and Tietz, D. F. 1998. Phylogeny of the Metazoa based on morphological and 18S ribosomal DNA evidence. *Cladistics* 14: 249–285.

3

The Early Cambrian Radiation of Mollusca

Pavel Yu. Parkhaev

The Early Cambrian interval is of great interest because of the explosive origin of abundant, diverse mineralized skeletons marking the beginning of a remarkable 500-million-year fossil record of metazoan evolution. The Lower Cambrian sequences contain representatives of most extant high-level taxa (e.g., sponges, molluscs, arthropods, brachiopods, echinoderms), as well as various extinct groups of organisms (archaeocyaths, trilobites, hyoliths, and various problematic forms). Molluscs—or, rather, the fossils that we at least believe to be the most ancient molluscs—first appeared just below the Precambrian-Cambrian boundary as a part of the mass skeletonization event (Runnegar 1982; McMenamin and McMenamin 1990; Bengtson and Conway Morris 1992; Fortey *et al.* 1996). For the most part they are found as members of diverse microfaunal assemblages, the so-called small shelly fossils (SSF; Matthews and Missarzhevsky 1975; for a review see Dzik 1994) that are commonly extracted from Cambrian rocks.

The earlier biota known from the Upper Proterozoic is the famous Ediacaran or Vendian fauna. It is composed of soft-bodied, possibly multicellular animals (Fedonkin 1987). Most of the finds are difficult to assign to any group of the younger Phanerozoic animals. However, *Kimberella quadrata* was interpreted as a primitive mollusc-like organism (Fedonkin and Waggoner 1997; Fedonkin 1998) due to the oval shape of the imprints (Figure 3.1A) with parts of some impressions suggesting a foot, a mantle edge, and a dorsal, non-mineralized "shell." The recent finds of *Kimberella* with traces of crawling (Figure 3.1B) prove the animal nature of this organism and favor its molluscan affinity (Ivantsov and Fedonkin 2001). However, to determine whether *Kimberella* is really a mollusc, specimens are needed that show a defined head with a ventral mouth. Also, more data on its shell morphology is needed. Only two Vendian fossils (i.e., tubular *Cloudina* and *Sinotubulus*) were apparently capable of biomineralization (Signor and Lipps 1992), and all others lacked hard mineralized tissues (Fedonkin 1992; Rozanov and Zhuravlev 1992: fig. 27; Ivantsov and Fedonkin 2001). This suggests that the radula, which is often a mineralized structure and a diagnostic feature of Mollusca, may not have been present or was

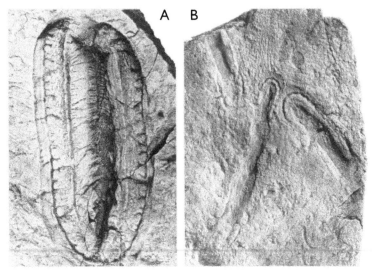

FIGURE 3.1. Soft-bodied mollusc-like fossil *Kimberella quadrata* from the Upper Vendian of the White Sea region. (A) PIN, no. 3993/5136, imprint of body, ×1.1; Winter Coast, Zimnegorsk Lighthouse, Mezen Formation. (B) PIN, no. 4853/9, 11, 12, imprints of bodies of three specimens with traces of crawling inside the sediment that buried the organisms, ×1.3; Summer Coast, Solza River, Ust-Pinega Formation. Photos courtesy of Andrey Yu. Ivantsov (Paleontological Institute, Russian Academy of Sciences, Moscow).

unmineralized in *Kimberella*. Consequently, the origin of molluscs may continue to be tied to the skeletonization event of the Nemakit-Daldynian–Tommotian age and with the development of the SSF faunas.

Chemical preparation, consisting of treating the rock with weak (usually acetic) acid, is to date the most efficient and widely used technique for the extraction of SSF. The insoluble residue contains original and secondary, silicified or phosphatized fossils. This technique has revealed numerous microscopic forms attributable to Mollusca, mostly cap-shaped or variously coiled, together with other diverse morphologies of problematic affinities.

The SSF molluscs range from 0.5 to a few millimeters. Although there have been larger specimens known from the Early Cambrian of Siberia (Valkov and Karlova 1984; Sundukov and Fedorov 1986; Dzik 1991; Gubanov and Peel 2000), Altaj, Russia (Rozanov *et al.* 1969; Missarzhevsky 1989), Mongolia (coll. E. A. Zhegallo, personal observation), and China (personal observation), these macroscopic forms are extremely rare.

Thus, most early molluscs were probably minute, rather than this distribution being simply an artifact of preservation or of the preparation technique, although why this is so, is unclear (Runnegar and Pojeta 1985; Gubanov and Peel 2000, but see Haszprunar 1992 for an opposing view). The prevalence of phosphatic internal molds in the residue is likely to be a direct consequence of the acid preparation resulting in the dissolution of the original carbonate shells, although, rarely, the shell is secondarily phosphatized or silicified (see Robison 1964; Runnegar and Jell 1976) and hence available for study.

Apart from the limpet-like and trochoid shell morphology, these fossils share with more modern molluscs characteristic shell microstructures. Thus, imprints of nacreous, prismatic, cross-lamellar, and other structures can be observed on the surface of internal molds (Runnegar 1983, 1985; Kouchinsky 1999, 2000; Parkhaev 2002b, 2006a; Feng and Sun 2003), and occasionally the original shell material is found replaced by phosphate. The shape of the shells and their microstructure are the only convincing evidence for their assignment to Mollusca, and some workers have disputed the molluscan affinities of several Early Cambrian forms (e.g., Yochelson 1975, 1978).

The most ancient probable molluscs are known from the terminal Precambrian of Siberia (uppermost Nemakit-Daldynian;[1] see Khomentovsky *et al.* 1990; Khomentovsky and Karlova 1993, 2002) and China (Lower Meishucunian) (Yu 1987; Qian and Bengtson 1989). These are the cap-shaped forms, *Purella*, *Anabarella*, and *Canopoconus*, and the spirally coiled *Latouchella*, and *Barskovia*. A range of univalved forms had evolved by the Tommotian, the basal stage of the Cambrian system (Figure 3.2). As the precise systematic position of these early forms remains controversial, a brief review of the main hypotheses follows.

Some "worm"-like organisms from the Burgess Shale (i.e., soft-bodied *Odontogriphus* and sclerite-bearing *Wiwaxia*) have been interpreted as stem-group molluscs on the basis of the interpretation of fossil structures, notably radula and ctenidia (Caron *et al.* 2006). However, interpretation of these putative structures seems doubtful, and possibly they represent other lophotrochozoans (Butterfield 2006).

A very rough estimate suggests that there are over 600 named species of Early-Middle Cambrian molluscs (personal observation). The most diverse faunas are found in China (about 250 nominal species, although over half of them are synonyms; see Parkhaev and Demidenko 2005) and Siberia (about 150 nominal species).

The Australian and Mongolian faunas include over 50 nominal species each. The rest of the species are distributed among such areas as North America, Morocco, Europe, Kazakhstan, and Iran. An estimate based on a current taxonomic revision of Chinese and Siberian material suggests that the number of valid species and genera is at least half as many as recognized in the literature, indicating an urgent need for revision of all Cambrian molluscs at the genus and species level.

BRIEF HISTORY OF THE STUDY OF ANCIENT MOLLUSCS AND THEIR SYSTEMATIC INTERPRETATION

PLACOPHORAN MOLLUSCS

The oldest fossil finds, more or less generally accepted as polyplacophoran molluscs, come from the Late Cambrian deposits (e.g., *Matthevia*, see Runnegar *et al.* 1979; or *Hemithecella* and *Elongata*, see Stinchcomb and Darrough 1995). However, a number of Early Cambrian forms have been repeatedly assigned to chitons by some authors. An especially diverse polyplacophoran fauna was reported from the Lower Cambrian (Meishucunian) of China (Yu 1987, 1989, 1990, 2001). These fossils are minute, isolated shell-like plates of different morphology. Strong doubts as to their polyplacophoran affinities have been expessed (Qian and Bengtson 1989; Runnegar 1996; Qian *et al.* 1999). These "microchitons" (e.g., *Stoliconus*, *Yangtzechiton*, *Luyanhaochiton*, *Meishucunchiton*, *Runnegarochiton*, *Tchangsichiton*) most likely represent different types of dorsal sclerites of problematic animals, similar to pairs of dorsal plates of the remarkable taxon *Halkieria evangelista* discovered from the Sirius Passet Lagerstätte of the Atdabanian of Greenland (Conway Morris and Peel 1990) and interpreted as a stem-group brachiopod (Conway Morris and Peel 1995). However, the latest comparative study of *Halkieria* and Recent chitons (Vinther and Nielsen 2005) favored a molluscan affinity of this enigmatic Cambrian fossil. The authors

---

1. The position of the Lower Cambrian boundary in the Fortune Head section, Burin Peninsula, Newfoundland, Canada (Landing 1994), cannot be reliably recognized within any other sections except the type one (Rozanov *et al.* 1997; Khomentovsky and Karlova 2005). In practice, geologists still use the regional schemes for the Precambrian-Cambrian interval (Peng *et al.* 2005) or use the Siberian stage standard (e.g., Gravestock *et al.* 2001; Khomentovsky and Karlova 2002, 2005). On the Siberian Platform, the base of the Cambrian system corresponds to the base of the Tommotian Stage (Rozanov and Sokolov 1984; Shergold *et al.* 1991; Khomentovsky and Karlova 2005).

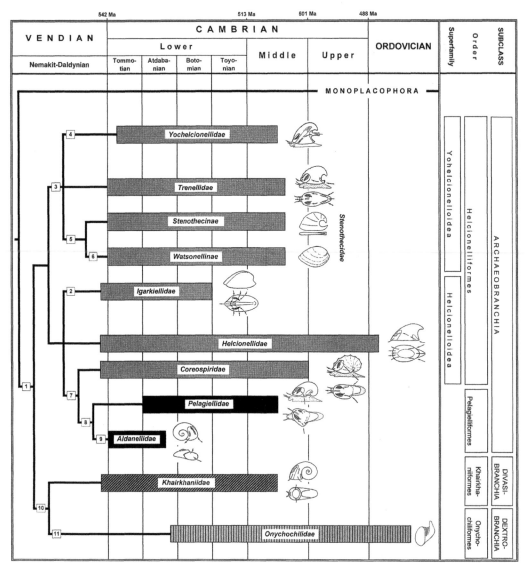

FIGURE 3.2. Phylogeny of the Cambrian univalved molluscs and ranges of the main higher taxa. Numbers indicate the origin of the following key features on each branch. 1, torsion; 2, anterior buttress; 3, posterior siphonal groove; 4, deep siphonal groove or snorkel; 5, strong lateral compression; 6, infaunal adaptations (internal plates, nonplanar aperture); 7, planispiral shell, spire whorls flattened, aperture elongated; 8, asymmetric shell; 9, turbospiral coiling with elevated spire; 10, planispiral shell, spire whorls and aperture circular, pallial cecum; 11, hyperstrophic shell (modified from Parkhaev 2002a; ages of stratigraphic boundaries from Geological Time Scale 2004). Note: Dextrobranchia and Divasibranchia are names introduced by Minichev and Starobogatov (1979) for groups of heterobranchs.

concluded that *Halkieria* may be a sister group for chitons and established a new class, Diplacophora, distinguished from Polyplacophora by the "posterior and anterior shell separated by elongate zone of scale-like sclerites, together surrounded by zones with other types of sclerites" (Vinther and Nielsen 2005: 87). While I generally support their opinion, separation of the halkieriids from polyplacophorans at a lower level (e.g., as a distinct subclass) may be advisable. By doing this, the general conception of polyplacophorans as molluscs with dorsal armoring composed of plate-like elements, varying in number, is retained. Finding a Carboniferous chiton, *Polysacos*, with 17 shell plates (Vendrasco et al. 2004), considerably extended the range

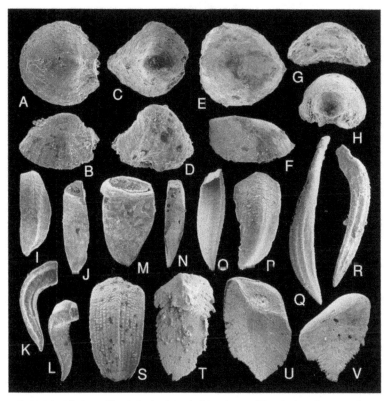

FIGURE 3.3. Sclerites of the Early Cambrian halkieriids, possible polyplacophoran molluscs. (A–H) Shell-like sclerites that can be interpreted as anterior and posterior dorsal shells of halkieriids. (I–V) blade-like sclerites of halkieriids (photographs from I through V courtesy of Yuliya E. Demidenko, Paleontological Institute of the Russian Academy of Sciences). (A, B, H) *Ocruranus finial*, Lower Cambrian, Meishucunian, Zhujiaqing Formation; Xiaowaitoushan, Meishucun, Yunnan, China: (A, B) PIN, no. 4552/1483, internal mold, ×21, Dahai Member: (A) dorsal view, (B) oblique lateral view. H, PIN, no. 4552/2123, shell, oblique apical view, ×20, Zhongyicun Member. (C, D) *Ocruranus trulliformis*, Lower Cambrian, Meishucunian, Zhujiaqing Formation, Zhongyicun Member; Xiaowaitoushan, Meishucun, Yunnan, China: C, PIN, no. 4552/2551, shell, dorsal view, ×23; D, PIN, no. 4552/2747, shell, lateral view, ×46. (E–G) *Eohalobia diandongensis*, Lower Cambrian, Meishucunian, Zhujiaqing Formation, Dahai Member; Xiaowaitoushan, Meishucun, Yunnan, China; PIN, no. 4552/1514, internal mold, ×20: E, dorsal view, F, oblique lateral view, G, oblique apical view. (I – L) *Siphogonuchites triangularis*, Lower Cambrian, Meishucunian, Zhujiaqing Formation, Zhongyicun Member; Xiaowaitoushan, Meishucun, Yunnan, China: I, PIN, no. 4552/2910, right sclerite, dorsal view, ×34; J, PIN, no. 4552/2741, left sclerite, ventral view, ×29; K, L, PIN, no. 4552/2200, left sclerite, ×38: K, lateral view, L, oblique ventral view. (M, N) *Halkieria parva*, Lower Cambrian, Botomian, Sellick Hill Formation; Sellick Hill, Fleurie Peninsula, South Australia: M, PIN, no. 4664/3104, palmate sclerite, ventral view, ×29; N, PIN, no. 4664/3176, cultrate sclerite, ventral view, ×18. (O–V) *Thambetolepis delicata*, Lower Cambrian, Botomian, Parara Limestone; Yorke Peninsula, South Australia: O, PIN, no. 4664/4230, cultrate sclerite, ventral view, ×31, bore-hole CD-2 (depth 12.56 m); P, PIN, no. 4664/4216, intermediate sclerite, between palmate and cultrate types, dorsal view, ×33, bore-hole CD-2 (depth 52.26 m); Q, PIN, no. 4664/4269, siculate sclerite, dorsal view, ×26, bore-hole CD-2 (depth 28.26 m); R, PIN, no. 4664/5338, siculate sclerite, dorsal view, ×33, bore-hole Minlaton-1 (depth 534.9 m); S, PIN, no. 4664/4006, palmate sclerite, dorsal view, ×26, bore-hole CD-2 (depth 28.82 m); T, PIN, no. 4664/4682, palmate sclerite, ventral view, ×26, bore-hole SYC-101 (depth 198.5 m); U, PIN, no. 4664/3013, palmate sclerite, ventral view, ×22, Horse Gully (HG0); V, PIN, no. 4664/3416, palmate sclerite, ventral view, ×23, Horse Gully (HG0).

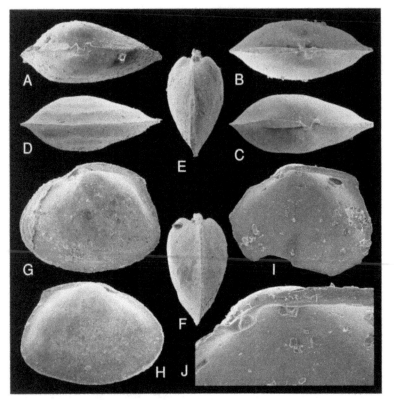

FIGURE 3.4. Early Cambrian bivalve *Pojetaia runnegari* from the Lower Cambrian, Botomian, Parara Limestone, Yorke Peninsula, South Australia. (A–C) Internal molds, views on a hinge composed of posteriorly placed ligament, two right and single left teeth: A, PIN, no. 4664/0396, ×42; B, PIN, no. 4664/0447, ×39; C, PIN, no. 4664/0473, ×35. (D) Internal mold viewed ventrally, PIN, no. 4664/0494, ×40. (E) internal mold viewed anteriorly, PIN, no. 4664/0451, ×39. (F) Internal mold viewed posteriorly, PIN, no. 4664/0450, ×33. (G) PIN, no. 4664/0417, internal mold, left lateral view, ×41. (H) PIN, no. 4664/0475, internal mold, left lateral view, ×41. (I, J) PIN, no. 4664/1613, right valve viewed interiorly: I, ×34, J, ×87.

from the normal eight. A reexamination of the Early Cambrian shell-like plates and sclerites (Figure 3.3) and reevaluation of their taxonomic affinities should provide useful information on early placophoran evolution (see Todt *et al.*, Chapter 4, for more discussion on this topic).

BIVALVED MOLLUSCS

The first bivalves, *Fordilla troyensis, F. sibirica,* and *Bulluniella borealis*, appear in the second half of the Tommotian and are only locally distributed (Barrande 1881; Pojeta *et al.* 1973; Pojeta 1975; Krasilova 1977; Jermak 1986). During the next Early Cambrian stage, the Atdabanian, the bivalve *Pojetaia runnegari* (Figure 3.4) was widely distributed geographically, being recorded from Australia, China, Mongolia, Transbaikalia, and Europe (Jell 1980; Runnegar and Bentley 1983; Gravestock *et al.* 2001; Parkhaev 2004b). Further bivalve genera (*Taurangia* MacKinnon, 1982, *Pseudomyonia* Runnegar, 1983, *Camya* Hinz-Schallreuter, 1995, and *Arhouriella* Geyer and Streng, 1998) have been described from the latest Early Cambrian and Middle Cambrian (see MacKinnon 1982, 1985; Runnegar 1983; Jermak 1986; Berg-Madsen 1987; Hinz-Schallreuter 1995; Geyer and Streng 1998). However, Early-Middle Cambrian taxa are difficult to link to the next bivalves that appear in the fossil record (in the Ordovician and later) because of a major gap in the Late Cambrian molluscan fossil record (Budd and

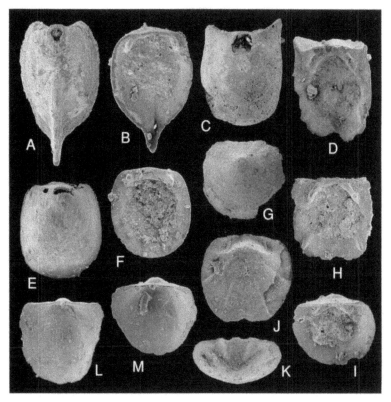

FIGURE 3.5. Early Cambrian problematic bivalved fossils. (A–K) Siphonoconchs. (L, M) *Aroonia*. A–D, *Apistoconcha siphonalis*: A, PIN, no. 4664/0019, valve of B-morphotype, exterior view, ×24; B, PIN, no. 4664/0094, valve of B-morphotype, interior view, ×43; C, PIN, no. 4664/0005, valve of A-morphotype, exterior view, ×36; D, PIN, no. 4664/0018, valve of A-morphotype, interior view, ×29; E, F, H, *Apistoconcha presiphonalis*, Lower Cambrian, Botomian, Parara Limestone; Horse Gully, Yorke Peninsula, South Australia: E, PIN, no. 4664/0065, valve of B-morphotype, exterior view, ×33; F, PIN, no. 4664/0058, valve of B-morphotype, interior view, ×47; H, PIN, no. 4664/0074, valve of A-morphotype, interior view, ×47; G, I–K, *Apistoconcha apheles*, Lower Cambrian, Botomian, Parara Limestone; Curramulka Quarry, Yorke Peninsula, South Australia: G, PIN, no. 4664/0042, valve of B-morphotype, exterior view, ×15; I, PIN, no. 4664/0053, valve of A-morphotype, interior view, ×36; J, K, PIN, no. 4664/0074, valve of B-morphotype, interior views, ×47; L, M, *Aroonia seposita* L, PIN, no. 4664/0035, tooth-valve, interior view, ×33; M, PIN, no. 4664/0036, pit-valve, interior view, ×38.

Jensen 2000). This hiatus casts some doubt on the bivalve affinity of *Fordilla* and *Pojetaia* (e.g., Yochelson 1978, 1981). Nevertheless, the presence of truly separate valves, a primitive hinge, adductor scars, and, most importantly, housing for what is assumed to be a ligament (Figure 3.4A–C), favor their bivalve nature. Hinz-Schallreuter (2000) and Pojeta (2000) recently reviewed all known bivalves reported from the Cambrian. In the latter publication, many bivalve taxa are reinterpreted as bra-chiopods or forms with doubtful affinities, but *Fordilla*, *Pojetaia*, *Taurangia*, and several others are confirmed as ancient Bivalvia.

Apart from bivalves, several Early Cambrian problematic groups, such as siphonoconchs (Figure 3.5A–K) and stenothecoids, are also charac-terized by their bivalved shells. Based on a peculiar type of bilateral symmetry (with a dorsal and a ven-tral valve) and some minor features of shell mor-phology, these groups were given class rank: the Siphonoconcha Parkhaev, 1998 and Stenothecoida

Yochelson, 1968 (= Probivalvia Aksarina, 1968) (see Aksarina 1968; Yochelson 1968, 1969; Pelman 1985; Parkhaev 1998). Their relationship to Mollusca is, however, questionable.

UNIVALVED MOLLUSCS

Following the first description of a Cambrian mollusc in 1847 (*Metoptoma? rugosa* Hall, 1847 = *Helcion subrugosus* d'Orbigny, 1850), the number of taxa had significantly increased by the end of the nineteenth century, as more publications appeared dealing with Cambrian fauna (e.g., Barrande 1867, 1881; Billings 1872; Shaler and Foerste 1888; Tate 1892; Matthew 1895, 1899). New species and genera of univalved molluscs were affiliated with extant families and higher taxa of gastropods; for example, cap-like forms were grouped together with patelloids, whereas those spirally coiled were grouped with trochids or other primitive groups. Such an approach to the systematics of Cambrian molluscs dominated studies until the middle of the twentieth century (e.g., Cobbold 1921, 1935; Cobbold and Pocock 1934; Kobayashi 1933, 1935, 1937, 1939, 1958; Resser 1938) when the situation was markedly changed because: (1) the introduction of Monoplacophora, initially as an order of fossil gastropods (Odhner in Wenz 1940:5 Wenz in Knight 1952: 46), and later as a class of molluscs following the discovery of living forms (Lemche 1957; Knight *et al.* 1960); and (2) the application of the chemical extraction of fossils from Cambrian rocks (also see previous discussion), resulting in the recognition of diverse, well-preserved molluscs (Rozanov and Missarzhevsky 1966; Rozanov *et al.* 1969).

Following the pioneer work by Rozanov and Missarzhevsky on the excellently preserved Cambrian fossils of Siberia, diverse molluscan assemblages from the Cambrian of Australia were interpreted in terms of functional morphology (Runnegar and Pojeta 1974, 1985; Runnegar and Jell 1976, 1980; Pojeta and Runnegar 1976; Runnegar 1981, 1983, 1985). Since then, the systematic position of the Cambrian univalved molluscs has been divided between three competing opinions (Table 3.1): (1) They share gastropod affinities;

(2) they share monoplacophoran affinities; (3) most of them represent a separate molluscan class.

GASTROPOD AFFINITY OF CAMBRIAN UNIVALVES

The placement of the ancient univalved molluscs in Gastropoda was accepted by almost all malacologists (e.g., Knight *et al.* 1960; Rozanov and Missarzhevsky 1966; Rozanov *et al.* 1969; but see Knight 1952[2]) before the discovery of extant untorted exogastric univalves, the monoplacophorans (Lemche 1957; Lemche and Wingstrand 1959). The limpet-like fossils *Scenella* and *Helcionella* were thought to be the oldest patelloids, whereas the spirally coiled forms *Pelagiella* and *Coreospira* were assigned to trochoids or pleurotomarioids. After the establishment of the class Monoplacophora, the cap-shaped taxa were excluded from Gastropoda (Runnegar and Pojeta 1974, 1985; Runnegar and Jell 1976; 1980; Pojeta and Runnegar 1976; Runnegar 1981, 1983; Missarzhevsky 1989). The most significant contributions during that period were the studies by Runnegar *et al.* (Runnegar and Pojeta 1974, 1985; Runnegar and Jell 1976; 1980; Pojeta and Runnegar 1976; Runnegar 1981, 1983) that supported the untorted nature of helcionelloideans[3] (see following).

Beginning from the late 1960s to early 1970s, the Russian school of malacologists undertook a morphological and taxonomic study of higher gastropod taxa, together with an evaluation of all Mollusca (Starobogatov 1970, 1976, 1977; Golikov and Starobogatov 1975; Minichev and

---

2. Knight (1952) proposed a classification of Gastropoda composed of two subclasses, that is, Isopleura (includes orders Monoplacophora, Polyplacophora, and Aplacophora) and Anisopleura (includes superorders Prosobranchia, Opisthobranchia, and Pulmonata). The Cambrian genera *Scenella* and *Helcionella* were treated in the family Triblidiidae of the order Monoplacophora, that is, they formally belong to gastropods in this classification, but were considered to be untorted forms.

3. Helcionelloideans, a group of Cambrian univalved molluscs with a bilaterally symmetric, cap-shaped, or coiled shell. In different publications its rank varies greatly, that is, from the subfamily Helcionellinae (Wenz 1938) to the class Helcionelloida (Peel 1991a, b). Here helcionelloideans are considered as the order Helcionelliformes within the gastropod subclass Archaeobranchia (Figure 3.2).

TABLE 3.1
Alternative Systematic Positions of Main Groups of Cambrian Univalved Molluscs

Helcionellids	Gastropods	Knight et al. 1960; Rozanov and Missarzhevsky 1966; Rozanov *et al.* 1969; Golikov and Starobogatov 1975; Starobogatov 1976; Minichev and Starobogatov 1979; Missarzhevsky and Mambetov 1981; Golikov and Starobogatov 1988; Parkhaev 1998, 2000, 2002a; Gravestock *et al.* 2001
	Monoplacophorans	Knight 1952; Runnegar and Pojeta 1974, 1985; Pojeta and Runnegar 1976; Runnegar and Jell 1976; 1980; Qian *et al.* 1979; Yu 1979, 1987; Runnegar 1981, 1983, 1985; Missarzhevsky 1989
	Separate molluscan class	Yochelson 1978; Peel and Yochelson 1987; Peel 1991a,b; Gubanov and Peel 2000; Yu Wen 1984, 1987; Geyer 1986, 1994
Pelagiellids	Gastropods	Rozanov and Missarzhevsky 1966; Rozanov *et al.* 1969; Runnegar and Pojeta 1974; Runnegar and Jell 1976, 1980; Qian *et al.* 1979; Yu 1979, 1987; Golikov and Starobogatov 1975; Starobogatov 1976; Minichev and Starobogatov 1979; Golikov and Starobogatov 1988; Missarzhevsky 1989; Parkhaev 1998, 2000, 2002a; Gravestock *et al.* 2001
	Advanced monoplacophorans	Runnegar 1981
	Separate molluscan class	Yochelson 1978; Linsley and Kier 1984; Geyer 1994
Aldanellids	Gastropods	Rozanov and Missarzhevsky 1966; Rozanov *et al.* 1969; Runnegar and Pojeta 1974, 1985; Golikov and Starobogatov 1975; Starobogatov 1976; Pojeta and Runnegar 1976, 1985; Runnegar and Jell 1976; 1980; Qian *et al.* 1979; Yu 1979, 1987; Runnegar 1981, 1983, 1985; Golikov and Starobogatov 1988; Missarzhevsky 1989; Parkhaev 1998, 2000, 2002a; Gravestock *et al.* 2001
	Separate molluscan class	Linsley and Kier 1984
	Non-molluscs	Yochelson 1975, 1978; Bockelie and Yochelson 1979

Starobogatov 1979; Starobogatov and Moskalev 1987). This study resulted in a new classification with Gastropoda being divided into eight subclasses (Golikov and Starobogatov 1988). The systematic position of some taxa, including several groups of Early Paleozoic molluscs, was revised, and the majority of Cambrian univalved molluscs were interpreted as gastropods. Arguments supporting helcionelloids as gastropods, however, were not entirely convincing, because these studies focused mainly on Recent taxa and barely touched the Cambrian forms.

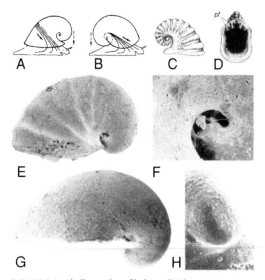

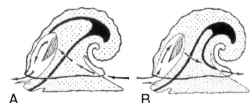

FIGURE 3.7. Two possible schemes of internal organization of the helcionelloid molluscs. (A) Torted. (B) Untorted. Arrows indicate supposed water current circulation in the mantle cavity (Parkhaev 2000).

FIGURE 3.6. Shell muscles of helcionelloids. (A, B) Position of muscle attachment area in respect to the type of shell coiling (Parkhaev 2000, modified from Starobogatov 1970): A, exogastric shell with, shell muscles attached to the peripheral wall; B, shell endogastric, with shell muscles attached to the columellar wall. (C, D) Shell of *Latouchella merino* with internal parietal folds (*pf*) supporting and dividing muscle threads (Parkhaev 2000, redrawn from photos in Peel, 1991a,b): C, lateral view, D, apertural view. (E, F) *Latouchella korobkovi*, internal mold, PIN, no. 4386/1411, Lower Cambrian, Tommotian, Kotui River, Anabar Region, Siberian Platform: E, left lateral view, ×15, F, "columellar" area with imprints of prismatic pallial myostracum, ×66. (G, H) *Anhuiconus microtuberus*, Lower Cambrian, Botomian, Yorke Peninsula, South Australia (Parkhaev 2002b), internal mold: G, PIN, no. 4664/1867, left lateral view, ×9, H, PIN, no. 4664/1738, "columellar" area with imprints of prismatic pallial myostracum, ×55.

Additional support for the hypothesis that helcionelloids are gastropods was provided by recent insights into possible functional aspects of the shell structures of Cambrian univalved molluscs (Parkhaev 2000, 2001, 2002a, b). It is possible to reconstruct details of the internal anatomy and paleoecology of Cambrian molluscs and to begin to resolve uncertainties relating to their systematics. Particular attention was given to the features determining water current circulation inside the shell (grooves, siphons, sinuses) and the patterns of shell muscle arrangement (Parkhaev 2000, 2002b, 2004a). Arguably, torsion appears to be the main apomorphy distinguishing gastropods from their monoplacophoran ancestors, but this character

cannot be directly recognized from shell morphology (Harper and Rollins 1982; Haszprunar 1988). Shell orientation in helcionelloids, therefore, can only be established from indirect evidence. Recently discovered muscle attachment scars in the columellar area of coiled helcionelloids (Parkhaev 2002b, 2004a), coupled with earlier suggestions that the internal columellar folds functioned to divide and support muscles in some helcionelloid shells (Parkhaev 2000), strongly argues for the endogastric[4] nature of these molluscs (Figure 3.6).

The position of the mantle cavity in helcionelloids is of great significance in interpreting whether or not the animal was torted. A lateral position is highly unlikely because their shells are strongly compressed laterally (Yochelson 1978). The posterior subapical region in various helcionelloids is significantly narrower than the opposite anterior side, so the most likely position of the mantle cavity is within the wider contra-apical area of the shell, that is, above the head. In such a position, the mantle cavity is unaffected by pressure exerted from the spire, while the shell muscles are housed in the posterior (parietal) side of the last whorl and do not fill a space in the area occupied by the mantle cavity (Parkhaev 2000). One can suggest two alternative reconstructions of helcionelloids corresponding to a torted (Figure 3.7A) and an untorted condition (Figure 3.7B), with water in these alternative schemes circulating in opposite

---

4. An endogastric shell has a posteriorly directed spire (Figure 3.6B), in contrast to an exogastric shell, which has an anteriorly directed spire (Figure 3.6A).

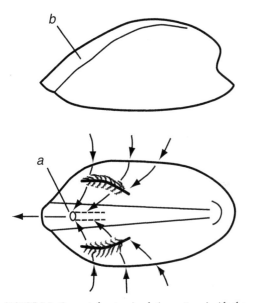

FIGURE 3.8. Suggested water circulation pattern inside the shell of the Igarkiellidae. *b* = anterior buttress; *a* = anus (after Parkhaev 2000).

directions. To either validate or disprove a particular reconstruction, parallels can be drawn from other molluscs, particularly those with special structures associated with inhalant and exhalant currents. The most striking form is *Yochelcionella*, with a tubular projection (called a "snorkel") on the subapical side of the shell (see Hinz-Schallreuter 1997 and Parkhaev 2004b for the latest revisions of the genus). The snorkel's morphology suggests that its function was inhalant (Parkhaev 2001). A similar and probably phylogenetically related form, *Eotebenna*, with a suggested semi-infaunal mode of life (Peel 1991a, b; Parkhaev 2002a), has a deep subapical sinus with a similar assumed inhalant function. Moreover, this genus morphologically links *Yochelcionella* to forms with shallow parietal sinuses or grooves. Finally, the helcionelloids with an anterior buttress (Figure 3.8) suggest the exhalation of water through the anterior sector of the aperture (Parkhaev 2000). Thus, the torted variant of helcionelloid organization (Figure 3.7A) appears to be a plausible hypothesis. In addition, the origin of asymmetry among ancient gastropods (Figure 3.9) can be explained in terms of water circulation pattern (Parkhaev 2001).

Finally, the recent discoveries (Parkhaev 2006b) of columellar muscles in turbospiral helcionelloids-aldanellids and the morphological similarity of their protoconch with larval shells of primitive modern gastropods support the position of helcionelloid molluscs within the Gastropoda.

MONOPLACOPHORAN AFFINITY OF CAMBRIAN UNIVALVES

Several Early Cambrian limpet-like forms were grouped with Monoplacophora (e.g., *Scenella* and *Helcionella*, originally assigned by Knight [1952] to an untorted gastropod family Tryblidiidae). In a subsequent publication (Knight *et al.* 1960), the monoplacophorans were treated as a separate class, and *Scenella* alone shared monoplacophoran affinities.

In the 1970s, a rich molluscan assemblage from the basal Middle Cambrian of Australia (Coonigan Formation, New South Wales, and Currant Bush Limestone, Queensland) was described (Runnegar and Pojeta 1974; Runnegar and Jell 1976, 1980). The excellent preservation of silicified shells from this locality and their taxonomic diversity provided the opportunity to study their functional morphology and to make assumptions about their ecology and systematic position. Helcionelloid molluscs with cap-like shells were thought to be monoplacophorans based on the interpretation of *Yochelcionella* with a snorkel on the subapical part of the shell by Runnegar *et al.* (Runnegar and Pojeta 1974; Runnegar and Jell 1976, 1980). These authors believed that the snorkel lay anteriorly and water flowed through it into the pallial cavity. With this interpretation, the shell would be exogastrically coiled, conforming to a monoplacophoran placement. This reconstruction of the general morphology of helcionelloids placed it at the base of the new phylogenetic scheme (Figure 3.10B) for Mollusca (Runnegar and Pojeta 1974, 1985; Runnegar and Jell 1976, 1980; Pojeta and Runnegar 1976).

Based on the study of Early Paleozoic bivalves and rostroconchs, the Mollusca were divided into two subphyla: the Cyrtosoma and Diasoma (Pojeta 1971; Pojeta and Runnegar

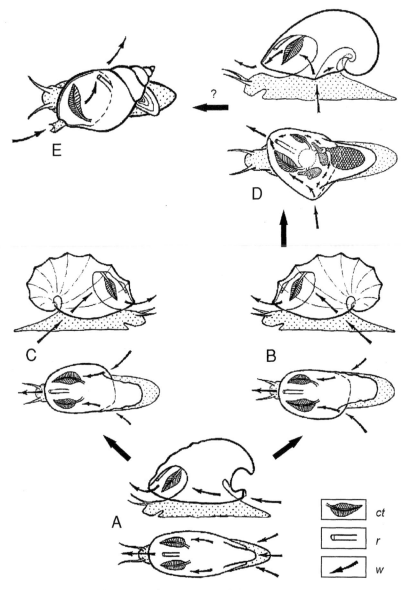

FIGURE 3.9. Suggested asymmetry formation among helcionelloids. (A) With symmetric shell and mantle cavity. (B) With slightly asymmetric (dextral) shell, mantle cavity still almost symmetric. (C) Same condition in slightly sinistral shell. (D) With considerably asymmetric (dextral shell), mantle cavity asymmetric. (E) Trochoideans and caenogastropods (*ct* = ctenidia; *r* = rectum; *w* = water currents). Condition A is proposed for most of helcionelloids; conditions B and C can be found among the Coreospiridae, and D in Pelagielliformes (see Parkhaev 2001 for detailed explanation).

1976; Runnegar 1996). In this model, the untorted, exogastric helcionelloids are regarded as ancestral to initially laterally compressed Diasoma (Rostroconchia + Bivalvia + Scaphopoda) and initially dorsoventrally elongated Cyrtosoma (Cephalopoda + Gastropoda). The origin of gastropods, with the advent of torsion, has been directly linked to the evolution of turbospiral coiling (Runnegar 1981, 1996). However, the recent discovery of columellar attachment areas of shell muscles, suggesting that coiled helcionelloid shells are endogastric (Parkhaev 2002b, 2004a), significantly weakened this hypothesis.

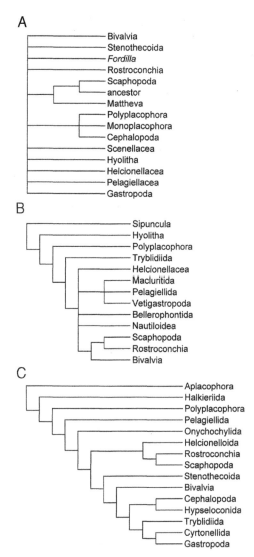

FIGURE 3.10. Alternative models of phylogenetic relationships of major mollusc taxa (from Runnegar 1996): (A) Based on Yochelson 1978. (B) Based on Runnegar and Pojeta 1974, 1975; Runnegar *et al.* 1975, 1979. (C) based on Peel 1991a.

SEPARATE CLASS-LEVEL LINEAGES

Yochelson controversially suggested that several taxa should be removed from Mollusca; in particular, the spirally coiled *Aldanella* Vostokova, 1962 was considered to be a worm tube (Yochelson 1975, 1978; Bockelie and Yochelson 1979), while the cap-shaped shells of *Scenella* or *Palaeacmaea* were interpreted as casts of a medusoid organism (Yochelson and Gil-Cid 1984; Webers and Yochelson 1999; Babcock and Robison 1988). Also, the morphology of

Yochelcionella was reinterpreted as endogastric and untorted with a posterior exhalant snorkel (Yochelson 1978). All other helcionelloids were reconstructed similarly with a posteriorly directed apex. Thus, Yochelson (1978) was the first to argue for a separate taxonomic position for helcionelloids, as embodied in his "two waves model" of molluscan evolution (see Runnegar 1996) with the first "blind" wave of Cambrian molluscan lineages and the second wave of modern classes appearing in the Ordovician.

Yochelson (1978) regarded *Pelagiella* as an asymmetric, endogastric but untorted mollusc, hence separating it from both the Gastropoda and Monoplacophora. A similar conclusion was reached later by Runnegar (1981), who argued that *Pelagiella* evolved an asymmetric shell but failed to complete torsion, and proposed it to be intermediate between Monoplacophora and Gastropoda. This line of thought was followed by Linsley and Kier (1984) in their study of the the the functional aspects of the morphology of Paleozoic hyperstrophic shells, and they also concluded that they were untorted. They removed several taxa from Gastropoda, including the Cambrian Onychochilidae, Pelagiellidae, and Aldanellidae, and established a new class, the Paragastropoda. Later, Peel further developed Yochelson's ideas (Peel and Yochelson 1987; Peel 1991a, b; Gubanov and Peel 1999, 2000, 2001) and described helcionelloid molluscs as endogastric but untorted, and separated them into a new class, the Helcionelloida (Peel 1991a, b), a move supported by some (Berg-Madsen and Peel 1987: fig. 1b; Geyer 1986, 1994). In addition, Geyer (1994) reconstructed the pelagiellids as exogastric, untorted molluscs and affiliated them into the class Amphigastropoda Simroth in Wenz (1940:9) (= Galeroconcha Salvini-Plaven, 1980, = Tergomya Peel, 1991b) along with the bellerophonts and monoplacophorans, including triblidiids.[5]

---

5. Triblidiida Lemche, 1957 is an order of the Early Paleozoic Monoplacophora with cap-shaped shell, anterior apex, and paired muscle scars.

Runnegar (1996) noted that the "viability of Peel's model depends ultimately on his interpretation of the snorkel of *Yochelcionella* and its possible homologues." Peel's (1991a, b) interpretation of this structure as exhalant was strongly criticized by Runnegar (1996) and Parkhaev (2001). The most convincing argument favoring the inhalant functioning of the snorkel comes from its morphology (Parkhaev 2001). The snorkel has a characteristic funnel-like flaring distally that is clearly visible on well-preserved specimens (Pei 1985: pl. 1, fig. 1; see also Runnegar and Jell 1976: fig. 11a-5, 1980: fig. 1; Missarzhevsky and Mambetov 1981: pl. 15, fig. 10; Bengtson *et al.* 1990: fig. 162A). Such a shape is diagnostic of "entry" structures (e.g., siphons of bivalves or funnel-shaped pores in archaeocyathian walls) because it increases the efficiency of water movement; that is, the velocity of water inside the tube is increased by the combining of velocity vectors in the inhalant funnel (Vogel 1988). In addition, the funnel-like shape of inhalant structures prevents the effect of jet contraction, which occurs in cylindric tubes (Butikov *et al.* 1989). If this interpretation is correct, the endogastric, untorted helcionelloid model is considerably weakened.

Chinese malacologists generally followed the earlier treatments of Cambrian molluscs, with coiled forms as gastropods and cap-shaped ones as monoplacophorans (e.g., Yu 1974, 1979, 1981, 1987; Luo *et al.* 1982; Xing *et al.* 1984; Qian *et al.* 1999). Yu (1984) introduced a new class-rank group, the Merismoconcha, for a few molluscan genera found in the Meishucunian of China. According to Yu, the univalved shell of these organisms suggests a metameric organization inherited from polyplacophorans, a group also identified among the vast amount of Early Cambrian Chinese SSF taxa (see previous discussion).

In summary, consideration of the hypotheses on the nature and systematic position of the Cambrian univalved molluscs suggests that the interpretation of helcionelloids as ancient gastropods (Parkhaev 2002a, b, 2004c, 2005) is the most plausible. However, comparison of

their shell morphology and reconstructed features of their internal anatomy suggests that helcionelloids differed from other gastropods. Here they are thought to be endogastric torted molluscs with a primarily symmetrical shell (cap-shaped or planispiral), having a symmetrical pallial complex with a primitive postero-anterior water circulation in the mantle cavity. Such a diagnosis justifies their placement in a separate subclass of gastropods, the Archaeobranchia (Gravestock *et al.* 2001; Parkhaev 2002a).

A number of new high-level taxa, ranging from class to order, have been introduced to accommodate Early Cambrian univalved molluscs. The concepts of those taxa are contrasted in Table 3.2 with the current concept of the Archaeobranchia. As shown in the table, no previously introduced name exactly corresponds to the Archaeobranchia, and no one has previously considered helcionelloid molluscs as the earliest gastropods.

Below is a synopsis of the families included in Archaeobranchia (Figure 3.2), with a summary of their main characteristics and place in early gastropod evolution. A phylogenetic scenario of Early Cambrian gastropod evolution follows this next section.

MAJOR GROUPS OF CAMBRIAN GASTROPODS

HELCIONELLIDAE Wenz, 1938 (Figures 3.11, 3.12). This family comprises numerous, diverse forms having the simplest morphology. The conical shell varies from low to high with a central, subcentral, or posterior apex. The aperture is simple, devoid of any grooves or deep sinuses. Posterior or anterior shallow notches occur in some advanced genera. The first members appeared in the terminal Precambrian (Nemakit-Daldynian) and were present until the Ordovician (Tremadocian). Forms with a comparatively low shell and a marginally placed apex are very similar to monoplacophorans, which could be ancestral to helcionelloids. For some genera treated here such as Helcionellidae (marked

TABLE 3.2

TABLE 3.2

Relation of the Subclass Archaeobranchia with Earlier Introduced High-Level Groups of the Early Cambrian Molluscs

HIGH-LEVEL TAXON	CONCEPT WITH REFERENCE	RELATIONSHIP WITH ARCHAEOBRANCHIA
Class Paragastropoda	Univalved untorted anisostrophically coiled molluscs (Linsley and Kier, 1984)	The families Aldanellidae and Pelagiellidae (i.e., order, Orthostrophina Linsley and Kier, 1984) considered as torted, i.e., gastropods, and assigned to the subclass Archaeobranchia as the order Pelagielliformes. The rest of the paragastropods (order Hyperstrophina Linsley and Kier, 1984) are placed in other gastropod subclasses.
Order Eomonoplacophora	Untorted exogastric molluscs (Missarzhevsky 1989)	All valid families are included in the order Helcionelliformes of the subclass Archaeobranchia, except the family Khairkhaniidae, which comprises the order Khairkhaniiformes included in the Heterobranchia.
Class Helcionelloida	Univalved untorted endogastric molluscs (Peel 1991a, b)	Peel declined to give the exact volume and hierarchical subdivision of the class, but the composition of assigned genera approximately corresponds to that of the order Helcionelliformes, subclass Archaeobranchia.
Class Rostroconchia	Bivalved molluscs with single-valved protoconch (Pojeta *et al.* 1972; Runnegar and Pojeta 1974; Pojeta and Runnegar 1976)	The genera *Watsonella* and *Eurkcapegma* are considered univalved and assigned to Archaeobranchia. Most of the remaining are Class Rostroconchia (Pojeta and Runnegar 1976), and some are Bivalvia (Starobogatov 1977).

by *), there is no information on internal or other shell morphology that distinguishes them from monoplacophorans. A thorough study of their protoconch morphology and microstructure could shed light on their affinities.

Valid genera included: *Absidaticonus, Aequiconus, Anuliconus, Asperconella, Bemella, Calyptroconus*, *Ceratoconus, Chabaktiella, Chuiliella, Codonoconus, Daedalia, Emarginoconus, Fenqiaronia, Hampilina, Helcionella, Igorella, Ilsanella, Lenoconus, Marocella*, *Miroconulus, Obtusoconus, Pararaconus, Prosinuites, Pseudoscenella*,

Randomia, Salanyella, Scenella, *Securiconus, Tannuella, Tichkaella, Truncatoconus*, *Tuoraconus,* and *Yangtzeconus.*

IGARKIELLIDAE   Parkhaev, 2001 (Figure 3.13). Overall shell morphology is very similar to the Helcionellidae, except for one peculiar apomorphic feature—a buttress running along the anterior field of a cap-like shell from the apex toward the anterior margin of the aperture and forming a groove on the inner surface of the shell (Figure 3.13). This structure is interpreted as a drainage sump in the mantle cavity, where

FIGURE 3.11. Representatives of the family Helcionellidae. (A, B) *Aegitellus placus*, Lower Cambrian, Tommotian, Zhujiaqing Formation, Zhongyicun Member; Xiaowaitoushan, Meishucun, Yunnan, China; PIN, no. 4552/1373, shell, ×30: A, left oblique view, B, dorsal view. (C, D) *Ilsanella atdabanica?*, Lower Cambrian, Tommotian; Selinde River, Uchur-Maya Region, Siberian Platform; PIN, no. 5083/0628, internal mold of immature (?) specimen, ×15: C, left oblique view, D, dorsal view. (E, F) *Truncatoconus campylurus*, Lower Cambrian, Tommotian, Zhujiaqing Formation, Dahai Member; Xiaowaitoushan, Meishucun, Yunnan, China; PIN, no. 4552/1506, internal mold, ×26: E, oblique right view, F, dorsal view. (G, H) *Bemella septata*, Lower Cambrian, Tommotian; Aldan River, Siberian Platform; PIN, no. 5083/0436, shell, ×12: G, left view, H, oblique dorsal view. (I) *Auricullina papulosa*, Lower Cambrian, Tommotian; Tiktirikteekh Creek, middle reaches of Lena River, Siberian Platform; PIN, no. 5083/0038, internal mold, dorsal view, ×32. (J) *Fenqiaronia proboscis*, Lower Cambrian, Botomian, Parara Limestone; bore-hole SYC-101 (depth 169.30 m); Yorke Peninsula, South Australia; PIN, no. 4664/1730, internal mold with shell fragments, ×13, (K, L) *Emarginoconus mirus*, Lower Cambrian, Tommotian, Zhujiaqing Formation, Zhongyicun Member; Xiaowaitoushan, Meishucun, Yunnan, China; PIN, no. 4552/1341, shell, ×34: K, dorsal view, L, oblique posterior view.

water accumulated before being exhaled with waste. The earliest members of the family are known from the Nemakit-Daldynian and basal Tommotian, and the group persisted until the Botomian or possibly later. The presence of the buttress suggests a water circulation pattern (Figure 3.8) similar to that of the Paleozoic bellerophontids, to which the igarkiellids could be ancestral. However, testing this latter assumption requires the study of material of late Middle

Cambrian–early Late Cambrian age, which has not been done to date because of an apparent gap in the fossil record.

Valid genera included: *Gonamella*, *Igarkiella*, *Mastakhella*, and *Protoconus*.

COREOSPIRIDAE Knight, 1947 (Figure 3.14). The family includes genera characterized by a planispiral shell with a compressed last whorl and an elongate aperture. It appeared at the base of the Tommotian and persisted until the Middle

FIGURE 3.12. Representatives of the family Helcionellidae. (A, B) *Igorella emeiensis*, Lower Cambrian, Tommotian, Zhujiaqing Formation, Dahai Member; Xiaowaitoushan, Meishucun, Yunnan, China specimen PIN, no. 4552/0136, internal mold, ×10: A, oblique right view, B, dorsal view. (C, D) *Igorellina monstrosa*, Lower Cambrian, Tommotian; Rassokha River, West Anabar Region, Siberian Platform; PIN, no. 5083/0053, internal mold, ×26: C, oblique apertural view, D, right view. (E, F) *Igorella maidipingensis*, Lower Cambrian, Tommotian, Zhujiaqing Formation, Dahai Member; Xiaowaitoushan, Meishucun, Yunnan, China; holotype PIN, no. 4552/0144, internal mold, ×14: E, left view; F, dorsal view.
(G, H) *Ceratoconus striatus*, Lower Cambrian, Tommotian; Kengede River, East Anabar Region, Siberian Platform; PIN, no. 5083/0091, internal mold, ×29: G, oblique posterior view, H, right view; (I) *Lenoconus sulcatus*, Lower Cambrian, Tommotian; Selinde River, Uchur-Maya Region, Siberian Platform; PIN, no. 5083/0514, internal mold, left view, ×31.
(J) *Obtusoconus honorabilis*, Lower Cambrian, Tommotian, Zhujiaqing Formation, Dahai Member; Xiaowaitoushan, Meishucun, Yunnan, China; PIN, no. 4552/1167, internal mold, left view, ×29. (K) *Anuliconus magnificus*, Lower Cambrian, Botomian, Parara Limestone; Horse Gully, Yorke Peninsula, South Australia; holotype PIN, no. 4664/0544, internal mold, right view, ×24. (L) *Obtusoconus brevis*, Lower Cambrian, Botomian, Sellick Hill Formation; Myponga Beach, Fleurieu Peninsula, South Australia; PIN, no. 4664/1337, internal mold, left view, ×29. (M) *Daedalia daedala*, Lower Cambrian, Botomian, Parara Limestone; Horse Gully, Yorke Peninsula, South Australia; holotype PIN, no. 4664/0511, internal mold, left view, ×40. (N) *Aequiconus zigzac*, Lower Cambrian, Botomian, Parara Limestone; Horse Gully, Yorke Peninsula, South Australia; holotype PIN, no. 4664/1507, internal mold, left view, ×29.

Cambrian. This group probably originated from helcionelloid ancestors, which include some forms with a strongly hooked apex and which appear to be intermediate between typical cap-shaped helcionelloids and coiled coreospirids. A slight deviation from symmetrical planispiral coiling occurs in some taxa. Such slight dextrality or sinistrality can occur simultaneously within the same genus and even species (along with normal, bilaterally symmetrical specimens). The variation in coiling in this group suggests that dextral or sinistral asymmetry of

FIGURE 3.13. Representatives of the family Igarkiellidae. (A–C) *Protoconus crestatus*, Lower Cambrian, Tommotian, Zhujiaqing Formation, Dahai Member; Xiaowaitoushan, Meishucun, Yunnan, China; PIN, no. 4552/1530, internal mold, ×26: A, dorsal view, B, left oblique view, C, oblique anterior view. (D, E) *Purella cristata*, Lower Cambrian, Tommotian; Selinde River, Uchur-Maya Region, Siberian Platform; PIN, no. 5083/0625, internal mold, ×33: D, oblique right view, E, dorsal view. (F–H) *Protoconus elegans*, Lower Cambrian, Tommotian, Zhujiaqing Formation, Dahai Member; Xiaowaitoushan, Meishucun, Yunnan, China; PIN, no. 4552/1467, internal mold, ×25: F, dorsal view, G, left oblique view, H, oblique anterior view. (I, J) *Igarkiella levis*, Lower Cambrian, Tommotian; Selinde River, Uchur-Maya Region, Siberian Platform; PIN, no. 5083/0144, internal mold, ×38: I, oblique anterior view, J, dorsal view. (K, L) *Gonamella rostrata*, Lower Cambrian, Tommotian; Fomich River West Anabar Region, Siberian Platform; PIN, no. 5083/0186, internal mold, ×35: K, oblique right view, L, dorsal view.

the body was not yet fixed (Minichev and Starobogatov, 1979).

Valid genera included: *Cambrospira, Coreospira, Kutanjia, Latouchella, Pseudoyangtzespira,* and *Tichkaella.*

TRENELLIDAE Parkhaev, 2001 (Figure 3.15). Members of the family are characterized by a shell with a distinct groove or arch developed at the posterior end of the aperture (Figure 3.15B, C, G, H), presumably to facilitate water intake (Parkhaev 2000). This apomorphic character, which separates them from their helcionelloid ancestors, appeared in taxa from the Early

Tommotian and persisted until the mid-Middle Cambrian. Further development of this structure into a tubular projection gave rise to the Yochelcionellidae.

Valid genera included: *Figurina, Horsegullia, Mackinnonia, Obscurella, Oelandia, Parailsanella, Perssuakiella, Prosinuites, Rugaeconus, Trenella, Tubatoconus,* and *Xianfengella.*

YOCHELCIONELLIDAE Runnegar and Jell, 1976 (Figure 3.16K–N). Among the Cambrian molluscs, representatives of this family have the most remarkable shell morphology. Its diagnostic feature is a tubular projection, the

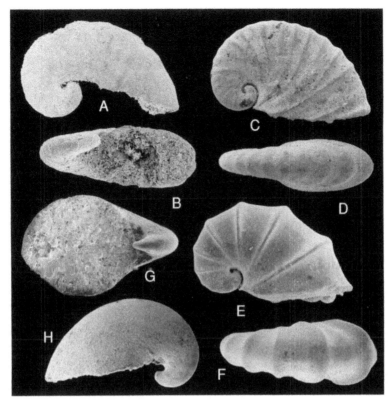

FIGURE 3.14. Representatives of the family Coreospiridae. (A, B) *Pseudoyangtzespira selindeica*, Upper Vendian, Nemakit-Daldynian; Selinde River, Uchur-Maya Region, Siberian Platform; PIN, no. 5083/0604, internal mold, ×23: A, right view, B, apertural view. (C, D) *Latouchella korobkovi*, Lower Cambrian, Tommotian, Zhujiaqing Formation; Xiaowaitoushan, Meishucun, Yunnan, China; internal molds: C, PIN, no. 4552/0147, right view, ×12, Zhongyicun Member; D, PIN, no. 4552/1554, dorsal view, ×21, Dahai Member. (E, F) *Latouchella memorabilis*, Lower Cambrian, Tommotian; West Anabar Region, Siberian Platform: E, PIN, no. 5083/0148, internal mold right view, ×17, Rassokha River; F, PIN, no. 5083/0182, internal mold, oblique dorsal view, ×24, Fomich River. (G, H) *Anhuiconus microtuberus*, Lower Cambrian, Botomian, Parara Limestone; Yorke Peninsula, South Australia; PIN, no. 4664/1867, internal mold, ×13: G, apertural view, H, left view.

snorkel, on the posterior shell surface, which is thought to be homologous to the parietal groove of the trenellids (Peel 1991a, b; Parkhaev 2001). A transitional phase from a groove to a snorkel is seen in Eotebenna, members of which possess a very deep groove with converging lower edges separated by a narrow slit. The earliest yochelcionellids are known from the Middle Tommotian, but they are most diverse from the Botomian through the early Middle Cambrian.

Valid genera included: *Eotebenna, Runnegarella, Yochelcionella,* and possibly, *Enigmaconus.*

STENOTHECIDAE Runnegar and Jell, 1980 (Figure 3.16A–J). Representatives have a trenellid-

like shell characterized by very strong lateral compression. The family first appeared in the uppermost Nemakit-Daldynian and Early Tommotian, achieved its maximum diversity in the late Atdabanian–Botomian, and persisted until the the mid-Middle Cambrian. Strong lateral compression of the shell is assumed (Parkhaev 2006c) to be an adaptation to new types of habitats—dense algal fields (subfamily Stenothecinae) and soft-sediment environments (subfamily Watsonellinae). The lateral compression is also thought to be an adaptation to an infaunal habit; this view is also favored by an increased curvature of the aperture and the appearance

FIGURE 3.15. Representatives of the family Trenellidae. (A) *Parailsanella murenica*, Lower Cambrian, Botomian, Parara Limestone; bore-hole SYC-101 (depth 205.60 m), Yorke Peninsula, South Australia; PIN, no. 4664/1656, internal mold, right view, ×33. (B, C) *Trenella bifrons*, Lower Cambrian, Botomian, Parara Limestone; Horse Gully, Yorke Peninsula, South Australia; holotype PIN, no. 4664/0665, internal mold, ×33: B, right view, C, oblique dorsal view. (D, E) *"Securiconus" costulatus*, Lower Cambrian, Tommotian; Kotui River, West Anabar Region, Siberian Platform; PIN, no. 5083/0007, internal mold, ×22: D, oblique posterior view, E, right view. (F) *Horsegullia horsegulliensis*, Lower Cambrian, Botomian, Parara Limestone; Horse Gully, Yorke Peninsula, South Australia; holotype PIN, no. 4664/1499, internal mold, left view, ×19. (G, H) *Mackinnonia rostrata*, Lower Cambrian, Botomian, Parara Limestone; Horse Gully, Yorke Peninsula, South Australia; G, PIN, no. 4664/0233, internal mold right view, ×20; H, PIN, no. 4664/0274, internal mold, posterior view, ×39. (I, J) *Leptostega hyperborea*, Lower Cambrian, Botomian; Anabar River basin, Siberian Platform; holotype PIN, no. 5083/0092, internal mold, ×33: I, left view, J, dorsal view.

of internal plates for the attachment of strong pedal musculature, suggestive of a digging foot (genera Eurekapegma and Watsonella), as supposed by some authors (MacKinnon 1985; Landing 1989; Peel 1991a, b).

Valid genera included: *Anabarella, Eurekapegma, Mellopegma, Stenotheca,* and *Watsonella.*

PELAGIELLIDAE Knight, 1952 (Figure 3.17). This family is characterized by a turbospiral, dextrally coiled shell with a somewhat triangular aperture having a drawn basal part. The basal angulation of the aperture (Figure 3.17B, D, F) is assumed to be an inhalant area homologous to the posterior groove or arch in coreospirids (Parkhaev 2001), which are thought to be the pelagiellid ancestors (Parkhaev 2002a). The first members of the family appeared in the earliest Atdabanian, were most diverse from the Late Atdabanian–Toyonian to the Middle Cambrian and persisted until the Middle Cambrian.

Valid genera included: *Costipelagiella, Pelagiella,* and *Tannuspira.*

FIGURE 3.16. Representatives of the families Stenothecidae (A–J) and Yochelcionellidae (K - N). (A, B) *Stenotheca drepanoida*, Lower Cambrian, Botomian, Parara Limestone; Yorke Peninsula, South Australia; A, PIN, no. 4664/1731, internal mold, left view, ×26, bore-hole SYC-101 (depth 168.80 m); B, PIN, no. 4664/0608, internal mold, dorsal view, ×30, Horse Gully. (C, D) *Watsonella crosbyi*, Lower Cambrian, Tommotian; Rassokha River, West Anabar Region, Siberian Platform; PIN, no. 5083/0047, internal mold, ×33: C, oblique posterior view, D, right view. (E, F) *Anabarella australis*, Lower Cambrian, Yorke Peninsula, South Australia: E, PIN, no. 4664/1780, shell, right view, ×29; Atdabanian, Kulpara Formation, bore-hole CD-2 (depth 32.66 m); F, PIN, no. 4664/0945, shell, apertural view, x54; Atdabanian, Kulpara Formation, Horse Gully. (G, H) *Mellopegma uslonica*, Lower Cambrian, Botomian, Bystraya Formation; Georgievka, East Transbaikalia; G, PIN, no. 2019/1047, internal mold, right view, ×34; H, PIN, no. 2019/1049, internal mold, dorsal view, ×31. (I, J) *Anabarella plana*, Lower Cambrian, Tommotian; Selinde River, Uchur-Maya Region, Siberian Platform; PIN, no. 5083/0634, internal mold, ×31: I, left oblique view; J, dorsal view. (K–M) *Yochelcionella crassa*, Lower Cambrian: K, PIN, no. 3302/5001, apical fragment of internal mold, right view, ×22, Botomian, Shingein-Nuruu, West Mongolia; L, PIN, no. 2019/1086, apical fragment of internal mold, oblique dorsal view, ×40; Atdabanian, Bystraya Formation, Georgievka, East Transbaikalia; M, PIN, no. 2019/1112, apical fragment of internal mold, left view, ×29; Atdabanian, Bystraya Formation, Georgievka, East Transbaikalia. (N) *Runnegarella americana*, Lower Cambrian, Newfoundland, Canada; specimen ex J.S. Peel, internal mold, left view, ×26.

ALDANELLIDAE Linsley and Kier, 1984 (Figure 3.18). The members of this family have a turbospiral, mostly dextral shell with a protruding spire and an elliptical aperture. Although this type of shell could be derived from a pelagiellid with the development of greater asymmetry through the extension of the spire, aldanellids first appear in the basal Tommotian, before the first appearance of pelagiellids, suggesting an independent origin from coreospirids. They persisted until the mid-Atdabanian.

Valid genera included: *Aldanella* and *Nomgoliella*.

FIGURE 3.17. Representatives of the family Pelagiellidae. (A, B) *Pelagiella adunca*, Lower Cambrian, Atdabanian; Chekurovka, lower reaches of Lena River, Siberian Platform; A, PIN, no. 5083/0300, internal mold, spire view, ×37; B, PIN, no. 5083/0298, internal mold, apertural view, ×35. (C, D) *Pelagiella subangulata*, Lower Cambrian, Botomian, Parara Limestone; Yorke Peninsula, South Australia: C, PIN, no. 4664/1708, internal mold, spire view, ×31, bore-hole SYC-101 (depth 194.45 m); D, PIN, no. 4664/1556, internal mold, apertural view, ×18, bore-hole SYC-101 (depth 216.35 m). (E, F) *Pelagiella madianensis*, Lower Cambrian, Botomian, Parara Limestone; Horse Gully, Yorke Peninsula, South Australia; E, PIN, no. 4664/1253, internal mold, spire view, ×20; F, PIN, no. 4664/1143, shell, apertural view, ×37. (G, H) *Tannuspira magnifica*, Lower Cambrian, Botomian, Sanashtyk-Gol Horizon, Eastern Tannu-Ola Range, Altai-Sayanian Folded Belt; holotype GIN, no. 3593/505, shell, ×3.3: G, spire view, H, apertural view.

KHAIRKHANIIDAE Missarzhevsky, 1989 (Figure 3.19 A–I). This family is characterized by a spirally coiled shell and includes both planispiral and slightly dextral and sinistral forms. Compared to the elongated cross section of the whorls in coreospirids, the whorls of khairkhaniid shells are almost circular in cross section. This family could represent another lineage derived from helcionelloids that appeared during the latest Nemakit-Daldynian–earliest Tommotian and persisted until the Middle Cambrian. If this interpretation is correct, the spirally coiled shell seen in this group has been independently derived from that of the Coreospiridae. The coiling of the relatively narrow tube into a tight spiral was accompanied by major transformations inside the mantle cavity and resulted in the origin of the pallial cecum (Parkhaev 2002a: text-fig. 4). Thus, this family composes a monotypic order Khairkhaniiformes and may be the most ancient member of the Heterobranchia (Parkhaev, in press).

Valid genera included: *Ardrossania, Barskovia, Khairkhania, Michniakia, Philoxenella, Protowenella*, and *Xinjispira*.

FIGURE 3.18. Representatives of the family Aldanellidae. (A, B) *Aldanella crassa*, Lower Cambrian, Tommotian; Kotui River, Anabar Region, Siberian Platform; A, PIN, no. 4386/1096, internal mold, spire view, ×14; B, PIN, no. 4386/1208, internal mold, apertural view, ×15. (C, D) *Aldanella attleborensis*, Lower Cambrian, Tommotian; Selinde River, Uchur-Maya Region, Siberian Platform; PIN, no. 5083/0479, internal mold, ×19: C, spire view, D, oblique apertural view. (E, F) *Aldanella utchurica*, Lower Cambrian, Tommotian; Maimakan River, Uchur-Maya Region, Siberian Platform; PIN, no. 5083/0463, internal mold, ×15: E, spire view, F, oblique apertural view. (G, H) *Aldanella operosa*, Lower Cambrian, Atdabanian; Achchagyi-Kyyry-Taas Creek, middle reaches of Lena River, Siberian Platform; G, PIN, no. 5083/0264, shell, spire view, ×31; H, PIN, no. 5083/0262, shell, apertural view, ×31. (I, J) *Aldanella rozanovi*, Lower Cambrian, Tommotian; Olenek River, Olenek Uplift, Siberian Platform; PIN, no. 5083/0460, internal mold, ×26: I, spire view, J, oblique apertural view. (K, L) *Aldanella golubevi*, Lower Cambrian, Tommotian; Kotui River, West Anabar Region, Siberian Platform; K, PIN, no. 4386/1523, internal mold, spire view, ×19; L, PIN, no. 4386/1520, internal mold, apertural view, ×16.

ONYCHOCHILIDAE Koken,1925 (Figure 3.19J, K). This family accommodates many of the Early Paleozoic taxa with hyperstrophic shells, along with a few earliest Cambrian forms. They are not included in the Archaeobranchia and are instead placed within extinct order Onychochiliformes (Figure 3.2). Its members are characterized by possessing a hyperstrophic, turbospiral shell, a mantle cavity with a pallial cecum, and they possibly retain only the right pallial organs (Starobogatov 1976; Golikov and Starobogatov 1988). This family may have originated from khairkhaniids, as suggested by a gradual transition from the almost planispiral Khairkhaniidae to the hyperstrophic Onychochilidae within the lineage *Protowenella*→*Xinjispira*→*Beshtashella*.

Main genera included: Apart from numerous Ordovician and younger genera, a single, variable genus occurs in the Early Cambrian: *Beshtashella* (= *Yuwenia*, = *Kistasella*).

FIGURE 3.19. Representatives of the families Khairkhaniidae (A–I) and Onychochilidae (J, K). (A, B) *Khairkhania rotata*, Lower Cambrian, Tommotian; Rassokha River, West Anabar Region, Siberian Platform; PIN, no. 5083/0153, internal mold with shell fragments, ×33: A, left view, B, oblique apertural view. (C, D) Khairkhaniidae gen. et sp. indet., Lower Cambrian, Tommotian; Chekurovka, lower reaches of Lena River, Siberian Platform; PIN, no. 5083/0279, internal mold, ×34: C, spire view, D, apertural view. (E, F) *Philoxenella spirallis*, Lower Cambrian, Tommotian; Rassokha River, West Anabar Region, Siberian Platform; PIN, no. 5083/0639, internal mold, ×17: E, spire view, F, oblique apertural view. (G, H) *Barskovia hemisymmetrica*, Lower Cambrian, Tommotian; Rassokha River, West Anabar Region, Siberian Platform; PIN, no. 5083/0132, internal mold, ×27: G, spire view, H, oblique apertural view. (I) *Ardrossania pavei*, Lower Cambrian, Botomian, Parara Limestone; bore-hole Cur-D1B (depth 278.35 m), Yorke Peninsula, South Australia; PIN, no. 4664/1535, internal mold with shell fragments, right view, ×29. (J, K) *Beshtashella tortilis*, Lower Cambrian, Botomian, Parara Limestone; Horse Gully, Yorke Peninsula, South Australia; J, PIN, no. 4664/1008, internal mold, dorsal view, ×39; K, PIN, no. 4664/1817, internal mold, apertural view, ×45.

THE CAMBRIAN RADIATION OF GASTROPODS

According to the interpretation presented above, gastropods appeared during the latest Precambrian (Nemakit-Daldynian) to Earliest Cambrian (Tommotian) (for an alternative interpretation see Frýda *et al.*, Chapter 10). They formed a paraphyletic group, the Archaeobranchia, which can be positioned at the base of the gastropod phylogenetic tree, and comprised eight family-level taxa arranged in two orders (Parkhaev 2002a, in press) (Figure 3.2). The position of two other families (Khairkhaniidae and Onychochilidae), are thought to be derived from the Archaeobranchia, is questionable. Most likely, they can be placed somewhere at the root of the Heterobranchia. The Archaeobranchia reached its maximum diversity during the Early to Middle Cambrian and extended into the Ordovician (Gubanov and Peel 2001). The most primitive members of this group had a simple, cap-shaped shell with a central or sub-central apex and an apertural margin devoid of any notches. The two major morphogical trends occurred at the base of the archaeobranchian radiation. The first was the development of structures enhancing the efficiency of water circulation, such as special sinuses, grooves, buttresses, or tubes. The second trend is the formation of a spirally coiled shell; this, as discussed previously, appeared independently in at least two lineages and led to the origin of coreospirids and khairkhaniids. The most advanced coiled archaeobranchians (pelagiellids and aldanellids) developed an asymmetrical, turbospiral shell.

The phylogenetic relationships of major Recent gastropod taxa have been disputed (e.g., contrast Golikov and Starobogatov 1975, 1988; Starobogatov 1970; Minichev and Starobogatov 1979; Salvini-Plawen 1980; Graham 1985; Haszprunar 1988; Bieler 1992; Ponder and Lindberg 1997), although there is now an increasingly accepted arrangement (Haszprunar 1988; Bieler 1992; Ponder and Lindberg 1997). Most of the major living taxa probably represent crown groups of different stems and possess rather advanced shell characters that are difficult to trace through the long history of gastropod evolution. In the earliest group, the archaeobranchians, a number of lineages can be identified, although these results are tentative because our knowledge of Cambrian molluscs remains very incomplete. The considerable gap in the fossil record between Cambrian and Ordovician gastropods, the latter more readily fitting within the commonly accepted groups, makes it especially difficult to establish relationships between the Cambrian and later taxa.

Possibly, the internally asymmetrical forms (Parkhaev 2000: text-fig. 3) with a turbospiral shell from the order Palegielliformes, gave rise to the Vetigastropoda. The position of the bellerophonts is of special interest. Golikov and Starobogatov (1988) included them as the most primitive order within their subclass Scutibranchia (= Vetigastropoda in part) because some possess an exhalant antero-median slit. Their supposed basal position is questionable, because shell slits in Recent gastropods can be convergent. Hence, it can be argued that the "shell slit, used alone, is of little phylogenetic importance" (Haszprunar 1988: 372). The order Bellerophontiformes is probably best kept separate from "Scutibranchia" and placed in Archaeobranchia on the basis of their planispiral shell coiling. As already suggested, the bellerophonts could arise from Igarkiellids, which have similar buttresses. Runnegar and Jell (1976) associated the Early to Middle Cambrian symmetrically coiled genus *Prowenella* with bellerophonts, allowing Dzik (1981: fig. 6) to hypothesize a relationship between bellerophonts and some of the Early Cambrian helcionelloids.

ADAPTIVE RADIATIONS

Adaptive radiation is usually considered as the acquisition of a number of new morphological, physiological, or behavioral features that allow the members of a group to occupy new ecological niches. Because the interpretation of hard-part morphology reflects only hints of some of these features, a study of adaptive radiation based on fossils is challenging and speculative. Such studies are facilitated by the presence of modern relatives of a group, or unrelated but convergent analogs, but are frustrated when there are no obvious living analogs.

The Cambrian molluscs are morphologically and phylogenetically very distant from extant groups, so hypotheses about their biology and structure are highly speculative and based on ambiguous interpretations. Consequently, such studies are virtually absent, with only a few publications that have touched upon this problem in a general way (Gubanov et al. 1999; Gubanov and Peel 1999; Kouchinsky 2001). Many specialists do not share the opinion espoused here, that most of the Early Cambrian univalved molluscs compose a phylogenetically uniform group (see

Table 3.1). However, the following discussion is based on the assumption that the helcionelloid molluscs do form a monophyletic group (Figure 3.2), with all families direct or indirect descendants of the root (paraphyletic) group, family Helcionellidae, which is, in turn, derived from monoplacophorans.

The most simple type of radula (docoglossate) is present among the most primitive Recent gastropods, the patellogastropods. It is similar in structure to the radula of chitons and monoplacophorans (i.e., stereoglossate; Golikov and Starobogatov 1988). We can suppose that archaeobranchians possessed a docoglossate radula adapted for grazing biofilms on relatively hard substrates. It is unlikely that suspension feeding or, especially, predation originated among the Early Cambrian univalve forms, as has been suggested in some publications (Burzin *et al.* 2001; Kouchinsky 2001), although these feeding types are present among more advanced gastropods (see Ponder and Lindberg 1997: table 4).

The prevalent shell form of the archaeobranchian basal family, Helcionellidae, is limpet-like, and this is considered plesiomorphic, inherited from monoplacophoran ancestors (Haszprunar 1988), although limpets have secondarily and independently evolved in various gastropod groups from coiled ancestors (Ponder and Lindberg 1997: table 6). Modern limpets inhabit a great variety of habitats, from intertidal rocks (most patellogastropods, Siphonariidae) to hard (hot-vent groups) or biogenic (Cocculinidae, Pseudococculinidae) substrates in abyssal depths or in fresh water (Acroloxidae, some Planorbidae [= Ancylidae]), where they occupy a wide range of habitats. Similarly, the Early Cambrian limpet-like molluscs may have occurred in a wide range of marine habitats. Several forms, such as large *Bemella* (Figure 3.11G, H), *Ilsanella, Tannuella*, and *Randomia*, appear to have inhabited shallow water shores, because they are sometimes associated with archaeocyath-algal bioherms (Sundukov and Fedorov 1986; Dzik 1991; Landing 1992). The smaller forms with depressed shells (Figure 3.11E, F, K, L) may have grazed

algae. Some members of the Helcionellidae have laterally compressed shells (Figure 3.12A–F, N), a possible adaptation for crevice or cave dwelling, life among dense algae, or even life on narrow stem-like algae, in much the same way as some modern Lottiidae live on seagrasses.

Some helcionellids had a highly conical shell (Figure 3.12G–L), unlike those of modern gastropods. An approximate analogous form is seen in caecids (e.g., *Caecum, Fartulum*, and *Brochina*), which live in a wide variety of shallow marine habitats including rock crevices, under stones, interstitially in gravel, or among *Zostera* fields (Golikov and Kusakin 1978; Rehder 1994).

The formation of a coiled shell may have led to increasing mobility. A compact, coiled shell not only allows a decrease in shell size while retaining the same volume, but also results in a smaller aperture, allowing a thin neck between the visceral mass and cephalopodium (Ponder and Lindberg 1997). This flexible neck and narrowed foot enabled more active behavior. The formation of spirally coiled shells among archaeobranchians probably occurred at least twice (see earlier discussion). It happened first in Coreospiridae, characterized by a large, longitudinally elongated aperture and relatively small spire (Figure 3.14). This group possibly arose from laterally compressed helcionellids, whose shell underwent gradual bending and coiling of the apex (Figure 3.12A–F). A similar, or possibly a little more active, lifestyle is suggested for coreospirids: that is, dwelling in fields of macroalgae or on their narrow stems. Possibly, the most similar modern analogs are certain (freshwater) Planorbidae usually associated with dense weed beds.

The second lineage in which spirally coiled shells formed, Khairkhaniidae, is characterized by the aperture diameter being considerably less than that of the shell (Figure 3.19A–I). This shell form probably originated from the ancestral helcionellids with highly conical shells (Figure 3.12G–I). Some are symmetric, others slightly dextral or slightly sinistral. Numerous Recent gastropods have shells of similar shape and size, such as Skeneopsidae (*Skeneopsis*) and Tornidae (*Circulus*), as well as freshwater

Valvatidae (*Valvata*) and many Planorbidae. Those forms inhabit a very wide range of habitats, ranging from filamentous algae to interstitial, and the marine taxa occur from the intertidal to the deep sea. Based on these analogies, it is not possible to predict the habitat of kairkhaniids with any certainty.

The formation of a hyperstrophic[6] shell of the family Onychochilidae (Figure 19J, K), which may have evolved from khairkhaniids, was probably accompanied by an increase in mobility, with the shell elongated along the animal's body axis. There are few species with hyperstrophic shells among Recent gastropods, but similar bulliform shells with long, narrow apertures are typical of many bulliform Opisthobranchia and some freshwater Pulmonata (e.g., Physidae, some Planorbidae). The former group includes mainly infaunal active predators as well as some herbivores, whereas the latter families inhabit various substrates and are herbivorous. Both epifaunal or infaunal lifestyles are possible for Cambrian onychochilids, and in the latter case they may have been detritus feeders.

The families Pelagiellidae and Aldanellidae developed a turbospiral shell, which probably enhanced mobility compared with its assumed ancestor Coreospiridae. Asymmetry of the shell was achieved by different transformations of shell shape: by the spire projection in aldanellids (Figure 3.18) and by the protrusion of the left (basal) part of the last whorl in pelagiellids (Figure 3.17). The posterior rotation of the spire was enabled in the ancestral Coreospiridae (see Parkhaev 2001: text-fig. 3c), resulting in the angle between the axis of coiling and the axis of the cephalopodium being less than 90° (see reconstructions in Parkhaev 2001: text-fig. 3g; Parkhaev, 2006b: text-fig. 1). Based on aldanellid shell morphometry, analogous shells among Recent gastropods are Tornidae (e.g., *Tornus*, *Pseudoliotia*, and *Pygmaeorota*), some Trochidae (*Margarites*), and Skeneidae (*Skenea*), whereas analogs for Pelagiellidae can be found among some members of the Trichotropidae

(*Lepistes*), Littorinidae (*Lacuna*), and Vanikoridae (*Vanikoro*). Considering the variety of habitats occupied by living analogs, the Cambrian forms could have occupied habitats ranging from algal thalli to crevices or caves. Species of *Aldanella* and *Pelagiella* are commonly encountered in Cambrian rocks, possibly because they had rather high population densities, analogous to some Recent *Lacuna* and *Margarites*, with population densities reaching 1,500 individuals per square meter (Skarlato 1987).

The formation of structures controlling water currents inside the pallial cavity is seen in two lineages of helcionelloids that arose from helcionellids. The first lineage, Igarkiellidae, has a groove (which looks like a buttress externally) inside the limpet-like shell, and it runs from the apex toward the anterior edge of the aperture (Figure 3.13). In spite of the obvious adaptive significance of the internal groove as a structure for the localization of the exhalant water current, and hence the increased efficiency of water circulation inside the pallial cavity (Figure 3.8), we have no data suggesting that igarkiellids developed new adaptive zones compared with their helcionellid ancestors. Judging from the low or even depressed shell, igarkiellids may have been slowly moving epifaunal dwellers, grazing biofilms from various substrates.

The shell was modified to facilitate water intake in Trenellidae and in its offshoot Yochelcionellidae. In these groups the posterior groove of some ancestral helcionellids was modified to form an arch (Figure 3.15B, D, H) or even a siphon (snorkel) (Figure 3.16K–N). In spite of those innovations, the general helcionellid shell shape was retained in trenellids and yochelcionellids, suggesting that these derived families may have remained in the same adaptive zone as their ancestors. A few members of the Yochelcionellidae (e.g., *Eotebena* and *Runnegarella*) are characterized by extreme lateral compression of the shell and a strongly arched apertural margin. These taxa may have been semi-infaunal (Peel 1991b: fig. 32), with the snorkel of *Runnegarella* (Figure 3.16N) and the deep posterior sinus of *Eotebenna* (Runnegar 1996:

---

6. See Frýda *et al.*, Chapter 10, for an alternative view.

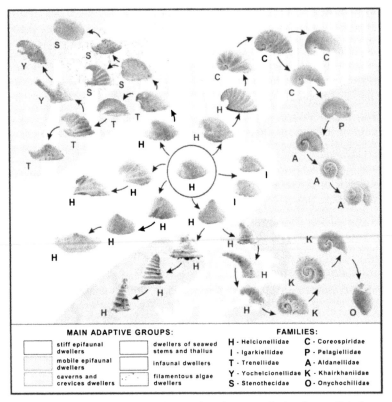

FIGURE 3.20. Hypthetical adaptive radiation of Cambrian univalved molluscs. All shells shown with anterior to the right. Arrows indicate general trends in shell morphogenesis and are not intended to indicate phylogenetic relationships (Parkhaev 2006c).

MAIN ADAPTIVE GROUPS:

stiff epifaunal dwellers

mobile epifaunal dwellers

caverns and crevices dwellers

dwellers of seawed stems and thallus

infaunal dwellers

filamentous algae dwellers

FAMILIES:

H - Helcionellidae   C - Coreospiridae
I - Igarkiellidae   P - Pelagiellidae
T - Trenellidae   A - Aldanellidae
Y - Yochelcionellidae   K - Khairkhaniidae
S - Stenothecidae   O - Onychochilidae

fig. 6.3) probably being adaptations for water intake above the sediment surface.

Members of the Stenothecidae, possibly derived from trenellids, also had laterally compressed shells. Typical stenothecids (*Stenotheca* and *Anabarella*) are characterized by an almost planar aperture (Figure 3.16A, B, E, F), whereas the members of the subfamily Watsonellinae (*Watsonella* and *Eurekapegma*) have strongly arched anterior and posterior apertural margins, possibly implying an infaunal mode of life in soft sediments. This assumption is supported by the finding of *Watsonella in situ* with the shell perpendicular to the sediment bedding (Landing 1989). The genus *Eurekapegma* has folds on the inner sides of the shell (MacKinnon 1985: fig. 6A; Runnegar 1996: fig. 6.3). which could be interpreted as additional attachment surfaces for a highly developed muscular system of a foot adapted for digging.

The morphological analogs for the Cambrian *Stenotheca* and *Anabarella* are absent from the modern malacofauna. The extent of shell compression of these fossil genera is high; that is, length/width ratio is 5–6 or even higher. A planar aperture suggests a mode of life crawling over rather firm substrate, possibly among macro or filamentous algae, similar to some highly compressed modern planorbids, many of which live in bushy macrophytes and filamentous algae.

The main trends of adaptive radiation of the Cambrian univalved molluscs surmised from the foregoing discussion are presented in Figure 3.20. Thus, the predominant ecological type at the beginning of the Cambrian were epifaunal grazers, feeding on biofilms, inhabiting mainly macroalgae, often in the shelter provided by rocks (under stones, or in caves, crevices, etc.), while the number and diversity of reef and open rock dwellers was relatively low. Infaunal or semi-infaunal taxa inhabited soft bottom environments (*Eotebenna*, *Runnegarella*, *Watsonella*, and *Eurekapegma*), and at least some of these taxa may have been detritus feeders. It should be emphasized that the majority of families, and hence, the main ecotypes, originate almost simultaneously, near the Precambrian-Cambrian boundary (Figure 3.2). This rapid radiation may possibly have resulted from new opportunities to explore a wider trophic space than previously available.

Early Cambrian bivalves (see previous discussion) are few, very small, and almost exclusively found with the valves closed (Kouchinsky 2001; personal observation), so isolated valves (Figure 3.4I, J) are the rare exceptions. These articulated finds have been interpreted as evidence for an infaunal lifestyle of *Pojetaia* and *Fordilla* (Runnegar and Bentley 1983; Ermak 1986; Kouchinsky 2001). However, absence of siphons and a weak ligament do not favor the burrowing ability of those clams. Possibly, the rapid postmortem phosphatization of the organic tissues of bivalves caused bonding of the valves and prevented them from separation. Rapid fossilization was common in the Cambrian and in some peculiar cases produced Lagerstätten with perfectly preserved tiny arthropods, known as "Orsten"-type localities (Walossek 2003).

In addition, numerous *Pojetaia runnegari* commonly co-occur with gastropods of the genera *Pelagiella* and *Anabarella* in the Late Atdabanian–Early Botomian of Australia (personal observation) and were possibly members of the same biocenosis. As suggested above, *Pelagiella* and *Anabarella* may have been epifaunal dwellers, inhabiting mainly algal substrates. Possibly *Pojetaia* and similar forms lived in dense algal mats as some minute marine and freshwater bivalves do today. Judging from the small size of Early Cambrian bivalves, they were probably deposit feeders, collecting particles with foot or labial palps. Filter-feeding, active burrowing, and the corresponding conquest of the infaunal soft bottom environment by later bivalves were a Late Cambrian–Early Ordovician phenomenon.

There are few data on the paleoecology of the Cambrian placophorans. The Late Cambrian forms described by Stinchcomb and Darrough (1995) were found in deposits with stromatolite remains. Late Cambrian *Matthevia* was also interpreted as a stromatolitic dweller (Runnegar *et al.* 1979). Articulated scleritomes of *Halkieria evangelista* were found in silts, which are assumed to accumulate in rather deep water and anoxic conditions (Conway Morris and Peel 1995). However, *Halkieria* finds may not be autochthonous and animals may have been transported from shallower environments. The numerous disarticulated sclerites of halkieriids in shallow-water assemblages of the Early Cambrian SSF suggest that they lived in those habitats, and may have had much the same lifestyle as most living polyplacophorans; grazing various types of algal and colonial animal substrates in sublittoral to intertidal habitats.

GAPS IN KNOWLEDGE

In spite of considerable progress during the last decades, our knowledge on the earliest molluscs, and hence the early evolution of the phylum, is still far from complete. The major gaps in knowledge and some directions for further studies are briefly outlined in the following paragraphs.

NEW DATA SETS

The global event of mass phosphate accumulation has a peak at the beginning of the Early Cambrian (Luvsandanzan and Rozanov 1984; Rozanov 1992), and consequently the phosphatized small shelly fauna comes mainly from that interval.

While there have been many studies on faunas from the Precambrian–Cambrian boundary and Lower Cambrian, there has been comparatively much less coverage of younger formations. To date, molluscan faunas from the latest Middle to Upper Cambrian are still poorly known, and this considerable gap is a serious impediment to our understanding of the early evolution of Mollusca. Thus, it is important that future research resolves the phylogenetic linkages

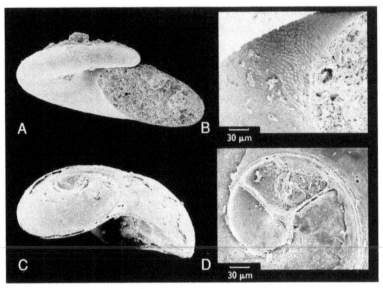

FIGURE 3.21. Recent discoveries in the morphology of ancient gastropods. (A, B) *Aldanella rozanovi*, PIN, no. 5083/1446, internal mold, Lower Cambrian, Tommotian; 0.5 km above mouth of Ary-Mas-Yuryakh Creek, Kotui River, West Anabar Region, Siberian Platform: A, apertural view, ×23, B, magnified fragment of columellar area showing polygonal microornamentation (replica from pallial myostracum). (C, D) *Aldanella operosa*, PIN, no. 5083/0264, shell, Lower Cambrian, Atdabanian; Achchagyi-Kyyry-Taas Creek, middle reaches of Lena River, Siberian Platform: C, oblique spire view, ×54; D, magnified fragment of apical area showing two septa.

between the bizarre Early Cambrian forms and the more familiar molluscs from the Ordovician and younger Paleozoic strata.

Another impotant problem is the accurate recognition of monoplacophorans among the early univalved molluscs. As it is commonly supposed that gastropods originated from untorted monoplacophoran ancestors, the remains of the latter should be present in the Cambrian fossil record. However, in practice it is very difficult to distinguish between a torted and untorted condition in Early Cambrian univalves (Harper and Rollins 1982; Parkhaev 2000a). Possibly, a study of the apical area of these fossils can assist, since the embryonic shells of monoplacophorans differ significantly from those of gastropods (Ponder and Lindberg 1997).

NEW APPROACHES

Microstructural studies of the shell and the micro-ornamentation of internal molds of ancient molluscs are probably the most promising avenues for future studies. Recent investigations in this area resulted in the discovery of muscle attachments in shells belonging to Cambrian helcionelloids (Parkhaev 2002b, 2004a, 2006b; Ushatinskaya and Parkhaev 2005). Additional studies on suitable material would certainly provide new data that would be of value in the resolution of ancient molluscan morphology and refining the existing models and reconstructions.

Studies of the protoconch and early ontogeny of ancient molluscs have been undeservedly deficient. The history of many molluscan lineages is inferred largely from analyses of early ontogeny (see Chapters 9–13), and Cambrian molluscs are arguably no exception. The latest investigations in that field (Parkhaev 2006b) discovered a protoconch of the genus *Aldanella* with a pair of septa, dividing it from the teleoconch (Figure 3.21C, D).

The presence of septation in the initial part of the shell is a common phenomenon among Recent and fossil gastropods (Yochelson 1971).

The septation maintains the watertightness of the apical part of the shell, which, being the oldest shell part, is subjected to the most prolonged corrosive effect of seawater, and being the most distal part of the shell, suffers from mechanical damage.

The presence of a protoconch and septa in the initial part of the aldanellid shell (Figure 3.21C, D) is characteristic of gastropods and rejects Yochelson's interpretation of aldanellids as sedentary polychaetes (Yochelson 1975, 1978; Bockelie and Yochelson 1979). In all 14 available specimens the formation of septa is very regular: the first septum is always 100 μm long, with a second one 150 μm long. The angle between the two septa is about 90°. The formation of the second septum occurred when the shell diameter reached 700–740 μm. The presence of columellar muscle scars in aldanellids (Figure 3.21B) and the morphological similarity of their protoconch with larval shells of primitive modern gastropods support the position of the family within one of the basal gastropod groups (Parkhaev 2006b). The discovery of additional protoconchs among the earliest univalve molluscs, and their comparative study with embryonic shells of younger fossil and modern molluscs, will be an extremely important and promising field for future researchers.

The recent consideration of *Halkieria* as a mollusc (Vinter and Nielsen 2005; see previous discussion) necessitates reconsideration of the whole group of sclerite-bearing fossils, including halkieriids and allied groups of coeloscleritophorans.

Bruce Runnegar (1996) wrote that his lifelong desire was to find a micromolluscan Lagerstätte that could provide data on the soft-body anatomy of ancient molluscs. In fact, all the known Cambrian localities of extraordinary preservation (such as Burgess Shale, Chengjiang, Sirius Passet, and Sinsk) yield representatives of various metazoan lineages (e.g., arthropods, hyoliths, brachiopods, sponges, problematic groups, and soft-bodied organisms), but, until recently, not conchiferan molluscs. The extraordinary preservation of a gastropod from the Silurian (Sutton *et al.* 2006) brings new hope to this endeavor.

It is no less necessary, and it is also the main objective of the present review, to accentuate the importance of collaborative studies in understanding the systematics and biology of the first molluscs.

ACKNOWLEDGMENTS

This paper has been supported by the Russian Foundation for Basic Research (project 03-04-48367), Grants of the President of the Russian Federation to Support Young Russian Scientists and Leading Scientific Schools (projects nos. NSh-974.2003.5, MK-723.2004.4 and MK-2836.2007.4, and the Programme of the Presidium of the Russian Academy of Sciences "Origin and Evolution of Biosphere." The author attendance on the 15th World Congress of Malacology was sponsored by the Russian Foundation for Basic Research (project 03-04-48367) and by the Museum of Paleontology, University of California, Berkeley, California, United States. I am thankful to D. R. Lindberg, W. F. Ponder, E. L. Yochelson, and another anonymous reviewer for valuable comments. I am greatly indebted to W. F. Ponder for language improvements.

REFERENCES

Aksarina, N. A. 1968. [Probivalvia—a new class of ancient molluscs]. In *Novye dannye po geologii i poleznym iskopaemum Zapadnoi Sibiri [New data on the geology and mineral products of the West Siberia]* 3: 77–86 [in Russian].

Babcock, L. E., and Robison, R. A. 1988. Taxonomy and paleobiology of some Middle Cambrian *Scenella* (Cnidaria) and Hyolithids (Mollusca) from Western North America. *Paleontological Contributions of the University of Kansas* 121: 1–22.

Barrande, J. 1867. *Système Silurien du centre de la Bohême.* ière Partie, *Recherches paléontologiques.* Vol. 3: *Classe des Mollusques, Ordre des Pteropodes.* Prague and Paris: J. Barrande and W. Waagen.

———. 1881. *Système Silurien du centre de la Bohême.* ière Partie, *Recherches paléontologiques.* Vol. 6, *Classe des Mollusques, Ordre des Acéphalés.* Prague and Paris: J. Barrande and W. Waagen.

Bengtson, S., and Conway Morris, S. 1992. Early radiation of biomineralizing phyla. In *Origin and early evolution of the Metazoa*. Edited by H. Lipps and W. Signor. New York: Plenum Press.

Bengtson, S., Conway Morris, S., Cooper, B. J., Jell, P. A., and Runnegar, B. 1990. Early Cambrian fossils from South Australia. *Association of Australasian Palaeontologists Memoirs* 9: 1–364.

Berg-Madsen, V. 1987. *Tuarangia* from Bornholm (Denmark) and similarities in Baltoscandian and Australasian late Middel Cambrian faunas. *Alcheringa* 11: 245–259.

Berg-Madsen, V., and Peel, J. S. 1987. *Yochelcionella* (Mollusca) from the late Middle Cambrian of Bornholm, Denmark. *Bulletin of the Geological Society of Denmark* 36: 259–261.

Bieler, R. 1992. Gastropod phylogeny and systematics. *Annual Review of Ecology and Systematics* 23: 311–338.

Billings, F. 1872. On some fossills from the primordial rocks of Newfoundland. *Canadian Naturalist* 6: 465–479.

Bockelie, T. G., and Yochelson, E. L. 1979. Variation in a species of «worm» from the Ordovician of Spitsbergen. *Saartrykk av Norsk Polarinsitutt* 167: 225–237.

Budd, G. E., and Jensen, S. 2000. A critical reappraisal of the fossil record of the bilaterian phyla. *Biological Reviews* 75: 253–295.

Butterfield, N. 2006. Hooking some stem-group "worms": fossil lophotrochozoans in the Burgess Shale. *BioEssays* 28: 1161–1166.

Burzin, M. B., Debrenne, F., Zhuravlev, and Yu, A. 2001. Evolution of shallow-water level bottom communities. In *The ecology of the Cambrian radiation*. Edited by A. Yu Zhuravlev and R. Riding. New York: Columbia University Press.

Butikov, E. I., Bykov, A. A., and Kondratev, A. S. 1989. Mekhanika zhidkostei [Liquid mechanics]. In *Fizika v primerakh i zadachakh [Physics in examples and exercises]*. Moscow: Nauka, pp. 143–165 [in Russian].

Caron, J. B., Scheltema, A., Schander, C., and Rudkin, D. 2006. A soft-bodied mollusc with radula from the Middle Cambrian Burgess Shale. *Nature* 442: 159–163.

Cobbold, E. S. 1921. The Cambrian Horizons of Comley (Shropshire) and their Brachiopoda, Pteropoda, Gastropoda, etc. *Quarterly Journal of the Geological Society of London* 76-4 (304): 325–386.

———. 1935. Lower Cambrian faunas from Hérault, France. *Annals and Magazine of Natural History, Series 10* 16: 25–49.

Cobbold, E. S., and Pocock, R. W. 1934. The Cambrian area of Rushton (Shropshire). *Philosophical Transactions of the Royal Society of London* (B) 223: 305–409.

Conway Morris, S., and Peel, J. S. 1990. Articulated halkieriids from the Lower Cambrian of north Greenland. *Nature* 345: 802–805.

———. 1995. Articulated halkieriids from the Lower Cambrian of North Greenland and their role in early protostome evolution. *Philosophical Transactions of the Royal Society of London* B 347: 305–358.

Dzik, J. 1981. Larval development, musculature, and relationships of *Sinuitopsis* and related Baltic bellerophonts. *Norsk Geologisk Tidskrift* 61: 111–121.

———. 1991. Is fossil evidence consistent with traditional views of the early metazoan phylogeny? In *The Early Evolution of Metazoa and the Significance of Problematic Taxa*. Edited by A. Simonetta and S. Conway Morris. Cambridge, UK: Cambridge University Press.

———. 1994. Evolution of "small shelly fossils" assemblages of the Early Paleozoic. *Acta Paleontologica Polonica* 39 (3): 247–313.

Ermak, V. V. 1986. [Early Cambrian fordillids (Bivalvia) from north of the Siberian Platform]. In Biostratigrafi ya i paleontologiya kembriya Severnoi Azii [Cambrian Biostratigraphy and Palaeontology of Northern Asia]. Edited by I. T. Zhuravleva. Trudy Instituta geologii i geofi ziki, Sibirskogo otdeleniya Akademii nauk SSS 669: 183–188 [in Russian].

Fedonkin, M. A. 1987. [Vendian non-skeletal fauna and its place in evolution of Metazoa]. *Trudy Paleontologicheskogo instituta Akademii Nauk SSSR* 226: 1–174 [in Russian].

———. 1992. Vendian Faunas and the Early Evolution of Metazoa. In *Origin and early evolution of the Metazoa*. Edited by J. H. Lipps and Ph. W. Signor. New York: Plenum Press.

———. 1998. [A second birth of *Kimberella*]. *Priroda [Nature]* 1998 (1): 3–10 [in Russian].

Fedonkin, M. A., and Waggoner, B. M. 1997. The Late Precambrian fossil *Kimberella* is a mollusc-like bilaterian organism. *Nature* 388 (28): 868–871.

Feng, W.-m., and Sun, W.-g. 2003. Phosphate replicated and replaced microstructure of molluscan shell from the earliest Cambrian of China. *Acta Paleontologica Polonica* 48 (1): 21–30.

Fortey, R. A., Briggs, D. E. G., and Wills, M. A. 1996. The Cambrian evolutionary "explosion": decoupling cladogenesis from morphological disparity. *Biological Journal of the Linnean Society* 57: 13–33.

Geyer, G. 1986. Mittelkambrische Molluscen aus Marokko und Spanien. *Senckenbergiana Lethaea* 67: 55–118.

———. 1994. Middle Cambrian molluscs from Idaho and early conchiferan evolution. Studies in stratigraphy and paleontology in honor of Donald W. Fischer. *Bulletin NY State Museum* 481: 69–86.

Geyer, G., and Streng, M. 1998. Middle Cambrian Pelecypods from the Anti-Atlas, Morocco. *Revista Española de Micropaleontolgía*. No. 0 (extr. homen. prof. Gonzalo Vidal): 83–96.

Golikov, A. N., and Kusakin, O. G. 1978. [Shelly gastropods of the intertidal zone of seas of the USSR] *Opredeliteli po faune SSSR* 116. Leningrad: Nauka [in Russian].

Golikov, A. N., and Starobogatov, Ya. I. 1975. Systematics of prosobranch gastropods. *Malacologia* 15 (1): 185–232.

———. 1988. [Questions of phylogeny and systematics of Prosobranch gastropods]. *Trudy Zoologicheskogo Instituta Akademii Nauk SSSR* 176: 4–77 [in Russian].

Graham, A. 1985. Evolution within the Gastropoda: Prosobranchia. In *The Mollusca, 10: Evolution.* Edited by E. R. Trueman and M. R. Clarke. Orlando: Academic Press:

Gravestock, D. I., Alexander, E. M., Demidenko, Yu. E., Esakova, N. V., Holmer, L. E., Jago, J. B., Lin, T.-r., Melnikova, L. M., Parkhaev, P. Yu., Rozanov, A. Yu., Ushatinskaya, G. T., Zang, W.-l., Zhegallo, E. A., and Zhuravlev, A. Yu. 2001. The Cambrian Biostratigraphy of the Stansbury Basin, South Australia. *Trudy Paleontologicheskogo Instituta Rossiiskoi Akademii Nauk* 282: 1–344.

Gubanov, A. P., Kouchinsky, A. V., and Peel, J. S. 1999. The first evolutionary-adaptive lineage within fossil mollusks. *Lethaia* 32: 155–157.

Gubanov, A. P., and Peel, J. S. 1999. *Oelandiella*, the Earliest Cambrian Helcionelloid Mollusc from Siberia. *Paleontology* 42 (2): 211–222.

———. 2000. Cambrian monoplacophoran molluscs (class Helcionelloidea). *American Malacological Bulletin* 15 (2): 139–145.

———. 2001. Latest helcionelloid molluscs from the Lower Ordovician of Kazakhstan. *Palaeontology* 44 (4): 681–694.

Harper, J. A., and Rollins, H. B. 1982. Recognition of Monoplacophora and Gastropoda in the fossil record: a functional morphological look at the bellerophont controversy. *Proceedings of the Third North American Paleontological Convention* 1: 227–232.

Haszprunar, G. 1988. On the origin and evolution of major gastropod groups, with special reference to the Streptoneura. *Journal of Molluscan Studies* 54: 367–441.

———. 1992. The first molluscs—small animals. *Bollettino di Zoologia* 59: 1–16.

Hinz-Schallreuter, I. 1995. Muscheln (Pelecypoda) aus dem Mittelkambrium von Bornholm. *Geschiebekunde aktuell* 11 (3): 71–84.

———. 1997. Einsaugstutzen oder Auspuff? Das Rästel um *Yochelcionella* (Mollusca, Kambrium). *Geschiebekunde aktuell* 13 (4): 105–140.

———. 2000. Middle Cambrian Bivalvia from Bornholm and a review of Cambrian bivalved Mollusca. *Revista Española de Micropaleontología* 32 (2): 225–242.

Ivantsov, A. Yu, and Fedonkin, M. A. 2001. [Traces of self-maintained movement—a final evidence of animal nature of Ediacaran organisms]. In *Evolyutsiya zhizni na Zemle [Life evolution on the Earth]. Materialy II Mezhdunarodnogo simpoziuma [Proceedings of the Second International Symposium],* Tomsk: 133–137 [in Russian].

Jell, P. A. 1980. Earliest known pelecypod on Earth—a new Early Cambrian genus from South Australia. *Alcheringa* 4 (3–4): 233–239.

Khomentovsky, V. V., and Karlova, G. A. 1993. Biostratigraphy of the Vendian-Cambrian beds and the lower Cambrian boundary in Siberia. *Geological Magazine* 130 (1): 29–45.

———. 2002. The Boundary between Nemakit-Daldynian and Tommotian Stages (Vendian–Cambrian) of Siberia. *Stratigraphy and Geological Correlation* 10 (3): 217–238. [Transl. from *Stratigrafiya. Geologicheskaya korrelyatsiya* 10 (3): 13–34].

———. 2005. The Tommotian Stage base as the Cambrian lower boundary in Siberia. *Stratigraphy and Geological Correlation* 13 (1): 21–34. [Transl. from *Stratigrafiya. Geologicheskaya korrelyatsiya* 13 (1): 26–40].

Khomentovsky, V. V., Valkov, A. K., and Karlova, G. A. 1990. [New data on the biostratigraphy of the transitional beds in the basin of Aldan River]. In *Pozdnii dokembrii i rannii paleozoi Sibiri: voprosy regionalnoi stratigrafii [Late Precambrian and Early Paleozoic of Siberia: questions of regional stratigraphy].* Edited by V. V. Khomentovsky and A. S. Gibsher. Novosibirsk: Nauka [in Russian].

Knight, B. J. 1952. Primitive gastropods and their bearing on gastropod classification. *Smithsonian Miscellaneous Collections* 117 (13): 1–56.

Knight, B. J., Cox, L. R., Keen, M. A., Smith, A. G., Batten, R. L., Yochelson, E. L., Ludbrook, N. H., Robertson, R., Yonge, C. M., and Moore, R. C. 1960. Mollusca 1. In *Treatise on Invertebrate Paleontology.* Lawrence, KS: University of Kansas Press and Geological Survey of America.

Kobayashi, T. 1933. Upper Cambrian of the Wuhutsui Basin, Liaotung, with special reference to the limit of the Chaumitian (or Upper Cambrian) of eastern Asia, and its subdivision. *Journal of the Faculty of Science, Imperial University of Tokyo* 11 (1–2): 55–155.

———. 1935. The Cambro-Ordovician formations and faunas of south Chosen. Palaeontology, Part 3: Cambrian faunas of south Chosen with a special study on the Cambrian trilobite genera and families. *Journal of the Faculty of Science, Imperial University of Tokyo, Section II* 4 (2): 49–344.

———. 1937. The Cambro-Ordovician shelly faunas of South America. *Journal of the Faculty of Science, Imperial University of Tokyo, Section II* 4 (4): 1–426.

———. 1939. Restudy on Lorenz's *Raphistoma broeggeri* from Shantung with a note on *Pelagiella*. *Jubilee Publication in Commemoration of Prof. H. Yabe's 60th birthday*: Sendai, Japan: Yabe kyōju kanreki kinen kai, pp. 283–288.

———. 1958. On some Cambrian gastropods from Korea. *Japanese Journal of Geology and Geography* 29 (1–3): 111–118.

Kouchinsky, A. V. 1999. Shell microstructures of the Early Cambrian *Anabarella* and *Watsonella* as new evidence on the origin of the Rostroconchia. *Lethaia* 32: 173–180.

———. 2000. Shell microstructures in Early Cambrian molluscs. *Acta Palaeontologica Polonica* 45 (2): 119–150.

———. 2001. Molluscs, hyoliths, stenothecoids and Coeloscleritophorans. In *The ecology of the Cambrian radiation*. Edited by A. Yu. Zhuravlev and R. Riding. New York: Columbia University Press.

Krasilova, I. N. 1977. [Fordillids (Bivalvia) from the Lower Paleozoic of the Siberian Platform]. *Paleontologicheskii Zhurnal* 1977 (2): 42–48 [in Russian].

Landing, E. 1989. Paleoecology and distribution of the Early Cambrian rostroconch *Watsonella crosbyi* Grabau. *Journal of Palaeontology* 63 (5): 566–573.

———. 1992. Lower Cambrian of southern Newfoundland: Epeirogeny and Lazarus faunas, lithofacies-biofacies linkages, and the myth of a global chronostratigraphy. In *Origin and Early Evolution of the Metazoa*. Edited by J. H. Lipps and P. W. Signor. New York: Columbia University Press.

———. 1994. Precambrian-Cambrian boundary global stratotype ratified and a new perspective of Cambrian time. *Geology* 22: 179–182.

Lemche, H. 1957. A new living deep-sea mollusc of the Cambro-Devonian class Monoplacophora. *Nature* 179: 413–416.

Lemche, H., and Wingstrand, K. G. 1959. The anatomy of *Neopilina galatheae* Lemche, 1957 (Mollusca, Tryblidiacea). *Galathea report* 3: 9–57. Copenhagen: Danish Science Press.

Linsley, R. M., and Kier, W. M. 1984. The Paragastropoda: a proposal for a new class of Paleozoic Mollusca. *Malacologia* 25 (1): 241–254.

Luo, H.-l., Jiang, Z.-w., Wu, X., Song, X.-l., Ouyang, L., et al. 1982. [*The Sinian-Cambrian Boundary in Eastern Yunnan, China*]. Yunnan Publishing House, Kunming [in Chinese].

Luvsandanzan, B., and Rozanov, A.Yu. 1984. [On the age of ancient phosphorites of Asia] *Doklady Akademii Nauk SSSR. Geologiya [Transactions of the Academy of Sciences of the USSR. Geology]* 277 (1): 164–167.

MacKinnon, D. I. 1982. *Taurangia paparua* n. gen. and n. sp., a late Middle Cambrian pelecypod from New Zealand. *Journal of Paleontology* 56 (3): 589–598.

———. 1985. New Zealand late Middle Cambrian molluscs and the origin of Rostroconchia and Bivalvia. *Alcheringa* 9 (1–2): 65–81.

Matthew, G. F. 1895. Notice of a new genus of Pteropods from the Saint John Group (Cambrian). *American Journal of Sciences and Arts, Series 3* 25 (178): 105–111.

———. 1899. The Etcheminian fauna of Smith Sound, Newfoundland. *Transactions of the Royal Society of Canada, Section 4* 5: 97–123.

Matthews, S. C., and Missarzhevsky, V. V. 1975. Small shelly fossils of late Precambrian and early Cambrian age: a review of recent work. *Journal of the Geological Society* 131: 289–304.

McMenamin, M. A. S., and McMenamin, D. L. S. 1990. *The Emergence of Animals: the Cambrian Breakthrough*. New York: Columbia University Press.

Minichev, Yu. S., and Starobogatov, Ya. I. 1979. [Subclasses of gastropods and their phylogenetic relationships]. *Zoologicheskii Zhurnal* 58 (3): 293–305 [in Russian].

Missarzhevsky, V. V. 1989. [The Oldest Skeletal Fossils and Stratigraphy of the Precambrian and Cambrian Boundary Strata]. *Trudy Geologicheskogo instituta Akademii Nauk SSSR* 443: 1–237 [in Russian].

Missarzhevsky, V. V., and Mambetov, A. M. 1981. [Stratigraphy and Fauna of the Cambrian and Precambrian Boundary Beds in Malyi Karatau]. *Trudy Geologicheskogo instituta Akademii Nauk SSSR* 326: 1–91 [in Russian].

Parkhaev, P. Yu. 1998. Siphonoconcha—a new class of Early Cambrian bivalved organisms. *Paleontological Journal* 32 (1): 1–15 [Transl. from *Paleontologicheskii Zhurnal* 1998 (1): 3–16].

———. 2000. The functional morphology of the Cambrian univalved molluscs—helcionellids. 1. *Paleontological Journal* 34 (4): 392–399 [Transl. from *Paleontologicheskii Zhurnal* 2000 (4): 32–39].

———. 2001. The functional morphology of the Cambrian univalved molluscs—helcionellids. 2. *Paleontological Journal* 35 (5): 470–475 [Transl. from *Paleontologicheskii Zhurnal* 2001 (5): 20–26].

———. 2002a. Phylogenesis and the System of the Cambrian univalved molluscs. *Paleontological Journal* 36 (1): 25–36 [Translated from *Paleontologicheskii Zhurnal* 2002 (1): 27–39].

———. 2002b. Muscle scars of the Cambrian univalved molluscs and their significance for systematics. *Paleontological Journal* 36 (5): 453–459 [Translated from *Paleontologicheskii Zhurnal* 2002 (5): 15–19].

———. 2004a. New data on the morphology of shell muscles in Cambrian helcionelloid molluscs. *Paleontological Journal* 38 (3): 254–256 [Translated from *Paleontologicheskii Zhurnal* 2004 (3): 27–29].

———. 2004b. Malacofauna of the Lower Cambrian Bystraya Formation of Eastern Transbaikalia. *Paleontological Journal* 38 (6): 590–608 [Translated from *Paleontologicheskii Zhurnal* 2004 (3): 9–25].

———. 2004c. The earliest stage of gastropod evolution—a Cambrian basement. *Abstract volume of the 32nd International Geological Congress, August 20–28, 2004, Florence, Italy*. Pt 1, Abstract. Number. 171-12: 804.

———. 2006a. On the genus *Auricullina* Vassiljeva, 1998 and the shell pores of the Cambrian helcionelloid molluscs. *Paleontological Journal* 40 (1): 20–33 [Translated from *Paleontologicheskii Zhurnal* 2006 (1): 20–32].

———. 2006b. New data on the morphology of ancient gastropods of the genus *Aldanella* Vostokova, 1962 (Archaeobranchia, Pelagielliformes). *Paleontological Journal* 40 (3): 244–252 [Translated from *Paleontologicheskii Zhurnal* 2006 (3): 15–21].

———. 2006c. Adaptive radiation of the Cambrian helcionelloid molluscs. In *Evolution of Biosphere and Biodiversity. On the 70th anniversary of A.Yu. Rozanov*. Moscow, KMK: pp. 282–296 [in Russian].

Parkhaev, P. Yu., and Demidenko, Yu. E. 2005. Taxonomy of the Cambrian molluscs from China. *Acta Micropalaeontologica Sinica* 22 Suppl.: 139–140.

Peel, J. S. 1991a. Functional morphology of the Class Helcionelloida nov., and the early evolution of the Mollusca. In *The Early Evolution of Metazoa and the Significance of Problematic Taxa*. Edited by A. Simonetta and S. Conway Morris. Cambridge, UK: Cambridge University Press.

———. 1991b. Functional morphology, evolution and systematics of Early Palaeozoic univalved molluscs. *Bulletin of the Grønlands Geologiske Undersøgelse* 161: 1–116.

Peel, J.S., and Yochelson, E. L. 1987. New information on *Oelandia* (Mollusca) from the Middle Cambrian of Sweden. *Bulletin of the Geological Society of Denmark* 36: 263–273.

Pei, F. 1985. [First discovery of *Yochelcionella* from the Lower Cambrian in China and its significance].

Acta Micropalaeontologica Sinica 2 (4): 395–400 [in Chinese].

Pelman, Yu. L. 1985. [New stenothecoids from the Lower Cambrian of Western Mongolia]. In *Problematiki pozdnego dokembriya i paleozoya [Problematic organisms of the Late Precambrian and Early Paleozoic]. Trudy Instituta Geologii I Geofiziki Sibirskogo otdeleniya Akademii Nauk SSSR* 632: 103–114.

Peng, S., Babcock, L. E., and Zhu, M. (eds.). 2005. *Cambrian System of China and Korea*. Hefei, Anhui Province, China: University of Science and Technology of China Press.

Pojeta, J. 1971. Review of Ordovician pelecypods. *United States Geological Survey Professional Paper* 695: 1–46.

———. 1975. *Fordilla troyensis* Barrande and early pelecypod phylogeny. *Bulletins of American Paleontology* 67 (287): 363–385.

———. 2000. Cambrian Pelecypoda (Mollusca). *American Malacological Bulletin* 15 (2): 157–166.

Pojeta, J., and Runnegar, B. 1976. The paleontology of rostroconch molluscs and the early history of the phylum Mollusca. *United States Geological Survey Professional Paper* 968: 1–88.

Pojeta, J., Runnegar, B., Morris, N.J., and Newell, N. D. 1972. Rostroconchia: a new class of bivalved mollusks. *Science* 177: 264–267.

Pojeta, J., Runnegar, B., and J. Kriz. 1973. *Fordilla troyensis* Barrande: the oldest known pelecypod. *Science* 180: 866–868.

Ponder, W. F., and Lindberg, D. R. 1997. Towards a phylogeny of gastropod molluscs: an analysis using morphological characters. *Zoological Journal of the Linnean Society* 119: 83–265.

Qian, Y., and Bengtson, S. 1989. Palaeontology and biostratigraphy of the early Cambrian Meishucunian Stage in Yunnan Province, South China. *Fossils and Strata* 24: 1–156.

Qian, Y., Chen, M., He, T.-g., Zhu, M.-y., Yin, G.-z., Feng, W.-m., Xu, J.-t., Jiang, Z.-w., Lio, D.-y., Li, G.-x., Ding, L.-f., Mao, Y.-q., and Xiao, B. 1999. *Taxonomy and biostratigraphy of small shelly fossils in China*. Beijing: Science Press. [in Chinese with expanded English summary].

Rehder, H. A. 1994. *National Audubon Society: Field Guide to North American Seashells*. New York: Chanticleer Press Inc.

Resser, C. E. 1938. Fourth contribution to nomenclature of Cambrian fossils. *Smithsonian Miscellaneous Collection* 97 (10): 1–43.

Robison, R.A. 1964. Late Middle Cambrian faunas from Western Utah. *Journal of Paleontology* 38 (3): 510–566.

Rozanov, A. Yu. 1992. [Once again about the ancient phosphorites of Mongolia] *Sovetskaya geologiya [Soviet Geology]* 1992 (1): 79–81 [in Russian].

Rozanov, A. Yu., and Missarzhevsky, V. V. 1966. [Biostratigraphy and Fauna of the Cambrian Lower Horizons]. *Trudy Geologicheskogo instituta Akademii Nauk SSSR* 148: 1–127 [in Russian].

Rozanov, A. Yu., Missarzhevsky, V. V., Volkova, N. A., Voronova, L. G., Krylov, I. N., Keller, B. M., Korolyuk, I. K., Lendzion, K., Michniak, R., Pykhova, N. G., and Sidorov, A. D. 1969. [The Tommotian Stage and the Problem of the Lower Boundary of the Cambrian]. *Trudy Geologicheskogo instituta Akademii Nauk SSSR* 206: 1–380 [in Russian]. (English translation edited by M. E. Raaben. 1981. New Dehli, India: Amerind Publishing Co.).

Rozanov, A. Yu., Semikhatov, M. A., Sokolov, B. S., Fedonkin, M. A., and Khomentovsky, V. V., 1997. The decision on the Precambrian-Cambrian boundary stratotype: a breakthrough or misleading action? *Stratigraphy and Geological Correlation* 5 (1):19–28. [Transl. from *Stratigrafiya. Geologicheskaya korrelyatsiya* 5 (1): 19–28].

Rozanov, A. Yu., and Sokolov, B. S. (eds.). 1984. *Yarysnoe raschlenenie nizhnego kembriya. Stratigraphiya [Lower Cambrian Stage subdivision. Stratigraphy]*. Moscow: Nauka [in Russian].

Rozanov, A. Yu., and Zhuravlev, A. Yu. 1992. The Lower Cambrian fossil record of the Soviet Union. In *Origin and early evolution of the Metazoa*. Edited by J. H. Lipps and Ph. W. Signor. New York: Plenum Press.

Runnegar, B. 1978. Origin and evolution of the Class Rostroconchia. *Philosophical transactions of the Royal Society of London* B 284: 319–333.

———. 1981. Muscle scars, shell form and torsion in Cambrian and Ordovician univalved molluscs. *Lethaia* 14 (4): 311–322.

———. 1982. The Cambrian explosion: animals or fossils? *Journal of the Geological Society of Australia* 29: 395–411.

———. 1983. Molluscan phylogeny revised. *Association of Australasian Palaeontologists Memoirs* 1: 121–144.

———. 1985. Shell microstructure of Cambrian molluscs replicated by phosphate. *Alcheringa* 9: 245–257.

———. 1996. Early evolution of the Mollusca: the fossil record. In *Origin and evolution of the Mollusca*. Edited by J. Taylor. Oxford: Oxford University Press.

Runnegar, B., and Bentley, C. 1983. Anatomy, ecology and affinity of Australian Early Cambrian bivalve *Pojetaia runnegari* Jell. *Journal of Paleontology* 57 (1): 3–92.

Runnegar, B., and Jell, P. A. 1976. Australian Middle Cambrian molluscs and their bearing on early molluscan evolution. *Alcheringa* 1 (2): 109–138.

———. 1980. Australian Middle Cambrian molluscs: corrections and additions. *Alcheringa*, 4 (1–2): 111–113.

Runnegar, B., and Pojeta, J. 1974. Molluscan phylogeny: the paleontological viewpoint. *Science* 186: 311–17.

———. 1985. Origin and diversification of the Mollusca. In *The Mollusca, 10: Evolution*. Edited by E. R. Trueman and M. R. Clarke. Orlando: Academic Press.

Runnegar, B., Pojeta, J., Morris, N. J, Taylor, J. D., Taylor, M. E., and McClung, G. 1975. Biology of the Hyolitha. *Lethaia* 8: 181–191.

Runnegar, B., Pojeta, J., Taylor M. E., and Collins, D. 1979. New species of the Cambrian and Ordovician chitons *Matthevia* and *Chelodes* from Wisconsin and Queensland: evidence for the early history of polyplacophoran mollusks. *Journal of Paleontology* 53: 1374–1394.

Salvini-Plawen, L. v. 1980. A reconsideration of systematics in the Mollusca (phylogeny and higher classification). *Malacologia* 19: 249–278.

Shaler, N. S., and Foerste, A. F. 1888. Preliminary description of North Attleborough fossils. *Bulletin of the Museum of Comparative Zoology* 16: 27–41.

Shergold, J. H., Rozanov, A. Yu., and Palmer, A. R. (eds.). 1991. *The Cambrian System on the Siberian Platform*. Trondheim: IUGS Publication:

Signor, Ph. W., and Lipps, J. H. 1992. Origin and early radiation of the Metazoa. In *Origin and early evolution of the Metazoa*. Edited by J. H. Lipps and Ph. W. Signor. New York: Plenum Press.

Skarlato, O. A. 1987. [Molluscs of the White Sea] *Opredeliteli po faune SSSR* 151. Leningrad: Nauka [in Russian].

Starobogatov, Ya. I. 1970. [To the systematics of the Early Paleozoic Monoplacophora. *Paleontologicheskii Zhurnal* 1970 (3): 6–16 [in Russian].

———. 1976. [On the subclasses of the gastropod class]. In: *Osnovnye problemy sistematiki zhivotnyh [Main problems of animal systematics.]* Edited by V. N. Shimansky, G. K. Kabanov, E. L. Dmitrieva, Ya. I. Starobogatov, and B. A. Trofimov. Moscow: Paleontologicheskii Institut Akademii Nauk SSSR [in Russian].

———. 1977. [Systematic position of conocardiids and the system of Paleozoic Septibranchia (Bivalvia)]. *Byulleten Moskovskogo obschestva ispytatelei prirody, Otdelenie geologii* 52 (4): 125–139 [in Russian].

Starobogatov, Ya. I., and Moskalev, L. I. 1987. [Systematics of monoplacophorans]. In *Mollyuski: Rezul'taty i perspektivy ikh izucheniya* [Molluscs: results and perspectives of their study]. Leningrad: Nedra [in Russian].

Stinchcomb, B. L., and Darrough, G. 1995. Some molluscan Problematica from the Upper Cambrian-Lower Ordovician of the Ozark Uplift. *Journal of Paleontology* 69: 52–65.

Sundukov, V. M., and Fedorov, A. V. 1986. [Paleontology and age of the beds containing algal-archaeocyathan bioherms of the Medvezhya River]. In Biostratigrafiya i paleontologiya kembriya Severnoi Azii [Cambrian Biostratigraphy and Palaeontology of Northern Asia]. Edited by I. T. Zhuravleva. *Trudy Instituta geologii i geofiziki, Sibirskogo otdeleniya Akademii nauk SSS* 669: 108–119 [in Russian].

Sutton, M. D., Briggs, D. E. G., Siveter, D. J., and Siveter D. J. 2006. Fossilized soft tissues in a Silurian platyceratid gastropod. *Proceedings of the Royal Society, Series B* 273: 1039–1044.

Tate, R. 1892. The Cambrian fossils of South Australia. *Transactions of the Royal Society of South Australia* 15: 183–189.

Ushatinskaya, G. T., and Parkhaev, P. Yu. 2005. Preservation of imprints and casts of cells of the outer mantle epithelium in shells of Cambrian brachiopods, mollusks and problematics. *Paleontological Journal* 39 (3): 251–263 [Transl. from *Paleontologicheskii Zhurnal* 2005 (3): 29–39].

Valkov, A. K., and Karlova, G. A. 1984. [Fauna from transitional Vend-Cambrian beds of the lower reaches of Gonam River]. In *Stratigrafiya pozdnego dokembriya I rannego paleozoya. Srednyaya Sibir [Stratigraphy of Late Precambrian and Early Paleozoic. Middle Siberia]*. Edited by V. V. Khomentovsky. Novosibirsk: Izdatelstvo IGiG [in Russian].

Vendrasco, M. J, Wood, T. E., and Runnegar, B. N. 2004. Articulated Paleozoic fossil with 17 plates greatly expands disparity of early chitons. *Nature* 429: 288–291.

Vinter, J., and Nielsen, C. 2005. The Early Cambrian *Halkieria* is a mollusc. *Zoologica Scripta* 34: 81–89.

Vogel, S. 1988. *Life's Devices: The Physical World of Animals and Plants*. Princeton, NJ: Princeton University Press.

Walossek, D. 2003. The "Orsten" window—a three dimensionally preserved Upper Cambrian Meiofauna and its contribution to our understanding of the evolution of Arthropoda. *Paleontological Research* 7 (1): 71–88.

Webers, G. F., and Yochelson, E. L. 1999. A revision of Palaeacmaea (Upper Cambrian) (?Cnidaria). *Journal of Paleontology* 73 (4): 598–607.

Wenz, W. 1938. Gastropoda. Allgemeiner Teil und Prosobranchia. In *Handbuch der Paläozoologie*. Edited by O. H. Schindewolf. 6 (1): Berlin: Verlag von Gebrüder Borntaeger, 1–240.

———. 1940. Ursprung und frühe Stammesgeschichte der Gastropoden. *Archiv für Molluskenkunde* 72: 1–109.

Xing, Y.-s., Ding, Q.-x., Luo, H.-l., He, T.-g., Wang, Y.-g., et al. 1984 [1983]. [The Sinian-Cambrian Boundary of China]. *Bulletin of the Institute of Geology of the Chinese Academy of Geological Science* 10: 1–262. Beijing: Geological Publishing House [in Chinese].

Yochelson, E. L. 1968. Stenothecoida, a proposed new class of Cambrian Mollusca. *Abstracts of the International Paleontological Union, August 20–26, Prague, Czechoslovakia*. Prague, 34.

———. 1969. Stenothecoida, a proposed new class of Cambrian Mollusca. *Lethaia* 2 (1): 49–62.

———. 1971. A new Late Devonian gastropod and its bearing on problems of open coiling and septation. *Smithsonian Contribution to Paleobiology* 3: 231–241.

———. 1975. Discussion of Early Cambrian "mollusks." *Journal of the Geological Society* 131 (6): 661–662.

———. 1978. An alternative approach to the interpretation of the phylogeny of ancient molluscs. *Malacologia* 17 (2): 165–191.

———. 1981. *Fordilla troyensis* Barrande: "the oldest known pelecypod" may not be a pelecypod. *Journal of Paleontology* 55:113–125.

Yochelson, E. L., and Gil-Cid, D. 1984. Reevaluation of the systematic position of *Scenella*. *Lethaia* 17: 331–340.

Yu W. 1974. [Cambrian Gastropoda]. In *Handbook of the stratigraphy and paleontology of southwest China*. Beijing: Science Press [in Chinese].

———. 1979. Earliest Cambrian monoplacophorans and gastropods from Western Hubei with their biostratigraphical significance. *Acta Palaeontologica Sinica* 18 (3): 233–270 [in Chinese with English abstract].

———. 1981. New earliest Cambrian monoplacophorans and gastropods from W. Hubei and E. Yunnan. *Acta Palaeontologica Sinica* 20 (6): 552–556 [in Chinese].

———. 1984. On merismoconchids. *Acta Palaeontologica Sinica* 23 (4): 432–446 [in Chinese with English summary].

———. 1987. Yangtze micromolluscan fauna in Yangtze region of China with notes on the Precambrian-Cambrian boundary. In *Stratigraphy and Palaeontology of Systemic Boundaries in China. Precambrian-Cambrian Boundary* 1:19–344. Nanjing: Nanjing University Publishing House.

———. 1989. *Did the shelled molluscs evolve from univalved to multivalved forms or vice versa? Developments in Geoscience (Contribution to 28th International Geological Congress, 1989, Washington, DC, USA)* pp. 235–244.

———. 1990. The first radiation of shelled molluscs. *Palaeontologia Cathayana* 5: 139–170.

———. 2001. The earliest Cambrian Polyplacophorans from China. *Records of the Western Australian Museum* 20: 167–185.

4

Solenogastres, Caudofoveata, and Polyplacophora

Christiane Todt, Akiko Okusu, Christoffer Schander, and Enrico Schwabe

The phylogenetic relationships among the molluscan classes have been debated for decades, but there is now general agreement that the most basal extant groups are the "aplacophoran" Solenogastres (= Neomeniomorpha), the Caudofoveata (= Chaetodermomorpha) and the Polyplacophora. Nevertheless, these relatively small groups, especially the mostly minute, inconspicuous, and deep-water-dwelling Solenogastres and Caudofoveata, are among the least known higher taxa within the Mollusca.

Solenogastres and Caudofoveata are marine, worm-shaped animals. Their body is covered by cuticle and aragonitic sclerites, which give them their characteristic shiny appearance. They have been grouped together in the higher taxon Aplacophora (e.g., Hyman 1967; Scheltema 1988, 1993, 1996; Ivanov 1996), but this grouping is viewed as paraphyletic by others (e.g., Salvini-Plawen 1972, 1980, 1981b, 1985, 2003; Salvini-Plawen and Steiner 1996; Haszprunar 2000; Haszprunar et al., Chapter 2).

SOLENOGASTRES

There are about 240 described species of Solenogastres (Figure 4.1 A–C), but many more are likely to be found (Glaubrecht et al. 2005). These animals have a narrow, ciliated, gliding sole located in a ventral groove—the ventral fold or foot—on which they crawl on hard or soft substrates, or on the cnidarian colonies on which they feed (e.g., Salvini-Plawen 1967; Scheltema and Jebb 1994; Okusu and Giribet 2003). Anterior to the mouth is a unique sensory region: the vestibulum or atrial sense organ. The foregut is a muscular tube and usually bears a radula. Unlike other molluscs, the midgut of solenogasters is not divided in compartments but unifies the functions of a stomach, midgut gland, and intestine (e.g., Todt and Salvini-Plawen 2004b). The small posterior pallial cavity lacks ctenidia. The smallest solenogasters measure less than a millimeter in body length (e.g., *Meiomenia swedmarki*, *Meioherpia atlantica*), whereas the largest species are more than 30 cm long (e.g., *Epimenia babai*) and

71

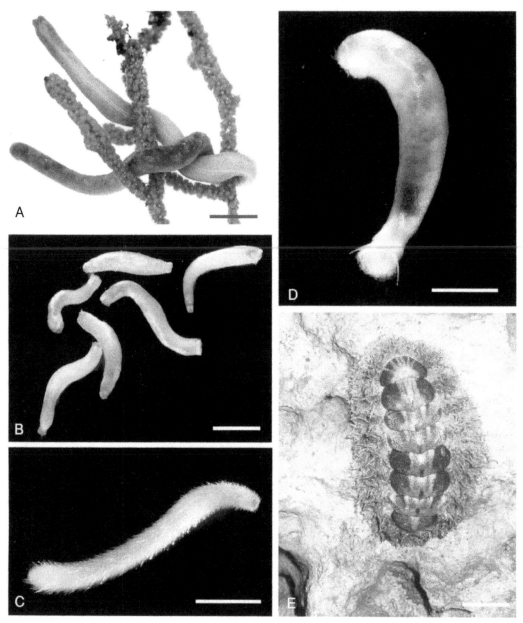

FIGURE 4.1. Living specimens of Solenogastres (A, B, C), Caudofoveata (D), and Polyplacophora (E). (A) *Epimenia* n. sp., from Japan, on its gorgonian prey, scale bar: 1 cm; there are blue patches on the dorsal mantle surface. From Okusu 2003. (B) Specimens of *Wirenia argentea* from Galicia (Spain) (micrograph by V. Urgorri), scale bar: 1.5 mm. (C) *Biserramenia psammobionta* from Galicia (Spain); note the long epidermal spicules of this interstitial animal (micrograph by V. Urgorri), scale bar: 0.25 mm. (D) *Prochaetoderma* sp. from Galicia, Spain; note the terminal knob with fringe of long, pointed sclerites at the posterium (micrograph by V. Urgorri), scale bar: 0.5 mm. (E) *Acanthopleura gemmata*, Sulawesi, Indonesia; with long hair-like projections covering the girdle; photo taken in the animal's natural habitat.

often colorful (Okusu 2003). There are a number of overviews of solenogaster morphology (e.g., Thiele 1913; Hoffmann 1930; Hyman 1967; Salvini-Plawen 1971, 1978, 1985) and microscopic anatomy (Scheltema *et al.* 1994), and there are some comprehensive studies that focused on the histology of the integument (Hoffman 1949) or the histology or physiology of the digestive tract (Baba 1940a; Salvini-Plawen 1967, 1981a; 1988; Scheltema 1981).

CAUDOFOVEATA

In Caudofoveata, a ventral groove and foot are lacking (Figure 4.1D). The mouth opening is partly or entirely surrounded by an oral shield (foot shield), an area covered by a thick layer of cuticle without sclerites. Caudofoveates are infaunal and feed on detritus or selectively on foraminiferans by burrowing in the mud with their oral shield. They have a muscular foregut bearing a radula, and their posterior midgut is divided into a dorsal tubular region (midgut duct) and a ventral midgut sac. The small, posterior pallial cavity bears a pair of true ctenidia. Caudofoveates range in length from a few millimeters (e.g., *Prochaetoderma raduliferum*; *Falcidens sterreri*) to 14 cm (e.g., *Chaetoderma productum*). Three major body regions are defined: anterium, trunk, and posterium. The latter may consist of a narrow shank and a terminal knob with characteristic elongate sclerites (typical for Prochaetodermatidae). With very few exceptions (e.g., *Chaetoderma rubrum*), caudofoveates are beige to brownish in color. About 120 species have been described so far (Glaubrecht *et al.* 2005). In caudofoveates, aside from the sometimes highly specialized radula, the variation in structure and arrangement of internal organs is limited, and knowledge of internal anatomy is mostly based on older studies (e.g.,Wirén 1892; Thiele 1913; Hoffmann 1930; van Lummel 1930; Hyman 1967; Salvini-Plawen 1971, 1975, 1985; Scheltema *et al.* 1994).

POLYPLACOPHORA

The monophyly of Polyplacophora has been well established (most recently Okusu *et al.* 2003), even if in a recent molecular analysis a species of monoplacophoran appears to be nested within the chitons (Giribet *et al.* 2006). The name Polyplacophora dates back to Gray (1821), but the term Placophora, which was first used by von Ihering (1876), is common, too, especially in German literature. The latter term is also used informally (e.g., Lindberg and Ponder 1996; Parkhaev, Chapter 3) to encompass the Aplacophora, Polyplacophora, and mollusc-like fossil taxa. The general morphology of Polyplacophora, with some information on histology, was described by Plate (1897, 1901), Hyman (1967), Kaas and Van Belle (1985), and Wingstrand (1985). Information on their microscopic anatomy was compiled by Eernisse and Reynolds (1994). These animals, commonly referred to as chitons, are dorsoventrally flattened, exclusively marine molluscs characterized by the presence of eight dorsal aragonitic shell plates (valves) and a broad ventral ciliated foot (Figures. 4.1E, 4.6A). The likewise ventrally positioned head is separated from the foot by a transverse groove. Surrounding the dorsal shell plates—or even completely engulfing them in some species—there is a thick marginal girdle (perinotum) covered by a chitinous cuticle. Embedded in this cuticle are calcium carbonate sclerites (Figure 4.6B), which are only occasionally lacking, and sometimes the cuticle additionally bears corneous processes (e.g., in *Chaetopleura*). The shell plates display a complex morphology and are composed of four layers: properiostracum, tegmentum, articulamentum, and myostracum. The articulamentum projects anteriorly and laterally beyond the tegmentum to form the sutural laminae and insertion plates. The shell plates characteristically bear so-called aesthetes, unique photo- and probably also mechano- and chemosensoric organs and in certain taxa ocelli (Figure 4.6C). The head in general lacks eyes and tentacles, but the mouth opening is laterally flanked by mouth lappets. Occasionally (e.g., in the genus *Placiphorella*) precephalic tentacles, which support the animal while feeding, may occur. The mantle cavity or pallial groove surrounds the foot and accommodates the terminal anal papilla, a multitude of laterally positioned ctenidia, the paired osphradium, and lateral sense organs. Chitons have complex muscle systems, including eight paired sets of dorsoventral muscle units that insert at the shell plates, the musculus rectus, which runs longitudinally underneath the shell plates, and a circular enrolling muscle. Usually chitons are grazers with a broad and exceptionally long stereoglossate radula (Figure 4.6D).

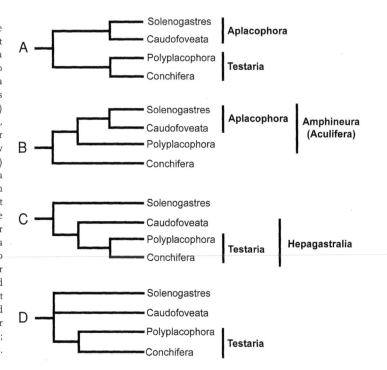

FIGURE 4.2. Alternative phylogenies for the placement of Solenogastres, Caudofoveata and Polyplacophora relative to Conchifera. (A). Aplacophora and Testaria as sister groups (after Waller 1998). (B) Aplacophora as Monophylum, Amphineura (Aculifera) sister group to Conchifera (after Ivanov 1996; Scheltema 1996). (C) Solenogastres and Caudofoveata as independent clades with Solenogastres branching earliest (after most parsimonious tree in Salvini-Plawen and Steiner 1996), additionally Caudofoveata grouping with Testaria into Hepagastralia (after Haszprunar 2000). (D) Solenogastres and Caudofoveata as independent clades with unresolved relationship to Testaria (after Salvini-Plawen and Steiner 1996; Salvini-Plawen 2003).

Their diet consists mainly of diatoms, detritus, and encrusting algae, but special feeding habits have been adopted by the carnivorous *Placiphorella* and *Lepidozona* (Latyshev *et al.* 2004), the xylophagous *Ferreiraella* (Sirenko 2004), or the true herbivorous *Stenochiton*. There are about 920 living species (Schwabe 2005), most living in the marine intertidal or sublittoral, with some deep-sea species also known (Kaas *et al.* 1998).

RELATIONSHIPS The two aplacophoran taxa and Polyplacophora were, and still are, considered by most morphologists as basal within Mollusca, preceding the conchiferan radiation, although their relative placement varies between proposed hypotheses (Figure 4.2A–D). In one scheme, Solenogastres and Caudofoveata have been incorporated in the phylum Aplacophora, the sister group of a clade Testaria consisting of Polyplacophora and Conchifera (Waller 1998). Alternatively, based on similarities in their nervous system, Polyplacophora was considered to be the sister group to Aplacophora, the two together forming the Amphineura (von Ihering

1876a, b; Spengel 1881; Hoffmann 1930), while a clade Aculifera was proposed for those groups having a cuticle with sclerites covering at least part of the mantle (e.g., Hatschek 1891; Scheltema 1988, 1996; Ivanov 1996). Other authors have argued that aplacophorans are paraphyletic with respect to a clade Testaria, comprising the remaining molluscs, within which Polyplacophora is a sister taxon to Conchifera (e.g., Wingstrand 1985; Salvini-Plawen 1980, 1985, 1990, 2003; Salvini-Plawen and Steiner 1996). Based on midgut morphology, Haszprunar (2000) additionally defines the clade Hepagastralia for Caudofoveata plus Testaria. A sister group relationship of Polyplacophora with Conchifera is often assumed in studies of conchiferan relationships (e.g., Giribet and Wheeler 2002), but it is also questioned (e.g., Lindberg and Ponder 1996).

Recent discoveries of sclerite-bearing fossils (see following discussion), additional developmental work with new techniques, and recent morphological and molecular studies have shed new light on molluscan origins and the evolution of Solenogastres, Caudofoveata, and

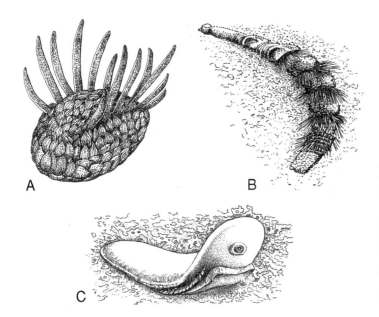

FIGURE 4.3. Fossils interpreted as early molluscs (drawings by C.-O. Schander). (A) *Wiwaxia corrugata* from the Burgess Shale. (B) The Silurian *Acaenoplax hayae*. Note the rows of acicular skeletal elements seen either as aculiferan-like sclerites or as polychaete-like chaetae. (C) The Cambrian *Odontogriphus omalus* from the Burgess Shale.

Polyplacophora. Here, we revisit some old hypotheses and review the newest findings for these fascinating groups.

THE FOSSIL RECORD AND MOLLUSCAN ORIGINS

Molluscan lineages probably extended at least as far back as the base of the Cambrian (543 Mya) (Glaessner 1969; Runnegar and Pojeta 1985; Bengtson 1992) or the Upper Precambrian, if trail-like impressions in the Ediacaran strata are correctly interpreted as traces left by the ventral muscular foot of *Kimberella quadrata*, a limpet-like animal with possibly molluscan affiliations (Fedonkin and Waggoner 1997; see also Parkhaev, Chapter 3). The known polyplacophoran fossil record extends from as early as the Upper Cambrian (Yochelson *et al.* 1965; Runnegar *et al.* 1979; Yates *et al.* 1992; Stinchcomb and Darrough 1995; Slieker 2000). No fossil aplacophorans are known, and thus there is no direct evidence of the time of origin of solenogasters or caudofoveates.

Recently, increasing numbers of problematic sclerite-bearing metazoans from the Early and Middle Cambrian have been discovered and assigned a taxonomic placement close to, or within, Mollusca. These animals have been compared to aplacophorans and polyplacophorans based on their sluglike appearance, muscular ventral foot, dorsal calcification patterns, gill arrangement, and the possible presence of a radula (e.g., Conway Morris and Peel 1995, Caron *et al.* 2006). Controversies still remain as to which extant Metazoa these fossils are most closely related, but, in any case, they assist in understanding the evolution of external calcification in Mollusca as well as in Metazoa in general.

The Middle Cambrian *Wiwaxia corrugata* from the Burgess Shale (Figure 4.3A) was considered "strikingly similar" to molluscs on the basis of body shape and the radular-like feeding apparatus (Conway Morris 1985). This radular-like structure was interpreted as homologous to the radula of an extant solenogaster *Helicoradomenia* (Scheltema 1998, Scheltema *et al.* 2003). Analyses of *Wiwaxia*'s sclerites, however, have led to doubts as to its molluscan affinity (Butterfield 1990, 1994; Watson Russel 1997) because the solid wiwaxiid sclerites were originally chitinous and had longitudinal ornamentation on their dorsal side, much like chrysopetalid polychaete paleae (specialized chaetae). The most recent and, so far, the most comprehensive study of the phylogenetic placement of *Wiwaxia*, was carried out by Eidbye-Jacobsen (2004). He found "no characters that could indicate any close relationship with Polychaeta or Annelida."

Thus, a molluscan affinity of *Wiwaxia* again seems plausible.

The Cambrian *Odontogriphus omalus* (Figure 4.3C) was originally interpreted as a lophophorate or with a possible connection to some Cambrian conodonts (Conway Morris 1976). This interpretation was based on a single specimen in rather poor condition. Newly discovered specimens from the Burgess Shale allowed a reinterpretation of this fossil, revealing several characters with possible molluscan affinity (Caron *et al.* 2006a; but see also Butterfield 2006 and Caron *et al.* 2006b). *Odontogriphus* was a dorsally-ventrally compressed, elongated animal with an oval body up to 12 mm long. It had a muscular foot lined by simple gills and a ventral mouth with a radula strikingly similar to that of *Wiwaxia*. There is also indications of a pair of salivary glands. It lacked a mineralized shell and sclerites. It was most likely a bacterial grazer feeding on the cyanobacterium *Morania*. The general shape of the body is suggestive of the Precambrian fossil *Kimberella* as well as the Cambrian *Wiwaxia*.

Another fossil repeatedly referred to as a mollusc is the Early Cambrian *Halkieria* from northern Greenland. The first entire articulated halkieriid to be discovered, *Halkieria evangelista*, is described as a long flat animal with a ventral creeping sole, dorsal sclerites, and two terminal valves (Conway Morris and Peel 1990, 1995). Although it possesses several mollusc-like characters, Conway Morris and Peel (1995) placed *Halkieria* either as the sister group to Annelida or within Brachiopoda. Halkieriid sclerites are filled with phosphate, leading to speculation that they originally were filled with tissue and formed by "external mineralization of a protruding organic template," like polychaete chaetae (Butterfield 1990; Bengtson 1992; Conway Morris and Peel 1995). The mode of sclerite formation was thought by Bengtson (1992, 1993) to be in contrast to that in aplacophorans and polyplacophorans, where sclerites are produced by invagination of a single cell or, in chitons, by invaginated groups of cells (Haas 1981; Eernisse and Reynolds 1994). However, there are certain hollow aplacophoran sclerites that form around an organic template protruding from the sclerite-producing cell (Hoffman 1949; Okusu 2002). Moreover, the terminal valves and sclerites of *Halkieria* may have been aragonitic and thus similar in their mineralogical composition to those of aplacophorans and polyplacophorans (Eernisse and Reynolds 1994; Vinther and Nielsen 2005). Similarities between the three distinct types of sclerites in halkieriids (siculates, cultrates, and palmates) and wiwaxiids (ventro-laterals, upper-laterals, and dorsals) and the zones of sclerites in aplacophorans and polyplacophorans have been noted (Conway Morris and Peel 1990; Bengtson 1992; Conway Morris and Peel 1995; Scheltema 1998). Scheltema and Ivanov (2002) also suggested that the serially clustered siculate sclerites in *Halkieria* are homologous to the seven transverse regions devoid of sclerites seen in a solenogaster postlarva. Most recently, Vinther and Nielsen (2005) convincingly demonstrated the molluscan affinity of *Halkieria*. They stressed a number of similarities between *Halkieria* and Polyplacophora, such as the overall morphology and sclerite arrangement, but the terminal valves of *Halkieria* are different from polyplacophoran valves and more similar to conchiferan shells in lacking a tegmentum layer and pore canals. The hollow siculate sclerites also show resemblance to those in the fossil polyplacophoran *Echinochiton* (Pojeta *et al.* 2003). Consequently, the new class Biplacophora for molluscs with two shell plates and a covering of sclerites was introduced (Vinther and Nielsen 2005), and the class Coeloscleritophora, a taxon that used to unify a number of fossils with hollow sclerites (Bengtson and Missarzhevsky 1981), was declared to be polyphyletic.

Intrepretations of *Halkieria* as a mollusc have raised further questions regarding the evolution of shell and sclerites among molluscs. It has been suggested that polyplacophoran and halkieriid valves are formed through coalescence of calcium carbonate sclerites (Pojeta 1980; Salvini-Plawen 1985; Eernisse and Reynolds 1994). The extant chiton, *Acanthochitona*, and the fossil ?coeloscleritophoran? *Maikhanella*, both have valves with scaly sculptures that appear to be composed of merged neighboring sclerites (Bengtson 1992). Other spicule-bearing fossils, however, lack those sculptures. The valves (shells) of *Maikhanella* have been suggested to grow by marginal accretion and their sclerites by interpolation, just as occurs in some molluscs and was suggested for halkieriids (Bengtson 1992; Conway, Morris, and Peel 1995), but new findings on a Recent vetigastropod, *Vacerrena kesteveni*, show that a scaly shell-surface may be a calcified periostracal sculpture (Ponder *et al.* 2007). Scheltema (1998) doubts that chiton valves are formed by coalescence of calcium carbonate sclerites and points out that seven transverse regions devoid of sclerites in a solenogaster postlarva (see preceding paragraph) may be homologous to the chiton larval shell fields. If chiton valves are not formed by coalescence of sclerites, they may have originated simply through modification of spicular calcification mechanisms (Carter and Hall 1990), necessitating only a simple step in the evolution of a shell from sclerites (Scheltema and Schander 2006).

The exceptionally well-preserved Silurian *Acaenoplax hayae* (Figure 4.3B) was thought to be related to aplacophorans (Sutton *et al.* 2001a, b, 2004). *Acaenoplax* is a vermiform fossil with about 18 iterated rows of ridges bearing needle-shaped sclerites similar to those in annelids, seven dorsal calcareous plates similar to those in chitons, a single posterior ventral plate, and posterior gills (Sutton *et al.* 2001a, b). Its seven dorsal valves and single ventral valve have been interpreted to be homologous with valves 1–6 and 8 of chitons and with the seven dorsal transverse regions free of sclerites in an aplacophoran postlarva (Scheltema and Ivanov 2002; see preceding paragraphs). Although this may seem to corroborate the Aculifera hypothesis, this placement was challenged by Steiner and Salvini-Plawen (2001), who suggested that an annelid affinity of *Acaenoplax* was just as likely because of its lack of explicit molluscan characters and overall similarity to some Recent tube-dwelling annelids.

Hoare and Mapes (1995) discussed the Devonian problematic taxon *Strobilepis* from the Moscow Formation in New York (United States) and introduced a new Carboniferous (Pennsylvanian) problematic genus *Diadeloplax*, from the Gene Autry Formation in Oklahoma (United States). These two genera were placed in the new family Strobilepidae and new class Multiplacophora whose phylum assignment remained uncertain. They noted that multiplacophorans characteristically have 12 plates that have diverse shapes and, at least in part, lack bilateral symmetry, and that small auxiliary plates are always associated with larger intermediate plates. The recent discovery of an exceptionally well-preserved specimen of another multiplacophoran, *Polysacos vickersianum*, from the Carboniferous of Indiana (United States), enabled a more accurate reconstruction of the body plan of this group (Vendrasco *et al.* 2004). The animal is very similar to a chiton in body shape and bears 17 shell plates and a lateral fringe of spines. Both plates and spines are most likely homologous to polyplacophoran valves, and the valves are articulated as in modern chitons. The oldest multiplacophoran fossils are Devonian and thus much younger than the oldest chitons (see following discussion). Vendrasco *et al.* (2004) place the multiplacophorans as an order within the Polyplacophora, implying an early divergence from the eight shell plate plan in Polyplacophora. It is possible that changes in the number and patterning of shell plates involved only small changes in homeobox genes, analogous to the changes that have occurred in the relative number of vertebrae in modern snakes (Cohn and Tickle 1999; Wiens and Slingluff 2001). Although it has been shown that homeobox genes are involved in the patterning and formation of modern chiton shell plates (Jacobs *et al.* 2000), details have not yet been investigated.

The oldest polyplacophoran fossils are known from the Upper Cambrian (Yates *et al.* 1992), and since then, with exception of multiplacophorans, their general body plan and valve morphology did not change significantly. This is confirmed by findings of numerous complete articulated specimens, such as *Glaphurochiton concinnus* from the Carboniferous of Illinois (United States) (Yochelson and Richardson 1979). Fossil plates, however, show that the occurrence of microaesthete structures must be interpreted as a post-Paleozoic innovation (Hoare 2000).

Smith and Hoare (1987) divided the Polyplacophora into three subclasses: Paleoloricata, Phosphatoloricata, and Neoloricata. Later (Sirenko 1997) followed Bergenhayn (1955) and Van Belle (1983) in accepting two lineages within the Polyplacophora: Paleoloricata and the more derived, articulamentum-bearing Neoloricata (or Loricata in Sirenko 1997). All extant chitons belong to Neoloricata, whereas fossil forms are classified within both groups. In the Neoloricata there are Cenozoic and Mesozoic taxa, while only Paleoloricata are known from the Paleozoic. Sirenko (1997) recognized four orders, including five suborders and 14 families from the Paleozoic. Hoare (2000) suggested minor changes in the system but otherwise accepted Sirenko's conclusions. Nevertheless, a few problems with uncertain affiliations to Polyplacophora still exist, such as *Luyanhaochiton* from the Lower Cambrium of China (Hoare 2000; see also Parkhaev 2007, Chapter 3).

DEVELOPMENT

Studies on the early embryology and development of aplacophoran molluscs are rare, and thus comparisons with other molluscan classes remain difficult (for review see Verdonk and Van den Biggelaar 1983; Buckland-Nicks *et al.* 2002). Knowledge of the development of Solenogastres is restricted to a few species, whereby the early studies of Pruvot (1890), Heath (1918), and Baba (1938, 1940b) were only recently added to by Okusu's (2002) work on the embryogenesis and development of *Epimenia babai*. This description of early embryogenesis revealed that cleavage is spiral, unequal, and holoblastic. Solenogastres are hermaphrodites with internal fertilization and

have free-swimming, lecithotrophic larvae with an enlarged swimming test (pericalymma) with differing numbers of rows of ciliated prototrochs. The apical test of *E. babai* larvae is completely ciliated with an apical tuft and a single prototroch composed of compound cilia (Figure 4.4A). It is lost during metamorphosis. The pericalymma test is often regarded as homologous to the enveloping test of protobranch bivalve larvae as well as to the velum of bivalve and gastropod veliger larvae (for review see Nielsen 2004). Homology of these structures remains uncertain, and either they are interpreted as similarly modified apical structures evolved from a basic trochophore specialized in swimming (Jaegersten 1972; Nielsen 1987, 2004) or the pericalymma test is seen as a primitive trait within the Mollusca (Salvini-Plawen 1972, 1980, 1988; Chaffee and Lindberg 1986). The trunk region of the larvae is unciliated and gives rise to definitive ectodermal structures, such as cuticle, epidermis, and epidermal sclerites. No external metameric iteration can be found at any stage, and there is no evidence of protonephridia.

Earlier findings (Nielsen 1995, 2004) have been recently supported by a thorough study on *Chaetoderma* employing electron microscopy and fluorescence staining of musculature (Nielsen *et al.* 2007). This study shows lecitotrophic (pseudo-)trochophore larvae with a prototroch and a telotroch and a pair of protonephridia. In the later stages, a ventral suture and seven dorsal transverse rows of spicules are present.

Chiton embryos, as studied to date, undergo equal cleavage in a typical spiralian pattern (Heath, 1899; Grave, 1932; Van den Biggelaar, 1996). The resulting trochophore larvae are lecithotrophic and possess a unique prototroch composed of two to three irregular rows of differentially ciliated trochoblasts, as shown for *Chiton polii* (see Kowalevsky 1883), *Ischnochiton rissoi* (see Heath 1899), *Lepidopleurus asellus* (see Christiansen 1954), and *Chaetopleura apiculata* (see Henry *et al.* 2004). The free-swimming larval stage ranges from a few minutes to a few days. After settlement, the apical tuft and the prototroch may persist for a while. Metamorphosis starts with a dorsoventral flattening of the body. A detailed summary of larval development in chitons was presented by Buckland-Nicks et al. (2002).

The relationship of larval shell formation with expression of the *engrailed* gene has been reported in various molluscs (Wray et al. 1995), as has the expression of this gene in cells adjacent to the shell fields in chiton larvae (Jacobs et al. 2000).

A first cell lineage study (Henry et al. 2004) pointed out that polyplacophoran epidermal sclerites arise from different, if overlapping, sets of cells than the shell plates and the conchiferan shell, an important finding for consideration of the evolution of molluscan shells. The same study demonstrated that the larval ocelli of *Chaetopleura apiculata* develop posttrochally from a unique set of cells not seen in other spiralians.

Detailed investigations of myogenesis using fluorescent markers during the early development of chitons showed that serial muscle structures and dorsal shell plates do not develop simultaneously (Friedrich et al. 2002; Wanninger and Haszprunar 2002). This indicates that hypotheses indicating a sister taxon relationship between molluscs and other segmented protostomes such as Annelida, based on the serial repetition of organs (e.g., Götting 1980; Ghiselin 1988; Nielsen 1995), are not supported.

PHYLOGENY AND SYSTEMATICS

SISTER GROUP RELATIONSHIPS

Although a number of attempts have been made to resolve molluscan phylogeny using both morphological and molecular sequence data, there has not yet been any consensus on the position of aplacophoran taxa and Polyplacophora within Mollusca (see Figure 4.2).

One problem with most phylogenetic studies is the lack of a representative taxon sampling for the basal clades (Ghiselin 1988; Winnepenninckx et al. 1994, 1996; Rosenberg et al. 1997;

Lydeard et al. 2000; Giribet and Wheeler 2002). There are only a few molecular analyses that have included representatives of Solenogastres (Okusu 2003; Okusu et al. 2003; Passamaneck et al. 2004; Giribet et al. 2006) and Caudofoveata (Winnepenninckx et al. 1994; Okusu 2003; Okusu et al. 2003; Passamaneck et al. 2004; Giribet et al. 2006). Obtaining DNA sequence data has been challenging for aplacophoran taxa because they are difficult to collect and because of contamination issues in Solenogastres (Okusu and Giribet 2003). In an investigation of molluscan phylogeny using large-subunit and small-subunit nuclear rRNA sequences of 33 molluscan taxa, including a solenogaster, a caudofoveate, and four chitons, neither the Aculifera hypothesis nor the Testaria hypothesis is supported (Passamaneck et al. 2004). In this study, Polyplacophora does not emerge as a basal clade, and it groups only in some of the analyses with Solenogastres and never with Caudofoveata. A recent analysis of five genes and gene fragments from 101 species representing all molluscan classes shows Solenogastres and Caudofoveata as independent clades near the base of the tree but Polyplacophora as more derived and forming a clade (Serialia) with Monoplacophora (Giribet et al. 2006).

Recent attempts to study chiton phylogenetic relationships using several combined genes (Okusu et al. 2003) resulted in a well-resolved phylogeny of chitons but could not resolve the placement of chitons relative to Solenogastres, Caudofoveata, and Conchifera.

The notion of a basal position of Solenogastres, Caudofoveata, and Polyplacophora was recently supported by Lundin and Schander's studies on the ultrastructure of locomotory cilia in Solenogastres (2001b), Caudofoveata (1999), and Polyplacophora (2001a). These cilia are of the common metazoan type, with paired ciliary rootlets orientated at almost 90° to each other and without an accessory centriole. Such paired ciliary rootlets do not occur in gastropods, bivalves, and monoplacophorans (Lundin and Schander 2001b), nor in scaphopods (Lundin and Schander, unpublished data).

SYSTEMATICS AND PHYLOGENY OF APLACOPHORAN MOLLUSCS

Histology has been the standard method used for species identification and classification in aplacophorans, mostly because of their small size, lack of a shell, and often poor preservation of sclerites in non-buffered fixatives. Thus, the morphological and histological data available for Solenogastres and Caudofoveata are surprisingly detailed compared to other molluscan taxa. External characters are sufficient for a species diagnosis in many caudofoveates but only in relatively few solenogasters. However, the addition of internal hard-part characters (radula and copulatory stylets), usually allows identification of members of both groups (Scheltema and Schander 2000), but knowledge of anatomical and histological characters is of great importance for systematics and phylogenetic analyses.

In both Solenogastres and Caudofoveata, classification is based on comprehensive publications by Salvini-Plawen (1975, 1978). Some recent additions have been made and doubts on the monophyly of certain clades raised (e.g., Scheltema 1999), but the general concepts remain unchallenged.

SOLENOGASTRES Solenogaster higher classification uses external characters, such as types of sclerites (solid elements versus hollow elements, flat scales versus rimmed or trough-like elements), thickness of the cuticle, and general characteristics of the lateroventral foregut glands. Four orders were recognized by Salvini-Plawen (1978) (see also Figure 4.8):

Pholidoskepia: Cuticle is thin, scerites are scales in one layer, lateroventral foregut glands are either endoepithelial (no glandular duct) or with duct and exoepithelial gland cells (e.g., Wireniidae, Dondersiidae, Lepidomeniidae).

Neomeniamorpha: Cuticle is thin; sclerites are scales, massive acicular elements, rimmed, trough-like, and harpoon-shaped elements; no lateroventral foregut glands present (e.g., Neomeniidae, Hemimeniidae).

Sterrofustia: Cuticle is thick, sclerites are solid acicular or scalelike elements, lateroventral foregut glands are diverse (e.g., Phyllomeniidae, Imeroherpiidae).

Cavibelonia: Cuticle is thick, sclerites are hollow acicular elements, additional solid elements may occur, lateroventral foregut glands are diverse and include tubular glands with intraepithelial glandular cells (e.g., Pararrhopaliidae, Rhopalomeniidae, Simrothiellidae, Epimeniidae).

Solenogaster phylogenetics still struggles with the great diversity of hard-part as well as soft-body characters among the families and with the lack of a general concept as to the plesiomorphic character states. Most phylogenetic analyses based on morphology (e.g., Scheltema and Schander 2000) have included only a limited number of taxa. A recent comprehensive study of solenogaster phylogeny based on morphological characters included all genera (Salvini-Plawen 2003). Although poor resolution was obtained, Cavibelonia was monophyletic and derived (see also Salvini-Plawen 2004), whereas Pholidoskepia emerged from a basal polytomy. Handl and Todt (2005) discussed the evolution of foregut glands in solenogasters and the so-called *Wirenia*-type lateroventral foregut glands (Figure 4.4B, a), without a duct or lumen, seen in the pholidoskepian Gymnomeniidae, were considered to be the most primitive exant type. *Pararrhopalia*-type glands (Figure 4.4B, c) occur in some Pholidoskepia and Cavibelonia taxa, while certain gland types (e.g. *Helicoradomenia*-type, Figure 4.4B, d; *Simrothiella*-type, Figure 4.4B, e) occur in Cavibelonia only.

Due to the ontogenetic change from solid sclerites to hollow needles seen in some species, the hollow epidermal sclerites are considered derived, thus ruling out Cavibelonia as a basal clade. Hollow needles, however, also occur in the Acanthomeniidae, a taxon closely related to pholidoskepian taxa, such as the Dondersiidae (Salvini-Plawen 2003; see also Scheltema 1999, Handl and Salvini-Plawen 2001). Thus the

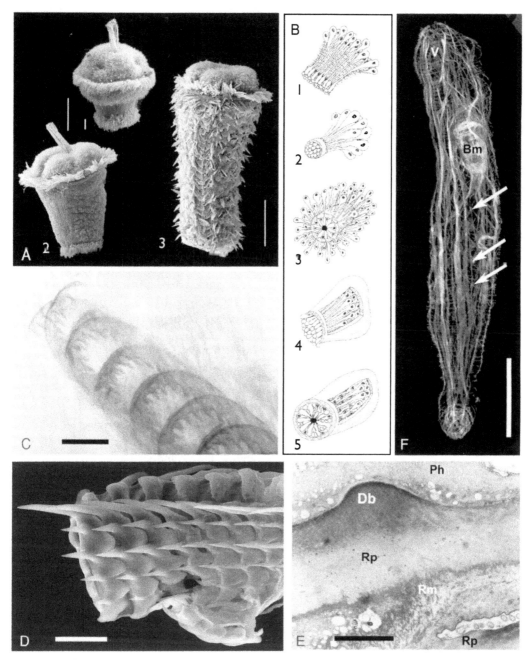

FIGURE 4.4. Solenogastres and Caudofoveata, development and important characters. (A) Larvae of *Epimenia babai* (Solenogastres) during the completion of metamorphosis, 1: 4–6 days old, 2, 3: 9–12 days old; scale bar: 100 µm. From Okusu 2002. (B) Examples for lateroventral foregut glands of Solenogastres, 1: *Wirenia*-type, 2: *Meioherpia*-type, 3: *Pararrhopalia*-type, 4: *Helicoradomenia*-type, 5: *Simrothiella*-type. (C) Radula of *Scutopus robustus* (Caudofoveata), light micrograph; scale bar: 50 µm. (D) Part of the right half of a radula of *Helicoradomenia* sp. (Solenogastres), scanning electron micrograph; scale bar: 20 µm. (E) Ultrathin section of a radular plate of *Helicoradomenia acredema*, transmission electron micrograph, Db = denticle base; Ph = pharynx lumen; Rm = radular membrane; Rp = radular plate. (F) Confocal scanning micrograph of Alexa-phalloidin stained *Meioherpia atlantica*, the arrows indicate spiral muscle fibers of the body wall; Bm = buccal musculature; V = vestibulum; scale bar: 0.1 mm.

homology of certain hollow sclerites may be questioned (Salvini-Plawen 2003).

Attempts toward a phylogeny of solenogasters by means of molecular methods has been hampered by technical problems (see Okusu and Giribet 2003), but refined techniques and intensified efforts should provide results in the near future.

CAUDOFOVEATA   This taxon is less diverse than Solenogastres, with only three or four families recognized, which are based on characters of the radula, mouth-shield, and body shape (Salvini-Plawen 1975, but see Ivanov 1981) (see Figure 4.8):

Limifossoridae: Radula is bipartite, of several transverse rows, without lateral supports; body is homogenously shaped; mouth shield is disk- or U-shaped posterior of mouth opening, or paired lateral to mouth opening.

Prochaetodermatidae: Radula is bipartite, in several transverse rows, with ventral and lateral supports; posterior body is tail-shaped, mouth shield is paired lateral to mouth opening.

Chaetodermatidae: Radula is generally represented by only one pair of teeth, with large ventral and lateral supports; body is homogenously shaped or posterior body is tail-shaped; mouth shield is U-shaped posterior to mouth opening or encircling mouth opening.

An additional family, Scutopodidae, was introduced by Ivanov (1981) but was rejected by Salvini-Plawen (e.g., 1992), who included *Scutopus* within Limifossoridae.

There are no modern phylogenetic analyses published for Caudofoveata. *Scutopus* and *Psilodens* are probably the most basal genera because some species have traces of a retained ventral suture innervated from the ventral nerve cords, as well as primitive radular (distichous pairs of teeth with median denticles; Figure 4.4C) (Salvini-Plawen 1975, 1985, 1988) and midgut configuration (Scheltema 1981; for *Psilodens* see Salvini-Plawen 1988, 2003). In contrast, Chaetodermatidae have a highly derived radula,

usually a single pair of teeth with prominent lateral and ventral supports, and the stomach has a gastric shield. The radula of Prochaetodermatidae appears to represent an intermediate state (Salvini-Plawen and Nopp 1974; Salvini-Plawen 1975: fig. 6, and slightly modified in 1988: fig. 1), but the phylogenetic relationship between Prochaetodermatidae and the other families is not well resolved (see Figure 4.8).

MORPHOLOGICAL CHARACTERS   Over the last few decades, modern techniques, such as scanning and transmission electron microscopy, have provided new insights into the morphology and histology of solenogasters and caudofoveates and helped to further define characters valuable for systematics and phylogeny. Some recent studies are summarized as follows.

Haszprunar (1986, 1987) supported the homology of the dorsoterminal sense organ (DTS) in Solenogastres and Caudofoveata with the usually paired osphradia of chitons and higher molluscs but suggested an independent origin of the unpaired condition of the DTS in the two aplacophoran taxa.

In both solenogastres and caudofoveates, the mantle sclerites exhibit extraordinary variability in size and shape, but certain sclerite types are characteristic at higher taxonomic levels (e.g., the hollow hooklike elements of Pararrhopaliidae). Information on sclerite thickness can be gained by the use of cross-polarized light or by scanning electron microscopy (Scheltema and Ivanov 2000, 2004). Because they vary according to their location, sclerites should be sampled from standardized body regions for taxonomic purposes (e.g., Scheltema 1976, 1985; Scheltema and Ivanov 2000).

Scheltema *et al.* (2003), in a review of the radula of basal molluscs, presented a theory on the nature of the primitive molluscan radula. Like Eernisse and Kerth (1988), she argued that the most basal type was the distichous or bipartite radula with rows of paired radular plates. This type is present in the solenogaster genus *Helicoradomenia* (Figure 4.4D) and the caudofoveate genus *Scutopus* (Figure 4.4C). In

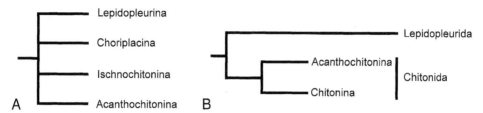

FIGURE 4.5. Alternative phylogenies for Recent Neoloricata. (A) Unresolved tree based on Kaas and Van Belle (1994) and Kaas *et al.* (1998). (B) Phylogenetic tree after Sirenko (1993, 1997).

contrast, Salvini-Plawen (1988, 2003; see also Sirenko and Minichev 1975), suggested that the monoserial radula type, consisting of rows of single teeth, was the most primitive. Wolter (1992) showed that radular formation in aplacophoran groups is like that in higher molluscs, although each tooth is continuous with the underlying membrane, there being no separate tooth base as in chitons or conchiferans (Figure 4.4E) and there is no subradular membrane. The radula is basically composed of a chitin-rich organic matrix (Peters 1972; Salvini-Plawen and Nopp 1974; Wolter 1992) with deposition of minerals (caudofoveates: Cruz *et al.* 1998; solenogasters: C. Todt, personal observation). As in chitons (see following), such studies promise additional phylogenetic characters.

In solenogaster systematics, foregut glands are among the most important characters, especially the multicellular lateroventral and dorsal glands (Salvini-Plawen 1972, 1978). Handl and Todt (2005) clarified the foregut gland terminology and modified Salvini-Plawen's (1978) classification system of the lateroventral glands. In addition, a number of ultrastructural studies showed the complexity of multicellular foregut glands, which are composed of up to five different types of glandular cells and nonglandular supporting cells (Todt and Salvini-Plawen 2004a, 2005; Todt, in press).

Attempts to apply modern fluorescence techniques to study musculature (Figure 4.4F) and nervous systems in Solenogastres are under way, and preliminary results have been presented as conference contributions (D. Eheberg and G. Haszprunar, R. Croll, and R. Hochberg, personal communication).

POLYPLACOPHORA SYSTEMATICS AND PHYLOGENY

Until recently, the higher classification of Polyplacophora has remained unsettled (Bergenhayn 1955; Smith 1960; Van Belle 1983; Eernisse 1984; Sirenko 1993, 1997; Buckland-Nicks 1995). Traditionally, classifications were based primarily on the morphology of shell plates (valves), spicules, and perinotum processes (e.g., Smith 1960; Van Belle 1983; Kaas *et al.*, 1998), the shell and spicules being the only characters available for fossil chitons (Smith 1960; Van Belle 1983). Of the four layers of the shell plates (properiostracum, tegmentum, articulamentum, myostracum) two are of highest taxonomic relevance: the often colorful and sculptured tegmentum and the articulamentum, which underlies the tegmentum and also forms the insertion plates (see previous discussion). All extant species (order Neoloricata) have been divided into three suborders (e.g., Bergenhayn 1930; Smith 1960; Kaas and Van Belle 1985; Van Belle 1983, 1985). Gowlett-Holmes (1987) reestablished the monotypic Choriplacina (for *Choriplax grayi*), and her proposal was followed by others (Kaas and Van Belle 1994; Kaas *et al.* 1998) (Figure 4.5A).

Lepidopleurina: Articulamentum may have unslit insertion plates or none; tegmentum is well developed; perinotum is narrow to wide, dorsally covered with elongate scale-like spicules, ventrally either naked or with scales.

Choriplacina: Articulamentum is well developed with large, unslit insertion plates; tegmentum is reduced; perinotum

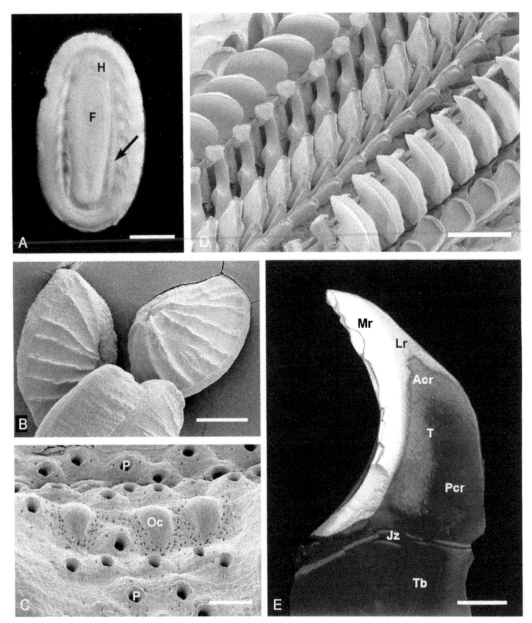

FIGURE 4.6. Polyplacophora; characters relevant for taxonomy and systematics. (A) Specimen of Ischnochitonidae, ventral view showing head (H), foot (F), and gills (arrow); scale bar: 2 mm. (B–D) Scanning electron micrographs of *Acanthopleura* spp. provided by L. Brooker. (B) Three girdle scales from *A. loochooana*, note the sculptured surface; scale bar: 100 μm. (C) Section of the lateral region of an intermediate valve of *A. brevispinosa* showing three ocelli and numerous apical and subsidiary pores of aesthetes; scale bar: 50 μm. (D) Radula of *A. echinata*, scale bar: 400 μm. (E) Back-scattered electron image of ground and polished resin-infiltrated major lateral tooth of *A. spinosa* composed of tooth base (Tb) and tooth proper (T) fused at a distinct junction zone (Jz); brightness of tooth compartments varies according to mineral contents: magnetite region (Mr), lepitocrocite-region (Lr), anterior cusp region (Acr), posterior cusp region (Pcr); scale bar: 50 μm (micrograph by L. Brooker).

A Gill placement

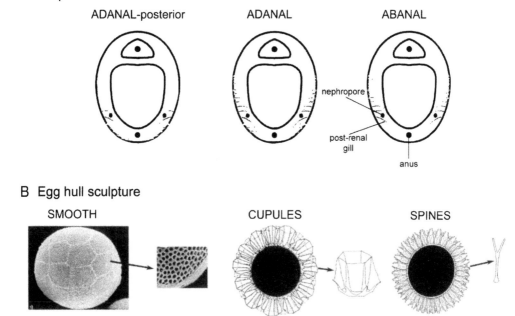

ADANAL-posterior   ADANAL   ABANAL

nephropore

post-renal gill

anus

B Egg hull sculpture

SMOOTH   CUPULES   SPINES

C Sperm morphology

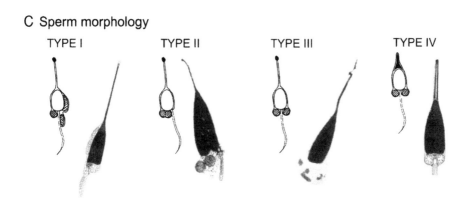

TYPE I   TYPE II   TYPE III   TYPE IV

FIGURE 4.7. Polyplacophora, characters important for phylogeny illustrating variations in gill placement (A), egg hull sculpture (B), and sperm morphology (C). A, from Okusu (2003), B, two left hand figures from Buckland-Nicks and Hodgson (2000), others from Sirenko (1993), C, schematic drawings from Okusu *et al.* (2003), others from Buckland-Nicks and Hodgson (2000). For further information see text.

is wide and fleshy, appears naked, dorsally with randomly distributed minute spicules.

Ischnochitonina: Articulamentum is well developed, generally with slits in all valves; teeth of insertion plates are pectinated or smooth; number of slits in the first valve generally higher than five; perinotum has

various types of elements (scales, hairs, spicules).

Acanthochitonina: Articulamentum is well developed with insertion plates in all valves; number of slits in the first valve does not exceed five; teeth of insertion plates never pectinated; perinotum wide and fleshy,

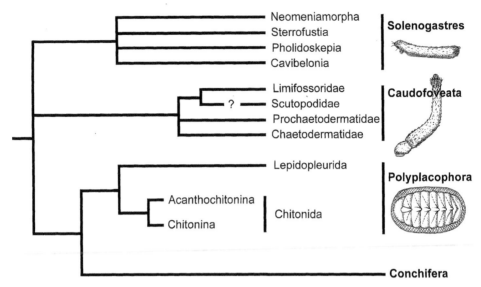

FIGURE 4.8. Tree diagram summarizing major clades within Solenogastres, Caudofoveata, and Polyplacophora with regard to recent knowledge. Within Caudofoveata, families are given as the highest taxonomic level because there are no orders defined. Note the lack of resolution in many positions and on different levels.

generally with spicules of different size, never scaly.

Some of the diagnostic characters in the preceding list have been criticized as being inappropriate for the higher classification of chitons. Although the tegmental structure of shell plates is of taxonomic relevance at the specific level (Haas 1972), the nature of the articulamentum is of interest for higher classification and reflects an evolutionary trend (Sirenko 1997). Therefore, an undeveloped articulamentum lacking insertion plates in the terminal and intermediate valves and with short and mainly unconnected apophyses is seen as the basal condition, and is still present in a few extant chitons (e.g., *Leptochiton*). The derived condition with either slit (e.g., *Ischnochiton*) or unslit insertion plates (e.g., *Choriplax*) is more common. Insertion plates with smooth teeth (e.g., *Ischnochiton*) are considered to be more primitive than those with pectinated teeth (e.g., *Chiton*).

Russell-Hunter (1988) discussed the importance of gill placement in chiton phylogeny, and recent work has shown a correlation among egg hull type, sperm morphology, and gill placement (Eernisse 1984; Sirenko 1993; Buckland-Nicks 1995; Okusu et al. 2003) (Figure 4.7). Chiton eggs have hull processes that are primarily secreted by the egg (Richter 1986) and seem to direct sperm to localized areas during fertilization (Buckland-Nicks 1993). The processes are typically either cup-shaped or spiny (Figure 4.8B) and show species-specific differences (Pearse 1979). Chitons with elaborate egg hulls also have sperm with asymmetrically arranged mitochondria and a long filamentous anterior extension of the nucleus (Pearse 1979), which has a reduced acrosomal vesicle at its tip (Type I and II sperm *sensu* Buckland-Nicks et al., 1990; Buckland-Nicks 1995) (Figure 4.8C). The ctenidia are positioned in characteristic numbers and arrangements along each side of the foot within the pallial cavity (for example, see Kaas and Van Belle 1985: fig. 3), even if variations in the exact number of ctenidia within species occur (Plate 1897, 1899, 1901). During ontogeny, the first ctenidial pair to appear is post-renal (immediately behind the nephridiopore) (Pelseneer 1899). In certain chitons, ctenidia are added exclusively anterior to the post-renal gill pair (abanal type), while in others they are added anteriorly and posteriorly (Eernisse 1984; Sirenko 1993; Eernisse and Reynolds 1994).

Sirenko (1993, 1997) updated the former classifications (Thiele 1909–1910; Bergenhayn 1930; Smith 1960; Van Belle 1983) and divided extant chitons into two orders, Lepidopleurida and Chitonida, the latter having two suborders, Chitonina and Acanthochitonina (Figure 4.5B). This is consistent with Buckland-Nicks' (1995) phylogenetic analysis, using 25 characters scored from egg hull, sperm, shell valves and ctenidia, of 10 polyplacophoran families (25 species examined in total), with two aplacophorans as outgroup taxa.

Lepidopleurida: Valve characters are presumably primitive, without slits in the insertion plates; ctenidia are adanal and restricted to the posterior region; sperm are ectaquasperm; eggs are smooth with extraordinary thick egg hulls.

Chitonida: Valve characters are presumably derived, with either slit or unslit insertion plates extending laterally into the girdle; ctenidia are of adanal or abanal type, always with a space between them and the anus papilla; sperm have a filamentous extension of the nucleus and reduced acrosome; there are elaborate egg hull processes.

The Chitonida was further divided into two suborders:

Chitonina: Ctenidial placement is adanal; a posterior extension of midpiece alters sperm shape; spiny, narrow-based egg hull projections; ocelli occur in some genera (e.g., *Onithochiton*).

Acanthochitonina: Ctenidial placement is abanal; the overall sperm shape differs from the preceding groups (for detailed descriptions see Buckland-Nicks 1995); egg hull with broadly based cupules that are not spiny; ocelli are absent.

This classification of Polyplacophora is corroborated by a recent molecular phylogenetic analysis of chiton relationships (Okusu *et al.* 2003), which included representatives of 28 species belonging to 13 families based on the combination of five genes (18S rRNA, 28S rRNA, 16S rRNA, COI, and histone 3). The resulting topology supports the two lineages, Lepidopleurida and Chitonida, but refutes monophyly of many classical taxonomical groups *sensu* Kaas and Van Belle. Okusu *et al.* (2003) further showed a strong correlation of egg hull morphology with the molecular phylogenetic trees. The study showed Lepidopleurida to be the more basal clade and Chitonida was divided into three lineages: taxa with simply round to weakly hexagonal cupules of the egg hull, abanal gills, and type I sperm (clade A in Okusu *et al.* 2003: fig. 8); taxa with egg hulls with strongly hexagonal cupules with flaps, abanal gills, and type I sperm (clade B in Okusu *et al.* 2003: fig. 8); and taxa with various shapes of spiny egg hulls, adanal gills, and type II sperm (clade C in Okusu *et al.* 2003, fig. 8; Chitonoidea *sensu* Sirenko 1997).

A number of additional characters, useful for systematics and phylogeny at different levels, have been investigated over the past few decades, and are summarized in the following paragraphs.

The position and morphology of osphradia vary among chiton taxa and may also be useful phylogenetically. According to ultrastructural data (Haszprunar 1986, 1987), a true osphradium is present only in Chitonida, while the more basal Lepidopleurida show branchial and lateral sense organs that do not appear to be homologous. However, some (if not all) genera of Lepidopleuridae have dark pigmentation under the mouth lappets, which may represent a true, anteriorly positioned osphradium (E. Schwabe, personal observation).

The occurrence and distribution of other sensory elements, including various types of aesthetes and ocelli in the shell plates (e.g., Fischer and Renner 1979; Currie 1992), as well as so-called ampullary cells and FMRF-amide-positive[1] neurons situated anteriorly underneath the apical ciliary tuft in chiton larvae (Haszprunar et al. 2002; Voronezhskaya et al. 2002), also appear to reflect phylogenetic relationships.

Radular characters have been used for chiton classification in the past (e.g., Thiele 1893, 1909–1910) but since then have been shown to be too homoplastic at the deeper levels (Eernisse 1984; Sirenko 1993, 1997; Eernisse and Reynolds 1994; Buckland-Nicks 1995; Okusu et al. 2003). Nevertheless, they are valuable at certain taxonomic levels (e.g., Bullock 1988; Saito 2004). Saito (2004), for example, points out that selected radular characters within the Cryptoplacoidea correlate with a reduction of the tegmentum within that group. Morphometric data such as the ratio of radular length to total body length, length of the radular cartilages to total radular length, and number of radular teeth rows to radular length may also be of phylogenetic relevance (E. Schwabe, personal observation). There is a wealth of data on radular mineralization in chitons (e.g., Macey et al. 1994; Macey and Brooker 1996; Macey et al. 1996; Lee et al. 1998; Brooker et al. 2003; Wealthall et al. 2005) and, according to Brooker and Macey (2001), specific traits in radular biomineralization can also be of systematic importance. With the help of light and scanning electron microscopy as well as energy-dispersive spectroscopy (see Figure 4.6E), they showed that iron levels in the teeth of some species only recently merged into Acanthopleura by Ferreira (1986) differ considerably from the traditional members of this taxon, including its type species.

Interesting information is also available on the karyotypes of chitons. Yum (1993) provided cytogenetic data for eight species and thus extented Nakamura's (1985) list to 22. The diploid chromosome number in chitons ranges from 12 to 26, with Acanthochitonidae showing a higher variability, ranging from 16 to 24, while Chitonidae are more uniform, ranging from 24 to 26 and all Ischnochitonidae are 24. In Ischnochitonidae, the chromosome arm morphology is meta- or submetacentric only, while additional telo- or subtelocentric arm morphologies occur in other chiton taxa.

The oxygen-binding protein hemocyanin has been found in chitons, cephalopods, protobranch bivalves, and gastropods. As its origin is calculated to be Precambrian it has been explored for its potential to resolve molluscan evolution (Lieb and Markl 2004). The importance of this protein for a species-level phylogeny and as a marker for evolutionary studies was demonstrated for basal gastropods by Streit et al. (2006), and attempts to reveal chiton phylogenetic relationships by means of this new molecular approach are in progress (B. Lieb, personal communication).

ADAPTIVE RADIATIONS

The most outstanding innovations of early molluscs in comparison to their putative predecessors, and to other exant spiralians with similar lifestyles, are the differentiation of a dorsal mantle completely covered in cuticle and sclerites, a ciliated ventral foot for locomotion, and the development of the radula as an effective feeding apparatus.

A protective cover composed of sclerites (scleritome) can be found in many of the earliest known putative molluscs (Wiwaxia, Halkieria) as well as in all three basal groups of extant molluscs (with additional shell plates in Polyplacophora) and thus may be viewed as a symplesiomorphy of modern Mollusca (most recently by Scheltema and Schander 2006). It is interesting to note, however, that the earliest fossils presumably belonging to the molluscan lineage (Kimberella, Odontogriphus) did not possess any shell or scleritome at all, indicating that these structures were derived within early molluscs. The evolutionary advantage of a scleritome composed of numerous small sclerites, such as in

---

1. FMRF-amide, a molluscan cardio-excitatory neurotransmitter, is a tetrapeptide composed of phenylalanine (F), methionine (M), arginine (R), and phenylalanine (F) residues, with the terminal acid group converted to an amide group.

solenogasters and caudofoveates, is obviously protection against predators and not so much against the physical impacts of tides and water currents. This probably accounts for aplacophorans being largely restricted to more sheltered habitats such as deep-water soft sediments and sublittoral hard bottoms. The very few shallow-water species (mostly Solenogastres) occur in coral reefs or are part of the subtidal meiobenthos. In contrast, many chitons inhabit the rocky intertidal, where they withstand strong physical forces protected by their tough cuticle and shell plates and are kept in place by their broad, highly muscular foot.

The model archimollusc of textbooks typically resembles a chiton or untorted limpet in body shape, and the vermiform shape of aplacophorans is usually viewed as a derived feature (Salvini-Plawen 1972, 1985, 2003; Scheltema 1993, 1996) or sometimes a plesiomorphic one (Haszprunar 2000; Haszprunar et al., Chapter 2). The expansion along the longitudinal body axis may be explained as an adaptation to epizoism (Solenogastres) or burrowing (Caudofoveata) (Salvini-Plawen 1972; but see Scheltema 1996 for a contrasting view). The complete reduction of a foot in caudofoveates, combined with the appearance of a mouth shield, is generally seen as connected to their burrowing lifestyle.

The radular morphologies of the three clades discussed herein, reflect divergent feeding habits. Based on fossil evidence, the most primitive radula (*Wiwaxia, Odontogriphus*) was used for algal mat grazing (Caron et al. 2006). Some authors argue that the primitive radula was used for either shoveling in detritus or grabbing large food items and was a broad structure consisting of several rows of wide, sclerotized teeth with denticles, the individual teeth connected by a flexible cuticle (e.g., Salvini-Plawen 2003; Scheltema et al. 2003). This type of radula is found in some caudofoveates (*Scutopus*; Figure 4.4C) and solenogasters (*Helicoradomenia*; Figure 4.4D). From this state pincer-like structures for picking up individual diatoms evolved within caudofoveates, while multiple rows of distichous hooks

with long and pointed denticles and a variety of other radular morphologies adapted for carnivory were developed in solenogasters. The extremely long radular ribbon of all modern chitons, which bears multiple sclerotized and sturdy teeth in part impregnated with metals, is, in contrast, a specialized tool for grazing on hard substrates.

GAPS IN KNOWLEDGE

As shown above, recent research in the fields of palaeontology, ultrastructure, and molecular biology has led to a better understanding of basal molluscs, their biology, and internal relationships. Although modern approaches, such as selective staining techniques for nervous tissues and musculature, *in situ* hybridization combined with tracing of gene expression in development, and multigene approaches for phylogenetic analyses, have already brought a wealth of important knowledge about Polyplacophora, such investigations are still largely lacking for the aplacophoran taxa. In Polyplacophora, however, information about more taxa needs to be added to the existing data matrices to strengthen phylogenetic concepts. This includes morphological data, such as sperm and egg hull structure, chromosome numbers, and radula characters, as well as molecular data. Comparative investigations of sense organs, excretory organs, and larval characters are needed to clarify the usefulness of these characters for phylogeny. The same is true for protein coding sequences, such as hemocyanin or ribosomal protein coding sequences, revealed by expressed sequence tag (EST) projects or selective analysis. For the aplacophoran taxa we still lack molecular studies that include a representative set of taxa. Even though our knowledge of the morphology of Solenogastres and Caudofoveata is extensive, the homology of certain characters between these taxa (mouth shield, vestibulum, foot; regions of the gonopericardial tract) and Polyplacophora (midgut regions; excretory system) is not yet well established. Moreover, additional cladistic analyses based on morphological characters are needed for both the aplacophoran taxa.

ACKNOWLEDGMENTS

The authors are grateful to Amélie Scheltema and Luitfried Salvini-Plawen for continuous help and support and for fruitful discussions. Photographic material was generously provided by Lesley Brooker and Victoriano Urgorri and the drawings of fossils by Carl-Otto Schander. This work was partly supported by a grant from the Swedish Research Council (to CS).

REFERENCES

Baba, K. 1938. The later development of a solenogastre, *Epimenia verrucosa* (Nierstrasz). *Journal of the Department of Agriculture of the Fukuoka University* 6: 21–40.

———. 1940a. The mechanisms of absorption and excretion in a solenogastre, *Epimenia verrucosa* (Nierstrasz). *Journal of the Department of Agriculture of the Fukuoka University* 6: 119–166.

———. 1940b. The early development of a solenogastre, *Epimenia verrucosa* (Nierstrasz). *Annotationes Zoologicae Japonenses* 19: 223–256.

Bengtson, S. 1992. The cap-shaped Cambrian fossil *Maikhanella* and the relationship between coeloscleritophorans and molluscs. *Lethaia* 25: 401–420.

———. 1993. The molluscan affinity of coeloscleritophorans—reply. *Lethaia* 26: 48.

Bengtson, S., and Missarzhevsky, V. V. 1981. Coeloscleritophora, a major group of enigmatic Cambrian metazoans. U.S. Geological Survey open-file report 81–743 (Short Papers from the Second International Symposium on the Cambrian System): 19–21.

Bergenhayn, J. R. M. 1930. Kurze Bemerkungen zur Kenntnis der Schalenstruktur und Systematik der Loricaten. *Kungliga Svenska Vetenskapsakademiens Handlingar* 9: 3–54.

———. 1955. Die fossilen schwedischen Loricaten nebst einer vorläufigen Revision des Systems der ganzen Klasse Loricata. *Kungliga Fysiografiska Sällskapets Handlingar* N.F. 66: 1–44.

Brooker, L. R., Lee, A. P., Macey, D. J., Bronswijk, W. v., and Webb, J. 2003. Multiple-front iron-mineralization in chiton teeth (*Acanthopleura echinata*: Mollusca: Polyplacophora). *Marine Biology* 142: 447–454.

Brooker, L. R., and Macey, D. J. 2001. Biomineralization in chiton teeth and its usefulness as a taxonomic character in the genus *Acanthopleura* Guilding, 1829 (Mollusca: Polyplacophora). *American Malacological Bulletin* 16: 203–215.

Buckland-Nicks, J. 1993. Hull capsules of chiton eggs: parachute structures and sperm focusing devices? *The Biological Bulletin* 184: 269–276.

———. 1995. Ultrastructure of sperm and sperm-egg interaction in Aculifera: Implications for molluscan phylogeny. In *Advances in spermatozoal phylogeny*. Edited by B.G.M. Jamieson, J. Ausió, and J. L. Justine. Paris: *Mémoires du Muséum National d'Histoire Naturelle* 166, pp. 129–153.

Buckland-Nicks, J., Chia, F. S., and Koss, R. 1990. Spermiogenesis in Polyplacophora, with special reference to acrosome formation (Mollusca). *Zoomorphology* 109: 179–188.

Buckland-Nicks, J. and Hodgson, A. N. 2000. Fertilization in Callochiton castaneus (Mollusca). *Biological Bulletin* 199: 59–67.

Buckland-Nicks, J., Gibson, G., and Koss, R. 2002. Phylum Mollusca: Polyplacophora, Aplacophora, Scaphopoda. In *Atlas of Marine Invertebrate Larvae*. Edited by C. M. Young. San Diego, San Francisco: Academic Press, pp. 245–259.

Bullock, R. C. 1988. The genus *Chiton* in the New World (Polyplacophora: Chitonidae). *The Veliger* 31: 141–191.

Butterfield, N. J. 1990. A reassessment of the enigmatic Burgess Shale fossil *Wiwaxia corrugata* (Matthew) and its relationship to the polychaete *Canadia spinosa* Walcott. *Paleobiology* 16: 287–303.

———. 1994. Burgess Shale type fossils from a Lower Cambrian shallow-shelf sequence in Northwestern Canada. *Nature* 369: 477–479.

———. 2006. Hooking some stem-group "worms": fossil lophotrochozoans in the Burgess Shale. *BioEssays* 28: 1161–1166,

Caron, J.-B., Scheltema, A. H., Schander, C., and Rudkin, D. 2006a. A soft-bodied mollusk with a radula from the Middle Cambrian Burgess Shale. *Nature* 442: 159–163.

———. 2006b. Reply to Butterfield on stem-group "worms": fossil lophotrochozoans in the Burgess Shale. *BioEssays* 29: 1–3.

Carter, J. G., and Hall, R. M. 1990. Polyplacophora, Scaphopoda, Archaeogastropoda and Paragastropoda (Mollusca). In *Skeletal Biomineralization: Patterns, Processes and Evolutionary Trends*. Edited by J. G. Carter. New York: Van Nostrand Reinhold, pp. 25–51.

Chaffee, C., and Lindberg, D. R. 1986. Larval biology of early Cambrian molluscs: the implications of small body size. *Bulletin of Marine Science* 39: 536–549.

Christiansen, M. E. 1954. The life history of *Lepidopleurus asellus* (Spengler) (Placophora). *Nytt Magasin für Zoologi* 2: 52–72.

Cohn, M. J., and Tickle, C. 1999. Developmental basis of limblessness and axial patterning in snakes. *Nature* 399: 474–479.

Conway Morris, S. 1976. A new Cambrian lophophorate from the Burgess Shale of British Columbia. *Palaeontology* 19: 199–222.

———. 1985. The Middle Cambrian metazoan *Wiwaxia corrugata* (Matthew) from the Burgess Shale and Ogygopsis Shale, British Columbia, Canada. *Philosophical Transactions of the Royal Society of London, Series B* 307: 507–586.

Conway Morris, S., and Peel, J. S. 1990. Articulated halkieriids from the Lower Cambrian of North Greenland. *Nature* 345: 802–805.

———. 1995. Articulated halkieriids from the Lower Cambrian of North Greenland and their role in early protostome evolution. *Philosophical Transactions of the Royal Society of London, Series B* 347: 305–358.

Cruz, R., Lins, U., and Farina, M. 1998. Minerals of the radular apparatus of *Falcidens* sp. (Caudofoveata) and the evolutionary implications for the phylum Mollusca. *The Biological Bulletin* 194: 224–230.

Currie, D. R. 1992. Aesthete channel morphology in three species of Australian chitons (Mollusca: Polyplacophora). *Journal of the Malacological Society of Australia* 13: 3–14.

Eernisse, D. J. 1984. *Lepidochitona* Gray, 1821 (Mollusca: Polyplacophora), from the Pacific Coast of the United States: Systematics and reproduction. Ph.D. Dissertation, University of California, Santa Cruz.

Eernisse, D. J., and Kerth K. 1988. The initial stages of radular development in chitons (Mollusca, Polyplacophora). *Malacologia* 28: 95–103.

Eernisse, D. J., and Reynolds, P. D. 1994. Polyplacophora. In *Microscopic Anatomy of Invertebrates*. Vol. 5, *Mollusca I*. Edited by F. W. Harrison. New York: Wiley-Liss, pp. 56–110.

Eidbye-Jacobsen, D. 2004. A reevaluation of *Wiwaxia* and the polychaetes of the Burgess Shale. *Lethaia* 37: 317–335.

Fedonkin, M. A., and Waggoner, B. M. 1997. The late precambrian fossil *Kimberella* is a mollusc-like bilaterian organism. *Nature* 388: 868–871.

Ferreira, A. J. 1986. A revision of the genus *Acanthopleura* Guilding, 1829 (Mollusca: Polyplacophora). *The Veliger* 28: 221–279.

Fischer, F. P., and Renner, M. 1979. SEM-Observations on the shell plates of three Polyplacophorans (Mollusca, Amphineura). *Spixiana* 2: 49–58.

Friedrich, S., Wanninger, A., Brückner, M., and Haszprunar, G. 2002. Neurogenesis in the mossy chiton, *Mopalia muscosa* (Gould) (Polyplacophora): evidence against molluscan metamerism. *Journal of Morphology* 253: 109–117.

Ghiselin, M. T. 1988. The origin of molluscs in the light of molecular evidence. *Oxford Surveys in Evolutionary Biology* 5: 66–95.

Giribet, G., and Wheeler, W. C. 2002. On bivalve phylogeny: a high-level analysis of the Bivalvia (Mollusca) based on combined morphology and DNA sequence data. *Invertebrate Biology* 121: 271–324.

Giribet, G., Okusu, A., Lindgren, A. R., Huff, S. W., Schrödl, M., and Nishiguchi, M. L. 2006. Evidence for a clade composed of molluscs with serially repeated structures: Monoplacophorans are related to chitons. *Proceedings of the National Academy of Sciences of the U.S.A.* 103: 7723–7728.

Glaessner, M. F. 1969. Decapoda. In *Treatise on Invertebrate Paleontology*, Part R: *Arthropoda 4*. Edited by R. C. Moore. Boulder, CO, and Lawrence, KS: Geological Society of America and the University of Kansas Press, pp. 399–566.

Glaubrecht, M., Maitas L., and Salvini-Plawen, L. v. 2005. Aplacophoran Mollusca in the Natural History Museum Berlin. An annotated catalogue of Thiele's type specimens, with a brief review of "Aplacophora" classification. *Mitteilungen des Museums für Naturkunde Berlin, Zoologische Reihe* 81: 145–166.

Götting, K. J. 1980. Origin and relationships of the Mollusca. *Zeitung für Zoologische Systematik und Evolutionsforschung* 18: 24–27.

Gowlett-Holmes, K. L. 1987. The suborder Choriplacina Starobogatov & Sirenko, 1975 with a redescription of *Choriplax grayi* (H. Adams & Angas, 1864) (Mollusca: Polyplacophora). *Transactions and Proceedings of the Royal Society of South Australia* 111: 105–110.

Grave, B. H. 1932. Embryology and life history of *Chaetopleura apiculata*. *Journal of Morphology* 54: 153–160.

Gray, J. E. 1821. A natural arrangement of Mollusca, according to their internal structure. *The London Medical Repository* 15: 229–239.

Haas, W. 1972. Untersuchungen über die Mikro- und Ultrastruktur der Polyplacophorenschale. *Biomineralisation Research Report* 6: 1–52.

———. 1981. Evolution of calcareous hardparts in primitive molluscs. *Malacologia* 21: 403–418.

Handl, C. H., and Salvini-Plawen, L. v. 2001. New records of Solenogastres-Pholidoskepia (Mollusca) from Norwegian fjords and shelf waters including two new species. *Sarsia* 86: 367–381.

Handl, C. H., and Todt, C. 2005. The foregut glands of Solenogastres (Mollusca): anatomy and revised terminology. *Journal of Morphology* 265: 28–42.

Haszprunar, G. 1986. Feinmorphologische Untersuchungen an Sinnesstrukturen ursprünglicher

Solenogastres (Mollusca). *Zoologischer Anzeiger* 217: 345–362.

———. 1987. The fine morphology of the osphradial sense organs of the Mollusca IV. Caudofoveata and Solenogastres. *Philosophical Transactions of the Royal Society of London, Series B* 315: 63–73.

———. 1992. The first molluscs—small animals. *Bolletitno de Zoologia* 59: 1–16.

———. 2000. Is the Aplacophora monophyletic? A cladistic point of view. *American Malacological Bulletin* 15: 115–130.

Haszprunar, G., Friedrich, S., Wanninger, A., and Ruthensteiner, B. 2002. Fine structure and immunocytochemistry of a new chemosensory system in the chiton larva (Mollusca: Polyplacophora). *Journal of Morphology* 251: 210–218.

Hatschek, B., ed. 1891. *Lehrbuch der Zoologie.* Jena: Gustav Fischer Verlag.

Heath, H. 1899. The development of *Ischnochiton*. *Zoologische Jahrbücher. Abteilung für Anatomie und Ontogenie der Tiere* 12: 567–656.

———. 1918. Solenogastres from the eastern coast of North America. *Memoirs of the Museum of Comparative Zoology at Harvard College* 45: 185–263.

Henry, J. Q., Okusu, A., and Martindale, M. Q. 2004. The cell lineage of the polyplacophoran, *Chaetopleura apiculata*: variation in the spiralian program and implications for molluscan evolution. *Developmental Biology* 272: 145–160.

Hoare, R. D. 2000. Considerations on Paleozoic Polyplacophora including the description of *Plasiochiton curiosus* n. gen. and sp. *American Malacological Bulletin* 15: 131–137.

Hoare, R. D., and Mapes, R. H. 1995. Relationships of the Devonian *Strobilepis* and related Pensylvanian problematica. *Acta Palaeontologica Polonica* 40: 111–128.

Hoffman, S. 1949. Studien über das Integument der Solenogastren. *Zoologiska Bidrag fran Uppsala* 27: 293–427.

Hoffmann, H. 1930. Amphineura. In *Bronn's Klassen und Ordnungen des Tier-Reiches* 3, 1. Abteilung, Nachträge. Leipzig: Akademische Verlagsgesellschaft, pp. 1–453.

Hyman, L. H. 1967. Class Aplacophora. In *The Invertebrates*. Vol. VI, *Mollusca I*. Edited by L. H. Hyman. New York: McGraw-Hill Book Company, pp. 13–70.

Ivanov, D. L. 1981. *Caudofoveatus tetradens* gen. et sp. n. and diagnosis of the subclass Caudofoveata (Mollusca, Aplacophora). *Zoologicheskij Zhurnal* 60: 18–28 [in Russian].

———. 1996. Origin of Aculifera and problems of monophyly of higher taxa in molluscs. In *Origin and evolutionary radiation of the Mollusca*. Edited by J. D. Taylor. Oxford: Oxford University Press, pp. 59–65.

Jacobs, D. K., Wray, C. G., Wedeen, C. J., Kostriken, R., DeSalle, R., Staton, J. L., Gates, R. D., and Lindberg, D. R. 2000. Molluscan engrailed expression, serial organization and shell evolution. *Evolution and Development* 2: 340–347.

Jaegersten, G. 1972. *Evolution of the metazoan life cycle*. A comprehensive theory. London, New York: Academic Press, pp. 1–282.

Kaas, P., and Van Belle, R. A. 1985. *Monograph of living chitons (Mollusca: Polyplacophora)*. Vol. 2, *Suborder Ischnochitonina, Ischnochitonidae: Schizoplacinae, Callochitoninae and Lepidochitoninae*. Leiden: E. J. Brill/W. Backhuys, 1–198.

———. 1994. *Monograph of living chitons (Mollusca: Polyplacophora)*. Vol. 5, *Suborder Ischnochitonina, Ischnochitonidae: Ischnochitoninae (continued)*. Leiden: E. J. Brill, pp. 1–464.

Kaas, P., Jones, A. M., and Gowlett-Holmes, K. L. 1998. Class Polyplacophora. In *Mollusca: The Southern Synthesis*. Vol. 5, *Fauna of Australia*. Edited by P. L. Beesley, G. J. B. Ross, and A. Wells. Melbourne: CSIRO Publishing, pp. 161–177.

Kowalevsky, M. A. 1883. Embryogénie du *Chiton polii* (Philippi) avec quelques remarques sur le développement des autres Chitons. *Annales du Muséum d'Histoire Naturelle, Marseilles* 1: 1–46.

Latyshev, N. A., Khardin, A. S., Kasyanov, S. P., and Ivanova M. B. 2004. A study on the feeding ecology of chitons using analysis of gut contents and fatty acid markers. *Journal of Molluscan Studies* 70: 225–230.

Lee, A. P., Webb, J., Macey, D. J., Bronswijk, W. v., Savarese, A. R., and Charmaine de Witt, G. 1998. In situ Raman spectroscopic studies of the teeth of the chiton *Acanthopleura hirtosa*. *Journal of Biological Inorganic Chemistry* 3: 614–619.

Lieb, B., and Markl, J. 2004. Evolution of molluscan hemocyanins as deduced from DNA sequencing. *Micron* 35: 117–119.

Lindberg, D. R., and Ponder, W. F. 1996. An evolutionary tree for the Mollusca: branches or roots? In *Origin and Evolutionary Radiation of the Mollusca*. Edited by J. Taylor. Oxford: Oxford University Press, pp. 67–75.

Lummel, L. v. 1930. Untersuchungen über einige Solenogastren. *Zeitschrift zur Morphologie und Ökologie der Tiere* 18: 347–383.

Lundin, K., and Schander, C. 1999. Ultrastructure of gill cilia and ciliary rootlets of *Chaetoderma nitidulum* Lovén, 1844 (Mollusca, Chaetodermomorpha). *Acta Zoologica* 80: 185–191.

———. 2001a. Ciliary ultrastructure of polyplacophorans (Mollusca, Amphineura, Polyplacophora). *Journal of Submicroscopic Cytology and Pathology* 33: 93–98.

———. 2001b. Ciliary ultrastructure of neomeniomorphs (Mollusca, Neomeniomorpha = Solenogastres). *Invertebrate Biology* 120: 342–349.

Lydeard, C., Holznagel, W. E., Schnare, M. N., and Gutell, R. R. 2000. Phylogenetic analysis of molluscan mitochondrial LSU rDNA sequences and secondary structures. *Molecular Phylogeny and Evolution* 15: 83–102.

Macey, D. J., Webb, J., and Brooker, L. R. 1994. The structure and synthesys of biominerals in chiton teeth. *Bulletin de l'Institut océanographique, Monaco,* 14: 191–197.

Macey, D. J., and Brooker, L. R. 1996. The junction zone: Initial site of mineralization in radula teeth of the chiton *Cryptoplax striata* (Mollusca: Polyplacophora). *Journal of Morphology* 230: 33–42.

Macey, D. J., Brooker, L. R., Webb, J., and St. Pierre, T. G. 1996. Structural organization of the cusps of the radular teeth of the chiton *Plaxiphora albida*. *Acta Zoologica* 77: 287–294.

Nakamura, H. K. 1985. A review of molluscan cytogenetic information based on the CISMOCH—computerized index system for molluscan chromosomes. Bivalvia, Polyplacophora and Cephalopoda. *Venus* 44: 193–225.

Nielsen, C. 1987. Structure and function of metazoan ciliary bands and their phylogenetic significance. *Acta Zoologica* 68: 205–262.

———. 1995. *Animal evolution; Interrelationships of the living phyla.* Oxford, UK: Oxford University Press.

———. 2004. Trochophora larvae: cell-lineages, ciliary bands, and body regions. 1. Annelida and Mollusca. *Journal of Experimental Zoology* 302: 35–68.

Nielsen, C., Haszprunar, G., Ruthensteiner, B., and Wanninger, A. 2007. Early development of the aplacophoran mollusc *Chaetoderma*. *Acta Zoologica* 88: 231–247.

Okusu, A. 2002. Embryogenesis and development of *Epimenia babai* (Mollusca Neomeniomorpha). *The Biological Bulletin* 203: 87–103.

———. 2003. Evolution of "early" molluscs: integrating phylogenetic, development, and morphological approaches. Ph.D. dissertation, Harvard University.

Okusu, A., and Giribet, G. 2003. New 18S rRNA sequences from neomenioid aplacophorans and the possible origin of persistent exogenous contamination. *Journal Molluscan Studies* 69: 385–387.

Okusu, A., Schwabe, E., Eernisse, D. J., and Giribet, G. 2003. Towards a phylogeny of chitons (Mollusca, Polyplacophora) based on combined analysis of five molecular loci. *Organisms Diversity and Evolution* 3: 281–302.

Passamaneck, Y. J., Schander, C., and Halanych, K. M. 2004. Investigation of molluscan phylogeny using large-subunit and small-subunit nuclear rRNA sequences, and analysis of rate variation across lineages. *Molecular Phylogenetics and Evolution* 32: 25–38.

Pearse, J. S. 1979. Polyplacophora. In *Reproduction of Marine Invertebrates.* Vol. 5. *Molluscs: Pelecypoda and lower classes.* Edited by A. C. Giese and J. S. Pearse. New York: Academic Press, pp. 27–85.

Pelseneer, P. 1899. Recherches morphologiques et phylogénétiques sur les mollusques Archaiques. *Memoires de l'Academie Royale des Sciences de Belgique* 57: 1–112.

Peters, W. 1972. Occurrence of chitin in Mollusca. *Comparative Biochemistry and Physiology* 41: 541–550.

Plate, L. H. 1897. Die Anatomie und Phylogenie der Chitonen. Fauna Chilensis 1 (1). *Zoologische Jahrbücher, Abteilung für Systematik, Ökologie und Geographie der Tiere* 1 Suppl. 4: 1–243.

———. 1899. Die Anatomie und Phylogenie der Chitonen. Fauna Chilensis 2 (1). *Zoologische Jahrbücher, Abteilung für Systematik, Ökologie und Geographie der Tiere* 2 Suppl. 5: 15–216.

———. 1901. Die Anatomie und Phylogenie der Chitonen. Fauna Chilensis 2 (2). *Zoologische Jahrbücher, Abteilung für Systematik, Ökologie und Geographie der Tiere* 3 Suppl. 5: 281–600.

Pojeta, J., Jr. 1980. Molluscan phylogeny. *Tulane Studies in Geology and Paleontology* 16: 55–80.

Pojeta, J., Jr., Eernisse, D. J., Hoare, R. D., and Henderson, M. D. 2003. *Echinochiton dufoei:* a new spiny Ordovician chiton. *Journal of Paleontology* 77: 646–654.

Ponder W. F., Parkhaev P. Yu., and Beechey D. L. 2007. A remarkable similarity in scaly shell structure in Early Cambrian univalved limpets (Monoplacophora; Maikhanellidae) and a Recent fissurellid limpet (Gastropoda: Vetigastropoda) with a review of Maikhanellidae. *Molluscan Research* 27: 129–139.

Pruvot, G. 1890. Sûr le développement d'un solenogastre. *Comptes Rendus de l'Académie des Sciences Paris* 114: 1211–1214.

Richter, H.-P. 1986. Ultrastructure of follicular epithelia in the ovary of *Lepidochitona cinerea* (L.) (Mollusca: Polyplacophora). *Development Growth and Differentiation* 28: 7–16.

Rosenberg, G., Tillier, S., Tillier, A., Kuncio, G. S., Hanlon, R. T., Masselot, M., and Williams, C. J. 1997. Ribosomal RNA phylogeny of selected major clades in the Mollusca. *Journal of Molluscan Studies* 63: 301–309.

Runnegar, B., Pojeta, J., Jr., Taylor, M. E., and Collins, D. 1979. New species of the Cambrian and Ordovician chitons *Matthevia* and *Chelodes* from Wisconsin and Queensland: evidence for the early history of polyplacophoran mollusks. *Journal of Paleontology* 53: 1374–1394.

Runnegar, B., and Pojeta, J., Jr. 1985. Origin and diversification of the Mollusca. In *The Mollusca*. Vol. 10, *Evolution*. Edited by E. R. Trueman and M. R. Clarke. London: Academic Press, pp. 1–57.

Russell-Hunter, W. D. 1988. The gills of chitons (Polyplacophora) and their significance in molluscan phylogeny. *American Malacological Bulletin* 6: 69–78.

Saito, H. 2004. Phylogenetic significance of the radula in chitons, with special reference to the Cryptoplacoidea (Mollusca: Polyplacophora). *Bollettino Malacologico* Suppl. 5 : 83–104.

Salvini-Plawen, L. v. 1967. Über die Beziehungen zwischen den Merkmalen von Standort, Nahrung und Verdauungstrakt von Solenogastres (Aculifera, Aplacophora). *Zeitschrift zur Morphologie und Ökologie der Tiere* 59: 318–340.

———. 1971. *Schild- und Furchenfüßer (Caudofoveata und Solenogastres), verkannte Weichtiere am Meeresgrund*. Ziemsen (Wittenberg): Die Neue Brehm-Bücherei 441, pp. 1–95.

———. 1972. Zur Morphologie und Phylogenie der Mollusken: Die Beziehungen der Caudofoveata und der Solenogastres als Aculifera, als Mollusca und als Spiralia. *Zeitschrift für wissenschaftliche Zoologie* 184: 205–304.

———. 1975. *Marine Invertebrates of Scandinavia 4: Mollusca Caudofoveata*. Oslo: Universittetsforlaget. pp. 1–55.

———. 1978. Antarktische und subantarktische Solenogastres (Eine Monographie 1889–1974). *Zoologica* 44: 1–315.

———. 1980. A reconsideration of systematics in the Mollusca (Phylogeny and higher classification). *Malacologia* 19: 249–178.

———. 1981a. The molluscan digestive system in evolution. *Malacologia* 21: 371–401.

———. 1981b. On the origin and evolution of the Mollusca. *Atti Convegni Lincei (Roma)* 49: 235–293.

———. 1985. Early evolution and the primitive groups. In *The Mollusca*. Vol. 10, *Evolution*. Edited by K. Wilbur. New York: Academic Press, pp. 59–150.

———. 1988. The structure and function of molluscan digestive systems. In *The Mollusca*. Vol. 11, *Form and Function*. Edited by K. Wilbur, New York: Academic Press, pp. 301–379.

———. 1990. Origin, phylogeny and classification of the phylum Mollusca. *Iberus* 9: 1–33.

———. 1992. On certain Caudofoveata from the VEMA-Expedition. In *Proceedings of the 9th int. Malacol. Congress* (Edinburgh 1986). Leiden: Unitas Malacologica, pp. 317–333.

———. 2003. On the phylogenetic significance of the aplacophoran Mollusca. *Iberus* 21: 67–97.

———. 2004. Contributions to the morphological diversity and classification of the order Cavibelonia (Mollusca: Solenogastres). *Journal of Molluscan Studies* 70: 73–93.

Salvini-Plawen, L. v., and Nopp, H. 1974. Chitin bei Caudofoveata (Mollusca) und die Ableitung ihres Radulaapparates. *Zeitschrift für Morphologie der Tiere* 77: 77–86.

Salvini-Plawen, L. v., and Steiner, G. 1996. Synapomorphies and plesiomorphies in higher classification of Mollusca. In *Origin and evolutionary radiation of the Mollusca*. Edited by J. D. Taylor. Oxford: Oxford University Press, pp. 29–51.

Scheltema, A. H. 1976. Two new species of *Chaetoderma* from off West Africa (Aplacophora, Chaetodermatidae). *Journal of Molluscan Studies* 42: 223–234.

———. 1981. Comparative morphology of the radulae and alimentary tracts in the Aplacophora. *Malacologia* 20: 361–383.

———. 1985. The aplacophoran family Prochaetodermatidae in the North American Basin, including *Chevroderma* n.g. and *Spathoderma* n.g. (Mollusca, Chaetodermomorpha). *Biological Bulletin* 169: 484–529.

———. 1988. Ancestors and descendants: Relationships of the Aplacophora and Polyplacophora. *American Malacological Bulletin* 6: 57–68.

———. 1993. Aplacophora as progenetic aculiferans and the coelomate origin of molluscs as the sister taxon of Sipuncula. *The Biological Bulletin* 184: 57–78.

———. 1996. Phylogenetic position of Sipuncula, Mollusca and the progenetic Aplacophora. In *Origin and evolutionary radiation of the Mollusca*. Edited by J. D. Taylor. Oxford: Oxford University Press, pp. 53–58.

———. 1998. Class Aplacophora. In *Mollusca: The Southern Synthesis*. Vol. 5, *Fauna of Australia*. Edited by P. L. Beesley, G. J. B. Ross, and A. Wells. Melbourne: CSIRO Publishing, pp. 145–159.

———. 1999. Two solenogaster molluscs, *Ocheyoherpia trachia* n.sp. from Macquarie Island and *Tegulaherpia tasmanica* Salvini-Plawen from Bass Strait (Aplacophora: Neomeniomorpha). *Records of the Australian Museum* 51: 23–31.

Scheltema, A. H., and Jebb, M. 1994. Natural history of a solenogaster mollusc from Papua New Guinea, *Epimenia australis* (Thiele) (Aplacophora, Neomeniomorpha). *Journal of Natural History* 28: 1297–1318.

Scheltema, A. H., and Ivanov, D. L. 2000. Prochaetodermatidae of the Eastern Atlantic Ocean and Mediterranean Sea (Mollusca: Aplacophora). *Journal of Molluscan Studies* 66: 313–362.

———. 2002. An aplacophoran postlarva with iterated dorsal groups of spicules and skeletal similarities to Paleozoic fossils. *Invertebrate Biology* 121: 1–10.

———. 2004. Use of birefringence to characterize Aplacophora sclerites. *The Veliger* 47: 153–156.

Scheltema, A. H., Kerth, K., and Kuzurian, A. M. 2003. The original molluscan radula: comparisons among Aplacophora, Polyplacophora, Gastropoda, and the Cambrian fossil *Wiwaxia corrugata*. *Journal of Morphology* 257: 219–245.

Scheltema, A. H., and Schander, C. 2000. Discrimination and Phylogeny of Solenogaster species through the morphology of hard parts (Mollusca, Aplacophora, Neomeniomorpha). *The Biological Bulletin* 198: 121–151.

———. 2006. Exoskeletons: Tracing Molluscan Evolution. *Venus* 65: 19–26.

Scheltema, A. H., Tscherkassky, M., and Kuzirian, A. M. 1994. Aplacophora. In *Microscopic Anatomy of Invertebrates*. Vol. 5 Mollusca 1. Edited by F. H. Harrison and A. J. Kohn. New York: Wiley-Liss, pp. 13–54.

Schwabe, E. 2005. A catalogue of recent and fossil chitons (Mollusca: Polyplacophora) Addenda. *Novapex* 6: 89–105.

Sirenko, B. I. 1993. Revision of the system of the order Chitonida (Mollusca: Polyplacophora) on the basis of correlation between the type of gill arrangement and the shape of the chorion processes. *Ruthenica* 3: 93–117.

———. 1997. The importance of the development of articulamentum for taxonomy of chitons (Mollusca, Polyplacophora). *Ruthenica* 7: 1–24.

———. 2004. The ancient origin and persistence of chitons (Mollusca, Polyplacophora) that live and feed on deep submerged land plant matter (xylophages). *Bollettino Malacologico* Suppl. 5: 111–116.

Sirenko, B. I., and Minichev, Y. 1975. Développement ontogénétique de la radula chez les polyplacophores. *Cahiers de Biologie Marine* 16: 425–433.

Slieker, F. J. A. 2000. *Chitons of the World*. L'Informatore Piceno, Ancona.

Smith, A. G. 1960. Amphineura. In *Treatise on invertebrate paleontology I*. Mollusca 1. Edited by R. C. Moore. Lawrence, Kansas: University of Kansas Press, pp. 141–176.

Smith, A. G., and Hoare, R. D. 1987. Paleozoic Polyplacophora: a checklist and bibliography. *Occasional Papers of the California Academy of Sciences* 146: 1–71.

Spengel, J. W. 1881. Die Geruchsorgane und das Nervensystem der Molluscen. Ein Beiträg zur Erkenntnis der Einhalt des Molluscentypus. *Zeitschrift für Wissenschaftliche Zoologie* 35: 333–383.

Steiner, G., and Salvini-Plawen, L. 2001. Invertebrate evolution (Communications arising): *Acaenoplax*—polychaete or mollusc? *Nature* 414: 601–602.

Stinchcomb, B. L., and Darrough, G. 1995. Some molluscan problematica from the upper Cambrian-Lower Ordovician of the Ozark uplift. *Journal of Paleontology* 69: 52–65.

Streit, K., Geiger, D. L., and Lieb, B. 2006. Molecular phylogeny and the geographic origin of Haliotidae traced by haemocyanin sequences. *Journal of Molluscan Studies* 72: 105–110.

Sutton, M. D., Briggs, D. E. G., Siveter, D. J., and Siveter, D. J. 2001a. An exceptionally preserved vermiform mollusc from the Silurian of England. *Nature* 410: 461–463.

———. 2001b. Invertebrate evolution. *Acaenoplax*—polychaete or mollusc? Reply. *Nature* 414: 602.

———. 2004. Computer reconstruction and analysis of the vermiform mollusc *Acaenoplax hayae* from the Herefordshire Lagerstätte (Silurian, England), and implications for molluscan phylogeny. *Palaeontology* 47: 293–318.

Thiele, J. 1893. Polyplacophora, Lepidoglossa, Schuppenzüngler. In *Das Gebiss der Schnecken* 2. Edited by F. H. Troschel. Berlin: Nicolaische Verlagsbuchhandlung, pp. 353–401.

———. 1909–1910. Revision des Systems der Chitonen. II. Teil. *Zoologica* 22: 71–132.

———. 1913. Solenogastres. In *Tierreich*. Berlin: R. Friedländer und Sohn, pp. 1–57.

Todt, C. 2006. Ultrastructure of multicellular foregut glands in selected Solenogastres (Mollusca). *Zoomorphology* 125: 119–134.

Todt, C., and Salvini-Plawen, L. v. 2004a. Ultrastructure and Histochemistry of the foregut in *Wirenia argentea* and *Genitoconia rosea* (Mollusca, Solenogastres). *Zoomorphology* 123: 65–80.

———. 2004b. Ultrastructure of the midgut epithelium in *Wirenia argentea* (Mollusca, Solenogastres). *Journal of Molluscan Studies* 70: 213–224.

———. 2005. The digestive tract of *Helicoradomenia* (Solenogastres), aplacophoran molluscs from the hydrothermal vents of the East Pacific Rise. *Invertebrate Biology* 124: 230–253.

Van Belle, R. A. 1983. The systematic classification of the chitons (Mollusca: Polyplacophora). *Informations de la Société Belge de Malacologie* 11: 1–179.

———. 1985. The systematic classification of the chitons (Mollusca: Polyplacophora). Addenda I (with the description of the genus *Incisiochiton* gen. n.). *Informations de la Société Belge de Malacologie* 13: 49–59.

Van den Biggelaar, J. A. M. 1996. The significance of the early cleavage pattern for the reconstruction of gastropod phylogeny. In *Origin and evolutionary radiation of the Mollusca*. Edited by J. Taylor. London: Oxford University Press, pp. 155–160.

Vendrasco, M. J., Wood, T. E., and Runnegar, B. N. 2004. Articulated Palaeozoic fossil with 17 plates greatly expands disparity of early chitons. *Nature* 429: 288–291.

Verdonk, N. H., and Van den Biggelaar, J. A. M. 1983. Early development and formation of the germ layers. In *The Mollusca*. Vol. 3, *Development*. Edited by N. H. Verdonk, J. A. M. van den Biggelaar, and A. S Tompa. New York: Academic Press, pp. 91–122.

Vinther, J., and Nielsen, C. 2005. The Early Cambrian *Halkieria* is a mollusc. *Zoologica Scripta* 34: 81–89.

Von Ihering, H. 1876a. Versuch eines natürlichen Systemes der Mollusken. *Jahrbücher der deutschen Malakozoologischen Gesellschaft* 3: 97–147.

———. 1876b. Beiträge zur Kenntnis des Nervensystems der Amphineuren und Arthrocochliden. *Morphologische Jahrbücher* 3: 155–178.

Voronezhskaya, E. E., Tyurin, S. A., and Nezlin, L. P. 2002. Neuronal development in larval chiton *Ischnochiton hakodadensis* (Mollusca: Polyplacophora). *The Journal of Comparative Neurology* 444: 25–38.

Waller, T. R. 1998. Origin of the molluscan class Bivalvia and a phylogeny of major groups. In *Bivalves: An Eon of Evolution*. Edited by P. A. Johnston and J. W. Haggart. Calgary: University of Calgary Press, pp. 1–45.

Wanninger, A., and Haszprunar, G. 2002. Chiton myogenesis: perspectives for the development and evolution of larval and adult muscle systems in molluscs. *Journal of Morphology* 251: 103–113.

Watson Russel, C. 1997. Patterns of growth and setal development in the deep sea worm, *Strepternos didymopyton* (Polychaeta: Chrysopetalidae). *Bulletin of Marine Science* 60: 405–426.

Wealthall, R. J., Brooker, L. R., Macey, D. J., and Griffin, B. J. 2005. Fine structure of the mineralized teeth of the chiton *Acanthopleura echinata* (Mollusca: Polyplacophora). *Journal of Morphology* 265: 165–175.

Wiens, J. J., and Slingluff, J. L. 2001. How lizards turn into snakes: A phylogenetic analysis of body-form evolution in anguid lizards. *Evolution* 55: 2303–2318.

Wingstrand, K. G. 1985. On the anatomy and relationships of recent Monoplacophora. *Galathea Report* 16: 1–94.

Winnepenninckx, B., and Backeljau, T. 1996. 18S rRNA alignments derived from different secondary structure models can produce alternative phylogenies. *Journal of Zoological Systematics and Evolutionary Research* 34: 135–143.

Winnepenninckx, B., Backeljau, T., and De Wachter, R. 1994. Small ribosomal subunit RNA and the phylogeny of Mollusca. *The Nautilus* Suppl. 2: 98–110.

Wirén, A. 1892. Studien über die Solenogastren. II. *Chaetoderma productum, Neomenia, Proneomenia acuminata. Svenska Vetenskapsakademiens Handlingar* 25: 1–100.

Wolter, K. 1992. Ultrastructure of the radula apparatus in some species of aplacophoran molluscs. *Journal of Molluscan Studies* 58: 245–256.

Wray, C. G., Jacobs, D. K., Kostriken, R., Vogler, A. P., Baker, R., and Desalle, R. 1995. Homologues of the engrailed gene from five molluscan classes. *FEBS Letters* 365: 71–74.

Yates, A. M., Gowlett-Holmes, K. L., and McHenry, B. J. 1992. *Triplicatella disdoma* Conway Morris, 1990, reinterpreted as the earliest known polyplacophoran. *Journal of the Malacological Society of Australia* 13: 71 (abstract).

Yochelson, E. McAllister, J. F., and Reso. A. 1965. Stratigraphic distribution of the late Cambrian mollusk *Matthevia* Walcott 1885. *U.S. Geological Survey Professional Paper* 525B: 73–78.

Yochelson, E., and Richardson, E. S., Jr. 1979. Polyplacophora molluscs of the Essex fauna (Middle Pennsylvanian, Illinois). In *Mazon Creek Fossilis*. Edited by H. Nitecki. New York: Academic Press, pp. 321–332.

Yum, S. 1993. *Systematic study of Korean Neoloricates (Polyplacophora, Mollusca) based on karyotype and character analyses*. M. S. thesis Sung Kyun Kwan University, pp. 1–231.

5

Monoplacophora (Tryblidia)

Gerhard Haszprunar

Dall (1893) first noted the similarity of shells of Silurian *Tryblidium* with Recent patellid taxa and warned that it was dangerous to conclude that *Tryblidium* anatomy would have been similar to that of living patellids—"it is almost inconceivable that the Silurian form should have any closely allied recent representative." Moreover, the symmetry of the adductor scars of the monoplacophoran fossils suggested to Dall "a peculiar disposition of the organs which might, indeed, have paralleled in some particulars the organization of some of the Chitons of that ancient time." The same argument was repeated about 50 years later by Odhner (in Wenz 1940) when introducing the taxon Monoplacophora as a theoretical concept for an extinct stem group of conchiferan molluscs and another 17 years before the recovery of living monoplacophorans confirmed Dall's insight. It was one of the zoological sensations of the twentieth century, when a living representative, *Neopilina galatheae,* was found in 1952 by the Danish *Galathea* expedition (Lemche 1957) and anatomically described in detail by Lemche and Wingstrand (1959).

Given the likely multiple paraphyly or even polyphyly of the fossils grouped as monoplacophorans and the differing use of the latter taxon (e.g., Knight and Yochelson 1960; Peel 1991; Parkhaev 2002), the name Tryblidia, introduced by Lindström (1884), is used here because it is a more precise name for the group that includes the extant monoplacophorans.

To date, 29 species have been formally described from deep or cold waters (Table 5.1). At present, details of external morphology are known from about 10 species, while detailed anatomy is available for four species. Only a single species has been studied by means of transmission electron microscopy (TEM). Most recently, Giribet *et al.* (2006) provided the first molecular data on the group.

SYSTEMATIC POSITION

EXTANT MONOPLACOPHORA (TRYBLIDIA)

The systematic position of the Tryblidia as the earliest extant offshoot of the Conchifera is generally accepted among morphologists. The analysis of a segment of 28S rDNA of *Laevipilina antarctica* within a multigene analysis has been interpreted by Giribet *et al.* (2006) to

TABLE 5.1
Current Taxonomy of the Tryblidia

Family Neopilinidae Lemche, 1957

Genus *Neopilina* Lemche, 1957

[a]*Neopilina (N.) galatheae* (Lemche 1957: 414, figs. 1–4): 37 mm; off Coasta Rica, 9°23´ N, 89°32´ W, 3,590–3,718 m.

Neopilina (N.) bruuni (Menzies 1968: 2; figs. 1c, 4A–E, 5): 15 mm; Milne-Edwards Deep, Peru-Chile Trench, 4,823–4,925 m.

Neopilina (Lemchephiala) rebainsi (Moskalev et al. 1983: 988, figs. 5, 7–9): 21 mm; SE off Falkland Islands, 56°29,0´ S, 50°51,1´ W, 4,660–5,630 m.

Genus *Vema* Clarke and Menzies, 1959

[a]*Neopilina (Vema) ewingi* (Clarke and Menzies 1959: 1026, figs. 1A–F): 33 mm; off Peru, 5,607–6,489 m.

Vema bacescui (Menzies 1968: p. 2, figs. 1a, 2A–C, 3): 28 mm; Milne-Edwards Deep, Peru-Chile Trench, 5,986–6,134 m.

? *Vema levinae* (Warén and Gofas 1997: 226, figs. 1A, 13A–G, 14A–C, 15B): 3.8 mm, south off SW Mexico, 1,058 m.

Vema occidua (Marshall 2006: 62, figs. 1C, G–I, 2E, F, 4): 5.4 mm; West Norfolk Ridge, northern New Zealand, 785–800 m.

Genus *Laevipilina* McLean, 1979

[a]*Vema (Laevipilina) hyalina* (McLean 1979: 11, figs. 1–11, 20–22): 2–3 mm, off California, 174–384 m.

Laevipilina rolani (Warén and Bouchet 1990: 450, figs. 1–8): 1.9 mm, off NW Spain, 985–1,000 m.

Laevipilina n. sp. (McLean, unpublished data; see Warén and Bouchet 1990): 2–3 mm; off Central America.

Laevipilina antarctica (Warén and Hain 1992: 167, figs. 2–5, 6–8, 10–16, 19, 27): 3.0 mm; Weddell and Lazarev Sea, Antarctica; 210–3,136 m (see also Schrödl *et al.* 2006)

Laevipilina cachuchensis (Urgorri et al. 2005: 59, figs. 1A–D, 2, 3A–D, 4–6): 1.9 mm; El Cuchuchu Bank, N off Asturias (Spain), 580–600 m.

Laevipilina theresae (Schrödl 2006: p. 225, figs. 1A, B): 2.5 mm; Antarctica, off Kapp Norvegia, 765–840 m.

Genus *Monoplacophorus* Moskalev, Starobogatov, and Filatova, 1983

[a]*Monoplacophorus zenkewitchi* (Moskalev *et al.* 1983: 993, figs. 5, 11–13): 4.8 mm; W of Hawaii, 20°41.7´ N, 170°52.9´ W, 2,000 m.

Genus *Rokopella* Starobogatov and Moskalev, 1987

Acmaea euglypta (Dautzenberg and Fisher 1897: 181, pl. 4, figs. 25, 26): 2.3 mm; Azores, 1,600 m.

[a]*Neopilina oligotropha* (Rokop 1972: 91, figs. 1–9): 3.0 mm; N off Hawai, Mid-Pacific, 6,065–6,079 m.

Neopilina goesi (Warén 1988: 676, figs. 1–3, 6–11): 1.8 mm; Caribbean, Virgin Islands, 360–540 m.

Rokopella brummeri (Goud and Gittenberger 1993: 74, figs. 1–10): 1.45 mm; E of Mid-Atlantic Ridge, 45°21.3´ N, 27°9.1´ W, 2,162 m.

Rokopella segonzaci (Warén and Bouchet 2001: 118, figs. 1a–e): 0.90 mm; hydrothermal vents 37°50.54´ N, 31°31.30´ W, 860–870 m.

Rokopella capulus (Marshall 2006: 62, figs. 1A, B, D–F, 4): 2.4 mm; northern edge of central Chatham Rise New Zealand, 900–970 m.

Genus *Veleropilina* Starobogatov and Moskalev, 1987

Tectura reticulata (Seguenza 1876: 264; see also Warén and Gofas 1997): 1.6 mm, Mediterranean/ Tyrrhenian Sea, 180–600 m.

Acmaea zografi (Dautzenberg and Fisher 1896: 101, pl. 22, figs. 16, 17; see also Warén and Gofas 1997): 5 mm; Mid-Atlantic Ridge and adjoining seamounts, 30°–38° N, 600–1,400 m.

TABLE 5.1

(continued)

---

[a]Neopilina (Neopilina) veleronis (Menzies and Layton 1963: 402, figs. 7A–F, 8G, 9, 10A–C): 2.6 mm
Mexico, Cedrus Islands, 2,730–2,769 m.

Veleropilina sp. (see Warén and Gofas 1997: 222, figs. 1D, 8A, B, 9E, 10A–C, 15A): 1.7 mm; off
southern Point of Baja California, 1,950 m.

Genus *Adenopilina* Starobogatov and Moskalev, 1987

[a]Neopilina adenensis (Tebble 1967: 663, fig. 1–3): 10.7 mm; Gulf of Aden, 3,000–3,950 m.

Family Micropilinidae Haszprunar and Schaefer, 1997

Genus *Micropilina* Warén, 1989

[a]Micropilina minuta (Warén 1989: 3, fig. 2): 1.1 mm; Iceland, 63°23′N, 13°25′W, 770–926 m.

Micropilina tangaroa (Marshall 1990: 107, figs. 2–5): 1.5 mm; N off New Zealand, 31°31′S, 172°50′E,
1,216–1,385 m.

Micropilina arntzi (Warén and Hain 1992: 173, figs. 9, 17, 18, 20–26, 28, 29): 0.91 mm; Lazarev Sea,
Antarctica, 191–765 m (see also Schrödl *et al.* 2006)

Micropilina rakiura (Marshall 1998: 53, figs. 1–9): 1.25 mm; S of New Zealand, 47°18.17′S,
165°49.9′E, 896–1,038 m.

Micropilina reinga (Marshall 2006: 64, figs. 2A–D, 4): 0.68 mm; NW of Cape Reinga northern
ealand, 394–400 m.

Micropilina wareni (Marshall 2006: 66, figs. 3A–I, 4): 1.05 mm; W of Cape Reinga, northern
New Zealand, 785–800 m.

Incertae Sedis

Neopilina sp. (Menzies 1968): 4.5 mm; off Peru, 3,909–6,354 m.

Neopilina sp. (Rosewater 1970): 2.3 mm; SE of Falkland Islands, 1,627–2,044 m.

Neopilina sp. (Moskalev *et al.* 1983): 5.1 mm; off North Chile, 4,600 m.

NOTE: Mainly after Warén (1989), Warén and Hain (1992), Warén and Gofas (1997), Urgorri *et al.* (2005), and Marshall (2006) including maximum shell length and distribution. As discussed in detail by Warén and Gofas (1997), the genera *Rokopella* and *Veleropilina* in particular are somewhat doubtful. Genera and species are chronologically arranged.
[a]Type species.

reflect a clade consisting of Polyplacophora and Tryblidia, but this hypothesis needs testing with further data.

FOSSIL MONOPLACOPHORANS

Concerning the fossil record, the situation is complex and is far from being clarified. Most authorities separate "tergomyan" (including the Tryblidia) from "cyclomyan" taxa; the latter (Helcionellida and others) are variously classified (e.g., as untorted Bellerophontida) (see recent reviews by Horný 1991; Peel 1991; Parkhaev 2002, Chapter 3). If so, Tergomya, and accordingly the Tryblidia, would indeed represent one of the earliest conchiferan offshoots and a very distinct lineage.

However, the tergomyan-cyclomyan distinctness, and the respective dichotomy, is only valid if the mantle retractors are ignored, the insertion areas of which form the pallial line in bivalves or limpets. If, based on extant representatives of both shell shapes, the pallial lines (i.e., the insertion dots of the mantle retractor muscles) are considered (Figure 5.1), the principal differences become reduced only to the relative position of the apex. Such variability is present among extant taxa in the Patellogastropoda and even within

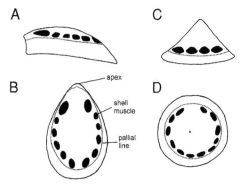

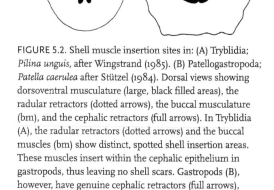

FIGURE 5.1. Comparison of tergomyan and cyclomyan shell muscle types (based on Horný 1991, Peel 1991). (A, B) Shell muscle insertions of a tryblidian (*Pilina*) in lateral and dorsal views. The circle of shell muscle scars does not contain the shell apex (tergomyan type), whereas the line of insertion areas of the mantle retractors (i.e., the pallial line) does encircle the apex. (C, D) A cup-shaped shell in which the circle of shell muscle scars is situated around the apex (cyclomyan type). According to the identical condition of the pallial line, there is no major difference between the tergomyan and cyclomyan type.

FIGURE 5.2. Shell muscle insertion sites in: (A) Tryblidia; *Pilina unguis*, after Wingstrand (1985). (B) Patellogastropoda; *Patella caerulea* after Stützel (1984). Dorsal views showing dorsoventral musculature (large, black filled areas), the radular retractors (dotted arrows), the buccal musculature (bm), and the cephalic retractors (full arrows). In Tryblidia (A), the radular retractors (dotted arrows) and the buccal muscles (bm) show distinct, spotted shell insertion areas. These muscles insert within the cephalic epithelium in gastropods, thus leaving no shell scars. Gastropods (B), however, have genuine cephalic retractors (full arrows), which are situated adjacent to the anterior portion of the dorsoventral muscles.

a single family, the Lepetidae. The majority of lepetid taxa are tergomyan with anteriorly overhanging apex, but certain genera (e.g., *Propilidium*) are cyclomyan and show a posterior apex. Accordingly, based on this analogy, the position of the apex ("tergomyan" versus "cyclomyan") alone is not valid to separate the Tryblidia from the remaining monoplacophoran lineages.

If one accepts the polyplacophoran shell plates as precursors of the conchiferan shell based on the position of the shell muscles (Wingstrand 1985), the cyclomyan condition is plesiomorphic for Conchifera. Accordingly, the probably apomorphic tergomyan condition could have evolved independently several times in early conchiferan evolution.

MONOPLACOPHORANS AND FOSSIL PATELLOGASTROPODS

A second problem exists in distinguishing Paleozoic tryblidians from patellogastropods— both usually preserved as internal molds (steinkerns)—because even extant forms have been repeatedly confused by earlier authors (see Table 5.1) and indeed are very similar in

shape and the general pattern of shell muscle insertion areas. Nacreous shell structure (e.g., MacClintock 1967, Erben *et al.* 1968; Hedegaard and Wenk 1998) and a broad, cap-shaped protoconch (e.g., Wingstrand 1985) are hallmarks of tryblidians but are usually not found in fossils because of poor preservation. The only other clear difference between tryblidian and patellogastropod fossils is found in the (granular) insertion area of the paired radular retractor (Tryblidia) versus the (homogeneous) one of the paired head retractor (Patellogastropoda), both of which show a similar position (Figure 5.2). All recent phylogenetic morphological analyses have the Patellogastropoda as the earliest gastropod clade (e.g., Haszprunar 1988, Ponder and Lindberg 1997), and the limpet shell shape might be plesiomorphic for Gastropoda (Haszprunar 1988, 1992), although this view is not shared by some (e.g., Ponder and Lindberg 1997). Accordingly, several taxa of formal Cambrian "Monoplacophora" or "Tergomya" or "Tryblidia," in which the characters mentioned previously could not be established or have been ignored, may be torted and could be members of

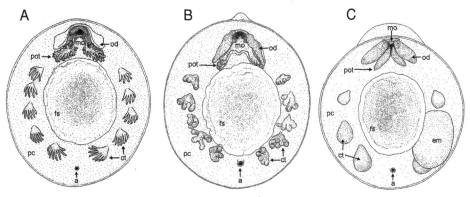

FIGURE 5.3. Ventral view of monoplacophorans representing three different size classes. (A) *Neopilina galatheae* (body length about 30 mm). (B) *Laevipilina antarctica* (body length about 3 mm). (C) *Micropilina arntzi* (body length about 0.8 mm). Abbreviations: a = anal opening; ct = ctenidia; em = brooded embryo; fs = foot sole; mo = mouth opening; od = oral disk lappets (so-called velum); pc = pallial cavity; pot = postoral tentacles.

the Patellogastropoda (see also Lindberg, Chapter 11).

SEGMENTATION (METAMERY) VERSUS SERIALITY

The main controversy around the Tryblidia has concerned their serial multiplication of several organ systems (shell muscles, lateropedal connectives, atria, nephridia, and ctenidia). This condition has been interpreted as evidence for a segmented molluscan ancestor (e.g., Lemche 1959a, b; Lemche and Wingstrand 1959; Götting 1980; Wingstrand 1985). However, many authors (e.g., Steinböck 1963; Vagvolgyi 1967; Salvini-Plawen 1972, 1981, 1985, 1991) have strongly argued against this interpretation and favored an independent evolution of seriality.

Lemche and Wingstrand (1959) described certain "typical annelid" characteristics such as a paired "dorsal coelom" being interconnected with the nephridia by segmental coelomoducts. However, the reinvestigation by Wingstrand (1985) on better-preserved material showed that the "paired dorsal coelomic cavities" were in fact esophageal pouches and that the connecting ducts between nephridia and the so-called dorsal coelomic cavities do not exist. Unfortunately, the latter "character"—although Wingstrand's findings have been confirmed by all subsequent studies (reviewed by Haszprunar

and Schaefer 1997a)—is still found in anatomical schemes in current textbooks.

With the discovery and detailed microanatomical and fine-structural investigations of two small representatives, *Laevipilina antarctica* (3 mm; Schaefer and Haszprunar 1997a, b) and *Micropilina arntzi* (0.9 mm; Haszprunar and Schaefer 1997b), substantial new data became available. In particular, the small (0.9 mm), partly paedomorphic *M. arntzi*, with only three pairs of ctenidia and nephridia and one pair of gonads, proved crucial for the segmentation versus seriality hypothesis (Figure 5.3). Meanwhile further progenetic species, *Rokopella segonzaci* (0.9 mm) from hydrothermal vents (Warén and Bouchet 2001) with three pairs of ctenidia, and *Micropilina minuta* (1.2 mm) from the Irish Sea with four pairs of ctenidia (personal observation), as well as a juvenile (with four pairs rather than the five seen in adults) of *Laevipilina antarctica*, confirmed that the serial tryblidian ctenidia develop from posterior to anterior as in extant Polyplacophora (except Lepidopleurida) and not the opposite way as found in the annelids. The same is true for the serial gonads: the paedomorphic *M. arntzi* has the most posterior one compared with larger species, while a male of *L. antarctica* showed an intermediate condition of the serial repetition of the gonad, where the most anterior pair (of three) is not yet fully separated from the others.

Also, the recent results on polyplacophoran ontogeny, particularly myogenesis and neurogenesis (Wanninger and Haszprunar 2002; Friedrich et al. 2002; Voronezhskaya et al. 2002), clearly suggest nonsegmentation in Mollusca. Accordingly, there remains little doubt that the multiplication of organ systems in the extant Tryblidia is an autapomorphic feature rather than a somewhat reduced or modified annelid-like segmentation. Also, the tryblidian conditions of the mantle folds are unique and thus probably not plesiomorphic for Conchifera (Schaefer and Haszprunar 1997b). In contrast, limpet-like shell shape (but not the type of mantle margin), the eight pairs of shell muscles, the cord-like nervous system with laterally placed statocysts, and the gononephridial type of gamete release reflect plesiomorphic conditions for the Conchifera (Schaefer and Haszprunar 1997a).

SYSTEMATICS OF EXTANT TRYBLIDIA

Until recently, all extant representatives were classified within a single family (Neopilinidae). The unique combination of plesiomorphic (regular esophageal pouches) and several apomorphic (lack of heart, brooding) characters in *Micropilina arntzi* was used by Haszprunar and Schaefer (1997a, b) to erect a second family, Micropilinidae. However, because a phylogenetic study of extant Tryblidia has yet to be undertaken, the status of this family remains untested.

The subdivision of Neopilinidae into genera is currently based on shell shape and structure and radular features. It has become clear that the number of ctenidia (and nephridia or gonads) is not a valid criterion as originally thought. It is not unlikely that molecular data will ultimately result in changes to the currently accepted taxonomy summarized in Table 5.1.

OUTLOOK

Without doubt, the Tryblidia is the least known molluscan class, as indicated by the nearly total lack of ecological, fine-structural, immunocytochemical, ontogenetic (including evo-devo) data.

The small molecular data set recently provided by Giribet et al. (2006) is possibly based on a mismatch of polyplacophoran and neopilinid samples (G. Steiner, personal communication). Each of these data sets are urgently needed to improve our current poor understanding of tryblidian biology and evolutionary history.

ACKNOWLEDGMENTS

I thank Mrs. Kühbandner (Zoologische Staatssammlung München) for doing the drawings. I am also grateful to the valuable comments and remarks of the reviewers and the editors.

REFERENCES

Bandel, K. 1982. Morphologie und Bildung der frühontogenetischen Gehäuse bei conchiferen Mollusken. *Facies (Erlangen)* 7: 1–198, pls. 1–22.

Clarke, A. H., and Menzies, R. J. 1959. *Neopilina (Vema) ewingi*, a second living species of the Paleozoic class Monoplacophora. *Science* 129: 1026–1027.

Dall, W. H. 1893. Phylogeny of the Docoglossa. *Proceedings of the Academy of Natural Sciences, Philadelphia, Annales and Magazing of Natural History* (6)12: 285–287.

Dautzenberg, P., and Fischer, H. 1896. Dragages effectués par l'*Hirondella* et par la *Princesse-Alice*, 1888–1896: 1. Mollusques Gastéropodes. *Mémoires de la Societé Zoologique de France* 9: 395–498.

———. 1897. Dragages effectués par l'*Hirondella* et par la *Princesse-Alice*, 1888–1896: *Mémoires de la Societé Zoologique de France* 10: 139–234, pls. 3–7.

Erben, H. K., Flass, G., and Siehl, A. 1968. Über die Schalenstruktur von Monoplacophoren. *Abhandlungen der Akademie der Wissenschaften und der Literatur, Mainz, Mathematisch-Naturwissenschaftliche Klasse* 1968: 1–24.

Friedrich, S., Wanninger, A., Brückner, M., and Haszprunar, G. 2002. Neurogenesis in the mossy chiton, *Mopalia muscosa* (Gould) (Polyplacophora): evidence versus molluscan metamerism. *Journal of Morphology* 253: 109–117.

Giribet, G., Okusu, A., Lindgren, A. R., Huff, S. W., Schrödl, M., and Nishiguchi, M. L. 2006. Evidence for a clade composed of molluscs with serially repeated structures: Monoplacophorans are related to chitons. *Proceedings of the National Academy of Sciences of the U.S.A.* 103: 7723–7728.

Götting, K.-J. 1980 Argumente für eine Deszendenz der Mollusken von metameren Antezedenten. *Zoologische Jahrbücher, Abteilung Anatomie* 103: 211–218.

Goud, J., and Gittenberger, E. 1993. *Rokopella brummeri* sp. nov., a new monoplacophoran species from the Mid-Atlantic Ridge in the northern Atlantic Ocean (Monoplacophora, Neopilinidae). *Basteria* 57: 71–78.

Haszprunar, G. 1988. On the origin and evolution of major gastropod groups, with special reference to the Streptoneura. *Journal of Molluscan Studies* 54: 367–441.

Haszprunar, G., and Schaefer, K. 1997a. Monoplacophora. In *Microscopic Anatomy of Invertebrates*. Vol. 6B, *Mollusca II*. Edited by F. W. Harrsion and A. J. Kohn. New York: Wiley-Liss, pp. 415–457.

———. 1997b. Anatomy and phylogenetic significance of *Micropilina arntzi* (Mollusca, Monoplacophora, Micropilinidae fam.nov.). *Acta Zoologica (Stockholm)* 77: 315–334.

Hedegaard, C., and Wenk, R. 1998. Microstructure and texture patterns of mollusc shells. *Journal Molluscan Studies* 64: 133–136.

Horný, R. J. 1991 ("1988"). Problems of classifying the cyclomyan molluscs (Mollusca, Monoplacophora): a historical review. *Casopis Národního Muzea Praze* 157: 13–32 [in Czech, English abstract].

Knight, J. B., and Yochelson, E. L. 1960. Monoplacophora. In: *Treatise on Invertebrate Paleontology*. Edited by R. C. Moore. Vol. I, *Mollusca 1*. Lawrence, KS: Geological Society of America and University of Kansas Press, pp. I77–I84.

Lemche, H. 1957. A new living deep-sea mollusc of the Cambrio-Devonian class Monoplacophora. *Nature* 179: 413–416.

———. 1959a. Molluscan phylogeny in the light of *Neopilina*. *Proceeding of the 15th International Congress on Zoology, London*, pp. 380–381.

———. 1959b. Protostomian interrelationships in the light of *Neopilina* (including discussion of Boettger). *Proceeding of the 15th International Congress on Zoology, London*, pp. 381–389.

Lemche, H., and Wingstrand, K. G. 1959. The anatomy of *Neopilina galatheae* Lemche, 1957. *Galathea Report* 3: 9–71, 56 pls.

Lindström, G. 1884. On the Silurian Gastropoda and Pteropoda of Gotland. *Kongliga Svenska Vetenskaps-Akademiens Handlingar, Stockholm* 19 (6).

MacClintock, C. 1967. Shell structure of patelloid and bellerophontoid gastropods (Mollusca). *Peabody Museum of Natural History (Yale University)* 22: 1–140, pls.1–32.

Marshall, B. A. 1990. *Micropilina tangaroa*, a new monoplacophoran (Mollusca) from Northern New Zealand. *The Nautilus* 104: 105–107.

———. 1998. A new monoplacophoran (Mollusca) from southern New Zealand. *Molluscan Research* 19: 53–58.

———. 2006. Four new species of Monoplacophora (Mollusca) from the New Zealand region. *Molluscan Research* 26: 61–68.

McLean, J. H. 1979. A new monoplacophoran limpet from the continental shelf of southern California. *Contributions of the Science Los Angeles County Museum* 307: 1–19.

Menzies, R. J. 1968. New species of *Neopilina* of the Cambrio-Devonian class Monoplacophora from the Milne-Edwards Deep of the Peru-Chile Trench, R/V Anton Bruun. *Marine Biological Association of India, Proceeding of the Symposium on Mollusca* 3: 1–9, pls. I–IV.

Menzies, R. J., and Layton, W., Jr. 1963. A new species of monoplacophoran molluscs, *Neopilina (Neopilina) veleronis* from the slope of the Cedros Trench, Mexico. *Annales and Magazines of Natural History* (13) 5: 401–406, pls. 7–10.

Moskalev, L. I., Starobogatov, Ya. I., and Filatova, Z. A. 1983. New data on the Monoplacophora of the abyssal of the Pacific and the Southern Atlantic Ocean. *Zoologicheskyi Zhournal* 62: 981–995 [in Russian, English abstract].

Parkhaev, P. Yu. 2002. Phylogenesis and the system of the Cambrian univalved mollusks. *Paleontological Journal* 36: 25–36.

Peel, J. S. 1991. The classes Tergomya and Helcionelloida, and early molluscan evolution. In *Functional morphology, evolution and systematics of early Paleozoic univalved molluscs*. Edited by Peel, J. S. *Grönlands Geologiske Undersuchungen. Bulletin* 161: 11–65.

Ponder, W. F, and Lindberg, D. R. 1997. Towards a phylogeny of gastropod molluscs: an analysis using morphological characters. *Zoological Journal of the Linnean Society* 119: 83–265.

Rokop, R. J. 1972. A new species of Monoplacophora from the Abyssal North Pacific. *The Veliger* 15: 91–95, 2 pls.

Rosewater, J. 1970. Monoplacophora in the South Atlantic Ocean. *Science* 167: 1485–1486.

Salvini-Plawen, L. v. 1972. Zur Morphologie und Phylogenie der Mollusken: Die Beziehung der Caudofoveata und Solenogastres als Aculifera, als Mollusca und als Spiralia. *Zeitschrift für wissenschaftliche Zoologie* 184: 205–394.

———. 1981. On the origin and evolution of the Mollusca. In *Origine dei Grande Phyla dei Metazoi. Atti dei Convegni Lincei (Roma)* 49: 235–293.

———. 1985. Early evolution and the primitive groups. In *The Mollusca*. Vol. 10, *Evolution*. Edited by E. R. Trueman and M. R. Clarke. London and New York: Academic Press, pp. 59–150.

———. 1991. Origin, phylogeny and classification of the phylum Mollusca. *Iberus* 9: 1–33.

Schaefer, K., and Haszprunar, G. 1997a. Anatomy of *Laevipilina antarctica*, a monoplacophoran limpet (Mollusca) from Antarctic waters. *Acta Zoologica (Stockholm)* 77: 295–314.

———. 1997b. Organisation and fine structure of the mantle of *Laevipilina antarctica* (Mollusca, Monoplacophora). *Zoologischer Anzeiger* 236: 13–23.

Schrödl, M. 2006. *Laevipilina theresae*, a new monoplacophoran species from Antarctica *Spixiana* 29: 225–227.

Schrödl, M., Linse, K, and Schwabe, E 2006. Review on the distribution and biology of Antarctic Monoplacophora, with first abyssal record of *Laevipilina antarctica*. *Polar Biology* 29: 721–727.

Seguenza, G. 1876. Studi stratigrafici sulla formazione Pliozena dell' Italia meridionale. Elenco dei Cirripedi e Molluschi dell Antico Plioceno. *Bollettino del regio Comitato Geologico d'Italia* 7: 259–271.

Starobogatov, Ya. I., and Moskalev, L. I. 1987. Systematics of the Monoplacophora. In: *Molluscs, Results and Perspectives of Investigation* [in Russian]. Edited by Ya. I. Starobogatov, A. N. Golikov, and I. M. Likarev. Leningrad, Institute of Zoology, Academy of Sciences of the SSSR, pp. 7–11.

Steinböck, O. 1963. Über die Metamerie und das Zölom der *Neopilina galatheae* Lemche, 1957. *Verhandlungen der Deutschen Zoologischen Gesellschaft 1962*: 385–403.

Stützel, R. 1984. Anatomische und ultrastrukturelle Untersuchungen an der Napfschnecke *Patella* L. unter besonderer Berücksichtigung der Anpassung an den Lebensraum. *Zoologica (Stuttgart)* 135: 1–54, pls. 1–35.

Tebble, N. 1967. A *Neopilina* from the Gulf of Aden. *Nature* 215: 663–664.

Urgorri, V., Garcia-Alvarez, O., and Luque, A. 2005. *Laevipilina cachuchensis*, a new neopilinid (Mollusca: Tryblidia) from off North Spain. *Journal of Molluscan Studies* 71: 59–66.

Vagvolvyi, J. 1967. On the origin of molluscs, the coelom, and coelomic segmentation. *Systematic Zoology* 16: 153–168.

Voronezhskaya, E. E., Tyurin, S. A., and Nezlin, L. P. 2002. Neuronal development in larval chiton *Ischnochiton hakodadensis* (Mollusca, Polyplacophora). *Journal of Comparative Neurology* 444: 25–38.

Wanninger, A., and Haszprunar, G. 2002. Chiton myogenesis: Perspectives for the development and evolution of larval and adult muscle systems in molluscs. *Journal of Morphology* 252: 103–113.

Warén, A. 1988. *Neopilina goesi*, a new Caribbean monoplacophoran mollusk dredged in 1869. *Proceedings of the Biological Society of Washington* 101: 676–681.

———. 1989. New and little known Mollusca from Iceland. *Sarsia* 74: 1–28.

Warén, A., and Bouchet, P. 1990. *Laevipilina rolani*, a new monoplacophoran from off Southwestern Europe. *Journal of Molluscan Studies* 56: 449–453.

———. 2001. Gastropoda and Monoplacophora from hydrothermal vents and seeps; new taxa and records. *The Veliger* 44: 116–231.

Warén, A., and Gofas, S. 1997. A new species of Monoplacophora, redescription of the genera *Veleropilina* and *Rokopella*, and new information on three species of the class. *Zoologica Scripta* 25 ("1996"): 215–232.

Warén, A., and Hain, S. 1992. *Laevipilina antarctica* and *Micropilina arntzi*, two new monoplacophorans from the Antarctic. *The Veliger* 35: 165–176.

Wenz, W. 1940. Ursprung und frühe Stammesgeschichte der Gastropoden. *Archiv für Molluskenkunde* 72: 1–10.

Wingstrand, K. G. 1985. On the anatomy and relationships of recent Monoplacophora. *Galathea Report* 16: 7–94, 12 pls.

6

Bivalvia

Gonzalo Giribet

Bivalves, the second largest class of molluscs, are aquatic (primarily marine), bilaterally symmetrical, and characterized by a laterally compressed body enclosed in a bivalved shell that articulates dorsally via a hinge and a ligament. Extant bivalves are abundant components of the marine fauna, from the intertidal—where mussels or oysters can be the dominant invertebrate—to the abyss—where protobranchiate bivalves can constitute an important component of the benthic biomass. These two examples demonstrate two of the extreme ecologies of bivalves: the infaunal sediment burrowers (Figure 6.1D) and the sessile epibenthic forms with cemented attachment (Figure 6.1B; Yonge 1979; Harper 1991) or attachment by way of a byssus (Figure 6.1A; Yonge 1962). Other sessile forms are borers in corals (Morton 1990a), soft rock or hard sediments, or wood. A few species of scallops (Pectinidae) and file clams (Limidae; Figure 6.1C) have evolved an escape mechanism by swimming using jet propulsion. The heterodont *Solecurtus strigilatus* uses jet propulsion to avoid capture by rapid escape, burrowing to more than 50 cm in depth (Bromley and Asgaard 1990). A few groups harvest symbiotic

bacteria (e.g., solemyids, bathymodiolines, vesicomyids, lucinids, and thyasirids; Reid 1990) or zooxanthellae, such as some fragines and tridacnines (Cardiidae; Fankboner and Reid 1990), *Tridacna* being one of the largest noncolonial marine invertebrates.

In general, bivalves have little motility as adults, and although some primitive taxa (most protobranchs) are deposit feeders, most bivalves have hypertrophied gills used for filter feeding. These bivalves—whether infaunal free-living, byssally attached, or cemented—obtain their food by filtering large amounts of water through their gills. Food particles trapped on thin mucus sheets on the gills are conducted anteriorly to the labial palps and mouth using ciliary mechanisms characteristic of the different bivalve lineages (Atkins 1936, 1937a, b, c, 1938a, b, c; Stasek 1963). A few members of the clade Anomalodesmata have lost their gills and feed by sucking their prey into the mantle cavity by means of modified siphons (Morton 1981). A few others obtain all or part of their food through symbiotic relationships with bacteria, such as the members of the family Solemyidae (most of which lack a gut), the deep-sea mussels (*Bathymodiolus*), and

FIGURE 6.1. Habitus of different bivalves. (A) *Pteria colymbus* from Bahamas attached to a mangrove root by its byssus (by). (B) *Anomia* sp. individuals from Bermuda attached to a rock. (C) *Ctenoides scaber* from Bermuda showing the mantle tentacles (mt) and the foot (fo). (D) *Tapes decussatus*, from Galicia, Spain, buried in sediment and showing the exhalant siphon (es) and inhalant siphon (is). (E) *Spondylus* sp. from the Caribbean (Panama) showing the mantle tentacles (mt) and complex pallial eyes (ey).

several members of Heterodonta, including lucinids, shipworms, and hydrothermal vent or hydrocarbon seep vesicomyids (e.g., Cavanaugh 1983; Distel 1998; Williams *et al.* 2004). Giant clams and some other bivalves, including some other cardiids, maintain a functional digestive tract but supplement their diets via symbiosis with zooxanthellae (Yonge 1980; Morton 2000).

DIVERSITY AND ABUNDANCE

Bivalves constitute an important portion of marine biomass. A study in a tropical Indo-Pacific 295-km² area on the west coast of New Caledonia found 2,738 species of marine molluscs, of which 519 were bivalves (Bouchet *et al.* 2002). Although this only represented 19% of

species diversity, the number of bivalve specimens constituted 35% of the total. Bivalves also make up a large percentage of the total molluscan diversity in temperate areas. Studies on the Garraf coast in the western Mediterranean showed that bivalves constituted 29% of the total molluscan diversity of 676 species (Giribet and Peñas 1997; Peñas and Giribet 2003). A slightly lower proportion of bivalves (20.7% of the total molluscan diversity of 655 species) has been registered for Alborán, also in the western Mediterranean (Peñas *et al.* 2006).

A study including 930 species of northeastern Pacific marine shelf bivalves shows a strong latitudinal diversity gradient (measured as number of species per degree of latitude) that is closely related to mean sea surface temperature (Roy *et al.* 2000). Protobranchiate bivalves are an exception to the strong latitudinal gradient seen in the data, perhaps because they have non-planktotrophic larvae. In addition to the latitudinal diversity gradient, the proportion of bivalves with respect to other molluscs generally increases with depth and latitude, perhaps because of the number of protobranchiate forms both in the deep sea and at higher latitudes. For example, bivalves were the dominant molluscan group (44% of species) in the Bransfield Strait (Antarctica), making up 83% of all mollusc specimens, and they were also the dominant species (Arnaud *et al.* 2001), with different protobranchiate species being well represented in most samples.

Little work has been done in evaluating the population structure (genetic diversity) of bivalve species through their marine (e.g., Benzie and Williams 1992, 1995, 1998; Macaranas *et al.* 1992; Ó Foighil and Jozefowicz 1999; Nikula and Väinölä 2003; Zardus *et al.* 2006) or continental (e.g., Machordom *et al.* 2003) ranges, despite the importance of these data for conservation. Data on the bathymetric distribution in four species of deep-sea protobranchiate bivalves indicate that genetic diversity decreases with depth (Chase *et al.* 1998; Etter *et al.* 2005; Zardus *et al.* 2006), perhaps indicating that the steep, topographically complex, and dynamic bathyal zone, which stretches as a narrow band along continental margins, may play a more important role in the evolutionary radiation of the deep-sea fauna than the much more extensive abyss (Etter *et al.* 2005), at least in the case of infaunal bivalves.

Paleobiological contributions have shown that, although sensitive to mesh size and environment, time-averaged death assemblages retain a strong signal of species' original rank orders, both in bivalves and in gastropods (Kidwell 2001). This implies that naturally accumulated death assemblages provide a reliable means of acquiring abundance data through time. It has also been shown that preservation of shells is not biased by shell composition in marine bivalves (Kidwell 2005). On the contrary, biases in the fossil record, measured as the percentage of bivalve genera and subgenera missing from the fossil record, are shown by animals that tend to have small body size, reactive shell structures, commensal or parasitic habit, deep-sea distribution, narrow geographic range, or recent date of formal taxonomic description in the neontological literature (Valentine *et al.* 2006). Most missing taxa show two or more of these features and tend to be concentrated in particular families.

While bivalves are mostly marine, several lineages have colonized freshwater environments. The most prominent are Unionoida (Palaeoheterodonta) and several families of heterodonts, namely Corbiculidae, Dreissenidae, and Sphaeriidae. Several members of other pteriomorphian and heterodont lineages live in brackish environments without entering the freshwater realm proper. The only freshwater pteriomorphians are the species of the arcoid genus *Scaphula* (e.g., Janaki Ram and Radhakrishna 1984). Freshwater bivalve diversity reaches its peak in North America, with more than 300 extant native species (e.g., McMahon and Bogan 2001), many of which are narrow-range endemics, often with endangered populations (e.g., Neves *et al.* 1997). Taxonomy within Unionoida is in clear need of revision (Graf 2000), but the six recognized families show interesting biogeographic patterns. Unionidae and Margaritiferidae are Laurasian, whereas the four other families (Hyriidae, Etheriidae, Mycetopodidae, and Iridinidae [= Mutelidae]) are Gondwanan in origin (Graf 2000; Graf and Ó Foighil 2000a).

ECONOMIC SIGNIFICANCE AND HUMAN IMPACTS

Bivalves are an important source of animal protein, with major fisheries including mussels (Mytilidae), arks (Arcidae), oysters (Ostreidae), scallops (Pectinidae), cockles (Cardiidae), venus shells (Veneridae), and razor clams (Solenidae). Pearl production is important in some nations, especially those in the Indo-Pacific (Landman *et al.* 2001). Although humans have consumed bivalves for millennia, the first documented historical record of bivalve aquaculture dates from the eighth century (Kurokura 2004). Bivalve production comprises a significant component of aquaculture in both value and weight; in Japan, for example, in 1990 aquaculture generated 260,000 metric tons of oysters and 18,000 metric tons of scallops. On the Pacific North American coast, 39,000 metric tons valued at $82 million dollars were produced in 1996 (Coan et al. 2000). Economically, the most important bivalves for the aquaculture industry are perhaps the blue mussels (*Mytilus edulis* and *M. galloprovincialis*), with a peak production of 500,000 annual metric tons in the late 1990s (data from the Food and Agriculture Organization of the United Nations [FAO] web site: http://www.fao.org). The total amount of marine bivalve catch (non-aquacultured) in 2002 was more than 2 million metric tons worldwide (data from FAO web site). Gosling (2003) provides detailed information on bivalve fisheries and aquaculture, including most common practices.

Because of their filter feeding, bivalves have the potential for accumulating heavy metals, bacteria, viruses, and toxins derived from marine dinoflagellates and diatoms. With the expansion of aquaculture and subsequent increase in shellfish consumption, the risk of bivalve-derived diseases constitutes an increasing concern for public health management (DePaola et al. 1990; Gosling 2003; Rehnstam-Holm and Hernoth 2005).

Teredinids, commonly known as shipworms, and xylophagine pholadids, are highly specialized bivalves adapted for boring into wood (Turner 1966), and as such they have an impact on wooden constructions and ships in all the oceans. They can invade new wood only during the mobile larval stage, and they enter the wood block by a small hole, which is only slightly enlarged during the life of the animal. However, the internal space grows quickly, and the damage in the wood remains mostly undetected from the outside until the interior is almost completely destroyed. Biofouling by mussels (or zebra mussels in freshwater environments) can also be of economic importance, and when settled on ships, they can have a tremendous impact in fuel efficiency. Many species of bivalves bore into limestone, dead corals (e.g., lithophagine mytilids, gastrochaenids, and petricolids), and even living corals, primarily aided by chemical processes (Valentich-Scott and Dinesen 2004), constituting important agents in bioerosion of corals.

Some marine and freshwater invasive species pose important ecological and economic threats. For example, the zebra mussel (*Dreissena polymorpha*) seriously affects native populations of freshwater unionoids in North America and colonizes all kinds of hard substrates, outcompeting most other benthic invertebrates. A huge economic impact has been documented in the United States on water intake systems for power stations, with estimated costs from 1989 to 1995 of $69 million (O'Neill 1997). This impact is also important in the intake systems for nuclear power stations in Spain (J. M. Giribet, personal communication) and other European countries. Other significant invasive bivalves include the freshwater Southeast Asian species of the genus *Corbicula* that have invaded North America, South America, and Europe (e.g., Araujo et al. 1993; Siripattrawan et al. 2000) and several marine species, especially venerids of commercial interest (e.g., Miller et al. 2002).

The larvae of unionoids (glochidia and lasidia) are parasitic on fishes and, in large numbers, can (although rarely) affect the health of the host. At the same time, the disappearance of the fish hosts in many river systems is contributing to the extinction of many freshwater mussels (e.g., Araujo et al. 2002).

Bivalves are hosts to many metazoan parasites, as well as viruses, bacteria, fungi, apicomplexans, and ciliates. Among the metazoan parasites, the most important are digenean trematodes, as larval flukes have been reported from virtually every marine bivalve species examined (see Lauckner 1983; Gosling 2003).

Two major aspects impact conservation issues on bivalves. Biological invasions, such as the ones discussed previously, have been documented in many freshwater and marine regions. In the case of marine environments, aquaculture has been an important factor in human-mediated introductions of several species of mussels and oysters. Perhaps some of the best-documented marine invasions are those of the Manila clam (*Venerupis philippinarum*) in the Mediterranean and North Atlantic (e.g., Pranovi *et al.* 2006) and the Pacific oyster in several parts of the Indo-Pacific (see Chapman *et al.* 2003). Bivalve invasions can have a tremendous impact in freshwater environments, where invader species such as the zebra and quagga (*Dreissena bugensis*) mussels or *Corbicula fluminea* and *C. leana* have dramatic economical and biological impact. The second conservation issue is related to habitat destruction and anthropogenic perturbations, which directly affect freshwater bivalve populations, notably unionoideans (Neves *et al.* 1997; Lydeard *et al.* 2004). Habitat degradation may also have other consequences; for example, water flow fragmentation caused by river regulation can affect host fish populations (Araujo and Ramos 2001) and drastically alter habitat characteristics.

BIVALVE MORPHOLOGY

In this section a brief account of bivalve morphology and anatomy is provided that is intended as an introduction to terms and characters important in bivalve systematics, both at higher levels and in the taxonomy of genera and species. This section is also intended to assist the reader in following the phylogenetic and evolutionary discussions later in the chapter. For more detailed information about bivalve morphology, including general anatomy (e.g., Figure 6.2), the reader is referred to Purchon (1968), Cox (1969), Trueman and Clark (1988), Oliver (1992), Morse and Zardus (1997), and Morton *et al.* (1998).

Bivalves are characterized by having two shell valves hinged dorsally and connected by an elastic ligament, and the valves are held together by adductor muscles (typically two), which attach to their inner surfaces. The valves are opened by the relaxation of the compressed ligament and closed by contraction of the adductor muscles. The adductor muscles (Figure 6.2) and other muscle attachments leave impressions in the inner part of each valve; these include the pedal muscles and the pallial retractor muscles attached along the pallial line (and pallial sinus in most of those species with siphons). These characteristics are easily observed in open shells (Figure 6.3A) in most bivalves. They have been used together with hinge morphology, the position of the ligament with respect to the umbones (the top of each valve; Figure 6.3B), the presence or absence of internal elements of the ligament, and the crystalline microstructure of the shell in systematics and taxonomy. These characters are outlined briefly below.

The hinge is formed by a series of interlocking teeth and sockets that prevent the valves from sliding against one another and has traditionally served to define major bivalve lineages. *Taxodont* dentition is found in most protobranchiates and many pteriomorphians, at least in the larval stage. The teeth that form this type of hinge are numerous, small, and of similar shape (Figure 6.4B). *Isodont* dentition is a reduced form of the taxodont type, consisting of a few teeth placed symmetrically at either side of the ligament. Mytiloids have an even more reduced hinge where there are no true teeth, only a few small denticles situated on either side of the ligament, this being *dysodont* dentition. Hinges with heterogeneous dentition formed by irregular larger teeth appeared most prominently in the heteroconch lineage (the subclasses Palaeoheterodonta and Heterodonta), where there are two main types: the *schizodont* (Figure 6.4D) and *heterodont* (Figure 6.3) hinges. Hinge reduction to produce an *edentulous* hinge has occurred in multiple lineages, including anomalodesmatans (Le Pennec 1980; Figure 6.4F). Accessory hinge structures such as chomata and secondary teeth are common in pteriomorphians (reviewed in Giribet and Wheeler 2002).

Adductor muscles (and their scars) have also served to define groups. The presence of two muscles of similar size (*isomyarian* condition;

A

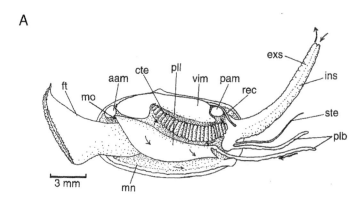

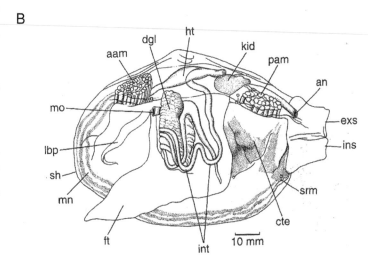

FIGURE 6.2. Gross anatomy of (A) the protobranchiate *Malletia obtusata* (after Yonge 1939) and (B) the heterodont *Mactra glauca*. *Abbreviations:* aam = anterior adductor muscle; an = anus; cte = ctenidium;, dgl = digestive gland; exs = exhalant siphon; ft = foot; ht = heart; ins =inhalant siphon; int = intestine; kid = kidney; lbp = labial palp; mn = mantle; mo = mouth; pam = posterior adductor muscle; plb = palp proboscis; pll = palp lamella; rec = rectum; ste = sensory tentacle; sh = shell; srm = siphonal retractor muscle; vim = visceral mass. Images from Beesley *et al.* (1998).

Figures 6.2 and 6.3A) seems to be the plesiomorphic state for bivalves. Certain bivalves are *anisomyarian*, with reduction of the anterior muscle, as in many pteriomorphians (Yonge 1953; Gilmour 1990). The anisomyarian condition has apparently progressed in some pteriomorphian lineages to the complete loss of the anterior adductor. An analogous loss of the anterior adductor has occurred in the giant clams *Tridacna* and *Hippopus*. In certain anisomyarian heterodonts, it is the posterior adductor muscle that has been reduced (e.g., Gastrochaenidae).

The *umbo* (umbones or umbos in plural) is the area surrounding the beak of the shell (Figure 6.3A, B), on the dorsal side. A line projecting ventrally from the umbones effectively divides the shell into an anterior and a posterior part. If the umbones are situated centrally, the valves are *equilateral*, but if the umbones are displaced toward either end, they are *inequilateral*. The umbones derive from the larval shell. If the beaks face each other across the dorsal margin, they are said to be *orthogyrate*, but more often they point anteriorly (*prosogyrate*) or posteriorly (*opisthogyrate*). Shell shape is also an important taxonomic character. Most species have a regular shape that changes only by increasing in size, but others have irregular shells that adapt to their substrate, as in oysters and many other pteriomorphians. The valves are typically of similar shape (*equivalve*) or dissimilar (*inequivalve*) in several groups of pteriomorphians (e.g., scallops) and some myioids and anomalodesmatans. The outline of the shell differs considerably, with some being circular, ovate (Figure 6.3A), elliptical, truncate (Figure 6.5D), rostrate, rectangular,

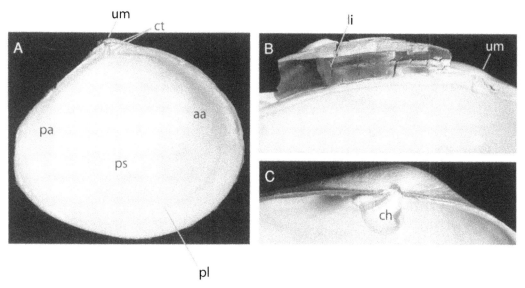

FIGURE 6.3. (A) Interior of left valve of *Tellina* illustrating the posterior adductor (pa), anterior adductor (aa), pallial line (pl), pallial sinus (ps), umbone (um) and cardinal teeth (ct). (B) Detail of the ligament (li) and umbonal region of *Calyptogena magnifica*. (C) Chondrophore (ch) of *Mya arenaria*.

triangular (Figure 6.5B), spatulate, and bialate, among others. In many cases bivalve taxonomists use terms to refer to shape that require prior taxonomic knowledge; such is the case of the mytiliform, modioliform, pteriiform (Figure 6.1A), or ensiform shape. Many bivalves have areas demarcated by angles or ridges, and two dorsal regions are most useful in taxonomy: the *lunule*, anterior to the umbones, and the *escutcheon*, posterior to the umbones and often depressed. Species with large siphons may have a large posterior gape (Figure 6.5D, E) whereas an anterior or antero-ventral gape, usually for the extension of the foot, is referred to as *pedal gape*. If a large byssus is present, there may also be a *byssal gape*, often in an antero-ventral position. Scallops, spiny oysters (Figure 6.5C), and other pteriomorphians have shells with earlike projections at both sides of the dorsal margin called *auricles* (Figure 6.5A), which in some cases are greatly expanded into wings, as in *Pteria* (Figure 6.1A). In scallops the anterior right auricle has a byssal notch accompanied by a series of small teeth on its inner margin, the *ctenolium*.

Shell sculpture is also of utmost importance in distinguishing species and includes various *radial* (Figure 6.4C–E) and *concentric* (Figure 6.4B) elements, with the concentric elements sometimes oblique to the commarginal line. Some present spines (Figure 6.5C) or squamate sculpture (Figure 6.4E). The nomenclature used to designate all types of sculpture is beyond the scope of this chapter. Radial structure may be composed of lines, threads, riblets or ribs, with different cross-sectional shapes. Concentric sculpture is likewise variable, including growth lines, lirations, ridges, undulations, lamellae, and frills. Other sculpture may not strictly correspond to radial or concentric elements, such as cross bars (often on ribs), tubercles, scales, or spines. The valves' margins are often smooth or can have a series of rugae or interlocking projections that may reflect the surface radial sculpture (Figure 6.4C–E). Such margins are termed serrate, denticulate, or crenulate.

Although most characteristics of the shell are applicable only to bivalves, the microstructure allows for phylogenetic comparison not only within bivalves but also among other molluscs. Bivalve shell microstructure is well known (Taylor *et al.* 1969, 1973; Carter 1990b).

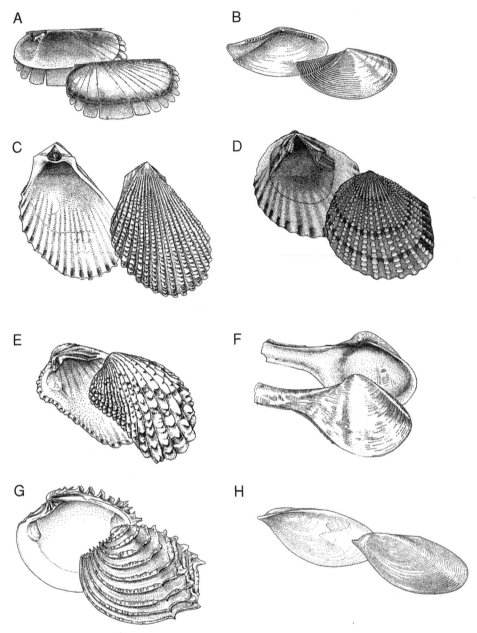

FIGURE 6.4. (A) Protobranchia, Solemyoidea: *Solemya australis*. (B) Protobranchia, Nuculanoidea: *Nuculana dohrni*. (C) Pteriomorphia, Limoidea: *Lima lima*. (D) Palaeoheterodonta, Trigonioidea: *Neotrigonia bednalli*. (E) Heterodonta, Carditoidea: *Cardita calyculata*. (F) Heterodonta, Anomalodesmata, Cuspidarioidea: *Cuspidaria angasii*. (G) Heterodonta, Veneroidea: *Bassina disjecta*. (H) Heterodonta, Gastrochaenoidea: *Gastrochaena cymbium*. All images from Beesley *et al.* (1998).

It includes mineralogical composition, number of shell layers, their structure, and the presence of accessory structures such as myostracal pillars, chalky lenses (in oysters), and the presence of a flexible shell margin resulting from an extension of the periostracum beyond the edge of the calcified shell, as in Solemyoidea.

The bivalve shell is made of crystalline calcium carbonate either deposited on, or embedded within, an organic matrix composed of glycoproteins. The whole shell is covered by

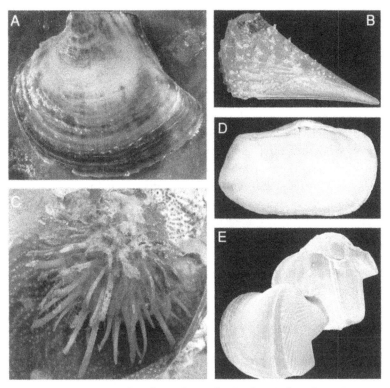

FIGURE 6.5. (A) *Isognomon alatus* from Bahamas showing the typical axe shape. (B) *Pinna carnea* from Bahamas. (C) *Spondylus* sp. from the Caribbean of Panama with prominent spines and a shell covered by sponges. (D) Internal view of the left valve of *Panopea glycymeris*, Galicia, Spain. (E) Scanning electron micrograph of the gaping shell of *Xylophaga dorsalis* from the Mediterranean.

an outer organic layer, the periostracum. All parts of the shell are secreted by the epithelial cells of the mantle. These cells are joined to the shell only at the outer fold of the mantle by the periostracum and at the points of attachment of the pallial, adductor, pedal, and other muscles (Figure 6.3A). Except at these points, the calcium carbonate and organic matrix of the shell are deposited from the extrapallial fluid. The periostracum is secreted at the extreme margin of the mantle and serves as a substrate for subsequent deposition of calcium carbonate and organic matrix. Under normal conditions of calcification, two polymorphs of calcium carbonate, calcite and aragonite, may be deposited. Some species contain aragonite alone; others, a combination of calcite and aragonite that always occur in discrete layers. The shell may consist of several layers, each formed by one

carbonate and organic matrix aggregate (Taylor *et al.* 1969). The mode of deposition of calcium carbonate results in seven basic shell structure types described by Taylor *et al.* (1969): nacreous, prismatic, foliated, crossed-lamellar, complex crossed-lamellar, homogeneous, and the myostracal layers associated with muscle attachment areas. The presence and arrangement of these layers has phylogenetic utility. Among the extant taxa, protobranchs, pteriomorphians, and palaeoheterodonts are wholly aragonitic and may present two or three shell layers, but most other bivalves have calcitic layers (Taylor *et al.* 1969, 1973).

The ligament (Figure 6.3B) is a mid-dorsal, largely uncalcified region that unites the two calcareous valves (Yonge 1978; Waller 1990). Ligament shape, position, and structure are important characters in bivalve systematics.

Many types occur in the different groups of bivalves (e.g., simple, duplivincular, alivincular, transverse, parivincular; Carter 1990a; Waller 1990). Ligaments have a fibrous central portion (the inner ligament) beneath an outer non-fibrous part (the outer ligament). They may be partially calcified (Taylor *et al.* 1969), often with fine aragonitic needles supporting the polymerized organic structure. *Amphidetic* ligaments are antero-posteriorly symmetrical (in equivalve and equilateral shells), but they are termed *opisthodetic* if located mostly posterior to the umbones. Ligaments often have associated shelly supports such as *nymphae*, *pseudonymphae*, or a *lithodesma*, a small calcareous plate or ossicle that divides the ligament into two compressive units that is found in many anomalodesmatans. The ligament may sit in a hollowed-out depression in the hinge plate known as a *resilifer*, located beneath the umbo. A spoon-shaped, projecting resilifer is termed a *chondrophore* (Figure 6.3C), found in mactrids and some myoids and anomalodesmatans (Giribet and Wheeler 2002).

The anatomy of bivalves differs considerably from that of other molluscs. There is no buccal mass, radula, salivary, or esophageal glands. The body is, in all but some highly specialized taxa, contained within the shell valves and is surrounded by the mantle cavity. The mantle lobes are either joined or free; if joined they can present different degrees of fusion. At least three lineages have developed posterior fusion of the mantle margin to form inhalant and exhalant siphons (Yonge 1939, 1957, 1982), although the primitive unfused condition is still observed in several groups, including many protobranchiates and pteriomorphians as well as trigonioids (Gould and Jones 1974; Morton 1987). All bivalves (except septibranchs) have a pair of ctenidia (gills) suspended laterally in the mantle cavity (see following discussion). In protobranchiate bivalves the labial palps collect food (Figure 6.2A), but in other bivalves this pair of complexly ridged and ciliated structures presort particles passed from the ctenidia before transferring it to the mouth.

Although the primitive mode of feeding in bivalves was deposit feeding, as presently found in nuculoids and nuculanoids, most others use their gills for feeding, in addition to respiration. Their gills have enlarged, folded, and extended anteriorly so that they lie laterally. In this position each ctenidium forms a large filtering apparatus divided into an inner and outer demibranch, each demibranch typically with ascending and descending lamellae. Water enters the mantle cavity (anteriorly in primitive groups, posteriorly in others), the current maintained by the action of powerful lateral cilia on each filament. Particles are sieved from the water by a film of mucus in conjunction with special cilia, and potential food particles are transferred to ciliated grooves on the margins of the lamellae. These food grooves direct the particles toward the labial palps and mouth.

The different grades of ctenidial organization have played a major role in bivalve systematics (Pelseneer 1889; Ridewood 1903; Atkins 1936, 1937a, b, c, 1938a, b, c). Detailed descriptions of the ciliation, food grooves, and ciliary currents are available for many species through the work of Atkins, Yonge, Morton, and many others. Protobranch bivalves have leaflike ctenidia, but most other bivalves have long, W-shaped filaments that form very large ctenidia. Each ctenidium, left and right, is arranged into two longitudinal lamellae, or *demibranchs*, inner and outer, although the outer demibranch is sometimes secondarily lost. If the individual filaments that compose the ctenidium are free, or loosely attached by ciliary disks, this is known as the *filibranch* condition, typical of many pteriomorphians. In the *pseudolamellibranch* condition there is permanent cross-fusion of gill filaments, as found in some pteriomorphians and palaeoheterodonts. The *eulamellibranch* condition, in which the lamellae are formed by perforated sheets of tissue, is found in most heterodonts. In addition to these primary conditions, the ctenidia are often classified into *homorhabdic*, when the filaments are alike, and *heterorhabdic*, when there are different types of filaments as a consequence of the folding (or plication) of the gill area, as it is found in the majority of higher bivalves. In the deep-sea predatory anomalodesmatans, the gill is reduced to a muscular plate or septum with small pores, a condition that is termed *septibranch*. Finally, in some lucinids and vesicomyids that live in sulfide-oxidizing environments, the gill filaments may be thickened to host chemoautotrophic bacteria.

The foot is extensible and varies from being elongated (Figure 6.1C) to spade-like (Figure 6.2B) and laterally compressed, but it can be reduced to absent in some sessile taxa. Its movements are controlled by hydrostatic pressure as well as by the contractions of the numerous pedal muscles, chiefly the pedal retractors, and sometimes pedal elevators and protractors. In byssate bivalves, a functional byssal gland is located generally in the posterior end of the foot or, sometimes, toward the middle.

The bivalve digestive system is characterized by a complex stomach and associated structures described in detail by Purchon (e.g., 1956, 1957, 1958, 1987b, 1990), Reid (1965), and Dinamani (1967). The stomach has been divided in five major types with phylogenetic significance (reviewed in Purchon 1987b). Digestion often occurs in two phases: an extracellular phase occurs in the stomach, whereas intracellular digestion is restricted to the digestive diverticula opening laterally from the stomach (Morton 1983). The ingested particles are moved inside the stomach by ciliary action, with further sorting occurring in a complex array of sorting areas. Rejected material is passed directly to the midgut for discharge with the feces. The principal organ of extracellular digestion is the crystalline style. This projects into the stomach from the style sac, which may be just a modified part of the midgut (as in Nuculanidae), a side sac united with the midgut (as in pteriomorphians), or, in the more specialized condition, from a sac separate from the midgut (as in Pholadidae, Purchon 1968). The style is rotated by cilia against a gastric shield lining the posterodorsal wall of the stomach. Its dissolution releases enzymes into the stomach, breaking down mucoid-bound food strings and causing primary extracellular digestion. Such material

is transferred into large cecal embayments in the stomach wall and thence via a duct system to the digestive diverticula (Morton *et al.* 1998). For further details on the digestive process see Morton (1983) or Morton *et al.* (1998). The anus is located posteriorly, and the intestine is typically convoluted.

The heart, within the pericardium, consists of a pair of auricles attached laterally to a single, median ventricle. In most bivalves, the ventricle is loosely wrapped around the rectum, which passes through the pericardial cavity (Morton *et al.* 1998). The blood constitutes 40% to 60% of the fresh tissue weight and serves as a transport medium as well as a hydrostatic skeleton, of importance for locomotion and siphon extension. Hemocyanin, the typical molluscan respiratory pigment, has been reported in several protobranchiate bivalves (e.g., Nuculidae and Nuculanidae; Terwilliger *et al.* 1988). Most other bivalves lack a respiratory pigment, but some heterodonts have hemoglobin, which can be found in muscle tissue, in hemocoelic erythrocytes, or dissolved in the hemolymph (Manwell 1963). In bivalves, hemoglobin has been reported for the families Arcidae, Astartidae, Carditidae, Crassatellidae, Lucinidae, and Vesicomyidae (Manwell 1963; Terwilliger and Terwilliger 1985; Taylor *et al.* 2005).

The bivalve excretory system is related to removal of nitrogenous waste, ionic and osmotic regulation, and detoxification of metals. It comprises the heart, the pericardial coelom, and a pair of conducting tubes leading from the renopericardial apertures to the nephridiopores opening into the suprabranchial chamber (Morton *et al.* 1998), often called kidneys or coelomoducts. Ultrafiltration takes place across either the walls of the pericardium or the auricles (Mangum and Johansen 1975). In the case of the giant clam *Tridacna*, where the excretory system is involved also in the uptake of nutrients, the system accounts for 10% of the animal's tissue weight (Yonge 1980).

Most bivalves are dioecious (or gonochoristic), although in a few cases hermaphroditism exists. Bivalves have paired gonads that open into the suprabranchial chamber via a gonopore, either close to the nephridiopore or united with it into a common urogenital aperture (Morton *et al.* 1998). There is usually no way to differentiate the sex of a bivalve without dissecting it, but females of some freshwater unionoids have more swollen shells than the males.

Ultrastructural studies in bivalves have also proved to have phylogenetic utility, particularly the studies of sense organs (e.g., Moir 1977; Rosen *et al.* 1978; Haszprunar 1983, 1985a, b, 1987), sperm and spermatogenesis (e.g., Popham 1974; Franzén 1983; Hodgson and Bernard 1986; Healy 1989, 1995a, b), ciliation (Lundin and Schander 2001), and larval structures (Cragg and Nott 1977; Zardus and Morse 1998).

DEVELOPMENT

Early larval development is well known for many bivalves, but early embryogenesis is only well understood in a few. Detailed studies on cell lineages are available for the pteriomorphian *Ostrea edulis* (Fujita 1929), the palaeoheterodont *Lasmigona complanata* (Lillie 1895), and two freshwater heterodonts: *Dreissena polymorpha* (Meisenheimer 1901) and *Sphaerium striatinum* (Woods 1931). Additional but less detailed early embryological data are available for several other species including the protobranchiate bivalves *Yoldia limatula* (Drew 1899), *Nucula delphinodonta* (Drew 1901), *Acila castrensis* (Zardus and Morse 1998), *Solemya reidi* (Gustafson and Reid 1986, 1988) and *S. velum* (Gustafson and Lutz 1992); the pteriomorphians *Placopecten magellanicus* (Drew 1906; as "*Pecten tenuicostatus*") and *Chlamys hastata* (Hodgson and Burke 1988); and the heterodonts *Codakia orbicularis* (Gros *et al.* 1997), *Lucinoma aequizonata* (Gros *et al.* 1999), *Anomalocardia brasiliana* (Mouëza *et al.* 1999), and *Chione cancellata* (Mouëza *et al.* 2006).

During early embryogenesis, fertilized eggs undergo typical spiral development. As in other molluscs, the embryo ectoderm is divided into a pretrochal and a post-trochal region by a band of ciliated cells, the so-called prototroch cells. The pretrochal region (the future head region)

originates from the first quartet of micromeres (1a–d), formed at third cleavage. In all molluscs except bivalves, this first quartet forms a "molluscan cross" (Verdonk and van den Biggelaar 1983), although a structure similar to a molluscan cross has been observed in *Solemya reidi* (Gustafson and Reid 1986) and *S. velum* (Gustafson and Lutz 1992). However, the true significance of the molluscan cross—a specific figure formed by cells $1a^{121}$–$1d^{121}$, $1a^{122}$–$1d^{122}$, and $2a^{11}$–$2a^{11}$—and whether it is a distinct pattern observed in some molluscs (and sipunculans), has been questioned in recent studies (Jenner 2003; Maslakova *et al.* 2004), as it is not seen in chitons, some gastropods, neomenioids, and most bivalves (Maslakova *et al.* 2004).

Most marine bivalves go through a trochophore stage before turning into a free-swimming veliger larva, which then develops into a pediveliger (e.g., Carriker 1990; Cragg 1996; Morse and Zardus 1997). An apical tuft is common in the free-swimming larvae of Bivalvia (Cragg 1996; Waller 1998), although it is absent in some members of Ostreidae (Waller 1981). In most bivalves, the swimming-crawling, bivalve-shelled pediveliger is a critical stage between planktonic and benthic existence in most bivalves. A review by Carriker (1990) reported a pediveliger stage in 31 families. This important larval stage possesses a two-valved, hinged, mineralized shell (*prodissoconch*), a strongly ciliated velum, and a densely ciliated foot (e.g., Yonge 1926; Hodgson and Burke 1988; Carriker 1990; Gros *et al.* 1997). The pediveliger larvae often have statocysts, and these can have a single statolith (in heterodonts) or several statoconia (in pteriomorphians; Cragg and Nott 1977; Giribet and Wheeler 2002). Pediveligers can also have pallial eyes that lie roughly at the center of each valve just beneath the larval shell. Each eye consists of a pigmented epithelial cup surrounding a central amorphous lens, which opens toward the exterior and with a nerve leading inward (Carriker 1990).

Protobranchiate bivalves form a test larva or pericalymma (e.g., Drew 1899; Gustafson and Reid 1986; Zardus and Morse 1998).

Unionoidea lack the normal trochophore, veliger, and pediveliger stages and instead go through a larval stage known as a glochidium or lasidium (e.g., Hoggarth 1999; Graf 2000). Rather than being free-swimming, these larvae attach to, and encyst in, fish, these hosts providing upstream dispersal opportunities.

TRADITIONAL TAXONOMY

One of the first widely used classifications of bivalves was that of Thiele (1929–1935),[1] who divided them into three orders: Taxodonta, Anisomyaria, and Eulamellibranchiata. The order Taxodonta included two "stirps" that correspond to the protobranchiate bivalves and to the order Arcoida. Anisomyaria contained five stirps of pteriomorphian bivalves. The largest order, Eulamellibranchiata, was divided into the four suborders, Schizodonta, Heterodonta, Adapedonta, and Anomalodesmata, with a total of 25 stirps that correspond roughly to the currently accepted superfamilies. Schizodonta is equivalent to the modern concept of Palaeoheterodonta, and Anomalodesmata remains a clade. Heterodonta includes many heteroconch families, but Adapedonta is a polyphyletic assemblage of "myoid" and several "veneroid" families.

Three decades after the publication of Thiele's *Handbook*, the higher bivalve lineages were mostly delimited (a compilation of alternative classification systems is given by Schneider 2001). A landmark paper by Newell (1965), later incorporated into the *Treatise on Invertebrate Paleontology* (Newell 1969; see also Vokes 1968, 1980), established the basis of the modern classification of bivalves with six subclasses and 36 superfamilies. The subclasses Palaeotaxodonta and Cryptodonta include the three superfamilies of protobranchiate bivalves, which had been considered monophyletic by several other authors, and the subclass Protobranchia (or Protobranchiata; e.g., Cox 1960; Stasek 1963;

Boss 1982). Cope (1996b) further proposed the subclass Lipodonta, mostly equivalent to Newell's Cryptodonta. Newell's two protobranchiate subclasses correspond to the subsequent orders Nuculoida and Solemyoida, respectively. His other four subclasses (Pteriomorphia, Palaeoheterodonta, Heterodonta, and Anomalodesmata) have remained mostly unchanged even in the most modern treatises (e.g., Beesley *et al.* 1998; Coan *et al.* 2000; Okutani 2000), whereas Bieler and Mikkelsen (2006) had Anomalodesmata as one of four orders within Heterodonta.

The superfamilies proposed by Newell (1965) have been retained in the most recent classifications, although later a new ordinal-level classification was proposed (e.g., see Vokes 1980; Beesley *et al.* 1998). In particular, Protobranchia was divided into Nuculoida and Solemyoida; Pteriomorphia was divided into Mytiloida, Arcoida, Pterioida, Limoida, and Ostreoida; Palaeoheterodonta was divided into Trigonioida and Unionoida; Heterodonta was divided into Veneroida and Myoida; and Anomalodesmata included the single order Pholadomyoida. Paleontologists continued using two subclasses for the protobranchiate bivalves (e.g., Vokes 1980).

Protobranchiate bivalves (Solemyoidea [Figure 6.4A], Nuculoidea, Nuculanoidea [Figure 6.4B]) are primitive, marine, infaunal bivalves with aragonitic shells with nacre (in solemyoids and nuculoids), or a homogeneous structure (in nuculanoids), and origins in the Lower Ordovician. They are characterized by the form of their ctenidia (or gills), which somewhat resemble those of some basal gastropods, consisting of broad bipectinate lamellae arising from a central septum. This arrangement contrasts with the narrow, V- or W-shaped gill filaments of most other bivalves (Reid 1998).

Pteriomorphians (Figures 6.1A–C, E, 6.4C, and 6.5A–C) are marine, mostly byssate, epifaunal forms, many of which are well known and economically important (e.g., mussels, arks, oysters, and scallops). Pteriomorphians have evolved asymmetries in the adductor muscles becoming heteromyarian or monomyarian, the foot has reduced (lacking in some groups) and

---

1. I have followed the English translation of Thiele (Bieler and Mikkelsen 1998).

tend to have inequilateral, asymmetric, or even highly distorted shells, generally composed of calcite and aragonite. Arcoids and some mytiloids have a fully aragonitic shell, whereas some other groups (e.g., some pectinids) may have fully calcitic shells.

Palaeoheterodonts include two very distinct orders, the marine relictual Trigonioida (Figure 6.4D) and the diverse freshwater Unionoida (freshwater mussels, pearl mussels). Their shells are composed of aragonite deposited in the form of simple prisms (nacre). Trigonioida is represented in the Recent fauna by the sole genus *Neotrigonia*, but trigonioids were diverse in Mesozoic seas. They have a characteristic schizodont hinge, consisting of two large, diverging, blade-like teeth in the right valve that interlock with two deep and narrow sockets in the left valve (Morton 1987). Unionoida includes several families of freshwater bivalves, especially abundant and diverse in North America. Like *Neotrigonia*, unionoids have prismatonacreous shells that are equivalve and inequilateral, parivincular and opisthodetic ligaments, and two adductor muscles with associated pedal retractors, among other characters (Prezant 1998a).

Heterodonta (Figures 6.1D, 6.2, 6.3, 6.4E, 6.5D–E) is the largest, most widely distributed and most diverse of the bivalve subclasses (Prezant 1998b), and perhaps not surprisingly, a paraphyletic group that includes Anomalodesmata (see following; Figure 6.4F), which is why it is difficult to define narrowly. While most heterodonts have aragonitic shells, calcite is found in Chamidae and in the extinct Hippuritoidea (Taylor *et al.* 1969). Aragonite is deposited in simple prisms (nacre) in anomalodesmatans, but crossed-lamellar/complex crossed-lamellar structures are found in the remainder heterodonts, with some exceptions that also include composite prisms, such as in lucinids and some tellinoids and veneroids (Taylor *et al.* 1973). Most heterodonts are marine, although at least three lineages have independently colonized fresh waters (Corbulidae, Dreissenidae, and Sphaeriidae; e.g., Giribet and Distel 2003).

Typically siphonate and dimyarian, heterodonts are mostly filter feeders with large eulamellibranch ctenidia and small palps. They include free-living, boring, and commensal species, and most members have a heterodont dentition, with relatively few, but complicated interlocking hinge teeth. Shell shape, microstructure, mineralogy, and associated structures are highly variable. Heterodont shells are abundant in the Mesozoic, although the group extends back into the Paleozoic (Prezant 1998b).

Anomalodesmata comprises the single order Pholadomyoida in most classifications, or two (Pholadomyoida and Septibranchia) in Coan *et al.* (2000). The group includes an array of strange and specialized bivalves, most of which are marine, some estuarine. Most species are small, nestling, or burrowing forms with prismatonacreous shells and either have modified eulamellibranchiate ctenidia or have reduced their ctenidia to a septibranch condition (e.g., Cuspidariidae; Figure 6.4F; Prezant 1998c). The septibranch families are fascinating deep-water carnivorous bivalves; others are remarkable adventitious tube dwellers (the "watering pot" shells or Clavagellidae; e.g., Morton 1985). Anomalodesmatans are known from the Ordovician (Runnegar 1974; Cope 1996a), with one extant family known since the Carboniferous (Harper *et al.* 2000a). Although the monophyly of Anomalodesmata has not been disputed, its status as a subclass has been challenged by virtually all molecular analyses, which unambiguously nest the family within Heterodonta (see references herein). From a morphological perspective the position of anomalodesmatans suggested by the molecular analyses may look hardly tenable, especially in relation to the prismatonacreous shell ultrastructure. However, a character often considered convergent in some of the (former) heterodonts and certain anomalodesmatans is the presence of a fourth pallial aperture between the incurrent siphon and the pedal gape (Harper *et al.* 2000a; Giribet and Wheeler 2002). Dreyer *et al.* (2003) listed putative anomalodesmatan apomorphies, in addition to the fourth pallial aperture, which include a shell that frequently bears spicules, a ligament with associated lithodesma and a suite of anatomical features such as radial arenophilic glands within the mantle, Type III gills (*sensu* Stasek 1963) and Type 4 stomachs (*sensu* Purchon 1956). However, none of these characters are exclusive to Anomalodesmata. For example, Morton (1980a) and Allen (2000) have reported a lithodesma in the ligament of two leptonids. Nor are any of the characters cited above present in all recognized anomalodesmatans (Morton 1985; Harper *et al.* 2000a, 2006); for example, Cuspidariidae, Thraciidae, and some Lyonsiidae have lost the prismatonacreous shell microstructure and the carnivorous "septibranch" taxa have Type 2 stomachs (Dreyer *et al.* 2003). Given the major morphological differentiation of this group, despite some homoplasy in many of the characters used to define them, further study of the evolution of the key morphological characters in the light of the seemingly stable molecular results should provide some fascinating insights into their evolutionary relationships.

THE FOSSIL RECORD AND FOSSIL BIVALVES IN PHYLOGENETIC STUDIES

Bivalves have a rich fossil record dating back to the Cambrian (Pojeta and Runnegar 1974; Runnegar and Bentley 1983; Runnegar and Pojeta 1992; see also Parkhaev, Chapter 3). There has been an interesting debate with respect to the affinities of early bivalves, with the stenothecid monoplacophorans and rostroconchs as the main contenders (e.g., Waller 1998; Carter *et al.* 2000). Rostroconchs diversified and radiated in the Early Ordovician, prior to the large radiation of bivalves in the Middle Ordovician (Pojeta 1978). As bivalves diversified, rostroconchs slowly decreased their diversity.

The earliest known bivalve is the Early Cambrian *Pojetaia runnegari* from Australia (Figure 6.6A), with other early taxa being the Middle Cambrian *Tuarangia gravgaerdensis* from New Zealand and the worldwide Upper

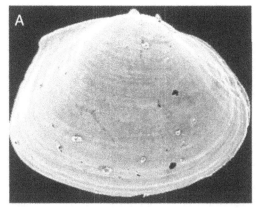

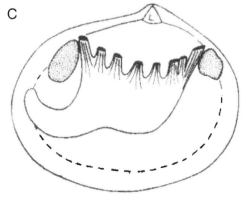

FIGURE 6.6. (A) *Pojetaia runnegari* from the Botomian Age (Early Cambrian) of southeast Gondwana (South Australia; image modified from www.palaeos.com/Invertebrates/Molluscs/Bivalvia/Bivalvia.html). (B) *Babinka prima* from the middle Ordovician of Bohemia (MCZ 113850, hypotype). (C) *Babinka prima* showing the eight pedal retractors and a reconstructed foot and pallial line (modified from Nielsen 2001).

Cambrian *Fordilla troyensis*. These three species formed the first three offshoots of the bivalve tree in the cladistic analysis of Carter *et al.* (2000), contradicting early ideas about affinities with crown group bivalves (Runnegar and

Bentley 1983; Pojeta and Runnegar 1985). By the Middle Ordovician the roots of all the extant bivalve subclasses were in place (Pojeta 1978) and all major bivalve feeding niches exploited, with the exception of the predatory lifestyle of septibranchs, which appeared later in the Triassic (e.g., Allen and Morgan 1981). Some major bivalve lineages became extinct, including rudists, a group of sessile epifaunal bivalves that flourished in low-latitude shallow seas from the Late Jurassic to the Cretaceous. Rudists contributed to the growth of reef platforms, especially in the late Cretaceous—just before going extinct (Skelton and Smith 2000).

The Ordovician lucinoid bivalve *Babinka prima* (Figure 6.6B and C) has been the center of discussions on molluscan evolution because of the eight sets of pedal retractor scars (McAlester 1965) thought homologous with the eight sets of pedal retractor muscles found in chitons and monoplacophorans (Nielsen 2001; Giribet et al. 2006) and have been interpreted as a possible plesiomorphic state in molluscs inherited from a putative segmented ancestor (e.g., Nielsen 2001). Despite the many phylogenetic schemes proposed by paleontologists, only a few so far are based on explicit numerical methodologies (e.g., Carter et al. 2000; Harper et al. 2000a).

Paleontologists recognized several lineages (e.g., Palaeoheterodonta; Newell 1965, 1969) much earlier than neontologists did (e.g., Healy 1989). In the case of Palaeoheterodonta, the hinge of primitive unionoids was of the schizodont type, like those of trigonioids (Figure 6.4D), consisting of two large diverging, blade-like teeth in the right valve that interlock with two deep and narrow sockets in the left valve (Morton 1987). This type of hinge is not found in most extant unionoids. Thus, while information from the many extinct bivalve lineages can be used in phylogenetic analyses, it is of necessity restricted to shell morphology or shell structure. Although bivalve anatomy is reflected in the shell more than in any other group of molluscs, a recent study on anomalodesmatans has shown that shell characters

alone are insufficient to resolve phylogeny (Harper et al. 2000a).

SISTER GROUP RELATIONSHIPS

Bivalves have a shell secreted by a discrete shell gland, as in gastropods, scaphopods, cephalopods, and probably in monoplacophorans. As such, bivalves form part of the clade known as Conchifera. Although the monophyly of Conchifera is more or less accepted (e.g., Salvini-Plawen and Steiner 1996; Waller 1998; but see Giribet et al. 2006 for an alternative position on Monoplacophora), the relationships within Conchifera are less than clear, and, in particular, the identity of the sister taxon of the Bivalvia is controversial. There are two common views. The first hypothesis, widely accepted for the last few decades, is the sister group relationship of bivalves to scaphopods. This was originally postulated by Lacaze-Duthiers (1858), based on anatomical and embryological data, and was followed by many other authors (e.g., Stasek 1972; Steiner 1992; Giribet 1998). Paleontologists also considered a relationship of scaphopods and bivalves to the extinct class Rostroconchia—a hypothesis known as the Diasoma concept (Runnegar and Pojeta 1974; Pojeta and Runnegar 1976, 1985; Runnegar 1978; Pojeta et al. 1987; this grouping was named Loboconcha by Salvini-Plawen 1980, 1985). Other authors also argued for monophyly of Scaphopoda + Bivalvia based on foot structure (Hennig 1979; Lauterbach 1984), and Hennig (1979) introduced the name Ancrypoda (= anchor foot) for this clade. This relationship between Scaphopoda and Bivalvia was also supported by some cladistic analyses (Götting 1980; Lauterbach 1983; Salvini-Plawen and Steiner 1996).

Lindberg and Ponder (1996) argued against the Diasoma concept and for a cephalopod-scaphopod-gastropod clade (as first suggested by Grobben 1886). Waller (1998) also excluded Scaphopoda from Loboconcha (= Diasoma), placing scaphopods as the sister group of Cephalopoda and corroborated by later cladistic analyses of morphological and molecular

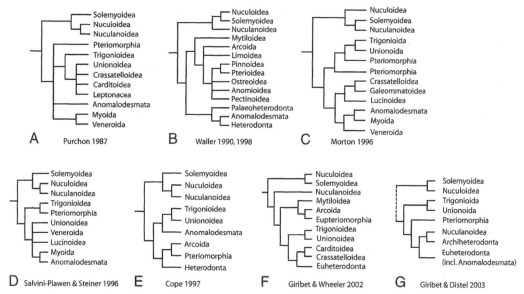

FIGURE 6.7. Phylogenetic hypotheses of higher bivalve relationships proposed by different authors. (A) Purchon (1987) based on phenetic analysis of morphological data. (B) Waller (1990, 1998) based on non-numerical cladistic analyses of morphology. (C) The proposed evolutionary tree of Morton (1996). (D) Salvini-Plawen and Steiner (1996) parsimony analysis of morphological data. (E) Suggested evolutionary tree by Cope (1997). (F) Giribet and Wheeler (2002) based on the simultaneous analysis of morphology and 2 molecular markers. (G) Giribet and Distel (2003) based on the parsimony analysis of molecular data.

data (Waller 1998; Haszprunar 2000; Giribet and Wheeler 2002). The monophyly of scaphopods, gastropods, and cephalopods has also been supported by recent developmental studies in scaphopods (Wanninger and Haszprunar 2002b).

One of the most important characters used to support the sister-group relationship of bivalves and scaphopods was the supposed presence of a "bivalved" larval shell in scaphopods and bivalves. While *engrailed* expression in the larval shells of bivalves clearly occurs in two fields (one per valve; Jacobs *et al.* 2000), the expression in scaphopods is actually restricted to a single shell gland (Wanninger and Haszprunar 2001) and is thus not bivalved.

Under the scenario that assumes monophyly of a clade including scaphopods, gastropods, and cephalopods, bivalves could well constitute the sister group of all remaining extant conchiferans, perhaps to the exclusion of Monoplacophora, which have been recently suggested to form

the clade Serialia together with Polyplacophora (Giribet *et al.* 2006).

THE PHYLOGENY OF LIVING BIVALVES

MORPHOLOGICAL PHYLOGENIES

Because bivalves present so many modifications from the plesiomorphic conchiferan condition, it is difficult to homologize (and polarize) structures such as the shell hinge, ligament, labial palps, elaborate gill ciliation, and specialized structures in the stomach, which are useful for bivalve taxonomy but are absent or much less modified in the other molluscan classes. In addition, paleontologists and neontologists have disputed the monophyly and phylogenetic position of many groups, such as Anomalodesmata, Protobranchia, or Palaeoheterodonta (e.g., compare Figure 6.7A–D). Those employing molecular sequence data have also openly questioned the monophyly of the traditional Heterodonta (e.g., Campbell

2000; Steiner and Hammer 2000; Giribet and Wheeler 2002; Giribet and Distel 2003; Taylor et al. 2005; Harper et al. 2006; Taylor and Glover 2006) and Protobranchia (Giribet and Wheeler 2002; Giribet and Distel 2003; Giribet et al. 2006), as well as Veneroida and Myoida (Campbell 2000; Steiner and Hammer 2000; Giribet and Wheeler 2002; Giribet and Distel 2003; Taylor et al. 2005; Harper et al. 2006; Taylor and Glover 2006).

Bivalves have been the focus of numerous systematic and phylogenetic treatments. Some classification systems focused on single organs or characters, such as the ctenidia (e.g., Pelseneer 1889; Atkins 1936, 1937a, b, c, 1938a, b, c), the stomach (Purchon 1956, 1957, 1958, 1959, 1960, 1963), the degree of mantle fusion (Yonge 1957), the labial palps (Stasek 1963), and shell microstructure (Taylor et al 1969, 1973; Carter 1990b). The hinge has also played an important role in establishing certain bivalve clades (e.g., Taxodonta, Schizodonta, Heterodonta), as has the structure of the ligament (e.g., Waller 1990).

Bivalve studies have pioneered the development of numerical systematics; one of the first numerical phenetic analyses for any group of organisms being that of Purchon (1978; see also Purchon 1987a). This study, based on the simultaneous analysis of multiple character systems, preceded many other morphological treatments of bivalve systematics (Waller 1990, 1998; Morton 1996; Salvini-Plawen and Steiner 1996; Cope 1997, 2000; Carter et al. 2000; Harper et al. 2000a; Giribet and Wheeler 2002 [see many of these hypotheses in Figure 6.7]). Many of these latter treatments are in general agreement with the classification proposed by Newell (1965; see previous discussion).

Morphological cladistic analyses of lower bivalve ranks (e.g., families or superfamilies) are beginning to be produced for several different clades, including Pectinoidea (Waller 2006), Pterioidea (Tëmkin 2006), Unionoidea (Graf 2000; Graf and Cummings 2006), Veneroidea (Mikkelsen et al. 2006), and Anomalodesmata (Harper et al. 2000a). Additional detailed stud-ies not based on numerical data analyses exist for arcoids (Oliver and Holmes 2006).

MOLECULAR PHYLOGENIES AND THEIR IMPACT

The widespread use of molecular data in systematics in the early 1990s was embraced by the bivalve community. These include studies encompassing all Bivalvia (Steiner and Müller 1996; Adamkewicz et al. 1997; Campbell et al. 1998; Hoeh et al. 1998; Giribet and Carranza 1999; Steiner 1999; Campbell 2000; Steiner and Hammer 2000; Giribet and Wheeler 2002; Giribet and Distel 2003) as well as major groups within the class such as Pteriomorphia (Canapa et al. 2000; Matsumoto and Hayami 2000; Steiner and Hammer 2000; Matsumoto 2003), Unionoida (Hoeh et al. 1999; Graf and Ó Foighil 2000b; Graf 2002; Huff et al. 2004; Graf and Cummings 2006), Heterodonta (Canapa et al. 1999, 2001, 2003; Park and Ó Foighil 2000; Campbell et al. 2004; Williams et al. 2004; Taylor et al. 2005; Taylor and Glover 2006)—including Anomalodesmata (Dreyer et al. 2003; Harper et al. 2006). Other studies focused on larger components of the class, or sampled representative taxa throughout the Bivalvia (e.g., Page and Linse 2002; Kirkendale et al. 2004; Campbell et al. 2005; Kappner and Bieler 2006; Mikkelsen et al. 2006).

Although by no means an exhaustive list, the publications listed in the preceding paragraph represent nearly a decade of molecular study of bivalve relationships. This trend continues to grow steadily, as seen in two recent meetings (Wells 2004; Malchus and Pons 2006), where roughly half of the bivalve presentations focused on molecular aspects of systematics and population genetics. Molecular techniques are currently at the forefront of biodiversity and evolutionary studies. Mollusc workers in general (see Lydeard and Lindberg 2003, and other chapters in this volume), and bivalve workers in particular (see previous citations), have taken full advantage of the possibilities that this source of data has to offer.

As of April 2006, GenBank, the sequence repository database of NCBI (National Center for Biotechnology Information, National Library of Medicine and National Institutes of Health, United States: www.ncbi.nlm.nih.gov) yielded more than 34,000 hits to the keyword Bivalvia. In theory this means that there are about 34,000 sequences of bivalves deposited in the database (not accounting for redundancy or other artifacts; for example, a paper with the keyword "Bivalvia" will have attributed to it all the new sequences published in that paper regardless of the taxonomy of the deposited sequence). Assuming that biases are equal for all molluscan classes, the same database contained only 52 hits for the term "Aplacophora" (following the taxonomy adopted by the database), 85 for Scaphopoda, and 284 for Polyplacophora, perhaps reflecting the greater diversity and abundance of bivalves. Within a year (from April 2005 to April 2006) the number of sequences deposited for Cephalopoda increased from 2,007 to 37,800. The most telling comparison is with gastropods, with almost an order of magnitude more diversity than that of bivalves, but the GenBank query results in slightly more than 41,100 hits.

The relatively healthy state of bivalve studies is also reflected in the numerous volumes published as result of special meetings and symposia. The symposium "Evolutionary Systematics of Bivalve Molluscs" organized by C. M. Yonge and T. E. Thompson (Cambridge, UK, 1978) was the first of a series of symposia that focused on the evolution and systematics of bivalves resulting in a collection of papers as part of a regular issue of the *Philosophical Transactions of the Royal Society of London B* (Vol. 284, pp. 199–436). A memorial symposium to honor Sir C. M. Yonge followed in Edinburgh in 1986 and resulted in a 355-page volume (Morton 1990b). The following symposia, in Drumheller in 1995 and in Cambridge, UK, in 1999, recruited molecular systematics into the tool kit of bivalve workers. These meetings resulted into two key volumes synthesizing much of the modern bivalve biology and systematics (Johnston and Haggart 1998; Harper *et al.* 2000b). Two recent landmarks in bivalve research were the International Marine Bivalve Workshop organized at Long

Key, Florida, in July 2002 that resulted in a special issue of *Malacologia* (Bieler and Mikkelsen 2004) and the bivalve symposium organized at the World Congress of Malacology in Perth in July 2004 that resulted in another special issue of the *Zoological Journal of the Linnean Society* (Bieler 2006), although these events, unlike the previous meetings, were restricted to invited participants.

To date, no complete bivalve genome has been sequenced or has been scheduled for production. Another important aspect in bivalve genetics is the discovery of paternally inherited mitochondrial genomes in bivalves (Zouros *et al.* 1994; Cao *et al.* 2004), a phenomenon that could be much more general than previously thought.

A NEW CLASSIFICATION OF BIVALVES

After decades of debate about the position of several bivalve lineages (e.g., see Figure 6.7 for differences in the position of Trigonioidea or Anomalodesmata in the studies of Salvini-Plawen and Steiner [1996], Cope [1997], and Waller [1998]), the addition of molecular data has assisted in stabilizing some hypotheses. This does not mean that all contentious issues in bivalve systematics have been solved by molecular data, and it certainly does not mean that all bivalve workers agree with the ideas presented here. Some of these other ideas are summarized in Figure 6.7 and in the taxonomic discussion provided previously. Consideration of molecular data has also introduced new debates, such as the paraphyly of protobranchiate bivalves (e.g., Giribet and Wheeler 2002), indicating that the protobranchiate gill is a plesiomorphic trait retained by more than one lineage.

Given the recent progress in bivalve systematics and evolution, it is time to propose a modified classification of bivalves that should be stable with the addition of data and therefore useful in the long term. The new classification is based mostly on the combined analysis of morphology and molecular data by Giribet and Wheeler (2002) but also takes into account results from other studies (Campbell 2000;

Steiner and Hammer 2000; Dreyer *et al.* 2003; Giribet and Distel 2003; Williams *et al.* 2004; Taylor *et al.* 2005; Bieler and Mikkelsen 2006). The classification is restricted so far to the higher lineages, for which evidence is stable or has high nodal support. Controversial issues that do not show enough support or stability (e.g., paraphyly of Protobranchia) are left for resolution at a later time. The proposed classification is a cladistic hypothesis, and ranks are not provided.

The new bivalve classification is introduced by discussing all major bivalve clades and by summarizing the evidence that supports such groups. The structure of this section will follow that of the classification outlined in Table 6.1, which is based on the cladogram shown in Figure 6.8.

Several higher groups (usually treated as subclasses) of bivalves are recognized by modern bivalve workers as outlined earlier in this chapter. Protobranchia has had its monophyly questioned in some recent studies (e.g., Campbell *et al.* 1998; Giribet and Wheeler 2002; Giribet and Distel 2003; see also Bieler and Mikkelsen 2006). It is composed of the clades Solemyoida, Nuculoida, and Nuculanoida (see Figure 6.7A–E). Most earlier studies (e.g., Purchon 1987a; Waller 1990, 1998; Morton 1996; Salvini-Plawen and Steiner 1996; Cope 1997) suggested that Nuculoida comprised Nuculoidea and Nuculanoidea (Figure 6.7A, D, E), but recent phylogenetic work supports a closer relationship of Nuculoidea to Solemyoidea (Waller 1998; Giribet and Wheeler 2002; Giribet and Distel 2003; see also Bieler and Mikkelsen 2006; Figure 6.7B, F, G). Protobranchia have their highest diversity in the deep sea, are all infaunal, and most deposit-feed (by way of their long labial palps). All have simple (protobranchiate) ctenidia (i.e., not modified for filter feeding) used solely for respiration. Nuculoid and nuculanoid forms are known from the Ordovician, and solemyoids have been recorded from the Devonian (Cox *et al.* 1969). Whereas Solemyoidea and Nuculoidea lack siphons, nuculanoids have developed posterior mantle fusion and, unlike

TABLE 6.1
The New Bivalve Classification

Bivalvia Linnaeus, 1758

Opponobranchia **new name**
Nuculoida Dall, 1889
Solemyoida Dall, 1889
Nuculanoida Carter, D. C. Campbell and
M. R. Campbell, 2000
Autolamellibranchiata Grobben, 1894
Pteriomorphia Beurlen, 1944
Heteroconchia Cox, 1960
Palaeoheterodonta Newell, 1965
Trigonioida Dall, 1889
Unionoida Stoliczka, 1871
Heterodonta Neumayr, 1883
Archiheterodonta **new name**
Euheterodonta Giribet and Distel, 2003
(including Anomalodesmata)

other protobranchs, lack nacre. Solemyoids have acquired bacterial endosymbionts and, as a consequence, have a much reduced gut. Some studies suggest Nuculanoidea are the sister group of the remaining bivalves (Autolamellibranchiata; Giribet and Wheeler 2002), or at least form a monophyletic group with them (Giribet and Distel 2003). The clade containing Nuculoidea and Solemyoidea is characterized by the opposite arrangement of gill filaments along the gill axis (Waller 1998), which results in a unique pattern of ciliary currents, designated the Type-A ciliary mechanism by Atkins (1937b). This clade has not been previously formally recognized and is here named Opponobranchia, referring to the opposite arrangement of the gill filaments in the protobranchiate gill. Whether Opponobranchia and Nuculanoidea form the clade Protobranchia (e.g., Waller 1998), or Opponobranchia is the sister clade to (Nuculanoidea + Autolamellibranchiata; e.g., Giribet and Wheeler 2002), is still contentious, and therefore I prefer to leave their relationships as unresolved (as in Figure 6.8).

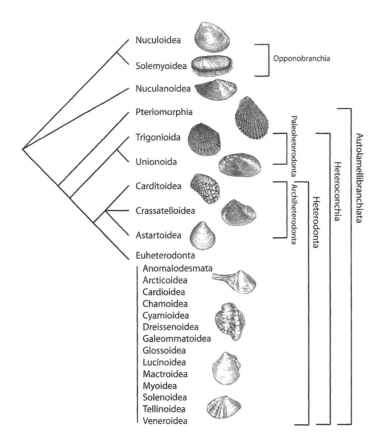

FIGURE 6.8. Hypothesized relationships among higher bivalve taxa following the classification proposed in this chapter. Resolution within Euheterodonta is, to say the least, contentious, but it is clear that it includes all members of the old order Myoida as well as the members of the monophyletic Anomalodesmata. All images from Beesley *et al.* (1998).

AUTOLAMELLIBRANCHIATA   This group includes all bivalves with their ctenidia modified for filter feeding (and subsequent modifications or reduction). They are primarily divided into Pteriomorphia and Heteroconchia. Pteriomorphia comprises mostly byssate epifaunal bivalves, including some of the most important commercial bivalves, such as mussels, arks, oysters, pearl oysters, and scallops. Pteriomorphians include 10 superfamilies (Mytiloidea, Arcoidea [including Limopsoidea following Steiner and Hammer 2000; Giribet and Distel 2003], Pterioidea, Pinnoidea, Limoidea, Ostreoidea, Plicatuloidea, Dimyoidea, Pectinoidea, and Anomioidea) of mostly monomyarian or strongly heteromyarian forms with great variation in shell form and with filibranch or pseudolamellibranch ctenidia (with exceptions). The oldest pteriomorphians are arcoid-like forms from the Lower Ordovician (Cox *et al.* 1969). The phylogenetic relationships of Pteriomorphia have been well studied using different sources of information (Waller 1978; Healy *et al.* 2000; Steiner and Hammer 2000; Giribet and Wheeler 2002; Giribet and Distel 2003; Matsumoto 2003; Malchus 2004; Tëmkin 2006), although a consensus has yet to be reached (see Matsumoto 2003).

HETEROCONCHIA   This group, comprising Palaeoheterodonta and Heterodonta, has been recognized in almost all analyses of bivalve relationships, from fossils to molecules (e.g., Waller 1998; Carter *et al.* 2000; Giribet and Wheeler 2002). Palaeoheterodonta includes a few extant Australian marine species of *Neotrigonia* (Trigonioida) and the freshwater mussels or naiads (Unionoida). Trigonioida is represented today by 6 or 7 species of the genus *Neotrigonia*, but its origins can be traced back to the early-middle Ordovician genus *Noradonta* (Cope 2000) with the group subsequently being diverse in the Mesozoic (Stanley 1978). Unionoida comprises a diverse group of freshwater bivalves with about 175 genera distributed in all continents except Antarctica (Roe and Hoeh 2003) and a fossil record extending back to the Upper Devonian (Cox *et al.* 1969). Recently phylogenetic studies on unionoids have flourished, in part driven by conservation considerations (Lydeard *et al.* 1996, 2000; Graf 2000, 2002; Graf and Ó Foighil 2000a, 2000b; Roe *et al.* 2001; Roe and Hoeh 2003; Serb *et al.* 2003; Huff *et al.* 2004). Although several authors have disputed the monophyly of Palaeoheterodonta (Purchon 1987a; Salvini-Plawen and Steiner 1996; Cope 2000), others support it (Newell 1965, 1969; Healy 1989; Morton 1996; Cope 1997; Waller 1998), and it is one of the best-corroborated clades in molecular analyses of bivalve relationships (Hoeh *et al.* 1998; Graf and Ó Foighil 2000a; Giribet and Wheeler 2002; Giribet and Distel 2003).

HETERODONTA This clade includes the majority of living bivalve lineages. Traditionally the Heterodonta included the members of Veneroida and Myoida. However, the non-monophyly of Myoida and Veneroida, as well as the paraphyly of Heterodonta with respect to Anomalodesmata (e.g., Campbell 2000; Steiner and Hammer 2000; Dreyer *et al.* 2003; Giribet and Wheeler 2002; Giribet and Distel 2003; Taylor *et al.* 2005; Taylor and Glover 2006), makes it necessary to redefine this clade. Heterodonta is considered here to be composed of two main lineages: one named here Archiheterodonta nov. and another, previously named Euheterodonta (Giribet and Distel 2003), including the remaining heterodont families.

ARCHIHETERODONTA NOV This lineage includes some of the oldest known bivalve fossils dating back to the Silurian (Allen 1985) or perhaps the Ordovician (Cox *et al.* 1969). The clade includes Astartoidea, Carditoidea, and Crassatelloidea, which are marine infaunal and some byssate forms, with an entire pallial line lacking a sinus, reflecting their nonsiphonate condition. Hemoglobin has also been reported in several members of this clade (e.g., Taylor *et al.* 2005). A reduction of one

of the two centrioles during spermiogenesis has been reported in members of the families Carditidae and Crassatellidae (Healy 1995b) and this may constitute a synapomorphy of Archiheterodonta. The antiquity of the group and its phylogenetic position are consistent with earlier hypotheses about the primitive status of its members relative to other heterodonts based on morphological data (Yonge 1969; Purchon 1987a). Additionally, support for the sister group relationship of Archiheterodonta to Euheterodonta is widely found in molecular analyses (Campbell 2000; Park and Ó Foighil 2000; Giribet and Wheeler 2002; Dreyer *et al.* 2003; Giribet and Distel 2003; Williams *et al.* 2004; Taylor *et al.* 2005; Harper *et al.* 2006; Taylor and Glover 2006), and to a certain degree based on fossil data (J. G. Carter, D. C. Campbell, and M. R. Campbell, 2006, unpublished results).

EUHETERODONTA This group (= *Unnamed node 3* of Giribet and Wheeler 2002) is the most speciose and widely distributed group of bivalves, characterized by infaunal groups, some of which present striking adaptations such as bacterial symbiosis (Lucinidae, Thyasiridae, Teredinidae), symbiotic zooxanthellae (Cardiidae: Fraginae and Tridacninae), or the secondary acquisition of a predatory lifestyle (Cuspidarioidea). Shell ultrastructure is variable (see previous discussion), but nacre is found only in the members of Anomalodesmata (see previous discussion), implying a possible secondary gain of nacreous shells in this clade. Shell morphology becomes highly plastic in some lineages, and Euheterodonta is difficult to define based on morphological characters alone. Euheterodonta includes the traditional "veneroidan" families not included in Archiheterodonta, the polyphyletic "Myoida," and the monophyletic Anomalodesmata, previously recognized as a subclass of its own (e.g., Newell 1965, 1969; Vokes 1968, 1980; Beesley *et al.* 1998). The group includes many lineages, including Anomalodesmata and the superfamilies Lucinoidea, Galeommatoidea, Cyamioidea, Tellinoidea, Gastrochaenoidea, Hiatelloidea,

Solenoidea, Arcticoidea, Glossoidea, Corbiculoidea, Sphaerioidea, Veneroidea, Cardioidea, Chamoidea, Mactroidea, Pholadoidea, Myoidea, and Dreissenoidea (Campbell 2000; Park and Ó Foighil 2000; Steiner and Hammer 2000; Giribet and Wheeler 2002; Dreyer et al. 2003; Giribet and Distel 2003; Taylor et al. 2005; Harper et al. 2006; Taylor and Glover 2006). Internal relationships within Euheterodonta are not well understood, although important progress has been made for Lucinoidea (Williams et al. 2004; Taylor et al. 2005; Taylor and Glover 2006) and Anomalodesmata (Harper et al. 2000a, 2006; Dreyer et al. 2003). Lucinoidea has been shown to be polyphyletic (Williams et al. 2004; Taylor et al. 2005; Taylor and Glover 2006), but not in Giribet and Distel (2003), who used a contaminated 28S rRNA sequence for Thyasira (J. D. Taylor, personal communication). The status of Arcticoidea is unstable to analytical parameter variation (Giribet and Wheeler 2002; Giribet and Distel 2003). The two groups previously included in Corbiculoidea are now treated as two unrelated clades, Corbiculoidea and Sphaerioidea (Park and Ó Foighil 2000; Giribet and Wheeler 2002; Giribet and Distel 2003). Cardioidea now includes the previously recognized superfamily Tridacnoidea (e.g., Beesley et al. 1998) as tridacnines within Cardiidae. Phylogenetic revision of the large superfamily Veneroidea using morphological and molecular data has suggested that Veneroidea is polyphyletic relative to the current classification of bivalves because Petricolidae and Neoleptonidae, the latter considered part of Cyamioidea (e.g, Beasley et al. 1998), have independent origins from Veneridae; and that Veneridae is paraphyletic with respect to the accepted families because it includes both Petricolidae and Turtoniidae (Mikkelsen et al. 2006). Reassessment of the boundaries of these taxa is clearly necessary. Monophyly of the other large superfamily, Tellinoidea, has not yet been properly investigated, but preliminary molecular data suggest monophyly (Giribet and Distel 2003; Taylor and Glover 2006; G. Giribet, unpublished sequence data).

The view of bivalve phylogenetics presented here heavily relies on numerical analyses of molecular and morphological data. It is acknowledged that the shell characters available in fossil taxa also play an important role in understanding the evolution of bivalves and may not be fully reflected in this hypothesis. However, it is also true that paleontological views have relied heavily on non-numerical interpretations. A global strategy to combine broad anatomical, paleontological, and molecular data was recently initiated in the International Congress on Bivalvia (Bellaterra, July 22–27, 2006) as the source for a new *Treatise on Invertebrate Paleontology* (Carter et al., work in progress). This broad initiative will attempt to reconciliate neontological and paleontological evidence for Bivalvia.

ADAPTIVE RADIATIONS

In the light of our revised phylogeny, we can assess the adaptive radiations of Bivalvia. Early bivalves were most probably infaunal with their first diversification during the Cambrian, when the genera *Fordilla*, *Pojetaia*, and *Tuarangia* appeared, but their diversity remained low. Following the hypertrophization of ctenidia and accompanying suspension feeding, the first increase in bivalve diversity and body plan disparity occurred during the Early Paleozoic, and by the Ordovician all extant higher lineages and feeding types were present (Pojeta 1978; Morton 1996). A structure that was an important factor in this diversification was the development of a larval byssus in the basal autolamellibranchs, absent in protobranchiate bivalves and later retained in some adult forms (Morton 1996). This enabled exploitation of hard substrates (e.g., rocks, coral), allowing the animals to become epibenthic, opening new adaptive niches. Retention of the byssus in the adult (notably pteriomorphians) also allowed bivalves to anchor themselves and live in more energetic systems, such as reef crests and exposed hard shores (Morton 1996). Cementation also evolved in Paleozoic oysters and scallops, with similar consequences. Pteriomorphians and archiheterodonts diversified during this next wave of bivalve diversification and colonized mostly shallow environments, leaving the deeper waters to the older protobranchiate lineages.

The next radiation of bivalves occurred during the Mesozoic, starting with a pool of bivalves already widely radiated, but depleted by the Permo-Triassic mass extinction. For example, shallow-water protobranchs were severely decimated, surviving mostly in the deep sea (Allen 1983). New lineages also arose and radiated, some to disappear quite quickly (Gould and Calloway 1980). Lucinoids and trigonioids diversified during this period, and it is also during this period that many "myoid" forms exploited their different niches, including byssate taxa, deep burrowers, and ultimately borers of coral (e.g., some Mytilidae, Gastrochaenidae), rock (e.g., some Mytilidae, Gastrochaenidae, Pholadidae), and wood (Pholadidae, Teredinidae).

Cementation of one of the valves to the substrate is an adaptation that originated independently in more than 16 bivalve families, including several marine pteriomorphians (e.g., Dimyidae, Ostreidae, Pectinidae, Placunidae, Plicatulidae, Spondylidae), marine heterodonts and anomalodesmatans (Chamidae, Clavagellidae, Cleidothaeridae, Myochamidae), and some freshwater unionoids (Etheriidae) and corbiculids (*Posostrea*) (Yonge 1979; Bogan and Bouchet 1998), whereas Anomiidae cement by way of a calcified byssus. Some of them, for example, oysters and the extinct Mesozoic rudists, have been extremely successful over geological time (Harper 1991). Cementation in bivalves is mostly a post-Paleozoic habit (Harper 1991), peaking in the Late Triassic and Jurassic, as a possible response to the Mesozoic Marine Revolution (Vermeij 1977), or the appearance of many predatory groups of marine animals during the Mesozoic. An experimental setting has demonstrated that major bivalve predators have preference for noncemented bivalves, indicating that cementation may have actually served to avoid predation (Harper 1991).

Anomalodesmatans pose an interesting case of a group that has diversified anatomically but that never showed much species diversity, perhaps because many occupy narrow, marginal niches (Morton 1985). They originated in the Middle Ordovician (Morton 1985), diversifying during the Middle and Late Paleozoic and Mesozoic. For most Anomalodesmata the advent of the Mesozoic was a period of declining importance, probably as a result of competition with the emerging "veneroid" forms, generalists that came to dominate shallow-water, soft substrates during the Mesozoic—and that still do (Morton 1981). The other lineages continued radiating subsequently, especially the epibyssate pteriomorphians that colonized hard substrates. In the deep sea, protobranchiate bivalves, arcoids, and propeamussiids also radiated widely, and the deep sea anomalodesmatans evolved into the septibranch lineages.

Another important episode in bivalve evolution is the colonization of freshwater environments by an ancestral unionoid during the Triassic, obtaining access to a bivalve-free ecosystem. Such radiation could have been triggered by the evolution of a novel mode of development using glochidium-type larvae with fish as intermediate hosts. Another parasitic larva, the lasidia, produced by the members of Etherioidea, is considered to be more derived (Graf 2000).

During the Cenozoic bivalves radiated into many of the modern forms while the current configuration of the oceans often resulted in taxon provincialism. During this time, soft, nutrient-rich sediments on continental margins allowed the diversification of shallowly burrowing, globular, strongly ribbed shells (e.g., cardiids), some of which exploited zooxanthellae symbiosis, or deep burrowers, with smooth blade-like shells (Solenoidea and Tellinoidea). Nonattached pectinids also become abundant during the Cenozoic, some of them becoming proficient swimmers (Morton 1980b).

During this time some bivalve lineages were able to exploit new niches that have resulted into unusual groups as a result of their diversity or peculiar modes of life. Among these, Tellinoidea is the largest bivalve superfamily, and most of its members show adaptations to their deep burrowing habits, including long and separate siphons, enlarged palps, and complex stomachs, and have been considered—according to some authors—to represent the culmination of the heterodont radiation (Willan 1998). The siphons are extremely extensible and bear distal papillae. The inhalant siphon ingests deposited organic matter from the sediment surface while water and feces are expelled through the exhalant siphon (Trevallion 1971). This novel feeding mode is often regarded as a key innovation that gave origin to the large diversity within the family Tellinidae. Galeommatoidea is a diverse group of generally small-sized, highly modified marine bivalves that often have commensal relationships with other invertebrates (e.g., Ponder 1998; Mikkelsen and Bieler 1992). Some species have an active foot used for crawling, such as *Galeomma turtoni* (Popham 1940), others such as *Ephippodonta* (Middelfart 2005) resemble crawling limpets, while yet others have largely lost their internal shells and become sluglike (e.g., *Phlyctaenachlamys* [Popham 1939]).

GAPS IN KNOWLEDGE

Numerous and extraordinary studies on functional morphology (see the many studies by B. Morton, R. D. Purchon, and C. M. Yonge, among others, e.g., Yonge 1939, 1957, 1980; Morton 1987, 2000; Purchon 1987b) and sperm ultrastructure (e.g., Healy 1995a, b, 1996; Healy *et al.* 2000; Keys and Healy 2000; Healy *et al.* 2006), as well as the many cladistic morphological and molecular studies, place bivalves in the forefront of molluscan research. All of these facets of bivalve research have contributed to resolving many of the branches of the bivalve tree, both at deep and shallow nodes. However, the recent focus on molecular studies, in particular, has impacted on important research in other areas such as anatomical, ultrastructural, and developmental studies. Other emerging disciplines such as *evo-devo*[2] have

---

2. Evo-devo is a common term that stands for a relatively new biological discipline that integrates evolution and development. See Chapter 16.

been largely neglected in bivalves while yielding important data in other groups of molluscs (e.g., van den Biggelaar *et al.* 1986; Wanninger *et al.* 1999; Wanninger and Haszprunar 2001, 2002a, b, 2003; Friedrich *et al.* 2002; Okusu 2002; Henry *et al.* 2004; see also Wanninger *et al.*, this volume, Chapter 16). Developmental studies on musculature (Wanninger *et al.* 1999; Wanninger and Haszprunar 2002a, b) and the nervous system (Friedrich *et al.* 2002; Wanninger and Haszprunar 2003) are currently being applied to various molluscan lineages and other metazoans and have aided in resolving long-standing controversies on character homology. However, no modern study has been published on the development of the musculature in a bivalve group. Gene expression data are equally important (e.g., Jacobs *et al.* 2000; Wanninger and Haszprunar 2001), and little information is yet available on any species of bivalve (see Jacobs *et al.* 2000), and only a few Hox and paraHox genes have been isolated for a few bivalves (Barucca *et al.* 2003; Iijima *et al.* 2006).

In a recent review on bivalve phylogenetics, Giribet and Distel (2003) concluded their chapter with a list of hot topics on bivalve systematics and evolution. These included major phylogenetic questions such as the status of the protobranchiate bivalves or the archiheterodont families; taxonomy of Arcoida, Mytilidae, and Heterodonta; a better understanding of the anomalies of the Anomalodesmata; and other evolutionary aspects such as the origins of carnivorism in septibranchs, colonization of freshwater environments or the deep sea, and the origins of rock- and wood-boring habits or algal and bacterial symbioses. Many of those remain unexplored and require the compilation of new data sets including more species and more characters (both morphological and molecular). A recent study on molluscan phylogenetics (Giribet *et al.* 2006) included sequence data for 101 molluscs (24 of which are bivalves) and five genes, totaling *ca.* 6.5 kb per complete taxon. It is to be expected that future bivalve studies include at least as many taxa and genes as this study on molluscan relationships. It should also be expected that new and more comprehensive morphological matrices than those currently available (e.g., Waller 1998; Carter *et al.* 2000; Harper *et al.* 2000; Giribet and Wheeler 2002) will also be assembled. Ideally, both sets of data matrices—the molecular and morphological ones—will be generated for largely overlapping sets of species to allow for the only sensible way to synthesize bivalve evolutionary relationships: the integration of all phylogenetic information.

More recently there have also been important developments in related scientific disciplines. Cell lineage studies using 4-D microscopy (Hejnol and Schnabel 2005), immunohistochemistry on musculature (e.g., Wanninger and Haszprunar 2002a, 2002b) and nervous systems (e.g., Wanninger and Haszprunar 2003), and developmental genetics (e.g., Jacobs *et al.* 2000) are the main unexplored areas in Bivalvia that could provide new insights into the evolutionary history of these fascinating animals. Even more exciting times are yet to come for bivalve workers.

ACKNOWLEDGMENTS

I dedicate this chapter to the memory of Lluis Dantart, former colleague and friend from the Departament de Biologia Animal, Universitat de Barcelona, who passed away as a consequence of a fatal car crash. He taught me about molluscs, SEM, and photography. Dave Lindberg and Winston Ponder are acknowledged for soliciting this chapter and for their continuous encouragement. Winston Ponder has provided numerous corrections on earlier versions and has dealt patiently with the many delays in delivering the different versions of this chapter. Rüdiger Bieler kindly provided final proofs of his edited volume *Bivalvia—A Look at the Branches*. Manuela Krakau provided literature and interesting discussions on bivalve parasites. Two anonymous reviewers are also acknowledged for their comments, which greatly helped to improve this chapter.

REFERENCES

Adamkewicz, S. L., Harasewych, M. G., Blake, J., Saudek, D., and Bult, C. J. 1997. A molecular phylogeny of the bivalve mollusks. *Molecular Biology and Evolution* 14: 619–629.

Allen, J. A. 1983. The ecology of deep-sea molluscs. In *The Mollusca*. Vol. 6, *Ecology*. Edited by W. D. Russell-Hunter. Orlando, FL: Academic Press, pp. 29–75.

———. 1985. The Recent Bivalvia: their form and evolution. In *The Mollusca*. Vol. 10, *Evolution*. Edited by E. R. Trueman and M. R. Clarke. Orlando, FL: Academic Press, pp. 337–403.

———. 2000. A new deep-sea species of the genus *Neolepton* (Bivalvia; Cyamioidea; Neoleptonidae) from the Argentine basin. *Malacologia* 42: 123–129.

Allen, J. A., and Morgan, R. 1981. The functional morphology of Atlantic deep water species of the families Cuspidariidae and Poromyidae (Bivalvia): an analysis of the evolution of the septibranch condition. *Philosophical Transactions of the Royal Society of London, Series B* 294: 413–546.

Araujo, R., Cámara, N., and Ramos, M. A. 2002. Glochidium metamorphosis in the endangered freshwater mussel *Margaritifera auricularia* (Spengler, 1793): a histological and scanning electron microscopy study. *Journal of Morphology* 254: 259–265.

Araujo, R., Moreno, D. and Ramos, M. A. 1993. The asiatic clam *Corbicula fluminea* (Müller, 1774) (Bivalvia: Corbiculidae) in Europe. *American Malacological Bulletin* 10: 39–49.

Araujo, R., and Ramos, M. A. 2001. Life-history data on the virtually unknown *Margaritifera auricularia*. In *Ecology and Evolution of the Freshwater Mussels Unionoida*. Edited by G. Bauer and K. Wächtler. Berlin: Springer, pp. 143–152.

Arnaud, P. M., Troncoso, J. S., and Ramos, A. 2001. Species diversity and assemblages of macrobenthic Mollusca from the South Shetland Islands and Bransfield Strait (Antarctica). *Polar Biology* 24: 105–112.

Atkins, D. 1936. On the ciliary mechanisms and interrelationships of lamellibranchs. Part I. New observations on sorting mechanisms. *Quarterly Journal of Microscopical Sciences* 79: 181–308.

———. 1937a. On the ciliary mechanisms and interrelationships of lamellibranchs. Part II: sorting devices on the gills. *Quarterly Journal of Microscopical Sciences* 79: 339–373.

———. 1937b. On the ciliary mechanisms and interrelationships of lamellibranchs. Part III: types of lamellibranch gills and their food currents. *Quarterly Journal of Microscopical Sciences* 79: 375–421.

———. 1937c. On the ciliary mechanisms and interrelationships of lamellibranchs. Part IV: cuticular fusion, with special reference to the fourth aperture in certain lamellibranchs. *Quarterly Journal of Microscopical Sciences* 79: 423–445.

———. 1938a. On the ciliary mechanisms and interrelationships of lamellibranchs. Part V: note on the gills of *Amusium pleuronectes*. *Quarterly Journal of Microscopical Sciences* 80: 321–329.

———. 1938b. On the ciliary mechanisms and interrelationships of lamellibranchs. Part VI: the pattern of lateral ciliated cells of the gill filaments of the Lamellibranchia. *Quarterly Journal of Microscopical Sciences* 80: 331–344.

———. 1938c. On the ciliary mechanisms and interrelationships of lamellibranchs. Part VII: latero-frontal cilia of the gill filaments and their phylogenetic value. *Quarterly Journal of Microscopical Sciences* 80: 346–436.

Barucca, M., Olmo, E., and Canapa, A. 2003. Hox and paraHox genes in bivalve molluscs. *Gene* 317: 97–102.

Beesley, P. L., Ross, G. J. B., and Wells, A., eds. 1998. *Mollusca: The Southern Synthesis.* Vol. 5, *Fauna of Australia.* Melbourne: CSIRO Publishing. viii + 1234 pp.

Benzie, J. A. H., and Williams, S. T. 1992. Genetic structure of giant clam (*Tridacna maxima*) populations from reefs in the Western Coral Sea. *Coral Reefs* 11: 135–141.

———. 1995. Gene flow among giant clam (*Tridacna gigas*) populations in Pacific does not parallel ocean circulation. *Marine Biology* 123: 781–787.

———. 1998. Phylogenetic relationships among giant clam species (Mollusca: Tridacnidae) determined by protein electrophoresis. *Marine Biology* 132: 123–133.

Bieler, R., ed. 2006. Bivalvia—a look at the branches. *Zoological Journal of the Linnean Society* 148: 221–552.

Bieler, R., and Mikkelsen, P. M., eds. 1998. *Handbook of Systematic Malacology.* Parts 3 and 4. Washington, DC: Smithsonian Institution Libraries. pp. 1193–1690.

———, eds. 2004. Bivalve studies in the Florida Keys. Proceedings of the International Marine Bivalve Workshop in Long Key, Florida, July 2002. *Malacologia* 46: 241–677.

———. 2006. Bivalvia—a look at the branches. *Zoological Journal of the Linnean Society* 148: 223–235.

Bogan, A., and Bouchet, P. 1998. Cementation in the freshwater bivalve family Corbiculidae (Mollusca: Bivalvia): a new genus and species from Lake Poso, Indonesia. *Hydrobiologia* 389: 131–139.

Boss, K. J. 1982. Mollusca. In *Synopsis and Classification of Living Organisms.* Edited by S. P. Parker. New York: McGraw-Hill Book Company, pp. 1092–1166.

Bouchet, P., Lozouet, P., Maestrati, P., and Heros, V. 2002. Assessing the magnitude of species richness in tropical marine environments: exceptionally high numbers of molluscs at a New Caledonia site. *Biological Journal of the Linnean Society* 75: 421–436.

Bromley, R. G., and Asgaard, U. 1990. *Solecurtus strigilatus*: a jet-propelled burrowing bivalve. In *The Bivalvia—Proceedings of a Memorial Symposium in Honour of Sir Charles Maurice Yonge, Edinburgh, 1986.* Edited by B. Morton. Hong Kong: Hong Kong University Press, pp. 313–320.

Campbell, D. C. 2000. Molecular evidence on the evolution of the Bivalvia. In *The Evolutionary Biology of the Bivalvia.* Edited by E. M. Harper, J. D. Taylor, and J. A. Crame. London: The Geological Society of London, pp. 31–46.

Campbell, D. C., Hoekstra, K. J., and Carter, J. G. 1998. 18S ribosomal DNA and evolutionary relationships within the Bivalvia. In *Bivalves: An Eon of Evolution—Palaeobiological Studies Honoring Norman D. Newell.* Edited by P. A. Johnston and J. W. Haggart. Calgary: University of Calgary Press, pp. 75–85.

Campbell, D. C., Serb, J. M., Buhay, J. E., Roe, K. J., Minton, R. L., and Lydeard, C. 2005. Phylogeny of North American amblemines (Bivalvia, Unionoida): prodigious polyphyly proves pervasive across genera. *Invertebrate Biology* 124: 131–164.

Campbell, M. R., Steiner, G., Campbell, L. D., and Dreyer, H. 2004. Recent Chamidae (Bivalvia) from the Western Atlantic Ocean. *Malacologia* 46: 381–415.

Canapa, A., Barucca, M., Marinelli, A., and Olmo, E. 2000. Molecular data from the 16S rRNA gene for the phylogeny of Pectinidae (Mollusca: Bivalvia). *Journal of Molecular Evolution* 50: 93–97.

———. 2001. A molecular phylogeny of Heterodonta (Bivalvia) based on small ribosomal subunit RNA sequences. *Molecular Phylogenetics and Evolution* 21: 156–161.

Canapa, A., Marota, I., Rollo, F., and Olmo, E. 1999. The small-subunit rRNA gene sequences of venerids and the phylogeny of Bivalvia. *Journal of Molecular Evolution* 48: 463–468.

Canapa, A., Schiaparelli, S., Marota, I., and Barucca, M. 2003. Molecular data from the 16S rRNA gene for the phylogeny of Veneridae (Mollusca: Bivalvia). *Marine Biology* 142: 1125–1130.

Cao, L., Kenchington, E., and Zouros, E. 2004. Differential segregation patterns of sperm mitochondria in embryos of the blue mussel (*Mytilus edulis*). *Genetics* 166: 883–894.

Carriker, M. R. 1990. Functional significance of the pediveliger in bivalve development. In *The Bivalvia—Proceedings of a Memorial Symposium in Honour of Sir Charles Maurice Yonge, Edinburgh, 1986*. Edited by B. Morton. Hong Kong: Hong Kong University Press, pp. 267–282.

Carter, J. G. 1990a. Evolutionary significance of shell microstructure in the Palaeotaxodonta, Pteriomorphia and Isofilibranchia (Bivalvia: Mollusca). In *Skeletal Biomineralization: Patterns, Processes and Evolutionary Trends, Vol. 1*. Edited by J. G. Carter. New York: Van Nostrand Reinhold, pp. 135–296.

———. 1990b. Shell microstructural data for the Bivalvia. In *Skeletal Biomineralization: Patterns, Processes and Evolutionary Trends, Vol. 1*. Edited by J. G. Carter. New York: Van Nostrand Reinhold, pp. 297–411.

Carter, J. G., Campbell, D. C., and Campbell, M. R. 2000. Cladistic perspectives on early bivalve evolution. In *The Evolutionary Biology of the Bivalvia*. Edited by E. M. Harper, J. D. Taylor, and J. A. Crame. London: The Geological Society of London, pp. 47–79.

Cavanaugh, C. M. 1983. Symbiotic chemoautotrophic bacteria in marine invertebrates from sulphide-rich habitats. *Nature* 302: 58–61.

Chapman, J. W., Miller, T. W., and Coan, E. V. 2003. Live seafood species as recipes for invasion. *Conservation Biology* 17: 1386–1395.

Chase, M. R., Etter, R. J., Rex, M. A., and Quattro, J. M. 1998. Bathymetric patterns of genetic variation in a deep-sea protobranch bivalve, *Deminucula atacellana*. *Marine Biology* 131: 301–308.

Coan, E. V., Valentich Scott, P., and Bernard, F. R. 2000. *Bivalve Seashells of Western North America. Marine Bivalve Mollusks from Arctic Alaska to Baja California*. Santa Barbara, CA: Santa Barbara Museum of Natural History. 764 pp.

Cope, J. C. W. 1996a. Early Ordovician (Arenig) bivalves from the Llangynog Inlier, South Wales. *Paleontology* 39: 979–1025.

———. 1996b. The early evolution of the Bivalvia. In *Origin and Evolutionary Radiation of the Mollusca*. Edited by J. D. Taylor. Oxford: Oxford University Press, pp. 361–370.

———. 1997. The early phylogeny of the class Bivalvia. *Palaeontology* 40: 713–746.

———. 2000. A new look at early bivalve phylogeny. In *The Evolutionary Biology of the Bivalvia*. Edited by E. M. Harper, J. D. Taylor, and J. A. Crame. London: The Geological Society of London, pp. 81–95.

Cox, L. R. 1960. Thoughts on the classification of the Bivalvia. *Proceedings of the Malacological Society of London* 34: 60–80.

———. 1969. General features of the Bivalvia. In *Treatise on Invertebrate Paleontology, Part I, Mollusca 1*. Edited by R. Moore. Lawrence, KS: University of Kansas Press and Geological Society of America, pp. N2–N129.

Cox, L. R., Newell, N. D., Boyd, D. W., Branson, C. C., Casey, R., Chavan, A., Coogan, A. H., Dechaseaux, C., Fleming, C. A., Haas, F., Hertlein, L. G., Kauffman, E. G., Myra Keen, A., LaRocque, A., McAlester, A. L., Moore, R. C., Nuttall, C. P., Perkins, B. F., Puri, H. S., Smith, L. A., Soot-Ryen, T., Stenzel, H. B., Trueman, E. R., Turner, R. D., and Weir, J. 1969. Part N. Bivalvia. In *Treatise on Invertebrate Paleontology, Part N, Mollusca 6*. Edited by R. C. Moore. Lawrence, KS: The Geological Society of America and The University of Kansas, pp. N2–N129.

Cragg, S. M. 1996. The phylogenetic significance of some anatomical features of bivalve veliger larvae. In *Origin and Evolutionary Radiation of the Mollusca*. Edited by J. D. Taylor. Oxford: Oxford University Press, pp. 371–380.

Cragg, S. M., and Nott, J. A. 1977. The ultrastructure of the statocysts in the pediveliger larvae of *Pecten maximus* (L.) (Bivalvia). *Journal of Experimental Marine Biology and Ecology* 27: 23–36.

DePaola, A., Hopkins, L. H., Peeler, J. T., Wentz, B., and McPhearson, R. M. 1990. Incidence of *Vibrio parahaemolyticus* in U. S. coastal waters and oysters. *Applied and Environmental Microbiology* 56: 2299–2302.

Dinamani, P. 1967. Variation in the stomach structure of the Bivalvia. *Malacologia* 5: 225–268.

Distel, D. L. 1998. Evolution of chemoautotrophic endosymbioses in bivalves. Bivalve-bacteria chemosymbioses are phylogenetically diverse but morphologically similar. *BioScience* 48: 277–286.

Drew, G. A. 1899. The anatomy, habits, and embryology of *Yoldia limatula*, Say. *Memoirs from the Biological Laboratory of the John Hopkins University* 4: 1–37 + 5 pl.

———. 1901. The life-history of *Nucula delphinodonta* (Mighels). *Quarterly Journal of Microscopical Sciences* 44: 313–392.

———. 1906. The habits, anatomy, and embryology of the giant scallop (*Pecten tenuicostatus*, Mighels). *The University of Maine Studies* 6: 1–71.

Dreyer, H., Steiner, G., and Harper, E. M. 2003. Molecular phylogeny of Anomalodesmata (Mollusca: Bivalvia) inferred from 18S rRNA sequences. *Zoological Journal of the Linnean Society* 139: 229–246.

Etter, R. J., Rex, M. A., Chase, M. R., and Quattro, J. M. 2005. Population differentiation decreases with depth in deep-sea bivalves. *Evolution* 59: 1479–1491.

Fankboner, P. V., and Reid, R. G. B. 1990. Nutrition in giant clams (Tridacnidae). In *The Bivalvia— Proceedings of a Memorial Symposium in Honour of Sir Charles Maurice Yonge, Edinburgh, 1986.* Edited by B. Morton. Hong Kong: Hong Kong University Press, pp. 195–209.

Franzén, Ä. 1983. Ultrastructural studies of spermatozoa in three bivalve species with notes on evolution of elongated sperm nucleus in primitive spermatozoa. *Gamete Research* 7: 199–214.

Friedrich, S., Wanninger, A., Brückner, M., and Haszprunar, G. 2002. Neurogenesis in the mossy chiton, *Mopalia muscosa* (Gould) (Polyplacophora): evidence against molluscan metamerism. *Journal of Morphology* 253: 109–117.

Fujita, T. 1929. On the early development of the common Japanese oyster. *Japanese Journal of Zoology* 2: 353–358.

Gilmour, T. H. J. 1990. The adaptive significance of foot reversal in the Limoida. In *The Bivalvia— Proceedings of a Memorial Symposium in Honour of Sir Charles Maurice Yonge, Edinburgh, 1986.* Edited by B. Morton. Hong Kong: Hong Kong University Press, pp. 247–263.

Giribet, G. 1998. "Molluscs as evolving constructions" and phylogenetic deconstructivism. *Iberus* 16: 123–128.

Giribet, G., and Carranza, S. 1999. What can 18S rDNA do for bivalve phylogeny? *Journal of Molecular Evolution* 48: 256–261.

Giribet, G., and Distel, D. L. 2003. Bivalve phylogeny and molecular data. In *Molecular Systematics and Phylogeography of Mollusks.* Edited by C. Lydeard and D. R. Lindberg. Washington, DC: Smithsonian Books, pp. 45–90.

Giribet, G., Okusu, A., Lindgren, A. R., Huff, S. W., Schrödl, M., and Nishiguchi, M. K. 2006. Evidence for a clade composed of molluscs with serially repeated structures: monoplacophorans are related to chitons. *Proceedings of the National Academy of Sciences of the U. S. A.* 103: 7723–7728.

Giribet, G., and Peñas, A. 1997. Malacological marine fauna from Garraf coast (NE Iberian Peninsula). *Iberus* 15: 41–93.

Giribet, G., and Wheeler, W. C. 2002. On bivalve phylogeny: a high-level analysis of the Bivalvia (Mollusca) based on combined morphology and DNA sequence data. *Invertebrate Biology* 121: 271–324.

Gosling, E. 2003. *Bivalve Molluscs. Biology, Ecology and Culture.* Oxford: Fishing News Books. 443 pp.

Götting, K.-J. 1980. Origins and relationships of the Mollusca. *Zeitschrift für zoologische Systematik und Evolutionsforschung* 18: 24–27.

Gould, S. J., and Calloway, C. B. 1980. Clams and brachiopods—ships that pass in the night. *Paleobiology* 6: 383–396.

Gould, S. J., and Jones, C. C. 1974. The pallial ridge of *Neotrigonia*: functional siphons without mantle fusion. *The Veliger* 17: 1–7.

Graf, D. L. 2000. The Etherioidea revisited: a phylogenetic analysis of hyriid relationships (Mollusca: Bivalvia: Paleoheterodonta: Unionoida). *Occasional Papers of the Museum of Zoology, University of Michigan* 729: 1–21.

———. 2002. Molecular phylogenetic analysis of two problematic freshwater mussel genera (*Unio* and *Gonidea*) and a re-evaluation of the classification of Nearctic Unionidae (Bivalvia: Palaeoheterodonta: Unionoida). *Journal of Molluscan Studies* 68: 55–64.

Graf, D. L., and Cummings, K. S. 2006. Palaeoheterodont diversity (Mollusca: Trigonioida + Unionoida): what we know and what we wish we knew about freshwater mussel evolution. In *Bivalvia—A Look at the Branches.* Edited by R. Bieler. *Zoological Journal of the Linnean Society* 148: 343–394.

Graf, D. L., and Ó Foighil, D. 2000a. Molecular phylogenetic analysis of 28S rDNA supports a Gondwanan origin of Australasian Hyriidae (Mollusca: Bivalvia: Unionoida). *Vie et Milieu* 50: 245–254.

———. 2000b. The evolution of brooding characters among the freshwater pearly mussels (Bivalvia: Unionoidea) of North America. *Journal of Molluscan Studies* 66: 157–170.

Grobben, C. 1886. Zur Kenntnis und Morphologie und der Verwandtschaftsverhältnisse der Cephalopoda. *Journal of Natural History* 33: 267–287.

Gros, O., Duplessis, M. R., and Felbeck, H. 1999. Embryonic development and endosymbiont transmission mode in the symbiotic clam *Lucinoma aequizonata* (Bivalvia: Lucinidae). *Invertebrate Reproduction and Development* 36: 93–103.

Gros, O., Frenkiel, L., and Mouëza, M. 1997. Embryonic, larval, and post-larval development in the symbiotic clam *Codakia orbicularis* (Bivalvia: Lucinidae). *Invertebrate Biology* 116: 86–101.

Gustafson, R. G., and Lutz, R. A. 1992. Larval and early post-larval development of the protobranch bivalve *Solemya velum* (Mollusca: Bivalvia). *Journal of the Marine Biological Association of the U.K.* 72: 383–402.

Gustafson, R. G., and Reid, R. G. B. 1986. Development of the pericalymma larva of *Solemya reidi* (Bivalvia: Cryptodonta: Solemyidae) as revealed by light and electron microscopy. *Marine Biology* 93: 411–427.

———. 1988. Larval and post-larval morphogenesis in the gutless protobranch bivalve *Solemya reidi* (Cryptodonta: Solemyidae). *Marine Biology* 97: 373–387.

Harper, E. M. 1991. The role of predation in the evolution of cementation in bivalves. *Palaeontology* 34: 455–460.

Harper, E. M., Dreyer, H., and Steiner, G. 2006. Reconstructing the Anomalodesmata (Mollusca: Bivalvia): morphology and molecules. In *Bivalvia— A Look at the Branches*. Edited by R. Bieler. *Zoological Journal of the Linnean Society* 148: 395–420.

Harper, E. M., Hide, E. A., and Morton, B. 2000a. Relationships between the extant Anomalodesmata: a cladistic test. In *The Evolutionary Biology of the Bivalvia*. Edited by E. M. Harper, J. D. Taylor, and J. A. Crame. London: The Geological Society of London, pp. 129–143.

Harper, E. M., Taylor, J. D., and Crame, J. A. (eds) 2000b. *The Evolutionary Biology of the Bivalvia*. London: The Geological Society of London.

Haszprunar, G. 1983. Comparative analysis of the abdominal sense organs of Pteriomorpha (Bivalvia). *Journal of Molluscan Studies* Suppl. 12A: 47–50.

———. 1985a. On the anatomy and fine-structure of a peculiar sense organ in *Nucula* (Bivalvia, Protobranchia). *The Veliger* 28: 52–62.

———. 1985b. The fine structure of the abdominal sense organs of Pteriomorpha (Mollusca, Bivalvia). *Journal of Molluscan Studies* 51: 315–319.

———. 1987. The fine morphology of the osphradial sense organs of the Mollusca. III. Placophora and Bivalvia. *Philosophical Transactions of the Royal Society of London B Biological Sciences* 315: 37–61.

———. 2000. Is the Aplacophora monophyletic? A cladistic point of view. *American Malacological Bulletin* 15: 115–130.

Healy, J. M. 1989. Spermiogenesis and spermatozoa in the relict bivalve genus *Neotrigonia*: relevance to trigonioid relationships, particularly with Unionoidea. *Marine Biology* 103: 75–85.

———. 1995a. Comparative spermatozoal ultrastructure and its taxonomic and phylogenetic significance in the bivalve order Veneroida. In *Advances in Spermatozoal Phylogeny and Taxonomy*. Edited by B. G. Jamieson, J. Ausió, and J.-L. Justine. Paris: Éditions du Muséum Paris, pp. 155–166.

———. 1995b. Sperm ultrastructure in the marine bivalve families Carditidae and Crassatellidae and its bearing on unification of the Crassatelloidea with the Carditoidea. *Zoologica Scripta* 24: 21–28.

———. 1996. Molluscan sperm ultrastructure: correlation with taxonomic units within Gastropoda, Cephalopoda and Bivalvia. In *Origin and evolutionary radiation of the Mollusca*. Edited by J. D. Taylor. Oxford: Oxford University Press, pp. 99–113.

Healy, J. M., Keys, J. L., and Daddow, L. Y. M. 2000. Comparative sperm ultrastructure in pteriomorphian bivalves with special reference to phylogenetic and taxonomic implications. In *The Evolutionary Biology of the Bivalvia*. Edited by E. M. Harper, J. D. Taylor, and J. A. Crame. London: The Geological Society of London, pp. 169–190.

Healy, J. M., Mikkelsen, P. M., and Bieler, R. 2006. Sperm ultrastructure in *Glauconome plankta* and its relevance to the affinities of the Glauconomidae (Bivalvia: Heterodonta). *Invertebrate Reproduction and Development* 49: 29–39.

Hejnol, A., and Schnabel, R. 2005. The eutardigrade *Thulinia stephaniae* has an indeterminate development and the potential to regulate early blastomere ablations. *Development* 132: 1349–1361.

Hennig, W. 1979. *Taschenbuch der speziellen Zoologie. Teil 1, Wirbellose I.* Zürich-Frankfurt: Verlag Harri Deutsch. 392 pp.

Henry, J. Q., Okusu, A., and Martindale, M. Q. 2004. The cell lineage of the polyplacophoran, *Chaetopleura apiculata*: variation in the spiralian program and implications for molluscan evolution. *Developmental Biology* 272: 145–160.

Hodgson, A. N., and Bernard, R. T. F. 1986. Ultrastructure of the sperm and spermiogenesis of three species of Mytilidae (Mollusca, Bivalvia). *Gamete Research* 15: 123–135.

Hodgson, C. A., and Burke, R. D. 1988. Development and larval morphology of the spiny scallop, *Chlamys hastata*. *Biological Bulletin* 174: 303–318.

Hoeh, W. R., Black, M. B., Gustafson, R. G., Bogan, A. E., Lutz, R. A., and Vrijenhoek, R. C. 1998. Testing alternative hypotheses of *Neotrigonia* (Bivalvia: Trigonioida) phylogenetic relationships using cytochrome c oxidase subunit I DNA sequences. *Malacologia* 40: 267–278.

Hoeh, W. R., Bogan, A. E., Cummings, K. S., and Guttman, S. I. 1999. Evolutionary relationships among the higher taxa of freshwater mussels (Bivalvia: Unionoida): inferences on phylogeny and character evolution from analyses of DNA sequence data. *Malacological Review* 31/32: 123–141.

Hoggarth, M. A. 1999. Description of some of the glochidia of the Unionidae (Mollusca: Bivalvia). *Malacologia* 41: 1–118.

Huff, S. W., Campbell, D., Gustafson, D. L., Lydeard, C., Altaba, C. R., and Giribet, G. 2004. Investigations into the phylogenetic relationships of the threatened freshwater pearl-mussels (Bivalvia,

Unionoidea, Margaritiferidae) based on molecular data: implications for their taxonomy and biogeography. *Journal of Molluscan Studies* 70: 379–388.

IIjima, M., Akiba, N., Sarashina, I., Kuratani, S., and Endo, K. 2006. Evolution of *Hox* genes in molluscs: a comparison among seven morphological diverse classes. *Journal of Molluscan Studies* 72: 259–266.

Jacobs, D. K., Wray, C. G., Wedeen, C. J., Kostriken, R., DeSalle, R., Staton, J. L., Gates, R. D., and Lindberg, D. R. 2000. Molluscan engrailed expression, serial organization, and shell evolution. *Evolution & Development* 2: 340–347.

Janaki Ram, K., and Radhakrishna, Y. 1984. The distribution of freshwater Mollusca in Guntur District (India) with a description of *Scaphula nagarjunai* sp. n. (Arcidae). *Hydrobiologia* 119: 49–55.

Jenner, R. A. 2003. Unleashing the force of cladistics? Metazoan phylogenetics and hypothesis testing. *Integrative and Comparative Biology* 43: 207–218.

Johnston, P. A., and Haggart, J. W. 1998. *Bivalves: An Eon of Evolution*. Calgary: University of Calgary Press. 461 pp.

Kappner, I., and Bieler, R. 2006. Phylogeny of venus clams (Bivalvia: Venerinae) as inferred from nuclear and mitochondrial gene sequences. *Molecular Phylogenetics and Evolution* 40: 317–331.

Keys, J. L., and Healy, J. M. 2000. Relevance of sperm ultrastructure to the classification of giant clams (Mollusca, Cardioidea, Cardiidae, Tridacninae). In *The Evolutionary Biology of the Bivalvia*. Edited by E. M. Harper, J. D. Taylor, and J. A. Crame. London: The Geological Society of London, pp. 191–205.

Kidwell, S. M. 2001. Preservation of species abundance in marine death assemblages. *Science* 294: 1091–1094.

———. 2005. Shell composition has no net impact on large-scale evolutionary patterns in mollusks. *Science* 307: 914–917.

Kirkendale, L., Lee, T., Baker, P., and Ó Foighil, D. 2004. Oysters of the Conch Republic (Florida Keys); a molecular phylogenetic study of *Parahyotissa mcgintyi*, *Teskeyostrea weberi* and *Ostreola equestris*. *Malacologia* 46: 309–326.

Kurokura, H. 2004. The importance of seaweeds and shellfishes in Japan: present status and history. *Bulletin of Fisheries Research Agency* Suppl. 1: 1–4.

Lacaze-Duthiers, F. J. H. 1858. *Histoire de l'organisation, du développement, des moeurs et des rapports zoologiques du Dentale*. Paris: Victor Masson. 287 pp.

Landman, N. H., Mikkelsen, P. M., Bieler, R., and Bronson, B. 2001. *Pearls: A Natural History*. New York: H. N. Abrams in association with the American Museum of Natural History and the Field Museum. 232 pp.

Lauckner, G. 1983. Diseases of Mollusca: Bivalvia. In *Diseases of Marine Animals*. Vol. II, *Introduction. Bivalvia to Scaphopoda*. Edited by O. Kinne. Chichester: John Wiley and Sons, pp. 477–961.

Lauterbach, K.-E. 1983. Erörterungen zur Stammesgeschichte der Mollusca, insbesondere der Conchifera. *Zeitschrift für zoologische Systematik und Evolutionsforschung* 21: 201–216.

———. 1984. Das phylogenetische System der Mollusca. *Mitteilungen der Deutschen Malakologischen Gesellschaft* 37: 66–81.

Le Pennec, M. 1980. The larval and post-larval hinge of some families of bivalve molluscs. *Journal of the Marine Biological Association of the U.K.* 60: 601–617.

Lillie, F. R. 1895. The embryology of the Unionidae. A study in cell-lineage. *Journal of Morphology* 10: 1–100.

Lindberg, D. R., And Ponder, W. F. 1996. An evolutionary tree for the Mollusca: branches or roots? In *Origin and Evolutionary Radiation of the Mollusca*. Edited by J. D. Taylor. Oxford: Oxford University Press, pp. 67–75.

Lundin, K., and Schander, C. 2001. Ciliary ultrastructure of protobranchs (Mollusca, Bivalvia). *Invertebrate Biology* 120: 350–357.

Lydeard, C., Cowie, R. H., Ponder, W. F., Bogan, A. E., Bouchet, P., Clark, S. A., Cummings, K. S., Frest, T. J., Gargominy, O., Herbert, D. G., Hershler, R., Perez, K. E., Roth, B., Seddon, M. B., Strong, E. E., and Thompson, F. G. 2004. The global decline of nonmarine mollusks. *BioScience* 54: 321–330.

Lydeard, C., and Lindberg, D. R., eds. 2003. *Molecular Systematics and Phylogeography of Mollusks*. Washington, DC: Smithsonian Books. 312 pp.

Lydeard, C., Minton, R. L., and Williams, J. D. 2000. Prodigious polyphyly in imperilled freshwater pearly-mussels (Bivalvia: Unionidae): a phylogenetic test of species and generic designations. In *The Evolutionary Biology of the Bivalvia*. Edited by E. M. Harper, J. D. Taylor, and J. A. Crame. London: The Geological Society of London, pp. 145–158.

Lydeard, C., Mulvey, M., and Davis, G. M. 1996. Molecular systematics and evolution of reproductive traits of North American freshwater unionacean mussels (Mollusca: Bivalvia) as inferred from 16S rRNA gene sequences. *Philosophical Transactions of the Royal Society of London B Biological Sciences* 351: 1593–1603.

Macaranas, J. M., Ablan, C. A., Pante, M. J. R., Benzie, J. A. H., and Williams, S. T. 1992. Genetic structure of giant clam (*Tridacna derasa*) populations from reefs in the Indo-Pacific. *Marine Biology* 113: 231–238.

Machordom, A., Araujo, R., Erpenbeck, D., and Ramos, M.A. 2003. Phylogeography and conservation genetics of endangered European Margaritiferidae (Bivalvia: Unionoidea). *Biological Journal of the Linnean Society* 78: 235–252.

Malchus, N. 2004. Constraints in the ligament ontogeny and evolution of pteriomorphian Bivalvia. *Palaeontology* 47: 1539–1574.

Malchus, N., and Pons, J.M. (eds.) 2006. *International Congress on Bivalvia. Scientific Program and Abstracts*. 87 pp. Barcelona: Universitat Autonoma de Barcelona.

Mangum, C.P., and Johansen, K. 1975. The colloid osmotic pressures of invertebrate body fluids. *Journal of Experimental Biology* 63: 661–671.

Manwell, C. 1963. The chemistry and biology of hemoglobin in some marine clams. I. Distribution of the pigment and properties of the oxygen equilibrium. *Comparative Biochemistry and Physiology* 16: 209–218.

Maslakova, S.A., Martindale, M.Q., and Norenburg, J.L. 2004. Fundamental properties of the spiralian developmental program are displayed by the basal nemertean *Carinoma tremaphoros* (Palaeonemertea, Nemertea). *Developmental Biology* 267: 342–360.

Matsumoto, M. 2003. Phylogenetic analysis of the subclass Pteriomorphia (Bivalvia) from mtDNA COI sequences. *Molecular Phylogenetics and Evolution* 27: 429–440.

Matsumoto, M., and Hayami, I. 2000. Phylogenetic analysis of the family Pectinidae (Bivalvia) based on mitochondrial cytochrome *c* oxidase subunit I. *Journal of Molluscan Studies* 66: 477–488.

McAlester, A.L. 1965. Systematics, affinities, and the life habits of *Babinka*, a traditional Ordovician lucinoid bivalve. *Paleontology* 8: 231–246.

McMahon, R.F., and Bogan, A.E. 2001. Mollusca: Bivalvia. In *Ecology and classification of North American freshwater invertebrates*, 2nd ed. Edited by J.H. Thorp and A.P. Covich. San Diego. Academic Press, pp. 3310–3429.

Meisenheimer, J. 1901. Entwicklungsgeschichte von *Dreissenia polymorpha*. *Zeitschrift für Wissenschaftliche Zoologie* 69: 1–137.

Middelfart, P. 2005. Review of *Ephippodonta sensu lato* (Galeommatidae: Bivalvia), with descriptions of new related genera and species from Australia. *Molluscan Research* 25: 129–144.

Mikkelsen, P.M., and Bieler, R. 1992. Biology and comparative anatomy of three new species of commensal Galeommatidae, with a possible case of mating behavior in bivalves. *Malacologia* 34: 1–24.

Mikkelsen, P.M., Bieler, R., Kappner, I., and Rawlings, T.A. 2006. Phylogeny of Veneroidea (Mollusca: Bivalvia) based on morphology and molecules. In *Bivalvia—A Look at the Branches*. Edited by R. Bieler. *Zoological Journal of the Linnean Society* 148: 439–521.

Miller, A.W., Hewitt, C.L., and Ruiz, G.M. 2002. Invasion success: does size really matter? *Ecology Letters* 5: 159–162.

Moir, A.J.G. 1977. On the ultrastructure of the abdominal sense organ of the giant scallop, *Placopecten magellanicus* (Gmelin). *Cell and Tissue Research* 184: 359–366.

Morse, M.P., and Zardus, J.D. 1997. Bivalvia. In *Microscopic anatomy of invertebrates*. Vol. 6A, *Mollusca II*. Edited by F.W. Harrison and A.J. Kohn. New York: Wiley-Liss, pp. 7–118.

Morton, B. 1980a. Some aspects of the biology and functional morphology (including the presence of a ligamental lithodesma) of *Montacutona compacta* and *M. olivacea* (Bivalvia: Leptonacea) associated with coelenterates in Hong Kong. *Journal of Zoology* 192: 431–455.

———. 1980b. Swimming in *Amusium pleuronectes* (Bivalvia: Pectinidae). *Journal of Zoology, London* 190: 375–404.

———. 1981. The Anomalodesmata. *Malacologia* 21: 35–60.

———. 1983. Feeding and digestion in Bivalvia. In *The Mollusca*. Vol. 5, *Physiology*, Part 2. Edited by A.S.M. Saleuddin and K.M. Wilbur. New York: Academic Press, pp. 65–147.

———. 1985. Adaptive radiation in the Anomalodesmata. In *The Mollusca*. Vol. 10, *Evolution*. Edited by E.R. Trueman and M.R. Clarke. Orlando, FL: Academic Press, pp. 405–459.

———. 1987. The functional morphology of *Neotrigonia margaritacea* (Bivalvia: Trigonacea), with a discussion of phylogenetic affinities. *Records of the Australian Museum* 39: 339–354.

———. 1990a. Corals and their bivalve borers—the evolution of a symbiosis. In *The Bivalvia—Proceedings of a Memorial Symposium in Honour of Sir Charles Maurice Yonge, Edinburgh, 1986*. Edited by B. Morton. Hong Kong: Hong Kong University Press, pp. 11–46.

———, ed. 1990b. *The Bivalvia—Proceedings of a Memorial Symposium in Honour of Sir Charles Maurice Yonge, Edinburgh, 1986*. Hong Kong: Hong Kong University Press.

———. 1996. The evolutionary history of the Bivalvia. In *Origin and Evolutionary Radiation of the Mollusca*. Edited by J.D. Taylor. Oxford: Oxford University Press, pp. 337–359.

———. 2000. The biology and functional morphology of *Fragum erugatum* (Bivalvia: Cardiidae) from

Shark Bay, Western Australia: the significance of its relationship with entrained zooxanthellae. *Journal of Zoology* 251: 39–52.

Morton, B. S., Prezant, R. S., and Wilson, B. 1998. Class Bivalvia. In *Mollusca: The Southern Synthesis. Fauna of Australia*. Vol. 5. Edited by P. L. Beesley, G. J. B. Ross, and A. Wells. Melbourne: CSIRO Publishing, pp. 195–234.

Mouëza, M., Gros, O., and Frenkiel, L. 1999. Embryonic, larval and postlarval development of the tropical clam, *Anomalocardia brasiliana* (Bivalvia, Veneridae). *Journal of Molluscan Studies* 65: 73–88.

———. 2006. Embryonic development and shell differentiation in *Chione cancellata* (Bivalvia, Veneridae): an ultrastructural analysis. *Invertebrate Biology* 125: 21–33.

Neves, R. J., Bogan, A. E., Williams, J. D., Ahlstedt, A. S., and Hartfield, P. W. 1997. Status of aquatic mollusks in the southeastern United States: a downward spiral of diversity. In *Aquatic fauna in peril: the southeastern perspective*. Edited by G. W. Benz and D. E. Collins. Chattanooga, TN: Southeast Aquatic Institute, pp. 43–85.

Newell, N. D. 1965. Classification of the Bivalvia. *American Museum Novitates* 2206: 1–25.

———. 1969. Classification of Bivalvia. In *Treatise on Invertebrate Paleontology*. Part N, *Mollusca 6*. Vol. 1, *Bivalvia*. Edited by R. Moore. Boulder, CO, and Lawrence, KS: Geological Society of America and University of Kansas, pp. N205–N244.

Nielsen, C. 2001. *Animal Evolution, Interrelationships of the Living Phyla*, 2nd ed. Oxford: Oxford University Press. 563 pp.

Nikula, R., and Väinölä, R. 2003. Phylogeography of *Cerastoderma glaucum* (Bivalvia: Cardiidae) across Europe: a major break in the Eastern Mediterranean. *Marine Biology* 143: 339–350.

Ó Foighil, D., and Jozefowicz, C. J. 1999. Amphi-Atlantic phylogeography of direct-developing lineages of *Lasaea*, a genus of brooding bivalves. *Marine Biology* 135: 114–122.

O'Neill, C. R., Jr. 1997. Economic impact of zebra mussels–results of the 1995 National Zebra Mussel Information Clearinghouse study. *Great Lakes Research Review* 3: 35–41.

Okusu, A. 2002. Embryogenesis and development of *Epimenia babai* (Mollusca Aplacophora). *Biological Bulletin* 203: 87–103.

Okutani, T., ed. 2000. *Marine Mollusks in Japan*. Takai University Press, Tokyo. 1173 pp.

Oliver, P. G. 1992. *Bivalved Seashells of the Red Sea*. Wiesbaden: Verlag Christa Hemmen.

Oliver, P. G., and Holmes, A. M. 2006. The Arcoidea (Mollusca: Bivalvia): a review of the current

phenetic-based systematics. In *Bivalvia—A Look at the Branches*. Edited by R. Bieler. *Zoological Journal of the Linnean Society* 148: 237–251.

Page, T. J., and Linse, K. 2002. More evidence of speciation and dispersal across the Antarctic Polar Front through molecular systematics of Southern Ocean *Limatula* (Bivalvia: Limidae). *Polar Biology* 25: 818–826.

Park, J. K., and Ó Foighil, D. 2000. Sphaeriid and corbiculid clams represent separate heterodont bivalve radiations into freshwater environments. *Molecular Phylogenetics and Evolution* 14: 75–88.

Pelseneer, P. 1889. Sur la classification phylogénétique des pélécypodes. *Bulletin Scientifique de la France et de la Belgique* 20: 27–52.

Peñas, A., and Giribet, G. 2003. Adiciones a la fauna malacologica del litoral del Garraf (NE de la Península Ibérica). *Iberus* 21: 177–189.

Peñas, A., Rolán, E., Luque, A. A., Templado, J., Moreno, D., Rubio, F., Salas, C., Sierra, A., and Gofas S. 2006. Moluscos marinos de la isla de Alborán. *Iberus* 24: 23–151.

Pojeta, J., Jr. 1978. The origin and early taxonomic diversification of pelecypods. *Philosophical Transactions of the Royal Society of London B Biological Sciences* 284: 225–243.

Pojeta, J., Jr., and Runnegar, B. 1974. *Fordilla troyensis* and the early history of pelecypod mollusks. *American Scientist* 62: 706–711.

———. 1976. The paleontology of rostroconch mollusks and the early history of the phylum Mollusca. *Geological Survey of Professional Papers (U.S.)* 968: 1–88.

———. 1985. The early evolution of diasome molluscs. In *The Mollusca*. Vol. 10, *Evolution*. Edited by E. R. Trueman and M. R. Clarke. New York: Academic Press, pp. 295–336.

Pojeta, J., Jr., Runnegar, B., Peel, J. S., and Gordon, M., Jr. 1987. Phylum Mollusca. In *Fossil Invertebrates*. Edited by R. S. Boardman, A. H. Cheetham, and A. J. Rowell. Oxford: Blackwell Scientific Publications, pp. 270–435.

Ponder, W. F. 1998. Superfamily Galeommatoidea. In *Mollusca: The Southern Synthesis. Fauna of Australia*. Vol. 5. Edited by P. L. Beesley, G. J. B. Ross, and A. Wells. Melbourne: CSIRO Publishing, pp. 316–318.

Popham, J. D. 1974. Proceedings: The ultrastructure of the acrosome reaction of the spermatozoon of, and gamete fusion in, the shipworm, *Bankia australis* (Bivalvia: Mollusca). *Journal of Anatomy* 118: 402–403.

Popham, M. L. 1939. On *Phlyctaenachlamys lysiosquillina gen.* and *sp. nov.*, a lamellibranch commensal in the burrows of *Lysiosquilla maculata*.

Great Barrier Reef Expeditions 1928–29. *Scientific Reports. British Museum (Natural History)* 6: 62–84.

———. 1940. The mantle cavity of some of the Erycinidae, Montacutidae and Galeommatidae with special reference to the ciliary mechanisms. *Journal of the Marine Biological Association of the U.K.* 24: 549–587.

Pranovi, F., Franceschini, G., Casale, M., Zucchetta, M., Torricelli, P., and Giovanardi, O. 2006. An ecological imbalance induced by a non-native species: the Manila clam in the Venice Lagoon. *Biological Invasions* 8: 595–609.

Prezant, R. S. 1998a. Subclass Palaeoheterodonta Introduction. In *Mollusca: The Southern Synthesis. Fauna of Australia*. Vol. 5. Edited by P. L. Beesley, G. J. B. Ross, and A. Wells. Melbourne: CSIRO Publishing, pp. 289–294.

———. 1998b. Subclass Heterodonta Introduction. In *Mollusca: The Southern Synthesis. Fauna of Australia*. Vol. 5. Edited by P. L. Beesley, G. J. B. Ross, and A. Wells. Melbourne: CSIRO Publishing, pp. 301–306.

———. 1998c. Subclass Anomalodesmata Introduction. In *Mollusca: The Southern Synthesis. Fauna of Australia*. Vol. 5. Edited by P. L. Beesley, G. J. B. Ross, and A. Wells. Melbourne: CSIRO Publishing, pp. 397–405.

Purchon, R. D. 1956. The stomach in the Protobranchia and Septibranchia (Lamellibranchia). *Proceedings of the Zoological Society of London* 127: 511–525.

———. 1957. The stomach in the Filibranchia and Pseudolamellibranchia. *Proceedings of the Zoological Society of London* 129: 27–60.

———. 1958. The stomach in the Eulamellibranchia; stomach types IV and V. *Proceedings of the Zoological Society of London* 131: 487–525.

———. 1959. Phylogenetic classification of the Lamellibranchia, with special reference to the Protobranchia. *Proceedings of the Malacological Society* 33: 224–230.

———. 1960. Phylogeny in the Lamellibranchia. In *Proceedings of the centenary and bicentenary congress of Biology, Singapore, Dec. 1958*. Edited by R. D. Purchon. Singapore: University of Malaya Press, pp. 69–82.

———. 1963. Phylogenetic classification of the Bivalvia, with special reference to the Septibranchia. *Proceedings of the Malacological Society* 35: 71–80.

———. 1968. *The Biology of the Mollusca*. Oxford: Pergamon Press.

———. 1978. An analytical approach to a classification of the Bivalvia. *Philosophical Transactions of the Royal Society of London, B* 284: 425–436.

———. 1987a. Classification and evolution of the Bivalvia: an analytical study. *Philosophical Transactions of the Royal Society of London, B* 316: 277–302.

———. 1987b. The stomach in the Bivalvia. *Philosophical Transactions of the Royal Society of London, B* 316: 183–276.

———. 1990. Stomach structure, classification and evolution of the Bivalvia. In *The Bivalvia–Proceedings of a Memorial Symposium in Honour of Sir Charles Maurice Yonge, Edinburgh, 1986*. Edited by B. Morton. Hong Kong: Hong Kong University Press, pp. 73–82.

Rehnstam-Holm, A.-S., and Hernoth, B. 2005. Shellfish and public health: A Swedish perspective. *AMBIO: A Journal of the Human Environment* 34: 139–144.

Reid, R. G. B. 1965. The structure and function of the stomach in bivalve molluscs. *Journal of Zoology, London* 147: 156–184.

———. 1990. Evolutionary implications of sulphide-oxidizing symbioses in bivalves. In *The Bivalvia—Proceedings of a Memorial Symposium in Honour of Sir Charles Maurice Yonge, Edinburgh, 1986*. Edited by B. Morton. Hong Kong: Hong Kong University Press, pp. 127–140.

———. 1998. Subclass Protobranchia. In *Mollusca: The Southern Synthesis. Fauna of Australia*. Vol. 5. Edited by P. L. Beesley, G. J. B. Ross, and A. Wells. Melbourne: CSIRO Publishing. Pp. 235–247.

Ridewood, W. G. 1903. On the structure of the gills of lamellibranchs. *Philosophical Transactions of the Royal Society of London, B* 194: 147–284.

Roe, K. J., Hartfield, P. D., and Lydeard, C. 2001. Phylogeographic analysis of the threatened and endangered superconglutinate-producing mussels of the genus *Lampsilis* (Bivalvia: Unionidae). *Molecular Ecology* 10: 2225–2234.

Roe, K. J., and Hoeh, W. R. 2003. Systematics of freshwater mussels (Bivalvia: Unionoida). In *Molecular Systematics and Phylogeography of Mollusks*. Edited by C. Lydeard and D. R. Lindberg. Washington, DC: Smithsonian Books, pp. 91–122.

Rosen, M. D., Stasek, C. R., and Hermans, C. O. 1978. The ultrastructure and evolutionary significance of the cerebral ocelli of *Mytilus edulis*, the Bay Mussel. *The Veliger* 21: 10–18.

Roy, K., Jablonski, D., and Valentine, J. W. 2000. Dissecting latitudinal diversity gradients: functional groups and clades of marine bivalves. *Proceedings: Biological Sciences* 267: 293–299.

Runnegar, B. 1974. Evolutionary history of the bivalve subclass Anomalodesmata. *Journal of Paleontology* 48: 904–939.

————. 1978. Origin and evolution of the Class Rostroconchia. *Philosophical Transactions of the Royal Society of London B Biological Sciences* 284: 319–330.

Runnegar, B., and Bentley, C. 1983. Anatomy, ecology, and affinities of the Australian Early Cambrian bivalve *Pojetaia runnegari* Jell. *Journal of Paleontology* 57: 73–92.

Runnegar, B., and Pojeta, J., Jr. 1974. Molluscan phylogeny: the paleontological viewpoint. *Science* 186: 311–317.

————. 1992. The earliest bivalves and their Ordovician descendants. *American Malacological Bulletin* 9: 117–122.

Salvini-Plawen, L. v. 1980. A reconsideration of systematics in the Mollusca. *Malacologia* 19: 249–278.

————. 1985. Early evolution and the primitive groups. In *The Mollusca*. Vol. 10, *Evolution*. Edited by E. R. Trueman and M. R. Clarke. Orlando, FL: Academic Press, pp. 59–150.

Salvini-Plawen, L. v., and Steiner, G. 1996. Synapomorphies and plesiomorphies in higher classification of Mollusca. In *Origin and evolutionary radiation of the Mollusca*. Edited by J. D. Taylor. Oxford: Oxford University Press, pp. 29–51.

Schneider, J. A. 2001. Bivalve systematics during the 20th century. *Journal of Paleontology* 75: 1119–1127.

Serb, J. M., Buhay, J. E., and Lydeard, C. 2003. Molecular systematics of the North American freshwater bivalve genus *Quadrula* (Unionidae: Ambleminae) based on mitochondrial ND1 sequences. *Molecular Phylogenetics and Evolution* 28: 1–11.

Siripattrawan, S., Park, J. K., and Ó Foighil, D. 2000. Two lineages of the introduced Asian freshwater clam *Corbicula* occur in North America. *Journal of Molluscan Studies* 66: 423–429.

Skelton, P. W., and Smith, A. B. 2000. A preliminary phylogeny for rudist bivalves: sifting clades from grades. In *The Evolutionary Biology of the Bivalvia*. Edited by E. M. Harper J. D. Taylor and J. A. Crame. London: The Geological Society of London, pp. 97–127.

Stanley, S. M. 1978. Aspects of the adaptive morphology and evolution of the Trigoniidae. *Philosophical Transactions of the Royal Society of London B* 284: 247–257.

Stasek, C. R. 1963. Synopsis and discussion of the association of ctenidia and labial palps in the bivalved Mollusca. *The Veliger* 6: 91–97.

————. 1972. The molluscan framework. In *Chemical Zoology*. Vol. 7, *Mollusca*. Edited by M. Florkin and B. T. Scheer. New York: Academic Press, pp. 1–44.

Steiner, G. 1992. Phylogeny and classification of Scaphopoda. *Journal of Molluscan Studies* 58: 385–400.

————. 1999. What can 18S rDNA do for bivalve phylogeny? Response. *Journal of Molecular Evolution* 48: 258–261.

Steiner, G., and Hammer, S. 2000. Molecular phylogeny of the Bivalvia inferred from 18S rDNA sequences with particular reference to the Pteriomorphia. In *The Evolutionary Biology of the Bivalvia*. Edited by E. M. Harper, J. D. Taylor, and J. A. Crame. London: The Geological Society of London, pp. 11–29.

Steiner, G., and Müller, M. 1996. What can 18S rDNA do for bivalve phylogeny? *Journal of Molecular Evolution* 43: 58–70.

Taylor, J. D., and Glover, E. A. 2006. Lucinidae (Bivalvia) – the most diverse group of chemosymbiotic molluscs. In *Bivalvia—A Look at the Branches*. Edited by R. Bieler. *Zoological Journal of the Linnean Society* 148: 421–438.

Taylor, J. D., Glover, E. A., and Williams, S. T. 2005. Another bloody bivalve: anatomy and relationships of *Eucrassatella donacina* from south western Australia (Mollusca: Bivalvia: Crassatellidae). In *The Marine Flora and Fauna of Esperance, Western Australia*. Edited by F. E. Wells, D. I. Walker, and G. A. Kendrick. Perth: Western Australian Museum, pp. 261–288.

Taylor, J. D., Kennedy, W. J., and Hall, A. 1969. The shell structure and mineralogy of the Bivalvia. Introduction. Nuculacea-Trigonacea. *Bulletin of the British Museum (Natural History), Zoology* Suppl 3: 1–125.

————. 1973. The shell structure and mineralogy of the Bivalvia II. Lucinacea—Clavagellacea, Conclusions. *Bulletin of the British Museum (Natural History), Zoology* 22: 255–294.

Tëmkin, I. 2006. Morphological perspective on the classification and evolution of Recent Pterioidea (Mollusca: Bivalvia). *Zoological Journal of the Linnean Society* 148: 253–312.

Terwilliger, N. B., Terwilliger, R. C., Meyhofer, E., and Morse, M. P. 1988. Bivalve hemocyanins—a comparison with other molluscan hemocyanins. *Comparative Biochemistry and Physiology B, Comparative Biochemistry* 89: 189–195.

Terwilliger, R. C., and Terwilliger, N. B. 1985. Molluscan hemoglobins. *Comparative Biochemistry and Physiology B, Comparative Biochemistry* 81B: 255–261.

Thiele, J. 1929–1935. *Handbuch der systematischen Weichtierkunde* (4 volumes). Jena, Germany: Gustav Fischer Verlag.

Trevallion, A. 1971. Studies on *Tellina tenuis* Da Costa. III. Aspects of general biology and energy flow. *Journal of Experimental Marine Biology and Ecology* 7: 95–122.

Trueman, E. R., and Clark, M. R., eds. 1988. *The Mollusca*. Vol. 11, *Form and Function*. New York: Academic Press.

Turner, R. D. 1966. A survey and illustrated catalogue of the Teredinidae. Cambridge, MA: The Museum of Comparative Zoology.

Valentich-Scott, P., and Dinesen, G. E. 2004. Rock and coral boring Bivalvia (Mollusca) of the Middle Florida Keys, U.S.A. In *Bivalve Studies in the Florida Keys. Proceedings of the International Marine Bivalve Workshop, Long Key, Florida, July 2002*. Edited by R. Bieler and P. M. Mikkelsen. *Malacologia* 46: 339–354.

Valentine, J. W., Jablonski, D., Kidwell, S., and Roy, K. 2006. Assessing the fidelity of the fossil record by using marine bivalves. *Proceedings of the National Academy of Sciences of the U S A* 103: 6599–6604.

van den Biggelaar, J. A. M., Kuhtreiber, W. M., Serras, F., Dorresteijn, A., Beekhuizen, H., and Schaap, D. 1986. Analysis of cell communication mechanisms involved in the induction of the stem cell of the mesodermal bands in embryos of *Patella vulgata* (Mollusca). *Acta Histochemica* Suppl. 32: 29–33.

Verdonk, N. H., and van den Biggelaar, J. A. M. 1983. Early development and formation of the germ layers. In *The Mollusca*. Vol. 3, *Development*. Edited by N. H. Verdonk, J. A. M. van den Biggelaar, and A. S. Tompa. New York: Academic Press, pp. 91–122.

Vermeij, G. J. 1977. The Mesozoic marine revolution: evidence from snails, predators and grazers. *Paleobiology* 3: 245–258.

Vokes, H. E. 1968. Genera of the Bivalvia: A systematic and bibliographic catalogue. Ithaca, NY: Paleontological Research Institution.

———. 1980. *Genera of the Bivalvia: A Systematic and Bibliographic Catalogue*, revised and updated. Ithaca, NY: Paleontological Research Institution.

Waller, T. R. 1978. Morphology, morphoclines and a new classification of the Pteriomorphia (Mollusca: Bivalvia). *Philosophical Transactions of the Royal Society of London B Biological Sciences* 284: 345–365.

———. 1981. Functional morphology and development of veliger larvae of the European oyster, *Ostrea edulis* Linné. *Smithsonian Contributions to Zoology* 328: 1–70.

———. 1990. The evolution of ligament systems in the Bivalvia. In *The Bivalvia—Proceedings of a Memorial Symposium in Honour of Sir Charles Maurice Yonge, Edinburgh, 1986*. Edited by B. Morton. Hong Kong: Hong Kong University Press, pp. 49–71

———. 1998. Origin of the molluscan class Bivalvia and a phylogeny of major groups. In *Bivalves: An Eon of evolution—Palaeobiological Studies Honoring Norman D. Newell*. Edited by P. A. Johnston and J. W. Haggart. Calgary: University of Calgary Press, pp. 1–45.

———. 2006. Phylogeny of families in the Pectinoidea (Mollusca: Bivalvia): importance of the fossil record. In *Bivalvia—A Look at the Branches*. Edited by R. Bieler. *Zoological Journal of the Linnean Society* 148: 313–342.

Wanninger, A., and Haszprunar, G. 2001. The expression of an engrailed protein during embryonic shell formation of the tusk-shell, *Antalis entalis* (Mollusca, Scaphopoda). *Evolution & Development* 3: 312–321.

———. 2002a. Chiton myogenesis: perspectives for the development and evolution of larval and adult muscle systems in molluscs. *Journal of Morphology* 251: 103–113.

———. 2002b. Muscle development in *Antalis entalis* (Mollusca, Scaphopoda) and its significance for scaphopod relationships. *Journal of Morphology* 254: 53–64.

———. 2003. The development of the serotonergic and FMRF-amidergic nervous system in *Antalis entalis* (Mollusca, Scaphopoda). *Zoomorphology* 122: 77–85.

Wanninger, A., Ruthensteiner, B., Dictus, W. J. A. G., and Haszprunar, G. 1999. The development of the musculature in the limpet *Patella* with implications on its role in the process of ontogenetic torsion. *Invertebrate Reproduction and Development* 36: 211–215.

Wells, F. E., ed. 2004. Molluscan Megadiversity: Sea, Land and Freshwater. *Proceedings of the World Congress of Malacology, Perth, Western Australia. 11–16 July 2004*. Perth: Western Australian Museum. 187 pp.

Willan, R. C. 1998. Superfamily Tellinoidea. In *Mollusca: The Southern Synthesis. Fauna of Australia*. Vol. 5. Edited by P. L. Beesley, G. J. B. Ross, and A. Wells. Melbourne: CSIRO Publishing, pp. 342–348.

Williams, S. T., Taylor, J. D., and Glover, E. A. 2004. Molecular phylogeny of the Lucinoidea (Bivalvia): non-monophyly and separate acquisition of bacterial chemosymbiosis. *Journal of Molluscan Studies* 70: 187–202.

Woods, F. H. 1931. History of the germ cells in *Sphaerium striatinum* (Lam.). *Journal of Morphology* 51: 545–595.

Yonge, C. M. 1926. Structure and physiology of the organs of feeding and digestion in *Ostrea edulis*. *Journal of the Marine Biological Association of the U.K.* 14: 295–386.

———. 1939. The protobranchiate Mollusca; a functional interpretation of their structure and evolution. *Philosophical Transactions of the Royal Society of London B Biological Sciences* 230: 79–147.

———. 1953. The monomyarian condition in the Lamellibranchia. *Transactions of the Royal Society of Edinburgh* 42: 443–478.

———. 1957. Mantle fusion in the Lamellibranchia. *Pubblicazioni della Stazione zoologica di Napoli* 29: 151–171.

———. 1962. On the primitive significance of the byssus in the Bivalvia and its effects in evolution. *Journal of the Marine Biological Association of the U.K.* 42: 113–125.

———. 1969. Functional morphology and evolution within the Carditacea (Bivalvia). *Proceedings of the Malacological Society of London* 38: 493–527.

———. 1978. Significance of the ligament in the classification of the Bivalvia. *Proceedings of the Royal Society of London Series B Biological Sciences* 202: 231–248.

———. 1979. Cementation in bivalves. In *Pathways in Malacology.*, Edited by S. van der Spoel A. C. van Bruggen and J. Lever. Utrecht: Bohn, Scheltema and Holkema, pp. 83–106.

———. 1980. Functional morphology and evolution in the Tridacnidae (Mollusca: Bivalvia: Cardiacea). *Records of the Australian Museum* 33: 735–777.

———. 1982. Mantle margins with a revision of siphonal types in the Bivalvia. *Journal of Molluscan Studies* 48: 102–103.

Zardus, J. D., Etter, R. J., Chase, M. R., Rex, M. A., and Boyle, E. E. 2006. Bathymetric and geographic population structure in the pan-Atlantic deep-sea bivalve *Deminucula atacellana* (Schenck, 1939). *Molecular Ecology* 15: 639–651.

Zardus, J. D., and Morse, M. P. 1998. Embryogenesis, morphology and ultrastructure of the pericalymma larva of *Acila castrensis* (Bivalvia: Protobranchia: Nuculoida). *Invertebrate Biology* 117: 221–244.

Zouros, E., Ball, A. O., Saavedra, C., and Freeman, K. R. 1994. Mitochondrial DNA inheritance. *Nature* 368: 818.

7

Scaphopoda

Patrick D. Reynolds and Gerhard Steiner

The Scaphopoda is the last class-level clade of the Mollusca to appear in the fossil record, around 360 Mya. This small group is currently recognized as having 816 extinct and 517 extant species (Steiner and Kabat 2004). They are commonly known as "tusk shells" for the consistent shell shape (Figure 7.1) and range in size from a few millimeters (mm) to several centimeters (cm) in length.

All scaphopods are marine, infaunal burrowers, globally distributed in the world's oceans from intertidal to abyssal depths. Despite their widespread distribution, scaphopods are only locally abundant and constitute minor components of most marine, soft-bottom, ecological assemblages. Detailed historical anatomical studies (the earliest monographs include Deshayes 1825; Clark 1849; Lacaze-Duthiers 1856–1857) use as a model system in early developmental studies (see Reverberi 1971; Reynolds 2002 for review), numerous reviews (e.g., Fischer-Piette and Franc 1969; Palmer and Steiner 1998; Steiner 1998b), and recent comprehensive treatments (Shimek and Steiner 1997; Reynolds 2002) have established

a sound, if generalized, knowledge of the biology of the group.

MAIN FEATURES

SHELL

As alluded to by the common name for the group (tusk shells), the shell of scaphopods is a conical tube, curved and open at both ends. While interpretation of the derivation of body axes has received some debate (Edlinger 1991; Steiner 1992b; Shimek and Steiner 1997; Waller 1998), convention usually names the larger aperture "anterior" and the convex side "ventral." The anterior aperture, usually the point of largest diameter of the shell, allows for extension of the foot and captacula, the smaller posterior aperture for expulsion of gametes (although oocytes exit via the anterior aperture in *Pulsellum lofotense* and *Cadulus subfusiformis* [Steiner 1993]) and feces; both allow for irrigation of the mantle, although the predominant pathway is seemingly through the posterior aperture (Clark 1849; Yonge 1937; Gainey 1972;

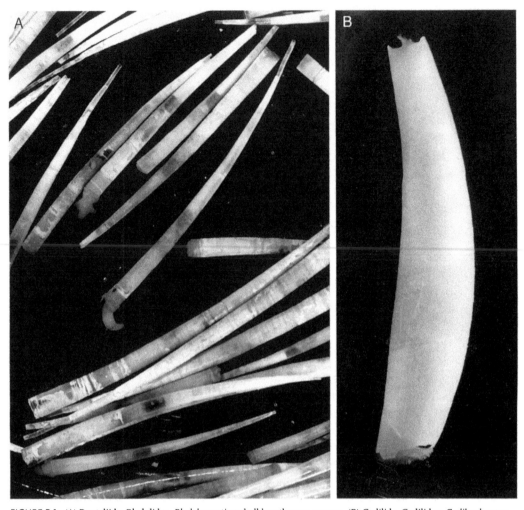

FIGURE 7.1. (A) Dentaliida, Rhabdidae, *Rhabdus rectius*; shell length ~30–50 mm. (B) Gadilida, Gadilidae, *Gadila aberrans*; shell length ~18 mm.

Taib 1980; Reynolds 1992b). The shell is added anteriorly in correspondence with the animal's growth and removed posteriorly (Reynolds 1992a) to periodically increase communication between the mantle cavity and the external environment. In several dentaliid genera (e.g., *Rhabdus, Episiphon*), a secondary, pipelike shell (Stasek and McWilliams 1973; Palmer 1974; Lamprell and Healy 1998) is secreted by the mantle at the posterior (apical) opening, with periodic decollation to increase the size of the posterior aperture (Reynolds 1992a).

Ornamentation, when present, consists of longitudinal or, in a few species, annular ribbing (e.g., Habe 1964; Palmer 1974; Scarabino 1995;

Lamprell and Healy 1998). Shell microstructure consists of a series of prismatic and crossed-lamellar layers, typically without a surviving periostracum (Steiner 1995; Reynolds and Okusu 1999).

MANTLE AND MANTLE CAVITY

The mantle cavity runs the length of the animal, created by a pair of lateral mantle folds that develop from the dorsal body wall, grow anteriorly and ventrally, and eventually fuse (Lacaze-Duthiers 1856–1857; Kowalevsky 1883; Wanninger and Haszprunar 2001). The anterior mantle margin is responsible for shell growth, although the posterior margin may secrete a secondary shell or "pipe," as well as dissolving or decollating the posterior shell (Reynolds 1992a). Glandular tissues are found within both mantle margins, and a cartilaginous ring supports the anterior margin and provides insertion for longitudinal and radial mantle muscles (Steiner 1991, 1992b).

Ctenidia, osphradia, and hypobranchial glands are absent. Simple sensory receptors are distributed around the anterior and posterior mantle openings (Steiner 1991, 1992b; Reynolds 1992a). A few to a few dozen transverse ciliary bands ring the midregion of the cavity (e.g., Lacaze-Duthiers 1856–1857; Yonge 1937; Steiner 1991; Shimek and Steiner 1997; Reynolds 2002) and contribute to mantle cavity circulation, and their high surface area and vascularization suggest a role in gas exchange (Reynolds 2002). Posterior to the ciliary bands, the anus opens atop a pulsatile anal bulb. The latter is flanked by an excretory pore from each kidney and slit-shaped epithelial openings from the hemocoel, which may accommodate rapid displacement of hemolymph during extreme foot contraction (Lacaze-Duthiers 1856–57; Plate 1892; Léon 1894; Boissevain 1904; Reynolds 1990b, 2002; Shimek and Steiner, 1997).

FEEDING

Most scaphopods are highly selective, infaunal-feeding microcarnivores, with foraminiferans constituting the majority of the diet (Dinamani 1964a, b; Shimek 1988, 1990; Steiner 1994; Langer et al. 1995; Gudmundsson et al. 2003; Glover et al. 2003). Some species exhibit greater omnivory, and some ingest significant quantities of inorganic material (Shimek 1990; Steiner 1994). Feeding behavior is characterized by the creation of a feeding cavity in the sediment by the burrowing action of the foot; the cavity may be spacious or minimal depending on the species and nature of the sediment (e.g., Poon 1987). All food items are manipulated in the first instance by the captacula; these cerebrally innervated tentacles number in the tens to hundreds and arise from a pair of dorsolateral head appendages, the captacula

shields (Shimek 1988). Captacula are extended by the locomotory activity of their ciliated, bulbous tips, whereas retraction is effected by longitudinal muscles (Shimek 1988; Byrum and Ruppert 1994; Shimek and Steiner 1997). The captacula ramify through the cavity and into the surrounding sediment. The bulbous tips of the captacula possess duo-gland capabilities, and particles are ultimately brought to the labial palps and mouth by adhesion, muscular contraction, and ciliary transport (reviewed in Reynolds 2002).

DIGESTIVE TRACT

The scaphopod gut is essentially U shaped, and most aspects of it are consistent throughout the class (Lacaze-Duthiers 1856–1857; Taib 1981a; Salvini-Plawen 1988; Shimek and Steiner 1997; Reynolds 2002). Ingested items are retained, for some time, in lateral buccal pouches before passing to the buccal cavity and being crushed by the relatively large radula.

The radula consists of five teeth per row, consisting of paired marginals, laterals, and a central rachidian tooth (e.g., Scarabino 1995; Lamprell and Healy 1998; Palmer and Steiner 1998). There is an unpaired dorsal supportive or protective "jaw" and a sensory subradular organ. The buccal cavity, with ciliated tracts, leads to an esophagus with large lateroventral esophageal glands. The esophagus opens to the stomach on the right-dorsal side. The stomach is partially lined with cuticle, is raised in places, and elsewhere is lined by ciliated tracts (Morton 1959; Salvini-Plawen 1981). One or paired digestive glands open into the stomach. Ventral to the entry from the esophagus, the intestine exits the stomach, from whence it coils before the rectum passes through the anal bulb, the anus opening to the mantle cavity (Salvini-Plawen 1988; Steiner 1994; Shimek and Steiner 1997; Reynolds 2002).

CIRCULATION AND EXCRETION

In contrast with the typical molluscan components of the digestive system, circulation and excretion in scaphopods are characterized by reduction and radical modification. There is no well-developed heart. The pericardium, located ventral to the stomach and posterior to the perianal sinus, has a muscular dorsal wall that performs regular pumping movements against the stomach and has been suggested as representing the ventricle (e.g., Plate 1891c, 1892). Alternatively, the perianal sinus with its pumping musculature may be the ventricle homolog (Fol 1889), as suggested by histological and ultrastructural studies (e.g., Lacaze-Duthiers 1856–1857; Reynolds 1990a, 2002). There are no auricles, their loss presumably associated with the absence of ctenidia. Pulsatile flow is initiated by both the perianal sinus and the dorsal pericardial wall, and circulation proceeds through a number of sinuses, including pedal and abdominal (reviewed in Shimek and Steiner 1997). The primary site of ultrafiltration in other molluscs (the auricular or pericardial border) is displaced in scaphopods, with podocytes occuring within the perianal sinus or pericardial border. The pericardium communicates with at least one kidney lumen, and both kidneys open to the mantle cavity (Reynolds 1990a, b, reviewed in 2002).

NERVOUS SYSTEM

The nervous system is typical of conchiferan tetraneury. The closely adjoined cerebral and pleural ganglia communicate by long, fused connectives to the pedal ganglia. The visceral connectives run from the pleural ganglia to the visceral ganglia and from hence to the pavilion ganglia, the latter being located near the dorsal apex. The buccal or stomatogastric system consists of paired subradular and radular ganglia. A variety of ciliated epithelial receptors have been identified from the anterior and posterior mantle edge (Steiner 1991; Reynolds 1992b), in addition to the subradular organ and captacula. A pair of statocysts are also present (Shimek and Steiner 1997; Reynolds 2002), with staticonia (Lacaze-Duthiers 1856–57). The early development of the nervous system was recently investigated by Wanninger and Haszprunar (2003).

REPRODUCTION AND DEVELOPMENT

Separate sexes (with rare hermaphrodites in some species; Reverberi 1971; D'Anna 1974) release eggs and sperm from the apical (posteriorly located) gonad to the exterior through the right kidney. Ultrastructural studies of gametogenesis and fertilization have been published by Reverberi (1969, 1970a, b, 1972), Van Dongen (1977), Dufresne-Dubé et al. (1983), Hou and Maxwell (1991), and Lamprell and Healy (1998), and reviewed in Reynolds (2002). Molluscan-pattern cleavage leads to a trochophore-like lecithotrophic stenocalymma, which develops into a single-shelled veliger-like larva. The tubular larval shell is eventually lost from the adult shell (Steiner 1995; reviewed in Reynolds 2002).

HISTORY OF CLASSIFICATION AND PHYLOGENY

Steiner and Kabat (2001, 2004) have reviewed the history of classification of the Scaphopoda, summarized here, in their recent and comprehensive specific and supraspecific catalogs of the group. Their classification is presented in Tables 7.1 and 7.2.

The class Scaphopoda was originally erected by Bronn (1862), and through general adoption, initially by Simroth (1894a,b) and Pilsbry and Sharp (1897–1898), it gained precedence over earlier names for the clade (i.e., Cirrhobranchiata de Blainville, 1824, Lateribranchiata Clark, 1851, and Solenoconques Lacaze-Duthiers, 1857). Both Starobogatov (1974) and Palmer (1974) proposed classifications that included the ordinal level, the former being accepted. Dentaliida and Gadilida now accommodate diversity within the clade. Gadilida is divided between two suborders: Entalimorpha and Gadilimorpha. At the time of the Steiner and Kabat (2001) publication, 14 families are distributed between the two orders (Table 7.1). Dentaliida comprise 276 species in 22 extant genera, Gadilida 241 extant species in 24 extant genera, with an additional 14 extinct valid genera, following the classification of Scarabino (1995). Recent departures from the Scarabino classification

TABLE 7.1
Classification of Dentaliida

Order Dentaliida da Costa, 1776

    Family Anulidentaliidae Chistikov, 1975

        Genus *Anulidentalium* Chistikov, 1975

        Genus *Epirhabdoides* Steiner, 1999

    Family Baltodentaliidae Engeser and Riedel, 1992

        Genus *Baltodentalium* Engeser and Riedel, 1992 (†)

    Family Calliodentaliidae Chistikov, 1975

        Genus *Calliodentalium* Habe, 1964

    Family Dentaliidae (Children, 1834)

        Genus *Antalis* H. and A. Adams, *1854*

        Genus *Coccodentalium* Sacco, 1896

        Genus *Compressidentalium* Habe, 1963

        Genus *Dentalium* Linnaeus, 1758

        Genus *Eodentalium* Medina and del Valle, 1985 (†)

        Genus *Eudentalium* Cotton and Godfrey, 1933

        Genus *Fissidentalium* Fischer, 1885

        Genus *Graptacme* Pilsbry and Sharp, 1897

        Genus *Paleodentalium* Gentile, 1974 (†)

        Genus *Paradentalium* Cotton and Godfrey, 1933

        Genus *Pictodentalium* Habe, 1963

        Genus *Plagioglypta* Pilsbry in Pilsbry and Sharp, 1898

        Genus *Schizodentalium* Sowerby, 1894

        Genus *Striodentalium* Habe, 1964

        Genus *Tesseracme* Pilsbry and Sharp, 1898

    Family Fustiariidae Steiner, 1991

        Genus *Fustiaria* Stoliczka, 1868

    Family Gadilinidae Chistikov, 1975

        Subfamily Episiphoninae Chistikov, 1975

            Genus *Episiphon* Pilsbry and Sharp, 1897

        Subfamily Gadilininae Chistikov, 1975

            Genus *Gadilina* Foresti, 1895

        Subfamily Lobantalinae Chistikov, 1975

            Genus *Lobantale* Cossmann, 1888 (†)

    Family Laevidentaliidae Palmer, 1974

        Genus *Laevidentalium* Cossmann, 1888

        Genus *Pipadentalium* Yoo, 1988 (†)

        Genus *Rhytiodentalium* Pojeta and Runnegar, 1979 (†)

        Genus *Scissuradentalium* Yoo, 1988 (†)

    Family Omniglyptidae Chistikov, 1975

        Genus *Omniglypta Kuroda and* Habe in Habe, 1953

    Family Prodentaliidae Starobogatov, 1974

        Genus: *Prodentalium* Young, 1942 (†)

    Family Rhabdidae Chistikov, 1975

        Genus *Rhabdus* Pilsbry and Sharp, 1897

    Dentaliida, *incertae sedis*

        Genus *Progadilina* Palmer, 1974 (†).

        Genus *Suevidontus* Engeser, Riedel, and Bandel, 1993 (†)

NOTE: Based on Steiner and Kabat 2001: Recent taxa from Table 1, fossil taxa (†) compiled from text. Scaphopoda, *incertae sedis*: Genus *Cyrtoconella* Patrulius, 1996 (†).

---

include the revision of Australian scaphopods by Lamprell and Healy (1998).

The relationship of scaphopods with other molluscan clades has been the source of considerable debate. Within the subphylum Conchifera, all major clades have been argued as sister taxa (reviewed in Steiner and Reynolds 2003). More recent morphological and molecular studies identify the Cephalopoda and Gastropoda as the most likely sister groups (Wanninger and Haszprunar 2001; Steiner and Dreyer 2003; reviewed in Steiner and Reynolds 2003; Passamaneck *et al.* 2004).

As might be expected for a clade with relatively limited diversity and abundance, wide geographic distribution and uncertain affinities within Mollusca, phylogeny of scaphopod taxa has received little attention. Among non-cladistic analysis, relationships among clades and character evolution were discussed by Emerson (1962)

TABLE 7.2

Classification of Gadilida

Order Gadilida Starobogatov, 1974

  Suborder Entalimorpha Steiner, 1992

    Family Entalinidae Chistikov, 1979

      Subfamily Bathoxiphinae Chistikov, 1983

        Genus *Bathoxiphus* Pilsbry and Sharp, 1897

        Genus *Rhomboxiphus* Chistikov, 1983

        Genus *Solenoxiphus* Chistikov, 1983

      Subfamily Entalininae Chistikov, 1979

        Genus *Entalina* Monterosato, 1872

      Subfamily Heteroschismoidinae Chistikov, 1982

        Genus *Costentalina* Chistikov, 1982

        Genus *Entalinopsis* Habe, 1957

        Genus *Heteroschismoides* Ludbrook, 1960

        Genus *Pertusiconcha* Chistikov, 1982

        Genus *Spadentalina* Habe, 1963

  Suborder Gadilimorpha Steiner, 1992

    Family Gadilidae Stoliczka, 1868

      Subfamily Gadilinae Stoliczka, 1868

        Genus *Bathycadulus* Scarabino, 1995

        Genus *Cadulus* Philippi, 1844

        Genus *Gadila* Gray, 1847

        Genus *Sulcogadila* Moroni and Ruggieri, 1981 (†)

      Subfamily Siphonodentaliinae Tryon, 1884

        Genus *Dischides* Jeffreys, 1867

        Genus *Polyschides* Pilsbry and Sharp, 1898

        Genus *Sagamicadulus* Sakurai and Shimazu, 1963

        Genus *Siphonodentalium* M. Sars, 1859

        Genus *Striocadulus* Emerson, 1962

    Gadilidae, *incertae sedis*

      Genus *Calstevenus* Yancey, 1973 (†)

      Genus *Gadilopsis* Woodring, 1925 (†)

    Family Pulsellidae Scarabino in Boss, 1982

      Genus *Annulipulsellum* Scarabino, 1986

      Genus *Pulsellum* Stoliczka, 1868

      Genus *Striopulsellum* Scarabino, 1995

    Family Wemersoniellidae Scarabino, 1986

      Genus *Chistikovia* Scarabino, 1995

      Genus *Wemersoniella* Scarabino, 1986

  Gadilida, *incertae sedis*

    Genus *Compressidens* Pilsbry and Sharp, 1897

    Genus *Megaentalina* Habe, 1963

NOTE: Based on Steiner and Kabat 2001: Recent taxa from Table 1, fossil taxa (†) compiled from text. Scaphopoda, *incertae sedis*: Genus *Cyrtoconella* Patrulius, 1996 (†).

and Chistikov (1975, 1978, 1979, 1984), with reference to shell, radular, and some anatomical characters. Emerson (1962) identified the genus *Entalina* as a link between the two main clades or orders of the class, because it exhibits characteristics of both, while *Plagioglypta* was later suggested as a common stem group by Chistikov (1984).

FOSSIL HISTORY

The fossil record of scaphopods, in large part represented by 816 extinct species (Steiner and Kabat 2004), has been studied only briefly to date. Summaries of stratigraphic ranges have been compiled by Emerson (1962) and Skelton

and Benton (1993), and Ludbrook (1960) commented upon a temporal increase in species diversity. The fluctuation in diversity over time received preliminary analysis (of only 242 species) by Reynolds (2002), who noted a number of significant putative extinction and origination periods.

Consideration of the earliest appearance of scaphopods in the fossil record has been fraught with uncertainty over the identification of material; serpulid polychaete tubes (e.g., Palmer *et al.* 2004) and nautiloid cephalopod shells (or portions thereof; e.g., Kues *et al.* 2006) being the main culprits. Several species that established Paleozoic origins for the group have been reassigned to non-scaphopod or non-molluscan taxa or are under scrutiny, such as *Plagioglypta iowensis* and *Rhytiodentalium kentuckyensis* from the Ordovician (Bretsky and Bermingham 1970; Pojeta and Runnegar 1979; see Engeser and Riedel 1996; Lamprell and Healy 1998; and Yochelson 1999 for discussion), and *Dentalium saturni*, *Dentalium antiquum*, and several other species from the Devonian (Ludbrook 1960; Emerson 1962; see Yochelson 1999, 2002; and Yochelson and Holland 2004 for discussion). Even Late Paleozoic records have recently been found to be unreliable (Kues *et al.* 2006). The earliest unequivocal record in the Dentaliida seemingly dates from the Mississippian (Carboniferous; 359.2 ± 2.5 Mya; Yochelson 1999). Similarly, the earliest gadilids date from the Paleogene (65.5 ± 0.3 Mya; Emerson 1962), with earlier records disproven following reexamination of material (Permian; see Yancey 1973; Yochelson 1999 for discussion).

MAJOR GROUPS

The two main clades of the Scaphopoda, the Dentaliida and Gadilida, are recognized at the ordinal level of classification. Several characters distinguish the two clades, many of which have been documented through cladistic analyses.

DENTALIIDA AND GADILIDA

MAIN SYNAPOMORPHIES
The tusk shape of the shell in Dentaliida is relatively consistent throughout the order, in comparison with that of the Gadilida (see following): a curved cone with the largest diameter at the anterior aperture consistent with an indeterminate maximum size and continuous growth. (Figure 7.1A). This is also the case in most members of the Gadilida, but members of one family (Gadilidae) have the maximum diameter about one-third of the body length back from the anterior edge, consistent with a defined adult size (Shimek 1989; Figure 7.1B).

Several features of the mantle distinguish the two orders. In all scaphopods, the anterior mantle margin secretes the primary shell and forms a substantial frontal epithelium that is in direct contact with the environment. In the Gadilida, this frontal epithelium possesses papillae, in contrast to the Dentaliida. Within the anterior mantle opening, a central fold surrounds an inner glandular area, whereas in the Dentaliida there is an additional, outer, glandular region, and the anterior opening of most species is lined by a ciliated band (Steiner 1991; Shimek and Steiner 1997).

The posterior mantle opening is a muscular valve-like mechanism, presumably to regulate water flow; in the dentaliids it is consistently found in a lateral orientation and in gadilids in a dorsoventral orientation (Steiner 1991; Shimek and Steiner 1997; Reynolds and Okusu 1999). The cylindrical "pavilion," which extends posteriorly from the valve, is open medially, and in gadilids this opening is ciliated (Steiner 1991).

In Dentaliida, the foot is characterized as being very muscular, with epipodial lobes surrounding a cone-shaped terminus, whereas gadilids differ in having the foot terminate in a terminal disk fringed with a series of epipodial papillae. All longitudinal pedal muscles are associated with the pedal wall in dentaliids, whereas a part of these muscles lies freely in the large pedal sinus in gadilids (Steiner 1992a; Shimek and Steiner 1997). Movement of the gadilid foot is more rapid than that of the dentaliids; retraction of the gadilid foot is by inversion, whereas in dentaliids the foot folds upon itself to retract into the mantle cavity (Steiner 1992a, b; Shimek and Steiner 1997; Reynolds and Okusu 1999).

The captacula possess longitudinal muscles that number 8 to 10 within dentaliid species and 5 to 7 within the Gadilida (Shimek 1988; Steiner 1998a; Reynolds and Okusu 1999).

Variation in a number of minor characteristics of the gut may prove to have order-level patterns. Salvini-Plawen (1988) noted that ciliated tracts in the anterior gut—dorsal in the buccal cavity, and both dorsal and ventral in the esophagus—of both dentaliids and gadilids were distinguished in the dentaliids by lateral ciliated folds, which run from the buccal cavity and fuse with the dorsal ciliated esophageal tracts.

Two organs within the digestive tract provide strong evidence for the evolutionary independence of these orders. First, the highly mineralized radula has long been used to distinguish between the two major clades of scaphopods. In the Dentaliida, the rachidian tooth is comparatively large and wider than high; lateral teeth have a narrow base and marginal teeth do not have a keel (also absent in one gadilid family, the Entaliinidae). In gadilids, the rachidian tooth is higher than wide, the lateral tooth base is broad, and the marginal tooth keel is present except for the Entaliinidae (e.g., Steiner 1998a, Reynolds and Okusu 1999). Radular movement in dentaliids is limited and maceration often incomplete, in contrast to the gadilids, where radular action is more vigorous and breaking down of the food more thorough (Shimek 1990; Shimek and Steiner 1997). Second, the digestive gland is single in the Gadilida and paired in the Dentaliida (reviewed in Shimek and Steiner 1997; Reynolds 2002).

The Dentaliida da Costa, 1776 comprise 32 genera (22 extant), distributed among 10 familes (3 subfamilies; Table 7.1). The Gadilia, with 4 families (5 subfamilies), consists of 27 (24 extant) genera (Table 7.2). One fossil putative scaphopod genus, *Cyrtoconella*, has uncertain affinities.

INNOVATIONS

Little work exists on the variation among these subtaxa. A few species, such as *Antalis dentalis*, *A. entalis*, and *A. vulgaris*, have been used as model taxa for histological (e.g., Deshayes 1825; Clark 1849; Lacaze-Duthiers 1856–57, 1885; Plate 1888, 1891a, b, c, 1892; reviewed in Shimek and Steiner 1997; Reynolds 2002) and developmental studies (e.g., Wilson 1904a, b), and organogenesis (e.g., Verdonk 1968a, b; Geilenkirchen *et al.* 1970, 1971; Verdonk *et al.* 1971; Van Dongen and Geilenkirchen 1974a, b, c, 1975; Van Dongen 1976a, b, c; Guerrier *et al.* 1978; Cather and Verdonk 1979; Jaffe and Guerrier 1981; Wanninger and Haszprunar 2001; reviewed in Reynolds 2002). A compilation of

biological data collected on disparate species is reviewed in Reynolds (2002). Because Scaphopoda is one of the "minor" classes of molluscs, and one whose members have a largely sparse and deeper-water distribution and whose form is apparently generally homogeneous, little comprehensive comparative data on biology or ecology is available. Reynolds (2002) presents a number of preliminary analyses on latitudinal and stratigraphic distribution on scaphopods, but only generalized for the class or for the Dentaliida.

Some generalities on biologically significant morphological innovations have been made but remain to be tested. For example, the smoother, anteriorly attenuated shell of some gadilids such as *Cadulus* and some other gadilids may enhance speed and depth of burrowing and, thus, capability for wide-ranging predation (Shimek 1989, 1990; reviewed in Reynolds 2002); adaptation to habitat may be reflected in the varied degrees of diet specialization among some species (Shimek 1990; reviewed Reynolds 2002).

PHYLOGENY

Phylogenetic analysis of scaphopod taxa has received considerable attention (Steiner 1992b, 1998a, 1999; Reynolds 1997; Reynolds and Okusu 1999; Steiner and Dreyer 2003; Steiner and Reynolds 2003), but the number of analyses is small, the available data is limited, and taxon sampling is poor. All strongly support monophyly of the Dentaliida and Gadilida, but resolution below that level varies widely.

MOLECULAR PHYLOGENIES

Molecular data available for scaphopods and their analysis has recently been reviewed by Steiner and Reynolds (2003); a summary of data available is given in Table 7.3. There are two published analyses, one using cytochrome *c* oxidase I (COI) mtDNA (Steiner and Reynolds 2003; Figure 7.2) and the other using 18S rDNA sequences (Steiner and Dreyer 2003; Figure 7.3). The COI analysis in Steiner and Reynolds (2003)

TABLE 7.3

Summary of morphological and molecular data published for scaphopod taxa. Morphology is compiled form coded data in Reynolds (1997), Reynolds & Okusu (1999), and Steiner (1992, 1998, 1999); molecular data from Steiner and Reynolds 2003, Steiner and Dreyer 2003, and GenBank (December 2006).

ORDER	FAMILY	SPECIES	MORPHOLOGY	18S	28S	16S	COI	OTHER
Dentaliida	Dentaliidae	*Antalis antillaris*					•	
		Antalis dentalis	•				•	
		Antalis entalis	•		•	•	•	histone3
		Antalis inaequicostata	•		•	•	•	histone3
		Antalis occidentalis	•					
		Antalis perinvoluta		•			•	
		Antalis pilsbryi		•	•		•	
		Antalis sp. BS660	•					
		Antalis sp. P 927 II	•					
		Antalis sp. P 930	•					
		Antalis sp. PR-2003					•	
		Antalis sp. Q917	•					
		Antalis vulgaris	•		•	•		histone3
		Dentalium austini		•				
		Dentalium laqueatum	•					
		Dentalium majorinum	•				•	
		Dentalium octangulatum		•	•			HOX genes
		Dentalium sp. MBB-1996					•	
		Fissidentalium candidum	•		•		•	
		Fissidentalium capillosum		•				
		Fissidentalium carduum	•					
		Fissidentalium megathyris	•					
		Fissidentalium zelandicum	•					
		Graptacme eborea				•	•	complete mt-genome; engrailed
	Rhabdidae	*Rhabdus dalli*	•					
		Rhabdus perceptum	•					
		Rhabdus rectius	•		•	•	•	histone3
	Fustiariidae	*Fustiaria rubescens*	•	•				
		Fustiaria stenoschiza	•					
	Calliodentaliidae	*Calliodentalium callipeplum*	•					
	Laevidentaliidae	*Laevidentalium eburneum*	•					
		Laevidentalium martyi	•					
		Laevidentalium sominium	•					
	Anulidentaliidae	*Epirhabdoides ivanovi*	•					
	Gadilinidae	*Episiphon kiachowwanense*	•					
		Episiphon sp.	•					
		Episiphon subtorquatum	•					
		Episiphon yamakawai					•	ITS2

TABLE 7.3
(continued)

ORDER	FAMILY	SPECIES	MORPHOLOGY	18S	28S	16S	COI	OTHER
		Gadilina insolita	•					
		Gadilina pachypleura	•					
	Entalinidae	*Bathoxiphus ensiculus*	•					
		Costentalina caymanica	•					
		Costentalina tuscarorae	•					
		Entalina platamodes	•					
Gadilida		*Entalina tetragona*	•	•			•	
		Heteroschismoides subterfissum	•	•				
	Pulsellidae	*Annulipulsellum euzkadii*	•					
		Pulsellum affine		•				
		Pulsellum lofotense	•					
		Pulsellum salishorum	•				•	
		Pulsellum sp. BS 904	•					
		Pulsellum sp. P 937 I	•					
		Striopulsellum sandersi	•					
	Wemersoniellidae	*Wemersoniella turnerae*	•					
	Gadilidae	*Cadulus artatus*	•					
		Cadulus cylindratus	•					
		Cadulus delicatulus	•					
		Cadulus fusiformis						engrailed
		Cadulus jeffreysi	•					
		Cadulus propinquus	•					
		Cadulus sp. 68	•					
		Cadulus sp. A		•				
		Cadulus sp. B		•				
		Cadulus sp. P927 I	•					
		Cadulus subfusiformis	•	•				
		Gadila aberrans	•				•	
		Gadila agassizii	•					
		Gadila fraseri	•					
		Gadila sp.	•					
		Gadila sp. ["kraeuteri"]	•					
	Siphonodentaliidae	*Polyschides carolinensis*					•	
		Polyschides olivi		•				
		Polyschides quadrifissatus	•					
		Siphonodentalium dalli	•					
		Siphonodentalium grandis	•					
		Siphonodentalium lobatum	•	•		•	•	complete mt-genome
		Siphonodentalium spectabilis	•					
	incertae sedis	*Compressidens platyceras*	•					

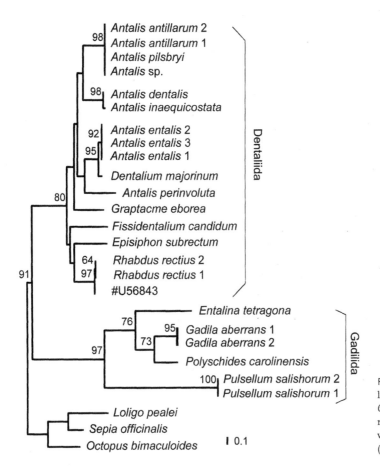

FIGURE 7.2. Maximum likelihood tree (–ln L = 6,153.28516), using COI mtDNA nucleotide sequences; bootstrap values >50% above branches (Steiner and Reynolds 2003).

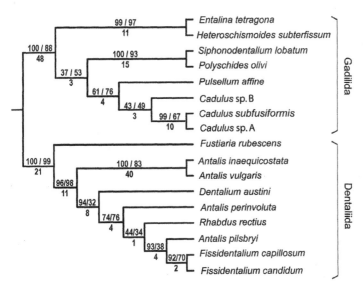

FIGURE 7.3. Strict consensus of three most parsimonious trees (L = 3,206; CI = 0.5221, RC = 0.3935), 18S. Bootstrap and puzzling values above branches, decay index values below (Steiner and Dreyer 2003).

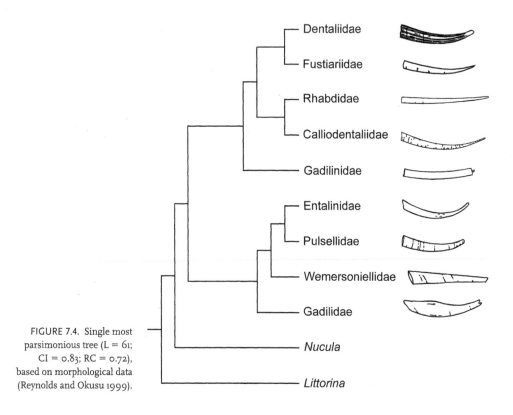

FIGURE 7.4. Single most parsimonious tree (L = 61; CI = 0.83; RC = 0.72), based on morphological data (Reynolds and Okusu 1999).

shows relatively strong support for relationships among the four species in four genera analyzed from within the Gadilida (Figure 7.2). Monophyly of Gadilidae (*Gadila aberrans* and *Polyschides carolinensis*) is supported, but not so the monophyly of the suborder Gadilimorpha (*Entalina tetragona,* suborder Entalimorpha, falling within the group). Taxon sampling within the Dentaliida is also limited, consisting of several species of Dentaliidae (mostly *Antalis*), with one representative from the Rhabdidae (*Rhaddus rectius*) and Gadilinidae (*Episiphon subrectum*). Resolution among family representatives is poor, with paraphyly of the Dentalidae indicated, whereas somewhat stronger support among Dentaliidae representatives renders *Antalis* paraphyletic.

Results from the 18S also support paraphyly of Dentalidae and *Antalis* (Figure 7.3), somewhat more strongly than in the COI analysis. Among Gadilida, there is strong support for the Entalimorpha (*Heteroschismoides subterfissum, Entalina tetragona*), less so for the Gadilimorpha. Monophyly of the gadilid

families represented by two or more terminals (Siphonodentaliidae, Gadilidae) is supported to varying degrees.

MORPHOLOGICAL PHYLOGENIES

Phylogenetic analysis based upon morphological data is as limited in studies, data, and taxon sampling as those based upon molecular data, although scaphopod morphology holds less promise for increasing character number. Character suites have revolved around shell shape, sculpture, and microstructure; the radula; mantle; foot and other musculature; and the gut (Steiner 1992b, 1998a, 1999; Reynolds 1997; Reynolds and Okusu 1999; for review see Steiner and Reynolds 2003).

Two representative trees from these analyses, which include differing data sets and taxon sampling, are presented in Figure 7.4 (Reynolds and Okusu 1999) and Figure 7.5 (Steiner 1998a). As is the case for the molecular data sets, monophyly of the two traditional order-level taxa, Dentaliida and Gadilida, is supported. Within the Gadilida, both support a basal position for

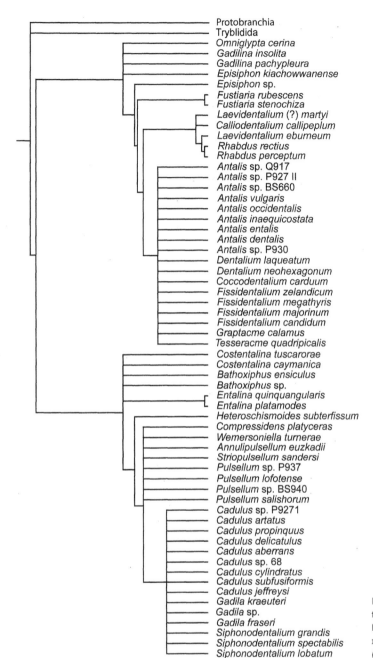

Protobranchia
Tryblidida
Omniglypta cerina
Gadilina insolita
Gadilina pachypleura
Episiphon kiachowwanense
Episiphon sp.
Fustiaria rubescens
Fustiaria stenochiza
Laevidentalium (?) *martyi*
Calliodentalium callipeplum
Laevidentalium eburneum
Rhabdus rectius
Rhabdus perceptum
Antalis sp. Q917
Antalis sp. P927 II
Antalis sp. BS660
Antalis vulgaris
Antalis occidentalis
Antalis inaequicostata
Antalis entalis
Antalis dentalis
Antalis sp. P930
Dentalium laqueatum
Dentalium neohexagonum
Coccodentalium carduum
Fissidentalium zelandicum
Fissidentalium megathyris
Fissidentalium majorinum
Fissidentalium candidum
Graptacme calamus
Tesseracme quadripicalis
Costentalina tuscarorae
Costentalina caymanica
Bathoxiphus ensiculus
Bathoxiphus sp.
Entalina quinquangularis
Entalina platamodes
Heteroschismoides subterfissum
Compressidens platyceras
Wemersoniella turnerae
Annulipulsellum euzkadii
Striopulsellum sandersi
Pulsellum sp. P937
Pulsellum lofotense
Pulsellum sp. BS940
Pulsellum salishorum
Cadulus sp. P9271
Cadulus artatus
Cadulus propinquus
Cadulus delicatulus
Cadulus aberrans
Cadulus sp. 68
Cadulus cylindratus
Cadulus subfusiformis
Cadulus jeffreysi
Gadila kraeuteri
Gadila sp.
Gadila fraseri
Siphonodentalium grandis
Siphonodentalium spectabilis
Siphonodentalium lobatum

FIGURE 7.5. Strict consensus tree (L = 91; CI = 0.527; RC = 0.444), based on morphological data (Steiner 1998a).

the Entalinidae (suborder Entalimorpha), if somewhat equivocal in Steiner (1998a). Relationships among gadilid taxa are not fully resolved, although a clade including Gadilidae, Pulsellidae, and Wemersoniellidae is consistent, as is the sister grouping of the Gadilidae subfamilies Gadilinae and Siphonodentaliinae. As indicated in analyses of molecular data, relationships among the Dentaliida are far from stable. Agreement exists in the basal position of the Gadilinidae, and the derived placement of a clade consisting of the Laevidentaliidae, Rhabdidae, and Calliodentaliidae. The position of the Fustiariidae, whether sister to the Dentaliidae or a larger group including the aforementioned clade, remains to be resolved.

All hypothesized relationships bear further examination using more complete taxonomic sampling and enhanced character development.

MORPHOLOGICAL CHARACTER EVOLUTION

Tracing the evolution of morphological characters depends in the first place upon the robustness of the phylogenetic hypothesis upon which it is based, and in scaphopods there is little congruence among analyses to date. Apart from the unambiguous result of monophyly of Dentaliida and Gadilida (the synapomorphies that define these two clades are discussed previously), few statements of relationships can be made with any confidence. In addition, many characters of interest (e.g., captacula, posterior mantle valves, mantle ciliated bands) do not lend themselves to outgroup analysis utilizing non-scaphopod clades. Our understanding of character evolution is hampered by incongruency, and, no doubt with greater resolution, several homoplastic characters will be confirmed.

The differentiation of the pedal longitudinal musculature is the only character providing a reliable inference of evolutionary transitions (Steiner 1992a, b). The dentaliids show the plesiomorphic condition with all muscles attached to the body wall. Among the gadilids, the Entalimorpha have central pedal retractor muscles detaching from the pedal wall at the midpiece of the foot. In the Gadilimorpha, the two massive central pedal retractors separate from the body wall much earlier, that is, at the level of the intestinal sinus, and traverse the entire foot from the base to the terminal disk. The entalimorph condition is, thus, intermediate between the plesiomorphic dentaliid and the more derived gadilimorph condition.

Shell sculpture, long used in taxonomy, offers little as most scaphopods are devoid of sculpture. Members of the Dentaliidae exhibit a wide variety of longitudinal ribbing (in number and height of ribs). Ribbing is otherwise only found in one family of Gadilida, the Entalinidae, which suggests this trait has evolved twice in the group. Members of the Omniglyptidae, and some Gadilinidae, have annular sculpture

(e.g., Habe 1964; Scarabino 1995; Reynolds and Okusu 1999), also suggesting that trait arose more than once.

Within the Dentaliida, the consistently basal position of the Gadilinidae allows some consideration of characters of interest. In all other dentaliids (members of the Calliodentaliidae, Dentaliidae, Fustiaridae, Laevidentaliidae, and Rhabdidae), there are two dorsoventral longitudinal retractor muscles, whereas there is only one in the Gadilinidae (*Gadilina* and *Episiphon*). Steiner (1998a), however, found incongruity in this character, a small second pair of dorsoventral muscles in an unidentified species of *Episiphon*. Dorsolateral ciliated slits, invaginations of the frontal mantle epithelium, which has abundant, with putatively sensory, ciliated receptors, are currently known only in members of the Rhabdidae, Calliodentaliidae, and some members of the Gadilinidae.

Within the Gadilida, possibly the most intriguing result to date is the relatively strong signal in the 18S analysis, suggesting a sister group relationship between the Pulsellidae and Gadilidae, to the exclusion of the Siphonodentaliidae (Steiner and Dreyer 2003). Although not supported by COI or morphological data sets, such a relationship would argue for the independent evolution of the attenuated anterior aperture in the latter two families (Steiner and Reynolds 2003). If indeed linked to deeper, faster burrowing, and a broadening of predatory range (see previous discussion), this would be an interesting evolutionary scenario for a functionally and ecologically significant character in the scaphopods. Such conjecture, in the absence of robust resolution using multigene or expanded morphological data sets, illustrates the current difficulties in assessing character evolution in the class.

ADAPTIVE RADIATION

The two main clades of scaphopods, Dentaliidae and Gadilidae, have been relatively equally successful in terms of species diversity, and members of both are distributed throughout the world's oceans. The comparative biogeography among scaphopod families remains to be analyzed.

As discussed previously, the main morphological differences between the Dentaliida and Gadilida, respectively, are the shape of the foot (thicker, more muscular, less active, S-shaped retraction vs. thinner, less muscular, more active, inversion), the number of digestive gland lobes (two vs. one), the shape and activity of the radula (larger, less active vs. smaller, more active), and the orientation of the posterior mantle valve (lateral vs. dorsoventral). Many gadilids have a noticably glossy, smooth shell in comparison to dentaliids, and some have an attenuated anterior aperture.

The evolution of these characters has been poorly studied to date, and few hypotheses of key innovations that may have driven diversification can be made. In general, however, it appears that members of the Gadilida are distinguished by a more active, predatory lifestyle. The foot and shell characteristics suggest faster and deeper burrowing activity in the clade (e.g., *Gadila aberrans*, Shimek 1989, 1990), which may be indicative of a distinction in the characteristics of adaptive radiation in these two groups.

FUTURE STUDIES

As can be gleaned from a comparison of the recognized taxonomic diversity in the class from Steiner and Kabat (2001, 2004; tables 7.1 and 7.2) with the level of taxonomic sampling in both the morphological and molecular phylogenetic analyses to date, much work remains to be done on scaphopod phylogeny. As a pragmatic proxy for phylogenetic distance, multispecies sampling across suprageneric taxa is still an unattained objective; an analysis with even half of the currently recognized genera is still some way off. Strategies for increasing taxonomic sampling require collaboration among malacologists and marine ecologists far beyond taxonomic interest. Commitments are needed to develop repositories of material suitable for molecular analysis, and collection efforts

that focus upon broad taxonomic coverage will assist in making headway in scaphopod phylogenetics.

REFERENCES

Blainville, H. M. D., de 1824. Mollusques, Mollusca. (Malacoz.). *Dictionnaire des Sciences naturelles* 32: 1–392.

Boissevain, M. 1904. Beiträge zur Anatomie und Histologie von *Dentalium*. *Jenaische Zeitschrift fürNaturwissenschaft* 38: 553–572.

Bretsky, P. W., and Bermingham, J. J. 1970. Ecology of the Paleozoic scaphopod genus *Plagioglypta* with special reference to the Ordovician of eastern Iowa. *Journal of Paleontology* 44: 908–924.

Bronn, H. G. 1862 (in 1862–1866). *Klassen und Ordnungen des Thier-Reichs, wissenschaftlich dargestellt in Wort und Bild. Dritter Band, Malacozoa.* Erste Abteilung. Leipzig and Heidelberg: C. F. Winter'sche Verlagshandlung, 1500 pp., 136 pls (1862: 523–650, pls 45–49).

Byrum, C. A., and Ruppert, E. E. 1994. The ultrastructure and functional morphology of a captaculum in *Graptacme calamus* (Mollusca, Scaphopoda). *Acta Zoologica* 75: 37–46.

Cather, J. N., and Verdonk, N. H. 1979. Development of *Dentalium* following removal of D-quadrant blastomeres at successive cleavage stages. *Wilhelm Roux's Archives of Developmental Biology* 187: 355–366.

Chistikov, S. D. 1975. Some problems of scaphopod taxonomy. In *U.S.S.R. Academy of Sciences Zoological Institute. Fifth Meeting on the Investigation of Molluscs. Molluscs: Their System, Evolution and Significance in Nature. Theses of communications.* Edited by I. M. Likharev. Leningrad: Nauka, pp. 21–23 [in Russian].

———. 1978. Some problems in the classification of the order Dentaliida Mollusca; Scaphopoda. *Malacological Review* 11: 71–73.

———. 1979. Phylogenetic relations of the scaphopods. In *U.S.S.R. Academy of Sciences Zoological Institute. Sixth Meeting on the Investigation of Molluscs. Molluscs: Main Results of Their Study. Abstracts of communications.* Edited by I. M. Likharev. Leningrad: Nauka, pp 20–22 [in Russian].

———. 1984. Phylogenetic relationships of the Scaphopoda. *Malacological Review* 17: 114–115.

Clark, W. 1849. On the animal of *Dentalium tarentinum*. *Annals and Magazine of Natural History* (2)4: 321–330.

———. 1851. On the classification of the British marine testaceous Mollusca. *Annals and Magazine of Natural History* (2)7: 469–481.

D'Anna, T. 1974. Ermafroditismo in *Dentalium enta-lis. Atti della Accademia Nazionale dei Lincei. Classe di Scienze Fisiche, Matematiche e Naturali. Rendiconti, Serie 8* 57: 673–677.

Deshayes, G. P. 1825. Anatomie et monographie du genre Dentale. *Mémoires de la Society d'histoire naturelle de Paris* 2: 321–378.

Dinamani, P. 1964a. Burrowing behaviour of *Dentalium. Biological Bulletin* 126: 28–32.

———. 1964b. Feeding in *Dentalium conspicuum. Proceedings of the Malacological Society of London* 36: 1–5.

Dufresne-Dube, L., Picheral, B., and Guerrier, P. 1983. An ultrastructural analysis of *Dentalium vulgare* (Mollusca, Scaphopoda) gametes with special reference to early events at fertilization. *Journal of Ultrastructural Research* 83: 242–257.

Edlinger, K. 1991. Zur Evolution der Scaphopoden-Konstruktion. *Natur und Museum* 121: 116–122.

Emerson, W. K. 1962. A classification of the scaphopod mollusks. *Journal of Paleontology* 36: 461–482.

Engeser, T. S., and Riedel, F. 1996. The evolution of the Scaphopoda and its implications for the systematics of the Rostroconchia Mollusca. *Mitteilungen aus dem Geologisch-Paläontologischen Institut der Universität Hamburg* 79: 117–138.

Fischer-Piette, E., and Franc, A. 1969. Classe de Scaphopodes. In *Traite de Zoologie, Mollusques, Gastéropodes et Scaphopodes*. Edited by P.-P. Grassé. Paris: Mason et Cie, pp. 987–1017.

Fol, H. 1889. Sur l'anatomie microscopique du Dentale. *Archives de Zoologie Experimentale et Générale, Deuxième Série* 7: 91–148. pls. 5–8.

Gainey, L. F. J. 1972. The use of the foot and the captacula in the feeding of *Dentalium* Mollusca: Scaphopoda. *Veliger* 15: 29–34.

Geilenkirchen, W. L. M., Timmermans, L. P. M., Van Dongen, C. A. M., and Arnolds, W. J. A. 1971. Symbiosis of bacteria with eggs of *Dentalium* at the vegetal pole. *Experimental Cell Research* 67: 477–478.

Geilenkirchen, W. L. M., Verdonk, N. H., and Timmermans, L. P. M. 1970. Experimental studies on morphogenetic factors localized in the first and the second polar lobe of *Dentalium* eggs. *Journal of Embryology and Experimental Morphology* 23: 237–243.

Glover, E. A., Taylor, J. D., and Whittaker, J. 2003. Distribution, abundance and foraminiferal diet of an intertidal scaphopod, *Laevidentalium lubricatum*, around the Burrup Peninsula, Dampier, Western Australia. In *The Marine Flora and Fauna of Dampier, Western Australia*. Edited by F. E. Wells, D. I. Walker, and D. S. Jones. Perth: Western Australian Museum, pp. 225–240.

Gudmundsson, G., Engelstad, K., Steiner, G., and Svavarsson, J. 2003. Diets of four deep-water scaphopod species (Mollusca) in the North Atlantic and Nordic Seas. *Marine Biology* 142: 1103–1112.

Guerrier, P., van den Biggelaar, J. A. M., Van Dongen, C. A. M., and Verdonk, N. H. 1978. Significance of the polar lobe for the determination of dorso ventral polarity in *Dentalium vulgare. Developmental Biology* 63: 233–242.

Habe, T. 1964. *Fauna Japonica. Scaphopoda Mollusca.* Tokyo: Biogeographical Society of Japan.

Hou, S. T., and Maxwell, W. L. 1991. Ultrastructural studies of spermatogenesis in *Antalis entalis* (Scaphopoda, Mollusca). *Philosophical Transactions of the Royal Society of London. B. Biological Sciences* 333: 101–110.

Jaffe, L. A., and Guerrier, P. 1981. Localization of electrical excitability in the early embryo of *Dentalium. Developmental Biology* 83: 370–373.

Kowalevsky, A. 1883. Étude sur l'embryogénie du Dentale. *Annales de la Museum d'Histoire Naturelle de Marseille* 1: 1–54.

Kues, B. S., Mapes, R. H., and Yochelson E. L. 2006. Nautiloid-scaphopod homeomorphy in the late Paleozoic of the United States. *Lethaia* 39: 91–93.

Lacaze-Duthiers, H. 1856–1857. Histoire de l'organisation et du développement du Dentale. *Annales des Sciences Naturelles, Quatrième Serie, Paris* Tome 6: 225–281, pls. 8–10; 319–385, pls. 11–13; Tome 7: 5–51, pls. 2–4; 171–255, pls. 5–9.; Tome 8: 18–44.

———. 1885. Note sur l'anatomie du Dentale. *Comptes rendus hebdomadaires des sceances de l'Académie des Sciences* 1885: 296–300.

Lamprell, K. L., and Healy, J. M. 1998. A revision of the Scaphopoda from Australian waters (Mollusca). *Records of the Australian Museum*, Suppl. 24: 1–189.

Langer, M. R., Lipps, J. H., and Guillermo, M. 1995. Predation on foraminifera by the dentaliid deep-sea scaphopod *Fissidentalium megathyris. Deep-Sea Research, Part I. Oceanographic Research Papers* 42: 849–857.

Léon, N. 1894. Zur Histologie des *Dentalium*-mantels. *Jenaische Zeitschrift Medizin und Naturwissenschaft* 29: 411–416.

Ludbrook, N. H. 1960. Scaphopoda. In *Treatise on Invertebrate Paleontology*. Edited by R. C. Moore. Lawrence, KS: Geological Society of America, pp I37–I41.

Morton, J. E. 1959. The habits and feeding organs of *Dentalium entalis. Journal of the Marine Biological Association of the United Kingdom* 38: 225–238.

Palmer, C. P. 1974. A supraspecific classification of the scaphopod Mollusca. *Veliger* 17: 115–123.

Palmer, C. P., Boyd, D. W., and Yochelson E. L. 2004. The Wyoming Jurassic fossil *Dentalium subquadratum* Meek, 1860 is not a scaphopod but a serpulid worm tube. *Rocky Mountain Geology* 392: 85–91.

Palmer, C. P., and Steiner, G. 1998. Class Scaphopoda. In *Mollusca: The Southern Synthesis*. Vol. 5. Edited by P. L. Beesley, G. J. B. Ross, and A. Wells. Melbourne: CSIRO Publishing, pp 431–450.

Passamaneck Y. J., Schander C., and Halanych, K. M. 2004. Investigation of molluscan phylogeny using large-subunit and small-subunit nuclear rRNA sequences. *Molecular Phylogenetics and Evolution* 32: 25–38.

Pilsbry, H. A., and Sharp, B. 1897–1898. Scaphopoda. *Manual of Conchology*. 17: 1–280, pls. 1–37.

Plate, L. H. 1888. Bemerkungen zur Organisation der Dentalien. *Zoologischer Anzeiger* 11: 509–515.

———. 1891a. Ueber einiger Organizations verhältnisse der Dentalien. *Sitzungsberichte der Gesellschaft zur Beförderung der gesammten Naturwissenschaften zu Marburg* 1890: 26–29.

———. 1891b. Über den Bau und die systematische Stellung der Solenoconchen. *Verhandlungen. Deutsche Zoologische Gesellschaft* 1: 60–66.

———. 1891c. Über das Herz der Dentalien. *Zoologischer Anzeiger* 14: 78–80.

———. 1892. Ueber den Bau und die Verwandtschaftsbeziehungen der Solenoconchen. *Zoologische Jahrbücher. Abteilung für Anatomie und Ontogenie der Tiere* 5: 301–386.

Pojeta, J., Jr., and Runnegar, B. 1979. *Rhytiodentalium kentuckyensis*, a new genus and new species of Ordovician scaphopod, and the early history of scaphopod mollusks. *Journal of Paleontology* 53: 530–541.

Poon, P. A. 1987. The diet and feeding behavior of *Cadulus tolmiei* Dall, 1898 (Scaphopoda: Siphonodentalioida). *Nautilus* 101: 88–92.

Reverberi, G. 1969. Il primo lobo polare dell'uovo di *Dentalium* al microscopio Elettronico. *Atti della Accademia Nazionale dei Lincei. Classe di Scienze Fisiche, Matematiche e Naturali. Rendiconti, Serie 8* 47: 557–560.

———. 1970a. The ultrastructure of *Dentalium* egg at the trefoil stage. *Acta Embryologiae Experimentalis* 1970: 31–43.

———. 1970b. The ultrastructure of the ripe oocyte of *Dentalium*. *Acta Embryologiae Experimentalis* 1970: 255–279.

———. 1971. *Dentalium*. In *Experimental embryology of marine and fresh-water invertebrates*. Edited by G. Reverberi. Amsterdam: North-Holland, pp. 248–264.

———. 1972. The fine structure of the ovaric egg of *Dentalium*. *Acta Embryologiae Experimentalis* 1972: 135–166.

Reynolds, P. D. 1990a. Functional morphology of the perianal sinus and pericardium of *Dentalium rectius* Mollusca: Scaphopoda with a reinterpretation of the scaphopod heart. *American Malacological Bulletin* 7: 137–149.

———. 1990b. Fine structure of the kidney and characterization of secretory products in *Dentalium rectius* Mollusca, Scaphopoda. *Zoomorphology* 110: 53–62.

———. 1992a. Mantle-mediated shell decollation increases posterior aperture size in *Dentalium rectius* Scaphopoda: Dentaliida. *Veliger* 35: 26–35.

———. 1992b. Distribution and ultrastructure of ciliated sensory receptors in the posterior mantle epithelium of *Dentalium rectius* Mollusca, Scaphopoda. *Acta Zoologica* 73: 263–270.

———. 1997. The phylogeny and classification of Scaphopoda Mollusca: an assessment of current resolution and cladistic reanalysis. *Zoologica Scripta* 26: 13–26.

———. 2002. The Scaphopoda. *Advances in Marine Biology* 42: 137–236.

Reynolds, P. D., and Okusu, A. 1999. Phylogenetic relationships among families in the Class Scaphopoda Phylum Mollusca. *Zoological Journal of the Linnean Society* 126: 131–154.

Salvini-Plawen, L. v. 1981. The molluscan digestive system in evolution. *Malacologia* 21: 371–401.

———. 1988. The structure and function of molluscan digestive systems. In *The Mollusca*. Vol. 11, *Form and Function*. Edited by E. R. Trueman and M. R. Clarke. New York: Academic Press, pp. 301–379.

Scarabino, V. 1995. Scaphopoda of the tropical Pacific and Indian Oceans, with description of 3 new genera and 42 new species. *Mémoires du Muséum National d'Histoire Naturelle, Paris* 167: 189–379.

Shimek, R. L. 1988. The functional morphology of scaphopod captacula. *Veliger* 30: 213–221.

———. 1989. Shell morphometrics and systematics: a revision of the slender, shallow-water *Cadulus* of the Northeastern Pacific Scaphopoda: Gadilida. *Veliger* 32: 233–246.

———. 1990. Diet and habitat utilization in a northeastern Pacific Ocean scaphopod assemblage. *American Malacological Bulletin* 7: 147–169.

Shimek, R. L., and Steiner, G. 1997. Scaphopoda. In *Microscopic Anatomy of Invertebrates*. Vol. 6B, *Mollusca II*. Edited by F. W. Harrison and A. J. Kohn. New York: Wiley-Liss, pp. 719–781.

Simroth, H. 1894a. I. Abteilung: Amphineura und Scaphopoda. In *Mollusca*. Edited by H. G. Bronn. Leipzig: Winter, pp. 356–467, pls. 15–22.

———. 1894b. Bemerkungen über dei Morphologie der Scaphopoden. *Zeitschrift für Naturwissenschaften* 67: 239–259.

Skelton, P. W., and Benton, M. J. 1993. Mollusca: Rostroconchia, Scaphopoda and Bivalvia. In *The Fossil Record*. Edited by M. J. Benton. London: Chapman and Hall, pp. 237–263.

Starobogatov, Y. I. 1974. Xenoconchias and their bearing on the phylogeny and systematics of some molluscan classes. *Paleontological Journal* 8: 1–13.

Stasek, C. R., and McWilliams, W. R. 1973. The comparative morphology and evolution of the molluscan mantle edge. *Veliger* 16: 1–19.

Steiner, G. 1991. Observations on the anatomy of the scaphopod mantle and the description of a new family, the Fustiariidae. *American Malacological Bulletin* 9: 1–20.

———. 1992a. The organisation of the pedal musculature and its connection to the dorsoventral musculature in Scaphopoda. *Journal of Molluscan Studies* 58: 181–197.

———. 1992b. Phylogeny and classification of Scaphopoda. *Journal of Molluscan Studies* 58: 385–400.

———. 1993. Spawning behavior of *Pulsellum lofotensis* (M. Sars) and *Cadulus subfusiformis* (M. Sars) (Scaphopoda, Mollusca). *Sarsia* 78: 31–33.

———. 1994. Variations in the number of intestinal loops in Scaphopoda Mollusca. *Marine Ecology* 15: 165–174.

———. 1995. Larval and juvenile shells of four North Atlantic scaphopod species. *American Malacological Bulletin* 11: 87–98.

———. 1998a. Phylogeny of Scaphopoda (Mollusca) in the light of new anatomical data on the Gadilinidae and some Problematica, and a reply to Reynolds. *Zoologica Scripta* 27: 73–82.

———. 1998b. Class Scaphopoda. In *Mollusca: The Southern Synthesis*. Vol. 5. Edited by P. L. Beesley, G. J. B. Ross, and A. Wells. Melbourne: CSIRO Publishing, pp 439–447.

———. 1999. A new genus and species of the family Annulidentaliidae Scaphopoda: Dentaliida and its systematic implications. *Journal of Molluscan Studies* 65: 151–161.

Steiner, G., and Dreyer, H. 2003. Molecular phylogeny of Scaphopoda Mollusca inferred from 18S rDNA sequences: support for a Scaphopoda–Cephalopoda clade. *Zoologica Scripta* 324: 343–356.

Steiner, G., and Kabat, A. R. 2001. Catalogue of supraspecific taxa of Scaphopoda Mollusca. *Zoosystema* 233: 433–460.

———. 2004. Catalog of species-group names of Recent and fossil Scaphopoda Mollusca. *Zoosystema* 264: 549–726.

Steiner G., and Reynolds, P. D. 2003. Molecular systematics of the Scaphopoda. In *Molecular Systematics and Phylogeography of Mollusks*. Edited by C. Lydeard and D. R. Lindberg. Washington, DC: Smithsonian Institution Press, pp. 123–139.

Taib, N. T. 1980. Some observations on living animals of *Dentalium entalis* L. *Journal. College of Science. University of Riyadh* 11: 129–144.

———. 1981a. Gross anatomy of the alimentary canal of *Dentalium entalis* L. Scaphopoda. *Journal. College of Science. University of Riyadh* 12: 139–145.

Van Dongen, C. A. M. 1976a. The development of *Dentalium* with special reference to the polar lobe. 5. Differentiation of the cell pattern in lobeless embryos of *Dentalium vulgare* Da Costa during late larval development. *Proceedings. Koninklijke Nederlandse Akademie van Wetenschappen. Series C. Biological and Medical Sciences* 79: 245–255, 9 figures, 2 tables.

———. 1976b. The development of *Dentalium* with special reference to the polar lobe. 6. Differentiation of the cell pattern in lobeless embryos of *Dentalium vulgare* Da Costa during late larval development. *Proceedings. Koninklijke Nederlandse Akademie van Wetenschappen. Series C. Biological and Medical Sciences* 79: 256–266.

———. 1976c. The development of *Dentalium* with special reference to the polar lobe. 7. Organogenesis and histogenesis in lobeless embryos of *Dentalium vulgare* da Costa as compared to normal development. *Proceedings. Koninklijke Nederlandse Akademie van Wetenschappen. Series C. Biological and Medical Sciences* 79: 454–465, 4 pls.

———. 1977. Mesoderm formation during normal development of *Dentalium dentale*. *Proceedings. Koninklijke Nederlandse Akademie van Wetenschappen. Series C. Biological and Medical Sciences)* 80: 372–376.

Van Dongen, C. A. M., and Geilenkirchen, W. C. M. 1974a. The development of *Dentalium* with special reference to the polar lobe. 1. Division chronology and development of the cell pattern in *Dentalium dentale* Scaphopoda. *Proceedings. Koninklijke Nederlandse Akademie van Wetenschappen. Series C. Biological and Medical Sciences* 77: 57–70.

———. 1974b. The development of *Dentalium* with special reference to the polar lobe. 2. Division chronology and development of the cell pattern in *Dentalium dentale* Scaphopoda. *Proceedings. Koninklijke Nederlandse Akademie van Wetenschappen. Series C. Biological and Medical Sciences* 77: 71–84.

———. 1974c. The development of *Dentalium* with special reference to the polar lobe. 3.

Division chronology and development of the cell pattern in *Dentalium dentale* Scaphopoda. *Proceedings. Koninklijke Nederlandse Akademie van Wetenschappen. Series C. Biological and Medical Sciences* 77: 85–100.

———. 1975. The development of *Dentalium* with special reference to the polar lobe. 4. Division chronology and development of the cell pattern in *Dentalium dentale* after removal of the polar lobe at first cleavage. *Proceedings. Koninklijke Nederlandse Akademie van Wetenschappen. Series C. Biological and Medical Sciences* 78: 358–375.

Verdonk, N. H. 1968a. The effect of removing the polar lobe in centrifuged eggs of *Dentalium*. *Journal of Embryology and Experimental Morphology* 19: 33–42.

———. 1968b. The relation of the two blastomeres to the polar lobe in *Dentalium*. *Journal of Embryology and Experimental Morphology* 20: 101–105.

Verdonk, N. H., Geilenkirchen, W. L. M., and Timmermans, L. P. M. 1971. The localization of morphogenetic factors in uncleaved eggs of *Dentalium*. *Journal of Embryology and Experimental Morphology* 25: 57–63.

Waller, T. R. 1998. Origin of the Molluscan Class Bivalvia and a phylogeny of major groups. In *Bivalves: An Eon of Evolution*. Edited by P. A. Johnston and J. W. Haggart. Calgary: University of Calgary Press, pp. 1–45.

Wanninger, A., and Haszprunar, G. 2001. The expression of an engrailed protein during embryonic shell formation of the tusk-shell, *Antalis entalis* Mollusca, Scaphopoda. *Evolution and Development* 35: 312–321.

———. 2003. The development of the serotonergic and FMRF-amidergic nervous system in *Antalis entalis* Mollusca, Scaphopoda. *Zoomorphology* 122: 77–85.

Wilson, E. B. 1904a. Experimental studies on germinal localization. *Journal of Experimental Zoology* 1: 1–72.

———. 1904b. Experimental studies in germinal localization. II. Experiments on the cleavage-mosaic in *Patella* and *Dentalium*. *Journal of Experimental Zoology* 1: 197–268.

Yancey, T. E. 1973. A new genus of Permian siphonodentalid scaphopods, and its bearing on the origin of the Siphonodentaliidae. *Journal of Paleontology* 47: 1062–1064.

Yochelson, E. L. 1999. Scaphopoda. In *Functional morphology of the Invertebrate Skeleton*. Edited by E. Savazzi. Chichester, UK: John Wiley and Sons, pp 363–367.

———. 2002. Restudy and reassignment of *Dentalium antiquum* Goldfuss, 1841 Middle Devonian. *Palaeontologische Zeitschrift* 762: 297–304.

Yochelson, E. L., and Holland, C. H. 2004. *Dentalium saturni* Goldfuss, 1841 Eifelian: Mollusca: complex issues from a simple fossil. *Palaeontologische Zeitschrift*, 781: 97–102.

Yonge, C. M. 1937. Circulation of water in the mantle cavity of *Dentalium entalis*. *Proceedings of the Malacological Society of London* 22: 333–337.

8

Cephalopoda

Michele K. Nishiguchi and Royal H. Mapes

Cephalopoda is one of the most intriguing and diverse classes of molluscs. Modern forms comprise the octopuses, squids, cuttlefish, and pearly nautilus (Figure 8.1). Cephalopods differ greatly from other molluscs—they are more active, fast-moving, intelligent carnivores, with highly advanced visual and nervous systems that allow them to be competitive and efficient predators. Their ability to sense their surrounding environment and adapt rapidly using camouflage or complex behavioral patterns, which have been observed during courtship, reproduction, and mating, demonstrates how complex these animals have become.

All fossil and modern taxa are marine, with a few found in estuarine habitats (as low as 15 ppt salinity). Fossil forms include the ammonoids, which became extinct 65 million years ago; the nautiloids (both orthoconic and coiled), of which *Nautilus* and *Allonautilus* are the only living descendants; and the Coleoidea, the order that accommodates all other living cephalopods.

Modern cephalopods have gained notoriety through being the subject of myths or science fiction (e.g., Verne 1896) and as an important food source. Importantly, they are used as model systems for a large variety of research studies (Hodgkin and Huyley 1952; Makman and Stefano 1984), including areas such as neurobiology, behavior, physiology, development, symbiosis, and growth (Arnold 1962; Hanlon *et al.* 1990; Gilly and Lucero 1992; Boletzky 2002; Forsythe *et al.* 2002; Boletzky 2003; Lee *et al.* 2003). There are close to 1,000 species of living cephalopods (Nesis 1987) found in all oceans, from polar seas to the tropics, with more in the Indo-West Pacific than elsewhere (Norman 2000). They inhabit a variety of marine ecosystems, including estuarine, benthic, pelagic, and the deep (>1,000 m) ocean. Because of their abundance and availability, they are economically important in many of the large fishing industries of Europe, Asia, Australia, New Zealand, and the Americas. It has been estimated that in the western United States alone, 2.7–3.6 million metric tons of squid, worth US $7 billion, are harvested annually (California Department of Fish and Game 2003). They represent a large percentage of the biomass in the ocean and are important in marine food webs, where they play significant roles as predators (mainly of crustaceans, fishes, and other molluscs) as

FIGURE 8.1. Cephalopod diversity. (A) *Nautilus pompilius* (M. Norman). (B) *Jeletzkya* (R. Johnson and E. Richardson, Jr.). (C) *Vampyroteuthis infernalis* (K. Reisenbichler). (D) *Pachyteuthis* sp. (D. Lindberg). (E) *Eutrephoceras* sp. (D. Lindberg). (F) *Octopus aegina* (M. Norman). (G) *Argonauta nodosa* (M. Nishiguchi). (H) *Euprymna tasmanica* (M. Norman). (I) *Sepioteuthis lessoniana* (M. Norman).

well as prey for other squids, fishes including sharks, seabirds, and marine mammals (Boyle and Boletzky 1996; Clarke 1996; Croxall and Prince 1996; Norbert and Klages 1996).

The cephalopod fossil record spans more than 450 million years, although this record is patchy because organic remains decrease in quality of preservation and information content with increasing time. Additionally, most fossils are only the remains of the hard, more durable skeletal material, which, in cephalopods as in most other molluscs, is calcium carbonate. Occasional cephalopod-bearing Lagerstätten (i.e., fossil bearing rock units with organisms having exceptional preservation, often including tissues and whole organs; Botter *et al.* 2002) have provided information not seen in typical fossils. In addition to the exceptionally

well-preserved cephalopod body fossils from the Konservat Lagerstätten of the Middle Jurassic (Callovian) of England and the Late Jurassic (Tithonian) lithographic limestone of Solenhofen, Germany, fossil cephalopods and their shells are also known from the Lower Carboniferous Bear Gulch Limestone (Hagadorn 2002) in Montana and the Upper Carboniferous Buckhorn Asphalt (Crick and Ottensman 1983; Squires 1973) in southern Oklahoma (both in the United States).

FOSSIL CEPHALOPODS

GROWTH

Like modern *Nautilus* and the coleoids, fossil cephalopods are believed to have determinate growth (i.e., the termination of growth and shell secretion presumably coincided with sexual maturity). Some fossil cephalopods reached large sizes; the largest nautiloid fossil recorded was four meters in length, whereas the largest ammonite was two meters in diameter (Lehmann 1981; Stevens 1988; Nixon and Young 2003). The smallest adult ammonoid shell we are aware of is *Maximites* from the Upper Carboniferous, with maturity attained at about 10 mm diameter (Frest *et al.* 1981).

SHELL MORPHOLOGY AND TERMINOLOGY

Almost always it is only the mineralized shell that is available for study in fossil cephalopods. The terminology for the hard parts of the shell is extensive; a good set of definitions is provided by Teichert (1964b). In general the shell is divided into a chambered phragmocone, used for buoyancy control, and the body chamber, which contained most of the bulk of the animal tissues and organs. In coleoids, the phragmocone and body chamber, if present, are partly or entirely covered by a rostrum or guard if a hard skeleton is present. Variations of these and other features provide the basis of the identification of different taxa. These morphological characters used in fossil cephalopods include the degree of coiling; chamber spacing; siphuncle position; septal neck shape; presence or absence of carbonate deposits within the chambers and, where present, the shape and placement of those deposits; protoconch shape and size; cross-sectional shape of the shell; rate of shell taper; suture pattern formed by the septa on the inside of the shell; mature shell modification of the body chamber that develops at presumed sexual maturity; shell ultrastructure (discussed next), rostrum composition and ultrastructure; guard shape; and numerous other features.

SHELL ULTRASTRUCTURE

The cephalopod shell is composed of aragonite deposited in two ultrastructural forms: nacreous plates (nacre) and prismatic needles (Bandel and Spaeth 1988). The external shell of living nautiloids, like that of most conchiferans, has an outer organic layer, the periostracum, which covers the outer prismatic needle layer and was presumably present in ammonoids and fossil nautiloids. The inner surface of the shell is composed of an aragonitic layer of nacre, as are structures such as septa and septal necks. Other shell layers can be present, depending on the ontogeny of the shell and the position of the outer whorls if the specimen is coiled. In some coleoids, because the mantle tissue covers the exterior of the shell, a layer of prismatic material (sometimes with organic material) is deposited on top of the original prismatic shell layer (the nacreous layer is missing) around the phragmocone or on the dorsal surface of the phragmocone and body chamber, forming the rostrum or guard seen in many fossil coleoids. The coleoid gladius, or proostracum, is composed of organic material or a combination of organic material and aragonite. In the Belemnoidea the rostrum can be composed of calcite, aragonite, and organic material (Bandel and Spaeth 1988).

Aragonite, the dominant building material of cephalopod shells, is an unstable mineral and can easily alter to calcite. The oldest unaltered cephalopod shell is from the Lower Carboniferous of Scotland (Hallam and O'Hara 1962). In the United States, the Upper Carboniferous Buckhorn Asphalt (Squires 1973) contains much unaltered cephalopod shell material (e.g., Kulicki *et al.* 2002). Other unaltered cephalopod shell occurrences are rare in the Paleozoic, but they are more common in younger units in the Mesozoic and Cenozoic.

BUOYANCY AND EQUILIBRIUM

The evolution of the morphology of the Cephalopoda was controlled, in large part, by the problems of shell equilibrium and the maintenance of neutral buoyancy while swimming, as described in the preceding section. Buoyancy problems persist despite most modern cephalopods lacking an external or internal shell (Figure 8.2).

Teichert (1988: table 1) listed 14 different mechanisms that modern and fossil cephalopods

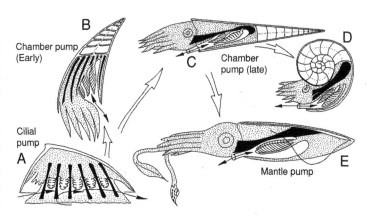

FIGURE 8.2. Diagram showing possible evolutionary directions in the methods of achieving more efficient respiratory currents in the history of the Cephalopoda from a hypothetical ancestor (A), through nautiloids (B–D) to modern coleoids (E) (from House 1988). The "chamber pump" alludes to the presence of a muscular hyponome or funnel.

have used to regulate buoyancy and equilibrium. Some of the most important are truncation of the posterior part of the phragmocone; endosiphuncular deposits in the phragmocone; cameral deposits in the phragmocone; replacing liquid-filling of the chambers with gas; shifting the gas-filled chambers to a position over the body chamber by overlap; shifting the liquid- and gas-filled chambers over the body chamber by coiling; using lighter-density chemicals in special tissues to reduce density; changing the shape or reducing the size and complexity of the phragmocone; thinning of the shell, septa, ornament, or other parts to reduce overall shell weight; construction of a rostrum on the phragmocone; and changing the configuration of the chamber by coiling the shell (Figure 8.2).

SEXUAL DIMORPHISM AND MATURITY

Many fossil cephalopods exhibit changes in shell morphology during growth, especially as the animal approaches and attains full size at maturity. Sexual dimorphism can be expressed by differences in shell size or modifications (Davis *et al.* 1996). Most research on maturity and sexual dimorphism in fossil cephalopods has been done on the Ammonoidea (see Davis *et al.* 1996 for references). Maturity and sexual dimorphism in fossil nautiloids, despite the numerous observations on living nautiloids (Collins and Ward 1987; Ward 1987), have received only modest study. This is also the case for the fossil coleoids.

From the Devonian (416–359 Mya) through the Cretaceous there are numerous cases in which mature ammonoid shells with nearly identical morphologies except for size are found in the same deposits; such cases are usually interpreted as different sexes rather than different species. Other shell morphology changes that have also been used as identifying maturity and sexual dimorphic characteristics include modification of the opening of aperture; septal approximation (i.e., septa becoming closer together); changes in the rate of coiling, ornament or in the cross-sectional shape of the body chamber; simplification of the suture pattern formed by the septa on the inside of the shell; and development of muscle scars (see Davis *et al.* 1996 for a more complete discussion). Additionally, there is approximation of transverse color bands and different patterns on the shells of some otherwise morphologically identical taxa that were presumably mature Triassic ammonoids (Mapes 1987a), which suggests that these specimens are different sexes.

INK

Buckland first described fossil ink from a Jurassic coleoid in 1836. Since then only modest numbers of coleoids from the Mesozoic have been recorded that contain ink (Doguzhaeva *et al.* 2004a) because the preservation of such material requires exceptional geologic conditions. The location and color of black or brown masses in the body chambers of fossil cephalopods is suggestive that the material could be fossil ink; however, proving that the material is ink has, until recently, been impossible without destroying the specimen, because identification of the melanin-based ink was previously dependent on destructive chemical analysis. The discovery of a new method of identifying fossil ink was made by Doguzhaeva et al. (2004a). Doguzhaeva independently discovered that ink from a Jurassic coleoid had a globular ultrastructure when viewed under the scanning electron microscope (SEM) at high magnification ($\times$10,000 to $\times$20,000) like that of modern coleoid ink from squid, octopus, and cuttlefish. Similar results were obtained on ink from Carboniferous coleoids (Doguzhaeva et al. 2004a; Doguzhaeva et al., in press, a). To date the oldest confirmed coleoid ink is from the Upper Carboniferous in the United States (Oklahoma and Illinois), indicating that the use of ink as a method of predator avoidance has been long utilized (Doguzhaeva et al. 2004a; Doguzhaeva et al., in press, a).

Although only some fossil coleoids are known to have ink, this is not surprising given the rarity of suitable preservation. It is possessed by most, but not all, modern coleoids, so some extinct coleoid taxa may also have not possessed it. No nautiloids are known to have ink, but there are unconfirmed reports that some Ammonoidea did possess it (Lehmann 1967; Doguzhaeva et al. 2004a). However, Lehmann (1988) reinterpreted his original material and concluded that his 1967 report was in error. Later Doguzhaeva et al. (2004b; in press, b) analyzed a Jurassic ammonoid using SEM techniques and identified preserved mantle tissue and the possible presence of ink. Thus, the debate as to whether some ammonoids had ink remains unresolved.

BEAK AND RADULAE

In general details, the beak (or mandibles, a modified jaw) and radulae of fossil cephalopods are known for most geological time periods, with many reports back to the Carboniferous (Mapes 1987). There are few reports of Devonian beaks

and none of radulae, and no reports of either structure prior to the Devonian. Reported occurrences of beaks are much more common than those of radulae; most are from ammonoids (Kennedy et al. 2002), with only a few nautiloid jaws having been recovered (Mapes 1987; Müller 1974). Jaws, in general, appear to be moderately conservative structures in regards to evolutionary innovations with all known Upper Paleozoic examples being chitinous. In the Mesozoic, some ammonoids replaced the chitinous lower jaw with two massive calcareous plates. The function of these plates is debated, with suggestions that either they functioned only as lower jaws or they had a dual function by acting as a lower jaw and as a protective operculum, equivalent to the hood on modern *Nautilus* (Lehmann 1981; Morton 1981; Seilacher 1993; Kennedy et al. 2002; Tanabe and Landman 2002). Radulae are known from several species of ammonoids (e.g., Saunders and Richardson 1979; Nixon 1996; and their citations) and coleoids (e.g., Saunders and Richardson 1979). These structures are similar to those found in extant coleoids. In general it is possible to separate nautiloid, ammonoid, and coleoid mandibles and radulae by their morphologies; however, it is seldom possible to separate them on a generic or specific level.

TISSUES AND ORGANS

Only a few reports have been published on the organs and soft tissues of ammonoids and nautiloids, with most being from fossil coleoids. Impressions of tentacle-bearing bodies assigned to octopods have been described from the Cretaceous and Jurassic of the Middle East and Europe (Kluessendorf and Doyle 2000; Haas 2003; Fuchs et al. 2003). The oldest impression of soft body tissues we are aware of is from the Mazon Creek Lagerstätte (Upper Carboniferous) in Illinois. This deposit has yielded several important coleoids. One is *Jeletzkya*, the famous ten-armed impression of a coleoid with arm hooks but without a well-preserved phragmocone (Figure 8.1). Another is an octopus-like form described as *Pohlsepia*, which has eight arms, two modified arms, a poorly defined head (including eyes, funnel, and beaks with a radula preserved between them) and fins; there is no evidence of any internal or external shell associated with the fossil. Other coleoids from this Lagerstätte are also known (Saunders and Richardson 1979; Allison 1987; Doguzhaeva *et al.*, in press, a). However, there are numerous reports of Mesozoic coleoids with preserved soft parts including arms with arm hooks, mantle tissue, gills, beaks, and radulae (e.g., Naef 1922; Doguzhaeva *et al.* 2002a; and citations therein). Gills are known only from Mesozoic coleoids (Bandel and Leich 1986; Mehl 1990).

Internal organs such as the stomach, crop, intestines, and circulatory system have been rarely reported in fossil cephalopods. In all reported cases involving the digestive system, the actual organs are not preserved. Instead, the undigested calcareous, chitinous, or phosphatic skeletal remains of meals are found clustered in specific areas of the body and are interpreted to mark the positions of the crop, stomach, or the intestine. In ammonoids, Lehman (1981) reported crinoid fragments, aptychi (lower beak/operculum) from smaller ammonoids, foraminiferans, and ostracods (Lehmann 1981; Nixon 1988). In orthoconic nautiloids, Quinn (1977) and Mapes and Dalton (2002) reported that the ammonoid clusters surrounding the body of large (1–3 meters in length) actinoceratid nautiloids from the Lower Carboniferous of Arkansas were the stomach contents of the nautiloid (Mapes and Dalton 2002). The oldest crop/stomach contents were discovered in an "orthoconic nautiloid" from the Lower Carboniferous Bear Gulch Lagerstätte (Landman and Davis 1988); these contents appear to be mostly composed of macerated fish scales. Subsequent study has shown that this specimen is a new coleoid (Mapes *et al.* 2007).

The circulatory system in fossil cephalopods is known only from impressions or grooves on the internal parts of mineralized structures such as the shell, cameral deposits, or rostrum and the siphuncle in fossil nautiloids

and ammonoids, which contained arteries and veins. In Permian ammonoids from Nevada, segments of the siphuncular tissues replaced by phosphate have been discovered and described (Tanabe *et al.* 2000).

COLOR PATTERNS

It is unknown whether fossil cephalopods had chromatophores like those seen in the mantle tissue of many modern coleoids. *Nautilus* and *Allonautilus* have reddish-colored transverse bands across the shell, and the tissue of the hood has reddish markings, which are not capable of changing in life (in contrast to modern coleoids). While it is unknown whether the head-foot region of ammonites and extinct nautiloids had similar coloration to modern nautiloids, the shells of different fossil cephalopod taxa exhibited a variety of different color patterns.

The actual color of the patterns on the fossil ammonoid and nautiloid shells is unknown because fossilization has destroyed the pigments, leaving patterns of gray in different shades (Teichert 1964a; Mapes and Evans 1995; Mapes and Davis 1996; Gardner and Mapes 2001); such patterns have not been discovered on fossil coleoid shells. Patterns include zigzag lines, transverse bands, longitudinal bands, and uniformly colored shells. Interestingly, while modern *Nautilus* and *Allonautilus* have transverse bands that do not follow the growth lines, all transverse bands on ammonoid forms do. Additionally, at maturity, in *Nautilus* and *Allonautilus* the transverse bands are confined to the umbilical sides of the shell, leaving the lateral and ventral sides of the shell uniform creamy white. In mature shells of ammonoids with transverse color bands, the color pattern extended entirely across the shell and covered the body chamber to the edge of the aperture at maturity.

In addition to the color patterns on nautiloids and ammonoids, a different kind of coloration was present in some fossils. This is not the product of a pigment but instead the result of scattering and differential absorption of light by the ultrastructure of the outer layers of the shell. Unaltered ammonoid shells from Poland,

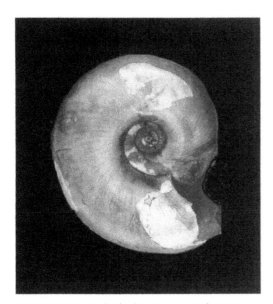

FIGURE 8.3. Longitudinal color pattern on *Cadoceras* sp., an ammonoid from Jurassic sediments near Luków, Poland. Except in the places where the shell is broken (white areas), the conch also exhibits an overall dark (reddish) iridescence; a longitudinal band of lighter iridescence can be seen on the darker (red) background around the umbilical region.

Russia, and Madagascar (all Mesozoic; precise ultrastructural and stratigraphic details are under study; Mapes, personal observation) with narrow longitudinal bands of red to reddish green on a darker red background have been observed (Figure 8.3). Such external colors would appear dark brown or dark gray at water depths below 10 meters. Interior shell layers appear to exhibit only deeper blue to purple colors; however, these colors would not be exposed during normal living conditions. The use of different refractions and reflections at different wavelengths of light to produce color patterns is not known in modern cephalopods, and indeed, we are not aware of any cases in modern taxa belonging to the phylum.

BIOLOGY OF LIVING CEPHALOPODS

Living cephalopods are divided into two major groups usually treated as subclasses: the Nautiloidea, containing *Nautilus* and *Allonautilus*, and the Coleoidea with all remaining taxa. Characteristics that distinguish these two subclasses are the presence of an external, many-chambered shell and multiple (60–90) suckerless tentacles in nautiloids, whereas coleoids have reduced (*Spirula*), internalized, or no shell, bearing eight to ten prehensile suckered appendages (arms and tentacles). Coleoids have a modified seven-element radula and a chitinous beak (similar to many ammonoids, see previous discussion), whereas nautiloids have a 13-element radula with a chitin and calcium carbonate beak. The nautiloid funnel has two separate folds, and they have two pairs of ctenidia (tetrabranchiate) and nephridia. Coleoids have a closed single tube for a funnel and one pair of ctenidia (dibranchiate) and nephridia. Coleoid eyes are more complex than the pinhole-camera eyes of nautiloids, and some even contain a cornea and lens (much like vertebrate eyes, an excellent example of convergent evolution). Modern nautiloids have no ink sac or chromatophores, whereas most coleoids have both. *Nautilus* has a simple type of statocyst, an oval cavity completely lined with hair cells (Young 1965). Coleoids have two types of statocysts: the "octobrachian" statocyst, which contains a spherelike sac with one gravity receptor system and an angular acceleration receptor system subdivided into nine segments (Young 1960; Budelmann *et al.* 1997), or the "decabrachian" statocyst, which is irregularly shaped with three gravity receptor systems and an angular acceleration receptor system subdivided into four segments (Budelmann *et al.* 1997). Both octobrachian and decabrachian statocysts are species specific in size and levels of organization and have been used in phylogenetic comparisons (Young 1984).

EGG LAYING

Laying of fertile eggs by modern *Nautilus* has been observed only in aquariums (Arnold 1987; Uchiyama and Tanabe 1999). Under these conditions single large (about 25 mm in diameter), yolky eggs are attached to a hard substrate. Embryo maturation takes 269 to 362 days, and the hatchling is a miniature of a full-sized adult.

Egg laying for modern coleoids is usually in two basic modes: those whose life histories are restricted to coastal and shelf habitats and those that are pelagic (Boyle and Rodhouse 2005). Cephalopods such as cuttlefish, most octopuses, and loliginid squids lay eggs that are attached to some type of substrate such as coral or kelp. Most of these squids ensheath their eggs with material from the nidamental glands and usually attach the eggs in areas where they are inconspicuous. Octopus eggs are more individually attached by a short stalk into strings or festoons, which are then attached to the substrate (Boyle and Rodhouse 2005). All octopod species brood their eggs, no matter whether or not they are attaching them to a specific substratum, and this characteristic has been used to delineate species complexes (i.e., *Octopus*; Boletzky, personal communication). For pelagic species, some ommastrephid squids extrude their eggs in a gelatinous mass, and some, such as the gonatid squids, are known to carry their eggs (Seibel *et al.* 2000). Young hatch directly from this gelatinous mass and are then fledged into the water column as paralarvae (see below). Other pelagic species, such as the Enoploteuthidae, release eggs singly into the water column (Young 1985), and the juvenile squids are then hatched in a pelagic state.

EARLY DEVELOPMENT

All cephalopods are direct developers; there is no larval stage (although juvenile cephalopods are referred to as paralarvae) or metamorphosis. This direct development is linked with the advent of swimming using jet propulsion (Boletzky 2003). Because coleoids lack an external shell, the development into a free-living, jet-propelled swimmer may have selected for direct development, rather than keeping the shell that would hinder this type of behavior. Cephalopod eggs are laden with yolk and vary in developmental time from a few days (O'Dor and Dawe 1998) to more than one year (Voight and Grehan 2000). The development time is dependent not only on species but on temperature (Boletzky 2003). Juvenile develop-ment is also variable with respect to arm crown morphologies among species. Hatchlings that develop from species with large eggs generally have fully developed arms, whereas species that produce smaller eggs have juveniles with short arms and tentacles with fewer suckers that cannot be immediately used like adult tentacles to capture prey (Boletzky 2003). Therefore, these juveniles use their arms to capture prey until the tentacles and suckers fully develop.

In *Nautilus*, the hatchlings have a shell diameter of about 25 mm. At maturity, shell diameters can exceed 210 mm diameter in the largest species, with estimates ranging from 2.5 to 15 years for *Nautilus* to achieve maturity depending on the species and the method of study (Landman and Cochran 1987). In *Nautilus* the sexes are separate and laying the single eggs continues for the life span of the female, which can be years or perhaps decades. In modern coleoids sexes are also separate and, while most species spawn their eggs only once, there are species that have several spawning events throughout their adult life span.

GROWTH

Modern cephalopods are the largest known invertebrates, reaching total lengths (mantle plus tentacles) ranging from 15 mm in pygmy squids (*Idiosepius*) to 20 m (*Architeuthis*). Some cephalopods grow rapidly, depending on temperature and age to senescence (Pecl and Moltschaniwskyj 1997; Semmens and Moltschaniwskyj 2000; Jackson and Moltschaniwskyj 2001; Moltschaniwskyj 2004). Often, in one season, there are two separate breeding cohorts that reach sexual maturity within a few months of one another (Moltschaniwskyj 2004), demonstrating that modern cephalopods can accelerate their growth rates depending on both environmental and physiological constraints.

MOVEMENT

Early cephalopods had a ventral apertural sinus that enabled jet propulsion for movement (House 1988). In all living cephalopods, the funnel enables movement through jet propulsion.

In squids and cuttlefishes, the funnel tube and collar close the entire length of the mantle. In some oegopsid squids (mainly Ommastrephidae), the funnel tube sits in a depression (funnel groove) mainly located in the lower head region. Certain aspects of the funnel groove (folds or ridges within the groove) can be used to distinguish subfamilies from one another. In myopsids, the groove is more like a depression and not as well defined as in Oegopsida, whereas in some octopuses the funnel tube is partially or fully embedded in the head tissues and is free only at the head end. In many of the fast-swimming squids, the anterior part of the funnel groove is bordered in front of the funnel aperture by a cuticular ridge. A pair of anterior adductor muscles are attached to the funnel anteriorly, and the posterior-lateral sides of the funnel are connected to the sac of the gladius or to the dorsal side of the mantle cavity by much stronger retractor muscles. For example, in the Jumbo Flying Squid, *Dosidicus,* the swimming velocity can increase from 4 to 14 knots (2–7.2 m/s) between becoming airborne and re-entering the sea as a result of the rapid expulsion of water from the mantle cavity (Packard 1985; House 1988). In most squids, cuttlefishes, and vampyromorphs, there is a funnel valve that supposedly strengthens or supports the funnel wall when the squid is swimming. There is no valve in octopuses or in some oceanic squids (Cranchiidae). The valve is also used to change the shape of the ink cloud when it is ejected through the funnel. Along with this, a funnel organ (Verrill's organ) is located on the dorsal side of the funnel just behind the valve. Its function is unknown, but it has been hypothesized that the mucus it secretes helps in maintaining the structure of the funnel or decreasing turbulence while swimming. The shape of this organ is species specific, and thus it is another useful character for taxonomy and systematics.

Fins are also used for locomotion in squids, even those that are not good swimmers. There are mainly five types of fin shapes, which can be used as a good taxonomic character for delineating families. These include marginal or fringing, rhomboidal or heart-shaped (cordate), kidney-shaped (reniform), round, and tonguelike fins (Nesis 1980). Smaller fins are usually less adapted to speed than larger ones but are better at maneuverability. It was previously thought that fins could be a good synapomorphy for Coleoidea (Young et al. 1998), and this may be true for modern forms, but they are preserved in very few fossils, so their distribution in fossil coleoids is uncertain.

FEEDING MODES

The mouth (buccal aperture) is highly adapted for predatory behavior. In the Devonian and Carboniferous and in most later fossil forms the mandibles (jaws) are a horny beak divided into an upper (shorter) and lower (longer) halves. In modern cephalopods, both mandibles contain pharyngeal plates and a frontal plate covering the lateral walls. These plates fuse, forming the rostrum and the cutting edges of the mandibles. Relative size and structure of the mandibles, as well as indentations along the cutting edge, differ greatly among cephalopods and are useful characters for the study of cephalopod evolution. In dibranchiate cephalopods, the radula consists of seven longitudinal rows of teeth, with a median (= central or rhachidian) tooth and the first, second, and third lateral teeth. Marginal plates may be present on either side but are not present in all species (they are absent in sepiolids and most oegopsids but are more developed in octopuses). Radulae are highly reduced in *Spirula* and finned octopuses. In other squids, such as *Gonatus*, the radula consists of five rows of teeth and the first lateral tooth is absent. Radulae have also been useful in determining differences among closely related species (Lindgren et al. 2005). The radula in most coleoids is used for gripping pieces of food torn by the beak and transferring it to the pharynx.

MODIFICATION OF ARMS AND TENTACLES

Cephalopods have the most modified foot among molluscs. The large muscular appendage has been divided into several appendages that can be used to manipulate and capture prey or be used for mating. The main characteristic that separates the Decabrachia from the Octobrachia is the number of arms, with the squids and cuttlefishes having ten arms, of which two are modified as retractable tentacles, and the octopuses having eight. There is debate on whether vampyromorphs are sister to the decabrachians (with their two reduced arms considered to be "squid-like") or whether they are more like octopuses. Arms and tentacles are attached to the outer lip surrounding the mouth by the buccal membrane. There are initially eight lappets or small triangular flaps that support the buccal membrane during development, at which point the first and sometimes fourth arms may merge together. Where the buccal lappet supports are attached is of great taxonomic significance for decabrachians, since in eight families of oegopsids, the supports are attached to the fourth arms dorsally, while in the remaining families of Oegopsida and Myopsida, they are attached ventrally (Nesis 1980).

Tentacles are always positioned between the third and fourth arms and not connected to the buccal funnel. In cuttlefishes, the tentacles are very elastic and can be retracted into special pockets, whereas in other squids the tentacles can be retracted slightly, but not entirely. Although all squids have tentacles (synapomorphic character), they may lose this feature as adults (as in the Octopoteuthidae and some Gonatidae), or obtain them later during juvenile development (Idiosepiidae). Arms have also been modified for functionality; not only are they used for capturing prey, but in the octopod *Argonauta* they are used for building a pseudo-shell (the "paper nautilus") to brood eggs.

The presence of arms or tentacles around the mouth is a synapomorphic character of the Cephalopoda. The oldest tentacle preservation is *Pohlsepia* from the Carboniferous period (Mazon Creek) (Klussendorf and Doyle 2000) and *Jeletzkya* (Johnson and Richardson 1968). No ammonoid or nautiloid tentacles or impressions made by tentacles are currently known. Suckers on the arms of squids and cuttlefishes are stalked and are hemispherical in shape, and the stalks can either be long or short, thin or thick, with outgrowths in the middle forming a cup (Nesis 1980). They are often arranged in two rows, but can also be found in four rows, particularly in the Sepiidae, Sepiolidae, and Gonatidae. The suckers can also increase in number toward the base of the arm and enlarged suckers may be found in both or one of the sexes. Suckers can also be absent or rearranged into one row. Squids with hectocotylzed arms (used for passing spermatophores to the female) have modifications of the suckers, with the larger ones found in the middle or near the tips and growing smaller towards the ends of the arms. Suckers also differ on the tentacular club and can vary in the number of rows (four and up to fifty, in some Sepiolidae and Mastigoteuthidae). Larger suckers are found in the center of the club (manus) and are smaller at the ends of the club (dactylus). Most suckers of squids and cuttlefishes are armed with horny rings but are smooth in Crahnchiidae or armed with teeth in some Loliginidae. Suckers are also modified into hooks (some Gonatidae and all Enoploteuthidae) and are either on median parts of the arms or only on a few arms. Hooks always develop from suckers by uneven elongation, bending, and longitudinal folding of the distal edge of the ring or by elongation of one or two teeth during the later stages of ontogenesis. Arms of octopuses and vampyromorphs have only suckers with neither stalks nor horny rings. Finned octopods and vampyromorphs have suckers flanked by a small row of cirri, which alternate between the suckers.

NERVOUS SYSTEM

Cephalopods have a more developed nervous system than any other invertebrate, with a highly developed brain and optic lobes (Budelmann *et al.* 1997). It has been previously noted that certain octopuses are capable of learning by observation (Fiorito and Scotto 1992) as well as by testing (Wells and Wells 1977). Of importance to neural biology was the discovery and subsequent use of the giant axon in *Loligo vulgaris* (J. Z. Young 1936, 1977). In addition, a highly advanced visual system, particularly in the coleoids, gives the capacity to recognize surroundings. *Nautilus* eyes are of the pinhole type, in which the lumen is filled with seawater. Both House (1988) and Lehmann (1985) speculated that this was probably the same for the ancestral Ammonoidea. Coleoids have a lens and a cornea, which improve their vision and is coupled with the ability to camouflage and signal using their chromatophores. This feature probably evolved within the endocochleate (having an internalized shell) Coleoidea, in which the development of chromatophores led to a change in defensive and offensive strategies for seeking or hiding from prey and predators (House 1988).

Not only is the visual system used to match surrounding habitats, but other adaptive features such as light organs, used for counter-illumination and signaling, exist in a number of squids and a few octopuses (Young and Roper 1976; R. E. Young 1977; Young *et al.* 1979b; Jones and Nishiguchi 2004; Nishiguchi *et al.* 2004). Cephalopods have either autogenic bioluminescence (luminescence produced by themselves with a eukaryotic luciferase) or bacteriogenic light organs (luminescence produced by symbiotic bacteria with prokaryotic luciferase). Squids are capable of visually detecting changes in light attenuation and can thereby mask their shadow so that predators or prey below cannot easily see them. This is especially important at night when many cephalopods are active and the only potential illumination is moonlight (Young *et al.* 1980; Jones and Nishiguchi 2004).

BEHAVIOR

Behavior is also presently used as a way to distinguish not only species, but populations as well (Hanlon 1988; Packard 1988; Hanlon and Messenger 1996). There is a multitude of behaviors that have been documented by cephalopod researchers and are related to the complexity of the nervous system and the brain. Some of these include camouflage and body

patterning, defense, communication, reproduction, and interspecies interactions (Hanlon and Messenger 1996). These intricate behaviors are not found in any of the other molluscan classes and have been key features that render the Cephalopoda unique within the phylum. Most of their behavioral features are ecologically similar to those of modern fishes, and probably the evolution of modern teleosts was largely responsible for the decline of cephalopods since the Mesozoic (Aronson 1991). Supposedly, many behavioral features were selected when fishes and reptiles living in coastal waters forced ectocochleate (having an external shell) cephalopods into deeper habitats (Packard 1972), but this is now refuted for modern cephalopods, with the exception of *Nautilus* (Aronson 1991).

Given such selective pressures, cephalopods have evolved a variety of behavioral traits linked to many of the senses (mechano-, chemo-, and photoreceptors) that allow the effectors (such as muscle, chromatophores, reflectors, photophores and the ink sac) to work in response to external stimuli. The brain, which has been mapped in a few key species (*Nautilus*, *Octopus*, and *Sepia*), forms a much more developed central nervous system and concentrated ganglia than in any other mollusc (Budelmann *et al.* 1997; Young 1988a), enabling the animal to efficiently organize the information received from all the sensory structures. This leads to the ability of matching habitat complexity, detecting and capturing prey, defense against predators (crypsis, flight, aggression), communication, learning, as well as complex mating rituals (which may be species specific) and subsequent brooding of eggs prior to hatching (Hanlon and Messenger 1996). Many of these behaviors can also be specific to species or even populations of individuals, which may then lead to subsequent sympatry and genetic differentiation between populations.

SHELL REDUCTION AND LOSS

Shell reduction (i.e., the change from having an external shell to having a reduced internal shell or no internal shell or supporting structure) was accomplished by the Carboniferous (*Pohlsepia*), and, based on the fossil record, the reduction did not appear to have a major impact on cephalopod evolution (Kluessendorf and Doyle 2000). Because of *Pohlsepia* (Kluessendorf and Doyle 2000) from the Upper Carboniferous, it is probable that shell reduction in the Coleoidea may have occurred at different times within different coleoid lineages. For example, a single evolutionary event giving rise to the Decabrachia as presented by Hass (2003) does not appear to be resolvable using a single lineage and the known fossil record.

In the *Pohlsepia* lineage, which is not well understood at this time, the complete loss of the internal chambered phragmocone, rostrum, and any kind of supporting structure (such as a pen), occurred much earlier in the overall evolution of the Coleoidea than had been expected given knowledge of the fossil record prior to 2000. With recognition of the early geologic age of this evolutionary internal shell reduction condition, it would appear that loss of buoyancy and equilibrium control by gas-filled chambers in the phragmocone or even a simple mineralized supporting structure did not provide a major evolutionary advantage to the coleoids. However, the coleoids did manage to survive three major extinction events (the Permian-Triassic, the Triassic-Jurassic, and the Cretaceous-Cenozoic), whereas the externally shelled ammonoids, which are the most abundant cephalopod group from the late Paleozoic to the end of the Mesozoic, survived only two of these extinctions. Interestingly, the ancestors of modern *Nautilus* and *Allonautilus* with their external shells managed to survive all three extinction events. The observation that the ammonoids were more abundant than the coleoids and that they were the dominant cephalopod group in the Late Paleozoic and Mesozoic is supported by the numerous marine Lagerstätten and other marine deposits around the world that preserve fossil cephalopods. Based on a nonquantifiable impression of collecting fossil cephalopods for decades and the understanding that externally shelled cephalopods were more

easily fossilized, coleoid fossils are considered to be very rare through the Late Paleozoic and Mesozoic; whereas, ammonoids are relatively common and are the dominant (in terms of abundance, diversity, and geographic distribution) fossil cephalopod group.

EVOLUTION AND PHYLOGENY OF CEPHALOPODS

SISTER GROUP RELATIONSHIPS

The two main competing hypotheses for cephalopod relationships with other molluscan classes have considered the placement of Cephalopoda with Gastropoda (as the Cyrtosoma) within the Conchifera (Haszprunar 1996; Salvini-Plawen and Steiner 1996; Haszprunar 2000) or whether the Scaphopoda are sister to the gastropod+cephalopod clade (Lindberg and Ponder 1996; Haszprunar 2000). Most recently, cephalopods have been widely accepted as sister to both Scaphopoda and Gastropoda based on molecular and morphological data (Waller 1998; Giribet and Wheeler 2002; Steiner and Dreyer 2003; Giribet *et al.* 2006).

FOSSIL GROUPS

In fossil cephalopods, the general higher-level details of the overall evolution and phylogeny of many of the nautiloid taxa (variably treated as orders, superorders, or even subclasses; Figure 8.4) appear to be moderately stable, though little research has been done since the 1970s (Orlov 1962; Teichert 1964a, 1967). Although most of the main details of cephalopod evolutionary trends appear to be well established, presumed ancestral links between many higher groups remain uncertain. Ammonoid evolution is known to be very complex because these animals have durable shells that quickly evolved complex features. These animals also had a worldwide distribution over a long span of geologic time, and they have received much more attention over the past two centuries because they are very useful in determining the age of different rock units (Moore 1957, 1964;

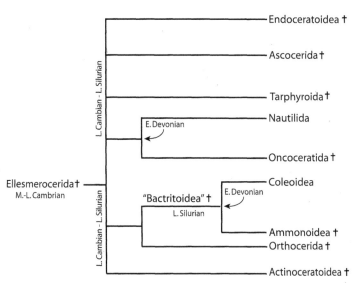

FIGURE 8.4. Generalized cephalopod phylogeny through time showing the general evolutionary trends from an Ellesmerocerida ancestor in Middle to Late Cambrian time to the extant Nautilida and Coleoidea in today's oceans. The ancestral molluscan that gave rise to the Ellesmerocerida was probably a monoplacophoran. The Ellesmerocerida lineage developed into a number of distinct subclasses, superorders, and orders, including the Orthocerida, between Late Cambrian and Late Silurian time. Only two genera with external shells (*Allonautilus* and *Nautilus*), which belong to the Nautilida, survive today. The Bactritoidea arose out of the Orthocerida in the Early Devonian. Slightly later in the Early Devonian, the Bactritoidea gave rise to the Ammonoidea, which became extinct at the end of the Cretaceous. Additionally, the Bactritoidea gave rise to the Coleoidea, which is the only other cephalopod group to survive today. The timing of the evolutionary origin of the Coleoidea from the Bactritoidea is presently unknown, with the majority of authorities suggesting a Devonian or Early Carboniferous timing for this significant evolutionary event. In order to clearly show the extant nautiloid and coleoid placements on the tree, some major groups are omitted from the diagram; all major cephalopod groups and their known geologic ranges are shown in Figure 8.7. Extinct taxa are designated with a dagger (†). E. = Early; L. = Late.

Orlov 1962; House 1981; Becker and Kullmann 1996; Page 1996; Wiedmann and Kullmann 1996; Kullmann 2002) (Figure 8.5). Even though the Nautiloidea have similar characteristics, the evolution of this group has received only modest attention because they evolved more slowly (Woodruff *et al.* 1987; Wray *et al.* 1995; Ward and Saunders 1997). To our knowledge only segments of the Ammonoidea phylogeny have been evaluated using cladistic analysis.

In contrast to the research pattern seen in the Nautiloidea, new coleoid material has been described during the past ten years that has significantly altered parts of the classifications proposed by pre-1995 coleoid researchers (e.g., Naef 1922; Jeletzky 1966; Donovan 1977). New

classifications, including some using cladistic methodologies, have been proposed (Engeser 1996; Pignatti and Mariotti 1996; Mariotti and Pignatti 1999; Haas 2002) (Figure 8.6). Despite this, the early phylogeny of the Coleoidea remains poorly understood.

The oldest cephalopod fossils are from the lower and middle part of the Yenchou Member of the Fengshan Formation (late Middle Cambrian) of northeast China (Chen and Teichert 1983). Only the genus *Plectronoceras* has been recovered from the lower part of this unit. This genus represents what is probably the best cephalopod archetype and presumably arose from ancient monoplacophorans with longiconic or breviconic shells similar to the

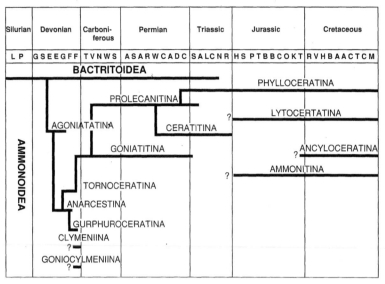

Silurian	Devonian	Carboni-ferous	Permian	Triassic	Jurassic	Cretaceous
L P	G S E E G F F	T V N W S	A S A R W C A D C	S A L C N R	H S P T B B C O K T	R V H B A A C T C M

BACTRITOIDEA

AMMONOIDEA

PHYLLOCERATINA

PROLECANITINA

AGONIATATINA

CERATITINA

LYTOCERATINA

GONIATITINA

ANCYLOCERATINA

AMMONITINA

TORNOCERATINA

ANARCESTINA

GURPHUROCERATINA

CLYMENIINA

GONIOCYLMENIINA

FIGURE 8.5. The evolutionary phylogeny of the Ammonoidea at the ordinal and subordinal level as derived from the Bactritoidea based on published data by Gordon 1966; Donovan *et al.* 1981; Glenister and Furnish 1981; Kullmann 1981; Tozer 1981; Wright 1981; House 1981, 1993; Wiedmann and Kullmann 1988, 1996; Becker and Kullmann 1996; Page 1996; and Kröger and Mapes 2007. Based on these works, there has been general agreement on most the phylogenetic relationships among most authorities, but some differences are unresolved. The overall relationships between the ammonoid suborders look deceptively simple; however, note that five of the 13 ammonoid suborders begin their origination with a question mark, indicating that the ancestral origins are presently unknown. There are also many unresolved ancestor and descendant relationships at the superfamily and family levels. A total number of ammonoid species is not yet available; however, Wiedmann and Kullmann (1996) indicated there are over 3,700 described ammonoid species in the Devonian, Carboniferous, and Permian. Given this impressive diversity and the fact that this cephalopod order has been studied for more than 250 years, it is surprising that subordinal ancestral-descendant relationships within the order are still unknown. This problem is even more complicated and extensive when one attempts to determine the phylogenies of the Bactritoidea, Nautiloidea, Coleoidea, and the other cephalopod orders.

Upper Cambrian *Hypseloconus* (Teichert 1988). During late Yenchou time, almost all nautiloid genera became extinct. This extinction was followed by a large evolutionary radiation seen in the overlying Wanwankou Member of the Fengshan Formation (Teichert 1988). By the Middle Ordovician, the nautiloids had diversified to the greatest morphological diversity recorded in the fossil record for ectocochleate cephalopods (Teichert and Matsumoto 1987). Almost all of the established major clades arose from Ellesmerocerida ancestors. These are the Intejocerida, Endocerida, Actinocerida, Discosorida, Ascocerida, Orthocerida, Barrandeocerida, Nautilida, Tarphycerida, and Oncoceratida (for details of these nautiloid groups see Moore

1964: fig. 7). Two were the ancestors to all the living cephalopods: the Nautilida, which gave rise to modern nautiloids (*Nautilus* and *Allonautilus*), the two living externally shelled cephalopod genera, and the Bactritoidea, which gave rise to the ammonoids and the coleoids (Chen and Teichert 1983) (Figures 8.5, 8.7).

The coleoids are generally thought to consist of eleven ordinal-level taxa, of which six are extinct (House 1988). The extinct groups with internal phragmocones are the Donovaniconida (Doguzhaeva *et al.*, in press, a), Aulacocerida, Phragmoteuthida, Belemnitida, Hematitida and Belemnoteuthida (Jeletzky 1966; Engeser and Reitner 1981; Teichert 1988; Figure 8.6, Table 8.1). The five living orders are the Octopoda, Sepiida,

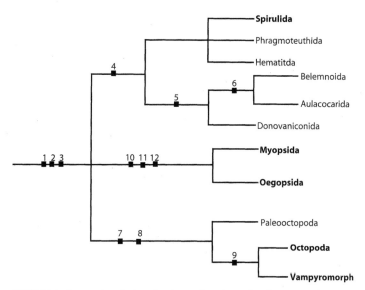

FIGURE 8.6. Proposed evolution of extinct and modern coleoids. Apomorphic characters include: (1) external shell; (2) 10 undifferentiated arms; (3) no suckers; (4) interior phragmocone; (5) arm hooks; (6) closing membrane in first chamber (suppression of prosiphon and cecum); (7) no internal shell; (8) modified arms; (9) presence of suckers; (10) arm 2 modified; (11) suppression of phragmocone; and (12) arm 4 modified. The ancestral bactritoid form probably had apomorphic characters (1), (2), and (3). This analysis evaluates most of the lineages of the Coleoidea. However, many other characters have not yet been evaluated, and we recommend that additional study of these evolutionary pathways is necessary to delineate the phylogeny of the Coleoidea more completely and accurately. Extant taxa are indicated in bold. See also Table 8.1.

Spirulida, Teuthida, and Vampyromorpha. Of these, the best documented in the fossil record is the Sepiida, since they retain all the morphological elements of the belemnitid shell (Jeletzky 1966, 1969). There are no known transitional taxa showing clear intermediate features between any of the higher coleoid groups.

Most characters used for classification of fossil coleoids include the phragmocone, rostrum or guard, and so forth. However, in general the Teuthida lack shelly hard parts but do have a modified proostracum (pen or gladius) made of organic material or a combination of organic and some carbonate material (Teichert 1988). These gladii typically do not fossilize well, and other hard parts such as phragmocones, although calcareous, are also rare as fossils. The Carboniferous coleoid taxon *Pohlsepia*, assigned to Palaeoctopoda, (Kluessendorf and Doyle 2000), lacks hard parts, and reveals the existence of coleoids without shells at that time. However,

from the Middle Carboniferous through the last of the late Paleozoic and through the early Mesozoic there has been no other recovery of coleoid material assignable to the Octopoda. Significantly, below the Middle Carboniferous there are no transitional forms from the primitive bactritoid stock that presumably gave rise to this important coleoid order.

New techniques to identify fossils as coleoids include the SEM analysis of fossil ink (see the section on ink above) and the identification of unique coleoid shell ultrastructures using SEM. For example, a Carboniferous "*Bactrites*" from Texas (Miller 1930) was reinterpreted as a coleoid belonging to the Spirulida (Doguzhaeva *et al.* 1999), extending the range of Spirulida from the Cretaceous (the oldest previously known spirulid) to the Carboniferous. Such discoveries, however, are rare (especially in the Paleozoic and early Mesozoic), given the generally poor fossilization potential of most coleoids.

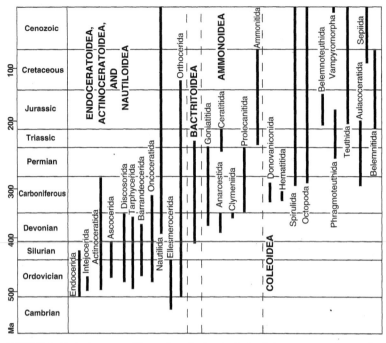

FIGURE 8.7. Geologic ranges of the major extinct and extant subclasses, superorders, and orders of the Cephalopoda (modified from House 1988; note that the terminology as to rank varies considerably in the literature). Some of the ancestor-descendant relationships within the Endoceratoidea, Actinoceratoidea, Nautiloidea, Bactritoidea, and Ammonoidea are known. However, the recent discovery of numerous new fossil coleoids from the Carboniferous has put to question some of the geologic ranges and the proposed origins of some of the fossil coleoid orders. Also, the fact that the origin of the earliest coleoids is unknown remains an unresolved evolutionary problem that complicates the development of a unified phylogeny for the Coleoidea.

The origin of the Coleoidea (Figure 8.4) is presently accepted as being from the Bactritoidea in the Paleozoic (Devonian and/or Carboniferous) (Figure 8.7). The Bactritoidea (Silurian to Triassic) have an egg- or ball-shaped protoconch in which the typical phragmocone of the genus *Bactrites* has a ventral marginal siphuncle with simple disk-shaped septa and an orthoconic (i.e., straight) shell (e.g., Mapes 1979). Primitive coleoids have phragmocones with these same characteristics. *Lobobactrites,* another bactritoidean, is similar in gross morphology to *Bactrites* except that the subspherical protoconch is twice as large in diameter and length as that of *Bactrites. Lobobactrites* is considered the probable ancestor of Ammonoidea because the earliest known ammonoids also have large, egg-shaped protoconchs (Erben 1964; Teichert 1988). The main difference between the Bactritoidea and the Ammonoidea is that all bactritoids have orthoconic shells, whereas those of the most primitive ammonoids are cyrtoconic (slightly curved) shells. By the end of the early Devonian, the ammonoid protoconch had generally begun to reduce in size and the shell became more tightly coiled. By the end of the Devonian, all ammonoids had a small, more or less spherical protoconch, and the shell was tightly coiled.

The ammonoid groups underwent several major extinctions and re-radiations from the early Devonian to the end of the Cretaceous (e.g., House 1988). The first extinction was near the end of the Devonian, the second at the Permo-Cenozoic boundary, and the third at the Triassic-Jurassic boundary. After surviving these three extinctions, ammonites became extinct at the Cretaceous-Cenozoic boundary (Figure 8.5). At

TABLE 8.1
Classification of Paleozoic and Modern Coleoids

Subclass Coleoidea Bather, 1888

Superorder Belemnoida Gray, 1849 (Carboniferous–Cretaceous)
Order Hematitida Doguzhaeva, Mapes, and Mutvei, 2002[a] (Carboniferous)
Family Hematitidae Gustomesov, 1976
Hematites (Flower and Gordon 1959): Upper Mississippian, Lower Eumorphoceras Zone (= Serpukhovan), Utah, Arkansas, United States
Bactritimimus (Flower and Gordon 1959): Upper Mississippian, Lower Eumorphoceras Zone (= Serpukhovan), Arkansas, United States
Paleoconus (Flower and Gordon 1959): Upper Mississippian, Lower Eumorphoceras Zone (= Serpukhovan), Arkansas, United States
Order Phragmoteuthida Jeletzky in Sweet, 1964 (Upper Permian–Jurassic)
Family Phragmoteuthididae Mojsisovics, 1882
Permoteuthis groenlandica Rosenkrantz, 1946: Upper Permian, Foldvik Creek Formation, Clavering Island, East Greenland
Order Donovaniconida Doguzhaeva, Mapes, and Mutvei[b] (Carboniferous)
Family Donovaniconidae Doguzhaeva, Mapes and Mutvei, 2002[c]
Donovaniconus (Doguzhaeva *et al.* 2002a): Upper Carboniferous, Desmoinesian, Oklahoma, United States
Saundersities (Doguzhaeva *et al.*, in press, b): Upper Carboniferous, Desmoinesian, Illinois, United States
Family Rhiphaeoteuthidae Doguzhaeva, 2002
Rhiphaeteuthis Doguzhaeva, 2002: Upper Carboniferous, Orenburgian, Southern Urals, Kazakhstan Republic (former USSR).
Family Uncertain
New genus: Lower Carboniferous, Bear Gulch Limestone, Montana, United States (Mapes *et al.* 2007)
Order Aulacoceratida Stolley, 1919 (Carboniferous–Jurassic)
Family Mutveiconitidae Doguzhaeva, 2002
Mutveiconites Doguzhaeva, 2002: Upper Carboniferous, Orenburgian, Southern Urals, Kazakhstan Republic (former Soviet Union).
Superorder Decembrachiata Winckworth, 1932 (Carboniferous–Holocene)
Order Spirulida Haeckel, 1896 (Carboniferous–Holocene)
Family Shimanskyidae Doguzhaeva, Mapes, and Mutvei, 1999
Shimanskya (Doguzhaeva *et al.* 1999): Upper Pennsylvanian, Virginian (= Stephanian); Texas, United States
Family Spirulidae Owen, 1836
Order Oegopsida Leach, 1917
Family Architeuthidae Pfeffer, 1900
Family Brachioteuthidae Pfeffer, 1908
Family Batoteuthidae Young and Roper, 1968
Family Chiroteuthidae Gray, 1849
Family Joubiniteuthidae Naef, 1922

TABLE 8.1
(continued)

Family Magnapinnidae Vecchione and Young, 1998

Family Mastigoteuthidae Verrill, 1881

Family Promachoteuthidae Naef, 1912

Family Cranchiidae Prosch, 1847

Family Cycloteuthidae Naef, 1923

Family Ancistrocheiridae Pfeffer, 1912

Family Enoploteuthidae Pfeffer, 1900

Family Lycoteuthidae Pfeffer, 1908

Family Pyroteuthidae Pfeffer, 1912

Family Gonatidae Hoyle, 1886

Family Histioteuthidae Verrill, 1881

Family Psychroteuthidae Thiele, 1920

Family Lepidoteuthidae Naef, 1912

Family Octopoteuthidae Berry, 1912

Family Pholidoteuthidae Voss, 1956

Family Neoteuthidae Naef, 1921

Family Ommastrephidae Steenstruup, 1857

Family Onychoteuthidae Gray, 1847

Family Thysanoteuthidae Keferstein, 1866

Order Myopsida Naef, 1916

Family Australiteuthidae Lu, 2005

Family Loliginidae Lesueur, 1821

Order Sepioidea Naef, 1916

Suborder Sepiida Keferstein, 1866

Family Sepiidae Keferstein, 1866

Suborder Sepiolida Naef, 1916

Family Sepiadariidae Fischer, 1882

Family Sepiolidae Leach, 1817

Superorder Vampyropoda Boletzky, 1992 (Carboniferous–Holocene)

Order Octopoda Leach, 1817 (Carboniferous–Holocene)

Family Palaeoctopodidae Dollo, 1912

Pohlsepia mazonensis (Kluessendorf and Doyle 2000): Middle Carboniferous, Desmoinesian, Francis Creek Formation, Illinois, United States

Suborder Cirrata Grimpe, 1916

Family Cirroteuthidae Keferstein, 1866

Family Stauroteuthidae Grimpe, 1916

Family Opisthoteuthidae Verrill, 1896

Suborder Incirrata Grimpe, 1916

Family Amphitretidae Hoyle, 1886

Family Bolitaenidae Chun, 1911

Family Octopodidae Orbigny, 1840, in Ferussac and Orbigny, 1834–1848

TABLE 8.1

(continued)

Family Vitreledonellidae Robson, 1932

Superfamily Argonautoida Naef, 1912

    Family Alloposidae Verrill, 1881

    Family Argonautidae Cantraine, 1841

    Family Ocythoidae Gray, 1849

    Family Tremoctopodidae Tryon, 1879

Order and family uncertain

    "Bactrites" woodi (Mapes 1979: pl. 18: figs. 8, 12): Upper Carboniferous, Missourian, Kansas, United States

    Undescribed Stark coleoids (see Doguzhaeva *et al.* 2002c): Upper Pennsylvanian, Missourian (= Kasimovian), Nebraska, United States

Superfamily Bathyteuthoida V

    Family Bathyteuthidae Pfeffer, 1900

    Family Chtenopterygidae Grimpe, 1922

    Family Idiosepiidae Fischer, 1882

Problematic specimens

    Boletzkya longa (Bandel *et al.* 1983): Devonian (Emsian), Hunsrückschiefer, Kaisergrube, Hunsrück, Germany

    Naefiteuthis breviphragmoconus (Bandel *et al.* 1983): Devonian (Emsian), Hunsrückschiefer, Kaisergrube, Hunsrück, Germany

    Protoaulacoceras longirostris (Bandel *et al.* 1983): Devonian, Hunsrückschiefer, Kaisergrube, Hunsrück, Germany

    Eoteuthis sp. (Termier and Termier 1971): Devonian of Morocco, North Africa

    Aulacoceras? sp. (de Konick 1843): Locality and age (?Devonian/Carboniferous)

    Eobelemnites caneyensis (Flower 1945): Unknown locality and age

    Jeletzkya douglassae (Johnson and Richardson 1968): Upper Carboniferous, Desmoinesian, Francis Creek Formation, Illinois, United States

    Unnamed coleoid from Czech Republic (Kostak *et al.* 2002): Early Carboniferous, Moravica Formation, Northern Moravia, Czech Republic

    Unnamed coleoid by Allison (1987): Upper Carboniferous, Desmoinesian, Francis Creek Formation, Illinois, United States

    Palaeobelemnopsis sinensis Chin, 1982:Upper Permian from China

NOTE: From Doyle 1993, Doyle *et al.* 1994, Pignatti and Mariotti 1995, Young *et al.* 1998, Doguzhaeva *et al.* 1999, 2002a,b, 2003, in press a, Haas 2002, and the web site maintained by T. Engeser (http://userpage.fu-berlin.de/~palaeont/fossilcoleoidea/hhierarchicalclassification.html) and by R. Young (http://tolweb.org/tree?group=Cephalopoda&contgroup=Mollusca).
[a]Doguzhaeva *et al.* 2002b.
[b]Doguzhaeva *et al.*, in press, b.
[c]Doguzhaeva *et al.* 2002a.
Taxa without stratigraphic ranges are Recent.

each extinction event, only one or a few genera survived, and it was those survivors that rapidly diversified into the empty ecological niches in the world's oceans. Thus, the Ammonoidea have an important place in invertebrate paleontology and stratigraphy, not only because of their diversity but through their pelagic development, which facilitated the attainment of worldwide distributions.

There are literature reports of putative coleoids from the Devonian (de Konick 1843),

Eoteuthis from Morocco (Termier and Termier 1971) and *Protoaulacoceras, Boletzkya,* and *Naefiteuthis* from Germany (Bandel and Boletzky 1988, Bandel *et al.* 1983). Unfortunately, none of these have been confirmed as being coleoids. De Konick's (1843) specimens were inadequately illustrated and are apparently lost, and the specimen identified by Termier and Termier (1971) has been determined to be a bactritoid (Doyle *et al.* 1994). The identification of specimens reported by Bandel *et al.* (1983) and reanalyzed by Bandel and Boletzky (1988) are questionable (Doyle *et al.* 1994), making the coleoid evolution of the Devonian period difficult to assess.

Of the known fossil coleoids, the oldest are *Hematites, Paleoconus,* and *Bactritimimus* (Flower and Gordon 1959; Gordon 1964) from the lower Carboniferous of Arkansas and Utah, United States, with only *Hematites* represented by numerous, well-preserved specimens. Doguzhaeva *et al.* (2002b) determined that *Hematites* and probably the other two genera were unique in many ways and established a new order, Hematitida. Some unique characteristics included the thick, blunt rostrum, which was partly calcified and organic. During early ontogeny, the animal initially grew the breviconic phragmocone with chambers containing internal cameral deposits and a short body chamber. The rostrum was secreted only as the animal neared maturity. Of over 100 *Hematites* specimens examined with rostrums, only one retained a visible protoconch,[1] described as relatively large and spherical in shape. A spherical protoconch and the characteristics of the phragmocone support the interpretation that the bactritoids were the ancestral stock of the coleoids, which probably originated in the earliest Carboniferous or perhaps the Devonian.

The Carboniferous is now known to contain several additional different coleoid genera. Many of these taxa have been described in the past ten years and have lead to the establishment of a number of new families (Table 8.1) complicating

previous hypotheses of the early coleoid phylogeny (e.g., Teichert and Moore 1964: K101, fig. 70; Teichert 1967: 198–199, fig. 20; Engeser and Bandel 1988; Pignatti and Mariotti 1996) (Figure 8.6). There has been general agreement that the origination of coleoids occurred in either the Early Carboniferous or the Devonian and that they were derived from a bactritoidean, with the implication that the group is monophyletic, arising from a paraphyletic Nautiloidea (Figure 8.4). Little consideration has yet been given to the possibility that several different taxa belonging to the Bactritoidea may have been involved in the origin of the Coleoidea at different times in the Devonian and Carboniferous. A polyphyletic origin of Coleoidea might help to explain the great diversity of different morphologies seen in the Carboniferous coleoids and why no closely related coleoid taxa have been conclusively documented from the Devonian.

Currently, there is little agreement among coleoid researchers as to how the higher groups are related to each other and how and when their originations occurred. In part, this lack of agreement is due to the lack of suitable fossils. Additionally, the understanding of the phylogeny of fossil Coleoidea may also have been confused by the premature application of cladistic analysis using data sets subject to major changes with the discovery of new fossils. This is not to suggest that cladistic analysis should not be applied to the coleoid data set, but rather that the results of such preliminary analyses should be treated with caution.

LIVING GROUPS

The first attempt at understanding cephalopod relationships among extant species began in the early 1800s, with the establishment of the subclasses Tetrabranchiata and Dibranchiata based on the number of ctenidia present (Owen 1836). Both Decapoda (Decabrachia, *sensu* Boletzky 2003) and Octopoda (Octobrachia, *sensu* Boletzky 2003) have been acknowledged since the works of d'Orbigny (d'Orbigny 1845) and are distinguished by a number of characters including the number of arms, presence or

---

1. Specimen now misplaced.

absence of chitinous sucker rings and sucker stalks, presence or absence of a buccal crown and lappets, a wide canal between the afferent and efferent vessels of the gill, reduced internal shell (common in both groups), broad neck fusion (found in both groups), and a medio-dorsal sac of the mantle cavity (Clarke 1988). These "superorders" were further subdivided into several groups usually treated as orders (Spirulida, Sepiida, Sepiolida, and Teuthida in the Decabrachia; Octopodida and Vampyromorphida in the Octobrachia) (e.g., Sweeney and Roper 1998). The majority of the earliest classifications were initiated by Naef (1921–1923, 1928), who used a large amount of detailed morphological, embryological, and paleontological data in his monographs. Although many of his groupings are still currently recognized today, some of his classification schemes have since been challenged. He did not use many of the adult morphological characters that are commonly used today, but nonetheless his observations provided a solid foundation and are some of the most influential and important works on cephalopod evolution.

Numerous characters have provided information in delineating family- to species-level differences (Clarke 1988), such as the gladii (Donovan and Toll 1988), mouth parts such as the buccal mass and beaks (Clarke and Maddock 1988a; Nixon 1988), statoliths (Clarke and Maddock 1988b; Young 1988b), ontogenetic development (Boletzky 1997, 2002, 2003), brain morphology (Budelmann et al. 1997; Nixon and Young 2003; Young 1988a) and photophores or the presence of a light organ (Young et al. 1979a; Young and Bennett 1988; Nishiguchi et al. 1998, 2004; Nishiguchi 2002). Since many of these characteristics are shared among sister taxa (e.g., bacteriogenic light organs—those that produce light by symbiotic luminous bacteria—are found only in two families of squids, the Loliginidae and the Sepiolidae), they provide additional information regarding the evolution of such structures and whether those features are derived, are synapomorphic, or have evolved independently several times.

EVOLUTION AND PHYLOGENETIC ANALYSIS OF RECENT CEPHALOPODS

CURRENT CLASSIFICATIONS

Overall, there has been strong support for the monophyly of Cephalopoda (Berthold and Engeser 1987; Salvini-Plawen and Steiner 1996; Lindgren et al. 2004; Passamaneck et al. 2004; Giribet et al. 2006). Most research in the past century has focused solely on describing groups, from levels of orders to species. In living cephalopods, Nautiloidea has one family (Nautilidae) and Coleoidea has two major groups (Octopodiformes and Decapodiformes). Several ordinal-group taxa have been recognized: Vampyromorpha (vampire squids) and Octopoda (shallow-water benthic and deep-water and pelagic octopus) within the Octopodiformes, and Oegopsida, Myopsida, Sepioidea (cuttlefish and bobtail squids) and Spirulida (ram's horn squid) (Young et al., 1998). There are many variations on this classification that differ in detail (number of orders, suborders, superfamilies recognized, and their rank). For example, the orders Cirroctopodida and Octopodida have also been classified as suborders Cirrata and Incirrata, using Octopoda as the order designation and Octopodiformes as the superorder designation (Young and Vecchione 1999; Norman 2000). An additional level of controversy exists when fossil taxa are incorporated in classifications (see also previous section) and different classifications result (Berthold and Engeser 1987) (Figure 8.6).

MORPHOLOGICAL DATA

The most extensive morphological phylogenetic analysis of coleoids is that of Young and Vecchione (1996), who examined 50 morphological characters of 24 species from 17 families to determine whether Cirrata, Incirrata, and Decabrachia were monophyletic and whether vampyromorphs were included within the Octobrachia or Decabrachia. Because of problems with character independence, lack of apomorphic characters, and possible presence of homoplasy among several key features, many characters were disregarded prior to analysis. Their analysis resulted in a better understanding of the relationships between incirrate and cirrate octopods, as well as the placement of Vampyromorpha as sister to the Octobrachia. Taxa within the Decabrachia were unresolved, although that group was not the primary focus of this analysis. In the ten years since their study, numerous Paleozoic coleoids have been described (see previous section and Table 8.1), which will undoubtedly have an impact on our understanding of early coleoid evolution. Other morphological studies focused on one or a few key characters for higher-level relationships (Roper 1969; Toll 1982; Boletzky 1987; Hess 1987; Nesis 1987; Voight 1997; Young et al. 1998; Vecchione et al. 1999; Haas 2002), but none were able to provide robust phylogenies for family-level relationships (e.g., within the Decabrachia).

MOLECULAR PHYLOGENETICS

SINGLE-GENE TREE PHYLOGENIES

The first attempt at using molecular data to determine family-level relationships of coleoids was by Bonnaud and co-workers (Bonnaud et al. 1994) using a 450–580 bp sequence of the 16S rDNA locus from 27 species of decabrachian squids, representing eight families. This study supported many of the higher order-level relationships resolved in some of the previous morphological studies (e.g., Young and Vecchione 1996; Young et al. 1998), yet it unsuccessfully delineated many of the family-level relationships. Following this, three more extensive molecular phylogenetic analyses were completed using loci from the mitochondrial genome; the cytochrome c oxidase subunit I (COI) and combined cytochrome c oxidase subunits II and III for 48 and 17 taxa respectively (Bonnaud 1995; Bonnaud et al. 1997; Carlini and Graves 1999). These studies demonstrated the monophyly of Coleoidea, Octobrachia, Vampyromorpha, and Decabrachia; that Vampyromorpha is sister to the Octobrachia; the polyphyly of Sepioidea; and the lack of resolution of lower-level taxa, particularly within the Decabrachia. Spirula, which was previously included within the Sepioidea (e.g., Voss 1977), was separated in both studies and grouped with the teuthoids. There were also some discrepancies between the analyses, with placement of Idiosepiidae either with Sepiadariidae (Carlini and Graves 1999), or within the Oegopsida (Bonnaud et al. 1997). A separate study used the actin gene family to determine coleoid phylogeny from 44 representative taxa (Carlini et al. 2000). The results recognized that multigene families of actin existed with the Cephalopoda (therefore producing gene trees more than taxon-specific trees), and the information had little resolution of decabrachian relationships, particularly within the orders of Teuthoidea, Sepioidea, and the teuthid suborder Oegopsida, but gave support to the Myopsida.

MULTILOCUS AND COMBINED PHYLOGENETIC ANALYSES

Concurrently, there has been a recognition that additional genes, or a combination of genes and morphology, was needed to provide more resolution, not only at the higher levels within coleoid cephalopods but also within "orders" of the Decabrachia. The first analysis using both molecular and morphological data sets (Carlini et al. 2001) reevaluated previous COI data (Carlini and Graves 1999) for octopod phylogeny in light of previous morphological evidence (Young and Vecchione 1996). Although a number of congruencies were supported using both data sets (monophyly of the Octopoda and of Cirrata), other discrepancies have not yet been resolved (monophyly of Incirrata). Studies using additional mitochondrial loci (Takumiya et al. 2005) or entire mitochondrial genomes (Yokobori et al. 2004; Akasaki et al. 2006) supported higher-level coleoid relationships but were still not able to resolve family-level hierarchies, particularly in the Decabrachia.

Because of conflict between molecular and morphological data sets, more recent analyses have used a combination of molecular data (using multiple genes) and morphology with a variety of analyses to determine both higher- and lower-level relationships within cephalopods (Lindgren et al. 2004; Strugnell et al. 2005). Morphology has been successfully used to define higher-level classification among the Octobrachia and Decabrachia, but

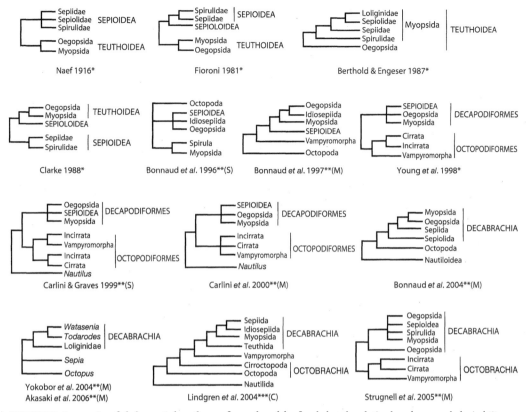

FIGURE 8.8. Summaries of phylogenetic hypotheses of several models of cephalopod evolution based on morphological (*), molecular (**), and combined data (***). Note the differences between many of the morphologically based phylogenies, as well as those using various molecular (single [S] or multiple [M] genes) and combined (molecular and morphology, [C]) data sets.

the lower-level relationships have been better resolved using molecular systematics. This has also been true for other metazoan groups where the diversity among classes is quite high (Giribet et al. 2001; Giribet and Edgecombe 2006). Incongruence between morphological and molecular data is not uncommon in other phylogenetic studies (Giribet 2003), and therefore, further investigation of the resolution between individual genes, synonymous and non-synonymous substitutions, and the use of coding genes (Strugnell et al. 2005) may help increase support for unresolved relationships. Inclusion of fossil data to "fill in" information regarding rapid radiations or extinctions may also help support nodes that contain problematic taxa (e.g., Vampyromorpha).

Comparisons of individual trees derived from a variety of analyses of single genes and morphological data have provided information

regarding which loci/characters have more or less resolution. This can be of use, since genes that evolve faster will have higher resolution at the family/species/population level, whereas slower-evolving or more conserved genes (such as the ribosomal genes) will have more resolution at the basal nodes of larger groups. For example, more conserved loci, such as 18S and 28S rDNA, provide information regarding the monophyly for Cephalopoda and Coleoidea but have little resolution among the lower-level relationships among orders and families (Lindgren et al. 2004, table 1). Combined analyses (Lindgren et al. 2004), along with several other studies (Bonnaud et al. 1994, 1997; Carlini and Graves 1999; Nishiguchi et al. 2004; Strugnell et al. 2005) have demonstrated that relationships among Sepiolida, Sepiida, Idiosepiida, and the Loliginidae are well supported by the addition of molecular data, while the position

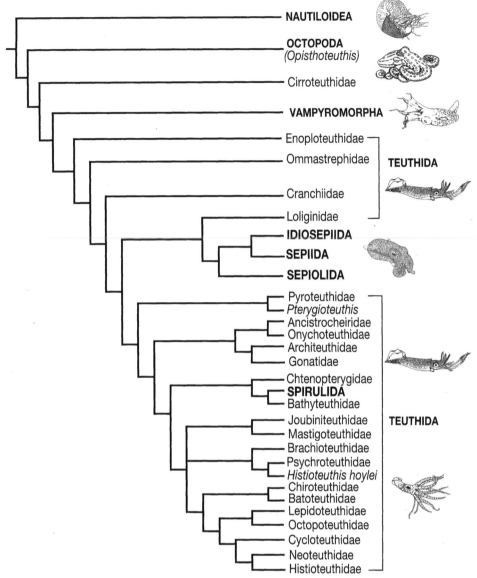

FIGURE 8.9. Schematic representation of modern cephalopod relationships based on the optimal parameter set using direct optimization (via parsimony) for the combined analysis of 101 morphological characters and molecular data from four loci (mitochondrial cytochrome *c* oxidase subunit I, nuclear 18S rDNA, the D3 expansion fragment of 28S rDNA, and histone H3). Taxa in capitals represent orders of cephalopods that appeared monophyletic in the analysis (from Lindgren *et al.* 2004). Drawings by G. Williams.

of *Spirula* within the sepioid groups remains to be resolved (Figures 8.8, 8.9). Although the molecular data has added increased resolution between the orders, it still does not resolve relationships between large family groups, such as the Oegopsida. This is probably because this order radiated rather quickly and, because sampling some of the deeper water families is difficult, they are under-represented in most molecular studies to date.

One of the most debatable relationships is the placement of Vampyromorpha within the Decabrachia (Figures 8.8, 8.9). Earlier studies have placed Vampyromorpha with the Octobrachia based on embryological and developmental data (Naef 1928, Young and Vecchione 1996;

Boletzky 2003) as well as several morphological, including ultrastructural, characters (Toll 1982, 1998; Healy 1989, Lindgren et al. 2004). However, molecular studies (Bonnaud et al. 1997; Carlini and Graves 1999; Lindgren et al. 2004) have shown different results depending on which molecular loci were tested and which outgroup taxa and parameter variations were used. The main conflict for this work was between the morphological and the molecular/combined data. Morphological data suggests a sister relationship between Vampyromorpha and the Octobrachia, whereas the molecular and combined data suggest a closer relationship between Vampyromorpha and the Decabrachia. Additional studies are needed to resolve these and other internal relationships, and the inclusion of more taxa (particularly from those groups that have been poorly sampled and therefore under-represented in previous results) will help provide a more detailed phylogeny of cephalopod evolution.

Additional molecular studies have recently been used to investigate decabrachian evolution with respect to the placement of *Spirula* using 18S rDNA (Warnke et al. 2003); lower-level relationships within families (16S rDNA for cirrate octopod relationships [Piertney et al. 2003] and microsatellite and 16S rDNA for loliginid relationships [Shaw et al. 1999; Anderson 2000; Reichow and Smith 2001]); identification of juvenile and adult gonatid squids (Seibel et al. 2000; Lindgren et al. 2005); as well as investigating the evolution of species complexes among octopods (Söller et al. 2000; Strugnell et al. 2004; Allcock et al. 1997; Guzik et al. 2005) and sepiolids (Nishiguchi et al. 2004), to name a few. Population genetic studies have also helped provide information regarding the migration of specific haplotypes and the phylogeography of sister species (Anderson 2000; Herke and Foltz 2002; Jones et al. 2006).

DEVELOPMENT

More recently, sophisticated techniques have improved our understanding of the development and neontology of cephalopods. Naef provided much of the foundation for the embryology of many cephalopod species, which is elegantly presented in his monograph (Naef 1928). This work provided cephalopod researchers with a large number of characters that were useful for morphological comparisons and, subsequently, cephalopod systematics and evolution. Recent techniques using *in situ* hybridization have allowed the expansion of a new field of research, namely evolutionary developmental biology, to help understand how specific genes are expressed among related taxa. This information can provide insight as to whether the same gene in a variety of organisms controls key developmental traits, or whether these genes have been co-opted to function for other developmental programs. For instance, master control genes such as *Hox* and *Pax6* have demonstrated the relationship of closely related molluscs and the conservative nature of these sequences (Halder et al. 1995, see also Wanninger et al., Chapter 16). Although the genetic information may not give the resolution needed to solve internal cephalopod nodes, the observation of protein expression can provide insights into the differentiation of cephalopod morphology and the co-option of genes for numerous functions (Tomarev et al. 1997; Callaerts et al. 2002; Hartmann et al. 2003; Lee et al. 2003).

Following developmental programs and combining them with phylogenetic information may also give insights as to the evolution of specific morphological features and provide evidence for relationships that are linked only by these features, such as bacteriogenic light organs in both sepiolids and loliginid squids (Foster et al. 2002; Nishiguchi et al. 1998, 2004) and symbiosis between squids and bacteria in accessory nidamental glands (Grigioni et al. 2000; Pichon et al. 2005). Elegant work has recently provided detailed maps of the neurodevelopment of myopsid, oegopsid, and idiosepiid squids (Shigeno et al. 2001a, b; Shigeno and Yamamoto 2002), exhibiting major differences among these squids. Protein regulation and gene expression has recently been examined in squids for use in diagnosing induction of tissue morphogenesis and apoptosis (Crookes et al. 2004; Doino and McFall-Ngai 1995; Foster and McFall-Ngai 1998; Montgomery and McFall-Ngai 1992; Small and McFall-Ngai 1998; Zinovieva et al. 1993). Finally, advanced tracer techniques, biochemical analyses, and direct in situ measurements of growth and other physiological attributes are providing information regarding the growth, life history strategies, and behavior of squids, both from the wild and in the laboratory (Forsythe et al. 2001, 2002; Hanlon et al. 1997; Huffard et al. 2005; Jones and Nishiguchi 2004; Kasugai 2000; Landman et al. 2004; Moltschaniwskyj 1994; Pecl and Moltschaniwskyj 1997; Shea 2005; Steer et al. 2004).

All these comparative studies have proven to be useful in providing additional information not only for delineating function and life history characteristics but also to delve into the possible evolutionary trajectories that may explain similar characteristics shared among distantly related cephalopods. Combining this information with molecular studies may help evolutionary biologists test homology hypotheses of morphological features or shared developmental patterns in gene expression, which may be evolutionarily derived. These can then be used as key innovations and included in cladistic analyses.

ADAPTIVE RADIATIONS

In fossil cephalopods, the focus of the few studies on adaptive radiation have been on the progressive changes of the external shell through geologic time as predators became more efficient (for an extended discussion of the theory of evolution and escalation see Vermeij 1987; also see Mapes and Chaffin 2003). Ward (1981) determined that the ornament on the shells of ammonoids became rougher through time (Devonian to Cretaceous). Signor and Brett (1984) analyzed the tarphycerids, barrandeocerids, and nautilids in part of the Paleozoic and determined that the origin of the durophagous (shell-crushing) predators in the Silurian and Devonian coincided with a gradually increased

robustness of the ornament in these nautiloid orders. As an additional observation on fossil cephalopod evolutionary radiations, Vermeij (1987) observed that the external shells of ancient nautiloids have not proven to be an effective counter to predation in the long run. The rationale behind this observation is that, given the evolutionary success of durophagous predators through time, and given that modern cephalopods without external shells (i.e., the Coleoidea) are so much more abundant, diverse, and worldwide in their distribution today as compared to modern Nautilus and Allonautilus with their external shells and limited geographic distribution, the externally shelled cephalopods were not competitive in the long run. This conclusion seems obvious given the then-known fossil record of ectocochleate cephalopods versus the abundance of endocochleate cephalopods living today. However, the externally shelled cephalopods (e.g., the Nautiloidea and Ammonoidea) had a long and diverse history (see Figure 8.5, for example, showing the complexity of ammonoid phylogeny) and that these lineages survived many extinction events and reradiated to refill and dominate the world's oceans. Notably, we now know that the evolution of shell-less coleoids occurred prior to the Upper Carboniferous and that, while these shell-less coleoids must have also survived the same extinction events in the Late Paleozoic and Mesozoic as the externally shelled cephalopods, they did not overwhelm them and become the dominant cephalopod group after those extinction events. It was only after almost all the externally shelled forms had become extinct (all the Ammonoidea and virtually all of the Nautiloidea) at the end of the Mesozoic that the coleoids radiated to become the dominant cephalopod group in the world's oceans.

Cephalopods, with their sensory/nervous/visual systems, are excellent predators, being entirely carnivorous on a large variety of prey species. Not only does their advanced nervous system allow for a great ability for capturing prey, but it also increased their ability to invade a multitude of niches. Based on observations of modern coleoids, the evolution of a complex visual system and chromatophores was probably related to the development of a well developed nervous system (Budelmann *et al.* 1997; Hanlon and Messenger 1996), which, through integration with a highly developed system of dermal chromatophores, enables most coleoid cephalopods to change color rapidly. Such color changes are used for camouflage; to display specific patterns toward predators, prey, other cephalopod species; or for intraspecific behavioral interactions, such as sexual displays (Hanlon and Messenger 1996), and presumably played a large part in the success of the group. The static presumed cryptic coloration of the shell and animal of living *Nautilus* is in marked contrast. The evolution from a shelled to a nonshelled coleoid probably began in the Devonian and/or early Carboniferous (see above) because more efficient and active predators evolved at that time. By shell reduction and the evolutionary selection of other advanced traits (i.e., complex brain driving other characters), the coleoids evolved elaborate sensory systems fulfilling tasks achieved by effectors (i.e., camouflage) to counter the evolutionary pressures created by efficient and more active predators (Nixon and Young 2003).

Along with major changes in behavior due to a highly advanced nervous system, cephalopods have evolved various mechanisms for controlling buoyancy and propulsion. As previously mentioned, buoyancy in cephalopods is controlled by the presence of gas spaces (as in taxa such as *Nautilus*, *Spirula*, and *Sepia*, which have a shell or cuttlebone). Cephalopods with no gas-filled compartments must rely on using jet propulsion to move continually through the water column to maintain buoyancy, or they achieve it by an increase in solutes such as ammonium chloride within the coelomic space or in vacuoles within their tissues (Boyle and Rodhouse 2005). The evolution of fins also helped to orient and maneuver the body during swimming, depending on the lifestyle of the squid (epipelagic fast swimmers versus slower, coastal species). Finally, the evolution of arms and tentacles enabled cephalopods to possess the ability to handle prey efficiently, allowing them to feed on a large variety of prey items. The arm crown, which is derived from the mouth of cephalopods (Packard 1972), is much more versatile for capturing and handling prey (Boyle and Rodhouse 2005). Along with arms and tentacles that possess suckers for holding and attaching onto prey, these adaptive characteristics contributed to the success of cephalopods as predators compared to other molluscan groups.

FUTURE STUDIES

The largest unresolved problems in the fossil Coleoidea are the problematic origins of the order and the relationships between the different families and genera that are assigned to the orders. To refine our understanding of the evolutionary events that gave rise to the Coleoidea will require new material, especially from Devonian-aged rock units. Efforts should focus on Lagerstätten where tissues can be preserved and there is excellent shell preservation. Paleoenvironments that had well-oxygenated water columns and anoxic conditions at the water/sediment interface are likely candidates for yielding well-preserved fossil coleoids.

There are also significant problems with obtaining living taxa. Many species do not survive commonly used collection techniques (such as deep-water trawls), suffering extensive damage and frustrating morphological studies. Similarly, there are difficulties in obtaining eggs or juveniles to study development with pelagic taxa and in obtaining rare or deep-water taxa for molecular studies.

The resolution of the problem of whether the coleoids have a monophyletic or paraphyletic origin will require additional research on existing specimens and the collection and analysis of new living and fossil material. Increased taxon sampling (both extinct and extant species) as well as the addition of larger molecular data sets (complete mitochondrial genomes, additional nuclear genes) will provide resolution both at the higher level relationships and those that have been particularly problematic (e.g., Decabrachia). Thus, we need to obtain and combine more fossil, morphological, developmental, and molecular data in order to increase our understanding of the interrelationships of these amazingly diverse and ecologically important molluscs.

ACKNOWLEDGMENTS

The authors would like to acknowledge the invaluable advice from S. v. Boletzky, whose expertise in many aspects of cephalopod evolution were greatly appreciated, as well as the three reviewers who provided additional insights for this chapter. The work is partially supported by NIH SO6 GM008136-31 and NSF DEB-0316516 to M. K. N. We would like to thank S. v. Boletzky, D. Lindberg, M. Norman, Conch Books, and *Science* magazine for reproduction of photographs used in this chapter.

REFERENCES

Akasaki, T., Nikaido, M., Tsuchiya, K., Segawa, S., Hasegawa, M., and Okada, N. 2006. Extensive mitochondrial gene rearrangements in coleoid Cephalopoda and their phylogenetic implications. *Molecular Phylogenetics and Evolution* 38: 648–658.

Allcock, A. L., Brierley, A. S., Thorpe, J. P., and Rodhouse, P. G. K. 1997. Restricted gene flow and evolutionary divergence between geographically separated populations of the Antarctic octopus *Pareledone turqueti*. *Marine Biology* 129: 97–102.

Allison, P. A. 1987. A new cephalopod with soft parts from the Upper Carboniferous Francis Creek Shale of Illinois, USA. *Lethaia* 20: 117–121.

Anderson, F. E. 2000. Phylogeny and historical biogeography of the loliginid squids (Mollusca: Cephalopoda) based on mitochondrial DNA sequence data. *Molecular Phylogenetics and Evolution* 15: 191–214.

Arnold, J. M. 1962. Mating behavior and social structure in *Loligo pealeii*. *Biological Bulletin* 123: 53–57.

———. 1987. Reproduction and embryology of *Nautilus*. In *Nautilus: The Biology and Paleobiology of a Living Fossil*. Edited by W. B. Saunders and N. H. Landman. New York: Plenum Press, pp. 13–16.

Aronson, R. B. 1991. Ecology, paleobiology and evolutionary constraint in the octopus. *Bulletin of Marine Science* 49: 245–255.

Bandel, K., and Boletzky, S. v. 1988. Features of development and functional morphology in the reconstruction of early coleoid cephalopods. In *Cephalopods Present and Past*. Edited by J. Wiedmann and J. E. Kullmann. Stuttgart: Schweizerbart'sche Verlagsbuchhandlung, pp. 229–246.

Bandel, K., and Leich, H. 1986. Jurassic Vampyromorpha (dibranchiate cephalopods). *Neues Jahrbuch für Geologie und Palaeontologie, Abhandlungen* 3: 129–148.

Bandel, K., Reitner, J., and Stürmer, W. 1983. Coleoids from the Lower Devonian Black Slate ("Hunsrück-Schiefer") of the Hunsrück (West Germany). *Neues Jahrbuch für Geologie und Palaeontologie, Abhandlungen* 165: 397–417.

Bandel, K., and Spaeth, C. 1988. Structural differences in the ontogeny of some belemnite rostra. In *Cephalopods Present and Past*. Edited by J. Wiedmann and J. E. Kullmann. Stuttgart: Schweizerbart'sche Verlagsbuchhandlung, pp. 247–261.

Becker, T. R., and Kullmann, J. 1996. Paleozoic ammonoids in time and space. In *Ammonoid Paleobiology*. Edited by N. H. Landman, K. Tanabe, and R. A. Davis. New York: Plenum Press, pp. 711–753.

Berthold, T., and Engeser, T. 1987. Phylogenetic analysis and systematization of the Cephalopoda. *Verhandlungen des Naturwissenschaftlichen Vereins Hamburg* 29:187–220.

Boletzky, S. v. 1987. Ontogenetic and phylogenetic aspects of the cephalopod circulatory system. *Experientia* 43: 478–483.

———. 1997. Developmental constraints and heterochrony: a new look at offspring size in cephalopod molluscs. *Geobios Memoire Special* 21: 267–275.

———. 2002. Yolk sac morphology in cephalopod embryos. In *Cephalopods Present and Past*. Edited by H. Summesberger and H. E. A. Daurer. Vienna: Abhandlungen der Geologischen Bundesanstalt, pp. 57–68.

———. 2003. Biology of early life stages in cephalopod molluscs. *Advances in Marine Biology* 44: 143–203.

Bonnaud, L. 1995. *Phylogenie des céphalopodes decapodes*. Paris: Museum National d'Histoire Naturelle.

Bonnaud, L., Boucher-Rodoni, R., and Monnerot, M. 1994. Phylogeny of decapod cephalopods based on partial 16S rDNA nucleotide sequences. *Comptes Rendus de l'Académie des Sciences de Paris* 317: 581–588.

———. 1997. Phylogeny of cephalopods inferred from mitochondrial DNA sequences. *Molecular Phylogenetics and Evolution* 7: 44–54.

Botter, D., Etter, W., Hagadorn, J. W., and Tang, C. M. 2002. *Exceptional Fossil Preservation: a Unique View on the Evolution of Marine Life.* New York: Columbia University Press.

Boyle, P. R., and Boletzky, S. v. 1996. Cephalopod populations: definitions and dynamics. *Philosophical Transactions of the Royal Society Series B: Biological Sciences* 351: 985–1002.

Boyle, P. R., and Rodhouse, P. 2005. *Cephalopods. Ecology and Fisheries.* Oxford: Blackwell.

Budelmann, B. U., Schipp, R., and Boletzky, S. v. 1997. Cephalopoda. In *Microscopic Anatomy of Invertebrates.* Edited by F. W. Harrison and A. J. Kohn. New York: Wiley-Liss, pp. 119–414.

California Department of Fish and Game. 2003. *Final Market Squid Fishery Management Plan.* California Department of Fish and Game. http://www.dfg.ca.gov/mrd/msfmp/index.html.

Callaerts, P., Lee, P. N., Hartmann, B., Farfan, C., Choy, D. W. Y., Zkeo, K., Fischbach, K.-F., Gehring, W. J., and Couet, H. G. d. 2002. *Hox* genes in the sepiolid squid *Euprymna scolopes*: Implications for the evolution of complex body plans. *Proceedings of the National Academy of Sciences of the United States of America* 99: 2088–2093.

Carlini, D. C., and Graves, J. E. 1999. Phylogenetic analysis of cytochrome *c* oxidase I sequences to determine higher-level relationships within the coleoid cephalopods. *Bulletin of Marine Science* 64: 57–76.

Carlini, D. C., Reece, K. S., and Graves, J. E. 2000. Actin gene family evolution and the phylogeny of coleoid cephalopods (Mollusca: Cephalopoda). *Molecular Biology and Evolution* 17: 1353–1370.

Carlini, D. C., Young, R. E., and Vecchione, M. 2001. A molecular phylogeny of the Octopoda (Mollusca: Cephalopoda) evaluated in light of morphological evidence. *Molecular Phylogenetics and Evolution* 21: 388–397.

Chen, J.-Y., and Teichert, C. 1983. Cambrian cephalopods. *Geology* 11: 647–650.

Clarke, M. R. 1988. Evolution of Recent cephalopods—a brief review. In *The Mollusca, Paleontology and Neontology of Cephalopods.* Edited by M. R. Clarke and E. R. Trueman. San Diego: Academic Press, pp. 331–341.

———. 1996. Cephalopods as prey. III. Cetaceans. *Philosophical Transactions of the Royal Society Series B: Biological Sciences* 351: 1053–1065.

Clarke, M. R., and Maddock, L. 1988a. Beaks of living cephalopods. In *The Mollusca, Paleontology and Neontology of Cephalopods.* Edited by M. R. Clarke and E. R. Trueman. San Diego: Academic Press, pp. 123–131.

———. 1988b. Statoliths from living species of cephalopods and evolution. In *The Mollusca,*

Paleontology and Neontology of Cephalopods. Edited by M. R. Clarke and E. R. Trueman. San Diego: Academic Press, pp. 169–184.

Collins, D., and Ward, P. D. 1987. Adolescent growth and maturity in *Nautilus.* In *Nautilus: The Biology and Paleobiology of a Living Fossil.* Edited by W. B. Saunders and N. H. Landman. New York: Plenum Press, pp. 421–432.

Crick, R. E., and Ottensman, V. M. 1983. Sr, Mg, Ca, and Mn, chemistry of skeletal components of a Pennsylvanian and Recent nautiloid. *Chemical Geology*: 147–163.

Crookes, W. J., Ding, L.-L., Huang, Q. L., Kimbell, J. R., Horwitz, J., and McFall-Ngai, M. J. 2004. Reflectins: The unusual proteins of squid reflective tissues. *Science* 303: 235–238.

Croxall, J. P., and Prince, P. A. 1996. Cephalopods as prey. I. Seabirds. *Philosophical Transactions of the Royal Society Series B: Biological Sciences* 351: 1023–1043.

d'Orbigny, A. 1845. *Mollusques vivants et fossiles, ou description de toutes les espèces de coquilles et de mollusques.* Paris: Adolphe Delahays.

Davis, R. A., Landman, N. H., Dommergues, J.-L., Marchand, D., and Bucher, H. 1996. Mature modifications and dimorphism in ammonoid cephalopods. In *Ammonoid Paleobiology.* Edited by N. H. Landman, K. Tanabe, and R. A. Davis. New York: Plenum Press, pp. 464–539.

de Konick, L. G. 1843. Notice sur une coquille fossile de terrains anciens de Belgique. *Bulletin de l'Académie Royale des Sciences, des Lettres et des Beaux-Arts de Belgique* 10: 207–208.

Doguzhaeva, L. A., Mapes, R. H., and Mutvei, H. 1999. A Late Carboniferous spirulid coleoid from the Southern Mid-continent (USA). In *Advancing Research on Living and Fossil Cephalopods.* Edited by F. Olóriz and Rodríguez-Tovar, New York/Boston: Kluwer Academic/Plenum Publishers, pp. 47–57.

———. 2002a. The coleoid with an ink sac and a body chamber from the Upper Pennsylvanian of Oklahoma, USA. *Berliner Paläobiologische Abhandlungen* 1: 34–38.

———. 2002b. Early Carboniferous coleoid *Hematites* Flower and Gordon, 1959 (Hematitida ord. nov.) from Midcontinent (USA). In *Cephalopods Present and Past.* Edited by H. Summesberger, K. Histon, and A. Daurer. Vienna: Abhandlungen Geologische Bundesanstalt. pp. 299–320.

———. 2004a. Occurrence of ink in Paleozoic and Mesozoic coleoids (Cephalopoda). *Mitteilungen des Geologisch-Paläontologischen Institutes der Universität Hamburg* 88: 145–156.

Doguzhaeva, L. A., Mutvei, H., Summesberger, H., and Dunca, E. 2004b. Bituminous soft body tissues in the body chamber of the Late Triassic ceratitid *Austrotrachyceras* from the Austrian Alps. *Mitteilungen des Geologisch-Paläontologischen Institutes der Universität Hamburg* 88: 37–50.

Doguzhaeva, L.A., Mapes, R.H., Mutvei, H., and Pabian, R.K. 2002c. The late Carboniferous phragmocone-bearing orthoconic coleoids with ink sacs: their environment and mode of life. In *Geological Society of Australia Abstracts*, number 68. Edited by G.A. Brock and J.A. Talent, First International Palaeontological Congress, July 2002, Macquarie University, NSW Australia: p. 200.

Doguzhaeva, L.A., Mapes, R., Summesberger, H., and Mutvei, H. In press, a. A late Carboniferous coleoid from the Mazon Creek (USA) with a radula, arm hooks, mantle tissues, and ink. In *Cephalopods Present and Past*. Edited by N.H. Landman, R.A. Davis, R.H. Mapes, and W.L. Manger. Berlin: Springer-Verlag.

———. In press, b. The preservation of soft tissues, shell, and mandibles in the ceratitid ammonoid *Austrotrachyceras* (Late Triassic), Austria. In *Cephalopods Present and Past*. Edited by N.H. Landman, R.A. Davis, R.H. Mapes, and W.L. Manger. Berlin: Springer-Verlag.

Doino, J.A., and McFall-Ngai, M.J. 1995. A transient exposure to symbiosis-competent bacteria induces light organ morphogenesis in the host squid. *Biological Bulletin* 189: 347–355.

Donovan, D.T. 1977. Evolution of the dibranchiate Cephalopoda. *Symposia of the Zoological Society of London* 38: 15–48.

Donovan, D.T., and Toll, R.B. 1988. The gladius in coleoid (Cephalopoda) evolution. In *The Mollusca, Paleontology and Neontology of Cephalopods*. Edited by M.R. Clarke and E.R. Trueman. San Diego: Academic Press, pp. 89–101.

Donovan, D.T., Callomon, J.H., and Howarth, M.K. 1981. Classification of the Jurassic Ammonitina. In *The Ammonoidea*. Edited by M.R. House and J.R. Senior. New York: Academic Press, pp. 101–155.

Doyle, P., Donovan, D.T., and Nixon, M. 1994. Phylogeny and systematics of the Coleoidea. *University of Kansas Paleontology Contributions, New Series* 5:1–15.

Engeser, T. 1996. The position of the Ammonoidea within the Cephalopoda. In *Ammonoid Paleobiology*. Edited by N.H. Landman, K. Tanabe, and R.A. Davis. New York: Plenum Press, pp. 3–19.

Engeser, T., and Bandel, K. 1988. Phylogenetic classification of coleoid cephalopods. In *Cephalopods: Present and Past*. Edited by J. Wiedmann and J. Kullmann. Stuttgart: Schweizerbart'sche Verlagsbuchhandlung, pp. 105–115.

Engeser, T., and Reitner, J. 1981. Beiträge zur Systematik von phragmokontragenden Coleoiden aus dem Untertithonium (Malm zeta, Solnhofener Plattenkalk) von Solnhofen und Eichstätt (Bayern). *Neues Jahrbuch für Geologie und Paläontologie, Monatshefte* 1981: 527–545.

Erben, H.K. 1964. Die Evolution der ältesten Ammonoidea (Lieferung I). *Neues Jahrbuch für Geologie und Paläontologie, Monatshefte* 120: 107–212.

Fiorito, G., and Scotto, P. 1992. Observational learning in *Octopus vulgaris*. *Science* 256: 545–547.

Flower, R.H. 1945. A belemnite from a Mississippian boulder of the Caney shale. *Journal of Paleontology* 19:490–503.

Flower, R.H., and Gordon, M., Jr. 1959. More Mississippian belemnites. *Journal of Paleontology* 33: 809–842.

Forsythe, J., Lee, P., Walsh, L., and Clark, T. 2002. The effects of crowding on the growth of the European cuttlefish, *Sepia officinalis* Linnaeus, 1758 reared at two temperatures. *Journal of Experimental Marine Biology and Ecology* 269: 173–185.

Forsythe, J.W., Walsh, L.S., Turk, P.E., and Lee, P.G. 2001. Impact of temperature on juvenile growth and age at first egg-laying of the Pacific reef squid *Sepioteuthis lessoniana* reared in captivity. *Marine Biology* 138:103–112.

Foster, J.S., Boletzky, S. v., and McFall-Ngai, M.J. 2002. A comparison of the light organ development of *Sepiola robusta* Naef and *Euprymna scolopes* Berry (Cephalopoda: Sepiolidae). *Bulletin of Marine Science* 70: 141–153.

Foster, J.S., and McFall-Ngai, M.J. 1998. Induction of apoptosis by cooperative bacteria in the morphogenesis of host epithelia tissues. *Development, Genes, and Evolution* 208: 295–303.

Frest, T.J., Glenister, B.F., and Frunish, W.M. 1981. Pennsylvanian-Permian cheiloceratacean ammonoid families Maximitidae and Pseudohaloritidae. *Journal of Paleontology Memoir* 11 55: 1–46.

Fuchs, D., Keupp, H., and Engeser, T. 2003. New records of soft parts of *Muensterella scutellaris* Muenster, 1842 (Coleoidea) from the Late Jurassic plattenkalks of Eichstätt and their significance for octobrachian relationships. *Berliner Paläobiologische Abhandlungen* 3: 101–111.

Gardner, G., and Mapes, R. 2001. Relationships of color patterns and habitat for Lower Triassic ammonoids from Crittenden Springs, Elko County, Nevada. *Revue de Paléobiologie Genève* 8: 109–122.

Gilly, W.F., and Lucero, M.T. 1992. Behavioral responses to chemical stimulation of the olfactory organ in the squid *Loligo opalescens*. *Journal of Experimental Biology* 162: 209–229.

Giribet, G. 2003. Stability in phylogenetic formulations and its relationship to nodal support. *Systematic Biology* 52: 554–564.

Giribet, G., and Edgecombe, G. D. 2006. Conflict between data sets and phylogeny of centipedes: an analysis based on seven genes and morphology. *Proceedings of the Royal Society of London Series B: Biological Sciences* 273:531–538.

Giribet, G., Edgecombe, G. D., and Wheeler, W. C. 2001. Arthropod phylogeny based on eight molecular loci and morphology. *Nature* 413: 157–161.

Giribet, G., Okusu, A., Lindgren, A. R., Huff, S. W., Schrodl, M., and Nishiguchi, M. K. 2006. Evidence for a clade composed of molluscs with serially repeated structures: Monoplacophorans are related to chitons. *Proceedings of the National Academy of Sciences of the United States of America* 103: 7723–7728.

Giribet, G., and Wheeler, W. C. 2002. On bivalve phylogeny: a high-level analysis of the Bivalvia (Mollusca) based on combined morphology and DNA sequence data. *Invertebrate Zoology* 121: 271–324.

Glenister, B. F., and Furnish, W. M. 1981. Permian Ammonoids. In *The Ammonoidea*. Edited by M. R. House and J. R. Senior. New York: Academic Press, pp. 49–64.

Gordon, M., Jr. 1964. Carboniferous cephalopods of Arkansas. *Geological Survey Professional Paper* 460.

Gordon, M., Jr. 1966. Classification of Mississippian coleoid cephalopods. *Journal of Paleontology* 40: 449–452.

Grigioni, S., Boucher-Rodoni, R., Demarta, A., Tonolla, M., and Peduzzi, R. 2000. Phylogenetic characterization of bacterial symbionts in the accessory nidamental glands of the sepioid *S. officinalis* (Cephalopoda: Decapoda). *Marine Biology* 136: 217–222.

Guzik, M. T., Norman, M. D., and Crozier, R. H. 2005. Molecular phylogeny of the benthic shallow-water octopuses (Cephalopoda: Octopodinae). *Molecular Phylogenetics and Evolution* 37: 235–248.

Haas, W. 2003. The evolutionary history of the eight-armed Coleoidea. *Abhandlungen der Geologischen Bundesanstalt, Wien* 57: 341–351.

Hagadorn, J. W. 2002. Bear Gulch: an exceptional Upper Carboniferous plattenkalk. In *Exceptional Fossil Preservation: a Unique View on the Evolution of Marine Life*. Edited by D. Botter, W. Etter, and C. M. Tang, New York: Columbia University Press, pp. 167–183.

Halder, G., Callaerts, P., and Gehring, W. J. 1995. Induction of ectopic eyes by targeted expression of the *eyeless* gene in *Drosophila*. *Science* 267: 1788–1792.

Hallam, A., and O'Hara, M. J. 1962. Aragonite fossils in the Lower Carboniferous of Scotland. *Nature* 195: 273–274.

Hanlon, R. T. 1988. Behavioral and body patterning characters useful in taxonomy and field identification of cephalopods. *Malacologia* 29:247–264.

Hanlon, R. T., Cooper, K. M., Budelmann, B. U., and Pappas, T. C. 1990. Physiological color change in squid iridophores. I. Behavior, morphology and pharmacology in *Lolliguncula brevis*. *Cell Tissue Research* 259: 3–14.

Hanlon, R. T., Claes, M. F., Ashcraft, S. E., and Dunlap, P. V. 1997. Laboratory culture of the sepiolid squid *Euprymna scolopes*: A model system for bacteria-animal symbiosis. *Biological Bulletin* 192: 364–374.

Hanlon, R. T., and Messenger, J. B. 1996. *Cephalopod Behaviour*. Cambridge, UK: Cambridge University Press.

Hartmann, B., Lee, P. N., Kang, Y. Y., Tomarev, S. I., de Couet, H. G., and Calherts, P. 2003. *Pax6* in the sepiolid squid *Euprymna scolopes*: evidence for a role in eye, sensory organ and brain development. *Proceedings of the National Academy of Sciences of the United States of America* 93: 13683–13688.

Haszprunar, G. 1996. The Mollusca: Coelomate turbellarians or mesenchymate annelids? In *Origin and Evolutionary Radiation of the Mollusca*. Edited by J. Taylor. Oxford: Oxford University Press, pp. 1–28.

———. 2000. Is the Aplacophoran monophyletic? A cladistic point of view. *American Malacology Bulletin* 15: 115–130.

Healy, J. M. 1989. Spermatozoa of the deep-sea cephalopod *Vampyroteuthis infernalis* Chun: ultrastructure and possible phylogenetic significance. *Philosophical Transactions of the Royal Society of London Series B: Biological Sciences* 323: 589–600.

Herke, S. W., and Foltz, D. W. 2002. Phylogeography of two squid (*Loligo pealei* and *L. plei*) in the Gulf of Mexico and northwestern Atlantic Ocean. *Marine Biology* 140: 103–115.

Hess, H. C. 1987. Comparative morphology, variability, and systematic applications of cephalopod spermatophores (Teuthoidea and Vampyromorpha). Ph.D. thesis, The University of Miami, Coral Gables, FL.

Hodgkin, A., and Huyley, A. 1952. Currents carried by sodium and potassium ions through the membrane of the giant squid axon of *Loligo*. *Journal of Physiology* 116: 449–472.

House, M. R. 1981. On the origin, classification and evolution of the early Ammonoidea. In *The Ammonoidea*. Edited by M. R. House and J. R. Senior. New York: Academic Press, pp. 3–36.

———. 1988. Major features of cephalopod evolution. In *Cephalopods Present and Past*. Edited by J. Wiedmann and J. Kullmann. Stuttgart: Schweizerbart'sche Verlagsbuchhandlung, pp. 1–16.

House, M. R. 1993. Fluctuations in ammonoid evolution and possible environmental controls. In *The Ammonoidea: Environment, Ecology, and Evolutionary Change*. Edited by M. R. House. Oxford, England: Clarendon Press, pp. 13–34.

Huffard, C. L., Boneka, F., and Full, R. J. 2005. Underwater bipedal locomotion by octopuses in disguise. *Science* 307: 1927.

Jackson, G. D., and Moltschaniwskyj, N. A. 2001. Temporal variation in growth rates and reproductive parameters in the small near-shore tropical squid *Loliolus noctiluca*; is cooler better? *Marine Ecology Progress Series* 218: 167–177.

Jeletzky, J. A. 1966. Comparative morphology, phylogeny, and classification of fossil Coleoidea. *University of Kansas Paleontology Contributions and Articles* 30: 1–162.

———. 1969. New or poorly understood Tertiary sepiids from southeastern United States and Mexico. *University of Kansas Paleontology Contributions and Articles* 41: 1–39.

Johnson, R. G., and Richardson Jr., E. S. 1968. Ten-armed fossil cephalopod from the Pennsylvanian of Illinois. *Science* 159: 526–528.

Jones, B. W., Lopez, J. E., Huttenberg, J., and Nishiguchi, M. K. 2006. Population structure between environmentally transmitted vibrios and bobtail squids using nested clade analysis. *Molecular Ecology* 15: 4317–4329.

Jones, B. W., and Nishiguchi, M. K. 2004. Counterillumination in the bobtail squid, *Euprymna scolopes* Berry (Mollusca: Cephalopoda). *Marine Biology* 144: 1151–1155.

Kasugai, T. 2000. Reproductive behavior of the pygmy cuttlefish *Idiosepius paradoxus* in an aquarium. *Venus, the Japanese Journal of Malacology* 59: 37–44.

Kennedy, W., Landman, N. H., Cobbin, A., and Larson, N. L. 2002. Jaws and radulae in *Rhaeboceras*, a late Cretaceous ammonite. In *Cephalopods Present and Past*. Edited by H. Summesberger, K. Histon, and A. Daurer. Vienna: Abhandlungen Geologische Bundesanstalt, pp. 113–132.

Kluessendorf, J., and Doyle, P. 2000. *Pohlsepia mazonensis*, an early "octopus" from the Carboniferous of Pennsylvanian of Illinois. *Paleontology* 43: 919–926.

Kostak, M., Marek, J., Neumann, P., and Pavel, M. 2002. An early Carboniferous coleoid (Cephalopoda Dibranchiata) fossil from the Kulm of Northern Moravia (Czech Republic). In *Berliner Paläobiologisch Abhandlungen*. Edited by K. Warnke. International Symposium "Coleoids Cephalopods Through Time." Program and Abstracts 1: 58–60.

Kröger, B., and Mapes, R. H. 2007. On the origin of bactritoids (Cephalopoda). In *Seventh International Symposium Cephalopods—Present and Past*, Hokkaido University, Sapporo, Japan, Abstracts Volume, pp. 9–10.

Kulicki, C., Landman, N. H., Heaney, M. J., Mapes, R. H., and Tanabe, K. 2002. Morphology of the early whorls of goniatites from the Carboniferous Buckhorn Asphalt (Oklahoma) with aragonite preservation. In *Cephalopods Present and Past*. Edited by H. Summesberger, K. Histon, and A. Daurer. Vienna: Abhandlungen Geologische Bundesanstalt, pp. 205–224.

Kullmann, J. 1981. Carboniferous goniatites. In *The Ammonoidea*. Edited by M. R. House and J. R. Senior. New York: Academic Press, pp. 37–48.

Kullmann, J. 2002. Ammonoid evolution during critical intervals before and after the Devonian-Carboniferous boundary and the mid-Carboniferous boundary. In *Cephalopods Present and Past*. Edited by H. Summesberger, K. Histon, and A. Daurer. Vienna: Abhandlungen Geologische Bundesanstalt, pp. 371–377.

Landman, N. H., Cochran, J. K., Cerrato, R., Mak, J., Roper, C. F. E., and Lu, C. C. 2004. Habitat and age of the giant squid (*Architeuthis sanctipauli*) inferred from isotopic analyses. *Marine Biology* 144: 685–691.

Landman, N. H., and Davis, R. A. 1988. Jaw and crop preserved in an orthoconic nautiloid cephalopod from the Bear Gulch Limestone (Mississippian, Montana). *New Mexico Bureau of Mines and Mineral Resources, Mem* 44: 103–107.

Landman, N. H., and Cochran, J. K. 1987. Growth and longevity of *Nautilus*. In *Nautilus: The Biology and Paleobiology of a Living Fossil*. Edited by W. B. Saunders and N. H. Landman. New York: Plenum Press, pp. 401–420.

Lee, P. N., Callaerts, P., de Couet, H. G., and Martindale, M. Q. 2003. Cephalopod *Hox* genes and the origin of morphological novelties. *Nature* 424: 1061–1065.

Lehmann, U. 1967. Ammoniten mit Tintenbeutel. *Paläontologische Zeitschrift* 41: 132–136.

———. 1981. *The Ammonites: Their Life and Their World*. New York: Cambridge University Press.

———. 1985. Zur Anatomie der Ammoniten: Teintenbeutel, Kieman, Augen. *Palaeontologische Zeitschrift* 59: 99–108.

———. 1988. On the dietary habits and locomotion of fossil cephalopods. In *Cephalopods Present and Past*. Edited by J. Wiedmann, and J. Kullmann. Stuttgart: E. Schweizerbart'sche Verlagsbuchhandlung, pp. 633–640.

Lindberg, D. R., and Ponder, W. F. 1996. An evolutionary tree for the Mollusca: Branches or roots? In *Origin and Evolutionary Radiation of the Mollusca*. Edited by J. Taylor. Oxford: Oxford University Press, pp. 135–154.

Lindgren, A. R., Giribet, G., and Nishiguchi, M. K. 2004. A combined approach to the phylogeny of Cephalopoda (Mollusca). *Cladistics* 20: 454–486.

Lindgren, A. R., Katugin, O. N., Amezquita, E., and Nishiguchi, M. K. 2005. Evolutionary relationships among squids of the family Gonatidae (Mollusca: Cephalopoda) inferred from three mitochondrial loci. *Molecular Phylogenetics and Evolution* 36: 101–111.

Makman, M. H., and Stefano, G. B. 1984. Marine mussels and cephalopods as models for study of neuronal aging. In *Invertebrate Models in Aging Research*. Edited by D. H. Mitchell and T. E. Johnson. Boca Raton, FL: CRC Press, pp. 165–189.

Mapes, R. H. 1979. Carboniferous and Permian Bactritoidea in North America. *University of Kansas Paleontological Contributions* 64:1–75.

———. 1987. Upper Paleozoic cephalopod mandibles: frequency of occurrence, modes of preservation, and paleoecological implications. *Journal of Paleontology* 61: 521–538.

Mapes, R. H., and Chaffin, D. T. 2003. Predation on cephalopods: a general overview with a case study from the upper Carboniferous of Texas. In *Predator-Prey Interactions in the Fossil Record*. Edited by P. H. Kelley, M. Kowalewski, and T. A. Hansen. New York: Kluwer Academic/Plenum Publishers, pp. 177–213.

Mapes, R., and Dalton, R. B. 2002. Scavenging or predation? Mississippian ammonoid accumulations in carbonate concretion halos around *Rayonnoceras* (Actinoceratoidea—Nautiloidea) body chambers from Arkansas. In *Cephalopods Present and Past*. Edited by H. Summesberger, K. Histon, and A. Daurer. Vienna: Abhandlungen Geologische Bundesanstalt, pp. 407–422.

Mapes, R. H., and Davis, R. A. 1996. Color patterns in ammonoids. In *Ammonoid Paleobiology*. Edited by N. H. Landman, K. Tanabe, and R. A. Davis. New York: Plenum Press, pp. 104–127.

Mapes, R., and Evans, T. S. 1995. The color pattern on a Cretaceous nautilid from South Dakota. *Journal of Paleontology* 69:785–786.

Mapes, R. H., Weller, E. A., and Doguzhaeva, L. A. 2007. An early Carboniferous coleoid cephalopod showing a tentacle with arm hooks and ink sac (Montana, USA). In *Seventh International Symposium Cephalopods – Present and Past*, Hokkaido University, Sapporo, Japan, Abstracts, pp. 123–124.

Mariotti, N., and Pignatti, J. S. 1999. The Xiphoteuthididae Bather, 1892 (Aulacocerida, Coleoidea). In *Advancing Research on Living and Fossil Cephalopods*. Edited by F. Olóriz and F. J. Rodríguez-Tovar. New York: Kluwer Academic/Plenum Publishers, pp. 161–170.

Mehl, J. 1990. Fossilerhaltung von Kiemen bei *Plesioteuthis prisca* (Rüppel 1829) aus untertithonen Plattenkalken der Altmühlalb. *Archaeopterix* 8: 77–91.

Miller, A. K. 1930. A new ammonoid fauna of Late Paleozoic age from western Texas. *Journal of Paleontology* 4: 383–412.

Moltschaniwskyj, N. A. 1994. Muscle tissue growth and muscle fibre dynamics in the tropical Loliginid squid *Photololigo* sp. (Cephalopoda: Loliginidae). *Canadian Journal of Fisheries and Aquatic Sciences* 51: 830–835.

———. 2004. Understanding the process of growth in cephalopods. *Marine and Freshwater Research* 55: 379–386.

Montgomery, M. K., and McFall-Ngai, M. J. 1992. The muscle-derived lens of a squid bioluminescent organ is biochemically convergent with the ocular lens. *Journal of Biological Chemistry* 267: 20999–21003.

Moore, R. C. 1957. *Treatise on Invertebrate Paleontology*. Lawrence, KS: Geological Society of America and University of Kansas Press.

———. 1964. *Treatise on Invertebrate Paleontology*. Lawrence, KS: Geological Society of America and University of Kansas Press.

Morton, N. 1981. Aptychi: the myth of the ammonite operculum. *Lethaia* 14: 57–61.

Müller, A. H. 1974. Über den Kieferapparat fossiler und rezenter Nautiliden (Cephalopoda) mit Bemerkungen zur Ökologie, Funktionsweise und Phylogenie. *Freiberger Forschungshefte* C 298: 7–17.

Naef, A. 1916. Ueber neue Sepioloiden aus dem Golf von Neapel. *Pubblicasioni della Stazione Zoologica di Napoli* 1: 1–10.

———. 1921–1923. *Cephalopoden. I. Teil. 1. Band: Systematik*. Fauna i Flora Golfo Napoli, Monograph 35-1 (English translation available from Smithsonian Institution Libraries, Washington, DC, United States).

———. 1922. *Die fossilen Tintenfische—eine paläozoologische Monographie* [English translation, Berliner Paläobiologische Abhandlungen, Band 5 (2004)].

———. 1928. *Die Cephalopoden (Embryologie)*. Fauna i Flora Golfo Napoli, 35(I-2) [English translation (2000) available from the Smithsonian Institution Libraries, Washington, DC, United States].

Nesis, K. N. 1980. Sepiids and loliginids: A comparative review of distribution and evolution of neritic cephalopods. *Zooogicheskii Zhurnal, Moscow* 59: 677–688.

———. 1987. *Cephalopods of the World.* Neptune City, NJ: T.F.H. Publications.

Nishiguchi, M. K. 2002. The use of physiological data to corroborate cospeciation events in symbiosis. In *Molecular Systematics and Evolution: Theory and Practice.* Edited by R. DeSalle, W. Wheeler, and G. Giribet. Basel: Birkhäuser, pp. 237–246.

Nishiguchi, M. K., Lopez, J. E., and Boletzky, S. v. 2004. Enlightenment of old ideas from new investigations: The evolution of bacteriogenic light organs in squids. *Evolution and Development* 6: 41–49.

Nishiguchi, M. K., Ruby, E. G., and McFall-Ngai, M. J. 1998. Competitive dominance among strains of luminous bacteria provides an unusual form of evidence for parallel evolution in sepiolid squid-*Vibrio* symbioses. *Applied and Environmental Microbiology* 64: 3209–3213.

Nixon, M. 1988. The feeding mechanisms and diets of cephalopods—living and fossil. In *Cephalopods Present and Past.* Edited by J. Wiedmann and J. Kullmann. Stuttgart: Schweizerbart'sche Verlagsbuchhandlung, pp. 641–952.

———. 1996. Morphology of the jaws and radula in ammonoids. In *Ammonoid Paleobiology.* Edited by N. H. Landman, K. Tanabe, and R. A. Davis. New York: Plenum Press, pp. 23–42.

Nixon, M., and Young, J. Z. 2003. *The Brains and Lives of Cephalopods.* New York: Oxford University Press.

Norbert, T., and Klages, W. 1996. Cephalopods as prey. II. Seals. *Philosophical Transactions of the Royal Society Series B: Biological Sciences* 351: 1045–1052.

Norman, M. 2000. *Cephalopods: a World Guide.* Hackenheim, Germany: Conch Books.

O'Dor, R. K., and Dawe, E. G. 1998. *Illex illecebrosus.* In *Squid Recruitment Dynamics—The Genus* Illex *as a Model, the Commercial* Illex *Species and Influences on Variability.* Edited by P. G. Rodhouse, E. G. Dawe, and R. K. O'Dor. FAO Fisheries Technical Paper 376. Rome: Food and Agriculture Organization of the United Nations, pp. 77–104.

Orlov, Y. A. 1962. Fundamentals of paleontology. *Mollusca–Cephalopoda* 5: 887.

Owen, R. 1836. Description of some new and rare Cephalopoda. *Proceedings of the Zoological Society of London* 4: 19–24.

Packard, A. 1972. Cephalopods and fish: the limits of convergence. *Biological Reviews* 47: 241–307.

———. 1985. A tale from the deep. *The Times* (London), October 26, 1985, No. 62,279: 11.

———. 1988. Visual tactics and evolutionary strategies. In *Cephalopods Present and Past.* Edited

by J. Wiedmann and J. Kullmann. Stuttgart: Schweizebart'sche Verlagsbuchhandlung, pp. 89–103.

Page, K. 1996. Mesozoic ammonoids in time and space. In *Ammonoid Paleobiology.* Edited by N. H. Landman, K. Tanabe, and R. A. Davis. New York: Plenum Press, pp. 755–794.

Passamaneck, Y. J., Schander, C., and Halanych, K. M. 2004. Investigation of molluscan phylogeny using large-subunit and small-subunit nuclear rRNA sequences. *Molecular Phylogenetics and Evolution* 32: 25–38.

Pecl, G. T., and Moltschaniwskyj, N. A. 1997. Changes in muscle structure associated with somatic growth in *Idiosepius pygmaeus*, a small tropical cephalopod. *Journal of Zoology London* 242: 751–764.

Pichon, D., Gaia, V., Norman, M. D., and Boucher-Rodoni, R. 2005. Phylogenetic diversity of epibiotic bacteria in the accessory nidamental glands of squids (Cephalopoda: Loliginidae and Idiosepiidae). *Marine Biology* 147: 1323–1332.

Piertney, S. B., Hudelot, C., Hochberg, F. G., and Collins, M. A. 2003. Phylogenetic relationships among cirrate octopods (Mollusca: Cephalopoda) resolved using mitochondrial 16S ribosomal DNA sequences. *Molecular Phylogenetics and Evolution* 27: 348–353.

Pignatti, J. S., and Mariotti, N. 1996. Systematics and phylogeny of the Coleoidea (Cephalopoda): a comment upon recent works and their bearing on the classification of the Aulacocerida. *Palaeopelagos* 5: 33–44.

Quinn, J. H. 1977. Sedimentary processes in *Rayonnoceras* burial. *Fieldiana Geology* 33: 511–519.

Reichow, D., and Smith, M. J. 2001. Microsatellites reveal high levels of gene flow among populations of the California squid *Loligo opalescens*. *Molecular Ecology* 10: 1101–1109.

Roper, C. F. E. 1969. Systematics and zoogeography of the worldwide bathypelagic squid *Bathyteuthis* (Cephalopoda: Oegopsida). *Bulletin of the United States National Museum* 291: 1–210.

Salvini-Plawen, L. v., and Steiner, G. 1996. Synapomorphies and plesiomorphies in higher classification of Mollusca. In *Origin and Evolutionary Radiation of the Mollusca.* Edited by J. Taylor. Oxford: Oxford University Press, pp. 29–51.

Saunders, W. B., and E. S. Richardson. 1979. Middle Pennsylvanian (Desmoinesean) (*sic*) Cephalopoda of the Mazon Creek fauna, northeastern Illinois. In *Mazon Creek Fossils.* Edited by M. H. Nitecki. New York: Academic Press, pp. 333–359.

Seibel, B. A., Hochberg, F. G., and Carlini, D. C. 2000. Life history of *Gonatus onyx* (Cephalopoda:

Teuthoidea): deep-sea spawning and post-spawning egg care. *Marine Biology* 137: 519–526.

Seilacher, A. 1993. Ammonite aptychi: how to transform a jaw into an operculum? *American Journal of Science* 293A: 20–32.

Semmens, J. M., and Moltschaniwskyj, N. A. 2000. An examination of variable growth in the loliginid squid *Sepioteuthis lessoniana*: a whole animal and reductionist approach. *Marine Ecology Progress Series* 193: 135–141.

Shaw, P. W., Pierce, G. J., and Boyle, P. R. 1999. Subtle population structuring within a highly vagile marine invertebrate, the veined squid *Loligo forbesi*, demonstrated with microsatellite DNA markers. *Molecular Ecology* 8: 407–417.

Shea, E. K. 2005. Ontogeny of the fused tentacles in three species of ommastrephid squids (Cephalopoda, Ommastrephidae). *Invertebrate Zoology* 124. 25–38.

Shigeno, S., Kidokoro, H., Tsuchiya, K., Segawa, S., and Yamamoto, M. 2001a. Development of the brain in the Oegopsid squid, *Todarodes pacificus*: An atlas up to the hatchling stage. *Zoological Science* 18: 527–541.

Shigeno, S., Tsuchiya, K., and Segawa, S. 2001b. Embryonic and paralarval development of the central nervous system of the loliginid squid *Sepioteuthis lessoniana*. *Journal of Comparative Neurology* 437: 449–475.

Shigeno, S., and Yamamoto, M. 2002. Organization of the nervous system in the pygmy cuttlefish, *Idiosepius paradoxus* Ortmann (Idiosepiidae, Cephalopoda). *Journal of Morphology* 254: 65–80.

Signor, P. W., and Brett, C. E. 1984. The mid-Paleozoic precursor to the Mesozoic marine revolution. *Paleobiology* 10: 229–245.

Small, A. L., and McFall-Ngai, M. J. 1998. A halide peroxidase in tissues interacting with bacteria in the squids *Euprymna scolopes*. *Journal of Cellular Biochemistry* 72: 445–457.

Söller, R., Warnke, K., Saint-Paul, U., and Blohm, D. 2000. Sequence divergence of mitochondrial DNA indicates cryptic biodiversity in *Octopus vulgaris* and supports the taxonomic distinctiveness of *Octopus mimus* (Cephalopoda: Octopodidae). *Marine Biology* 136: 29–35.

Squires, R. L. 1973. Burial environment, diagenesis, mineralogy, and Mg and Sr contents of skeletal carbonates in the Buckhorn Asphalt of Middle Pennsylvanian age, Arbuckle Mountains, Oklahoma. PhD. thesis, Department of Paleontology, California Institute of Technology, Pasadena.

Steer, M. A., Moltschaniwskyj, N. A., Nichols, D. S., and Miller, M. 2004. The role of temperature and maternal ration in embryo survival: using the dumpling squid *Euprymna tasmanica* as a model. *Journal of Experimental Marine Biology and Ecology* 307: 73–89.

Steiner, G., and Dreyer, H. 2003. Molecular phylogeny of Scaphopoda (Mollusca) inferred from 18S rDNA sequences: support for a scaphopod-cephalopod clade. *Zoologica Scripta* 32: 343–356.

Stevens, G. R. 1988. Giant ammonites: a review. In *Cephalopods Present and Past*. Edited by J. Wiedmann and J. Kullmann. Stuttgart: Schweizerbart'sche Verlagsbuchhandlung, pp. 140–166.

Strugnell, J., Norman, M. D., Drummond, A. J., and Cooper, A. 2004. The octopuses that never came back to earth: neotenous origins for pelagic octopuses. *Current Biology* 18: R300–R301.

Strugnell, J., Norman, M., Jackson, J., Drummond, A. J., and Cooper, A. 2005. Molecular phylogeny of coleoid cephalopods (Mollusca: Cephalopoda) using a multigene approach; the effect of data partitioning on resolving phylogenies in a Bayesian framework. *Molecular Phylogenetics and Evolution* 37: 426–441.

Sweeney, M. J., and Roper, C. F. E. 1998. Classification, type localities, and type repositories of Recent Cephalopoda. In *Systematics and Biogeography of Cephalopods*. Edited by N. A. Voss, M. Vecchione, R. B. Toll, and M. J. Sweeney. Smithsonian Contributions to Zoology 586. Washington, DC: Smithsonian Institution, pp. 561–599.

Takumiya, M., Kobayashi, M., Tsuneki, K., and Furuya, H. 2005. Phylogenetic relationships among major species of Japanese coleoid cephalopods (Mollusca: Cephalopoda) using three mitochondrial DNA sequences. *Zoological Science* 22: 147–155.

Tanabe, K., and Landman, N. H. 2002. Morphological diversity of the jaws of Cretaceous Ammonoidea. In *Cephalopods Past and Present*. Edited by H. Summesberger, K. Histon, and A. Daurer. Vienna: Abhandlungen Geologische Bundesanstalt, pp. 157–165.

Tanabe, K., Mapes, R., Sasaki, T., and Landman, N. H. 2000. Soft-part anatomy of the siphuncle in Permian ammonoids. *Lethaia* 33: 83–91.

Teichert, C. 1964a. Glossary of morphological terms used for nautiloids. In *Treatise on Invertebrate Paleontology*. Edited by R. C. Moore. Lawrence, KS: Geological Society of America and University of Kansas Press, pp. K54–K59.

———. 1964b. Morphology of hard parts. In *Treatise on Invertebrate Paleontology*. Edited by R. C. Moore. Lawrence, KS: Geological Society of America and University of Kansas Press, pp. K13–K53.

———. 1967. Major features of cephalopod evolution. In *Essays of Paleontology and Stratigraphy*. Edited by C. Teichert and E. L. Yochelson. Lawrence: University of Kansas Press, pp. 162–210.

———. 1988. Main features of cephalopod evolution. In *The Mollusca*. Edited by M. R. Clarke and E. R. Trueman. San Diego: Academic Press, pp. 11–79.

Teichert, C., and Matsumoto, T. 1987. The ancestry of the genus *Nautilus*. In *Nautilus: The Biology and Paleobiology of a Living Fossil*. Edited by W. B. Saunders and N. H. Landman. New York: Plenum Press, pp. 25–32.

Teichert, C., and Moore, R. C. 1964. Classification and stratigraphic distribution. In *Treatise on Invertebrate Paleontology*. Edited by R. C. Moore, Lawrence, KS: Geological Society of America and University of Kansas Press, pp. K94–K106.

Termier, H., and Termier, G. 1971. Les Prebelemnitida: un nouvel ordre des céphalopodes. *Annales de la Sociétéé Géologique du Nord* 90: 109–112.

Toll, R. B. 1982. The comparative morphology of the gladius in the order Teuthoidea (Mollusca: Cephalopoda) in relation to systematics and phylogeny. Ph.D. thesis, University of Miami, Coral Gables, FL.

———. 1998. The gladius in cephalopod systematics. In *Systematics and Biogeography of Cephalopods*. Edited by N. A. Voss, M. Vecchione, R. B. Toll, and M. J. Sweeney. Smithsonian Contributions to Zoology 586. Washington, DC: Smithsonian Institution, pp. 55–68.

Tomarev, S. I., Callaerts, P., Kos, L., Zinovieva, R., Halder, G., Gehring, W., and Piatigorsky, J. 1997. Squid *Pax-6* and eye development. *Proceedings of the National Academy of Sciences of the United States of America* 94: 2421–2426.

Tozer, E. T. 1981. Triassic Ammonoidea: classification, evolution and relationship with the Permian and Jurassic forms. In *The Ammonoidea*. Edited by M. R. House and J. R. Senior. New York: Academic Press, pp. 49–100.

Uchiyama, K., and Tanabe, K. 1999. Hatching of *Nautilus macromphalus* in the Toba Aquarium, Japan. In *Advancing Research on Living and Fossil Cephalopods*. Edited by F. Olóriz and F. J. Rodríguez-Tovar. New York: Kluwer Academic/Plenum Publishers. pp. 13–16.

Vecchione, M., Young, R. E., Donovan, D. T., and Rodhouse, P. G. 1999. Reevaluation of coleoid cephalopod relationships based on modified arms in the Jurassic coleoid Mastigophora. *Lethaia* 32: 113–118.

Vermeij, G. 1987. *Evolution and Escalation. An Ecological History of Life*. Princeton, NJ: Princeton University Press.

Verne, J. 1896. *Twenty Thousand Leagues under the Sea*. New York: Grosset and Dunlap.

Voight, J. R. 1997. Cladistic analysis of the octopods based on anatomical characters. *Journal of Molluscan Studies* 63: 311–325.

Voight, J. R., and Grehan, A. J. 2000. Egg brooding by deep-sea octopuses in the North Pacific Ocean. *Biological Bulletin* 198: 94–100.

Voss, G. L. 1977. Appendix II: Classification of recent cephalopods. *Symposium Zoological Society of London* 38: 575–579.

Waller, T. R. 1998. Origin of Molluscan class Bivalvia and a phylogeny of major groups. In *Bivalves: An Eon of Evolution*. Edited by P. A. Johnston, and J. W. Haggart. Calgary, Alberta, Canada: University of Calgary Press, pp. 1–45.

Ward, P. D. 1981. Shell sculpture as a defensive adaptation in ammonoids. *Paleobiology* 7: 96–100.

———. 1987. *The Natural History of Nautilus*. Boston: Allen and Unwin.

Ward, P. D., and Saunders, W. B. 1997. *Allonautilus*: A new genus of living nautiloid cephalopod and its bearing on phylogeny of the Nautilida. *Journal of Paleontology* 71: 1054–1064.

Warnke, K., Plötner, J., Santana, J. I., Rueda, M. J., and Llinas, O. 2003. Reflections on the phylogenetic position of *Spirula* (Cephalopoda): preliminary evidence from the 18S ribosomal RNA gene. In *Coleoid cephalopods through time*. Edited by K. Warnke, H. Keupp, and S. v. Boletzky. Berlin: Berliner Paläobiologische Abhandlungen, pp. 253–260.

Wells, M. J., and Wells, J. 1977. Cephalopoda: Octopoda. In *Reproduction of Marine Invertebrates*, Vol. 4: *Molluscs: Gastropods and Cephalopods*. New York: Academic Press, pp. 291–336.

Wiedmann, J. and Kullmann, J. 1988. Crises in ammonoid evolution. In *Ammonoid Paleobiology*. Edited by N. H. Landman, K. Tanabe, and R. A. Davis. New York: Plenum Press, pp. 795–813.

Wiedmann, J., and J. Kullmann. 1996. Crises in ammonoid evolution. In *Ammonoid Paleobiology*. Edited by N. H. Landman, K. Tanabe, and R. A. Davis. New York: Plenum Press, pp. 795–813.

Woodruff, D. S., Carpenter, M. P., Saunders, W. B., and Ward, P. D. 1987. Genetic variation and phylogeny in *Nautilus*. In *Nautilus: The Biology and Paleobiology of a Living Fossil*. Edited by W. B. Saunders, and N. H. Landman. New York: Plenum Press, pp. 65–83.

Wray, C. G., Landman, N. H., Saunders, W. B., and Bonacum, J. 1995. Genetic diversity and geographic diversification in *Nautilus*. *Paleobiology* 21: 220–228.

Wright, C. W. 1981. Cretaceous Ammonoidea. In *The Ammonoidea*. Edited by M. R. House and J. R. Senior. New York: Academic Press, pp. 157–174.

Yokobori, S.-I., Fukuda, N., Nakamura, M., Aoyama, T., and Oshima, T. 2004. Long-term conservation of six duplicated structural genes in Cephalopod mitochondrial genomes. *Molecular Biology and Evolution* 21: 2034–2046.

Young, J. Z. 1936. The giant nerve fibers and epistellar body of cephalopods. *Quarterly Journal of Microscopy Sciences* 78: 367–386.

———. 1960. The statocysts of *Octopus vulgaris*. *Philosophical Transactions of the Royal Society of London Series B: Biological Sciences* 152: 3–29.

———. 1965. The central nervous system of *Nautilus*. *Philosophical Transactions of the Royal Society of London Series B: Biological Sciences* 249: 1–25.

———. 1977. Brain, behavior, and evolution of cephalopods. *Symposium of the Zoological Society of London* 38: 377–434.

———. 1984. The statocysts of cranchid squids (Cephalopoda) *Journal of Zoology London* 203: 1–21.

———. 1988a. Evolution of the cephalopod brain. In *The Mollusca, Paleontology and Neontology of Cephalopods*. Edited by M. R. Clarke and E. R. Trueman. San Diego: Academic Press, pp. 215–228.

———. 1988b. Evolution of the cephalopod statocyst. In *The Mollusca, Paleontology and Neontology of Cephalopods*. Edited by M. R. Clarke and E. R. Trueman. San Diego: Academic Press, pp. 229–239.

Young, R. E. 1977. Ventral bioluminescence countershading in mid-water animals: evidence from living squid. *Science* 191: 1046–1048.

———. 1985. The common occurrence of oegopsid squid eggs in near-surface oceanic waters. *Pacific Science* 39: 359–366.

Young, R. E., and Bennett, T. M. 1988. Photophore structure and evolution within the Enoploteuthinae (Cephalopoda). In *The Mollusca, Paleontology and Neontology of Cephalopods*. Edited by M. R. Clarke and E. R. Trueman. San Diego: Academic Press, pp. 241–251.

Young, R. E., Kampa, E. M., Maynard, S. D., Mencher, F. M., and Roper, C. F. E. 1980. Counterillumination and the upper depth limits of midwater animals. *Deep-Sea Research* 27A: 671–691.

Young, R. E., and Roper, C. F. E. 1976. Intensity regulation of bioluminescence during countershading in living midwater animals. *Fishery Bulletin* 75: 239–252.

Young, R. E., Roper, C. F. E., Mangold, K., Leisman, G., and Hochberg, F. G. 1979a. Luminescence from non-bioluminescent tissues in oceanic Cephalopods. *Marine Biology* 53: 69–77.

Young, R. E., Roper, C. F. E., and Walters, J. F. 1979b. Eyes and extraocular photoreceptors in midwater cephalopods and fishes: their roles in detecting down-welling light for counterillumination. *Marine Biology* 51: 371–380.

Young, R. E., and Vecchione, M. 1996. Analysis of morphology to determine primary sister-taxon relationships within coleoid cephalopods. *American Malacological Bulletin* 12: 91–112.

———. 1999. *Octopodiformes*. Tree of Life web site. Tucson: The University of Arizona College of Agriculture and Life Sciences. http://tolweb.org/Octopodiformes.

Young, R. E., Vecchione, M., and Donovan, D. T. 1998. The evolution of coleoid cephalopods and their present biodiversity and ecology. In *Cephalopod Biodiversity, Ecology, and Evolution*. Edited by A. I. Payne, M. R. Lipinski, M. R. Clarke, and M. A. C. Roeleveld. *South African Journal of Marine Science* 20: 393–420.

Zinovieva, R. D., Tomarev, S. I., and Piatigorsky, J. 1993. Aldehyde dehydrogenase-derived omega-crystallins of squid and octopus. Specialization for lens expression. *Journal of Biological Chemistry* 268: 11449–11455.

9

Gastropoda

AN OVERVIEW AND ANALYSIS

*Stephanie W. Aktipis, Gonzalo Giribet, David R. Lindberg,
and Winston F. Ponder*

Gastropoda is the largest molluscan class and the second most speciose animal class, and it contains the greatest diversity of described marine species. Gastropods are characterized by having a single shell and an operculum, at least in the larval stage, and by the larva having undergone torsion. The group has radiated enormously in comparison to other molluscan classes and shows great disparity in external form, anatomy, behavior, and physiology, in part responsible for the enormous diversity of the clade. The adult body is generally covered with a single shell, which is often coiled and torted (usually dextrally) but may also be limpet-like (e.g., Patellogastropoda, Figure 9.1A; Cocculoidea; Fissurellidae, Figure 9.1B; Lepetodrilidae; Peltospiridae; Neomphalidae; Phenacolepadidae; Calyptraeidae; Hipponicidae; Trimusculidae) or even tubular (Caecidae, Figure 9.1D) or bivalved (Juliidae). Many have reduced or entirely lost their shell and have become slugs (e.g., Opisthobranchia, Systellommatophora, Arionidae, Cystopeltidae, Testacellidae, Succineidae, Athoracophoridae). Gastropods range in size from less than 1 mm (e.g., Limacinidae, Scissurellidae) to almost a meter in length (Turbinellidae).

Members of this ancient group have adapted to a variety of environments and are found in almost every ecosystem. Gastropods are present in every marine environment, including extreme ones such as hydrothermal vents (e.g., McLean 1989; Warén 2001; Warén et al. 2003). Terrestrial and freshwater gastropods have invaded these environments on multiple occasions and currently occur on every continent except Antarctica. Furthermore, gastropods are the only molluscan group to have invaded the land.

Estimates of the numbers of extant gastropod species range widely, with the minimum estimate around 40,000 and the maximum approaching 150,000 (Lindberg et al. 2004). They surpass other mollusc classes in species abundance and diversity in most marine environments. In a survey of a New Caledonian coral reef lagoon, almost twice as many gastropods were collected as bivalves, the next most abundant group, with these samples comprising 2,738 species of molluscs (not including cephalopods), 2,187 (almost 80%) of which were gastropods (Bouchet et al. 2002), many representing undescribed species. There may still be some gastropod families such as Eulimidae and

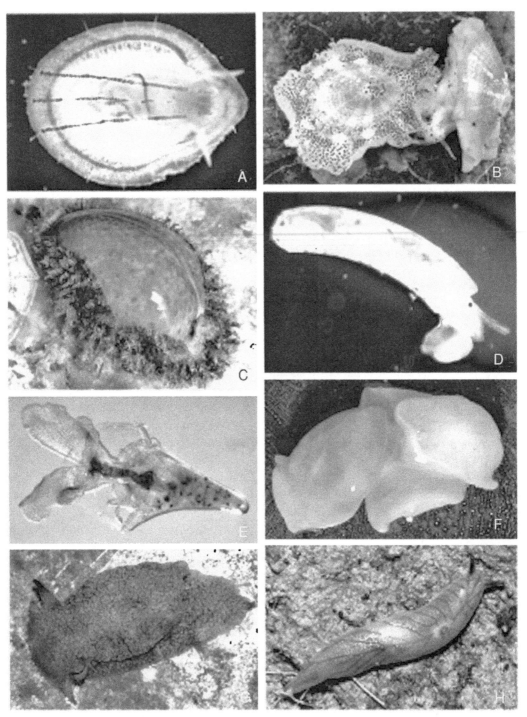

FIGURE 9.1. (A) Patellogastropod (Patellidae), *Helcion pellucidus*, Fiskebäckskil, Sweden (photograph by G. W. Rouse).
(B) Vetigastropoda (Fissurellidae), *Lucapina* sp., Bocas del Toro, Panama (photograph by G. Giribet). (C) Caenogastropoda
(Cypraeidae), *Erosaria spurca*, Cabo de Palos, Spain (photograph by G. Giribet). (D) Caenogastropoda (Caecidae), *Caecum*
sp., Fiskebäckskil, Sweden (photograph by G. W. Rouse). (E) Heterobranchia, Opisthobranchia (Cavolinidae), *Creseis* sp. off
Massachusetts, USA (photograph by A. Schulze). (F) Heterobranchia, Opisthobranchia (Philinidae), *Philine aperta*, Tjärnö, Sweden
(photograph by G. Giribet). (G) Heterobranchia, Opisthobranchia (Pleurobranchidae), *Pleurobranchaea meckeli*, Banyuls sur Mer,
France (photograph by G. Giribet). (H) Heterobranchia, Pulmonata (Athoracophoridae), *Pseudoaneitea* sp., near Rotorua,
New Zealand (photograph by G. Giribet).

Turridae with up to 80% of their species undescribed (Bouchet et al. 2004).

The higher groups of gastropods are now generally accepted as Patellogastropoda (the true limpets), Vetigastropoda (e.g., abalone, top shells, keyhole limpets), Neritimorpha (the nerites), Cocculinida (cocculinoids), Caenogastropoda (periwinkles, whelks, augers, cowries, strombs, cones, etc.) and Heterobranchia (including sea slugs, bubble shells, and many land snails and slugs) (Figure 9.1).

SIGNIFICANCE TO HUMANS

Some gastropods are of direct significance to humans as food sources, jewelry, vectors of disease, commercial pests, or pets. Abalone (Haliotidae), giant conch (Strombidae), periwinkle (Littorinidae), whelks (Buccinidae, Muricidae), escargot (mainly Helicidae), and others such as the apple snails (Ampullariidae), turbinids, trochids and muricids are important food sources in many countries. Gastropods are also harvested for their shells for ornaments, collector's items, and nacre. In historical times, secretions of Mediterranean muricid whelks were used to make the "royal purple" dye by the Romans.

Gastropods have been used in medicine since antiquity (Bonnemain 2005). As early as 450 BCE, snail mucus was used to treat pain related to burns, abscesses, and other skin disorders. In the eighteenth century, various gastropod preparations were used for dermatological disorders and for symptoms associated with tuberculosis and nephritis. In the nineteenth century there was renewed interest in the pharmaceutical and medical use of snails, and this continued into the twentieth century as well. For example, heterobranchs have been useful for investigations of Alzheimer disease (Alkon et al. 1998, Leong et al. 2001) and synthetic drugs derived from peptides found in Conus venom (e.g., ziconotide) have been recently approved for the treatment of chronic pain.

Some freshwater gastropods act as primary hosts to parasitic trematodes such as liver flukes (Fasciola and Opisthorchis), lung flukes (Paragonimus), or blood flukes (Schistosoma). These infections are particularly severe in Southeast Asia and China, where it is estimated that more than 40 million people may be infected by flukes (World Health Organization 2004). Some snails and slugs (mainly pulmonates) are serious agricultural pests. These and other invasive species, such as the freshwater snail Pomacea canaliculata, the giant African snail Achatina fulica, and the temperate common garden snail Cantareus aspersus, are voracious feeders capable of seriously damaging communities into which they have been introduced. Predation by the carnivorous gastropod Euglandina rosea introduced to "control" Achatina has resulted in the extinction of many endemic land snails on Pacific islands (Cowie 1992).

GASTROPOD MORPHOLOGY

Gastropod morphology can vary dramatically, but some general traits are shared among other molluscan groups and within the class. Like most other molluscan clades, they have a shell (nearly always single if present) secreted by the mantle, an open circulatory system that also serves as a hydrostatic skeleton, a radula used during feeding, a large muscular foot for locomotion, and a head. Like that of cephalopods, the head has a pair of eyes, but, uniquely to gastropods, it also bears a pair of cephalic tentacles.

Gastropods are distinguished from all other molluscs by undergoing torsion during development, generally during the late veliger larval stage. Their visceral mass rotates up to 180° with respect to the head and foot as a result of a combination of muscular action and differential growth. Although all gastropods tort in their development, several groups—particularly some opisthobranchs, which may only tort 90° (e.g., Page 1995)—secondarily untwist as postlarvae and are partially to fully detorted as adults.

The internal anatomy of gastropods has been highly modified as a result of the process of torsion. The most obvious modification is the post-torsional anterior placement of the mantle cavity. This, combined with the independent effects of anal-pedal flexure (Lindberg 1985; Ponder and Lindberg 1997) and shell coiling, has resulted in a cascade of evolutionary events that, in most groups, has resulted in rearranged, reduced, or absent organs on one side of the body (generally those on the post-torsional right side) (Lindberg and Ponder 1996; Ponder and Lindberg 1997). Specifically, the most obvious of these changes are the modifications to the plesiomorphic arrangement of mantle cavity organs in gastropods (paired ctenidia, osphradia, and hypobranchial glands), resulting in the independent loss of the right structure in most lineages (for more information see Haszprunar 1988c; Ponder and Lindberg 1996, 1997; Lindberg and Ponder 2001).

The gastropod ctenidium is primitively bipectinate (ctenidial leaflets on both sides of the axis) and is modified to monopectinate (ctenidial leaflets placed only on one side) in caenogastropods and a few vetigastropods. Paired bipectinate ctenidia are only found in some vetigastropods (collectively called Zeugobranchia in some early classifications). Other vetigastropods, neritimorphs, and some patellogastropods, possess a single pretorsional left bipectinate ctenidium, whereas caenogastropods have a single left monopectinate ctenidium. Some patellogastropods (Patellina) lack ctenidia entirely; gas exchange instead occurs through secondary gills within a groove (pallial groove) surrounding the foot. Heterobranchs all lack a true ctenidium (Haszprunar 1985c) but many have one or more secondary gills. Terrestrial gastropods lack ctenidia; respiration instead occurs though the highly vascularized mantle cavity or, in slugs, primarily through the body wall.

The visceral loop of the nervous system of torted gastropods is twisted into a figure eight, a condition known as streptoneury, resulting in the plesiomorphic displacement of the right and left pleural ganglia beneath the esophagus. Secondarily detorted gastropods exhibit the euthyneuran condition, in which the visceral loop is not twisted. In some of these, and most caenogastropods, the pleural ganglia are situated next to, or fused with, their respective cerebral ganglia in an apomorphic condition termed epiathroid. In the plesiomorphic condition (hypoathroid) the pleural ganglia lie nearer, or abut, the pedal ganglia (Fretter and Graham 1962; Haszprunar 1988c).

All living gastropods possess only one gonad—the pretorsional right gonad. Methods of reproduction vary widely; most are dioecious, but simultaneous or sequential hermaphrodites are found in some Patellogastropoda and Caenogastropoda and in all Cocculinoidea and Heterobranchia. Some freshwater gastropod families, such as Hydrobiidae, Viviparidae, and Thiaridae, have parthenogenetic members. Both external and internal fertilization occur, with internally fertilizing gastropods having a wide variety of egg encapsulation methods to protect the developing embryos. These range from hard to leathery egg capsules in neritimorphs and caenogastropods to jelly mucopolysaccharide masses, ribbons, or strings as in many opisthobranchs. Terrestrial snails, notably pulmonates, produce eggs surrounded by a protective (sometimes shelly) coating to reduce desiccation. Some terrestrial, freshwater, and marine gastropods brood embryos internally and give birth to live juveniles. Complex precopulatory behaviors have been observed in some gastropods, mainly pulmonates. In contrast to most other molluscan classes, only patellogastropods and externally fertilizing vetigastropods have a free-swimming trochophore larva that develops into a veliger larva with ciliated swimming lobes. In other gastropods the free-swimming trochophore stage is bypassed and, if present, is contained within the egg capsule, with the veliger the only free-swimming larval stage. Veliger larvae may be planktotrophic or lecithotrophic, or, in those species with direct development, the larval stage may be passed through in the egg capsule or bypassed altogether, the young hatching as juveniles. Nearly all freshwater gastropods and all terrestrial species have direct development, as do many marine gastropods. Various methods are used to furnish the embryo with nutrients during intracapsular development, including yolk, albumen, unfertilized eggs, and other embryos. Brooding may occur in different locations, depending on the group—externally on the shell, within the pallial part of the oviduct, within the mantle cavity, or in a special chamber in the head cavity. Reviews of gastropod reproduction are provided by Fretter and Graham (1962) and by Giese and Pearse (1977), Thompson (1976), Tompa et al. (1984), and Barker (2001b).

The gut morphology is very varied, often reflecting the wide range of feeding habits and diets. Perhaps most obvious, radular morphology varies considerably within gastropods. For example, patellogastropods have stiff, rasping docoglossate radulae, similar to those found in Polyplacophora. Vetigastropods and neritimorphs have a broomlike rhipidoglossate radula with many teeth in each row. The radula is reduced to seven teeth per row (taenioglossate) in many caenogastropods, although some, notably Epitoniidae and Janthinidae, have multiple fang-like teeth in each row (ptenoglossate type). Many of the carnivorous neogastropods have serrated rachiglossate radulae made up of three teeth per row, whereas others, such as Conidae, have highly modified harpoon-like toxoglossate radulae. The radular morphology in the heterobranchs is very diverse, with some even closely related families showing very different patterns. Some pulmonates and opisthobranchs have a broad radula with numerous similar teeth in each row, while in some opisthobranchs the radula is reduced to a few or even a single row of teeth or is lost entirely. Radular morphology may be adapted to a particular food type, but a diverse range of feeding habits can be associated with similar radulae. Notably, the taenioglossan type is found in scrapers, browsers, detritus feeders, filter feeders, grazing herbivores and carnivores, and predators. Some caenogastropods and opisthobranchs are suctorial feeders and typically lose the radula.

SUMMARY OF MAJOR GROUPS

EOGASTROPODA PATELLOGASTROPODA (= DOCOGLOSSA)

All living patellogastropods are limpets with cap-shaped shells. Many of the character states of patellogastropods appear to be plesiomorphic, such as the docoglossan radula, the simple gonad-renal tract of the reproductive system, and tubular protoconch morphology. Synapomorphies include the foliated shell structure, horseshoe-shaped muscles constricted into multiple bundles, the presence of retractile mantle tentacles and sensory strips along the pallial grooves, the lack of a propodium, the lack of heart penetration by the rectum, and two to five pairs of odontophoral cartilages (Ponder and Lindberg 1997; Sasaki 1998).

Patellogastropods range in size from about 3 mm to over 200 mm in length and are distributed throughout the world, where they typically occur on intertidal rock substrates. In addition, some species utilize algae and plants as substrates. In the deep sea, patellogastropods are also found at both cold seeps and hydrothermal vent sites. While most are marine, a few live in estuaries, and one species has invaded the brackish lower reaches of rivers (Lindberg 1990). The earliest patellogastropod verified by shell microstructure is from the Triassic of Italy (Hedegaard et al. 1997). However, it is likely that the group is descended from yet unrecognized, coiled Paleozoic group(s) (Ponder and Lindberg 1997).

Morphological phylogenetic analyses typically place patellogastropods at the base of the gastropod tree as the sister taxon to all other gastropods (Orthogastropoda), and, together with their assumed coiled ancestors, they constitute the clade Eogastropoda.

Patellogastropods have been traditionally divided into two major suborders: the Patellina and Acmaeina. However, the possibility that Patellogastropoda is a paraphyletic taxon and actually represents two distinct limpet groups that do not share a common patelliform ancestor has recently been raised (see Lindberg, Chapter 11).

ORTHOGASTROPODA

This group includes all the remaining gastropods that differ from patellogastropods by having hypobranchial gland(s), a propodium and anterior pedal gland, paired jaws, and a substantial reduction in the extent of the pallial nerves.

Much of the success of orthogastropods was probably due to the evolution of the rhipidoglossate radula. Although radular tooth number and disparity would be again reduced in later taxa, the evolution of the rhipidoglossate condition in conjunction with enhanced flexoglossate bending was a major innovation in gastropod evolution. This coupling of diverse tooth morphologies (especially those of the marginal teeth) with the longitudinal bending plane enabled the radular teeth to adopt new functions as shears, brooms, and knives (Guralnick and Smith 1999). It is likely that this new, highly versatile structure provided the early orthogastropods with highly efficient and diverse feeding modes.

VETIGASTROPODA

Vetigastropods are a large and diverse group of marine gastropods. Some maintain bilateral symmetry in organ systems other than the gut, kidneys, and gonad, and the shells of these taxa typically have slits or other openings. Synapomorphies include the pallial position of the renal glands, an epipodium often bearing tentacles, and the presence of unique sensory structures: ctenidial bursicles and sensory papillae (Ponder and Lindberg 1997; Geiger et al., Chapter 12).

At adult size, vetigastropods range from less than 1 mm to more than 300 mm in length, and they are found in most marine habitats, from the intertidal to the deep sea. They occur on rocky substrates, both on and in soft sediments and in the deep sea, including some on exotic biogenic habitats such as plant debris (e.g., waterlogged wood), egg cases of sharks and skates, and whale bones.

Although most living vetigastropods form a well demarcated group (see Geiger et al., Chapter 12) placed near the base of the gastropod tree, some Paleozoic clades (notably Bellerophonta and Euomphalina) are difficult to place, with differing interpretations as to their inclusion or exclusion from Vetigastropoda (see Frýda et al., Chapter 10). Impressive radiations have occurred at deep-sea hydrothermal vents and cold seeps, some of which, the "hot-vent taxa," are also problematic in their relationships and are not unequivocally part of Vetigastropoda (see subsequent discussion and Geiger et al., Chapter 12). Geiger et al. (Chapter 12) consider the clades Haliotidae, Scissurellidae, Pleurotomariidae, Fissurelloidea, Lepetelloidea, and Trochoidea to be unequivocal members of Vetigastropoda.

NERITIMORPHA (= NERITOPSINA)

Neritimorpha are largely a marine group with some terrestrial and freshwater taxa. Plesiomorphic characters include the rhipidoglossate radula, single bipectinate gill (which lacks skeletal rods), and three pairs of radular cartilages. Synapomorphies include the eccentric operculum, which typically has a peg on the inner surface, a unique accretionary protoconch (but see Frýda et al., Chapter 10), characteristic sperm ultrastructure, several unique odontophoral muscles, posterior esophageal glands, and a diaulic female reproductive system (Ponder and Lindberg 1997; Sasaki 1998; see Lindberg, Chapter 11, for details).

Neritimorphs have a relatively small size range: 2 to 40 mm in length. One taxon (Titiscaniidae) lacks a shell. In marine and freshwater realms neritimorphs are typically associated with hard substrates. Multiple terrestrial (Helicinidae, Hydrocenidae) and freshwater invasions have occurred (Neritidae), with some freshwater taxa still having an estuarine or marine larval phase. The earliest unequivocal record of Neritimorpha is Triassic (Bandel and Frýda 1999; Frýda et al., Chapter 10). Earlier Paleozoic records are based on teleoconch and protoconch similarities shared between extinct taxa, including some platycerids, and later neritimorph taxa. However, the recent discovery of fossilized soft parts in a platyceratid (Sutton et al. 2006) suggests that they may have shared anatomical similarities with patellogastropods rather than neritimorphs (see also Lindberg 2007, Chapter 11).

Although neritimorphs were noted as being very distinct from other "archaeogastropods" in the early twentieth century (e.g., Bourne 1908; Yonge 1947, Figure 9.2B) it is only in the last 30 years that this view has been broadly accepted. Their placement among the other gastropod groups has been problematic, although some now consider them to be the sister taxon of Apogastropoda (Caenogastropoda + Heterobranchia) (e.g., McArthur and Harasewych 2003).

COCCULIN = OIDEA (COCCULINIFORMIA IN PART)

All living members of this small, mainly deep-sea group have white, cap-shaped shells. Cocculinoid gills are vestigial and do not resemble the typical molluscan ctenidium; they may consist of a large folded or small papillate pseudoplicate gill or are further reduced to just respiratory leaflets on the dorsal surface of the mantle cavity (Strong et al. 2003). All cocculinoids are simultaneous hermaphrodites, producing eggs and sperm in a single gonad that opens into the mantle cavity through a glandular oviduct (Haszprunar 1988b); a single or paired seminal receptacle is present. Male reproductive anatomy includes a penis typically associated with the right cephalic tentacle. A pallial brood chamber is present in most taxa.

Cocculinoidean shells range in size from about 2 mm to 15 mm in length. They are distributed throughout the world's oceans at bathyal and abyssal depths, where they live on biogenic substrates including cephalopod beaks, waterlogged wood, and whale bones. They are likely descended from coiled snails, but their ancestors have not yet been identified in the fossil record. The earliest occurrence of cocculinoids in the fossil record is in the Cretaceous of Madagascar (Kiel 2006).

Their placement among the other gastropods is problematic; recent analyses placed them near the base of the gastropod tree (thereby representing an early offshoot in gastropod evolution), as sister to Neritimorpha

or a separate group near the vetigastropods. A detailed account of this group and the remarkably similar vetigastropod limpets that were together treated as the Cocculiniformia is provided by Haszprunar (1988b). Although the name Cocculiniformia continues to be used by some workers, subsequent anatomical and molecular analyses (Ponder and Lindberg 1997; Sasaki 1998; McArthur and Harasewych 2003; Colgan et al. 2000, 2003) have confirmed the paraphyletic nature of the "Cocculiniformia," and continued usage is unwarranted (see also Lindberg, Chapter 11).

CAENOGASTROPODA

Caenogastropods constitute the most diverse gastropod group living today and include almost every gastropod shell form and are successful in every major habitat. Synapomorphies that support the caenogastropod clade include the presence of a single monopectinate gill with skeletal rods, a well-developed osphradium, differentiated renal lamellae, presence of parasperm in many clades, one or two marginal radular teeth, reduced intestinal looping and anterior anus position, an epiathroid nerve ring and many anterior pedal nerves (Ponder and Lindberg 1997).

Caenogastropods include some of the largest and smallest gastropods, ranging in adult size from about 1 mm to over 900 mm (Ponder et al., Chapter 13). Most are marine, but several groups thrive in freshwater or terrestrial environments. The earliest caenogastropods appear in Late Silurian and Early Devonian rocks (see Frýda et al., Chapter 10) (i.e., from about 418 million years ago), with one major group, Neogastropoda, first appearing during the Cretaceous. For further details see Ponder et al., 2007, Chapter 13.

HETEROBRANCHIA

Like caenogastropods, Heterobranchia are a diverse gastropod group and include a wide array of gastropod shell forms, although many, such as the nudibranch and pulmonate slugs, have lost their shells. The heterobranch clade is supported by numerous synapomorphies, including a sinistral larval shell, pigmented mantle organ (PMO) in basal members, lack of a true ctenidium, simple oesophagus, lack of true odontophoral cartilages, medial position of the eyes in many taxa, and distinctive sperm ultrastructure characters (Hazprunar 1985c Ponder and Lindberg 1997).

Heterobranchs range in adult size from about 2 mm to over 750 mm. Two major groups—opisthobranchs (see Wägele et al., Chapter 14) and pulmonates (see Mordan and Wade 2007, Chapter 15) are recognized, as well as several basal groups (Valvatoidea, Rissoelloidea, Architectonicoidea, Pyramidelloidea). Most opisthobranchs are found in the marine environment, whereas the Pulmonata are predominately terrestrial with some marine and limnic groups. Like Caenogastropoda, heterobranchs first appear in Late Silurian and Early Devonian rocks from about 416 Mya (Frýda et al. 2007, Chapter 10).

GASTROPODS IN THE FOSSIL RECORD

The placement of some Cambrian and other lower Paleozoic univalve fossils within Gastropoda is questionable (e.g., see differing interpretations in Parkhaev, Chapter 3, and Frýda et al., Chapter 10). As detailed analyses of putative gastropods through the Cambrian (Parkhaev, Chapter 3) and undoubted gastropods through the rest of the Paleozoic (Frýda et al., 2007, Chapter 10) is given elsewhere in this volume, only a brief outline is provided here. Undoubted gastropods first appear during the Upper Cambrian, with members of one early lineage (Pleurotomariidae) still found in our oceans today (Harasewych 2002).

Gastropods experienced several radiations after their origination in the Cambrian. The initial diversification occurred during the Ordovician, with later major extinction/ origination events occurring between the Permian and Triassic and between the Triassic and Jurassic. These radiations occurred extremely rapidly (Erwin 1990), a fact that further complicates attempts to understand the timing of these divergences as well as the relationships of extinct and extant gastropod clades.

Significant early works concerning the phylogeny of extinct gastropods were published by Knight (1952) and Knight et al. (1954, 1960), and there have been a few recent attempts to undertake large-scale phylogenetic analyses using fossils (see also Frýda et al., Chapter 10, and Ponder et al., Chapter 13). Bandel (1997) attempted to determine the relationship of fossils to other extant gastropods, although he did not conduct a numerical parsimony analysis. Wagner (2002) performed an extensive cladistic analysis of early Paleozoic fossils based on an exemplar approach (*sensu* Prendini 2001), in which each of the 295 terminal taxa were treated as individual species, thus preventing the formulation of *a priori* assumptions of familial classification. Wagner's resulting phylogenetic hypothesis provided information concerning the evolution of early gastropods, but with equivocal results regarding the origin of modern groups.

PREVIOUS CLASSIFICATIONS

Influential phylogenetic schemes proposed for the Gastropoda in the last century are shown in Figure 9.2. Although many classification schemes based upon specific organ systems or characters were proposed in the nineteenth century (see Bieler 1992 and Ponder and Lindberg 1997 for overviews), modern gastropod classification begins with Thiele's (1929–1935) comprehensive taxonomic work in the early twentieth century (Figure 9.2A). This arrangement continued relatively unmodified through most of the twentieth century (Wenz 1938–1944; Cox 1960; Taylor and Sohl 1962; Morton and Yonge 1964).

A scheme proposed by Golikov and Starobogatov (1975) (Figure 9.2c) promoted much discussion, although it was neither the first nor the only paper published by Russian malacologists revising the phylogeny of "prosobranch" gastropods (Kay et al. 1998). However, it is one of the few revisions published in English and as such is most commonly cited in the literature. This controversial and complicated gastropod classification splits "prosobranchs" into three new subclasses, resurrecting many "traditional" orders and families or establishing new ones. Overall, 26 new gastropod clades were identified. However, the majority of these clades have never been widely accepted.

The 1980s marked the start of a renewed interest in gastropod phylogeny. Salvini-Plawen (1980) separated gastropods into four subclasses: Prosobranchia, Pulmonata, Gymnomorpha, and Opisthobranchia. Systematic studies of ultrastructural characters (Haszprunar 1985a, b; Healy 1988) led to the establishment of new concepts for grouping taxa. In particular, Haszprunar (1985c) established Heterobranchia, a clade composed of some taxa previously treated as "prosobranchs" (i.e., caenogastropods) as well as including Pulmonata and Opisthobranchia. In a separate analysis, Salvini-Plawen and Haszprunar (1987) used characters of the nervous system and sense organs to test the monophyly of Vetigastropoda (proposed by Salvini-Plawen (1980) as one of the main clades of "prosobranchs") as well as other higher gastropod clades including Caenogastropoda and Pentaganglionata (= Euthyneura). The families Campanilidae, Valvatidae, Rissoellidae, Omalogyridae, and the clade Allogastropoda were recognized as separate groups located between Caenogastropoda and Pentaganglionata. Finally, Lindberg (1988) formally established the clade Patellogastropoda based on morphological characters.

A later "clado-evolutionary" analysis of the streptoneuran (=prosobranch) gastropods divided the group into two paraphyletic grades: Archaeogastropoda and Apogastropoda (Haszprunar 1988a, c) (Figure 9.2D). Euthyneura was proposed as a clade composed of a monophyletic or paraphyletic Opisthobranchia and monophyletic Pulmonata. This study, although very influential, received criticism for its analytical methodology as well as acceptance of paraphyletic groups in a formal classification (Hickman 1988; Bieler 1990; Ponder and Lindberg 1997).

NUMERICAL PHYLOGENETIC ANALYSES

MORPHOLOGICAL ANALYSES

Ponder and Lindberg (1996) were the first to perform a numerical cladistic analysis for the class Gastropoda. In their complete analysis (1997), 117 morphological characters were coded for 40 taxa, and this analysis (Figure 9.2F) has become the basis of current gastropod classification. Haszprunar's use of paraphyletic taxa was rejected, and Gastropoda were instead split into two monophyletic clades: Eogastropoda (patellogastropods and their relatives) and Orthogastropoda (all remaining gastropods). Some well-supported clades recovered in the analysis include Patellogastropoda, Vetigastropoda, Neritimorpha, Caenogastropoda, and Heterobranchia.

Technological advancements in developmental biology as well as microscopy have allowed for the utilization of additional morphological characters in recent phylogenetic analyses. The use of cell lineage data (van den Biggelaar and Haszprunar 1996; Lindberg and Guralnick 2003) have been shown to be useful for reconstructing relationships among various gastropod clades. Some authors focusing on groups within Gastropoda have successfully used novel morphological characters to elucidate further the relationships of the various gastropod clades (Sasaki 1998; Strong 2003; Barker 2001b; Wägele and Willan 2000), while others have added to our understanding of the levels of character homoplasy among specific gastropod groups (Gosliner 1985, 1991).

Salvini-Plawen and Steiner (1996) (Figure 2.9E) provided several cladistic analyses of molluscan relationships, including two data matrices for streptoneuran and euthyneuran gastropods. In these analyses they used higher taxa as terminals (e.g., Cocculiniformia, Vetigastropoda, Caenogastropoda) and therefore could not test the monophyly of many of these taxa. Nonetheless they found

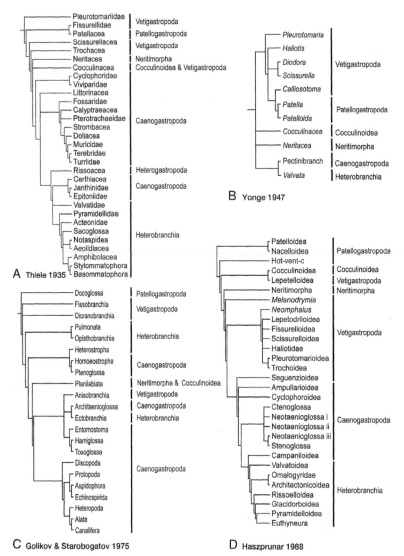

FIGURE 9.2. Phylogenetic hypotheses of gastropod relationships. Terminal taxa follow the authors' original treatment. To the right of each tree the terminal taxa are allocated to the six major gastropod groups as currently recognized. (A) Tree reconstructed following Thiele's (1929–1935 (1935)) discussion of molluscan phylogeny. (B) Tree reconstructed from Yonge's (1947) diagram "illustrating the possible course of evolution within the Prosobranchia." (C) Tree redrawn from Golikov and Starobogatov's "Scheme of the evolution and phylogeny of the Gastropoda." (D) Redrawn from Haszprunar (1988c, fig. 5).

some clades in common with the analyses of Ponder and Lindberg (1996, 1997), such as Orthogastropoda and Heterobranchia. Within Euthyneura they found monophyly of Pulmonata and Opisthobranchia.

MOLECULAR ANALYSES

Molecular analyses of gastropod relationships have varying degrees of success at identifying

the placement of gastropods among the Mollusca and the monophyly of Gastropoda. Early analyses by Rosenberg et al. (1994, 1997) could not establish gastropod monophyly. Winnepenninckx et al. (1994) found monophyly of the class but only fragments of two gastropod sequences were used. A later study (Winnepenninckx et al. 1996) using 18S rRNA complete sequences found monophyly of Gastropoda but

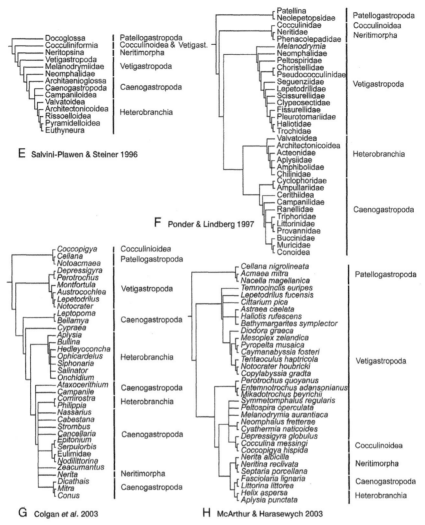

FIGURE 9.2 (cont). (E) Salvini Plawen and Steiner 1996. (F) Ponder and Lindberg 1997. (G) Colgan *et al.* 2000. (H) McArthur and Harasewych 2003. D–H were originally published as cladograms. G and H are molecular analyses, the rest based on morphology.

with low bootstrap support. A separate study using partial 18S rRNA sequences of 27 gastropods by Harasewych *et al.* (1997) also failed to establish the monophyly of Gastropoda.

Establishing supported and stable relationships of groups within Gastropoda has been slightly more successful. Using partial 18S rRNA and cytochrome *c* oxidase subunit I data, Harasewych *et al.* (1997, 1998) found monophyly of higher gastropod groups such as Apogastropoda, Heterobranchia, Caenogastropoda, Patellogastropoda, Neritimorpha, and Vetigastropoda. Colgan *et al.* (2000) analyzed

partial 28S rRNA and histone H3 sequence data and found support for the monophyly of Patellogastropoda and Euthyneura, but failed to recover with strong support other gastropod clades such as Caenogastropoda, Heterobranchia, or Vetigastropoda. The placement of Patellogastropoda was unstable. In a later analysis using six segments from four genes (Colgan *et al.* 2003), only Euthyneura was found monophyletic, and Patellogastropoda was placed in a derived clade with Vetigastropoda (Figure 9.2G). Colgan *et al.*'s (2000, 2003) analyses (Figure 9.2G) failed to recover some higher clades traditionally obtained with morphological data. A recent Bayesian analysis by McArthur and Harasewych (2003) (Figure 9.2H) using full and partial 18S rRNA sequences from 81 taxa found monophyly of Heterobranchia, Caenogastropoda, Apogastropoda, Neritimorpha, and Patellogastropoda. These results can only be considered preliminary because a non-gastropod out group was not included and there was not adequate sampling of the tree space; only 150,000 generations were examined by the Markov chain Monte Carlo (MCMC) sampling procedure, from which 100,000 were discarded as burn-in. Remigio and Hebert (2003) failed to recover any gastropod clade as monophyletic while exploring the effectiveness of cytochrome *c* oxidase subunit I in recovering deep divergences in gastropod phylogeny. Colgan *et al.* (2007), using molecular data from six genes, supported the monophyly of Caenogastropoda, Hypsogastropoda, and Heterobranchia.

McArthur and Harasewych (2003) also performed two supertree analyses, recoding trees from 11 different morphological and molecular data sets. The majority-rule consensus tree from the first analysis, including all 11 studies, recovered only the clade Apogastropoda (= Caenogastropoda + Heterobranchia) with high nodal support, but the use of supertrees in systematics is a contentious issue, especially given the loss of information (e.g., see opposing sides of the debate in Bininda-Emonds *et al.* 2003; Gatesy *et al.* 2004). After the non-cladistic analyses by Thiele (1929–1935), Golikov and Starobogatov (1975), and Haszprunar (1988c) were removed, the majority-rule consensus tree from their reduced analysis recovered three well-supported monophyletic clades: Apogastropoda, Caenogastropoda, and Heterobranchia. The root of this tree was unexpected, placing Neritimorpha as the sister group to all other gastropods, and there was an overall lack of support for the placement of vetigastropods, neritimorphs, patellogastropods, cocculinoids, and hot-vent taxa.

Another recent analysis on molluscan relationships included five markers and up to 6.5 Kb of sequence data per complete taxon for 32 gastropods (Giribet *et al.* 2006). This analysis failed to recover gastropod monophyly but monophyletic Patellogastropoda, Orthogastropoda, Neritimorpha, Caenogastropoda, Heterobranchia were recovered, but not Vetigastropoda.

In contrast to the results of higher-level gastropod phylogenies, many molecular analyses have focused on relationships within gastropod clades (e.g., Harasewych *et al.* 1997; Lydeard *et al.* 2001; Williams *et al.* 2003; Nakano and Ozawa 2004; Vonnemann *et al.* 2005), and have been more successful at recovering well-supported phylogenies, possibly because of their more recent divergence times. This lends support to the idea that failure to recover stable relationships between the different gastropod clades may be due either to deep divergences within Gastropoda or to taxon or character sampling deficiencies. A combined analysis of morphological and molecular characters should help recover deeper divergences in gastropod evolution, because morphology may overlap with the sequence data along a broad range of levels of resolution (Giribet 2002; Giribet and Wheeler 2002) and because having all sources of evidence maximizes explanatory power (Kluge 1989; Nixon and Carpenter 1996). Empirical studies using complete genomes in yeast have also shown the importance of combining all evidence even when individual genes may show certain levels of conflict (Rokas *et al.* 2003). However it has also been argued that not all systematic problems may be able to be resolved, especially in the case of the Cambrian explosion (e.g., Rokas *et al.* 2005; Rokas and Carroll 2006).

COMBINED ANALYSES

Although simultaneous analyses of molecules and morphology are common in many animal phyla, and some are available for other molluscan classes such as Bivalvia (Giribet and Wheeler 2002) and cephalopods (Lindgren *et al.* 2004), no combined analysis incorporating both molecular and morphological characters has been previously published for gastropods as a whole. We combined an updated version of the morphological data set from Ponder and Lindberg (1997) with *ca.* 3.2 Kb of molecular data for 36 gastropod taxa and 13 outgroups; molecular data from 24 of the gastropod taxa have been used in the broader molluscan analysis by Giribet *et al.* (2006).

MATERIALS AND METHODS

SPECIMENS USED IN MOLECULAR ANALYSIS

Details of the 36 gastropod specimens (representing 30 families) analyzed in this study are given in Appendices 1 and 2. The included taxa include representatives of all the major gastropod clades described earlier in this chapter. Thirteen outgroup taxa, representing four molluscan classes, were also included. Sequence data for four ingroup and four outgroup taxa were obtained from GenBank (see Table 9.2), but all other sequences were obtained from material available to the authors, although some had been published previously (Giribet and Wheeler 2002; Giribet and Distel 2003; Okusu *et al.* 2003; Lindgren *et al.* 2004; Giribet *et al.* 2006). In situations where only few loci were available from one species, sequence data from other species within the same genus or the same family were merged. Although this practice has been criticized by some authors because of the many possible pitfalls of using composite terminals (e.g., Maila *et al.* 2003), we used it only in those cases where there is high confidence in the monophyly of the composite taxon.

DNA SEQUENCE DATA

Genomic DNA was extracted from the 96% ethanol (EtOH)-preserved tissue using the DNeasy™ Tissue Kit from QIAGEN. Five molecular loci were polymerase chain reaction (PCR) amplified from the purified DNA template. The complete 18S rRNA gene was PCR-amplified in three overlapping fragments (1F–4R, 3F–18Sbi, and 18Sa2.0–9R), of size approximately 950, 900, and 850 bp, respectively. Other internal primers (4F, 5F, 7F, 7R) were used in samples that were challenging to amplify. Primers used in amplification and sequencing have been described elsewhere (Giribet *et al.* 1996; Whiting *et al.* 1997). The length of the 18S rRNA gene for the ingroup taxa ranged between 1,767 and 2,239 bp in length. The 28S rDNA D3 fragment was amplified and sequenced using primers 28Sa and 28Sb (Whiting *et al.* 1997), measuring 277–389 bp for the ingroup taxa. The nuclear protein-coding gene histone H3 (327 bp for all taxa) was amplified and sequenced using primers H3a F and H3a R (Colgan *et al.* 1998). A mitochondrial gene fragment of cytochrome *c* oxidase subunit I (COI hereafter) was amplified using primer pair LCO1490 and HCO2198 (Folmer *et al.* 1994), measuring 654–657 bp. A fragment of the mitochondrial ribosomal gene 16S rRNA of 409 to 597 bp was amplified using primers 16Sar and 16Sb (Xiong and Kocher 1991). In total, up to 3.2 Kb of unaligned sequence data was produced per complete terminal. PCR amplification, purification, and sequencing were performed according to standard protocols for molluscs, described elsewhere (Giribet and Wheeler 2002; Lindgren *et al.* 2004; Giribet *et al.* 2006). The only modification to these protocols occurred if initial PCR amplification failed. In those cases, a less stringent reaction using an additional 10 mM $MgCl_2$ (for a total of 25 mM $MgCl_2$) in the 50 μL reaction was performed.

Chromatograms obtained from the automated sequencer were viewed and fragments (contigs) assembled using the sequence-editing software Sequencher™ 4.2.2. Complete sequences were then edited in MacGDE: Genetic Data Environment for MacOSX (Linton 2005), where external primer regions were removed and the sequences split according to primer-delimited regions and secondary structure features following Giribet (2001), when necessary. To account for length variability among the sequences, 18S rRNA was partitioned into twenty-three fragments, 28S rRNA into three, 16S rRNA into seven, and COI into five. Hypervariable regions in 18S rRNA (fragments 3, 10, and 19) and 28S rRNA (fragment 2) were excluded from further analyses because such regions do not present obvious base-to-base correspondences and, hence, homology assignment is highly speculative (e.g., Giribet *et al.* 2000). All new sequences have been deposited in GenBank (Appendix 1).

MORPHOLOGICAL DATA

Forty-nine additional characters were added to the original 117 characters of Ponder and Lindberg's (1997) matrix for a total of 166 morphological characters (see Appendix 3). These additional characters were extracted from data matrices that have appeared since 1997 (e.g., Sasaki 1998; Guralnick and Smith 1999; Lindberg and Guralnick 2003; Strong 2003), derived from non-phylogenetic studies (e.g., Okoshi and Ishii 1996), suggested by G. Haszprunar (characters 120, 123–125), or constitute original characters from Ponder's studies (126–129). Additional characters analyzed by Barker (2001a, b) were not included because of our limited taxon sampling within the Heterobranchia. Taxa matched the species-level taxa used in the molecular analysis but were coded in the morphological data at family level, most being attributable to the families used in the Ponder and Lindberg (1997) data set.

ANALYTICAL METHODS

MORPHOLOGICAL DATA

Morphological data were analyzed under the parsimony criterion in TNT v. 1. 0 (Goloboff *et al.* 2003) with 1,000 random addition replicates (RAS) followed by tree bisection and reconnection (TBR) branch swapping, holding 10 trees per replicate, with a subsequent round using a hold value of 10,000 trees. Nodal support was determined using jackknifing (Farris *et al.* 1996); values were calculated from 1,000 replicates using TBR with a single starting point in TNT.

MOLECULAR AND COMBINED ANALYSIS

All data files used in this analysis as well as the output and standard error files generated from each analysis can be obtained from http://www.ucmp.berkeley.edu/science/archived_data.php Molecular and combined data were analyzed using the direct optimization method (Wheeler 1996) implemented in POY v 3.0.11 (Wheeler *et al.* 2004). Seven partitions were analyzed independently, including each of the five molecular loci and two combinations: (1) all of the molecular data and (2) all molecular and morphological data analyzed simultaneously. Although two loci are protein-coding (COI and H3), all molecular data were examined on a DNA level. Processes were executed in parallel using PVM (Parallel Virtual Machine) (commands –parallel–dpm–jobspernode 2), and preliminary tree space searched with 20 random addition replicates for the combined molecular and morphological analysis, or 10 random addition replicates for all other analyses (commands –multirandom–random *n*). Subtree pruning and regrafting (SPR), followed by tree bisection and reconnection (TBR) and tree fusing, were used as branch-swapping algorithms in all replicates. Nodal support was calculated with 100 jackknife replicates (Farris *et al.* 1996).

A parameter space of two variables (gap/change ratio and transversion/transition ratio) was explored (Wheeler 1995; Giribet 2003) for each of the partitions. A total of nine parameter sets were analyzed per partition; gap/change ratio values of 1, 2, and 4 were explored ("change" refers to the highest value for a base transformation, i.e., the transversions), as well as transversion/transition ratios of 1 (equal weights), 2 (transversions receive twice as much weight as transitions), and 4. Specifically, this resulted in the analysis of the nine parameter sets 111, 121, 141, 211, 221, 241, 411, 421, and 441. In order to prevent the molecular signal from overwhelming the morphological signal in the analysis, the morphological data set was weighed according to the highest value for a base transformation (e.g., indel cost) in the molecular analysis. Tree lengths for all the analyses are summarized in Table 9.1.

After the shortest trees from all initial searches under all parameters were gathered,

TABLE 9.1

Tree Lengths for the Individual and Combined Data Sets at Different Parameter Values, with ILD Values

	18S	28S	H3	16S	COI	MOR	TOT	ILD
111	3,466	279	1,054	3,822	3,314	464	13,393	0.0742
121	**5,422**	**384**	**1,505**	**6,067**	**5,068**	**928**	**20,900**	**0.0730**
141	9,204	577	2,373	10,323	8,396	1,856	35,425	0.0761
211	4,306	303	1,057	4,548	3,350	928	15,755	0.0802
221	7,037	417	1,505	7,408	5,114	1,856	25,437	0.0826
241	12,345	639	2,372	12,904	8,486	3,712	44,345	0.0877
411	5,734	335	1,058	5,604	3,366	1,856	19,816	0.0940
421	9,827	486	1,506	9,427	5,148	3,712	33,413	0.0990
441	17,920	772	2,370	16,846	8,544	7,424	60,208	0.1052

NOTE: Individual data sets: 18S: 18S rRNA; 28S: 28S rRNA; H3: histone H3; 16S: 16S rRNA; COI: cytochrome *c* oxidase subunit I; MOR: morphology. Combined data set: TOT = [18S + 28S + H3 + 16S + COI + MOR]). Bold row reflects the parameter set that minimizes incongruence among data sets.

a sensitivity analysis tree-fusing (SATF) search was performed in order to search the tree space more thoroughly. This method of analysis performs tree fusing on a collection of trees retained from previous searches to improve tree length further and has been used successfully to obtain shorter trees in other analyses (D'Haese 2003; Boyer *et al.* 2005). The shortest trees retained from all parameters of the preliminary round of combined molecular and total-evidence analyses were used as the starting trees in the SATF analyses for the second round of combined molecular and total-evidence analyses.

Congruence was used as an optimality criterion for choosing the combined analysis that minimized overall incongruence among molecular and morphological partitions (Wheeler 1995). A modified version of the incongruence length difference (ILD) metric was used to measure character congruence (Mickevich and Farris 1981; Farris *et al.* 1995). The ILD value was determined for each parameter by subtracting the sum of the lengths of the individual trees from the length of the combined tree and dividing that result by the length of the combined tree to determine the ILD value (see Table 9.1). The parameter with the lowest ILD score is the one maximizing congruence among all the partitions. Nodal stability (Giribet 2003) under

the nine different parameter sets was explored, and an alternative, more conservative estimation of the phylogenetic hypothesis was shown by the strict consensus of the shortest trees found under all explored parameter sets for the combined analyses of all molecular data and for the total evidence analyses.

RESULTS

MORPHOLOGICAL ANALYSIS

The tree search performed in TNT generated 72 shortest trees at 464 steps (CI = 0.61; RI = 0.88), which were hit 734 times. The strict consensus of these 72 cladograms (Figure 9.3) was consistent with the monophyly of all molluscan classes represented except Polyplacophora. Nodal support for the monophyly of each class, including Gastropoda, was found in more than 90% of jackknife replicates (jackknife frequency; JF hereafter), again with the exception of Polyplacophora and Bivalvia, in which nodal support was found in only 78% of jackknife replicates. Gastropoda were resolved into Patellogastropoda (100% JF) and Orthogastropoda (78% JF). Monophyly of Patellogastropoda, Caenogastropoda, Neritimorpha, and Heterobranchia was supported with at least 92% JF. Opisthobranchia and Vetigastropoda

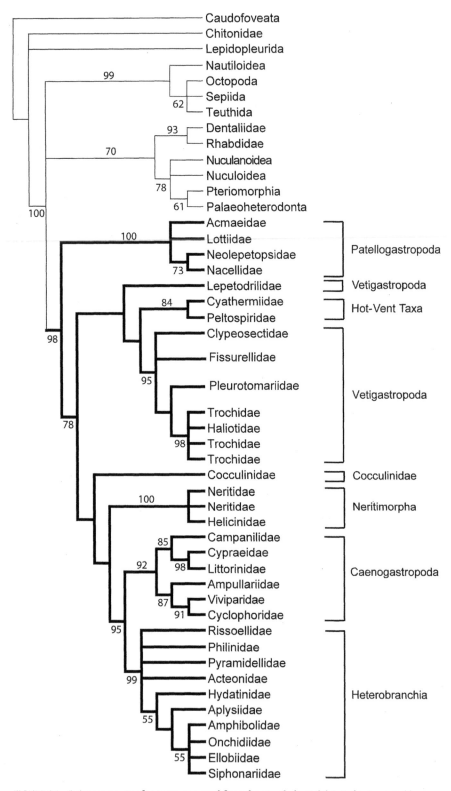

FIGURE 9.3. Strict consensus of 72 trees generated from the morphological data (464 steps). Bold branches indicate gastropod taxa. Numbers on branches indicate jackknife support of greater than 50% calculated in PAUP. See text for further details.

were not monophyletic, but a Caenogastropoda + Heterobranchia clade (Apogastropoda) was recovered with 95% JF.

Monophyly of the Patellogastropoda was supported by 100% JF. Monophyly of the "hot-vent taxa" (see above under Vetigastropoda) was supported with 84% JF, but their position within Vetigastropoda received a JF below 50%. The relationship of *Lepetodrilus* as sister to the rest of Vetigastropoda + hot-vent taxa also obtained a JF below 50%. Resolution within the remaining Vetigastropoda (95% JF) yielded a clade composed of Pleurotomariidae + aliotidae + Trochidae (but JF < 50%), and Fissurelloidea was not monophyletic. The Haliotidae + Trochidae clade was recovered in 98% of jackknife replicates, but its internal relationships were unresolved.

A clade containing Cocculinidae + Neritimorpha + Caenogastropoda + Heterobranchia was recovered, although the placement of Cocculinidae as sister to Neritimorpha + Apogastropoda received less than 50% JF. Two main clades were recovered within Caenogastropoda: Architaenioglossa (87% JF) and Sorbeoconcha (85% JF). Within Architaenioglossa, Cyclophoroidea + Viviparidae had 91% JF. Within Sorbeoconcha, Hypsogastropoda (represented by Littorinidae + Cypraeidae) also had 98% JF, and they were the sister clade to Campanilidae. Note that specimens from neogastropod families were not included in this analysis. Within Heterobranchia (99% JF), Pulmonata (= Amphibolidae + Onchidiidae + Siphonariidae + Ellobiidae) was monophyletic (55% JF), but Opisthobranchia was not.

CONGRUENCE ANALYSIS

Overall congruence for the simultaneous analysis of all data was maximized under parameter set 121, where the ratio between indels/transversions is 1:1 and the ratio between transversions and transitions is 2:1 (hence the absolute indel cost = 2, tv = 2, ts = 1). This parameter set resulted in an ILD score of 0.0730 (Table 9.1). Parameter set 111 received the immediate closest ILD score of 0.0742. The lowest ILD value for the combined molecular-only

analysis was also found under parameter set 121 and received an ILD score of 0.0666.

COMBINED ANALYSES OF ALL MOLECULAR DATA

Analysis of the combined molecular data set under the optimal parameter set yielded a single tree of 19,763 weighted steps. When rooted with Caudofoveata, this tree showed monophyly for Cephalopoda, Scaphopoda, Bivalvia and Polyplacophora, but not for Gastropoda (Fig. 9.4A). Gastropoda formed two clades; Heterobranchia appeared as sister to Scaphopoda, while the remaining Gastropoda were sister to a clade that includes Polyplacophora and Bivalvia. Jackknife support for any of these relationships was below 50%, and such nodes collapsed under the strict consensus of the analyses for the entire parameter space (Figure 9.4B).

Heterobranchia was recovered in 100% jackknife replicates and it was found under all parameter sets. Within Heterobranchia, *Onchidella* + *Ophicardelus* was the only clade with JF greater than 50%. Caenogastropoda, Neritimorpha, Patellogastropoda, and the hot-vent taxa were each monophyletic and supported with more than 50% JF, and Caenogastropoda, Neritimorpha, Neritidae, Pleurotomariidae, *Diodora*, Clypeosectidae + Haliotidae, *Clanculus* + *Calliostoma*, and the hot-vent taxa were stable to parameter set variation. The monophyly of Vetigastropoda was not found under any parameter set. Under the optimal parameter set, Pleurotomariidae grouped outside the remaining Vetigastropoda + Patellogastropoda. A clade of Vetigastropoda excluding Pleurotomariidae was supported with 77% JF, but the vetigastropod groups Fissurelloidea (including Fissurellidae and Clypeosectidae) and Trochidae sensu lato were not monophyletic. The monophyletic hot-vent taxa were sister to Cocculinidae, but this result had JF < 50%. Neritimorpha was sister to the Caenogastropoda, again with JF < 50%. Across the entire parameter space, Helicinidae was sister group to Neritidae (i.e., comprising Neritimorpha), and this relationship received 60% jackknife support. Relationships within

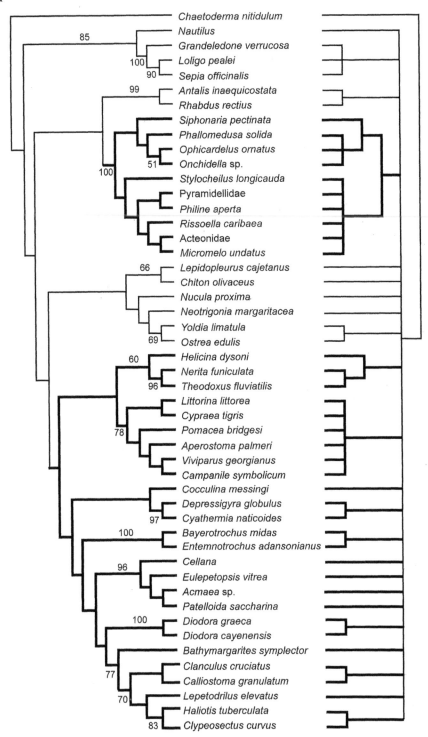

A

B

Chaetoderma nitidulum
Nautilus
Grandeledone verrucosa
Loligo pealei
Sepia officinalis
Antalis inaequicostata
Rhabdus rectius
Siphonaria pectinata
Phallomedusa solida
Ophicardelus ornatus
Onchidella sp.
Stylocheilus longicauda
Pyramidellidae
Philine aperta
Rissoella caribaea
Acteonidae
Micromelo undatus
Lepidopleurus cajetanus
Chiton olivaceus
Nucula proxima
Neotrigonia margaritacea
Yoldia limatula
Ostrea edulis
Helicina dysoni
Nerita funiculata
Theodoxus fluviatilis
Littorina littorea
Cypraea tigris
Pomacea bridgesi
Aperostoma palmeri
Viviparus georgianus
Campanile symbolicum
Cocculina messingi
Depressigyra globulus
Cyathermia naticoides
Bayerotrochus midas
Entemnotrochus adansonianus
Cellana
Eulepetopsis vitrea
Acmaea sp.
Patelloida saccharina
Diodora graeca
Diodora cayenensis
Bathymargarites symplector
Clanculus cruciatus
Calliostoma granulatum
Lepetodrilus elevatus
Haliotis tuberculata
Clypeosectus curvus

FIGURE 9.4. Cladograms based on the analyses of the combined molecular data. (A) Cladogram is single shortest tree (19,763 weighted steps) under the optimal parameter set (121). See text for further details and Table 9.2 for family designations. Bold branches indicate gastropod taxa. Numbers on branches indicate jackknife support values above 50%. (B) Cladogram is a strict consensus of all trees obtained under all the parameter sets explored for the total molecular analysis.

Caenogastropoda had low JF, but Littorinidae + Cypraeidae were monophyletic under most of the parameter space. JF > 50% were not found for any deep nodes within Gastropoda, and the lack of resolution in the strict-consensus cladogram of all parameter sets indicates a lack of stability of these relationships. Gastropod clades supported under all analytical parameter sets for the combined analysis of all molecular data were Neritimorpha, Caenogastropoda, Heterobranchia, and the hot-vent taxa.

COMBINED MOLECULAR AND MORPHOLOGICAL ANALYSES

When morphological and molecular data were combined in a simultaneous analysis under the optimal parameter set, a single shortest tree of 20,900 weighted steps was obtained. This tree is shown in Figure 9.5A, while Figure 9.5B illustrates the strict consensus of all trees obtained under all explored analytical parameters. When rooted with Caudofoveata, all molluscan classes included in the analyses were monophyletic across the entire parameter set, with the exception of Polyplacophora, which was not monophyletic under three parameter sets (411, 421, and 441). Although Gastropoda was monophyletic under all parameter sets in the total evidence analysis, its relationship to outgroup taxa was unstable and showed low JF. Under the optimal parameter set, Gastropoda was divided into two unsupported clades: ((Cocculinidae + hot-vent taxa) (Patellogastropoda + Vetigastropoda)) and (Neritimorpha (Caenogastropoda + Heterobranchia)). Patellogastropoda, Vetigastropoda, Neritimorpha, Caenogastropoda, and Heterobranchia were recovered under the entire parameter space, and JF for the optimal parameter set ranged from 57% to 100% (Figure 9.5A). In addition, the total-evidence analysis resolved the sister group relationship between Caenogastropoda and Heterobranchia (= Apogastropoda) under all parameter sets (52% JF). However, the relationship between all these clades (with the exception of Caenogastropoda + Heterobranchia) was unstable and received JF < 50% under the most congruent parameter set.

Patellogastropoda was supported by 100% JF, but the relationship of the patellogastropod families represented in this analysis found low support and stability to parameter variation. Patellogastropoda was the sister clade to Vetigastropoda only under two parameter sets (121 and 221); using other weighing schemes, Patellogastropoda is sister to Cocculinidae, ((Cocculinidae + hot-vent taxa) Vetigastropoda), or Orthogastropoda (trees not shown). Vetigastropoda was monophyletic under all parameter sets (57% JF), with Pleurotomariidae as the sister group to all other vetigastropods (87% JF), which were in turn divided into Fissurellidae and the remaining vetigastropods. This branching pattern was stable to parameter set variation (Figure 9.5B). Pleurotomarioidea, Trochidae, and Fissurelloidea were not monophyletic. The hot-vent taxa were monophyletic in 94% of jackknife replicates and under all parameter sets. Their sister group relationship to Cocculinidae was recovered under the optimal parameter set (66% JF).

Neritimorpha was sister to Apogastropoda (= Caenogastropoda + Heterobranchia) under all parameter sets except 141. The monophyly of the Neritimorpha and Neritidae (92% and 83% JF, respectively) was stable to parameter changes. Apogastropoda, a clade obtained under all parameter sets explored, received 52% JF. While Caenogastropoda and Heterobranchia were both well supported (85% and 98% JF, respectively) and stable to parameter change, under the optimal parameter set only one internal node (Pulmonata, 51% JF) was recovered in more than 50% of jackknife replicates. There was some internal stability to parameter variation; Architaenioglossa was monophyletic under all parameter sets, with an internal topology of ((Ampullariidae (Cyclophoroidea + Viviparidae)) recovered under all remaining parameter sets except 111 (Figure 9.6). Littorinidae + Cypraeidae were sister taxa under all parameter sets except for the optimal one and 241. Campanilidae was sister to Architaenioglossa under all parameter sets except for 111, 141, and 421, where it was sister to (Littorinidae + Cypraeidae). Within

A B

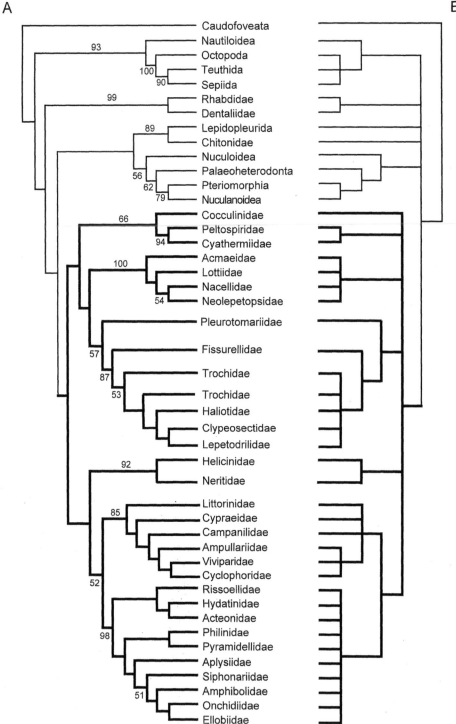

FIGURE 9.5. Cladograms based on the analyses of the combined morphological and molecular data.
(A) Cladogram is strict consensus of the shortest tree (20,900 weighted steps) for combined molecular and morphological analysis under the optimal parameter set (121). See text for further details. Bold branches indicate gastropod taxa. Numbers on branches indicate jackknife support values above 50%. (B) Cladogram is a strict consensus of all trees obtained under all the parameter sets explored for the combined molecular and morphological analysis.

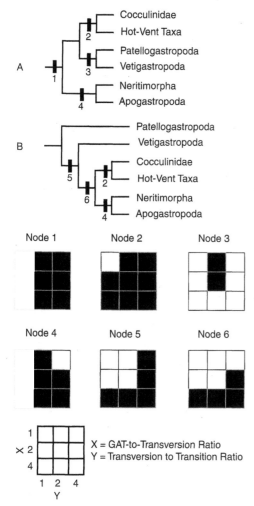

FIGURE 9.6. Two topologies of higher-level relationships among Gastropoda as derived from analyses of combined morphology and molecular data. (A) Topology generated under most congruent parameter set 121 (ILD = 0.0730). (B) Topology generated under parameter set 241 (ILD = 0.0877). Numbers under select nodes correspond with graphic plots of sensitivity analyses (Navajo Rugs) for the combined morphological and molecular data. Black squares indicate monophyly for a given parameter set, while white squares indicate nonmonophyly.

Heterobranchia, Pulmonata was monophyletic under all parameter sets except 421 and 441, but Opisthobranchia was not monophyletic.

DISCUSSION

This study is the first to use multiple molecular loci (*ca*. 3.2 Kb) and 166 morphological characters in a simultaneous analysis to provide one of the most comprehensive analyses of internal relationships among major gastropod clades. In this study, hypotheses of gastropod phylogeny based on the analysis of morphological or molecular data sets alone differed dramatically. Furthermore, results for individual molecular partitions were unstable because alignment parameters were varied, and only when the morphological data were included did the results become more stable. The monophyly of Gastropoda was not obtained under any molecular partitions, but after the inclusion of morphological characters a monophyletic Gastropoda was recovered. Additionally, the jackknife support for most other clades increased in the combined analysis of all data; morphology provided additional support to the deepest nodes, while molecules in general added support to the shallower nodes of the tree. This result supports the arguments of proponents of the total-evidence approach (see Kluge 1989; Nixon and Carpenter 1996); the most reliable evolutionary hypothesis for a particular group can be generated only by using all available data in the analysis.

Previous analyses of gastropods have shown monophyly of the group under morphology (Haszprunar 1988c; Ponder and Lindberg 1996, 1997) but failed to recover its monophyly in molecular studies (Rosenberg *et al.* 1994; Colgan *et al.* 2000, 2003). Although the monophyly of gastropods was not recovered in more than 50% of jackknife replicates in the combined analyses of all data (for parameter set 121), this clade was stable to parameter variation (Figures 9.5A, 9.6). The recovery of a monophyletic Gastropoda after the inclusion of morphological characters is due to the morphological data providing information about deep splits in gastropod lineages. Although to some this may look like conflict between morphology and molecules, the reality is that none of the clades supported in the combined analysis is contradicted by strongly supported clades appearing in the molecular-only analyses (also, see Rokas *et al.* 2003).

Hypotheses of relationships among the major gastropod lineages and the support found across the nine explored parameter sets are illustrated in Figure 9.6. Interestingly, most of the depicted relationships were stable to parameter variation, but jackknife support for the optimal parameter set was below 50% for any of the deep nodes obtained. Decoupling between nodal support and stability has been reported before for other molluscs (Giribet 2003), and it is often caused by having a low number of characters—although uncontradicted—supporting a node. This contrasts with other situations where high JF is coupled with low-stability support, as in the case of long-branch attraction.

Two major topologies were recovered under the different parameter sets in the combined analysis (Figure 9.6). In five of nine parameter sets, Gastropoda was split into the two main lineages recognized by Ponder and Lindberg (1996): Eogastropoda and Orthogastropoda. Under the most congruent parameter set and another parameter set, Vetigastropoda + Patellogastropoda was obtained, with the placement of Neritimorpha, Cocculinidae, and the hot-vent taxa fluctuating among Apogastropoda, Vetigastropoda, and Patellogastropoda as parameter sets were changed.

PATELLOGASTROPODA

Patellogastropoda was recovered as monophyletic in the morphological analysis and for every parameter set in the total-evidence analysis, but it was unstable to parameter change in the molecular-only analysis. The placement of the patellogastropods within Gastropoda also changed depending on the type of data used in the analysis. As in other morphological analyses (Golikov and Starobogatov 1975; Haszprunar 1988c; Ponder and Lindberg 1997), patellogastropods consistently formed the sister group to the rest of the gastropods (= Orthogastropoda) when only morphological characters were considered. As in other molecular analyses (Tillier *et al.* 1994; Harasewych *et al.* 1997; Rosenberg *et al.* 1997; Colgan *et al.* 2000, 2003; McArthur and Harasewych 2003; Giribet *et al.* 2006), the placement of Patellogastropoda was not consistent when only molecular loci were considered.

In the total-evidence analysis its relationship to other gastropods was variable under parameter set changes.

VETIGASTROPODA

Vetigastropoda, excluding the hot-vent taxa, was monophyletic only when all data, including morphology, were analyzed simultaneously, and it remained stable to parameter set variation (Figure 9.5). Pleurotomariidae was the sister group to the remaining vetigastropods under all analytical parameters (Fig. 9.5B), a placement contradicting previous morphological analyses (Haszprunar 1988c; Ponder and Lindberg 1997; Sasaki 1998) but corroborating results from molecular-based analyses (Harasewych *et al.* 1997; Geiger and Thacker 2005; Geiger *et al.*, Chapter 12).

Although the monophyly of Zeugobranchia (Pleurotomarioidea + Haliotoidea + Fissurelloidea + Scissurelloidea) cannot be fully tested without including representatives from Scissurelloidea in the analysis, Haliotoidea, Fissurelloidea and Pleurotomarioidea never formed a clade (but see Geiger *et al.*, Chapter 11). As recognized by Geiger and Thacker (2005), Fissurelloidea and Trochidae were not recovered as monophyletic in any analyses (Figure 9.7); Clypeosectidae did not group as sister to Fissurellidae, and *Haliotis* or *Lepetodrilus* fell within Trochidae depending on the analysis.

COCCULINOIDEA + HOT-VENT TAXA

It is beyond the scope of this study to address the monophyly of Cocculiniformia because of the lack of representation of Lepetelloidea. Haszprunar (1988b) considered Lepetelloidea as sister to Cocculinoidea within Cocculiniformia, a result that Ponder and Lindberg (1997) and Colgan *et al.* (2000, 2003) later contradicted. The monophyly of Cocculinoidea (Strong *et al.* 2003) also cannot be addressed because of the absence of Bathysciadiidae, but the relationship of Cocculinidae (represented by *Cocculina*) to other gastropod groups can be. Its placement contradicted the Cocculinoidea + Neritimorpha clade found by Golikov and Starobogatov (1975) and Ponder and Lindberg (1997) based on their

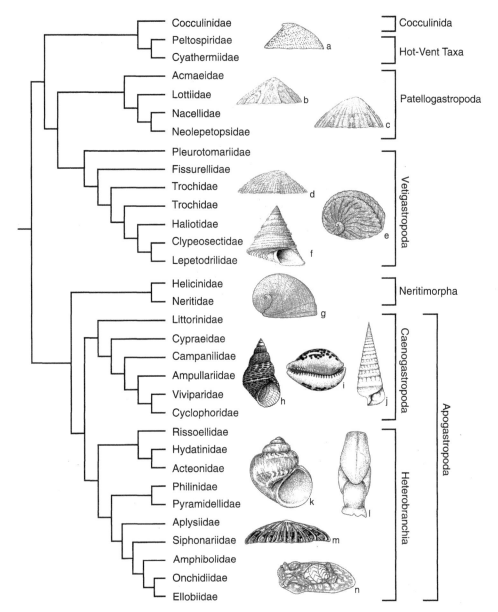

FIGURE 9.7. Schematic representation of gastropod relationships based on the optimal parameter set (121) for the combined analysis of morphological and molecular data. See text for further details. Taxa represented in the figure are (a) *Coccopigya hispida* (Cocculinidae); (b) *Patelloida saccharina* (Lottiidae); (c) *Cellana tramoserica* (Nacellidae); (d) *Diodora lineata* (Fissurellidae); (e) *Haliotis ruber* (Haliotidae); (f) *Calliostoma* sp. (Trochidae); (g) *Nerita atramentosa* (Neritidae); (h) *Littoraria luteola* (Littorinidae); (i) *Cypraea tigris* (Cypraeidae); (j) *Campanile symbolicum* (Campanilidae); (k) *Phallomedusa solida* (Amphibolidae); (l) *Philine columnaria* (Philinidae); (m) *Siphonaria atra* (Siphonariidae); (n) *Micromelo undatus* (Hydatinidae). All images from Beesley *et al.* (1998).

morphological analyses. Morphological data from this study supported a placement of Cocculinidae as the sister group to a clade containing Neritimorpha + Caenogastropoda + Heterobranchia. However, under the optimal parameter set in

both the molecular and combined morphological and molecular analyses, Cocculinidae formed a clade with the hot-vent taxa, as recognized in previous molecular analyses (McArthur and Harasewych 2003). The relationship between these two clades may suggest common ancestry between Cocculinidae and some hydrothermal vent taxa. The relationships of this problematic group are further discussed Chapters 11 and 12.

HOT-VENT TAXA

Cyathermiidae and Peltospiridae, two of three gastropod taxa recognized by Ponder and Lindberg (1997) as hot-vent taxa, were included in this analysis and formed a clade. Their third member of the hot-vent group, *Melanodrymia*, has many apomorphic characters, which make it difficult to classify (Fretter 1989; Haszprunar 1989). Limited molecular information has been collected for this taxon, resulting in unclear placements in other phylogenetic analyses (McArthur and Tunnicliffe 1998; McArthur and Koop 1999; McArthur and Harasewych 2003). Resolution of the relationships between the hot-vent taxa may be difficult until more morphological and molecular data are collected for a larger diversity within the group. However, this analysis strongly suggested common ancestry between Cyathermiidae and Peltospiridae. The grouping of these taxa outside the vetigastropods in our molecular analyses is at variance with the results obtained in some other analyses (e.g., Colgan *et al.* 2003; Strong *et al.* 2003). For further discussion see Geiger *et al.*, Chapter 12.

NERITIMORPHA

The monophyly of Neritimorpha has been widely recognized (Salvini-Plawen and Haszprunar 1987; Haszprunar 1988c; Ponder and Lindberg 1997; Harasewych *et al.* 1998; Sasaki 1998; McArthur and Harasewych 2003; Lindberg, Chapter 11). Results from this analysis did not differ; Neritimorpha (represented here by Neritidae and Helicinidae) was recovered as a well-supported clade in the morphological, molecular, and combined analyses.

The position of Neritimorpha within the gastropod tree was more contentious. The sister group relationship between Neritimorpha and Apogastropoda was recovered in the morphological and some total-evidence analyses. Relationships suggested in previous phylogenetic analyses dif-

fer; a Neritimorpha + Cocculinoidea clade was recovered in one morphological analysis with a bootstrap support value below 50 (Ponder and Lindberg 1997), while Harasewych *et al.* (1998) recovered Neritimorpha + Apogastropoda in a molecular analysis. Placement in analyses by Colgan *et al.* (2000; 2003) varied, with Neritimorpha recovered as sister to Apogastropoda, sister to all other gastropods, and within Caenogastropoda.

APOGASTROPODA

In the morphological and combined analyses Apogastropoda, as defined by Ponder and Lindberg (1997), was recovered. Although Apogastropoda has been recognized in previous molecular analyses (Tillier *et al.* 1994; Harasewych *et al.* 1998; Colgan *et al.* 2003), Heterobranchia was recovered as separate from Caenogastropoda in the molecular analysis for this study (Figure 9.4). JF for Apogastropoda was 95% in the morphological analysis, but only 52% of jackknife replicates recovered the clade under the most optimal parameter set for the combined analysis, illustrating a putative source of major incongruence between morphology and the selected molecular characters. With the combined analysis, this clade was, however, extremely stable to parameter variation, being recovered under every parameter (Figure 9.7).

Caenogastropoda, represented in this analysis by Cyclophoroidea, Ampullariidae, Viviparidae, Campanilidae, Littorinidae, and Cypraeidae, was monophyletic under all parameter sets in all analyses. Success at recovering a monophyletic Caenogastropoda in other molecular analyses has varied; Harasewych *et al.* (1998), McArthur and Harasewych (2003) and Colgan *et al.* (2007) recovered the clade, whereas Colgan *et al.* (2003) recovered it only in some analyses and Colgan *et al.* (2000) did not. The results of our study agreed with Ponder and Lindberg (1997), Strong (2003), Colgan *et al.* (2007), and most of Colgan *et al.*'s (2003) results by recovering Architaenioglossa within Caenogastropoda. However, the architaenioglossans as a group were weakly supported, not recovered under all data sets, and unstable to parameter set variation. The clade was monophyletic under the optimal parameter set in the total-evidence analysis but not in the molecular one. *Viviparus* was sister to *Aperostoma* in the morphological analysis and the optimal parameter set for the combined analysis (Figures 9.3, 9.5). A second caenogastropod clade, Sorbeoconcha, was monophyletic in only the morphological analysis. In the molecular and combined analyses, *Campanile* grouped with Architaenioglossa. Given the lack of inclusion of Cerithioidea, this agrees with previous results in Colgan *et al.* (2003; 2007). Although Caenogastropoda was a well-supported and stable clade in this analysis, there was little resolution of the internal nodes, a not unexpected result given the very limited sampling of this taxon (but see Ponder *et al.*, Chapter 13, for more comprehensive analyses of Caenogastropoda).

Heterobranchia was one of the best-supported clades in this study. It was first established by Haszprunar (1985c) and was recovered in later analyses (e.g., Ponder and Lindberg 1997; McArthur and Tunnicliffe 1998; McArthur and Koop 1999; Thollesson 1999; Dayrat and Tillier 2002; Grande and Templado 2004). Groups within Heterobranchia were unstable with low nodal support, and only Pulmonata was recovered in all the analyses. The monophyly of Euthyneura has varied in different analyses. Colgan *et al.* (2000; 2003; 2007) recovered this clade as monophyletic, but other morphological and molecular analyses (Thollesson 1999; Dayrat and Tillier 2002; Grande and Templado 2004) failed to recover a monophyletic Euthyneura. In this analysis, neither Opisthobranchia nor Euthyneura were monophyletic, likely because of the limited representation of clades within Heterobranchia. "Lower" heterobranchs such as Rissoelloidea and Pyramidelloidea were not recovered together in any of the analyses. In the morphological analysis, *Rissoella*, Pyramidellidae, and two cephalaspideans (Acteonidae and *Philine*) are unresolved in the strict consensus, but the third cephalaspidean (*Micromelo*) appears as sister group to the remaining heterobranchs. Under the optimal parameters for the

molecular and combined analyses, Pyramidellidae was again sister to *Philine*, and *Rissoella* was sister to the clade formed by the two cephalaspideans Acteonidae + *Micromelo*. However, these relationships were not stable to parameter variation in both the molecular and combined analyses. The relationships of euthyneuran taxa are explored in detail in chapters 14 and 15.

ADAPTIVE RADIATIONS

The adaptive radiation of gastropods was discussed in some detail by Ponder and Lindberg (1997) and in the relevant chapters in this volume that deal with the subgroups of gastropods, so we provide only a brief summary here.

Changes in developmental pathways via heterochronic processes are often invoked to explain the evolution of new morphologies and perhaps set the stage for adaptive radiations (Gould 1977; McKinney and MacNamara 1988; and references therein). These developmental changes can act on both larvae and adults, and multiple examples are provided by Ponder and Lindberg (1997), notably with respect to the evolution of many heterobranch and caenogastropod traits. In addition, with the recent revolution in molecular techniques associated with evolution development (see Wanninger *et al.*, Chapter 16), it is now possible to make direct observations of these and other processes.

One of the most important innovations related to feeding—the move from plesiomorphic grazing on surface films of diatoms, bacteria, and other microorganisms (see Caron *et al.* 2006), a habit persisting in many modern gastropods, to omnivorous and then to carnivorous grazing on sessile animals—precipitated some of the most important adaptive radiations through dietary specialization. A move to different feeding behaviors is often also accompanied by significant changes in habit and body form. Several groups of vetigastropods, heterobranchs, and caenogastropods feed exclusively on colonial animals, while some heterobranchs and caenogastropods have moved to mobile animal prey. A few groups have become pelagic neuston (the caenogastropod Janthini-dae and the nudibranch Glaucidae), and graze on floating cnidarians, whereas Gymnosomata (Heterobranchia) and Pterotracheoidea (Caeno-gastropoda) have become permanent active swimmers feeding on zooplankton and exhibit major morphological adaptations. Suspension feeding independently evolved in a few vetigas-tropods (Trochoidea), several caenogastropod families, and the pelagic heterobranch Thecoso-mata. Some other normally benthic gastropods can also swim, but for relatively short distances. On the adoption of a carnivorous diet sev-eral groups underwent significant radiations, including neogastropods, several other groups within caenogastropods, opisthobranchs, and "lower" heterobranchs (notably pyramidelloide-ans). Shell drilling evolved independently in a few caenogastropod families (Naticidae, Muri-cidae, Marginellidae, Buccinidae, Cassidae, and Capulidae) and two heterobranch genera: the opisthobranch *Okadaia* and the pulmonate *Aegopinella* (Kabat 1990).

A major innovation includes the modifica-tions to the gastropod mantle cavity in gastropod evolution. In particular, a switch in focus from exhalant flows from the mantle cavity to inhalant flows (Lindberg and Ponder 2001) maximized the potential of the chemosensory osphradium located in a forward-oriented opening (enabled by torsion). This opened up new adaptive pos-sibilities not only in feeding but also in predator avoidance and the location of mates.

The evolution of internal fertilization also opened up many new opportunities; notably the ability to encapsulate the eggs after fertilization. This allowed not only protection via a capsule or brood pouch to at least the veliger stage but also abbreviation of the life cycle (through "direct development"). The main clades where this occurs (Neritimorpha, Caenogastropoda, Het-erobranchia) have all undergone radiations into nonmarine habitats. Stylommatophoran pulmo-nates are by far the dominant terrestrial clade, but there are several other small non-marine groups of pulmonates. There are also freshwater and terrestrial Neritimorpha, the architaenio-glossan terrestrial Cyclophoroidea, and fresh-water Ampullarioidea and Viviparoidea and a number of other non-marine caenogastropod groups (mainly some Cerithioidea, Littorinoidea, and Rissooidea).

Convergences in body form are common in gastropods, but these do not necessarily relate to obvious adaptive situations. The limpet form has evolved many times and in all major gastropod clades, but limpets occupy many habitats that are often very divergent (coastal rock platforms, seagrass, lakes, hot vents, biogenic substrates in the deep sea, parasites on starfish). Similarly, the reduction and eventual loss of the shell (resulting in a slug) has occurred in some groups, notably in opisthobranchs and pulmonates, although it is rare or uncommon in other groups, including wormlike parasitic "slugs" in the family Eulimi-dae. Major changes in body form such as in these two examples will have enabled exploitation of different niches at the time when they evolved. Generalized conclusions to explain away such convergences are fraught with difficulty.

Significant changes in body morphology can provide access to new adaptive zones, but there is a generally a lack of obvious correlation with body form and shell morphology with habitat.

CONCLUSIONS AND FUTURE RESEARCH DIRECTIONS

This study provides an overview of the morphol-ogy and evolutionary history of gastropods as well as a fresh look at deep gastropod relationships by combining data from both morphological and molecular sources for the first time. The results of this analysis indicate that the combination of multiple sources of information and their simul-taneous analysis can show the conflict existing among data partitions as well as allow the hid-den signal within the combined data to emerge via congruence. Furthermore, this analysis is the first to suggest that neither morphology nor molecules alone are powerful enough to solve the deepest gastropod nodes. The extinc-tion of ancient gastropod clades, or the erosion of the molecular signal in this ancient group of animals, may be the cause of such failure. However, these results suggest that additional molecular, ultrastructural, developmental, and anatomical data may assist in resolving deep divergences although, it is clear that very little information currently exists for many critical taxa. The inclusion of additional data sources, such as sequence data from more nuclear protein-coding genes, mitochondrial gene order data, or genome content information, may also provide novel sources of information useful in elucidating gastropod relationships.

ACKNOWLEDGMENTS

We thank Annie Lindgren and Michele Nishiguchi for allowing us to use previously unpublished 16S rRNA sequences of *Grandeledone verrucosa* and *Sepia officinalis*, Jerry Harasewych for providing several tissues employed in this study, Amanda Bates for *Depressigyra globulus* specimens, Terry Gosliner for *Helicina dysoni* and *Stylochelus longicauda* specimens, Fred Thompson for the *Aperostoma palmeria* specimen, Don Colgan for DNA samples from *Pomacea bridgesi, Miralda ligata, Phallomedusa solida,* and *Ophicardelus ornatus* and Till Bayer for the previously unpublished H3 sequence of *Pomacea bridgesi* and COI sequence of *Miralda ligata*. Gerhard Haszprunar's assistance with morphological characters and states is also greatly appreciated. Greg Rouse and Anja Shulze provided images for Figure 9.1. We also thank Jerry Harasewych and two other reviewers for their comments that helped to improve this manuscript. This material is partly based upon work supported by the National Science Foundation under Grant No. 0334932 to G. G. and Grant No. DEB-0089624 to D. R. L.

APPENDIX 1: TAXON SAMPLING USED IN THE ANALYSES

Sampling, with MCZ DNA catalog number and GenBank accession codes, appears in Table 9.2.

APPENDIX 2: VOUCHER DATA FOR SPECIMENS USED IN MOLECULAR ANALYSES

Data for specimens represented by GenBank sequences are omitted. See Table 9.2 for family assignments. Abbreviations: CAS: California Academy of Sciences, MCZ: Museum of Comparative Zoology, Harvard University, USNM: U.S. National Museum, Smithsonian Institution.

Patelloida saccharina Linnaeus, 1758

Amakusa (Japan), July 1998; A. Okusu

Cellana nigrolineata Reeve, 1854

Minabe (Japan); M. G. Harasewych; MCZ DNA100662; Voucher specimens USNM 888623

Cellana cf. *tramoserica* Holten, 1802

Coogee Beach (Sydney, Australia), March 19, 2000; G. Giribet; MCZ DNA101183

"Acmaea" sp.

Collection data not available.

Eulepetopsis vitrea McLean, 1990

9°50.335′ N, 104°17.476′ W; 9°N East Pacific Rise (Mussel Bed), November 30, 2002; DSV *Alvin* Dive 3842; S. Aktipis; MCZ DNA100846

Cocculina messingi McLean and Harasewych, 1995

26°37.30′ N, 78°58.55′ W, on deployed wood (Bahamas), 418 m; supplied by M. G. Harasewych; MCZ DNA100663 (*ex* USNM 888655)

Helicina dysoni Pfeiffer, 1849

Nueva Esparta, (Isla de Margarita,Venezuela) Parque Nacional Cerro El Copey, near summit in low-growth succulent vegetation (similar to timberline); elevation 3,000 ft; donated by T. Gosliner; CAS 113077; MCZ DNA101386

Theodoxus fluviatilis (Linnaeus, 1758)

Font de Solla (Sóller, Mallorca, Spain), January 17, 1996; G. Giribet; MCZ DNA100668

Nerita funiculata Menke, 1851

31°20′37′′ N, 113°38′38′′ W; Sea of Cortez, Cholla Bay (Rocky Point, Mexico), March 27, 2003; G. Giribet; MCZ DNA100938 and MCZ DNA101206

Cyathermia naticoides Warén and Bouchet, 1989

9°50.60′ N, 104°17.49′ W; 9°N East Pacific Rise (Mussel Bed), December 11, 2002; DSV *Alvin* Dive 3853; S. Aktipis; MCZ DNA100855

Depressigyra globulus Warén and Bouchet, 1989

Marshmallow ASHES (Juan de Fuca Ridge); 1,525 m; Ropos Dive number R740-03-01; A. Bates; MCZ DNA101123

TABLE 9.2
Taxon Sampling Used in the Analyses

FAMILY	TERM FILE NAME	VOUCHER NUMBER	18S RRNA	28S RRNA	16S RRNA	HISTONE H3	COI
Lottiidae	Patelloida saccharina	Not available	AF046051[a]	DQ093445	DQ093466	DQ093492	
Nacellidae	Cellana[b]	MCZ DNA100662[c] (MCZ DNA 101183)[d]	DQ093425	DQ093446	DQ093467	(DQ093493)	DQ093515
Acmaeidae	Acmaea sp.	Not available	DQ093426	DQ093447		DQ093494	
Neolepetopsidae	Eulepetopsis vitrea	MCZ DNA100846	DQ093427	DQ093448	DQ093468	DQ093495	DQ093516
Cocculinidae	Cocculina messingi	MCZ DNA100663	AF120508	AY377696	AY377624	AY377777	AY377731
Helicinidae	Helicina dysoni	MCZ DNA101386	DQ093428	DQ093449	DQ093469	DQ093496	
Neritidae	Theodoxus fluviatilis	MCZ DNA100668	AF120515	AF120573	DQ093470		AF120633
Neritidae	Nerita funiculata	MCZ DNA10093 DNA101206	DQ093429	DQ093450	DQ093471	DQ093497	DQ093517
Cyathermiidae	Cyathermia naticoides	MCZ DNA100855	DQ093430	DQ093451	DQ093472	DQ093498	DQ093518
Peltospiridae	Depressigyra globulus	MCZ DNA101123	DQ093431	DQ093452	DQ093473	DQ093499	DQ093519
Pleurotomariidae	Bayerotrochus midas	MCZ DNA100666	AF120510	DQ093453	DQ093474	DQ093500	AY296820
Pleurotomariidae	Entemnotrochus adansonianus	MCZ DNA100665	AF120509	AY377694	AY377621	AY377774	
Lepetodrilidae	Lepetodrilus elevatus	MCZ DNA100930	DQ093432	DQ093454	DQ093475	DQ093501	DQ093520
Clypeosectidae	Clypeosectus curvus	GenBank	AF534990[a]				
Fissurellidae	Diodora graeca	MCZ DNA100114	AF120513	AF120572	DQ093476	DQ093502	AF120632
Fissurellidae	Diodora cayenensis	Not available	AY377659	AY377695	AY377623	AY377776	AY377730
Haliotidae	Haliotis tuberculata	MCZ DNA100110	AF120511	AF120570	AY377622	AY070145	AY377729
Trochidae sensu lato	Bathymargarites symplector	MCZ DNA101220	DQ093433	DQ093455	DQ093477	DQ093503	DQ093521
Trochidae sensu lato	Clanculus cruciatus	MCZ DNA100664	AF120514	DQ093456			
Trochidae sensu lato	Calliostoma granulatum	MCZ DNA100086	DQ093434	DQ093457	DQ093478	DQ093504	DQ093522
Cyclophoridae	Aperostoma palmeri	MCZ DNA101457	DQ093435	DQ093458	DQ093479	DQ093505	DQ093523
Ampullariidae	Pomacea bridgesi	DNA from DC	DQ093436		DQ093480	DQ093506	DQ093524
Viviparidae	Viviparus georgianus	MCZ DNA100112	AF120516	AF120574	AY377626	AY377779	AF120634
Campanilidae	Campanile symbolicum	GenBank	AF055648[a]		AY010507[a]	AF033683[a]	AY296828[a]
Littorinidae	Littorina littorea	MCZ DNA101389	DQ093437	DQ093459	DQ093481	DQ093507	DQ093525
Cypraeidae	Cypraea tigris	GenBank	AF055654[a]		AY161488[a]		AY161721[a]

Pyramidellidae	Pyramidellidaeᵇ	GenBankᶜ (DC)ᶠ	AY145367ᵃ		AF355163ᵃ		(DQ093526)
Rissoellidae	Rissoella caribaea	MCZ DNA100667	AF120520	DQ093460			DQ093527
Acteonidae	Acteonidaeᵇ	GenBankᵍ GenBankʰ	(AY427516)ᵃ		AF355186ᵃ		
Philinidae	Philine aperta	MCZ DNA101268	DQ093438	DQ093461	DQ093482	DQ093508	DQ093528
Aplysiidae	Stylocheilus longicauda	MCZ DNA101392	DQ093439		DQ093483	DQ093509	DQ093529
Hydatinidae	Micromelo undatus	MCZ DNA101398	DQ093443		DQ09348	DQ093513	
Amphibolidae	Phallomedusa solida	DNA from DC	DQ093440	DQ093462	DQ093484	DQ093510	
Onchididae	Onchidella sp.	MCZ DNA101393	DQ093441	DQ093463	DQ093485	DQ093511	
Siphonariidae	Siphonaria pectinata	MCZ DNA100660	X91973ᵃ	AF120578	AY377627	AY377780	AF120638
Ellobiidae	Ophicardelus ornatus	DNA from DC	DQ093442	DQ093464	DQ093486	DQ093512	DQ093530
Outgroups							
Caudofoveata	Chaetoderma nitidulum	MCZ DNA100838	AY377658	AY377692	AY377612	AY377763	AY377726
Lepidopleurida	Lepidopleurus cajetanus	MCZ DNA100108	AF120502	AF120565	AY377585	AY377735	AF120626
Chitonidae	Chiton olivaceus	MCZ DNA100157	AY377651	AY377682	AY377605	AY377755	AY377716
Dentaliida	Rhabdus rectius	Not available	AF120523	AF120580	AY377619	AY377772	AF120640
Dentaliida	Antalis inaequicostata	MCZ DNA101022	DQ093444	DQ093465		DQ093514	DQ093531
Nuculoidea	Nucula proxima	Not available	AF120526	AF120583	AY377617		AF120641
Nuculanoidea	Yoldia limatula	MCZ DNA100119	AF120528	AF120585		AY070149	AF120642
Pteriomorphia	Ostrea edulis	MCZ DNA100130	L49052ᵃ	AF137047	DQ093488	AY070151	AF120651
Palaeoheterodonta	Neotrigonia margaritacea	MCZ DNA100031	AF411690	AF411689	DQ093489	AY070155	U56850ᵃ
Nautiloidea	Nautilusᵇ	NMSUⁱ NMSUʲ	AY557455	(AF120567)	AY377628	(AF033704)	AY557514
Octopoda	Grandeledone verrucosa	NMSU	AY557468	AY557557	DQ093490	AY557413	AF000032
Sepiida	Sepia officinalis	NMSU	AY557471	AY557560	DQ093491	AY557415	AF000062
Teuthida	Loligo pealei	NMSU	AY557479	AY557565	AF421958	AY557423	AF000052

NOTE: NMSU indicates New Mexico State University.

ᵃObtained from GenBank.

ᵇMultiple specimens were used for the terminal taxon, and sequences or voucher numbers in parentheses are the alternative specimens.

ᶜCellana nigrolineata.

ᵈCellana cf. tramoserica.

ᵉBoonea seminuda.

ᶠHinemoa ligata.

ᵍActeon tornatilis.

ʰPupa solidula.

ⁱNautilus pompilius.

ʲNautilus scrobiculatus.

Bayerotrochus midas Bayer, 1965

Goulding's Cay, New Province Island (Bahamas); *ca.* 750 m; Johnson-Sea-Link I submersible (dive 3658); MCZ DNA100666; Voucher specimens USNM 888645

Entemnotrochus adansonianus Crosse and Fischer, 1861

Guadeloupe; Supplied by M. G. Harasewych; MCZ DNA100665 (*ex* USNM 888647)

Lepetodrilus elevatus McLean, 1988

9°50.60′ N 104°17.49′ W; 9°N East Pacific Rise (Mussel Bed), December 11, 2002; DSV *Alvin* Dive 3853; S. Aktipis; MCZ DNA100930

Diodora graeca Linnaeus, 1758

Tossa de Mar (Girona, Spain), August 5, 1997; G. Giribet and C. Palacín; MCZ DNA100114

Diodora cayenensis Lamarck, 1822

Sebastian Inlet, Florida (United States); M. G. Harasewych; Voucher specimens USNM 888660

Bathymargarites symplector Warén and Bouchet, 1989

9°N East Pacific Rise; 10 December 2003; DSV *Alvin* Dive 3951; T. Harmer; MCZ DNA101220

Haliotis tuberculata Linnaeus, 1758

Tossa de Mar (Girona, Spain), June 6, 1997; G. Giribet; MCZ DNA100110

Clanculus cruciatus Linnaeus, 1758

Tossa de Mar (Girona, Spain), June 6, 1997; G. Giribet; MCZ DNA100664

Calliostoma granulatum Born, 1778

Banyuls sur Mer (Languedoc-Roussillon, France); 30–60 m; July 2000; G. Giribet; MCZ DNA100086

Aperostoma palmeri Bartsch and Morrison, 1942

21°14.6′ N, 98°51.5′ W; 4 km NE of Taman, 12 km SW of Tamazunchale (San Luis Potosi, Mexico), January 27, 1998; Donated by F. G. Thompson; FGT 5802; UF 268155; F. G. Thompson and S. P. Christman,; MCZ DNA101457

Pomacea bridgesi Reeve, 1856

Brookfield (West of Brisbane, Queensland, Australia); voucher number C.333043; DNA supplied by D. Colgan

Viviparus georgianus Lea, 1834

Lake Talquin, Tallahassee (Florida, United States); supplied by M. G. Harasewych; MCZ DNA100112

Littorina littorea Linnaeus, 1758

58°12′ N, 11°19′ E; Kristineberg Marine Station, Fiskebäckskil (Sweden), July 24, 2004; AToL Protostome Expedition; MCZ DNA101389

Miralda ligata Angas, 1877

North end of Edwards Beach (Balmoral Sydney Harbour, Australia); March 30, 2002; W. F. Ponder; voucher number C. 408354; DNA supplied by D. Colgan

Rissoella caribaea Rehder, 1943

Fiesta Key (Florida, United States); supplied by J. Harasewych; MCZ DNA100667

Philine aperta Linnaeus, 1767

Tjärnö (Sweden); July 29, 2004; AToL Protostome Expedition; MCZ DNA101387

Stylocheilus longicauda Quoy & Gaimard, 1825

Guam; September 14, 1993; donated by T. Gosliner; MCZ DNA101392

Phallomedusa solida Von Martens, 1878

Tilligerry Creek (Port Stephens, New South Wales, Australia); DNA supplied by D. Colgan; voucher number C203229

Onchidella sp.

Santo Domingo, Barrio Los Angeles (California, United States), October 30, 1993; donated by T. Gosliner; MCZ DNA101393

Siphonaria pectinata Linnaeus, 1758

El Puerto de Santa María (Cádiz, Spain), April 27, 1993; G. Giribet leg.; MCZ DNA100660

Ophicardelus ornatus Férussac, 1821

Tilligerry Creek (Port Stephens, New South Wales, Australia); 25 January 1994; W. F. Ponder and P. Eggler; DNA supplied by D. Colgan

Micromelo undatus Bruguiére, 1792

Station 15, September 5, 2001; donated by T. Gosliner; MCZ DNA101398

Outgroups

Chaetoderma nitidulum Lovén, 1845

Kristineberg (Sweden), January 1998; A. Okusu and A. Scheltema; MCZ DNA100838

Lepidopleurus cajetanus Poli, 1791

Tossa de Mar (Girona, Spain), June 6, 1997; G. Giribet; MCZ DNA100108

Chiton olivaceus Spengler, 1797

Tossa de Mar (Girona, Spain), June 6, 1997; G. Giribet; MCZ DNA100157

Antalis inaequicostatum Dautzenberg, 1891

Banyuls sur Mer (Languedoc-Roussillon, France), 30 m, July 10, 2003; G. Giribet; MCZ DNA101022

Nucula proxima Say, 1822

Beaufort (North Carolina, United States); DNA from D. Campbell

Yolida limatula Say, 1831

Woods Hole (Massachusetts, United States), December 1997; Marine Biological Laboratory; MCZ DNA100119

Ostrea edulis Linnaeus, 1758

Barcelona (Spain); from fish market; G. Giribet; MCZ DNA100130

Neotrigonia margaritacea (Lamarck, 1804)

D'Entrecasteau channel (Tasmania, Australia), April 19, 2000; The Marine Discovery Centre (contacted through L. Turner, Tasmanian Museum); MCZ DNA1000031

Grandeledone verrucosa Verril, 1881

Voucher information in Lindgren *et al.* 2004

Sepia officinalis Linnaeus, 1758

Voucher information in Lindgren *et al.* 2004

APPENDIX 3: LIST OF ADDITIONAL CHARACTERS AND STATES USED IN THE MORPHOLOGICAL DATA SET

CHARACTERS 118–122: NEURAL

118—Cephalic tentacle epithelium characterized by ensheathing sensory cells (Künz and Haszprunar 2001). 0 = absent, 1 = present.

119—Muscular hydrostat system (Kier 1988). 0 = absent, 1 = present.

120—Giant nerve cells. 0 = absent, 1 = present.

121—Lateral nerve (Haszprunar 1988c). 0 = absent, 1 = present.

122—Cerebral ganglia. 0 = absent, 1 = present.

CHARACTERS 123–129: ALIMENTARY SYSTEM

123—Position of gastric shield. 0 = dorsal, 1 = ventral.

124—Gastric pouches. 0 = absent, 1 = left, 2 = right.

125—Pyloric ceca. 0 = absent, 1 = present.

126—Hood. 0 = present, 1 = fused, 2 = absent.

127—Mouth scales. 0 = absent, 1 = present.

128—Shape of mouth opening. 0 = round, 1 = triangular, 2 = vertical slit, 3 = horizontal slit.

129—Lateral expansion of mouth. 0 = not expanded, 1 = expanded.

CHARACTERS 130–132: RADULAR CARTILAGES

Data derived from Sasaki (1998) and Guralnick and Smith (1999). Numbers in parentheses refer to Sasaki's original character numbers.

130—Posterior cartilage (54). 0 = present, 1 = absent.

131—Anterolateral cartilage (55). 0 = present, 1 = absent.

132—Median cartilage (56). 0 = present, 1 = absent.

CHARACTERS 133–134: RESPIRATORY SYSTEM

133—Pneumostome. 0 = absent, 1 = present.

134—Gill filament shape (Lindberg and Ponder 2001). 0 = round, 1 = elongate-triangular.

CHARACTER 135: LARVAL

135—Velum. Ring = 0, lobed = 1.

CHARACTERS 136–143: MUSCULATURE

These characters are directly derived from Sasaki (1998). Sasaki's original character numbers are in parentheses.

136—Retractable mantle tentacles (2). 0 = absent, 1 5 present.

137—Dorsal protractor muscles of the odontophore (27). 0 = absent, 1 = present.

138—Anterior levator muscle of the odontophore (28). 0 = absent, 1 = present.

139—Posterior depressor muscle of odontophore (31). 0 = absent, 1 = present.

140—Postdorsal buccal tensor muscle (32). 0 = absent, 1 = present.

141—Retractor muscles of the subradular membranes (39). 0 = divided, 1 = fused.

142—Postmedian retractor muscles of the radular sac (41). 0 = absent, 1 = present.

143—Median tensor muscle of the radular sac (42). 0 = absent, 1 = present.

CHARACTERS 144–152: ELEMENTAL RADULAR COMPOSITION

Okoshi and Ishii (1996) measured the concentration of 17 elements in the radulae of 24 molluscan species; 10 of which represent family rank taxa used in our analysis. Although more sampling and study of radular elemental composition is needed, these preliminary data provide the first opportunity to explore and compare these characters across the Gastropoda. Moreover, current sampling is sufficient to rule out widespread convergence in taxa with similar feeding modes. For example, although patellogastropods and some vetigastropods and caenogastropods rasp rock surfaces for food, only patellogastropods show high concentrations of iron in their radular teeth. In contrast, some vetigastropods and the caenogastropod rock-rasping Littorinidae and carnivorous Nassariidae have higher concentrations of silicon than patellogastropods.

To determine whether elemental concentration carried phylogenetic signal, a 17×24 matrix was constructed from the data and an F-test performed to estimate the statistical significance of the variance present in the matrix. Because the distribution of the variance in the data was statistically significant ($p = 0.000$), Tukey's test (Hsu 1996) was used to identify those elements with statistically significant differences. Nine of the 17 elements were found to have statistically significant distributions: calcium, iron, magnesium, manganese, phosphorus, strontium, zinc, silicon, and copper. The elemental measurements were log transformed to reduce variance prior to gap coding, and the following character states were gap coded by graphing the distribution of values within each character and determining gaps or trend changes by eye. Obvious, isolated groupings and breaks in trends were recognized and assigned discrete character states.

144—Calcium (Ca). 0 = high, 1 = low.

145—Iron (Fe). 0 = high, 1 = low.

146—Magnesium (Mg). 0 = high, 1 = low.

147—Manganese (Mn). 0 = high, 1 = low, 2 = moderate.

148—Phosphorus (P). 0 = high, 1 = low.

149—Strontium (Sr). 0 = high, 1 = low.

150—Zinc (Zn). 0 = high, 1 = low, 2 = moderate.

151—Silicon (Si). 0 = high, 1 = low, 2 = moderate.

152—Copper (Cu). 0 = high, 1 = low, 2 = moderate.

CHARACTERS 153–166: DEVELOPMENTAL CELL LINEAGE TIMING

These character states represent the cell number of the developing embryo at which specific cell lineages originate. Complete character analysis is present in Lindberg and Guralnick (2003), and a discussion of character independence of cell lineage characters can be found in Guralnick and Lindberg (2002).

153—Median number of cells present at formation of $1q^1$ cell lineage (4th division). 0: 16, 1: 21.

154—Median number of cells present at formation of $1q^{11}$ cell lineage (5th division). 0: 24, 1: 28, 2: 32, 3: 38–40.

155—Median number of cells present at formation of $1q^{112}$ cell lineage (6th division). 0: 44–54, 1: 68–99.

156—Median number of cells present at formation of $1q^{121}$ cell lineage (6th division). 0: 36–49, 1: 55–72.

157—Median number of cells present at formation of $1q^{22}$ cell lineage (5th division). 0: 32, 1: 44–49, and 2: 60.

158—Median number of cells present at formation of $1q^{222}$ cell lineage (6th division). 0: 60, 1: 80, and 2: 123–132.

159—Median number of cells present at formation of $2q$ cell lineage (4th division). 0: 16, 1: 12.

160—Median number of cells present at formation of $2q^1$ cell lineage (5th division). 0: 32, 1: 24.

161—Median number of cells present at formation of $2q^{11}$ cell lineage (6th division). 0: 60, 1: 50–52, 2: 44, 3: 41, 4: 37–38, 5: 32.

162—Median number of cells present at formation of $3q$ cell lineage (5th division). 0: 20, 1: 23–24, and 2: 32.

163—Median number of cells present at formation of $3q^1$ cell lineage (6th division). 0: 54, 1: 60–61, 2: 48, 3: 40–41, 4: 36.

164—Median number of cells present at formation of $4Q$ cell lineage (6th division). 0: 85, 1: 63–64, 2: 48–49, 3: 42–44.

165—Median number of cells present at formation of 4d cell lineage (6th division). 0: 64–73, 1: 41–48, and 2: 25–32.

166—Median number of cells present at formation of 4d¹ cell lineage (7th division). 0: 90–130, 1: 39–49.

REFERENCES

Alkon, D. L., Favit, A., and Nelson, T. 1998. Evolution of adaptive neural networks: the role of voltage-dependent K⁺ channels. *Otolaryngology—Head and Neck Surgery* 119: 204–211.

Bandel, K. 1997. Higher classification and pattern of evolution of the Gastropoda. *Courier Forschungs Institut Senckenberg* 201: 57–81.

Bandel, K., and Frýda, J. 1999. Notes on the evolution and higher classification of the subclass Neritimorpha (Gastropoda) with the description of some new taxa. *Geologica et Palaeontologica* 33:219–235.

Barker, G. M. 2001a. Gastropods on land: Phylogeny, diversity and adaptive morphology. In *The Biology of Terrestrial Molluscs*. Edited by G. M. Barker. New York: CAB International, pp. 1–130.

———, ed. 2001b. *The Biology of Terrestrial Molluscs*. New York: CAB International.

Beesley, P. L., Ross, G. J. B., and Wells, A., eds. 1998. *Mollusca: The Southern Synthesis*. Fauna of Australia Vol. 5. Melbourne: CSIRO Publishing.

Bieler, R. 1990. Haszprunar's "clado-evolutionary" classification of the Gastropoda—a critique. *Malacologia* 31: 371–380.

———. 1992. Gastropod phylogeny and systematics. *Annual Reviews of Ecology and Systematics* 23: 311–338.

Bininda-Emonds, O. R., Jones, K. E., Price, S. A., Grenyer, R., Cardillo, M., Habib, M., Purvis, A., and Gittleman, J. L. 2003. Supertrees are a necessary not-so-evil: a comment on Gatesy *et al*. *Systematic Biology* 52: 724–729.

Bonnemain, B. 2005. *Helix* and drugs: snails in Western health care from antiquity to the present. *Evidence-Based Complementary and Alternative Medicine* 2: 25–28.

Bouchet, P., Lozouet, P., Maestrati, P., and Heros, V. 2002. Assessing the magnitude of species richness in tropical marine environments: exceptionally high numbers of molluscs at a New Caledonia site. *Biological Journal of the Linnean Society* 75: 421–436.

Bouchet, P., Sysoev, A., and Lozouet, P. 2004. An inordinate fondness for turrids. In *Molluscan Megadiversity: Sea, Land and Freshwater. World Congress of Malacology, 11–16 July, Perth, Western Australia*. Perth: Western Australian Museum, p. 12.

Bourne, G. C. 1908. Contributions to the morphology of the group Neritacea of aspidobranch gastropods, part I. The Neritidæ. *Proceedings of the General Meetings for Scientific Business of the Zoological Society of London* 1908: 810–887.

Boyer, S. L., Karaman, I., and Giribet, G. 2005. The genus *Cyphophthalmus* (Arachnida, Opiliones, Cyphophthalmi) in Europe: a phylogenetic approach to Balkan Peninsula biogeography. *Molecular Phylogenetics and Evolution* 36: 554–567.

Caron, J.-B., Scheltema, A., Schander, C., and Rudkin, D. 2006. A soft-bodied mollusc with radula from the Middle Cambrian Burgess Shale. *Nature* 442: 159–163.

Colgan, D. J., McLauchlan, A., Wilson, G. D. F., Livingston, S. P., Edgecombe, G. D., Macaranas, J., Cassis, G., and Gray, M. R. 1998. Histone H3 and U2 snRNA DNA sequences and arthropod molecular evolution. *Australian Journal of Zoology* 46: 419–437.

Colgan, D. J., Ponder, W. F., Beacham, E., and Macaranas, J. M. 2003. Molecular phylogenetic studies of Gastropoda based on six gene segments representing coding or non-coding and mitochondrial or nuclear DNA. *Molluscan Research* 23: 123–148.

———. 2007. Molecular phylogenetics of Caenogastropoda (Gastropoda: Mollusca). *Molecular Phylogenetics and Evolution* 42: 717–737.

Colgan, D. J., Ponder, W. F., and Eggler, P. E. 2000. Gastropod evolutionary rates and phylogenetic relationships assessed using partial 28S rDNA and histone H3 sequences. *Zoologica Scripta* 29: 29–63.

Cowie, R. H. 1992. Evolution and extinction of Partulidae, endemic Pacific Island land snails. *Philosophical Transactions of the Royal Society of London Series B: Biological Sciences* 335: 167–191.

Cox, L. R. 1960. Thoughts on the classification of the gastropoda. *Proceedings of the Malacological Society of London* 33: 239–261.

D'Haese, C. A. 2003. Sensitivity analysis and tree-fusing: faster, better. *Cladistics* 19: 150–151.

Dayrat, B., and Tillier, S. 2002. Evolutionary relationships of euthyneuran gastropods (Mollusca): a cladistic re-evaluation of morphological characters. *Zoological Journal of the Linnean Society* 135: 403–407.

Erwin, D. H. 1990. Carboniferous-Triassic gastropod diversity patterns and the Permo-Triassic mass extinction. *Paleobiology* 16: 187–203.

Farris, J. S., Albert, V. A., Källersjö, M., Lipscomb, D., and Kluge, A. G. 1996. Parsimony jackknifing outperforms neighbor-joining. *Cladistics* 12: 99–124.

Farris, J.S., Källersjö, M., Kluge, A.G., and Bult, C. 1995. Constructing a significance test for incongruence. *Systematic Biology* 44: 570–572.

Folmer, O., Black, M., Hoeh, W., Lutz, R., and Vrijenhoek, R.C. 1994. DNA primers for amplification of mitochondrial cytochrome *c* oxidase subunit I from diverse metazoan invertebrates. *Molecular Marine Biology and Biotechnology* 3: 294–299.

Fretter, V. 1989. The anatomy of some new archaeogastropod limpets (Superfamily Peltospiracea) from hydrothermal vents. *Journal of Zoology, London* 218: 123–169.

Fretter, V., and Graham, A. 1962. *British Prosobranch Molluscs. Their Functional Anatomy and Ecology.* London: Ray Society.

Gatesy, J., Baker, R.H., and Hayashi, C. 2004. Inconsistencies in arguments for the supertree approach: supermatrices versus supertrees of Crocodylia. *Systematic Biology* 53: 342–355.

Geiger, D.L., and Thacker, C.E. 2005. Molecular phylogeny of Vetigastropoda reveals non-monophyletic Scissurellidae, Trochoidea, and Fissurelloidea. *Molluscan Research* 25: 47–55.

Giese, A.C., and Pearse, J.S., Eds. 1977. *Reproduction of Marine Invertebrates.* Vol. IV. *Molluscs: Gastropods and Cephalopods.* New York: Academic Press.

Giribet, G. 2001. Exploring the behavior of POY, a program for direct optimization of molecular data. *Cladistics* 17: S60–S70.

———. 2002. Current advances in the phylogenetic reconstruction of metazoan evolution. A new paradigm for the Cambrian explosion? *Molecular Phylogenetics and Evolution* 24: 345–357.

———. 2003. Stability in phylogenetic formulations and its relationship to nodal support. *Systematic Biology* 52: 554–564.

Giribet, G., Carranza, S., Baguñà, J., Riutort, M., and Ribera, C. 1996. First molecular evidence for the existence of a Tardigrada + Arthropoda clade. *Molecular Biology and Evolution* 13: 76–84.

Giribet, G., and Distel, D.L. 2003. Bivalve phylogeny and molecular data. In *Molecular Systematics and Phylogeography of Mollusks.* Edited by C. Lydeard and D.R. Lindberg. Washington, DC: Smithsonian Books, pp. 45–90.

Giribet, G., Distel, D.L., Polz, M., Sterrer, W. and Wheeler, W.C. 2000. Triploblastic relationships with emphasis on the acoelomates and the position of Gnathostomulida, Cycliophora, Plathelminthes, and Chaetognatha: A combined approach of 18S rDNA sequences and morphology. *Systematic Biology* 49: 539–562.

Giribet, G., Okusu, A., Lindgren, A.R., Huff, S.W., and Schrödl, M. 2006. Evidence for a clade composed of molluscs with serially repeated structures: Monoplacophorans are related to chitons. *Proceedings of the National Academy of Sciences* 103: 7723–7728.

Giribet, G., and Wheeler, W.C. 2002. On bivalve phylogeny: a high-level analysis of the Bivalvia (Mollusca) based on combined morphology and DNA sequence data. *Invertebrate Biology* 121: 271–324.

Golikov, A.N., and Starobogatov, Y.I. 1975. Systematics of prosobranch gastropods. *Malacologia* 15: 185–232.

Goloboff, P.A., Farris, J.S., and Nixon, K. 2003. TNT: Tree analysis using New Technology. Version 1.0. Ver. Beta test v. 0.2. Program and documentation available at www.zmuc.dk/public/phylogeny/TNT.

Gosliner, T.M. 1985. Parallelism, parsimony, and the testing of phylogenetic hypotheses: the case of opisthobranch gastropods. In *Species and Speciation.* Edited by E.S. Vrba. Pretoria: Transvaal Museum, pp. 105–107.

———. 1991. Morphological parallelsim in opisthobranch gastropods. *Malacologia* 32: 313–327.

Gould, S.J. 1977. *Ontogeny and Phylogeny.*, Cambridge: Belknap Press.

Grande, C., and Templado, J. 2004. Molecular phylogeny of euthyneura (Mollusca: Gastropoda). *Molecular Biology and Evolution* 21: 303–313.

Guralnick, R.G., and D.R. Lindberg. 2002. Cell lineage data and spiralian: A reply to Nielsen and Meier. *Evolution* 56: 2558–2560.

Guralnick, R.G., and Smith, K. 1999. Historical and biomechanical analysis of integration and dissociation in molluscan feeding, with special emphasis on the true limpets (Patellogastropoda: Gastropoda). *Journal of Morphology* 241: 175–195.

Harasewych, M.G. 2002. Pleurotomarioidean gastropods. *Advances in Marine Biology* 42: 235–293.

Harasewych, M.G., Adamkewicz, S.L., Blake, J., Saudek, D., Spriggs, T., and Bult, C.J. 1997. Phylogeny and relationships of pleurotomariid gastropods (Mollusca: Gastropoda): an assessment based on partial 18S rDNA and cytochrome *c* oxidase I sequences. *Molecular Marine Biology and Biotechnology* 6: 1–20.

Harasewych, M.G., Adamkewicz, S.L., Plassmeyer, M., and Gillevet, P.M. 1998. Phylogenetic relationships of the lower Caenogastropoda (Mollusca, Gastropoda, Architaenioglossa, Campaniloidea, Cerithioidea) as determined by partial 18S sequences. *Zoologica Scripta* 27: 361–372.

Haszprunar, G. 1985a. The fine morphology of the osphradial sense organs of the Mollusca. Part 1: Gastropoda Prosobranchia. *Philosophical Transactions of the Royal Society of London Series B: Biological Sciences* 307: 457–496.

———. 1985b. The fine morphology of the osphradial sense organs of the Mollusca. Part 2: Allogastropoda (Architectonicidae and Pyramidellidae). *Philosophical Transactions of the Royal Society of London Series B: Biological Sciences* 307: 497–505.

———. 1985c. The Heterobranchia—a new concept of the phylogeny and evolution of the higher Gastropoda. *Zeitschrift für zoologische Systematik und Evolutionsforschung* 23: 15–37.

———. 1988a. A preliminary phylogenetic analysis of streptoneurous Gastropoda. *Malacological Review* Suppl. 4: 7–16.

———. 1988b. Comparative anatomy of cocculiniform gastropods and its bearing on archaeogastropod systematics. In Prosobranch Phylogeny. Edited by W. F. Ponder. *Malacological Review* Suppl. 4: 64–84.

———. 1988c. On the origin and evolution of major gastropod groups, with special reference to the Streptoneura. *Journal of Molluscan Studies* 54: 367–441.

———. 1989. The anatomy of *Melanodrymia aurantiaca* Hickman, a coiled archaeogastropod from the East Pacific hydrothermal vents (Mollusca, Gastropoda). *Acta Zoologica Stockholm* 70: 175–186.

Healy, J. M. 1988. Sperm morphology and its systematic importance in the Gastropoda. In Prosobranch Phylogeny. Edited by W. F. Ponder. *Malacological Review* Suppl. 4: 251–266.

Hedegaard, C., Lindberg, D. R., and Bandel, K. 1997. A patellogastropod limpet from the Triassic St. Cassian Formation of Italy. *Lethaia* 30: 331–335.

Hickman, C. S. 1988. Archaeogastropod evolution, phylogeny, and systematics: a re-evaluation. In Prosobranch Phylogeny. Edited by W. F. Ponder. *Malacological Review* Suppl. 4: 17–34.

Hsu, J. C. 1996. *Multiple Comparisons: Theory and Methods.* New York: Chapman and Hall.

Kabat, A. R. 1990. Predatory ecology of naticid gastropods with a review of shell boring predation. *Malacologia* 32: 155–193.

Kay, E. A., Wells, F. E., and Ponder, W. F. 1998. Class Gastropoda. In *Mollusca: The Southern Synthesis.* Fauna of Australia . Edited by P. L. Beesley, G. J. B. Ross, and A. Wells. Melbourne: CSIRO Publishing, pp. 565–604.

Kiel, S. 2006. New and little-known gastropods from the Albian of the Mahajanga Basin, Northwestern Madagascar. *Journal of Paleontology* 80: 455–476.

Kier, W. 1988. The arrangement and function of molluscan muscle. In *The Mollusca. Vol. 11: Form and Function.* Edited by E. Trueman and M. Clarke. London: Academic Press, pp. 211–252.

Kluge, A. G. 1989. A concern for evidence and a phylogenetic hypothesis of relationships among

Epicrates (Boidae, Serpentes). *Systematic Zoology* 38: 7–25.

Knight, J. B. 1952. Primitive fossil gastropods and their bearing on gastropod classification. *Smithsonian Miscellaneous Collections* 117: 1–55.

Knight, J. B., Batten, R. L., and Yochelson, E. L. 1954. Status of invertebrate paleontology, 1953. V. Mollusca: Gastropoda. *Bulletin of the Museum of Comparative Zoology* 112: 173–179.

Knight, J. B., Cox, L. R., Keen, A. M., Batten, R. L., Yochelson, E. L., and Robertson, R. 1960. Systematic descriptions. In *Treatise on Invertebrate Paleontology.* Part I. *Mollusca I.*. Edited by R. C. Moore. Lawrence, KS: Geological Society of America, pp. 1169–1310.

Künz, E., and Haszprunar, G. 2001. Comparative ultrastructure of gastropod cephalic tenacles: Patellogastropoda, Neritaemorphi and Vetigastropoda. *Zoologischer Anzeiger* 240: 101–210.

Leong, C. C., Syed, N. I., and Lorscheider, F. L. 2001. Retrograde degeneration of neurite membrane structural integrity of nerve growth cones following *in vitro* exposure to mercury. *Neuroreport* 12: 733–737.

Lindberg, D. R. 1985. Aplacophorans, monoplacophorans, polyplacophorans and scaphopods: the lesser classes. In *Mollusks. Notes for a Short Course.* Edited by T. W. Broadhead. University of Tennessee, Department of Geological Science Studies in Geology 13. Knoxville, TN: University of Tennessee, pp., 230–247.

———. 1988. The Patellogastropoda. In Prosobranch Phylogeny. Edited by W. F. Ponder. *Malacological Review* Suppl. 4: 35–63.

———. 1990. The systematics of *Potamacmaea fluvatilis* Blandford: A brackishwater patellogastropod (Patelloidinae: Lottiidae). *Journal of Molluscan Studies* 56: 309–316.

Lindberg, D. R., and Guralnick, R. P. 2003. Phyletic patterns of early development in gastropod molluscs. *Evolution and Development* 5: 494–507.

Lindberg, D. R., and Ponder, W. F. 1996. An evolutionary tree for the Mollusca: branches or roots? In *Origin and Evolutionary Radiation of the Mollusca.* Edited by J. Taylor. Oxford: Oxford University Press, pp. 67–75.

———. 2001. The influence of classification on the evolutionary interpretation of structure—a re-evaluation of the evolution of the pallial cavity in gastropod molluscs. *Organisms, Diversity and Evolution,* 1: 273–299.

Lindberg, D. R., Ponder, W. F., and Haszprunar, G. 2004. The Mollusca: relationships and patterns from their first half-billion years. In *Assembling the Tree of Life.* Edited by J. Cracraft and M. J. Donoghue. New York: Oxford University Press, pp. 252–278.

Lindgren, A. R., Giribet, G., and Nishiguchi, M. K. 2004. A total evidence phylogeny of the Cephalopoda (Mollusca). *Cladistics* 20: 454–486.

Linton, E. W. 2005. MacGDE: Genetic Data Environment for MacOS X. Ver. 2.0. Software available at http://www.msu.edu/~lintone/macgde/.

Lydeard, C., Holznagel, W. E., Glaubrecht, M., and Ponder, W. F. 2001. Molecular phylogeny of a circum-global, diverse gastropod superfamily (Cerithioidea: Mollusca: Caenogastropoda): pushing the deepest phylogenetic limits of mitochondrial LSU rDNA sequences. *Molecular Phylogenetics and Evolution* 22: 399–406.

Maila, M. J. J., Lipscomb, D. L., and Allard, M. W. 2003. The misleading effects of composite taxa in supermatrices. *Molecular Phylogenetics and Evolution* 27: 522–527.

McArthur, A. G., and Harasewych, M. G. 2003. Molecular systematics of the major lineages of the Gastropoda. In *Molecular Systematics and Phylogeography of Mollusks*. Edited by C. Lydeard and D. R. Lindberg. Washington, DC: Smithsonian Books, pp. 140–160.

McArthur, A. G., and Koop, B. F. 1999. Partial 28S rDNA sequences and the antiquity of hydrothermal vent endemic gastropods. *Molecular Phylogenetics and Evolution* 13: 255–274.

McArthur, A. G., and Tunnicliffe, V. 1998. Relics and antiquity revisited in the modern vent fauna. In *Modern Ocean Floor Processes and the Geological Record*. Edited by R. A. Mills and K. Harrison. London: Geological Society Special Publications. pp 271–291.

McKinney, M. L., and MacNamara, K. J. 1988. *Heterochrony. The Evolution of Ontogeny*. New York: Plenum Press.

McLean, J. H. 1989. New archaeogastropod limpets from hydrothermal vents: new family Peltospiridae, new superfamily Peltospiracea. *Zoologica Scripta* 18: 49–66.

Mickevich, M. F., and Farris, J. S. 1981. The implications of congruence in *Menidia. Systematic Zoology* 27: 143–158.

Morton, J. E., and Yonge, C. M. 1964. Classification and structure of the Mollusca. In *Physiology of the Mollusca*. Edited by K. M. Wilbur. New York: Academic Press, pp. 1–58.

Nakano, T., and Ozawa, T. 2004. Phylogeny and historical biogeography of limpets of the order Patellogastropoda based on mitochondrial DNA sequences. *Journal of Molluscan Studies* 70: 31–41.

Nixon, K. C., and Carpenter, J. M. 1996. On simultaneous analysis. *Cladistics* 12: 221–241.

Okoshi, K., and Ishii, T. 1996. Concentrations of elements in the radular teeth of limpets, chitons and other marine molluscs. *Journal of Marine Biotechnology* 3: 252–257.

Okusu, A., Schwabe, E., Eernisse, D. J., and Giribet, G. 2003. Towards a phylogeny of chitons (Mollusca, Polyplacophora) based on combined analysis of five molecular loci. *Organisms, Diversity & Evolution* 3: 281–302.

Page, L. R. 1995. The ancestral gastropod larval form is best approximated by hatching stage opisthobranch larvae: evidence from comparative developmental studies. In *Reproduction and development of marine invertebrates*. Edited by H. W. Wilson, S. A. Stricker, and G. L. Shinn. Baltimore, Johns Hopkins University Press, pp. 206–223.

Ponder, W. F., and Lindberg, D. R. 1996. Gastropod phylogeny—challenges for the 90s. In *Origin and Evolutionary Radiation of the Mollusca*. Edited by J. Taylor. Oxford: Oxford University Press, pp. 135–154.

———. 1997. Towards a phylogeny of gastropod molluscs: an analysis using morphological characters. *Zoological Journal of the Linnean Society* 119: 83–265.

Prendini, L. 2001. Species or supraspecific taxa as terminals in cladistic analysis? Groundplans versus exemplars revisited. *Systematic Biology* 50: 290–300.

Remigio, E. A., and Hebert, P. D. N. 2003. Testing the utility of partial COI sequences for phylogenetic estimates of gastropod relationships. *Molecular Phylogenetics and Evolution* 29: 641–647.

Rokas, A., and Carroll, S. B. 2006. Bushes in the Tree of Life. *PLoS Biology* 4: e352.

Rokas, A., Krüger, D., and Carroll, S. B. 2005. Animal evolution and the molecular signature of radiations compressed in time. *Science* 310: 1933–1938.

Rokas, A., Williams, B. L., King, N., and Carroll, S. B. 2003. Genome-scale approaches to resolving incongruence in molecular phylogenies. *Nature* 425: 798–804.

Rosenberg, G., Davis, G. M., Kuncio, G. S., and Harasewych, M. G. 1994. Preliminary ribosomal RNA phylogeny of gastropod and unionoidean bivalve mollusks. *The Nautilus* Suppl. 2: 111–121.

Rosenberg, G., Tillier, S., Tillier, A., Kuncio, G. S., Hanlon, R. T., Masselot, M., and Williams, C. J. 1997. Ribosomal RNA phylogeny of selected major clades in the Mollusca. *Journal of Molluscan Studies* 63: 301–309.

Salvini-Plawen, L. v. 1980. A reconsideration of systematics in the Mollusca (phylogeny and higher classification). *Malacologia* 19: 249–278.

Salvini-Plawen, L. v., and Haszprunar, G. 1987. The Vetigastropoda and the systematics of streptoneurous

Gastropoda (Mollusca). *Journal of Zoology, London* 211: 747–770.

Salvini-Plawen, L. v., and Steiner, G. 1996. Synapomorphies and plesiomorphies in higher classification of Mollusca. In *Origin and Evolutionary Radiation of the Mollusca*. Edited by J. D. Taylor. Oxford: Oxford University Press, pp. 29–51.

Sasaki, T. 1998. Comparative anatomy and phylogeny of the Recent archaeogastropoda (Mollusca: Gastropoda). *The University Museum The University of Tokyo Bulletin* 38: 1–223.

Strong, E. E. 2003. Refining molluscan characters: morphology, character coding and a phylogeny of the Caenogastropoda. *Zoological Journal of the Linnean Society* 137: 447–554.

Strong, E. E., Harasewych, M. G., and Haszprunar, G. 2003. Phylogeny of the Cocculinoidea (Mollusca, Gastropoda). *Invertebrate Biology* 122: 114–125.

Sutton, M. D., Briggs, D. E. G., and Siveter, D. J. 2006. Fossilized soft tissues in a Silurian platyceratid gastropod. *Proceedings of the Royal Society of London Series B: Biological Sciences* 273: 1039–1044.

Taylor, D. W., and Sohl, N. F. 1962. An outline of gastropod classification. *Malacologia* 1: 7–32.

Thiele, J. 1929–1935. *Handbuch der systematischen Weichtierkunde*. Jena: Gustav Fischer Verlag,.

Thollesson, M. 1999. Phylogenetic analysis of Euthyneura (Gastropoda) by means of the 16S rRNA gene: use of a "fast" gene for "higher-level" phylogenies. *Proceedings of the Royal Society of London Series B: Biological Sciences* 266: 75–83.

Thompson, T. E. 1976. *Biology of Opisthobranch Molluscs*. Vol. I. London: The Ray Society.

Tillier, S., Masselot, M., Guerdoux, J., and Tillier, A. 1994. Monophyly of major gastropod taxa tested from partial 28S rRNA sequences, with emphasis on Euthyneura and hot-vent limpets Peltospiroidea. *The Nautilus* Suppl. 2: 122–140.

Tompa, A. S., Verdonk, N. H., and van den Biggelaar, J. A. M., eds. 1984. *The Mollusca*. Vol. 7. *Reproduction*. Orlando, FL: Academic Press.

van den Biggelaar, J. A. M., and Haszprunar, G. 1996. Cleavage patterns and mesentoblast formation in the Gastropoda: an evolutionary perspective. *Evolution* 50: 1520–1540.

Vonnemann, V., Schrödl, M., Klussmann-Kolb, A., and Wägele, H. 2005. Reconstruction of the phylogeny of the Opisthobranchia (Mollusca: Gastropoda) by means of 18S and 28S rRNA gene sequences. *Journal of Molluscan Studies* 71: 113–125.

Wägele, H., and Willan, R. C. 2000. On the phylogeny of the Nudibranchia. *Zoological Journal of the Linnean Society* 130: 83–181.

Wagner, P. J. 2002. Phylogenetic relationships of the earliest anisostrophically coiled gastropods.

Smithsonian Contributions to Paleobiology 88: 1–152.

Warén, A. 2001. Gastropoda and Monoplacophora from hydrothermal vents and seeps; new taxa and records. *The Veliger* 44: 116–231.

Warén, A., Bengtson, S., Goffredi, S. K., and Van Dover, C. L. 2003. A hot-vent gastropod with iron sulfide dermal sclerites. *Science* 302: 1007.

Wenz, W. 1938–1944. Gastropoda, Teil 1: Allgemeiner Teil und Prosobranchia. In *Handbuch der Paläozoologie, 6*. Edited by O. H. Schindewolf. Berlin: Gebrüder Bornträger, pp. 1–240.

Wheeler, W. C. 1995. Sequence alignment, parameter sensitivity, and the phylogenetic analysis of molecular data. *Systematic Biology* 44: 321–331.

———. 1996. Optimization alignment: the end of multiple sequence alignment in phylogenetics? *Cladistics* 12: 1–9.

Wheeler, W. C., Gladstein, D., and De Laet, J. 2004. POY version 3.0. Program and documentation available at ftp.amnh.org/pub/molecular. American Museum of Natural History.

Whiting, M. F., Carpenter, J. M., Wheeler, Q. D., and Wheeler, W. C. 1997. The Strepsiptera problem: phylogeny of the holometabolous insect orders inferred from 18S and 28S ribosomal DNA sequences and morphology. *Systematic Biology* 46: 1–68.

Williams, S. T., Reid, D. G., and Littlewood, D. T. L. 2003. A molecular phylogeny of the Littoriniae (Gastropoda: Littorinidae): unequal evolutionary rates, morphological parallelism, and biogeography of the Southern Ocean. *Molecular Phylogenetics and Evolution* 28: 60–86.

Winnepenninckx, B., Backeljau, T., and De Wachter, R. 1994. Small ribosomal subunit RNA and the phylogeny of Mollusca. *The Nautilus* Suppl. 2: 98–110.

———. 1996. Investigation of molluscan phylogeny on the basis of 18S rRNA sequences. *Molecular Biology and Evolution* 13: 1306–1317.

World Health Organization. 2004. Report: Joint WHO/FAO workshop on foodborne trematode infections in Asia. World Health Organization Regional Office for the Western Pacific. Manila, Philippines, 26–28 November 2002. Available at: http://www.wpro.who.int/NR/rdonlyres/045FE9E3-78AD-42DF-868D-5C08D2755977/0/FBT.pdf.

Xiong, B., and Kocher, T. D. 1991. Comparison of mitochondrial DNA sequences of seven morphospecies of black flies (Diptera: Simuliidae). *Genome* 34: 306–311.

Yonge, C. 1947. The pallial organs in the aspidobranch Gastropoda and their evolution throughout the Mollusca. *Philosophical Transactions of the Royal Society of London Series B: Biological Sciences* 232: 443–518.

10

Paleozoic Gastropoda

Jiří Frýda, Alex Nützel, and Peter J. Wagner

Gastropods have a rich fossil record stretching back to the Cambrian. However, determining the relationships of those early taxa to each other and (especially) to extant taxa entails major challenges. Fossil gastropod shells typically offer only a limited number of characters and only occasionally opercular and muscle scar characters. Although numerous studies suggest that teleoconchs retain strong phylogenetic signal among closely related species and even genera, severe architectural restrictions result in a near absence of character states diagnosing large clades. Features that might diagnose large gastropod clades, such as shell mineralogy and protoconch morphology, are preserved only infrequently in early fossils. Thus, most Paleozoic fossils lack obvious markers that might link them to extant taxa.

As a result of these factors, gross shell form, coupled with the then-current models of gastropod phylogeny, are the basis for traditional ideas about Paleozoic gastropod phylogeny (e.g., Wenz 1938–1944; Knight *et al.* 1960; Pchelintsev and Korobkov 1960). Recently, detailed phylogenetic analyses focusing on specific rather than gross shell features, coupled with

new data on protoconch and shell mineralogy, have seriously challenged views about the relationships of Paleozoic gastropods both to each other and to extant taxa. Here, we review and evaluate ideas about the early evolution of gastropods, ranging from their origins to the evolution of the first undoubted members of living major extant gastropod taxa. We discuss inferences drawn from teleoconch and protoconch morphologies as well as possible relationships of extinct Paleozoic gastropod groups to living gastropods. Finally, we summarize macroevolutionary trends of their teleoconchs and protoconchs, and we discuss some ongoing controversies and adaptive radiations of the Paleozoic gastropods.

THE EARLIEST GASTROPODS AND THEIR RELATIONSHIPS TO OTHER MOLLUSCS

We cannot directly observe torsion on fossils, so it is difficult to demonstrate that any fossil is a gastropod (Yochelson 1967; Peel 1991a, b; see later discussion of the bellerophont controversy). The biological meaning (if any) of torsion is debated (e.g., Morton 1958; Runnegar

1981; Bandel 1982; Pennington and Chia 1985; Goodhart 1987; Geyer 1994; Page 1997), which leaves us with no certain expectations about the earliest gastropod teleoconch. Some neontological analyses imply that the earliest gastropods were limpets (Haszprunar 1988), but this is unlikely, given pervasive convergence to a limpet form from coiled ancestors (Ponder and Lindberg 1997). Thus, the earliest gastropods likely were coiled. Moreover, a recurring trend in gastropod evolution concerns altering of the mantle cavity in ways that accommodate the effects of torsion (Wagner 1996; Lindberg and Ponder 2001). The deep sinus of *Strepsodiscus* and *Schizopea*, placing the anus well behind the inferred locations of the gills and mouth, represents an obvious way to do this. One might question this interpretation of a sinus because similar features appear on taxa with tergomyan muscle scars (Horný 1991; Horný and Peel 1996), and sinuses even occur on some cephalopods. However, this only means that there are likely multiple reasons for possessing a sinus and does not alter the expectation that early gastropods should have had one. Thus, taxa such as *Strepsodiscus* probably represent the oldest gastropods (for an alternative view, see Parkhaev, Chapter 3).

A related issue is the place of the earliest gastropods in Cambrian molluscan phylogeny and which taxa do *not* fall within the gastropod clade. Wagner (1995a, 1997, 1999b) analyzed numerous Cambrian molluscs as possible outgroups for early gastropods and rostroconchs. This suggested that most Cambrian taxa once considered gastropods represent either a paraphyletic assemblage relative to gastropods, tergomyans, rostroconchs, and bivalves (e.g., *Helcionella*; see Runnegar and Pojeta 1974; Peel 1991b; Geyer 1994; but see Parkhaev, Chapter 3, for an alternative view) or distant relations of gastropods (e.g., *Pelagiella* and the Mimospirina; see following section on paragastropods). The earliest anisostrophic shells with deep sinuses, a single selenizone keel (but no slit), and very nearly bilaterally symmetrical lenticular apertures, such as *Schizopea*, form a clade with early

bellerophonts bearing the same sinus, selenizone, and apertural morphologies (e.g., *Strepsodiscus*), with this clade nested within tergomyans (see Wagner 1999b: fig. 7).

Because sample intensity relative to rates of homoplasy affects the accuracy of phylogenetic inference (e.g., Wagner 2000a), denser sampling of Cambrian molluscs in these analyses would be desirable. However, Wagner's iterative outgroup analyses with random selections of 75% of the outgroup taxa repeatedly reconstructed *Schizopea*'s place in molluscan phylogeny. Also, these analyses (Wagner 1999b) place taxa that might be relatives of early cephalopods (e.g., *Knightoconus*) elsewhere in the tergomyan clade. Thus, the implied relationships among classes (bivalves as an outgroup to tergomyans, cephalopods, and gastropods) are consistent with neontological reconstructions. Still, even this corroboration is tentative, as the Cambrian antecedents of cephalopods are also disputed.

RELATIONSHIPS INFERRED FROM TELEOCONCH MORPHOLOGY

Knight *et al.* (1960) presented a consensus view of early gastropod phylogeny in the *Treatise on Invertebrate Paleontology* (see also Knight 1952). This scheme (Figure 10.1) relied little on teleoconch characters. Instead, it interpreted Paleozoic teleoconchs in the context of neontologically derived phylogenetic models (in particular Yonge 1947). In these models (Figure 10.1), nearly symmetrical pleurotomarioids with a sinus and selenizone gave rise to "advanced archaeogastropods" (e.g., patellogastropods, neritoids, and trochoids) as well as the forbearers of caenogastropods and heterobranchs (including opisthobranchs and pulmonates). The classification proposed by Knight *et al.* (1960) lumped Paleozoic taxa into extant taxa whenever possible: for example, trochiform taxa (e.g., platyceratids) were trochines, most limpets were patelloids, and high-spired taxa lacking selenizones (e.g., loxonematoids and subulitoids) were caenogastropods. The phylogenetic positions of extinct morphologies were made

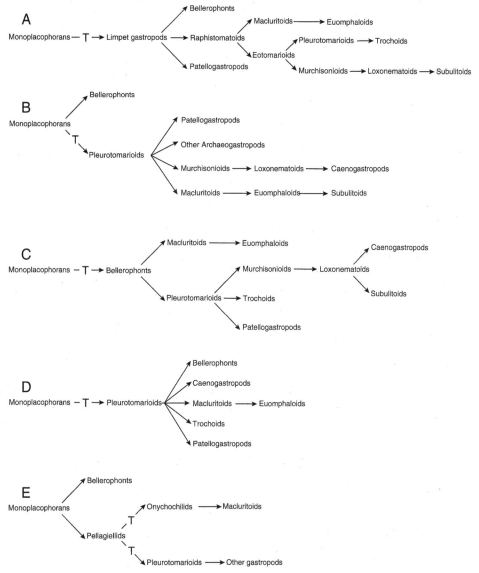

FIGURE 10.1. Summary of traditional paleontological hypotheses on phylogeny of the Gastropoda. (A) Ulrich and Scofield (1897), (B) Wenz (1938–1944), (C) Knight (1952), (D) Yochelson (1967, 1984), (E) Runnegar (1981) (modified from Figure 1, Wagner 1999b) T = advent of torsion.

consistent with Yonge's model. Thus, bilaterally symmetrical bellerophonts linked other gastropods to untorted molluscs. High-spired, sinus- and selenizone-bearing murchisonioids linked medium-spired, sinus- and selenizone-bearing pleurotomarioids and sinus-bearing but selenizone-less loxonematoids. Because Knight *et al.* (1960) were not adverse to polyphyletic taxa, their actual phylogenetic scenarios were not as simple as their classifications. However, this meant repetition of Yonge's (1947) scheme rather than challenges to it (see also Lindberg and Ponder 2001).

Proposed modifications to the *Treatise* consensus typically were based on hypothetical interpretations of internal anatomy rather than the study of teleoconchs (Figure 10.1D, E). Examples include the suggestion that bellerophontoids

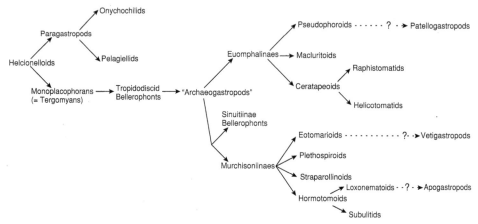

FIGURE 10.2. Phylogenetic hypothesis on phylogeny of the Paleozoic gastropods inferred from teleoconch morphology based on analysis of Wagner (1999b) (modified from Figure 37, Wagner 1999b).

were not gastropods (Wenz 1940; Runnegar 1981), the removal of many genera classified as gastropods to helcionelloids and paragastropods (e.g., Linsley and Kier 1984; Peel 1991b), and the linking of many so-called palaeotrochids[1] with euomphaloids (Morris and Cleevely 1981). Attempts to link extant taxa to Paleozoic forms were also based on inferences about internal anatomy (e.g., *Neomphalus* to euomphalids; McLean 1981) or from features such as shell mineralogy (e.g. fissurelloids with euphemitid bellerophonts; MacClintock 1968; McLean 1984; or, patelloids to platyceratoids; Ponder and Lindberg 1997).

In some groups, phylogenetic analyses using teleoconch characters began to challenge the *Treatise* consensus. Erwin's (1990b) preliminary cladistic analyses of *Treatise* families implied not only different relationships, but also different relationships among families depending on which generic exemplars were used. During the last decade, generic as well as family-level classification of the *Treatise* was criticized by several authors (see Tracey *et al.* 1993, Bandel and Frýda 1996; Bandel and Geldmacher 1996). These results corroborate the view that teleoconch characters are not informative for higher (e.g., family or higher) level analyses. Indeed,

many workers claim that teleoconch characters have little or no phylogenetic signal (e.g., Kool 1993; Bandel 2002a). However, low (i.e., species or genus)-level studies show similar patterns among teleoconch, morphologic, and even molecular characters (e.g., Haasl 2000; Vermeij and Carlson 2000; Papadopoulos *et al.* 2004). Although likelihood tests do indicate that teleoconch characters are more homoplastic than are soft-tissue characters (Wagner 2001a; Schander and Sundberg 2001), hierarchical structure among shell characters is not significantly lower than that among skeletal characters used in low-level analyses of vertebrates and echinoderms, and significantly greater than characters used in low-level trilobite studies (Wagner 2000b). Thus, teleoconch characters are likely akin to third-codon DNA data: informative at low levels but evolving too quickly and lacking sufficiently unique states to diagnose large clades.

Wagner (1999a, b) estimated relationships among nearly 400 Cambrian-Silurian species (Figures 10.1, 10.2). This work suggested that bellerophontiforms gave rise to "*Schizopea*-like ophiletiforms" (Wagner 1999a, b). Ophiletiforms in turn gave rise to two major clades. One retained ophiletiforms in basal members but gave rise to independently derived euomphaliforms, including the Macluritidae, derived Ophiletidae, and Helicotomidae. Euomphaliforms gave rise to the majority of

---

1. Palaeotrochidae is a poorly defined group containing a disparate group of taxa.

the trochiform taxa of the Ordovician and Silurian, including the Holopeidae, Pseudophoridae, Elasmonematidae, and Palaeotrochidae. Wagner (1999b) also suggested that the Platyceratoidea belonged to this clade, but based that argument on protoconch and shell mineralogy data, not teleoconch characters. However, the Euomphalidae proper evolved from one group of trochiform euomphalines in the early Silurian. The second began with murchisoniiforms (primarily species of *Hormotoma*). At least two classic "pleurotomarioid" clades evolved independently from *Hormotoma*: the Eotomarioidea and the Trochonematoidea (which include the Lophospiridae). The former clade includes the bulk of vetigastropod-like taxa, although this morphotype does not become prevalent until the Late Silurian, and the clade did yield occasional euomphaliforms (e.g., *Pleuronotus*). Lophospirids include two separate lines traditionally classified in the Trochonematoidea as well as trochiform taxa such as true *Gyronema* and some murchisoniiform taxa. Trochiforms such as the Straparollinidae also evolved from murchisoniiforms. Murchisoniiforms within this second clade also gave rise to true murchisonioids, which in turn gave rise to caenogastropod-like taxa (perhaps including the actual precursors of caenogastropods; see following discussion) at least twice in the Ordovician (Loxonematidae and Subulitidae). A second subulitiform clade, including species lumped in *Macrochilina*, arose within loxonematoids in the Silurian. Murchisoniiforms generated at least one other loxonemiform clade (e.g., taxa such as *Sinuspira*) in the Silurian. Unpublished analyses of Cambrian-Devonian species suggest that the earliest possible neritoids (taxa such as *Naticopsis*) arose from *Macrochilina*-like species and corroborate the idea that taxa such as the Palaeozygopleuridae are derived loxonematids (see also Horný 1955; Frýda 1993).

Because neontological estimates of phylogeny make only vague predictions about the early history of teleoconchs, it is difficult to contrast Wagner's (1999b) results with neontological expectations. However, the general pattern is consistent with several scenarios: Ponder and Lindberg's (1997) suggested precursors to patelloids (platyceratoids) come from the euomphaline clade (but is in conflict with protoconch data, see following discussion) whereas taxa suggested by some workers as precursors to vetigastropods, neritoids, and caenogastropods (see Wagner 1999b) belong to the murchisoniine clade, with vetigastropod precursors an outgroup to neritoid + caenogastropod precursors. The implied times of divergence even corroborate some molecular clock estimates (e.g., patelloids and other gastropods diverging ~500 Mya; Peterson 2004).

RELATIONSHIPS INFERRED FROM PROTOCONCH MORPHOLOGY

The gastropod protoconch I is the juvenile organic shell, which is usually mineralized by calcium carbonate during later ontogeny. The protoconch is produced by gastropods prior to hatching and metamorphosis (i.e., prior to the adult stage). Like many marine invertebrates, gastropods commonly have a biphasic lifecycle; they generally hatch with an embryonic shell (protoconch I, built within the egg) and subsequently have a planktonic larval stage. Gastropod larvae either feed on plankton (planktotrophic), during which time they build a larval shell (protoconch II), or do not feed but may add some additional shell material. A biphasic life cycle with a nonfeeding larva is supposedly the original state in gastropods and other marine invertebrates, and planktotrophy is considered to have been acquired subsequently (e.g., Haszprunar 1995). Different developmental modes (i.e., direct development, planktotrophy, lecithotrophy) can occur in closely related species (e.g., even within some genera of caenogastropods). However, other clades, such as the Vetigastropoda and Patellogastropoda, never produce planktotrophic larvae.

Protoconchs are important for three reasons. First, there is a rich fossil record of protoconchs extending back to the Paleozoic. Second, protoconchs reflect life histories to a high

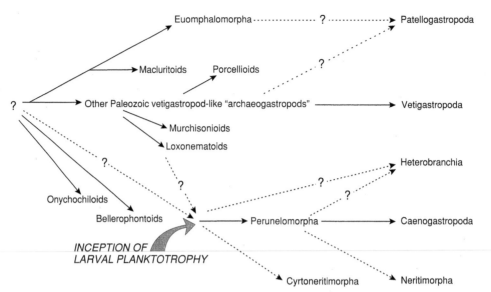

FIGURE 10.3. Phylogenetic hypothesis on phylogeny of the Paleozoic gastropods inferred from protoconch morphology based mainly on Frýda and Manda (1997), Frýda and Bandel (1997), Bandel and Frýda (1998, 1999), Bandel (1997, 2002a, b), Frýda and Blodgett (1998, 2001, 2004), Frýda (1999a, d, 2001), Nützel et al. (2002), Nützel (2002), and Frýda and Rohr (2004, 2006) (see text for discussion).

degree and thus provide paleoecological data. Finally, and most importantly for this discussion, there is empirical support for the notion that basic protoconch characters evolve less frequently than do teleoconch characters (Bandel 1997; Ponder and Lindberg 1997; Sasaki 1998; Frýda 1999a; Riedel 2000; Nützel et al. 2000; Haasl 2000) and that they compare favorably to anatomical characters in their phylogenetic utility (e.g., Riedel 2000). Unfortunately, good protoconch preservation is rare in the Paleozoic. However, during the last 25 years, protoconchs of many fossil and living gastropod groups have been reported, and these have provided crucial additional information.

Modern members of basal (e.g., patellogastropods and vetigastropods) and some more advanced (e.g., Cocculinoidea) clades feature only a coiled embryonic shell (protoconch I) with a relatively late calcification, followed by a teleoconch (e.g., Bandel 1982, 1997). The protoconchs of a majority of advanced gastropods (Neritimorpha, Caenogastropoda, and Heterobranchia) potentially consist of protoconch I and a protoconch II. The distributions of these general types on phylogenies based on soft anatomy

(e.g., Haszprunar 1988; Ponder and Lindberg 1997) or molecules (e.g., Rosenberg et al. 1994; Colgan et al. 2000, 2003) suggest that possessing a coiled protoconch I alone is plesiomorphic.

Use of fossil data reveals a more complicated evolutionary scenario than analyses based solely on living taxa due to extinction of several gastropod clades (Figure 10.3). Many Paleozoic (especially Early Paleozoic) gastropods had open-coiled protoconchs (e.g., Hynda 1986; Frýda 1999a; Dzik 1994, 1999; Nützel and Frýda 2003; Frýda and Rohr 2004, 2006; Nützel et al. 2006, and references therein), which are known only among a few highly derived pteropods today (see Bandel and Hemleben 1995). Indeed, many of these protoconchs are nearly orthoconic, although most are simply widely umbilicate. Open-coiled protoconchs likely represent the plesiomorphic condition, given that it is observed among several apparently distantly related taxa (Frýda 1998a, 1999a). Thus, it was also likely lost several times (see following discussion of trends). Notably, open coiling extends across taxa with only protoconch I (i.e., the Ordovician-Permian Euomphalomorpha and Ordovician Macluritoidea) and some with protoconch I and

II (i.e., the Ordovician-Permian Cyrtoneritimorpha and Ordovician-Carboniferous Perunelomorpha; see Frýda and Rohr 2004, and Frýda et al. 2006 for references).

Close-coiled protoconchs of the "vetigastropod" type are known since the Silurian, and they have been frequently recorded for Middle and Late Paleozoic taxa with trochiform and pleurotomariiform teleoconchs (e.g., Yoo 1988, 1994), as well as such taxa as the Devonian Murchisonioidea (Frýda and Manda 1997). The Ordovician-Permian Euomphalomorpha (Figure 10.5A–C, G, H) also bear only protoconch I, but in contrast to the latter groups it is open coiled (e.g., Yoo 1994; Bandel and Frýda 1998; Nützel 2002; Frýda et al. 2006). Possession of a close-coiled protoconch I unites several Paleozoic groups, but it is possible that this reflects a driven trend rather than common ancestry (see following discussion). More detailed analyses might reveal more specific synapomorphies within this "plesiomorphic" complex and determine whether protoconch characters unite any of these taxa.

Possession of protoconch II unites neritimorphs, caenogastropods, and heterobranchs, which corroborates recent neontological studies (see following discussion) but a homeostrophic, open-coiled, protoconch II occurs also in the Ordovician-Permian Cyrtoneritimorpha (see later discussion of the origin of Neritimorpha) and in the Ordovician-Carboniferous Peruneloidea (Perunelomorpha; see discussion of the origin of Caenogastropoda). The cyrtoneritimorphs share some teleoconch features with neritimorphs (see Bandel and Frýda 1999), but their relationships to other gastropods are unclear (Frýda and Heidelberger 2003). On the other hand, the peruneloids represent either basal caenogastropods or their sister group (Frýda 1999a; Frýda and Rohr 2004, and references therein). This strongly suggests that "advanced" gastropods evolved close-coiled protoconchs independently from "primitive" gastropods, in contradiction to the expectations of neontological studies.

Protoconch data also might corroborate the idea that caenogastropods are more closely related to neritimorphs than they are to heterobranchs. The unique highly convolute, homeostrophic protoconchs of extant neritimorphs do not appear until the Triassic (Bandel and Frýda 1999; Bandel 2000). However, possible Paleozoic precursors of neritimorphs possess a closely coiled (but not convolute) homeostrophic protoconch II that is similar to that in caenogastropods (e.g., Yoo 1994; Frýda 2001; Nützel and Mapes 2001, Bandel 2002a). Moreover, the earliest heterobranch protoconchs, featuring a different coiling axis of the larval shell than the teleoconch, are also tightly coiled and also appear in the Devonian (see discussion of the origin of Heterobranchia). Thus, it is possible that the homeostrophic protoconch II of caenogastropods, perunelomorphs, and cyrtoneritimorphs is plesiomorphic relative to the heterostrophic protoconch of heterobranchs. The function (if any) of shell heterostrophy (Figures 10.3, 10.4), which has been documented also in living and fossil vetigastropods (Hadfield and Strathmann 1990; Frýda and Blodgett, 2001; Frýda and Farrell, 2005) and the Ordovician Macluritoidea (Frýda and Rohr, 2006), is unknown. Frýda and Blodgett (2001, 2004) discussed shell heterostrophy in the Agnesiinae (Porcellioidea, Vetigastropoda) and concluded that its development in the "Archaeogastropoda" and Heterobranchia is not homologous.

It should be noted that Early and Middle Paleozoic Clisospiroidea (Mimospirina) have a protoconch II similar to that of advanced gastropods (Dzik 1983; Frýda 1989, 1993; Frýda and Rohr 1999, 2004), which is remarkable given that they are probably not even gastropods (see previous discussion, following discussion on Paragastropoda, and Parkhaev, Chapter 3, for an alternative view).

In summary, we are only just beginning to fully use protoconch and teleoconch characters in phylogenetic analyses of Paleozoic gastropods. Because of numerous autapomorphies and uncertain polarity, many of protoconch characters simply unite groups rather than offer hierarchical information about relationships among those groups. We expect this to change in the near future.

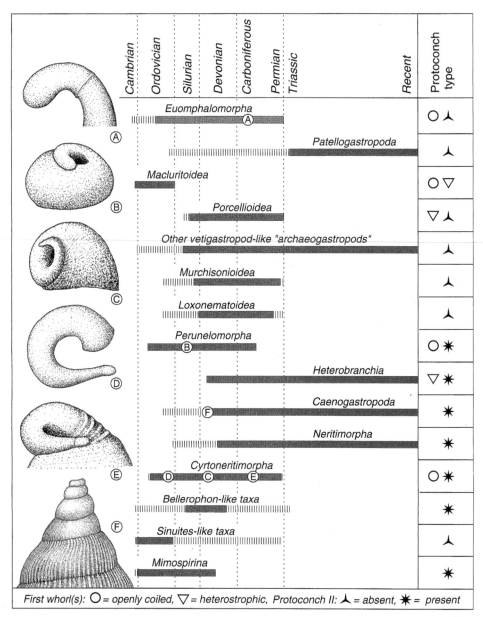

FIGURE 10.4. Diagram illustrating the stratigraphic ranges of main gastropod groups based on protoconch morphology (gray bars). Shaded bars show stratigraphic ranges inferred from teleoconch features (based on Frýda [1999a, 2005a] and Frýda and Rohr [2004, 2006]). Compare stratigraphic ranges of the Paleozoic gastropods with presumed phylogenetic relationships inferred from teleoconch and protoconch morphologies (Figures 10.2 and 10.3). Characteristic protoconchs are drawn on left side. (A) Euomphalomorph protoconch of the Early Carboniferous *Serpulospira*. (B) Late Silurian perunelomorph larval shell. (C) Early Devonian cyrtoneritimorph *Vltaviela*. (D) Late Ordovician cyrtoneritimorph larval shell. (E) Larval shell of the Carboniferous *Orthonychia*. (F) Early Devonian subulitid larval shell. (Sketches made according to Bandel and Frýda 1999, Dzik 1994, Frýda and Manda 1997, Yoo 1994, Frýda 1999a, 2001.)

FIGURE 10.5. Euomphaloidea (A–C, G–H) and Macluritoidea (D–F). (A) Carboniferous *Schizostoma crateriforme*, ×1. (B, C) Carboniferous *Euomphalus pentangulatus*, type species of *Euomphalus*, ×1.2. (D, E) Early Ordovician *Macluritella stantoni*, ×3.5. (F) Detail view of early whorls in *Macluritella stantoni* showing a distinct dextral coiling, ×20. (G) Aragonitic crossed-lamellar structure in the Carboniferous *Amphiscapha catilloides*. (H) Open-coiled first whorl of Devonian *Odontomaria* sp., boundary between the embryonic shell and teleoconch is indicated by white arrows, ×50.

MACROEVOLUTIONARY TRENDS AND PATTERNS

TELEOCONCH MORPHOLOGY

Active trends (*sensu* Fisher 1986) have eliminated several common morphotypes of the earliest snails. Prominent among these is the decreased diversity and then loss of near-planispiral euomphaliforms (Cain 1977). This trend appears to be due to extinction (Wagner 1996; Frýda and Rohr 2004), as convergent euomphaliforms evolve significantly more frequently than expected through the Devonian (Wagner and Erwin 2006). Wagner also found a driven trend (i.e., biased transition; McShea 1994) for sinus dimensions, with the large sinuses replaced by deep slits in some groups and by high asymmetry in others. The sinus trend is also consistent with trends in mantle evolution noted by Lindberg and Ponder (2001).

Rates of morphologic evolution were significantly greater among Early Ordovician gastropods than among later ones (Wagner 1995a). Notably, neontological trees suggest elevated rates of anatomical change given either Haszprunar's (1988) or Ponder and Lindberg's (1997) data sets (Wagner 1998) and even among mitochondrial genes (Colgan *et al.* 2000, 2003).

PROTOCONCH MORPHOLOGY

Nützel and Frýda (2003) documented a probable driven trend toward decreasing proportions

FIGURE 10.6. Bellerophontiform molluscs. (A) Early Devonian cyrtonellid *Kolihadiscus tureki*, ×15. (B, C) Middle Devonian *Bellerophon vasulites*, type species of *Bellerophon*, ×1.5. (D) Larval shell of Silurian *Bellerophon* sp., ×65. (E, F) Early Devonian *Ladamarekia miranda*, ×8. (G) Detail view of larval shell in Silurian *Bellerophon* sp., ×160.

of open-coiled protoconchs over the Paleozoic based on Frýda's (1998a, 1999a) observation that open-coiled protoconchs were apparently lost in multiple, unrelated lineages. However, we cannot yet rule out differential extinction as a contributing factor. Frýda (2004) noted an accompanying trend toward smaller embryonic shells. A characteristic feature of Early Paleozoic gastropods is the development of large embryonic shells, reflecting their lecithotrophic larval strategy. This changed dramatically during the Silurian and Devonian, during which small, close-coiled embryonic shells came to predominate. This coincided with the inception of larval planktotrophy and was followed by the Late Paleozoic radiation of Neritimorpha, Caenogastropoda, and Heterostropha. Thus, Frýda (2004, 2005b) suggested that fundamental changes in biogeochemical evolution of the Paleozoic oceans, linked to a pronounced increase in nutrient input to sea surface waters during the eutrophication episodes, triggered both the inception of larval planktotrophy and the diversification of groups with such larva.

ONGOING CONTROVERSIES

ARE BELLEROPHONTIFORM MOLLUSCS GASTROPODS?

Workers have debated whether the bilaterally symmetrical bellerophontiform molluscs (Figure 10.6) are gastropods for over 50 years. Opinions typically center around muscle scar patterns found on some steinkerns. Wenz (1940) documented segmented dorsal muscle scars on a Devonian cyrtonelloid bellerophontiform (Figure 10.6A), which fit expectations for an untorted mollusc. Knight (1947) documented unsegmented umbilical scars on bellerophontoids, which fit the expectations for an anatomically primitive gastropod (e.g., a vetigastropod). This led to a consensus view that bellerophontiforms include both "monoplacophorans" and gastropods (Knight *et al.* 1960; Yochelson 1967; Peel 1991a, b, 2001; Wahlman 1992; Horný 1991; Horný and Peel 1996), either because of parallel evolution of shell form (Yochelson 1967; Wahlman 1992) or because gastropods evolved amid untorted bellerophonts (Knight 1952).

Two primary challenges to this consensus exist. Runnegar (1981) inferred that bellerophonts were a monophyletic group only distantly related to gastropods. In this model, the gastropod muscle scars of some bellerophontiforms reflect a parallelism that allowed untorted bellerophonts to retract deep into the shell. This model also assumes that torsion represents an adult adaptation pertinent only to anisostrophic shells. Potential problems with this model include the possibility that torsion is a larval adaptation and that it is in conflict with the fossil record (i.e., the hypothesized order in which morphotypes evolved is at odds with their chronology in the fossil record).

Harper and Rollins (2000) argued that bellerophontiforms were a clade of gastropods. They noted that the larval muscles inducing torsion do not become the adult attachment muscles (Wanninger et al. 2000; contra Crofts 1955) and thus that the position of adult muscle did not place themselves at the coiling axis by inducing torsion. Wanninger et al. (2000) further noted that the bundled muscle attachments of patelloids create segmented muscle scars and that some phylogenetic models (e.g., Haszprunar 1988) hypothesized a limpet ancestor for gastropods. Thus, Harper and Rollins proposed that the patelloid muscle pattern is primitive for gastropods, with gastropod muscle scars evolving in parallel among bellerophonts and anisostrophic gastropods. Like the other monophyletic model, the order of morphotype evolution in this model is at odds with the order of morphotype appearance. Also, many workers (e.g., Ponder and Lindberg 1997, see following discussion) dispute the assumption that a limpet was the last common ancestor for gastropods. Moreover, the bundled muscle pattern bears only a superficial resemblance to the segmented muscle scar patterns of cyrtonelloids. Finally, and most critically, the assumed independence of attachment muscles and torsion confuses causation with correlation. Neither the nervous system nor the alimentary tract induces torsion, but torsion strongly modified both those systems and thus is correlated with the final anatomical patterns.

In other words, even though attachment muscles do not affect torsion, we still expect torsion to affect attachment muscles.

Wagner (2002) corroborated the idea that "monoplacophoran" and "gastropod" muscle scars separate two bellerophontiform clades. Using teleoconch characters, his study showed that the best trees with a single origin of the bellerophont shell are significantly less likely than are the best trees with two originations. That study also suggested that the group with gastropod muscle scars is paraphyletic relative to anisostrophic gastropods (Figure 10.2). Parsimony analyses using teleoconch characters separate species with typical monoplacophoran muscle scars from those with typical gastropod muscle scars (Wagner 2001b), although in those analyses parsimony implausibly suggested that monoplacophorans are a derived clade nested within gastropods.

The most important events influencing the higher-level taxonomic position of sinuitids were discoveries of paired muscle scars in some from the Ordovician-Devonian. These discoveries were interpreted as evidence for untorted, exogastrical orientation in these genera (e.g., Peel 1980; Runnegar 1981, 1983; Horný 1991; Wahlman 1992; Horný and Peel 1996), and sinuitids were transferred to the class Monoplacophora. However, Frýda and Gutiérrez-Marco (1996) pointed out that secondary shell deposits in some Ordovician sinuitids and Devonian-Carboniferous euphemitids (having typical gastropod muscle scars) are similar in position of the perinductura-inductura[2] boundary and in their form and ornamentation (see also Horný 1996). The same geometry of the

---

2. Moore (1941) described secondary shell deposits in *Euphemites* in detail and used the terms *inductura*, *perinductura*, and *coinductura*. The inductura is a secondary shell layer extended from the inner side of the aperture over the parietal region, columellar lip, and part or all of the outer shell surface. The perinductura is a secondary shell layer assumed to be secreted by a mantle flap reflected back over the outer apertural lip. This shell layer obscures the growth lines and is the lowest of three outer shell layers. The coinductura is a secondary shell layer extending over the inner lip within the aperture and covering only a small part of the inductura.

secondary shell deposits was later documented in Devonian plectonotids (Frýda 1999b). Frýda and Gutiérrez-Marco (1996) pointed out that these secondary shell deposits were secreted by homologous mantle flaps. Thus, Sinuitidae and Euphemitidae had probably similar soft-body organization, even though they were placed in different molluscan classes, that is, Gastropoda and Monoplacophora (see previous discussion).

Frýda (1999a) suggested that the presence or absence of a larval shell (protoconch II) diagnosed separate derivations of bellerophontiform gastropods. Bellerophontiforms such as *Bellerophon* (Figure 10.6B, C) possess a multiwhorled protoconch with a small diameter of the first whorl. This protoconch type was interpreted as a true larval shell (protoconch II; Figure 10.6D, G; Frýda 1999a). *Bellerophon* shares this feature with anisostrophic gastropods (e.g., neritimorphs, caenogastropods, and heterobranchs) and some Devonian cyrtonellids (e.g., *Cyclocyrtonella;* Horný 1993; Figure 10.6A). However, Ordovician sinuate bellerophontiform molluscs (*Sinuitopsis* and *Modestospira*) possess a relatively large, symmetrical protoconch formed only by the primary embryonic shell (protoconch I; e.g., Dzik 1981). This absence of a secondary larval shell is shared with patellogastropods, vetigastropods, and related Paleozoic taxa (including Euomphalomorpha (Figure 10.5A–C) with aragonitic crossed-lamellar shell structure, Figure 10.5G) and some Devonian limpets (e.g., *Ladamarekia,* Figure 10.6E, F).

MacClintock (1968) found that some Late Carboniferous bellerophonts possess complex crossed-lamellar inner layers with no nacre, which is very similar to the condition of the bilaterally symmetrical fissurelloids (see also McLean 1984). This is very difficult to reconcile with neontological data unless these bellerophontiforms are secondarily derived Vetigastropoda. This also raises the possibility that bellerophonts are polyphyletic within gastropods (Frýda 1999d; Wagner 2002; see also Parkhaev, Chapter 3, for an alternative view).

In summary, different models about the evolution of torsion and the nature of the earliest

gastropods make contradictory inferences about whether any or all bellerophonts were gastropods and whether the group was mono-, para-, or polyphyletic. Several lines of empirical evidence suggest polyphyly. Teleoconch data suggest that early bellerophonts were diphyletic, with clade membership coinciding with different muscle scar patterns. General protoconch characters link some Silurian and Devonian bellerophontids to more advanced gastropod groups but at least some Ordovician sinuitid bellerophonts have a protoconch similar to that seen in patellogastropods, euomphaloids, and vetigastropods, corroborating the idea of polyphyly. Alternatively, these data also corroborate an idea that the active trend toward advanced protoconchs (Nützel and Frýda 2003) was a driven trend (see previous discussion). Finally, shell mineralogy and microstructure link at least some Late Paleozoic bellerophonts to derived vetigastropods.

ORIGIN OF PATELLOGASTROPODA

There are two primary issues concerning the origins of patellogastropods: the nature of their last common ancestor with other gastropods, and which (if any) Paleozoic taxa are eogastropods (*sensu* Ponder and Lindberg 1997) and hence ancestral to patellogastropods.

Traditionally, workers thought patellogastropods to be derived from pleurotomarioids (*e.g.,* Knight 1952). However, based on soft anatomy, Golikov and Starobogatov (1975) and Lindberg (1988) suggested that they were the sister taxon to all other gastropods. Haszprunar (1988) further hypothesized that the limpet shell is a plesiomorphy retained by patellogastropods. However, Ponder and Lindberg (1997; also Lindberg and Ponder 2001) argued that this was a parallelism and that the last common ancestor of extant gastropods was coiled, given that different basal limpets share only superficial similarities because asymmetries in their anatomy indicate coiled ancestry.

The highly reduced teleoconch morphology of most limpets (including patellogastropods) leaves only shell microstructure, protoconchs, and maybe also muscle scars as sources of information about relationships to coiled taxa. The oldest limpets with patellogastropod shell structure are from Triassic strata (Hedegaard *et al.* 1997), but unfortunately, no information exists on shell structure in any Paleozoic limpets, which leaves us with no mineralogical evidence for or against the idea that some might be patellogastropods. Ponder and Lindberg (1997: 197) note that patellogastropods share a calcitic foliated structure with at least some platyceratids (see Carter and Hall 1990), and therefore suggest platyceratids to be possible coiled precursors for patelloids. However, there are some doubts whether this structure is really identical in the both groups (see Bandel and Geldmacher 1996). In addition, the latter hypothesis is also in conflict with protoconch data. Protoconchs of Silurian-Permian platyceratids (e.g., Bandel and Frýda 1999) consist of an embryonic as well as true larval shells (see subsequent discussion of the origin of Neritimorpha), in contrast to those of patellogastropods. Wagner (1999b) suggests that platyceratids were derived euomphalines, given that early platyceratoids share an orthoconic protoconch with his early euomphalines, such as *Pararaphistoma* (Dzik 1994), and calcitic shells with other early euomphalines, such as *Ophiletina* (see Rohr and Johns 1990). Given that the plausible ancestors for the Orthogastropoda reside in Wagner's definition of the murchisoniine clade, this in turn would predict that patellogastropods are the sister group of all other gastropods (Figures 10.2 and 10.3).

As noted previously, patellogastropods, like several other living and fossil groups (see also the discussion of Vetigastropoda below), lack protoconch II. Protoconch morphologies are known for only three Paleozoic taxa that include limpets. Limpets of the Middle and Late Paleozoic Cyrtoneritimorpha (Figure 10.8; see also *Origin of Neritimorpha*) and Middle Paleozoic Pragoscutulidae (Figure 10.10E; Frýda 2001, Cook *et al.*, in press, and references therein) possess a true larval shell (protoconch II). Thus, it is unlikely that either taxon is related closely to patellogastropods. Middle

Paleozoic members of the Ladamarekiidae (Figure 10.6E, F) have smooth, bowl-shaped protoconchs (Frýda, unpublished data) resembling those of monoplacophorans. Ponder and Lindberg (1997) noted a slight sinistral offset of the patellogastropod protoconchs and thus suggested that patelloids had a sinistrally coiled ancestor (but see discussion in Bandel 1982; Sasaki 1998). Recently Lindberg (2004) suggested that patellogastropods might even represent a diphyletic group.

It is not known whether crown group patellogastropods appeared in the Paleozoic. This uncertainty is likely due to the assumed coiled ancestor not being recognized or may be exacerbated by the adaptation to intertidal environments shown by living patellogastropods given the (relatively) poor fossil record of such environments.

THE VETIGASTROPODA

Because anatomical and external head–foot characters diagnose the Vetigastropoda, it is difficult to recognize this taxon reliably among fossils (see Salvini-Plawen and Haszprunar 1987; Ponder and Lindberg 1997; Geiger *et al.*, Chapter 12). However, features such as the presence of a nacreous shell layer and the absence of protoconch II also typify vetigastropods (see Bandel 1982, 1997). The latter character is assumed to be plesiomorphic because it is shared with some hot-vent taxa (Warén and Bouchet 2001, and references therein), Patellogastropoda (see Ponder and Lindberg 1997; Sasaki 1998), and fossil Euomphalomorpha (see previous discussion and Frýda *et al.* 2006 for references), as well as some bellerophontiform molluscs (see also the previous section on relationships inferred from protoconch data). However, the shape of protoconch I (i.e., left or right offset, close coiling versus open coiling, etc.) differs among these taxa (see previous discussion on patellogastropods; also Bandel 1982; Ponder and Lindberg 1997; Bandel and Frýda 1998; Nützel 2002). Whether the similarities represent symplesiomorphic or synapomorphic homologies or even parallelisms is still unclear. Basal vetigastropods appear to be diagnosed by teleoconch features such as a narrow sinus, deep slits, and bilineate selenizones.

In traditional classifications the Vetigastropoda include two prominent groups of Paleozoic Gastropoda: the Pleurotomarioidea and Trochoidea. These classifications imply that these vetigastropods appeared by the Late Cambrian and are the most common Paleozoic gastropod group (Knight *et al.* 1960; Tracey *et al.* 1993; Wagner 1999b; Frýda and Rohr 2004). However, numerous teleoconch, protoconch, and shell mineralogy studies suggest that the traditional Paleozoic definitions of both taxa are highly polyphyletic (e.g., Bandel and Frýda 1996; Bandel and Geldmacher 1996; Wagner 1999b, and references therein). Moreover, as Wagner (1999b) noted, the vetigastropod teleoconch is just as derived relative to early Ordovician gastropods as is the teleoconch of caenogastropods or neritimorphs. Teleoconchs consistent with extant pleurotomarioids and coiled scissurelloids do not appear until the Silurian, and they are not especially diverse until the Devonian (Wagner 1999a, b). Several gastropod groups with preserved shell structure (most nacreous) and typical vetigastropod-type protoconchs were reported from the Carboniferous Buckhorn Asphalt Deposit of Oklahoma, United States (Bandel *et al.* 2002). Although vetigastropod-style teleoconchs were diverse in the Late Paleozoic, it is still unclear whether crown group members of the Pleurotomarioidea, Fissurelloidea, Scissurelloidea, and Trochoidea diverged during the Paleozoic.

The oldest known vetigastropod-type protoconchs come from the Silurian (Frýda, unpublished data). Early Devonian taxa possess this style of protoconch, whether or not they have selenizones (Figure 10.7; Frýda and Rohr 2004, and references therein). Batten (1972) documented nacreous structures in Carboniferous "eotomarioids" similar to that of Devonian species known to have vetigastropod-type protoconchs (Figure 10.7; see references in Frýda and Rohr 2004). The same protoconch type was also documented in high-spired,

selenizone-bearing Early and Middle Devonian Murchisonioidea (Frýda and Manda 1997; Figures 10.7A, B, 7D–F, I), which were interpreted as a link among medium-spired, sinus- and selenizone-bearing pleurotomarioids and sinus-bearing but selenizone-less loxonematoids (e.g., Knight *et al.* 1960; see previous discussion and Figure 10.10B, C, G). Middle and Late Paleozoic taxa with vetigastropod-type protoconchs also possess the distinctive deep slit, reduced sinus, and bilineate selenizone prominent among pleurotomarioids and even coiled scissurelloids that might diagnose stem group vetigastropods.

Whether nacreous structures occurring in some living and extinct Vetigastropoda are derived or plesiomorphic is a matter of contention. X-ray diffraction studies by Hedegaard (1997) and Chateigner *et al.* (2000) suggested that gastropods and cephalopods derived nacreous structures independently. However, ongoing study of crystallographic textures and microstructures of molluscan shells suggests that nacre is homologous between living and fossil vetigastropods and cephalopods (Frýda *et al.* 2004, 2006, and unpublished data). If so, then the feature is plesiomorphic for gastropods, and vetigastropod-type nacre thus might have evolved long before vetigastropods did. Thus, nacre was probably lost multiple times, and its presence suggests a basal position of the vetigastropods, as do some recent molecular studies, in contrast to Ponder and Lindberg (1997)'s placement of the Vetigastropoda above Neritimorpha.

In summary, paleontological as well as neontological data now imply that vetigastropods do not represent the ancestral stock for all other gastropods, as was long supposed. Still, vetigastropods retain primitive features such as a sinus and nacreous shell structure, which is consistent with a fairly early divergence from other gastropods.

ORIGIN OF NERITIMORPHA

Neritimorpha (= Neritaemorphi and Neritopsina; see Bouchet and Rocroi 2005) are characterized

FIGURE 10.7. Paleozoic vetigastropod-like taxa: Murchisoniidae (A, B, D–F, I), Gyronematinae (C), Eotomariidae (G, J), and Porcelliidae (H). (A) Early Devonian *Murchisonia pragensis*, ×15. (B) Middle Devonian *Murchisonia coronata*, ×1. (C) Middle Devonian *Kitakamispira armata*, ×2. (D) Middle Devonian *Murchisonia bilineata*, ×1. (E, F) Juvenile shell of Early Devonian *Diplozone* sp., ×50. (G) Early Devonian *Zlichomphalina* sp., ×3. (H) Lateral view of Early Devonian *Alaskiella medfraensis* showing heterostrophic coiling, ×15. (I) Detail view of Early Devonian *Murchisonia holynensis* showing protoconch, ×40. (J) Early Devonian *Zlichomphalina* sp., boundary between the embryonic shell and teleoconch is indicated by white arrow, ×60.

by several anatomical apomorphies (Ponder and Lindberg 1997), distinct cleavage patterns (van den Biggelaar and Haszprunar 1996; Lindberg and Guralnick 2003), and molecular sequence data (Kano *et al.* 2002). Post-Paleozoic marine neritimorphs with planktotrophic larval development share a unique protoconch morphology (Bandel 1982) typically featuring highly convolute (protoconch II; Figure 10.8C, F). Their relationships to other gastropods were long enigmatic (see Ponder and Lindberg 1997), but recent publications of embryological (van den Biggelaar and Haszprunar 1996; Lindberg and Guralnick 2003) and even molecular data (see Colgan *et al.* 2003; Lindberg, Chapter 11) suggest that neritimorphs are more closely related to caenogastropods than to traditional archaeogastropods.

The oldest undoubted examples of the neritimorph protoconch are from Triassic strata (Bandel and Frýda 1999; Bandel 2000, and unpublished data). Frýda (1998a; 1999a) pointed out that presumed Paleozoic neritimorphs (Figure 10.8) have two different larval shells: one closely coiled (but not convolute),

FIGURE 10.8. Neritimorpha (A, C, D, F) and Cyrtoneritimorpha (B, E, G–I). (A) Middle Devonian *Paffrathopsis subcostata* showing shell color pattern, ×15. (B) Middle Devonian vltaviellid *Eifelcyrtus blodgetti*, ×20. (C, F) Late Triassic *Pseudorthonychia alata*. (C) Detail view of neritimorph larval shell, ×80, (F) ×6. (D) Middle Devonian *Paffrathia lotzi*, ×3. (E) Larval shell of Carboniferous *Orthonychia parva*, boundary between the embryonic shell and protoconch II is indicated by white arrow, ×60. (G) *Orthonychia parva*, boundary between the larval shell and teleoconch is indicated by white arrow, ×20. (H) Cyrtoneritimorph larval shell of Early Devonian *Vltaviella reticulata*, ×25. (I) Early Devonian *Vltaviella reticulata*, ×15.

and the other beginning with nearly orthoconic growth followed by an open-coiled, fishhook-like protoconch II (Figure 10.8). The first group, the Cycloneritimorpha includes "Platyceratidae," Plagiothyridae, Naticopsidae, Nerrhenidae, and "Oriostomatoidea" (Yoo 1994; Bandel and Frýda 1999; Bandel and Heidelberger 2001; Nützel and Mapes 2001), and displays a tightly coiled homeostrophic larval shell that is little different from the larval protoconch II of caenogastropods (see *Origin of Caenogastropoda* below). Frýda (1998b; 1999a; also Bandel and Frýda 1999) interpreted the second protoconch group, the Cyrtoneritimorpha (e.g., Vltaviellidae and Orthonychiidae; Figure 10.8B, E, G–I), as stem

group neritimorphs based on Middle and Late Paleozoic specimens. However, teleoconchs of the Silurian and Devonian cyrtoneritimorphs led Frýda and Heidelberger (2003; also, Frýda, unpublished data) to question whether cyrtoneritimorphs are stem group neritimorphs. Thus, the systematic position of Cyrtoneritimorpha is uncertain, and they may represent an independent clade of Paleozoic gastropods (Figure 10.3).

In summary, fossil data corroborate a close relationship between neritimorphs and caenogastropods (see discussion in *Relationships inferred from protoconch morphology*). However, when neritimorphs diverged from other

FIGURE 10.9. Paleozoic Perunelomorpha (A–G) and Caenogastropoda (H, I). (A, G) Silurian peruneloid larval shells, ×60. (B) Early Devonian chuchlinid *Havlicekiela parva*, ×30. (C) Apical view of Early Devonian *Zenospira pragensis*, ×25. (D) Early Devonian *Chuchlina minuta*, ×15. (E) *Zenospira pragensis*, ×14. (F) Apical view of Early Devonian *Havlicekiela parva*, ×25. (H) Larval shell of Early Devonian "caenosubulitids" *Balbiniconcha cerinka*, ×25. (I) Larval shell of Carboniferous *Cerithioides* sp., ×80.

gastropods obviously is a matter of contention given the divergent views on the stem members of the clade. If platyceratids are related to neritimorphs (but see previous discussion), then the oldest possible neritimorphs are Middle Ordovician (Wagner 1999b; Frýda and Rohr 2004). At latest, they had diverged from caenogastropods (and any closer relatives to caenogastropods) prior to the first appearance of caenogastropods in the Devonian (see following discussion). Regardless, there is no fossil evidence that crown group neritimorphs diverged during the Paleozoic.

ORIGIN OF CAENOGASTROPODA

Knight *et al.* (1960) linked five families that appear in the Ordovician to caenogastropods. However, these links are problematic, because general teleoconchs (high-spired to fusiform, eventually with siphonal canals) typifying caenogastropods likely evolved several times. Indeed,

teleoconch data alone suggest that putative caenogastropods such as the Subulitidae, Loxonematidae, Murchisoniidae, and Plethospiridae are polyphyletic, whereas taxa such as the Craspedostomatidae appear to be unrelated to those taxa (Wagner 1999b; Nützel *et al.* 2000). Also, protoconch data of the middle Paleozoic members of the Subulitidae (Frýda 1999c, 2001), Loxonematidae (see discussion in Frýda and Blodgett 2004), and Murchisoniidae (Frýda and Manda 1997) reveal their polyphyly. Caenogastropods do have diagnostic protoconchs with multiple whorls and well-separated, orthostrophic volutions (Figure 10.9I; see also *Relationships inferred from protoconch morphology*). Well-preserved Paleozoic caenogastropod-type protoconchs are first known from the Middle Paleozoic (Figure 10.9H; Frýda 2001), and they are well documented from the Late Paleozoic (e.g., Yoo 1988, 1994; Bandel 2002a; Bandel *et al.* 2002; Pan and Erwin 2002).

Paleozoic caenogastropods include several distinct clades. One is the highly diverse Zygopleuroidea (Knight 1930), which are probably related to some extant "ptenoglossans" (Bandel 1991; Nützel 1998), although that group is likely polyphyletic (see Ponder et al., Chapter 13). Another caenogastropod clade present in the Paleozoic is the Cerithioidea, which are commonly considered basal caenogastropods. Cerithioids are likely closely related to the slit-bearing, Murchisonia-like Goniasmatidae and Pithodeidae as well as the slitless Orthonematidae and Palaeostylidae (Nützel 1998; Nützel and Bandel 2000; Bandel et al. 2002). In contrast to slit-bearing, Murchisonia-like Goniasmatidae and Pithodeidae, the Middle Paleozoic Murchisonioidea (including Murchisonia) are not Caenogastropoda (see previous discussion and Figure 10.7). Cerithioidea were placed in "Palaeo-cacnogastropoda" together with Littorinoidea and Rissooidea (Bandel 1993, 2002a). However, Ponder and Lindberg (1997) suggest that this assemblage is polyphyletic. The Procaenogastropoda (Bandel 2002a) represents another problematic taxon of high rank that includes Paleozoic caenogastropods. Bandel (2002a) suggested that an indistinct transition between protoconch and teleoconch diagnoses the procaenogastropods, but Nützel and Pan (2005) suggested that this transition is probably a preservational artifact because well-preserved specimens show an abrupt transition, and they placed them in Caenogastropoda.

Subulitiform gastropods from the Middle and Late Paleozoic (Figure 10.9B–E, H) form a polyphyletic assemblage of caenogastropods and peruneloids (Frýda and Bandel 1997; Frýda 2001; Nützel et al. 2000; Bandel et al. 2002; Nützel and Cook 2002). However, most of these taxa share only superficial similarities with the Ordovician Subulites and are probably not directly related (Wagner 1999b; Nützel et al. 2000). On the other hand, some Middle Paleozoic subulitiform gastropods (e.g., Prokopiconcha and Balbiniconcha, Figure 10.9H) are closely related to the late Paleozoic "caenogastropod-like subulitids" (Frýda 2001). The

links (if any) of caenogastropod-like subulitids to extant groups are unclear.

Late Paleozoic caenogastropods, according to Bandel (2002a), include the nonmarine families Anthracopupidae, Dendropupidae, and "Palaeocyclophoridae" (Order Procyclophorida; Bandel 2002a). Although many analyses place the Architaenioglossa either within Caenogastropoda or as sister group to all other caenogastropods, there are no unequivocal Paleozoic architaenioglossans. Thus, the fossil record does not corroborate a basal divergence of nonmarine architaenioglossans from marine caenogastropods.

Neritimorphs and caenogastropods share planktotrophy and have homeostrophic larval shells. Paleozoic Neritimorpha possess a larval shell similar to that of caenogastropods, although post-Paleozoic neritimorphs have highly convolute larval shells with resorbed inner walls, and the teleoconchs generally have a higher whorl expansion rate. This suggests that the neritimorphs may be the sister taxon of caenogastropods. If so, neritimorphs and caenogastropods clearly diverged by the Early Devonian and likely last shared a common ancestor in the Silurian (Figure 10.4). However, fossil data also suggest another evolutionary scenario. The Ordovician-Carboniferous subulitiform Peruneloidea (Perunelomorpha) possesses homeostrophic protoconch II but with an open-coiled first whorl (Figure 10.9A, C, F, G). The teleoconch morphologies of the Middle Paleozoic peruneloids belonging to Chuchlinidae (e.g., Chuchlina, Zenospira, and Havlicekiela; Figure 10.9B–F; Frýda and Bandel 1997; Frýda 1999b) and Late Paleozoic Imoglóbidae (Nützel et al. 2000; Bandel 2002a; Nützel and Cook 2002) are similar to those in Late Paleozoic caenogastropod-like subulitids. Differences in the morphology of their protoconchs (i.e, open-coiled first whorl of peruneloids versus close-coiled first whorl of "caeno-subulitids") may reflect an active trend toward decreasing proportions of open-coiled protoconchs during the Paleozoic (see previous discussion). Thus, the Ordovician-Carboniferous Peruneloidea (Perunelomorpha) may represent basal caenogastropods or their sister group

(Frýda 1998a, 1999a, 2005a; Bouchet and Rocroi 2005). If so, then caenogastropods last shared a common ancestor with another gastropod group in the Ordovician (Frýda and Rohr 2004, and references therein).

In summary, true caenogastropods appear much later than implied by traditional works, and gross teleoconch characters linking Ordovician taxa to caenogastropods probably represent convergence. However, it is very possible that one of those groups includes the ancestors of caenogastropods and peruneloids. Caenogastropods almost certainly had diverged from neritimorphs and heterobranchs by the Devonian, given protoconch data. Moreover, the absence of appropriate material from earlier strata means that even older divergences are possible. Finally, fossils do corroborate ideas that caenogastropods are closely related to neritimorphs and maybe also heterobranchs (see Figures 10.2 and 10.3, and following discussion).

ORIGIN OF HETEROBRANCHIA

The Heterobranchia (= Heterostropha of Bandel 1994; but not of Ponder and Warén 1988) are diagnosed by several anatomical features (see Ponder and Lindberg 1997) as well as by a protoconch that coils in a different direction from the teleoconch. As noted previously, neontological studies typically suggest a close relationship between heterobranchs and caenogastropods, but differ over whether heterobranchs are most closely related to caenogastropods or to a caeno + neritimorph clade.

The majority of Paleozoic heterobranchs are high spired with slender apertures. The oldest known likely heterobranchs have been found in the Middle Paleozoic (Bandel 1994; Nützel 1998; Frýda 2000; Frýda and Blodgett 2001, 2004; Bandel and Heidelberger 2002), and the group is fairly diverse and abundant in the Late Paleozoic (e.g., Bandel 2002b, and references therein).

Paleozoic heterobranchs include the Streptacidoidea, Stuoraxidae (Architectonicoidea), as well as possible Devonian stem taxa such as *Palaeocarboninia* and *Kuskokwimia* (Figure 10.10A)

(Frýda and Blodgett 2001; Bandel 2002b; Bandel and Heidelberger 2002; Nützel *et al.* 2002), although their relationships to other heterobranchs are uncertain. Streptacidoids are relatively highly diverse in Carboniferous communities (see Knight *et al.* 1960; Bandel 2002b, and references therein). *Heteroloxonema* from the Givetian (late Middle Devonian) represents probably the oldest known streptacidoidean genus (Frýda 2000; also Bandel 1994; Nützel 1998; Figure 10.10D). Streptacidioids are also known from the Early Mesozoic (e.g., Bandel 1994) and it is likely that there are even Recent representatives (e.g., Bandel 1996, 2005).

The oldest known stuoraxids are from Permian strata (Pan and Erwin 2002; see discussion in Bandel 2002b; Nützel 2002; Frýda and Blodgett 2004). Thus, the Architectonicoidea diverged from other heterobranchs by the Late Paleozoic.

We currently know of no opisthobranchs or pulmonates from the Paleozoic. Although workers traditionally classified Paleozoic taxa such as *Girtyspira* and *Acteonina* in the Opisthobranchia (see Kollmann and Yochelson 1976), it is now known that these taxa had homeostrophic protoconchs (Nützel *et al.* 2000; Bandel 2002b) and are better included in Caenogastropoda. Morris' (1990) suggestion that opisthobranchs are derived subulitoids stemmed in part from the assumption that taxa such as *Girtyspira* were opisthobranchs. Nevertheless, both Wagner (1999b) and Nützel *et al.* (2000) indicate polyphyly among subulitiforms. Moreover, early heterobranchs such as *Kuskokwimia* (Figure 10.10A) and *Palaeocarboninia* possess subulitiform teleoconchs. Thus, it is distinctly possible that heterobranchs (and thus also opisthobranchs) arose from some subulitiform group.

Ponder and Lindberg (1997: 203) speculated that the ancestors of Heterobranchia arose from hyperstrophic dextral ancestors such as onychochilids. However, as noted previously, onychochilids likely represent a distinct molluscan clade only distantly related to gastropods (see following discussion and Parkhaev,

FIGURE 10.10. Paleozoic Heterobranchia (A, D, F), Loxonematoidea (B, C, G) , and Pragoscutulidae (E). (A) Early Devonian *Kuskokwimia moorei*, ×9. (B) Early Devonian *Pragozyga costata*, ×14. (C) Early Devonian palaeozygopleurid *Palaeozygopleura bohemica*, ×10. (D) Middle Devonian *Heteroloxonema moniliforme*, ×3. (E) Early Devonian *Pragoscutula wareni*, boundary between the larval shell and teleoconch is indicated by white arrow, ×20. (F) Heterostrophic larval shell of Carboniferous streptacidoidean, ×70. (G) Protoconch of Early Devonian loxonematoidean *Bohemozyga kettneri*, ×20.

Chapter 3, for an alternative view). Moreover, there is no evidence that onychochilids were hyperstrophic (Frýda and Rohr 2004). Macluritoids are little better as an alternative. Morris's (1991) functional analyses suggest that that macluritoids were orthostrophic, and Frýda and Rohr (2006) present evidence for sinistral heterostrophy in macluritids (Figure 10.5D–F). Finally, macluritoids likely went extinct at the end of the Ordovician (Wagner 1999b, Frýda and Rohr 2004, and reference therein), which would leave an improbable sampling gap in the fossil record or require a very different intermediate ancestor.

In summary, heterobranchs clearly diverged from other gastropods by the Devonian. However, the relationships of Paleozoic heterobranchs to extant ones are still unclear. Fossil data are consistent with a generally close relationship between heterobranchs and caenogastropods, as suggested by most neontological studies (see Lindberg and Guralnick 2003), but fossil data do not resolve specific scenarios.

PARAGASTROPODA—AN INDEPENDENT MOLLUSCAN CLASS?

Linsley and Kier (1984) used functional analyses to propose that many anisostrophic molluscs with near-planispiral, hyperstrophic, or sinistral coiling (e.g., the Clisospiroidea, Macluritoidea, and possibly Euomphaloidea) formed a class of untorted molluscs, which they called the Paragastropoda. They also included Pelagielloidea in this class, although they explicitly stated pelagielloids represented a separated derivation of the paragastropod condition. This led to two debates: Were some or all of these taxa untorted, and were these taxa actually a clade? Dzik (1983) separated the Clisospiridae and Onychochilidae from the Macluritoidea and Euomphaloidea by placing the former in the Mimospirina, but considered mimospirines to be "archaeogastropods." Wagner (1999b) linked Cambrian mimospirines to pelagielloids and found that this clade was only distantly related to gastropods. The same study further suggested that mimospirine-like macluritoids (e.g., *Palliseria*) are highly derived rather than the phylogenetic intermediates between macluritoids and onychochilids as assumed in other models (e.g., Runnegar 1981). Finally, Morris's (1991) water flow experiments found that gastropods inhabiting *Maclurites*-like shells could utilize Bernoulli effects, whereas untorted molluscs inhabiting the same shells could not. However, Morris (1991) corroborated Linsley and Kier's (1984) hypothesis that mimospirines were untorted.

Protoconch data also fail to unite the Paragastropoda. Onychochilids and clisospirids share sinistrally coiled protoconchs with large embryonic shells (Dzik 1983; Frýda 1992; Frýda and Rohr 1999), which were interpreted as larval shells formed during a non-planktotrophic larval stage (Frýda and Rohr 2004). In contrast to the Mimospirina, early members of the "Euomphaloidea" possess orthoconic protoconchs (Dzik 1994), whereas Devonian-Permian Euomphaloidea (including the type species of *Euomphalus*; Figure 10.5A–C) possess open-coiled early whorls lacking a true larval shell (Yoo 1994; Bandel and Frýda 1998; Nützel 2002). Finally, Frýda and Rohr (2006) document sinistral heterostrophy in a phylogenetically basal, Early Ordovician macluritoid (the oldest example of heterostrophy among gastropods).

In summary, teleoconch data and functional analyses both suggest paragastropods are diphyletic, with one group gastropods and the other non-gastropod (see Wagner 1999b, Figure 7; and here Figures 10.2 and 10.3; see also Parkhaev, Chapter 3, for an alternative view). Protoconch data unite onychochilids and clisospirids but fail to unite that group with macluritids or euomphalids.

DIVERSITY DYNAMICS

It seems as though a relatively steady increase in "extinction resistance" (Erwin and Signor 1990) has made gastropods one of the most diverse metazoan clades (Sepkoski 1981; Sepkoski and Hulver 1985; Bambach 1985; Signor 1985). As a consequence, gastropods increase in importance from the Paleozoic to the modern fauna (Sepkoski 1981). For future research, a particularly interesting question is which characteristics made gastropods so "extinction resistant?" Like all organisms, gastropods are products of evolution, and every species and each individual has adapted to its environment. Increases in diversity may or may not be called "adaptive radiations" but when working with fossil gastropods, it is rarely possible to infer to what a taxon was adapted.

There are many unresolved questions about phylogeny, stratigraphic range, and adequacy of the fossil record. Nevertheless, with almost 200 years of research on fossil gastropods, both Paleozoic and post-Paleozoic, we have abundant data on the diversity dynamics of this group. However, relatively few such studies are primarily concerned with Paleozoic gastropod diversity, and most were primarily based on Sepkoski's (1982, 2002) compilations (Signor 1985; Dmitriev 2005; and others, see following discussion).

Cambrian univalves, including early gastropods or their closest relatives, might have suffered an early extinction event during the Cambrian and subsequent turnover in the Ordovician. As with most other invertebrate clades, gastropod diversity increased sharply during the Early Ordovician radiation (Frýda and Rohr 2004).

The Ordovician was a remarkable period in the evolution of many metazoans, including gastropods. There are numerous hypotheses for the cause(s) of these radiations (Miller 2004). Regardless, several independent studies

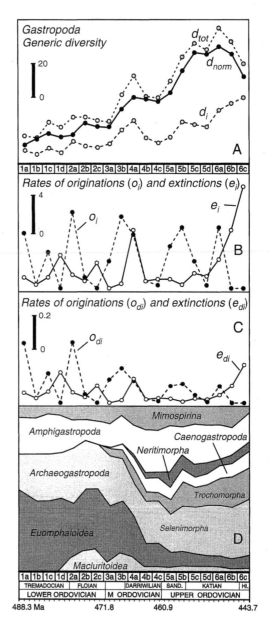

FIGURE 10.11. (A–C) Generic diversity (d_{tot}, d_{norm}, and d_i) of the Ordovician gastropods and their rates of originations (o_i and o_{di}) and extinctions (e_i and e_{di}). Total diversity (d_{tot}) is defined as the total number of genera that are recorded from the time unit; normalized diversity (d_{norm}) is defined as the number of gastropod genera ranging through the time unit plus half the number of genera confined to the unit or ranging beyond the time unit, but originating or ending within it; and genera per My. (d_i) is defined as the number of gastropod genera present within the time unit divided by its duration. Rate of originations (o_i) per My (or rate of extinction $[e_i]$ per My) is defined as the total number of genera originating (or going extinct) within the time unit divided by its duration, and mean per capita origination (o_{di}) (or extinction, e_{di}) rates is defined as the number of originations (or extinctions), divided by the total number of genera present and by the duration of the time unit in My. (D) Relative richness of the major gastropod taxa. Abbreviations: SAND. = Sandbian; HI. = Hirnantian. Modified from Frýda and Rohr (2004).

(Sepkoski 1995; Wagner 1995b, 1999b; Novack-Gottshall and Miller 2003; Frýda and Rohr 2004) showed that diversity of Ordovician gastropods was increasing from the earliest Ordovician until the Late Ordovician, when, close to the Ordovician-Silurian boundary, it dropped drastically (Figure 10.11A). Frýda and Rohr (2004) found two peaks of high diversity in Ordovician gastropods: one in the middle Late Ordovician (as shown in other studies) and another in the lowermost Darriwilian (late Middle Ordovician). Extinction in the Ordovician peaked close to the Ordovician-Silurian boundary (Sepkoski 1995; Wagner 1995b; Frýda and Rohr 2004), but the exact time of the highest extinction rate for each gastropod group seems to differ slightly, perhaps reflecting multiple crises during the uppermost Ordovician (Copper 2001). Wagner (1996) showed that taxa with large sinuses were preferentially eliminated at this time. Another strong peak of extinction was found at the beginning of the Darriwilian (Figure 10.11B; Frýda and Rohr 2004), which preferentially eliminated low-spired taxa (Wagner 1996).

Patterns of origination might be more complex than those of extinction. Although Wagner (1995b) and Novack-Gottshall and Miller (2003) suggested generally declining origination rates following a logistic pattern, Frýda and Rohr (2004) suggested that origination rates show four distinct pulses (Figure 10.11B, C). The first coincides with the base of the Floian ("Arenig," late Early Ordovician), during which both the vetigastropod-like taxa and Euomphaloidea diversified. Barnes *et al.* (1995) noted that a first-order regression/transgression couplet close to the Tremadocian-Arenig (Early Ordovician) boundary could be responsible for the extinction event, which was followed by a period of adaptive radiation during rapid transgression. Shallow seas attained a higher level of oxygenation during the transgression, which allowed the rapid radiation of organisms with a (relatively) high metabolism, such as gastropods. Novack-Gottshall and Miller (2003) showed that Ordovician gastropods were most diverse in shallower, carbonate-rich settings in more equatorial paleocontinents (as they are today), suggesting that the initial radiation of gastropods reflected, at least in part, the expansion of their preferred habitats. The interval close to the lower boundary of the Darriwilian exhibits high turnover rates, coinciding with the most distinct faunal change within Ordovician gastropod faunas (Frýda and Rohr 2004). The Euomphaloidea and Macluritoidea reach their highest diversity close to the lower boundary of the Darriwilian, and subsequently their diversities markedly decreased, with the Macluritoidea probably becoming extinct during the mass extinction event close to the Ordovician-Silurian boundary (Figure 10.11D). A third pulse in the Sandbian (early "Caradoc," Late Ordovician) saw a marked increase in the radiation of most gastropod groups (Frýda and Rohr 2004). Finally, a late spike in origination rates affecting Clisospiroidea and many groups of vetigastropod-like taxa occurs in the Katian (late "Ashgill," Late Ordovician).

The Silurian was a period of increasing diversity of most gastropod clades. This might have been connected with an increase in their ecological adaptation to specific environments. The proportion of species with high shells (mainly Loxonematoidea, Murchisonioidea, and Subulitoidea) continued to increase. In addition, slit-bearing taxa became diverse for the first time (Wagner 1999b). Platyceratid gastropods also diversify during this time; given their apparent specializations for parasitizing echinoderms, this might qualify as a true adaptive radiation (see Erwin 1992; Schluter 2000); unfortunately, evaluating this suggestion is difficult because of the difficulty in delimiting species and because the calcitic shells of platyceratids greatly increased their preservation rates (and thus their expected sampled richness) relative to most other gastropod taxa.

The Devonian was a time of distinct changes in marine gastropod communities (McGhee 2005; Frýda 2005a). Some Ordovician-Silurian groups became extinct (Mimospirina), new groups (e.g., Heterobranchia) appeared (Frýda and Blodgett 2001, 2004), and many groups (Caenogastropoda and Neritimorpha) underwent rapid radiation and specialization (Frýda 2001, Bandel and Frýda 1999; Ponder *et al.*, Chapter 13). The Devonian was also the time when the protoconch morphology of several gastropod groups underwent considerable change (see discussion on macroevolutionary trends relating to protoconch morphology above). These changes as well as Middle and Late Paleozoic radiations of Neritimorpha, Caenogastropoda, and Heterostropha are likely linked with fundamental changes in the biogeochemical evolution of Paleozoic oceans (Frýda 2005b). Due to a very poor Frasnian-Fammenian (Late Devonian) record, the effects of the end-Devonian extinction on gastropod diversity patterns are currently are not well known; however, Late Paleozoic and Devonian gastropod faunas share many taxa, implying that the extinction was not drastic for gastropods.

Erwin (1990a) reported a steep increase in gastropod generic diversity from the Fammenian to the Visean (Mississippian, Early Carboniferous) with high origination rates in the Tournaisian and Visean (Mississippian, Early Carboniferous). Diversity was apparently relatively stable from the Visean to the Latest Permian without major extinction events or radiations (Erwin 1990a). Well-preserved Late Paleozoic gastropod faunas are particularly well known from soft-bottom communities of the United States and Russia. Basal clades present in these faunas are the bellerophontids, euomphalids, slit-bearing vetigastropods, and naticopsids (see Knight et al. 1960 and Pchelintsev and Korobkov 1960 for a list of families and references) are frequently abundant. Moreover, the Late Paleozoic seems to have experienced a continued expansion of high-spired caenogastropods, such as Pseudozygopleuridae, Goniasmatidae, Orthonematidae, and Meekospiridae (see Ponder et al., Chapter 13). The first certain pseudozygopleurids occur in the Tournaisian, with the last occurring in the latest Permian (Yoo 1994; Kues and Batten 2001; Pan and Erwin 2002; Nützel 2005). They are highly diverse with several genera and more than 100 nominate species so that Knight (1930) used the term "explosive evolution" for the Late Paleozoic pseudozygopleurid expansion. Nützel (1998) speculated that they were parasitic, similar to some Recent small high-spired ptenoglossans (such as eulimids) of high diversity and low disparity (see also Ponder et al., Chapter 13). If so, then the pseudozygopleurid radiation might qualify as truly adaptive. The "subulitoid" family Soleniscidae likely originates in the Devonian or even earlier and is highly diverse and globally distributed in the late Paleozoic. Given that these taxa were probably predators, this too might qualify as an adaptive radiation. Notably, the earliest appearance of likely predators (true subulitids in the early Ordovician) did not result in a similar radiation. This is consistent with ideas that prominent adaptations need not lead to radiations (Fürsich and Jablonski 1984) or that predatory lifestyles were more viable in the late Paleozoic than in the Early Paleozoic (Vermeij 1987; Bambach 1993). Caenogastropods display another interesting pattern associated with major radiations such as that of the angiosperms (see Wing et al. 1993): that is, although they comprise most of the richness in some large Pennsylvanian gastropod faunas, they make up only about one-third of the individuals (see Ponder et al., Chapter 13), indicating a decoupling of diversification and ecological domination. Late Paleozoic basal Heterobranchia are mainly represented by the small high-spired Streptacididae, including the diverse, cosmopolitan genus Donaldina.

A recently discovered Latest Permian (Changhsingian) gastropod fauna from South China shows that most of the previously mentioned groups of Late Paleozoic gastropods were still present at the Permian-Triassic mass extinction (Pan and Erwin 2002; Nützel 2005), suggesting that gastropod extinction was sudden. Erwin (1990a) reported a sharp Late Permian decrease of gastropod diversity and high extinction rates, and his factor analysis suggests a clear turnover of Middle/Late Paleozoic and Triassic gastropods. Nützel (2005) compared the richest Early Triassic gastropod fauna (Sinbad Limestone, Utah; Batten and Stokes 1986; Nützel and Schulbert 2005) with typical Late Paleozoic gastropod faunas and found major differences in composition, stating that if the Sinbad fauna was found in the Paleozoic, it would stand out as highly unusual. Taxa that probably became extinct include the euomphaloids, pseudozygopleurids, orthonematids, and platyceratids. Taxa that trickled through include bellerophontoids and soleniscids, although these two groups might have had very different ultimate fates: bellerophontoids were a "dead clade walking" (sensu Jablonski 2002; see also Nützel 2005 for review), whereas soleniscids might include the precursors of some Mesozoic caenogastropod clades. Thus, gastropods were clearly affected by the end-Permian mass extinction event, and their subsequent rebound and recovery no doubt played a major role in shaping modern gastropod diversity.

The study of radiations and extinctions of gastropods is a rich field for future research. Available data and phylogenetic analyses still need much improvement. Moreover, description of new taxa and faunas from all parts of the world are needed. For example, although only 25% of nominate species and 15% of nominate families were described in the last 25 years, nearly one-third of Paleozoic gastropod genera have been named in the last 25 years. Nevertheless, the analysis of the data available now has resulted in interesting patterns that begin to provide an understanding of the evolutionary history of gastropods.

GAPS IN KNOWLEDGE

The following challenges can considerably improve our knowledge of the Paleozoic gastropods:

To find well-preserved pre-Devonian gastropod material with protoconchs and shell structures.

To place the Late Cambrian/Ordovician diversification of gastropods into the context of the initial Phanerozoic diversification of molluscs.

To link early Paleozoic faunas with Middle/Late Paleozoic ones.

To link of these Paleozoic faunas with Mesozoic to modern lineages.

Better understanding of phylogenetic relationships and biodiversity of the Paleozoic gastropods will require species-level teleoconch and (whenever possible) protoconch and shell structure data as well as exact stratigraphic and paleogeographic information. In addition, better clues from neontological data about the expected ancestral forms of many "basal" gastropod groups (e.g., many hot vent taxa) as well as improved resolution of thorny phylogenetic issues (e.g., the relationships of neritimorphs and heterobranchs to other extant gastropods) offer the potential to corroborate these results. Disagreements between neontological and paleontological analyses will be more difficult to evaluate, as these might reflect error in neontological inferences (which misled

evaluation of early paleontological phylogenies), paleontological inferences, or the assumptions about how the two should be compared (e.g., that fossil taxon A belongs to extant taxon B). Nevertheless, such disagreements indicate that some generalizations are incorrect, and thus facilitate the progress of research.

ACKNOWLEDGMENTS

We dedicate this work to the memories of Ellis L. Yochelson and Robert M. Linsley. Not only did both contribute greatly to our knowledge of Paleozoic snails, but they made these extinct animals interesting and left much food for thought for future generations of evolutionary biologists. This study was supported by grant KJB307020602 from the Grant Agency of the Czech Academy of Sciences to J. F. A. N. acknowledges grants from the Deutsche orschungsgemeinschaft (NU96/3-1, 6-1, and 10-1).

REFERENCES

Bambach, R. K. 1985. Classes and adaptive variety: the ecology of diversification in marine faunas through the Phanerozoic. In *Phanerozoic Diversity Patterns*. Edited by J. W. Valentine. Princeton, NJ, and San Francisco: Princeton University Press, pp. 191–253.

Bambach, R. K. 1993. Seafood through time: changes in biomass, energetics, and productivity in the marine ecosystem. *Paleobiology* 19: 372–397.

Bandel, K. 1982. Morphologie und Bildung der frühontogenetischen Gehäuse bei conchiferen Mollusken. *Fazies* 7: 1–198.

———. 1991. Über triassische "Loxonematoidea" und ihre Beziehungen zu rezenten und palaeozoischen Schnecken. *Paläontologische Zeitschrift*, 65, 3–4: 239–268.

———. 1993. Trochomorpha (Archaeogastropoda) aus den Cassian Schichten (Dolomiten, Mittlere Trias). *Annalen des Naturhistorischen Museums in Wien* 95: 1–99.

———. 1994. Triassic Euthyneura (Gastropoda) from St. Cassian Formation (Italian Alps) with a discussion on the evolution of the Heterostropha. *Freiberger Forschungsheft* C 452: 79–100.

———. 1996. Some heterostrophic gastropods from Triassic St. Cassian Formation with a discussion of the classification of the Allogastropoda. *Paläontologische Zeitschrift* 70: 325–365.

————. 1997. Higher classification and pattern of evolution of the Gastropoda. *Courier Forschungsinstitut Senckenberg* 201: 57–81.

————. 2000. The new family Cortinellidae (Gastropoda, Mollusca) connected to a review of the evolutionary history of the subclass Neritomorpha. *Neues Jahrbuch für Geologie und Paläontologie, Abhandlungen* 217: 111–129.

————. 2002a. Reevaluation and classification of Carboniferous and Permian Gastropoda belonging to the Caenogastropoda and their relation. *Mitteilungen Geologie-Paläontologie Institut der Universität Hamburg* 86: 81–188.

————. 2002b. About the Heterostropha (Gastropoda) from the Carboniferous and Permian. *Mitteilungen des Geologiisch-Paläontologischen Instituts der Universität Hamburg* 86: 45–80.

————. 2005. Living fossils among tiny Allogastropoda with high and slender shell from the reef environment of the Gulf of Aquba with remarks on fossil and recent relatives. *Mitteilungen des Geologisch-Paläontologischen Instituts der Universität Hamburg* 89: 1–24.

Bandel, K., and Frýda, J. 1996. *Balbinipleura*, a new slit bearing archaeogastropod (Vetigastropoda) from the Early Devonian of Bohemia and the Early Carboniferous of Belgium. *Neues Jahrbuch für Geologie und Paläontologie, Monatshefte* 6: 325–344.

————, J. 1998. Position of Euomphalidae in the system of the Gastropoda. *Senckenbergiana Lethaea* 78: 103–131.

————. 1999. Notes on the evolution and higher classification of the subclass Neritimorpha (Gastropoda) with the description of some new taxa. *Geologica et Palaeontologica* 33: 219–235.

Bandel, K., and Geldmacher, W. 1996. The structure of the shell of *Patella crenata* connected with suggestions to the classification and evolution of the Archaeogastropoda. *Freiberger Forschungsheft* C 464: 1–71.

Bandel, K., and Heidelberger, D. 2001. The new family Nerrhenidae (Neritimorpha, Gastropoda) from the Givetian of Germany. *Neues Jahrbuch für Geologie und Paläontologie, Abhandelungen* 2001: 705–718.

————. 2002. A Devonian member of the subclass Heterostropha (Gastropoda) with valvatoid shell shape. *Neues Jahrbuch für Geologie und Paläontologie, Abhandelungen* 2002 (9): 533–550.

Bandel, K., and Hemleben, C. 1987. Jurassic heteropods and their modern counterparts (planctonic Gastropoda, Mollusca). *Neues Jahrbuch fuer Geologie und Palaeontologie* 174: 1–22.

————. 1995. Observations on the ontogeny of thecosomatous pteropods (holoplanktic Gastropoda) in the southern Red Sea and from Bermuda. *Marine Biology* 124: 225–243.

Bandel, K., Nützel, A., and Yancey, T. E. 2002. Larval shells and shell microstructures of especially well-preserved Late Carboniferous gastropods from the Buckhorn Asphalt deposit (Oklahoma, USA). *Senckenbergiana Lethaea* 82: 639–689.

Barnes, C. R., Fortey, R. A., and Williams, S. H. 1995. The pattern of global bio-events during the Ordovician period. In *Global Events and Event Stratigraphy*. Edited by O. H. Walliser. New York: Springer, pp. 139–172.

Batten, R. L. 1972. The ultrastructure of five common Pennsylvanian pleurotomarian gastropod species of eastern United States. *American Museum Novitates* 2501: 1–34.

Batten, R. L., and Stokes, W. L. 1986. Early Triassic gastropods from the Sinbad Member of the Moenkopi Formation, San Rafael Swell, Utah. *American Museum Novitates* 2864: 1–33.

Bouchet, P. and Rocroi, J. P., eds. 2005. Classification and nomenclator of gastropod families. *Malacologia* 47: 1–368.

Cain, A. J. 1977. Variation in the spire index of some coiled gastropod shells, and its evolutionary significance. *Proceedings of the Royal Society of London. Series B, Biological Sciences* 277: 377–428.

Carter, J. G., and Hall, R. M. 1990. Part 3. Polyplacophora, Scaphopoda, Archaeogastropoda and Paragastropoda (Mollusca). In *Skeletal Biomineralisation: Patterns, Processes and Evolutionary Trends*. Vol. II, *Atlas and index*. Edited by J. G. Carter. New York: Van Nostrand Reinhold, pp. 101.

Chateigner, D., Hedegaard, C., and Wenk, H.-R. 2000. Mollusc shell microstructures and crystallographic textures. *Journal of Structural Geology* 22: 1723–1735.

Colgan, D. J., Ponder, W. F., Beacham, E., and Macarana, J. M. 2003. Gastropod phylogeny based on six segments from four genes representing coding or non-coding and mitochondrial or nuclear DNA. *Molluscan Research* 23: 123–148.

Colgan, D. J., Ponder, W. F., and Eggler, P. E. 2000. Gastropod evolutionary rates and phylogenetic relationships assessed using partial 28S rDNA and histone H3 sequence. *Zoologica Scripta* 29: 29–63.

Cook, A., Nützel, A., and Frýda, J. In press. Two Mississippian caenogastropod limpets from Australia and their meaning for the ancestry of the Caenogastropoda. *Journal of Paleontology*.

Copper, P. 2001. Reefs during the multiple crises towards the Ordovician-Silurian boundary: Anticosti Island, eastern Canada, and worldwide. *Canadian Journal of Earth Sciences* 38: 153–171.

Crofts, D. R. 1955. Muscle morphogenesis in primitive gastropods and its relation to torsion. *Proceedings of the Zoological Society of London* 125: 711–750.

Dmitriev, V. Y. 2005. Taxonomic diversification of marine gastropods. *Paleontological Journal* 39: 236–247.

Dzik, J. 1981. Larval development, musculature, and relationships of *Sinuitopsis* and related Baltic bellerophonts. *Norsk Geologisk Tidsskrift* 61: 111–121.

———. 1983. Larval development and relationships of *Mimospira*; a presumably hyperstrophic Ordovician gastropod. *Geologiska Foereningen i Stockholm Foerhandlingar* 104: 231–239.

———. 1994. Evolution of "small shelly fossils" assemblages. *Acta Palaeontologica Polonica* 39: 247–313.

———. 1999. Evolutionary origin of asymmetry in early metazoan animals. In *Advances in BioChirality*. Edited by G. Pályi, C. Zucchi, and L. Caglioti. Amsterdam: Elsevier, pp. 153–190.

Erwin, D. H. 1990a. Carboniferous-Triassic gastropod diversity patterns and the Permo-Triassic mass extinction. *Paleobiology* 16: 187–203.

———. 1990b. A phylogenetic analysis of major Paleozoic gastropod clades. *Geological Society of America—Abstracts with Program* 22: A265.

———. 1992. A preliminary classification of evolutionary radiations. *Historical Biology* 6: 25–40.

Erwin, D. H., and Signor, P. W. 1990. Extinction in an extinction-resistant clade: the evolutionary history of the Gastropoda. In *The Unity of Evolutionary Biology*, Vol. 1. Edited by E. C. Dudley. Proceedings of the Fourth International Congress Systematics and Evolutionary Biology. Portland, OR: Dioscorides Press, pp 152–160.

Fisher, D. C. 1986. Progress in organismal design. In *Patterns and Processes in the History of Life*. Edited by D. M. Raup and D. Jablonski. Berlin: Springer-Verlag, pp. 99–117.

Frýda, J. 1989. A new species of *Mimospira* (Clisospiridae, Gastropoda) from the Late Ordovician of Bohemia. *Bulletin of the Geological Survey* 64: 237–241.

———. 1992. Mode of life of a new onychochilid mollusc from the Lower Devonian of Bohemia. *Journal of Paleontology* 66: 200–205.

———. 1993. Oldest representative of the family Palaeozygopleuridae (Gastropoda) with notes on its higher taxonomy. *Journal of Paleontology* 67: 822–828.

———. 1998a. Did the ancestors of higher gastropods (Neritimorpha, Caenogastropoda, and Heterostropha) have an uncoiled shell? In *Abstracts, World Congress of Malacology*. Edited by R. Bieler and P. M. Mikkelsen. Chicago: UNITAS Malacologica, p. 107.

———. 1998b. Higher classification of the Paleozoic gastropods inferred from the early shell ontogeny, 108. In *Abstracts, World Congress of Malacology*. Edited by R. Bieler and P. M. Mikkelsen. Chicago: UNITAS Malacologica, p. 108.

———. 1999a. Higher classification of Paleozoic gastropods inferred from their early shell ontogeny. *Journal of the Czech Geological Society* 44: 137–152.

———. 1999b. Secondary shell deposits in a new plectonotid gastropod genus (Bellerophontoidea, Mollusca) from the Early Devonian of Bohemia. *Journal of the Czech Geological Society* 44: 309–315.

———. 1999c. Shape convergence in gastropod shells: an example from the Early Devonian *Plectonotus* (*Boucotonotus*)–*Palaeozygopleura* community of the Prague Basin (Bohemia). *Mitteilungen aus dem Geologisch-Paläontologischen Institut der Universität Hamburg* 83: 179–190.

———. 1999d. Suggestions for polyphyletism of Paleozoic bellerophontiform molluscs inferred from their protoconch morphology. In *Abstracts, 65th Annual Meeting, American Malacological Society*. Pittsburgh: American Malacological Society. p. 30.

———. 2000. Some new Givetian (Late Middle Devonian) gastropods from the Paffrath area (Bergisches Land, Germany). *Memoirs of the Queensland Museum* 45: 359–374.

———. 2001. Discovery of a larval shell in Middle Paleozoic subulitoidean gastropods with description of two new species from the Early Devonian of Bohemia. *Bulletin of the Czech Geological Survey* 76: 29–37.

———. 2004. Phylogeny of Paleozoic gastropods and origin of larval planktotrophy. In *Abstracts, World Congress of Malacology*. Edited by F. E. Wells. Perth: Western Australian Museum, pp. 42–43.

———. 2005a. Gastropods. In *Encyclopedia of Geology, 3*. Edited by R. C. Selley, L. R. M. Cocks, and I. R. Plimer. Amsterdam and Boston: Elsevier Academic, pp. 378–388.

———. 2005b. Inception of larval planktotrophy in the class Gastropoda as a consequence of eutrophication of middle Paleozoic sea-surface waters. In *Contributions to International Conference "Devonian terrestrial and marine environments: from continent to shelf."* Edited by E. A. Yolkin, N. G. Izokh, O. T. Obut, and T. P. Kipriyanova. Geo Novosibirsk, p. 63.

Frýda, J., and Bandel, K. 1997. New Early Devonian gastropods from the *Plectonotus* (*Boucotonotus*)–*Palaeozygopleura* community in the Prague Basin (Bohemia). *Mitteilungen Geologie-Paläontologie Institut der Universität Hamburg* 80: 1–57.

Frýda, J., and Blodgett, R. B. 1998. Two new cirroidean genera (Vetigastropoda, Archaeogastropoda) from the Emsian (late Early Devonian) of Alaska with notes on the early phylogeny of Cirroidea. *Journal of Paleontology* 72: 265–273.

———. 2001. The oldest known heterobranch gastropod, *Kuskokwimia* gen. nov., from the Early Devonian of west-cetnral Alaska, with notes on the early phylogeny of higher gastropods. *Bulletin of the Czech Geological Survey* 76: 39–53.

———. 2004. New Emsian (Late Early Devonian) gastropods from Limestone Mountain, Medfra B-4 Quadrangle, west-central Alaska (Farewell Terrane), and their paleobiogeographic affinities and evolutionary significance. *Journal of Paleontology* 78: 111–132.

Frýda, J., and Farrell, J. R. 2005. Systematic position of two Early Devonian gastropods with sinistrally heterostrophic shells from the Garra Limestone, Larras Lee, New South Wales. *Alcheringa* 29: 229–240.

Frýda, J., and Gutiérrez-Marco, J. C. 1996. An unusual new sinuitid mollusc (Bellerophontoidea, Gastropoda) from the Ordovician of Spain. *Journal of Paleontology* 70: 598–605.

Frýda, J., and Heidelberger, D. 2003. Systematic position of Cyrtoneritimorpha within the class Gastropoda with description of two new genera from Siluro-Devonian strata of Central Europe. *Bulletin of the Czech Geological Survey* 78: 35–39.

Frýda, J., Heidelberger, D., and Blodgett, R. B. 2006. Odontomariinae, a new Middle Paleozoic subfamily of slit-bearing euomphaloidean gastropods (Euomphalomorpha, Gastropoda). *Neues Jahrbuch für Geologie und Paläontologie, Monatshefte* 4: 225–248.

Frýda, J., and Manda, S. 1997. A gastropod faunule from the *Monograptus uniformis* graptolite Biozone (Early Lochkovian, Early Devonian) in Bohemia. *Mitteilungen Geologie-Paläontologie Institut der Universitat Hamburg* 80: 59–121.

Frýda, J., Rieder, M., Klementová, M., Weitschat, W., and Bandel, K. 2004. Discovery of gastropod-type nacre in fossil cephalopods: a tale of two crystallographic textures. In *Abstracts, World Congress of Malacology*. Edited by F. E. Wells. Perth: Western Australian Museum, p. 43.

Frýda, J., Rieder, M., Weitschat, W., Týcová, P., and Haloda, J. 2006. Evolution of nacre: a new evidence for close phylogenetic relationship of classes Cephalopoda and Gastropoda. In *Ancient life and modern approaches, Abstract of The second International Palaeontological Congress*, Edited by Q. Yan, Y. Wang, and E. A. Weldon. Beijing, pp. 51–52.

Frýda, J., and Rohr, D. M. 1999. Taxonomy and paleobiogeography of the Ordovician Clisospiridae and Onychochilidae. *Acta Universitatis Carolinae-Geologica* 43: 405–408.

———, D. M. 2004. Gastropods. In *The Great Ordovician Biodiversification Event*. Edited by B. D. Webby, F. Paris, M. L. Droser, and I. G. Percival. New York: Columbia University Press, pp. 184–195.

———, D. M. 2006. Shell heterostrophy in Early Ordovician *Macluritella* Kirk, 1927, and its implications for phylogeny and classification of Macluritoidea (Gastropoda). *Journal of Paleontology* 80: 264–271

Fürsich, F. T., and Jablonski, D. 1984. Late Triassic naticid drillholes: carnivorous gastropods gain a major adaptation but fail to radiate. *Science* 224: 78–80.

Geyer, G. 1994. Middle Cambrian mollusks from Idaho and early conchiferan evolution. *New York State Museum Bulletin* 481: 69–86.

Golikov, A. N., and Starobogatov, Y. I. 1975. Systematics of prosobranch gastropods. *Malacologia* 15: 185–232.

Goodhart, C. B. 1987. Garstang's hypothesis and gastropod torsion. *Journal of Molluscan Studies* 53: 33–36.

Haasl, D. M. 2000. Phylogenetic relationships among nassariid gastropods. *Journal of Paleontology* 74: 839–852.

Hadfield, M. G., and Strathmann, M. F. 1990. Heterostrophic shells and pelagic development in trochoideans: implications for classification, phylogeny and palaeoecology. *Journal of Molluscan Studies* 56: 239–256.

Harper, J. A., and Rollins, H. B. 2000. The bellerophont controversy revisited. *American Malacological Bulletin* 15: 147–156.

Haszprunar, G. 1988. On the origin and evolution of major gastropod groups, with special reference to the Streptoneura. *Journal of Molluscan Studies* 54: 367–441.

———. 1995. On the evolution of larval development in the Gastropoda, with special reference to larval planktotrophy. *Notiziario C.I.S.M.A.* 16 ("1994"): 5–13.

Hedegaard, C. 1997. Shell structures of the Recent Vetigastropoda. *Journal of Molluscan Studies* 63: 369–377.

Hedegaard, C., Lindberg, D. R., and Bandel, K. 1997. Shell microstructure of a Triassic patellogastropod limpet. *Lethaia* 30: 331–335.

Horný, R. J. 1955. Palaeozygopleuridae nov. fam. (Gastropoda) ze stredoceskeho devonu. *Sbornik Ustredního Ústavu Geologického, oddíl Paleontologicky* 21: 17–74.

———. 1991. Shell morphology and muscle scars of *Sinuitopsis neglecta* Perner (Mollusca, Monoplacophora). *Casopis Národního Muzea, Rada prírodovedná* 157: 81–105.

———. 1993. Shell morphology and mode of life of the Lower Devonian cyclomyan *Neocyrtolites* (Mollusca, Tergomya). *Casopis Národního Muzea, Rada prírodovedná* 162: 57–66.

———. 1996. Secondary shell deposits and presumed mode of life in *Sinuites. Acta musei Nationalis Pragae, Serie B* 51: 89–103.

Horný, R. J., and Peel, J. S. 1996. *Carcassonnella*, a new Lower Ordovician bellerophontiform mollusc with dorsally located retractor muscle attachments (Class Tergomya). *Bulletin of the Czech Geological Survey* 71: 305–331.

Hynda, V. A. 1986. *Melkaya bentosnaya fauna ordovika yugo-zapada Vostochno-Evropeyskoy platformy.* Kiev: Naukova Dumka.

Jablonski, D. 2002. Survival without recovery after mass extinctions. *Proceedings of the National Academy of Sciences U S A* 99: 8139–8144.

Kano, Y., Chiba, S., and Kase, T. 2002. Major adaptive radiation in neritopsine gastropods estimated from 28S rRNA sequences and fossil records. *Proceedings of the Royal Society of London. Series, Biological Sciences* 269 (1508): 2457–2465.

Knight, J. B. 1930. The gastropods of the St. Louis, Missouri, Pennsylvanian outlier: the Pseudozygopleurinae. *Journal of Paleontology Memoir* 4(4): 1–88.

———. 1947. Bellerophont muscle scars. *Journal of Paleontology* 21: 264–267.

———. 1952. Primitive fossil gastropods and their bearing on gastropod classification. *Smithsonian Miscellaneous Collections* 117: 1–56.

Knight, J. B., Cox, L. R., Batten, R. L., and Yochelson, E. L. 1960. Systematic descriptions. In *Treatise on Invertebrate Paleontology. Part I, Mollusca 1.* Edited by R. C. Moore. Lawrence, KS: University of Kansas Press, pp. I169–I1324.

Kollmann, H. A., and Yochelson, E. L. 1976. Survey of Paleozoic gastropods possibly belonging to the subclass Opisthobranchia. *Annalen des Naturhistorischen Museums Wien* 80: 207–220.

Kool, S. P. 1993. Phylogenetic analysis of the Rapaninae (Neogastropoda: Muricidae). *Malacologia* 35: 155–259.

Kues, B. S., and Batten, R. L. 2001. Middle Pennsylvanian gastropods from the Flechado Formation, north-central New Mexico. *Journal of Paleontology* 75 Suppl. 1: 1–95.

Lindberg, D. R. 1988. The Patellogastropoda. In *Prosobranch Phylogeny.* Edited by W. F. Ponder. *Malacological Review* Suppl. 4: 35–64.

———. 2004. Are the living patellogastropods sister taxa? In *Abstracts, World Congress of Malacology.* Edited by F. E. Wells. Perth: Western Australian Museum, p. 86.

Lindberg, D. R., and Guralnick, R. P. 2003. Phyletic patterns of early development in gastropod molluscs. *Evolution and Development* 5: 494–507.

Lindberg, D. R., and Ponder, W. F. 2001. The influence of classification of the evolutionary interpretation of structure—a reevaluation of the evolution of the pallial cavity of gastropod molluscs. *Organisms, Diversity and Evolution* 1: 273–299.

Linsley, R. M., and Kier, W. M. 1984. The Paragastropoda: a proposal for a new class of Paleozoic Mollusca. *Malacologia* 25: 241–254.

MacClintock, C. 1968. Shell structure of patelloid and bellerophontoid gastropods (Mollusca). *Peabody Museum of Natural History Bulletin* 22: 1–140.

McGhee, G. R. 2005. Devonian. In *Encyclopedia of Geology, 2.* Edited by R. C. Selley, L. R. M. Cocks, and I. R. Plimer. Amsterdam and Boston: Elsevier Academic, pp. 194–200.

McLean, J. H. 1981. The Galapagos rift limpet *Neomphalus*: relevance to understanding the evolution of a major Paleozoic-Mesozoic radiation. *Malacologia* 21: 291–336.

———. 1984. A case for derivation of the Fissurellidae from the Bellerophontacea. *Malacologia* 25: 3–20.

McShea, D. W. 1994. Mechanisms of large-scale evolutionary trends. *Evolution* 48: 1747–1763.

Miller, A. I. 2004. The Ordovician radiation: toward a new global synthesis. In *The great Ordovician Biodiversification Event.* Edited by B. D. Webby, F. Paris, M. L. Droser, and I. G. Percival. New York: Columbia University Press, pp. 380–388.

Moore, R. C. 1941. Upper Pennsylvanian gastropods from Kansas. *Kansas State Geological Bulletin* 38: 121–164.

Morris, N. J. 1990. Early radiation of the Mollusca. In Major evolutionary radiations. Edited by P. D. Taylor and G. P. Larwood. Oxford: Clarendon Press, pp. 73–90.

Morris, N. J., and Cleevely, R. J. 1981. *Phanerotinus cristatus* (Phillips) and the nature of euomphalacean gastropods, Molluscans. *Bulletin of the British Museum of Natural History (Geology)* 35: 195–212.

Morris, P. J. 1991. Functional morphology and phylogeny: an assessment of monophyly in the Kingdom Animalia and Paleozoic nearly-planispiral snail-like mollusks. Ph.D. dissertation, Harvard University.

Morton, J. E. 1958. Torsion and the adult snail; a reevaluation. *Proceedings of the Malacological Society of London* 33: 2–10.

Novack-Gottshall, P. M., and Miller, A. I. 2003. Comparative geographic and environmental diversity dynamics of gastropods and bivalves during the Ordovician radiation. *Paleobiology* 29: 576–604.

Nützel, A. 1998. Über die Stammesgeschichte der Ptenoglossa (Gastropoda). *Berliner Geowissenschaftliche Abhandlungen. Reihe E. Paläobiologie* 26: 1–225.

———. 2002. An evaluation of the recently proposed Palaeozoic gastropod subclass Euomphalomorpha. *Palaeontology* 45: 259–266.

———. 2005. Recovery of gastropods in the Early Triassic. In *The Biotic Recovery from the End-Permian Mass Extinction*. Edited by D. Bottjer and J.-C. Gall. *Comptes Rendus Palevol* 4: 501–515.

Nützel, A., and Bandel, K. 2000. Goniasmidae and Orthonemidae: two new families of the Palaeozoic Caenogastropoda (Mollusca, Gastropoda). *Neues Jahrbuch für Geologie und Paläontologie Abhandelungen* 2000 (9): 557–569.

Nützel, A., and Cook, A.G. 2002. *Chlorozyga*, a new caenogastropod genus from the Early Carboniferous of Australia. *Alcheringa* 26: 151–157.

Nützel, A., Erwin, D.H., and Mapes, R.H. 2000. Identity and phylogeny of the Late Paleozoic Subulitoidea (Gastropoda). *Journal of Paleontology* 74: 575–598.

Nützel, A., and Frýda, J. 2003. Paleozoic plankton revolution: evidence from early gastropod ontogeny. *Geology* 31: 829–831.

Nützel, A., Lehnert, O., and Frýda, J. 2006: Origin of planktotrophy—evidence from early molluscs. *Evolution and Development*, 8: 325–330.

Nützel, A., and Mapes, R.H. 2001. Larval and juvenile gastropods from a Carboniferous black shale: palaeoecology and implications for the evolution of the Gastropoda. *Lethaia* 34: 143–162.

Nützel, A., and Pan, H.-Z. 2005. Late Paleozoic evolution of the Caenogastropoda: larval shell morphology and implications for the Permian/Triassic mass extinction event. *Journal of Paleontology* 79: 1175–1188.

Nützel, A., Pan H.-Z., and Erwin, D.H. 2002. New taxa and some taxonomic changes of a latest Permian gastropod fauna from South China. *Documenta Naturae* 145: 1–10.

Nützel, A., and Schulbert, C. 2005. Gastropod lagerstätten in the aftermath of the end-Permian mass extinction—diversity and facies of two major Early Triassic occurrences. *Facies* 51: 480–500.

Page, L.R. 1997. Ontogenetic torsion and protoconch form in the archaeogastropod *Haliotis kamtschatkana*: evolutionary implications. *Acta Zoologica* 78: 227–245.

Pan, H.-Z., and Erwin, D.H. 2002. Gastropods from the Permian of Guangxi and Yunnan Provinces, South China. *Journal of Paleontology Memoir* 56 Suppl. 1: 1–49.

Papadopoulos, L.N., Todd, J.A., and Michel, E. 2004. Adulthood and phylogenetic analysis in gastropods: character recognition and coding in shells of Lavigeria (Cerithioidea, Thiaridae) from Lake Tanganyika. *Zoological Journal of the Linnean Society* 140: 223–240.

Pchelintsev, V.F., and Korobkov, I.A. 1960. Mollusca - Gastropoda. In *Osnovy paleontologii*. Edited by Y.A. Orlov. Moscow: Izdatel'stvo Akademia Nauk, pp. 1–360.

Peel, J.S. 1980. A new Silurian retractile monoplaocphoran and the origin of the gastropods. *Proceedings of Geological Association* 91: 91–97.

———. 1991a. Functional morphology of the Class Helcionelloida nov., and the early evolution of the Mollusca. In *The Early Evolution of the Metazoa and the Significance of Problematic Taxa*. Edited by A. Simonetta and S. Conway Morris. Cambridge, UK: Cambridge University Press, pp. 157–177.

———. 1991b. The Classes Tergomya and Helcionelloida, and early molluscan evolution. *Grønlands Geologiske Undersøgelse Bulletin* 161: 11–65.

———. 2001. Musculature and asymmetry in a Carboniferous pseudo-bellerophontoidean gastropod (Mollusca). *Palaeontology* 44: 157–166.

Pennington, J.T., and Chia, F.S. 1985. Gastropod torsion: a test of Gastrang's hypothesis. *Biological Bulletin* 169: 391–396.

Peterson, K.J. 2004. Dating the origins of marine invertebrate larvae with a molecular clock. *Geological Society of America—Abstracts with Program* 36: A-295.

Ponder, W.F., and Lindberg, D.R. 1997. Towards a phylogeny of gastropod molluscs: an analysis using morphological characters. *Zoological Journal of the Linnean Society* 119: 83–265.

Ponder, W.F., and Warén, A. 1988. Classification of the Caenogastropoda and Heterostropha—a list of the family-group names and higher taxa. In *Prosobranch Phylogeny*. Edited by W.F. Ponder. *Malacological Review* Suppl. 4: 288–326.

Riedel, F. 2000. Ursprung und Evolution der "höheren" Caenogastropoda. *Berliner geowissenschaftliche, Abhandlungen* E 32, 240 pp.

Rohr, D.M., and Johns, R.A. 1990. First occurrence of *Oriostoma* (Gastropoda) from the Middle Ordovician. *Journal of Paleontology* 64: 732–735.

Rosenberg, G., Kuncio, G.S., Davis, G.M., and Harasewych, M.G. 1994. Preliminary ribosomal RNA phylogeny of gastropod and unionoidean bivalve mollusks. *Nautilus* Suppl. 2: 111–121.

Runnegar, B. 1981. Muscle scars, shell form and torsion in Cambrian and Ordovician univalved molluscs. *Lethaia* 14: 311–322.

———. 1983. Molluscan phylogeny revisited. *Association of Australasian Palaontologists Memoir* 1: 121–144.

Runnegar, B., and Pojeta, J. 1974. Molluscan phylogeny: the paleontological viewpoint. *Science* 186: 311–317.

Salvini-Plawen, L. v., and Haszprunar, G. 1987. The Vetigastropoda and the systematics of streptoneurous gastropods (Mollusca). *Journal of Zoology A* 211: 747–770.

Sasaki, T. 1998. Comparative anatomy and phylogeny of the Recent Archaeogastropoda (Mollusca: Gastropoda). *The University Museum, The University of Tokyo, Bulletin* 38: 1–223.

Schander, C., and Sundberg, P. 2001. Useful characters in gastropod phylogeny: soft information or hard facts? *Systematic Biology* 50: 136–141.

Schluter, D. 2000. Ecological character displacement in adaptive radiation. *The American Naturalist* 156 Suppl.: S4–S16.

Sepkoski, J. J., Jr. 1981. A factor analytic description of the Phanerozoic marine fossil record. *Paleobiology* 7: 36–53.

———. 1982. A compendium of fossil marine families. Milwaukee Public Museum *Contributions in Biology and Geology* 51: 1–125.

———. 1995. The Ordovician radiations: diversification and extinction shown by global genus-level taxonomic data. In *Ordovician Odyssey: Short Papers for the Seventh International Symposium on the Ordovician System*. Edited by J. D. Cooper, M. L. Droser, and S. C. Finney. Fullerton, CA: SEPM, pp. 393–396.

———. 2002. A compendium of fossil marine animal genera. *Bulletins of American Paleontology* 363: 1–563.

Sepkoski, J. J., Jr., and Hulver, M. L. 1985. An atlas of Phanerozoic clade diversity diagrams. In *Phanerozoic Diversity Patterns*. Edited by J. W. Valentine. Princeton, NJ, and San Francisco: Princeton University Press, pp. 11–39.

Signor, P. W. 1985. Gastropod evolutionary history. In *Mollusks, Notes for a Short Course*. Edited by T. W. Broadhead. Knoxville: TN: University of Tennessee; Department of Geological Sciences, Studies in Geology 13. Knoxville, TN: University of Tennessee, pp. 157–173.

Tracey, S., Todd, J. A., and Erwin, D. H. 1993. Mollusca: Gastropoda. In *The fossil record*. Edited by M. J. Benton. London: Chapman and Hall, pp. 131–167.

van den Biggelaar, Jo A. M., and Haszprunar, G. 1996. Cleavage patterns and mesentoblast formation in the Gastropoda: an evolutionary perspective. *Evolution* 50: 1520–1540.

Ulrich, E. O., and Scofield, W. H. 1897. The Lower Silurian Gastropoda of Minnesota. In *The Geology of Minnesota*. Vol. 3, Part 2, *Paleontology*. Minneapolis, MN: Harrison and Smith, pp. 813–1081.

Vermeij, G. J. 1987. *Evolution and Escalation—An Ecological History of Life*. Princeton, NJ: Princeton University Press, pp. 1–544.

Vermeij, G. J., and Carlson, S. J. 2000. The muricid gastropod subfamily Rapaninae: phylogeny and ecological history. *Paleobiology* 26: 19–46.

Wagner, P. J. 1995a. Testing evolutionary constraint hypotheses with early Paleozoic gastropods. *Paleobiology* 21: 248–272.

———. 1995b. Diversity patterns among early gastropods—contrasting taxonomic and phylogenetic descriptions. *Paleobiology* 21: 410–439.

———. 1996. Contrasting the underlying patterns of active trends in morphologic evolution. *Evolution* 50: 990–1007.

———. 1997. Patterns of morphologic diversification among the Rostroconchia. *Paleobiology* 23: 115–150.

———. 1998. Anatomical disparity over time as inferred from modern gastropods: contrasting neontological and paleontological patterns. In *Abstracts, World Congress of Malacology*. Edited by R. Bieler and P. M. Mikkelsen. Chicago: UNITAS Malacologica, p. 348.

———. 1999a. Phylogenetics of Ordovician-Silurian Lophospiridae (Gastropoda: Murchisoniina): the importance of stratigraphic data. *American Malacological Bulletin* 15: 1–31.

———. 1999b. Phylogenetics of the earliest anisotrophically coiled gastropods. *Smithsonian Contributions to Paleobiology* 88: 1–132.

———. 2000a. The quality of the fossil record and the accuracy of phylogenetic inferences about sampling and diversity. *Systematic Biology* 49: 65–86.

———. 2000b. Exhaustion of cladistic character states among fossil taxa. *Evolution* 54: 365–386.

———. 2001a. Rate heterogeneity in shell character evolution among lophospiroid gastropods. *Paleobiology* 27: 290–310.

———. 2001b. Gastropod phylogenetics: progress, problems and implications. *Journal of Paleontology* 75: 1128–1140.

———. 2002. Likelihood tests of general phylogenetic hypotheses: how many times did bellerophont molluscs evolve? *Geological Society of America—Abstracts with Program* 34: 261.

Wagner, P. J., and Erwin, D. H. 2006. Patterns of convergence in general shell form among Paleozoic gastropods. *Paleobiology* 32: 315–336.

Wahlman, G.P. 1992. Middle and Upper Ordovician symmetrical univalved molluscs (Monoplacophora and Bellerophontina) of the Cincinnati Arch region. *United States Geological Survey Professional Paper* 1066-O: O1–O213.

Wanninger, A., Ruthensteiner, B., and Haszprunar, G. 2000. Torsion in *Patella caerulea* (Mollusca, Patellogastropoda): ontogenetic process, timing, and mechanisms. *Invertebrate Biology* 119: 177–187.

Warén, A., and Bouchet, P. 2001. Gastropoda and Monoplacophora from hydrothermal vents and seeps; new taxa and records. *The Veliger* 44: 116–231.

Wenz, W. 1938–1944. *Gastropoda*. Vol. Band 6, Teil 1–7, *Handbuch der Paläozoologie*. Edited by O.H. Schindewolf. Berlin: Bontraeger, pp. 1–1639.

———. 1940. Urspung und frühe Stammesgeschichte der Gastropoden. *Archiv für Molluskenkunde* 72: 1–10.

Wing, S.L., Hickey, L.J., and Swisher, C.C. 1993. Implications of an exceptional fossil flora for Late Cretaceous vegetation. *Nature* 363: 342–344.

Yochelson, E.L. 1967. Quo vadis, *Bellerophon?* In *Essays in Paleontology and Stratigraphy*. Edited by C. Teichert and E.L. Yochelson. Lawrence, KS: University of Kansas Press, pp. 141–161.

———. 1984. Historic and current considerations for revision of Paleozoic gastropod classification. *Journal of Paleontology* 58: 259–269.

Yonge, C.M. 1947. The pallial organs in aspidobranch Gastropoda and their evolution throughout the Mollusca. *Philosophical Transactions of the Royal Society of London Series B* 232B: 443–518.

Yoo, E.K. 1988. Early Carboniferous Mollusca from Gundy, Upper Hunter, New South Wales. *Records of the Australian Museum* 40: 233–264.

———. 1994. Carboniferous Mollusca from the Tamworth Belt, New South Wales, Australia. *Records of the Australian Museum* 46: 63–120.

11

Patellogastropoda, Neritimorpha, and Cocculinoidea

THE LOW-DIVERSITY GASTROPOD CLADES

David R. Lindberg

The Patellogastropoda, Cocculinoidea, and Neritimorpha represent three distinct gastropod clades. However, their placement on the gastropod tree, and therefore their relationships to each other as well as to other gastropod groups, have been and remain problematic (Ponder and Lindberg 1997; Colgan *et al.* 2003; McArthur and Harasewych 2003; Geiger and Thacker 2005; Aktipis *et al.*, Chapter 9; Geiger *et al.*, Chapter 12).

All three taxa are low in species diversity relative to other gastropod clades (Vetigastropoda, Caenogastropoda, and Heterobranchia), but there are no obvious common morphological or ecological features that might account for their diminished diversity. For example, while the patellogastropods and cocculinoids are exclusively limpets, neritimorph morphological diversity includes limpets, slugs, and coiled trochiforms similar to those seen across the Gastropoda. Patellogastropods and cocculinoids are also almost exclusively marine, but the Neritimorpha include marine, freshwater, and terrestrial taxa. All three groups have deep-sea representatives, but only the Patellogastropoda and Neritimorpha have intertidal members, and it is in this arduous habitat that these two groups have achieved their only substantial radiations.

The Patellogastropoda, Cocculinoidea, and Neritimorpha appear to be intriguing fragments of earlier gastropod radiations. The assumed coiled sister taxa of the patellogastropods and cocculinoids (Ponder and Lindberg 1997) are extinct and have yet to be identified in the fossil record. Plesiomorphic elements of the morphology of the living representatives are undoubtedly confounded by apomorphies that have appeared since their divergence from their respective last common ancestor with other gastropods. The occurrence of fossil neritimorph taxa provides a more complete picture, but the low diversity and marginal habitats of the basal members also suggests substantial extinction. The role of extinction in exacerbating long-branch attraction in both morphological and molecular data is well documented (e.g., Donoghue *et al.* 1989; Wiens 2005, 2006), and, given current sampling and methodologies, an obvious solution to our incomplete understanding of these groups is not readily apparent.

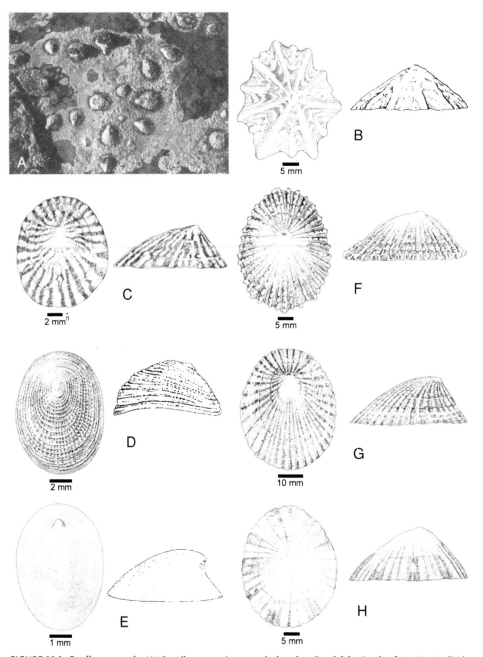

FIGURE 11.1. Patellogastropoda. (A) *Scutellastra* species on rocky bench at Brookdale, South Africa. (B) *Patelloida saccharina* (Lottiidae). (C) *Notoacmea petterdi* (Lottiidae). (D) *Pectinodonta kapalae* (Acmaeidae). (E) *Propilidium tasmanicum* (Lepetidae). (F) *Scutellastra peronii* (Patellidae). (G) *Nacella macquariensis*. (H) *Cellana tramoserica* (Nacellidae). (Figures 11.1B–H from Lindberg [1998] in *Mollusca—The Southern Synthesis* [Beesley *et al.* 1998]).

For more detailed treatment of the morphology and composition of these taxa, see the summaries of Haszprunar (1998), Lindberg (1998), Ponder (1998), and Scott and Kenny (1998) and references therein.

PATELLOGASTROPODA

The Patellogastropoda (Figure 11.1) or "true limpets" are some of the most ubiquitous gastropod components of marine littoral hard-substrate communities (Figure 11.1A) and include some of the best studied intertidal gastropods in the disciplines of ecology, physiology, behavior, and reproduction (see following discussion).

Living patellogastropods are divisible into three major groups: the Patellina, Acmaeina, and Lottiidae. The Patellina includes the Patellidae (e.g., *Patella, Scutellastra, Helcion*; Figure 11.1A, F) and the Nacellidae (e.g., *Cellana, Nacella*; Figure 11.1G, H). The Acmaeina includes the Lepetidae (e.g., *Lepeta, Cryptobranchia, Iothia*; Figure 11.1E), Pectinodontinae (e.g., *Pectinodonta, Serradonta*; Figure 11.1D), and the Acmaeidae (*Acmaea*), while the Lottiidae includes the Patelloidinae (*Patelloida*) and Lottiinae (e.g., *Lottia, Scurria*; Figure 11.1B, C). There are also less diverse taxa, including the deep-water *Eulepetopsis* and *Bathyacmaea*, the *Asteracmaea*, and members of the genus *Erginus*, whose affinities with the Acmaeidae and Lottiidae remain unknown. Because of their simple shell morphology and anatomy, patellogastropod classifications have tended to under-estimate their diversity. The application of molecular techniques is proving useful in resolving the evolutionary history of this group (see following discussion), and future systematic and nomenclatural revisions are certain.

Patellogastropoda are marine, although a few live in brackish habitats and one lottiid (*Potamacmaea fluviatilis*) may extend from brackish waters into the major rivers that drain into the Bay of Bengal in Southeast Asia (Lindberg 1990). Patellogastropods live on most hard substrates, including rock, wood, and shells of other molluscs and invertebrates, where they indiscriminately graze on diatoms, blue-green algae, and algal spores. Some lottiid and patellid species are associated with marine angiosperms and algae, and several subtidal taxa are restricted to calcareous algae substrates (e.g., *Acmaea, Erginus*; Lindberg 1998).

The Patellogastropoda are predominantly an intertidal and shallow subtidal group. However, some taxa such as the Lepetidae and Acmaeidae are mostly subtidal, while others such as the *Pectinodonta* and *Bathyacmaea* are deep-water, occurring to depths in excess of 4,000 m

(Sasaki *et al.*2005; Sasaki *et al.* 2006b). At those depths, the limpets utilize food such as bacteria-rich waterlogged wood and detritus, rather than algae and plants. Also in the deep sea, *Eulepetopsis* species are found living in cold seeps and in the vicinity of hydrothermal vents (McLean 1990).

Patellogastropods range in adult size from about 3 mm to over 200 mm in length. The smallest and largest species are typically found in the lowest intertidal zone or subtidally. Most intertidal species average between 20 and 40 mm in length. Subtidal species are typically white or pink in color, and intertidal species are typically drab browns and grays with white spots and radial rays. The shell color can be similar to that of the substrate on which the limpet occurs because of the incorporation of plant compounds into the shell (Lindberg and Pearse 1990).

NERITIMORPHA

The Neritimorpha[1] (Figure 11.2) have long been recognized as a separate and deep branch within the Gastropoda (Bourne 1908; Yonge 1947), and their evolutionary history is a microcosm of the radiation of the Gastropoda. Members of the Neritimorpha live in habitats ranging from tropical intertidal beach rock to cold seeps and hydrothermal vents at depths in excess of 2,500 m and in the lightless recesses of subtidal sea caves. They have invaded both freshwater and terrestrial habitats, become arboreal, evolved limpet- and slug-grade morphologies, and have some of the most elaborate reproductive systems known among gastropods.

The living Neritimorpha include the Neritopsidae, Neritidae, Neritiliidae, Helicinidae, Hydrocenidae, Phenacolepadidae, and Titiscaniidae. The Neritopsidae is represented by only two living species of *Neritopsis* (Figure 11.2C); however, fossil taxa referred to this group are present in the Late Paleozoic and are especially

1. An alternative name used for Neritimorpha (originally Neritaemorphi Koken, 1896) is Neritopsina Cox and Knight, 1960.

FIGURE 11.2. Neritimorpha. (A) *Nerita atramentosa* with the patellogastropod (*Cellana tramoserica*) and the vetigastropod (*Austrocochlea procata*), Collaroy, NSW, Australia. (B) *Phenacolepas osculans* on coralline algal substrate, Moorea, French Polynesia. (C) *Neritopsis radula* (Neritopsidae). (D) *Nerita plicata* (Neritidae). (E) *Neritina violacea* (Neritiliidae). (F) *Titiscania limacine* (Titiscaniidae). (G) *Pleuropoma draytonensis* (Helicinidae). (H) *Phenacolepas miriabilis* (Phenacolepadidae). (Figure 11.2A image by A. C. Miller © Australian Museum; Figure 11.2B image provided and © by D. Geiger; and Figure 11.2C–H from Scott and Kenny [1998] in *Mollusca— The Southern Synthesis* [Beesley *et al.* 1998]).

abundant in the Jurassic and Cretaceous (Kaim and Sztajner 2004). Members of the Neritidae include most shallow-water marine species (*Nerita, Smaragdia*; Figure 11.2A, D) as well as brackish and freshwater representatives (*Neritina, Theodoxus*); one neritid taxon (*Neritodryas*) is partially terrestrial and sometimes arboreal, occurring on trees and vegetation along fresh and brackish bodies of water (Little 1990). The Neritiliidae include both freshwater (*Neritilia*; Figure 11.2E) and sea cave taxa (*Pisulina*; Kano and Kase 2002), and the Helicinidae (Figure 11.2G) and Hydrocenidae are fully terrestrial groups. The extinct Paleozoic Dawsonellidae are also

considered to have been fully terrestrial (Solem and Yochelson 1979). The Phenacolepadidae are limpets and occur in both shallow water (*Phenacolepas*; Figure 11.2B, H) and in the deep sea at seeps and vents (*Bathynerita, Shinkailepas*; Kano *et al.* 2002; Sasaki *et al.*, 2006a). The Titiscaniidae are marine shell-less slugs (Figure 11.2F); only two living species are known (Scott and Kenny 1998).

Neritimorphs are conservative in their diets, most being grazers on algal spores, diatoms, and detritus. Some freshwater species such as *Theodoxus* are also omnivorous, including insect larvae in their diet (e.g., Pavlichenko 1977).

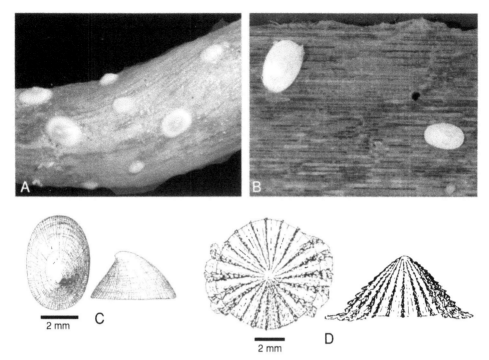

FIGURE 11.3. Cocculinoidea. (A) *Cocculina* sp. on waterlogged wood off Balicasag Island, Bohol, Philippines. Bathyal. (B) *Cocculina* sp. cf. *suruganensis* on waterlogged wood Kumano Basin, off Mie Prefecture, Japan. 757–777 m. (C) *Cocculina japonica* D. *Bathypelta pacificum*. (Figure 11.3A image provided and © by E. Guillot de Suduiraut; Figure 11.3B image provided and © by T. Sasaki; Figure 11.3C from Dall [1925]; and Figure 11.3D from Haszprunar [1998] in *Mollusca—The Southern Synthesis* [Beesley *et al.* 1998]).

Terrestrial representatives feed on detritus, algal spores, moss, and lichens. Seep and vent taxa are likely detritivores as well, feeding on the abundant bacterial mats that occur in these habitats.

Neritimorphs are found worldwide, but they reach their greatest diversity in tropical and warm temperate environments. They are generally small- to medium-sized gastropods between 2 and 40 mm in length, a relatively small size range compared with other major gastropod groups. Some aquatic taxa have complex color patterns. Some terrestrial neritimorphs also have bright-colored markings and glossy shells, although others are more subdued, mottled, and match their habitats well.

COCCULINOIDEA

The Cocculinoidea (Figure 11.3) are a group of small (2–15 mm) white limpets that occur on waterlogged wood (Figure 11.3A, B), whale bone, and cephalopod beaks in the deep sea

worldwide. While this habitat would appear to be incredibly sparse or rare given the vastness of the deep-sea floor, the fact that another group of gastropods (the Lepetelloidea in the Vetigastropoda; Ponder and Lindberg 1997; Geiger *et al.*, Chapter 12) have also evolved to utilize similar substrates in the deep sea suggests that this first impression is incorrect. All known Cocculinoidea have cap-shaped shells, but they have likely descended from coiled snails. As with patellogastropods, possible coiled ancestors have not yet been identified in the fossil record.

Living cocculinoid limpets were unknown until the advent of deep-sea exploration in the 1800s. At the time a variety of substrates were recovered by dredging, including cephalopod beaks, waterlogged wood, bones of fish and whales, and the egg cases of sharks and skates, and all of these substrates were found to have small white limpets living on their surfaces. Until recently all of these limpets were thought to share a common ancestor and were grouped in the Cocculiniformia (Haszprunar 1987). However, subsequent morphological (Ponder and Lindberg 1997) and molecular analysis (Colgan *et al.* 2000, 2003; McArthur and Harasewych 2003) suggest that the shared habitats and morphology are the result of convergence rather than common ancestry.

The Cocculinoidea includes two major groups—the Cocculinidae and the Bathysciadiidae. The Bathysciadiidae (*Bathysciadium* [Figure 11.3D] and *Teuthirostria*) occur on squid beaks, while the Cocculinidae (*Coccopigya, Cocculina* [Figure 11.3C] and *Macleaniella*) are known predominantly from wood and occasionally on whale bone (Haszprunar 1988a, 1998; McLean 1992).

SIGNIFICANCE

PATELLOGASTROPODA

Because of their large size, ease of manipulation, niche specializations, high diversity, and relative abundance on rocky shores, patellogastropods have been the subject of extensive ecological investigations (Underwood 1979; Branch 1981, and references therein). However, long before their role as intertidal guinea pigs was discovered, some of these same characteristics made intertidal patellogastropod species important components of aboriginal diets for over 150,000 years (Speed 1969), and even today, size is reduced through human predation in many areas (e.g., Hockey and Bosman 1986; Keough *et al.* 1993; Kyle *et al.* 1997; Lindberg *et al.* 1998). Patellogastropods are also used in monitoring the ecological health of rocky shore communities because of their ubiquitous presence and ecological role in rocky shore systems (Navrot *et al.* 1974; Crump *et al.* 1999; Roy *et al.* 2003). Otherwise, there appears to be little economic interest in most patellogastropods, although some species of *Cellana* are a significant fishery in Hawaii (Kay *et al.* 1982), and patellids have been substantially overharvested at the Azores in recent years (Ferraz *et al.* 2001).

Only two patellogastropods are listed by the IUCN Red List (IUCN 2006)—*Lottia edmitchelli* and *Lottia alveus*. However, the extinction of these two North American species does not appear to have been related to human activities (Lindberg 1984; Carlton *et al.* 1991). In contrast, local declines of the larger patellogastropods (such as in the Azores and Hawaiian Islands) can usually be associated with overharvesting.

Most recently, the complete nuclear genome of the patellogastropod *Lottia gigantea* has been sequenced (http://www.jgi.doe.gov/sequencing/why/CSP2005/limpet.html), making this the first complete molluscan genome available to researchers and promising to revolutionize the study of molluscan biology and evolution (see Simison and Boore, Chapter 17).

NERITIMORPHA

As with patellogastropods, the occurrence and abundance of neritimorph species in the intertidal has made them well suited for ecological studies (e.g., Ruwa and Jaccarini 1988; Cushman 1989; Praveenkumar and Devi 1998; Underwood 2004), as well as indicator species for environmental monitoring (Annapurna and Bhavanarayana 1993; Foster and Cravo 2003). Freshwater, and possibly some marine species, can serve as intermediate hosts for significant trematode parasites (e.g., Chernogorenko *et al.* 1978).

Most marine neritimorphs are quite common on open shores, and larger marine species have been used as food items by subsistence gatherers since prehistoric times (e.g., Anderson *et al.* 2001; Kirch and O'Day 2003). Freshwater and terrestrial species are often more restricted in their distribution and therefore are potentially more susceptible to disturbance. Habitat destruction is the largest threat to these species, with deforestation directly affecting terrestrial species. Increased sedimentation, pollution, water diversion, and reservoir projects impact freshwater species (e.g., *Septaria, Theodoxus*) (Lill 1993; Gortz 1998).

COCCULINOIDEA

Because of their occurrence in the deep sea, the Cocculinoidea have no significant interaction with humans. However, human interactions with their substrates could threaten or enhance this group. For example, increased logging and near-shore construction during the last 500 years has undoubtedly increased the amount of wood entering the ocean and subsequently the deep sea, whereas the advent of commercial whaling about 200 years ago likely caused a decrease in whale fall habitat. In addition, overexploitation of squid stocks could jeopardize the habitat of the Bathysciadiidae by reducing the number of squid beaks that accumulate on the ocean bottom.

MAIN FEATURES

PATELLOGASTROPODA

All living patellogastropods have cap-shaped shells with the apex typically situated at the center of the shell or slightly anterior. The protoconch is simple, cecum-like, and usually lost by the time the limpet reaches 1 mm in length. The operculum is present in the veliger stage but is lost at metamorphosis and is absent in the adult. The shell aperture is typically oval, although it can be modified to reflect the substrate on which the limpet lives; elongated apertures with parallel sides are found in species that live on algae or marine grasses with flat stipes and blades. Species living on algae with round stipes have elongated apertures with raised anterior and posterior edges, and the apertures of species that form home depressions are often highly crenulate. The exterior shell sculpture can consist of fine riblets, heavy ribs, concentric growth lines, or combinations of these elements. Subtidal species tend to lack heavy ribbing, and strong cancellate sculpture is known only in the living deep-sea pectinodontids.

Shell color patterns have both genetic and environmental components. Genetic color patterns are expressed regardless of the food source or type, whereas environmental coloration is determined by plant pigments and other compounds present in the diet (Lindberg and Pearse 1990). Thus, color patterns can change when limpets change diets (Sorenson and Lindberg 1991). Common color patterns are variegated, mottled, radial or rayed, tessellated, or solid. Combinations of these basic components are more common than simple patterns.

The inner surface of the shell bears a horseshoe-shaped muscle scar that opens anteriorly behind the head, and the mantle edge surrounding the aperture is studded with retractile sensory tentacles. The patellogastropod radula has a few robust brown "teeth" impregnated with iron compounds—the docoglossate condition. Unlike most gastropods, patellogastropods have several gill configurations (Powell 1973; Lindberg 1998). In the Patellina secondary gill leaflets are located around the edge of the foot (a "pallial gill"); in some taxa the gill is complete, while in other taxa it is interrupted in front of the head. In the Acmaeidae and Lottiidae the single left gill (ctenidium) is located over the head, as it is in many other gastropods. In some lottiids both a pallial gill and a ctenidium are present, while in the Lepetidae, *Neolepetopsis*, and species of *Erginus* gills are absent.

The crystalline microstructure of the patellogastropod shell distinguishes this group from all other living gastropods (MacClintock 1967). The layering and types of crystals that make up the patellogastropod shell are complex and varied. The primitive shell structures have both foliated and crossed-lamellar components, whereas in the vetigastropods the shell structures have combinations of intersected crossed-platy, nacreous, and prismatic structures (Hedegaard 1990, 1997). Some patellogastropod lineages have emphasized foliated structures, while others have emphasized crossed-lamellar structures (Lindberg 1988a; Fuchigami and Sasaki 2005). There are no shell structure groups associated with specific habitats, and although there are geographic restrictions of some shell structure groups in

Recent taxa, these patterns are not maintained when the fossil record is examined (Lindberg 1988a; Kase and Shigeta 1996; Fuchigami and Sasaki 2005).

As in the monoplacophorans, the patellogastropod gonad lies on the ventral surface of the visceral mass against the dorsal surface of the foot, in contrast to the dorsal position of the gonad in other gastropods and chitons (Lemche and Wingstrand 1959; Lindberg 1985, 1988b; Ponder and Lindberg 1997). The gonad opens into the right excretory organ in most species, and the gametes are expelled from the right excretory organ through the right excretory pore.

Brooding species are known in the genera *Erginus* and *Rhodopetala* from the boreal Pacific (Golikov and Kussakin 1972; Lindberg 1981, 1983). In some brooding species the excretory organs are rotated 90°, placing the left excretory organ above the rectum and the right excretory organ below; in several of these species the dorsal left excretory organ is modified into a brood chamber (Lindberg 1988b). Copulatory structures located under the right cephalic tentacle are found in some species of *Erginus* (Golikov and Kussakin 1972).

NERITIMORPHA

The typical neritimorph shell differs from that of other coiled gastropods in the internal whorls of the shell being absorbed as the snail grows, permitting the snail's body to be more limpet-like irrespective of the external shell morphology. This also means that a central shell axis—or columella—is missing. Only *Neritopsis* does not absorb the internal whorls, leaving the columella intact and contributing to its putative position as the basal living member of the Neritimorpha. In addition to absorbing the internal whorls, most coiled neritimorphs thicken the inner surface of the aperture, and many have apertural barriers. The operculum is typically thickened and calcareous and has a small peg or apophysis on the inner surface.

Neritimorph shell sculpture is lacking or is dominated by strong spiral cords; growth lines may also be prominent, giving rise to small beads at their intersection with the cords. Small species (< 5 mm) and freshwater species tend to be smoother, and spines may be present along the periphery of the penultimate whorl in some freshwater taxa. Color patterns are highly varied, but variegated, tessellate, and zigzag patterns predominate in the group, although some solid colors (dark and light) also occur. The Neritimorpha often have vivid colors, including red, yellow, blue, and green hues; freshwater species tend to be drabber in color.

The neritimorph radula is rhipidoglossate with a central tooth (lacking in *Neritopsis*), two to three inner lateral teeth (present as plates in *Neritopsis*), a large, robust outer lateral, and numerous marginal teeth. A single left ctenidium is present in the mantle cavity of marine and freshwater taxa, whereas in terrestrial groups the ctenidium is absent and respiration occurs in the vascularized mantle cavity (Scott and Kenny 1998).

Shell microstructure in the Neritimorpha is conservative and varies little regardless of habitat (i.e., marine, freshwater, or terrestrial). The group is diagnosed by an outer calcitic homogeneous layer, while the inner layers are composed of comarginal crossed-lamellar, cone complex crossed-lamellar layers, or both (Hedegaard 1990; Hedegaard and Lindberg, unpublished data).

Neritimorph reproductive systems are highly variable and complex (Bourne 1908, 1911; Andrews 1937; Haynes 2005). Female reproductive systems are referred to as monaulic, diaulic, or triaulic denoting the number of genital openings into the mantle cavity. The genital tracts are adorned with numerous glands and structures including the bursa copulatrix, spermatophore sac, and seminal receptacle (Scott and Kenny 1998). Male reproductive anatomy includes a penis located near the right cephalic tentacle (lacking in terrestrial taxa) and unique dimorphic spermatozoa (Healy 1988); sperm are transferred from males to females via spermatophores (Haynes 2005).

COCCULINOIDEA

All living cocculinoids have cap-shaped shells. The apex of the shell is typically situated at the center or nearer the posterior end of the shell. The protoconch is simple, rarely cecum-like, and symmetrically coiled for at least one-quarter of a whorl; it is often maintained on the adult shell. The shell aperture is typically oval but can also reflect the shape of the substrate to which the limpet is affixed. Shells are sculptured with concentric growth lines, and in some species additional fine radial threads extend from the apex to the shell margin. When the growth lines are strong, beads are sometimes formed at the intersections with the radial sculpture; in some species the strong growth lines and radial sculpture give the shell surface a can-cellate appearance. The shells are white and typically covered by a thick, sometimes spiny periostracum.

The inner surface of the shell bears a horse-shoe-shaped muscle scar that opens anteriorly. The cocculinoid radula is rhipidoglossate as in the Vetigastropoda and Neritopsina. Typically there is a central tooth, four inner lateral teeth, a large robust outer lateral, and numerous marginal teeth. The gills of cocculinoids are vestigial and do not resemble the typical mol-luscan gill or ctenidium (Haszprunar 1988a). Gill morphology is variable and may consist of a large folded or small papillate pseudoplicate gill or be further reduced to just respiratory leaflets on the dorsal surface of the mantle cavity (Strong *et al.* 2003).

Cocculinoid shell microstructure consists of an outer aragonitic homogeneous or simple prismatic layer overlying two distinct layers of cone complex crossed-lamellar microstructure (Hedegaard 1990; Hedegaard and Lindberg, unpublished data).

All cocculinoids are simultaneous her-maphrodites, producing eggs and sperm in a single gonad that opens into the mantle cavity through a glandular oviduct (Haszprunar 1988a); a single or paired seminal receptacle is present. Male reproductive anatomy includes a penis typically associated with the right cephalic tentacle. A pallial brood chamber is present in most taxa (Haszprunar 1987; Strong *et al.* 2003).

CLASSIFICATION

PATELLOGASTROPODA

Fifty years ago all patellogastropods were con-tained within a handful of genera in three families—Lepetidae, Acmaeidae, and Patellidae (e.g., Powell, 1973). Even 20 years ago most of the Lottiidae were placed in the single genus *Acmaea*, and the entire "order" was consid-ered an unusual superfamily of archaeogastro-pod snails (= Vetigastropoda, in part) (but see Golikov and Starobogatov 1975). However, in the last 20 years the morphological and molecu-lar distinctiveness of the Patellogastropoda has become more apparent and better documented (Golikov and Starobogatov 1975; Graham 1985; Lindberg 1988a; Haszprunar 1988b; Ponder and Lindberg 1997; Sasaki 1998; McArthur and Harasewych 2003; Nakano and Ozawa 2004).

This recent trend is actually a return to the perspective espoused by some early malacolo-gists. Initially, the Patellogastropoda were consid-ered to be quite different from other gastropods, and several higher taxa were proposed. Cuvier (1817) placed the patellogastropods with the polyplacophorans in the Cyclobranchia. Troschel (1866–93) proposed Docoglossa in recognition of the differences between the patellogastropod radula and that of other gastropods. Sars (1878) coined Onychoglossa on the basis of the clawlike morphology of the lateral teeth, while Lancaster (1883) proposed Phylliobranchia because of the structure of the pallial gill in *Patella*, and Ber-nard (1890) proposed Heterocardia based on the patellogastropod heart.

In a reversal of these earlier trends to distinguish the patellogastropods from other gastropod groups, Thiele (1925) placed them in his order Archaeogastropoda, ignoring the unique character combinations documented by previous workers (Figure 11.4A). The inclusion of patellogas-tropods within archaeogastropods assumed that they were an aberrant group of rhipidoglossates—a scenario for which there was no evidence—and they remained there (Figure 11.4B) until the 1970s.

The early proliferation of patellogastropod genera was reversed in the middle 1900s with

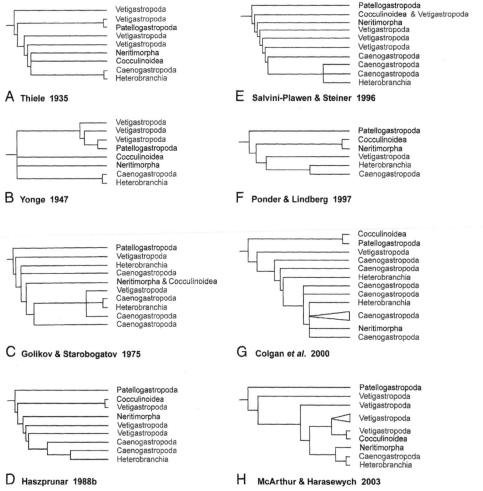

FIGURE 11.4. Placement of the Patellogastropoda, Neritimorpha, and Cocculinoidea in previous phylogenetic classifications of the Gastropoda. Branches have been collapsed to reflect the six major gastropod clades. A–F are morphology-based trees; F–G are molecular-based trees. A was reconstructed from Thiele's (1935) "Discussion of the Phylogeny of the Molluscs," B from Yonge (1947, Figure 31) "Diagrams illustrating the possible course of evolution within the Prosobranchia," and C is redrawn from Golikov and Starobogatov (1975, Figure 6) "Scheme of the evolution and phylogeny of the Gastropoda, ..." The remaining trees (D–H) were originally published as cladograms.

the merging of many regional genera based on overall similarities of gill, radular, and shell characters. However, new characters derived from shell microstructure, anatomy, development (Figure 11.4C–F), and molecular data (Figure 11.4G, H) suggest widespread convergence in the traditional characters. In addition, the fossil record suggests that some clades (e.g., *Scutellastra* and *Patelloida*) are relatively old (> 90 million years; Akpan *et al.* 1982; Lindberg and Vermeij 1985; Kase and Shigeta 1996; Hedegaard *et al.* 1997) while others (e.g., *Lottia*, *Scurria*) are rela-

tively young (< 50 million years; Lindberg 1988a; Espoz *et al.* 2004). Thus the emerging view of the world's patellogastropod faunas is one of few regional endemic taxa, but rather mosaics of different clades with unique regional histories of speciation and extinction (Lindberg 1988a, 1998).

NERITIMORPHA

Like the patellogastropods, the Neritimorpha were included in the Archaeogastropoda by Thiele (1925), although earlier anatomical studies by Bourne (1908, 1911) had suggested that there were substantial differences between them and other "archaeogastropod" groups. Yonge (1947) also recognized the uniqueness of the "Neritacea" (Figure 11.4B), and later, with Morton (Morton and Yonge 1964), formally removed "Neritacea" from the Archaeogastropoda. Golikov and Starobogatov (1975: 200; Figure 11.4C) used the taxon Neritimorpha, considering it "equal to the old superfamily Neritacea." Using morphological criteria Haszprunar (1988a; Figure 11.4D), Salvini-Plawen and Steiner (1996; Figure 11.4E), and Ponder and Lindberg (1997; Figure 11.4F) have all reported a distinct neritimorph clade. Similarly, Colgan *et al.* (2000, 2003; Figure 11.4G) and McArthur and Harasewych (2003; Figure 11.4H; Aktipis *et al.*, Chapter 9.) have also identified a similar clade based on molecular characters.

Partitioning taxa within the Neritimorpha has been relatively straight-forward. Traditionally, Neritimorpha (until recently as Neritacea) has included the Neritidae, Neritopsidae, Helicinidae, Hydrocenidae, and Titiscaniidae. Although overall shell and anatomical characters are similar between groups, shell, radular, opercular, and reproductive characters are often used in the diagnoses of individual taxa. For example, unlike most neritimorphs, the Neritopsidae do not reabsorb the internal shell whorls, and the radula lacks central and inner lateral teeth. Phenacolepadidae are limpet-like, but their anatomy is almost identical with that of neritids (Fretter 1984) and, although a shell-less slug, the Titiscaniidae are anatomically almost identical to Neritopsidae (Kano *et al.* 2002).

Explorations to deep sea vents and cold seeps have recently added several new neritimorph genera, including *Bathynerita, Shinkailepas*, and *Olgasolaris* (Sasaki *et al.* 2003). However, these taxa show little deviation from the neritimorph bauplan and are easily placed within the existing higher taxa (Kano *et al.* 2002).

COCCULINOIDEA

The Cocculinoidea have been recognized as a distinct group since their discovery and description in the late 1800s by Dall (1882). Subsequent anatomical studies in the early 1900s (Thiele 1903) placed them among the rhipidoglossate groups and ultimately in the Archaeogastropoda *sensu* Thiele (1925; Figure 11.4A). More recently, the Cocculinoidea have been placed as the sister taxon of Neritimorpha based on morphological data (Figure 11.4C, F) or as the sister taxon of the Vetigastropoda based on both morphological or molecular data (Figure 11.4D, F, H) or Patellogastropoda (Figure 11.4G).

Haszprunar (1987) united Cocculinoidea with Lepetelloidea in the taxon Cocculiniformia. The inclusion of these two groups in a single taxon was problematic because there were few characters, other than them being white, deep-water, rhipidoglossate limpets, that supported this action (see section on phylogeny below). Ponder and Lindberg (1996) suggested that the Cocculiniformia was diphyletic, and subsequent studies have mostly supported this position (see subsequent section on phylogeny and discussion in Geiger *et al.*, Chapter 12). Lindberg *et al.* (2004) treated the Cocculinida (i.e., Cocculinoidea) as a major group of orthogastropods, further distancing them from the vetigastropod lepetelloidians.

In the last 20 years there has been a proliferation of generic-rank taxa within the Cocculinoidea (e.g., *Coccocrater, Coccopigya, Macleaniella, Paracocculina*), but higher taxa have been relatively stable since the early 1900s, with a straight-forward division into the Cocculinidae and Bathysciadiidae (Haszprunar 1988a; Strong *et al.* 2003).

FOSSIL HISTORY

PATELLOGASTROPODA

There are numerous limpet taxa in the Paleozoic fossil record; however, none possess unique characters that convincingly place these taxa within Patellogastropoda. The Ordovician *Floripatella rousseaui* described by Yochelson (1988) has been argued by him to represent the earliest patellogastropod. However, the Y-shaped structure on the posterior margin of the steinkern appears to be an efferent mantle vessel, indicating that the living animal was untorted and therefore was not a gastropod but rather a cyclomyan monoplacophoran (Lindberg in Yochelson 1988). After reviewing muscle scar and protoconch data for Paleozoic limpets, Frýda et al., Chapter 10) conclude that it is currently not known whether crown group patellogastropods appeared in the Paleozoic fossil record.

Putative coiled patellogastropod ancestors have not yet been identified (Lindberg 1988a, 1998; Frýda et al., Chapter 10). Ponder and Lindberg (1997) noted that the only gastropods that share foliated shell microstructure are members of the extinct Paleozoic taxon Platyceratoidea and patellogastropods. Bandel (2000) argued, based on protoconch morphology, that late Paleozoic platyceratoids are more closely related to Neritimorpha. However, Sutton et al. (2006) have documented fossilized partial internal anatomy in a Silurian platycerid that they suggested has patellogastropod rather than neritimorph affinities.

The earliest patellogastropod limpet verified by shell microstructure is from the Triassic of Italy (Hedegaard et al. 1997), but it is in the late Cretaceous and Cenozoic periods that many of the higher crown taxa have their first occurrences in the fossil record (Apkan et al. 1982; Lindberg and Vermeij 1985; Kollman and Peel 1983; Lindberg 1988a; Lindberg and Marincovich 1988; Lindberg and Squires 1990; Kase and Shigeta 1996).

NERITIMORPHA

Neritimorph origination appears to be in the Late Paleozoic or Early Mesozoic. The earliest unambiguous occurrence of shells referable to the Neritidae and Neritopsidae is in the Triassic (Knight et al. 1960; Tracey et al. 1993). Earlier Paleozoic records (Silurian-Devonian) of neritimorphs (Frýda 2001; Frýda and Blodgett 2001) were based on supposed similarities of protoconch morphology that are now suspect; the first occurrence of undisputed neritimorph protoconchs is also in the Triassic (see Frýda et al. , Chapter 10). However, Kaim and Sztajner (2004), using shell and opercular evidence, have proposed an evolutionary scenario that includes

the Carboniferous *Naticopsis* in the neritimorph lineage. A Carboniferous or earlier origination of the group is also suggested by the first occurrences of the putative terrestrial neritimorph taxon—*Dawsonella* (Solem and Yochelson 1979), in the Upper Carboniferous of Illinois and Indiana (Kano et al. 2002).

Fossil Helicinidae and Hydrocenidae are known from the Cretaceous and Mesozoic, respectively (Kano et al. 2002). The occurrence of Hydrocenidae on all continents (except Antarctica) led Bandel (2000) to propose a Pangean origin for the group that would date their origin in the Jurassic. This timing is not falsified by the relationships of Helicinidae and Hydrocenidae, as presented in the molecular phylogeny of Kano et al. (2002). Based on the fossil record, the remaining major neritimorph taxa have Cenozoic originations. Both shallow- and deep-water Phenacolepadidae first appear in the Middle Eocene, and the marine and freshwater Neritiliidae also have their first occurrence in the Eocene (Kano et al. 2002). However, Kano et al. (2002) have argued that the neritiliids are likely older, but confused with Paleocene, and possibly Mesozoic, neritids.

COCCULINOIDEA

The earliest reported occurrence of Cocculinoidea is in the Cretaceous (lower Albian) of Madagascar (Kiel 2006). Surprisingly well-preserved specimens, some still associated with wood, have been described from the Cenozoic of New Zealand (Marshall 1985).

PHYLOGENY

PATELLOGASTROPODA

Most evolutionary and phylogenetic analyses place patellogastropods as the sister taxon to all other gastropods (Golikov and Starobogatov 1975; Haszprunar 1988b; Ponder and Lindberg 1997; McArthur and Harasewych 2003; Figure 11.4C–F, H). In a recent study (Giribet et al. 2006), using five gene fragments from both nuclear and mitochondrial genomes (18S rRNA, 28S rRNA, H3, COI, 16 S rRNA), found Gastropoda to be polyphyletic, with the patellogastropods as the sister of Solenogastres and nested within a clade of Cephalopoda and Scaphopoda. The remaining gastropod taxa (Orthogastropoda) were nested between higher bivalves and a clade composed of more basal bivalves + Polyplacophora and Monoplacophora. Although a sister relationship with the Solenogastres is surprising, the placement of the Patellogastropoda within a clade including the Cephalopoda and Scaphopoda has been suggested by others (Lindberg and Ponder 1996; Waller 1998; Haszprunar 2000; Steiner and Dreyer 2002). The exclusion of the remaining Gastropoda from this clade may indicate a significant genomic change between the Eogastropoda (patellogastropods + lineage(s) of extinct coiled gastropod ancestors) and the Orthogastropoda (see Ponder and Lindberg, Chapter 1).

Relationships within Patellogastropoda have been investigated using morphological characters by Ridgway et al. (1998; Patellidae only), Lindberg (1998), Sasaki (1998), and Guralnick and Smith (1999); molecular analyses include Koufopanou et al. (1999: 16S and 12S rRNA, Patellidae only), Harasewych and McArthur (2000: 18S rRNA), and Nakano and Ozawa (2004: 16S and 12S rRNA). While these analyses have not produced congruent results, there is some consensus, especially within character suites (Figure 11.5). All three morphological analyses recovered a Lottiidae + Acmaeidae clade (Figure 11.5A–C) and two of the analyses (Figure 11.5A, C) also recover a Patelloidea + Nacelloidea clade. Another difference between the three morphological trees is the placement of Lepetidae; they are sister to the Lottiidae + Acmaeidae clade in Lindberg (1998) and Guralnick and Smith (1999) analyses (Figure 11.5B, C), but sister to Lottiidae + Acmaeidae, and Patelloidea + Nacelloidea, in Sasaki's (1998) analysis (Figure 11.5A). Last, while some trees (Figure 11.5A, C) are relatively symmetrical, one is more asymmetric and comblike (Figure 11.5B). Nakano and Ozawa's (2004) molecular analysis (Figure 11.5E) is most similar to the morphological trees. Their

analysis recovers the Lottiidae + Acmaeidae and Patelloidea + Nacelloidea clades as do some morphological trees (Figure 11.5A, C), the major difference between them being the presence of a paraphyletic Lottiidae in the molecular tree where the Acmaeidae nest within the Lottiidae (Figure 11.5E). Although the molecular tree of Harasewych and McArthur (2000; Figure 11.5D) is like the morphological tree of Lindberg (1998; Figure 11.5B) in being asymmetric and comblike, none of the groupings are present in the other analyses.

Thus, most morphological and molecular analyses suggest that there are two distinct lineages of Patellogastropoda, the Lottiidae + Acmaeidae and the Patelloidea + Nacelloidea. These two taxa shared a common ancestor that was probably a coiled snail (Lindberg 1988a, 1998; Ponder and Lindberg 1997), although Haszprunar 1988b suggested that limpet-shaped shells were primary for gastropods based on the shell and muscle morphology of the outgroup Monoplacophora (see also Haszprunar, Chapter 5). However, a putative sister taxon has yet to be confidently identified (see Frýda et al., Chapter 10).

Recent reconsideration of patellogastropod ancestry (Lindberg 2004), spurred in part by the paraphyletic nature of the Cocculiniformia (see following discussion), has raised questions regarding the origins of patellogastropods, including whether the living taxa are paraphyletic (Figure 11.6D) and whether there was a single ancestral coiled lineage (Figure 11.6A) or multiple coiled lineages (Figure 11.6C) in the Paleozoic that gave rise to living patellogastropods.

To a large extend the supposed sister taxon relationship of the two major patellogastropod lineages (Lottiidae + Acmaeidae and Patelloidea + Nacelloidea) has been predetermined by their shared limpet shape. Many of the other characters that serve to unite these two groups are plesiomorphic (e.g., docoglossate radula, protoconch morphology, renopericardial configuration). In contrast, a substantial number of anatomical, molecular, and ecological characters require difficult explanations and evolutionary scenarios when they are forced to be derived from

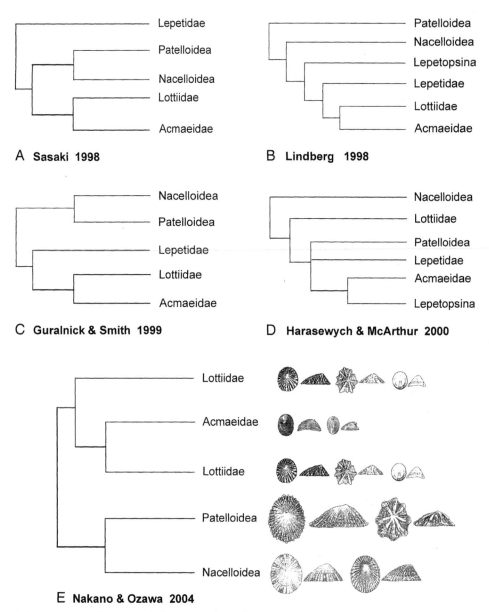

A **Sasaki 1998**

B **Lindberg 1998**

C **Guralnick & Smith 1999**

D **Harasewych & McArthur 2000**

E **Nakano & Ozawa 2004**

FIGURE 11.5. Trees from recent phylogenetic analyses of the Patellogastropoda. (A–C) Morphological analyses of Lindberg (1998), Sasaki (1998), and Guralnick and Smith (1999), respectively. (D–E) Molecular analyses of Harasewych and McArthur (2000) and Nakano and Ozawa (2004). Representative morphologies in each clade are presented next to Tree E and have been reproduced from Lindberg (1998) in *Mollusca—The Southern Synthesis* (Beesley *et al.* 1998).

a single shared common ancestor. For example, neural anatomy and nuchal cavity configuration of the Patelloidea + Nacelloidea suggests that the ancestral lineage was bilaterally symmetrical; however, in the Lottiidae + Acmaeidae both the nervous system and nuchal cavity configuration show the characteristic loss and reduction of structures on the post-torsional right side of the body—as seen in most asymmetrically coiled gastropod groups. Ecologically, Patelloidea + Nacelloidea taxa are restricted to intertidal and shallow subtidal habitats, while the Lottiidae + Acmaeidae taxa range from the upper reaches of the intertidal to depths in excess of 4,000 m.

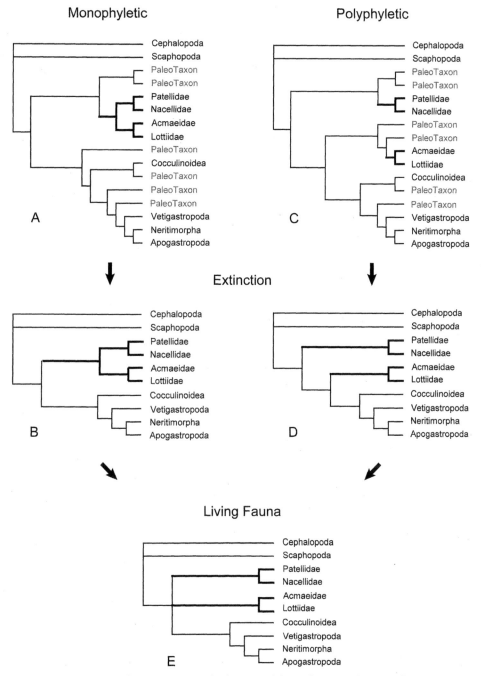

FIGURE 11.6. Current and alternative scenarios for the origin of the Patellogastropoda. (A) Patellogastropods are monophyletic, and (B) extinction of coiled sister paleo taxa consigns the living patellogastropods to the base of the living gastropod tree. (C) Alternative scenario in which the living "Patellogastropoda" have a polyphyletic origin, and (D) extinction of paleo taxa produces a grade of living patellogastropods. (E) Strict-consensus tree for the two scenarios (Figure 11.6B, D).

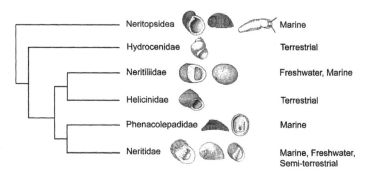

FIGURE 11.7. Tree from molecular phylogenetic (28 rRNA) analysis of the Neritimorpha (Kano *et al.* 2002). Representative morphologies and major associated biomes are presented for each taxon; morphological images from Scott and Kenny (1998) in *Mollusca—The Southern Synthesis* (Beesley *et al.* 1998), except for Neritiliidae by C. Huffard.

Lindberg (2004) suggested that the apomorphies seen in living patellogastropod clades may represent characters possessed by different lineages of coiled Paleozoic taxa that independently became limpet-like (Figure 11.6C) rather than subsequent divergence in the clades after the achievement of limpet morphology by a common ancestor (Figure 11.6A). With the extinction of these coiled brethren (Figure 11.6B, D), the two surviving lineages can easily be mistakenly grouped together based on plesiomorphic characters. For example, the strict-consensus tree of the relationships produced by these two scenarios (Figure 11.6E) may have been resolved by simply overweighting the limpet shell morphology as the primary synapomorphy of the taxon (Figure 11.1B–H)—a character that is well known to be homoplastic in gastropods (e.g., Graham 1985; Haszprunar 1988b; Ponder and Lindberg 1997; cf. Figure 11.1–3, 8). Limpet shell morphology also substantially contributed to the confusion surrounding the Cocculiniformia (see following discussion). Currently hypothesized relationships among these small white limpets (see following discussion) are more similar to the pattern resulting from a polyphyletic origin (Figure 11.6C) than to a monophyletic origin (Figure 11.6A). A similar situation may also apply to the Architaenioglossa, which can be argued as either a grade or a clade (see Ponder *et al.*, Chapter 13).

Plesiomorphic characters shared with coiled eogastropod ancestors would have included the docoglossate radula, pericardium-gonad-excretory organ configuration, foliated crossed-lamellar shell microstructure, protoconch morphology, and paired ctenidia. Subsequent divergence in the coiled ancestral lineages may have established different shell, radular, anatomical, and genomic apomorphies in these lineages prior to their giving rise to limpet-like taxa. Following the extinction of the coiled ancestral taxa, the surviving remnants of this early radiation would be the lineages of living patellogastropods.

NERITIMORPHA

Views on the phylogenetic position of Neritimorpha have been widely divergent. They have been considered to represent an early branch of gastropod evolution, grouping near the patellogastropods and cocculinoideans (Haszprunar 1988b; Figure 11.4D), the most basal taxon in the rhipidoglossate clade (Sasaki 1998), or as a transitional group between the traditional "archaeogastropod" taxa and the Caenogastropoda (Yonge 1947; Figure 11.4B). They have also been thought to be closely related to Cocculinoidea (Thiele 1903; Moskalev 1971; Golikov and Starobogatov 1975; Ponder and Lindberg 1997; also see previous discussion).

Ponder and Lindberg (1997) identified seven possible synapomorphies that united the Neritimorpha and Cocculinoidea as sister taxa—a divided shell muscle, absence of salivary glands, jaw structure, sublingual glands, enlarged fourth lateral teeth, a shift in the position of the anus from the posterior half to the anterior half of the pallial cavity, and absence of the right kidney. Sasaki (1998) and Sasaki *et al.* (2006a) have argued that these characters do not support this relationship, and new morphological data for the Cocculinoidea (Strong *et al.* 2003) further argues against a Cocculinoidea + Neritimorpha clade.

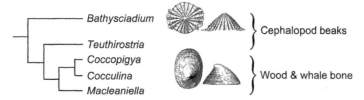

FIGURE 11.8. Tree from morphological phylogenetics analysis of the
Cocculinoidea (Strong *et al.* 2003). Representative morphologies and deep-
sea substrates used by each group are presented to the right of each taxon.
Bathyscadiid image from Haszprunar (1998) in *Mollusca—The Southern Synthesis*
(Beesley *et al.* 1998) and cocculinid image from Dall (1925).

Some recent morphological and molecular
analyses (Harasewych and McArthur 2000;
Colgan *et al.* 2003; Aktipis *et al.*, Chapter 9)
place the Neritimorpha as the sister taxon of the
Apogastropoda (Ponder and Lindberg 1997),
as suggested by Yonge (1947). Developmental
data also supports this relationship (van den
Biggelaar and Haszprunar 1996; Lindberg
and Guralnick 2003). In contrast, Colgan *et al.*
(2000; Figure 11.4G) resolved the Neritimorpha
as sister taxa to all other gastropods or nested
within the caenogastropods. The caenogastropod placement was also found in some of the
analyses of Colgan *et al.* (2003).

Phylogenetic relationships within the Neritimorpha have recently been investigated by
Kano *et al.* (2002), who identified six clades
based on 28S rRNA sequences—Neritopsidea
(including Neritopsidae and Titiscaniidae),
Hydrocenidae, Neritiliidae, Helicinidae, Neritidae, and Phenacolepadidae (Figure 11.7). The
topology of their tree suggests that terrestriality
has evolved three times (Neritidae, Hydrocenidae, and Helicinidae) and that the Neritiliidae
are a distinct clade not related to the Neritidae
as first suggested by Holthuris (1995).

COCCULINOIDEA

Placement of the Cocculinoidea on the gastropod tree remains problematic. After almost
70 years of unquestioned inclusion in the
Archaeogastropoda, new material and subsequent anatomical studies renewed interest in
the phylogenetic relationship of this disparate group to other gastropods. However, no
unambiguous placement has yet emerged,

and their position has fluctuated from being
the most basal group of the Orthogastropoda
(Haszprunar 1988b) to being the sister taxa of
the Neritimorpha (see previous discussion).

Recent molecular analyses have also failed
to agree on their placement. The analyses of
Colgan *et al.* (2000; Figure 11.4G) resolved the
Cocculinoidea either as the sister taxon of the
patellogastropods or at the base of the vetigastropods. Colgan *et al.* (2003) also resolved the
Cocculinoidea at the base of several vetigastropod taxa or alternatively within a clade that
included both caenogastropods and heterobranchs. McArthur and Harasewych (2003)
placed the Cocculinoidea with the hot-vent
clade Neomphalina (Figure 11.4H), Geiger and
Thacker (2005) place the Cocculinoidea at the
base of the Vetigastropoda, and the combined
analysis of Aktipis *et al.* (Chapter 9) place them
as the most basal group of gastropods.

Haszprunar (1987) united most deep-water
rhipidoglossate limpets in the taxon Cocculiniformia. However, the inclusion of the Cocculinoidea with Lepetelloidea in a single clade
was problematic from its inception because it
required difficult transformations in shell muscle, gill, gonopericardial, and alimentary characters (Haszprunar 1988b). Ponder and Lindberg
(1996, 1997) and Lindberg and Ponder (1996)
reviewed and analyzed these and other characters and concluded that Cocculiniformia were
paraphyletic—the lepetelloids were confidently
placed within the Vetigastropoda, while the cocculinoids were outside of the vetigastropods.
Recent molecular analyses (Colgan *et al.*
2000, 2003; McArthur and Harasewych 2003;

Figure 11.4G, H) have also confirmed the paraphyletic nature of the Cocculiformia suggesting continued debate regarding the validity of a single cocculiniform clade or its usage is unwarranted (see also Geiger *et al.*, Chapter 12).

Phylogenetic relationships within the Cocculinoidea have recently been investigated by Strong *et al.* (2003) using morphological characters (Figure 11.8). Their study upheld the division of the Cocculinoidea into Cocculinidae and Bathysciadiidae (Moskalev 1976) as well as the uniqueness of the genera *Teuthirostria* and *Fedikovella*.

ADAPTIVE RADIATIONS

PATELLOGASTROPODA

The patellogastropods are likely candidates to have produced adaptive radiations. As noted previously, they are found in diverse habitats from deep-sea hydrothermal vents to the highest reaches of the intertidal and brackish water. They occur as commensals on other invertebrate species and on marine plants, and their species diversity in some regions is well documented (Branch 1975; Ponder and Creese 1980; Lindberg 1988a; Sasaki and Okutani 1993; Lindberg 1998). However, while certain clades of patellogastropods have undergone extensive radiations within the last three million years (e.g., Lottiinae; Lindberg 1988a), others appear to have maintained morphological stasis for tens of millions of years. For example, a Tethyan clade comprised of *Patelloida* species, first suggested by Christiaens (1975) and later delimited morphologically and temporally by Lindberg and Vermeij (1985) and molecularly by Kirkendale and Meyer (2004), has a 99 million-year history in which shell morphology, habitat, and radular morphology (see Akpan *et al.* 1982) appear unchanged. In contrast, during this same period of time more than 50% of extant caenogastropods originated and radiated (Sepkoski 2002).

In the speciose lottiid groups there is still little evidence that adaptive radiations have taken place. For example, along the California coast there may be as many as 10–15 co-occurring

species, but there is little evidence of a radiation associated with the exploitation of new niches (Lindberg 2007a). The vast majority of these species occur in the assumed ancestral habitat—on exposed rock substrate, and although they are vertically segregated by tidal height, their diversity within each zone is comparable. The remaining species occur in more unique habitats such as commensals on algae, marine angiosperms and other gastropods. Multiple lineages have converged on these apomorphic habitats, but none have produced any subsequent radiations (Lindberg 2007a). Thus, while there is no question that there is a high diversity of limpets in a variety of diverse habitats, there is no evidence that this was produced by an adaptive radiation. Similar patterns are present in Australia (Ponder and Creese 1980; Lindberg 1998), South Africa (Lindberg 2007b), and Chile (Espoz *et al.* 2004).

More commonly, patellogastropod faunas and the lineages that produce them are aggregates of different clades that have accumulated in a region through geological time and have been shaped by the vagaries of colonization, origination, radiation, and extinction. Japan, Australia, and southern Africa have diverse patellogastropod faunas with co-occuring representatives of patellid, nacellid, and lottiid lineages. In Chile only nacellid and lottiid lineages are present; in Europe only lottiid and patellid taxa, and in the western United States and Caribbean only lottiid taxa, occur (Lindberg 1988a). However, nacellid taxa were present in the Eocene of the western United States (Lindberg and Hickman 1985), and patellid taxa occurred in the Caribbean during the Pliocene (Lindberg 2007b). Thus, higher-level patterns, like species-level patterns, appear to have been shaped by numerous biotic and physical factors, and interpretations of the evolutionary processes producing the current composition must be rigorously examined.

NERITIMORPHA

The neritimorphs appear similar to the patellogastropods in their habitat diversification history. In addition to the increased morphological diversity seen in the neritimorphs as compared to the living patellogastropods, there is also an increase in habitat diversity, with neritimorphs also successfully colonizing both freshwater and terrestrial biomes. However, as in patellogastropods, no clear-cut adaptive radiations appear to have occurred, with the possible exception of one of the terrestrial groups (see following discussion).

The plesiomorphic globose coiled shell is the common shell form in marine intertidal habitats, and there appears to be little modification to its basic morphology in species associated with seagrasses (*Smaragdia*) or in the freshwater taxa (*Neritina*, *Theodoxus*, Neritiliidae). Limpetlike forms have evolved in both freshwater (*Septaria*) and marine habitats (Phenacolepadidae), but neither morphological transformation nor the habitat-plus-morphological transition in *Septaria* has produced any notable subsequent radiation. In the Opisthobranchia the evolution of a shell-less, marine slug followed by subsequent key character transformations produced incredible radiations in several groups, including the Doridoidea (1,720 species) and Aeolidioidea (560 species; Wägele 2004; Wägele *et al.*, Chapter 14). In contrast, the evolution of the ancestor of the neritimorph slug taxon *Titiscania* has produced, at the most, two living species. In addition, lineage divergence in *Theodoxus* associated with post-Tethyan marine basin development in Europe and the Middle East has produced only six major lineages in this freshwater taxon despite repeated regional and local vicariance events in the late Neogene (Bunje and Lindberg 2007). Deep-sea hydrothermal vent and troglobite taxa also show low diversity although they have successfully invaded depauperate, albeit marginal, habitats. Last, the ancestors of two living taxa colonized the terrestrial biome—Helicinidae and Hydrocenidae.

The Helicinidae may represent an exception to the general pattern outlined herein because helicinids are often speciose in many tropical and subtropical environments (Richling 2004), especially on islands (e.g., Jamica, Rosenberg and Muratov 2006). Helicinids occupy a wide range of habitats and include arboreal species. Access to some of these niches, such as the leaf litter community,

is determined by body size, which can vary dramatically between faunas (Emberton 1995). In contrast, Hydrocenidae are not as diverse as Helicinidae. Thus, even the colonization of a new biome, which was likely achieved independently by the two taxa (Kano *et al.* 2002), appears to have had two very different diversification outcomes in the living terrestrial neritimorph taxa.

COCCULINOIDEA

Indications of an adaptive radiation in the Cocculinoidea have been largely dependent on whether they are included or excluded from the Cocculiniformia (see previous discussion). If included, the Cocculinoidea appear as outliers with a unique anatomy and as a second major radiation onto biogenic substrates—the other group being the putative sister taxon Lepetelloidea (Haszprunar 1988a). Recognition of the Cocculiniformia as a polyphyletic taxon recasts the Cocculinoidea as a relatively small clade with slight anatomical variation and little habitat diversification. They appear to be convergent with the Pseudococculinidae in their limpet-shaped shells and their occurrence on biogenic substrates rather than sharing a common ancestor that colonized these distinctive deep-sea habitats.

Thus, the pattern of diversification of the Cocculinoidea is similar to both Patellogastropoda and Neritimorpha, with low species and morphological diversity and minimal habitat diversity—cephalopod beaks and waterlogged wood, with a single species known from whale bone. Cephalopod beaks appear to be the first deep-sea substrate utilized by the living taxa (Figure 11.8). Temporally, beaks would also have been the first of the three habitats available in the deep sea to putative ancestors (beaks > wood > whale bone). In comparison, the Lepetelloidea occur on the same habitats as cocculinoideans plus five additional substrates, including polychaete tubes, algal and plant debris, crustacean carapaces, Chondrichthyes egg cases, and fish bone (Haszprunar 1988a, 1998).

Although low, the species and habitat diversity of living cocculinoideans does not negate the possibility that the original colonization of the deep sea by their common ancestor was an adaptive radiation, although its remnants are sparse and undistinguished today. The subsequent extinction of a putative coiled sister taxon makes resolution of this question difficult, if not impossible, and very similar to the situation now faced in the interpretation of the early evolutionary history of patellogastropods.

In summary, there is little evidence for adaptive radiations in the Patellogastropoda, Neritimorpha, or Cocculinoidea, with the single exception of the terrestrial Helicinidae. Because of their occurrence in deep time, it is possible that numerous adaptive radiations have occurred in the past, but the signals have been removed by subsequent extinctions and imperfections in the fossil record (Figure 11.6). The surviving lineages are either successful solutions to strong selection at these assumed extinction events, chance products of contingencies, or, more likely, the sum of both. It is interesting to note that all three of these taxa also show little modification to their feeding apparatus or mode—predominantly detritivores or herbivores. There are no active predators, suspension or deposit feeders, suctorial feeders, or ectoparasites on other animals in these groups (Purchon 1968; Ponder and Lindberg 1997), and this lack of feeding diversity may represent a potential constraint that has limited their niche options.

FUTURE RESEARCH

The Patellogastropoda, Cocculinoidea, and Neritimorpha appear to be intriguing remnants of early gastropod evolution. The Patellogastropoda, and perhaps the Cocculinoidea, represent some of the earliest gastropod lineages, while the Neritimorpha were probably the first molluscan lineage to move from the marine realm into freshwater and onto land.

One would think that the accessibility of patellogastropod and neritimorph taxa would have increased our basic knowledge of these groups. In some cases this has been true; we

know a great deal about patellogastropod biology and ecology, but the same cannot be said for the neritimorphs, and biological and ecological studies of the deep-sea cocculinoideans are virtually nonexistent. Moreover, even where members of these groups have been well studied, the studies were seldom done in a comparative way. Also, with the exception of detailed studies on the Patellidae of South Africa (Branch 1981 and references therein), there are few regions in which the basic ecological and life history data for most members of the group has been obtained, compared, and synthesized. In some areas, virtually no data exists despite the accessibility of taxa, such as developmental studies of marine and terrestrial neritimorph species. It is therefore ironic that one of the first developmental studies of a neritimorph species has been accomplished for a deep-water hydrocarbon seep taxon—*Bathynerita naticoidea* (Van Gaest, 2006).

In the molecular arena there is a need for more taxon sampling and multigene analyses. Again, for the patellogastropod and neritimorph taxa, this should not be difficult. But the deep-sea habitat of the Cocculinoidea makes this charge more challenging. The serendipitous snagging of deep-sea wood and other suitable substrates does not lend itself to systematic sampling or guarantee that the appropriate preservation of tissues for molecular studies will occur. Collaborative and creative partnerships with fishermen, other deep-sea researchers, or even a directed research effort should be explored.

Future research directions should also include the rigorous exploration of the Paleozoic gastropod record in search of the coiled sister taxa of patellogastropods and cocculinoideans. Moreover, in the patellogastropods, these searches must include the possibility of multiple lineages giving rise to the living taxa rather than a single precursor lineage. The extinction of the coiled sister taxa has left the living patellogastropods and cocculinoideans as the only surviving remnants of these earlier diversifications, and if this evolutionary scenario its correct, we face two tasks similar to, and as challenging as, trying to reconstruct the history of Caenogastropoda from only the living, limpet-like taxa (e.g., *Hipponix, Crepidula, Thyca, Concholepas*).

Future research programs will require greater integration of both paleontological and neontological approaches as argued by Hickman and Lindberg (1985) as well as the practice of more contemporary paleontological approaches (see Frýda *et al.*, Chapter 10). The discovery of future Lagerstätten, coupled with new approaches to research, will undoubtedly play an important role, as demonstrated by the Sutton *et al.* (2006) reconstruction of portions of a Silurian gastropod's anatomy. The complete nuclear genome of the patellogastropod *Lottia gigantea* will provide a new comparative molecular framework previously available only in model organisms such as *Drosophilia, Xenopus,* and *Homo,* and the addition of more molluscan genomes (*Aplysia, Octopus, Biomphalaria*) that are currently under way will provide the next generation of researchers with data sets that will undoubtedly shed light on the currently obscure deep history of these groups as well as the entire Gastropoda.

Like most other molluscan groups, the Patellogastropoda, Cocculinoidea, and Neritimorpha are works in progress. While previous studies have provided insights into their morphology and putative relationships, a new understanding of their place in gastropod evolution is just beginning to unfold. The emerging patterns and scenarios are much more complicated than the previous ones. For example, the patellogastropods may not be monophyletic; terrestriality in the neritimorphs likely evolved a minimum of three times; and not all white limpets on deep-sea biogenic substrates shared a common ancestor. The lineages that make up these taxa extend through both deep and shallow time, and today their living members may live side by side at a hydrothermal vent, on a waterlogged piece of wood, or in the intertidal zone. As a mosaic of lineages at multiple scales this pattern presents serious challenges to systematists, ecologists, paleontologists, and evolutionary biologists, who must appropriately frame their questions in their attempts to gain

an understanding of these diverse groups of gastropods.

ACKNOWLEDGMENTS

I am indebted to D. Geiger, W. F. Ponder, and T. Sasaki for their criticism and comments on earlier drafts of the manuscript. D. Geiger (Figure 11.2B), E. Guillot de Suduiraut (Figure 11.3A), A. Miller (Figure 11.2B), and T. Sasaki (Figure 11.3B) generously provided images and permission for their use. I am also grateful to P. Beesley and the Australian Biological Resource Study (ABRS) for permission to reuse images from *Mollusca—The Southern Synthesis* (Beesley *et al.* 1998). This is contribution No. 1963 from the University of California Museum of Paleontology.

This chapter is dedicated to the late Ellis Yochelson in appreciation of his scientific contributions, controversial and challenging ideas, appreciation of deep and shallow history, marvelous wit, and friendship.

REFERENCES

Anderson, A., Highman, T., and Wallace, R. 2001. The radiocarbon chronology of the Norfolk Island archaeological sites. *Records of the Australian Museum* Suppl. 27: 33–42.

Andrews, E. A. 1937. Certain reproductive organs in the Neritidae. *Journal of Morphology* 61: 525–563.

Annapurna, C., and Bhavanarayana, P. V. 1993. Some effects of crude oil in the marine gastropod *Nerita albicilla* (Linnaeus). *Geobios* 20: 132–135.

Akpan, E. B., Farrow, G. E., and Morris, N. 1982. Limpet grazing on Cretaceous algal-bored ammonites. *Palaeontology* 25: 361–367.

Bandel, K. 2000. The new family Cortinellidae (Gastropoda, Mollusca) connected to a review of the evolutionary history of the subclass Neritimorpha. *Neues Jahrbuch für Geologie und Palaontologie. Abhandlungen* 217: 111–129.

Beesley, P.L., Ross, G.J.B., and Wells, A. 1998. *Mollusca: The Southern Synthesis.* Vol. 5, *Fauna of Australia* (Part B). Melbourne: CSIRO Publishing.

Bernard, F. 1890. Recherches sur les organes palléaux des Gastéropodes Prosobranches. *Annales des Sciences Naturelles. Zoologie* (7)9: 89–404.

Bourne, G. C. 1908. Contributions to the morphology of the group Neritacea of aspidobranch gastropods, part I. The Neritidæ. *Proceedings of* the General Meetings for Scientific Business of the Zoological Society of London 1908: 810–887.

———. 1911. Contributions to the morphology of the group Neritacea of aspidobranch gastropods, part II. The Helicinidæ. *Proceedings of the General Meetings for Scientific Business of the Zoological Society of London* 1911: 759–809.

Branch, G. M. 1975. Mechanisms reducing intraspecific competition in *Patella* spp.—migration, differentiation and territorial behavior. *Journal of Animal Ecology* 44: 575–600.

———. 1981. The biology of limpets: physical factors, energy flow, and ecological interactions. *Oceanography and Marine Biology Annual Review* 19: 235–380.

Bunje, P. M. E., and Lindberg, D. R. 2007. Molecular phylogeny of a freshwater snail clade reveals lineage divergence associated with post-Tethyan marine basin development. *Molecular Phylogenetics and Evolution* 42: 373–387.

Carlton, J. T., Vermeij, G. J., Lindberg, D. R., Carlton D. A., and Dudley, E. C. 1991. The first historical extinction of a marine invertebrate in an ocean basin: the demise of the eelgrass limpet *Lottia alveus. Biological Bulletin* 180: 72–80.

Chernogorenko, M. I., Komarovova, T. I., and Kurandina, D. P. 1978. Life cycle of the trematode, *Plagioporus skrjabini* Kowal, 1951 (Allocreadiata, Opecoelidae)]. *Parazitologiia* 12: 479–86.

Christiaens, J. 1975. Revision provisoire des mollusques marins recents de la famille des Acmaeidae (seconde partie). *Informations de la Société Belge de Malacologie* 4: 49–116.

Colgan, D. J., Ponder, W. F., Beacham, E., and Macaranas, J. M. 2003. Gastropod phylogeny based on six segments from four genes representing coding or non-coding and mitochondrial or nuclear DNA. *Molluscan Research* 23: 123–148.

Colgan, D. J., Ponder, W. F., and Eggler, P. E. 2000. Gastropod evolutionary rates and phylogenetic relationships assessed using partial 28S rDNA and histone H3 sequences *Zoologica Scripta* 29: 29–63.

Crump, R. G., Morley, H. S., and Williams, A. D. 1999. West Angle Bay, a case study. Littoral monitoring of permanent quadrats before and after the Sea Empress oil spill. *Field Studies* 9: 497–511.

Cushman, J. H. 1989. Vertical size gradients and migratory patterns of two *Nerita* species in the northern Gulf of California, Mexico. *The Veliger* 32: 147–151.

Cuvier, G. 1817. *Le regne animal, distribute d'après son organization.* Vol. 2. Les Reptiles, les poissons, les Mollusques et les Annelids. Paris: Déterville 532 pp.

Dall, W. H. 1882. On certain limpets and chitons from the deep waters off the eastern coast of the United States. *Proceedings of the United States National Museum* 4: 400–414.

———. 1925. Illustrations of unfigured types of shells in the collection of the United States National Museum. *Proceedings of the United States National Museum* 66(17): 1–41.

Donoghue, M. J., Doyle, J. A., Gauthier, J., Kluge, A. G., and Rowe T. 1989. The importance of fossils in phylogeny reconstruction. *Annual Review of Ecology and Systematics* 20: 431–60.

Emberton, K. C. 1995. Land-snail community morphologies of the highest-diversity sites of Madagascar, North America, and New Zealand, with recommended alternatives to height-diameter plots. *Malacologia* 36: 43–66.

Espoz, C., Lindberg D. R., Simison, W. B., and Castilla J. C. 2004. Los patelogastrópodos intermareales endémicos a la costa rocosa de Perú y Chile. *Revista Chilena de Historia Natural* 77: 257–283.

Ferraz, R. R., Menezes, G. M., and Santos, R. S. 2001. Limpet (*Patella* spp.) (Mollusca: Gastropoda) exploitation in the Azores, during the period 1993–1998. *Arquipélago Life and Marine Sciences* Suppl. 2 Part B: 59–65.

Foster, P., and Cravo, A. 2003. Minor elements and trace metals in the shell of marine gastropods from a shore in tropical East Africa. *Water Air and Soil Pollution* 145: 53–65.

Fretter, V. 1984. The functional anatomy of the neritacean limpet *Phenacolepas omanensis* Biggs and some comparison with *Septaria. Journal of Molluscan Studies* 50: 8–18.

Frýda, J. 2001. Discovery of the larval shell in Middle Paleozoic subulitoidean gastropods with description of two new species from the early Devonian of Bohemia. *Bulletin of the Czech Geological Survey* 76: 29–37.

Frýda, J., and Blodgett, R. B. 2001. The oldest known heterobranch gastropod, *Kuskokwimia* gen. nov., from the Early Devonian of west-central Alaska, with notes on the early phylogeny of higher gastropods. *Bulletin of the Czech Geological Survey* 76: 39–53.

Fuchigami, T., and Sasaki, T. 2005. The shell structure of the Recent Patellogastropoda (Mollusca: Gastropoda) *Palaeontological Research* 9: 143–168.

Geiger, D. L., and Thacker, C. E. 2005. Molecular phylogeny of Vetigastropoda reveals non-monophyletic Scissurellidae, Trochoidea, and Fissurelloidea. *Molluscan Research* 25: 47–55.

Giribet, G., Okusu, A., Lindgren, A. R., Huff, S. W., and Schrödl, M. 2006. Evidence for a clade composed of molluscs with serially repeated structures: monoplacophorans are related to chitons. *Proceedings of the National Academy of Sciences of the United States of America* 103: 7723–7728.

Golikov, A. N., and Kussakin, O. G. 1972. Sur la biologie de la réproduction des patelles de la famille Tecturidae (Gastropoda: Docoglossa) et sur la position systématique des ses subdivisions. *Malacologia* 11: 287–294.

Golikov, A. N., and Starobogatov, Y. I. 1975. Systematics of prosobranch gastropods. *Malacologia* 15: 185–232.

Gortz, P. 1998. Effects of stream restoration on the macroinvertebrate community in the River Esrom, Denmark. *Aquatic Conservation* 8: 115–130.

Graham, A. 1985. Evolution within the Gastropoda: Prosobranchia. In *The Mollusca*. Vol. 10, *Evolution*. Edited by E. R. Trueman and M. R. Clark. New York: Academic Press, pp. 151–186.

Guralnick, R. P., and Smith, K. 1999. Historical and biomechanical analysis of integration and dissociation in molluscan feeding, with special emphasis on the true limpets (Patellogastropoda: Gastropoda). *Journal of Morphology* 241: 175–195.

Harasewych, M. G., and McArthur, A. G. 2000. A molecular phylogeny of the Patellogastropoda (Mollusca: Gastropoda). *Marine Biology* 137: 183–194.

Haszprunar, G. 1987. Anatomy and affinities of cocculinid limpets (Mollusca, Archaeogastropoda). *Zoologica Scripta* 16: 305–324.

———. 1988a. Comparative anatomy of cocculiniform gastropods and its bearing on archaeogastropod systematics. In *Prosobranch Phylogeny*. Edited by W. F. Ponder. *Malacological Review* Suppl. 4: 64–84.

———. 1988b. On the orgin and evolution of major gastropod groups, with special reference to the Streptoneura (Mollusca). *Journal of Molluscan Studies* 54: 367–441.

———. 1998. Superorder Cocculiniformia. In *Mollusca: The Southern Synthesis*. Vol. 5, *Fauna of Australia* (Part B). Edited by P. L Beesley, G. J. B. Ross, and A. Wells. Melbourne: CSIRO Publishing, pp. 653–664.

———. 2000. Is the Aplacophora monophyletic? A cladistic point of view. *American Malacological Bulletin* 15: 115–130.

Haynes, A. 2005. An evaluation of members of the genera *Clithon* Montfort, 1810 and *Neritina* Lamarck 1816 (Gastropoda: Neritidae). *Molluscan Research* 25: 75–84.

Healy, J. M. 1988. Sperm morphology and its systematic importance in the Gastropoda. In *Prosobranch Phylogeny*. Edited by W. F. Ponder. *Malacological Review*, Suppl. 4: 251–266.

Hedegaard, C. 1990. Shell structures of the Recent Archaeogastropoda. Thesis, Department of Ecology and Genetics, University of Aarhus, Denmark. Vol. 1, 154 pp, Vol. 2 Atlas, 78 pls.

———. 1997. Shell structures of the recent Vetigastropoda. *Journal of Molluscan Studies* 63: 369–377.

Hedegaard, C., Lindberg D. R., and Bandel, K. 1997. A patellogastropod limpet from the Triassic St. Cassian Formation of Italy. *Lethaia* 30: 331–335.

Hickman, C. S., and Lindberg, D. R. 1985. Perspectives on molluscan phylogeny. In *Mollusks. Notes for a Short Course*. Edited by T. W. Broadhead. Knoxville: University of Tennessee, Department of Geological Science, pp. 13–16

Hockey, P. A. R., and Bosman, A. L. 1986. Man as an intertidal predator in Transkei: disturbance, community convergence, and management of a natural food resource. *Oikos* 46: 3–14.

Holthuis, B. V. 1995 Evolution between marine and freshwater habitats: a case study of the gastropod Neritopsina. Ph.D. dissertation, University of Washington, Seattle.

IUCN. 2006. *2006 IUCN Red List of Threatened Species*. <http://www.iucnredlist.org>.

Kaim, A., and Sztajner, P. 2004. The opercula of neritopsid gastropods and their phylogenetics importance. *Journal of Molluscan Studies* 71: 211–219.

Kano, Y., Chiba S., and Kase, T. 2002. Major adaptive radiation in neritopsine gastropods estimated from 28S rRNA sequences and fossil records. *Proceedings of the Royal Society of London. Series B, Biological Sciences* 269: 2457–2465.

Kano, Y., and Kase, T. 2002. Anatomy and systematics of the submarine-cave gastropod *Pisulina* (Neritopsina: Neritiliidae). *Journal of Molluscan Studies* 68: 365–384.

Kase, T., and Shigeta, Y. 1996. New Species of Patellogastropoda (Mollusca) from the Cretaceous of Hokkaido, Japan and Sakhalin, Russia. *Journal of Paleontology* 70: 762–771.

Kay, E. A., Corpus G. C., and Magruder, W. H. 1982. *Opihi: Their Biology and Culture*. Aquaculture Development Program, State of Hawaii, Honolulu: Department of Land and Natural Resources.

Keough, M. J., Quinn, G. P., and King, X. 1993. A correlation between human collecting and intertidal mollusc populations on rocky shores. *Conservation Biology* 7: 378–390.

Kiel, S. 2006. New and little-known gastropods from the Albian of the Mahajanga Basin, northwestern Madagascar. *Journal of Paleontology* 80: 455–476.

Kirch, P. V., and O'Day, S. J. 2003. New archaeological insights into food and status: a case study from pre-contact Hawaii. *World Archaeology* 34: 484–497.

Kirkendale, L. A., and Meyer, C. P. 2004. Phylogeography of the *Patelloida profunda* group (Gastropoda: Lottiidae): diversification in a dispersal-driven marine system. *Molecular Ecology* 13: 2749–2762.

Knight, J. B., Cox, L. R., Keen, A. M., Batten, R. L., Yochelson, E. L., and Robertson, R. 1960. Systematic descriptions. In *Treatise on Invertebrate Paleontology*. Part I, *Mollusca 1*. Edited by R.C. Moore. Lawrence, KS: Ecological Society of America and University of Kansas Press, pp. I169–I310

Kollman, H.A., and Peel, J.S. 1983. Paleocene gastropods from Nûgssuaq, West Greenland. *Grønlands Geologiske Undersøgelse Bulletin* 146: 1–115.

Koufopanou V., Reid, D.G., Ridgway, S.A., and Thomas, R.H. 1999. A molecular phylogeny of the patellid limpets (Gastropoda: Patellidae) and its implications for the origins of their antitropical distribution. *Molecular Phylogenetics and Evolution* 11: 138–156.

Kyle, R., Pearson, B., Fielding, P.J., Robertson, W.D., and Birnie, S.L. 1997. Subsistence shellfish harvesting in the Maputal and marine reserve in northern KwaZulu-Natal, South Africa: rocky shore organisms. *Biological Conservation* 82: 183–192.

Lankester, E.R. 1883. Mollusca. In *Encyclopedia Britannica*, 9th ed. Edinburgh: A & C Black, pp. 95–158.

Lemche, H., and Wingstrand, K.G. 1959. The anatomy of *Neopilina galatheae* Lemche, 1957. *Galathea Reports* 3: 9–71.

Lill, K. 1993. On surviving populations of *Theodoxus fluviatilis* (Linnaeus 1758) (Gastropoda: Prosobranchia: Neritidae) from the middle course of the Weser River near Nieburg: with a water quality assessment based on physicochemical parameters of water quality. *Mitteilungen der Deutschen Malakozoologischen Gesellschaft* 50/51: 41–48.

Lindberg, D.R. 1981. Rhodopetalinae, a new subfamily of Acmaeidae from the boreal Pacific: anatomy and systematics. *Malacologia* 20: 291–305.

———. 1983. Anatomy, systematics, and evolution of brooding Acmaeid limpets. Ph.D. dissertation, Biology, University of California, Santa Cruz.

———. 1984. A Recent specimen of *Collisella edmitchelli* from San Pedro, California (Mollusca: Acmaeidae). *Bulletin of the Southern California Academy of Sciences* 83: 148–151.

———. 1985. Aplacophorans, monoplacophorans, polyplacophorans and scaphopods: the lesser classes. In *Mollusks. Notes for a Short Course*. Edited by T.W. Broadhead. Knoxville: University of Tennessee, Department of Geological Science, pp. 230–247.

———. 1988a. The Patellogastropoda. In *Prosobranch Phylogeny*. Edited by W.F. Ponder. *Malacological Review* Suppl. 4: 35–63.

———. 1988b. Gastropods: The neontological view. In *Heterochrony in Evolution: An Interdisciplinary Approach*. Edited by M. McKinney. New York: Plenum Press, pp. 197–216.

———. 1990. The systematics of *Potamacmaea fluvatilis* Blandford: a brackishwater patellogastropod (Patelloidinae: Lottiidae). *Journal of Molluscan Studies* 56: 309–316.

———. 1998. Order Patellogasropoda. In *Mollusca: The Southern Synthesis*. Vol. 5, *Fauna of Australia* (Part B). Edited by P. L Beesley, G.J.B. Ross, and A. Wells. Melbourne: CSIRO Publishing, pp. 639–652.

———. 2004. Are the living patellogastropods sister taxa? *Molluscan Megadiversity: Sea, Land, and Freshwater*. World Congress of Malacology, Edited by F.E. Wells. Perth: Western Australian Museum. 86.

———. 2007a. Patellogastropoda. In *Light and Smith's Manual. Intertidal Invertebrates of the Central California Coast*. Edited by J.T. Carlton. Berkeley: University of California Press, pp. 753–761.

———. 2007b. Reproduction, ecology and evolution of the Indo-Pacific limpet *Scutellastra flexuosa* (Quoy and Gaimard, 1834). *Bulletin of Marine Science* 81: 219–234.

Lindberg, D.R., Estes, J.A., and Warheit, K.I. 1998. Human influences on trophic cascades along rocky shores. *Ecological Applications* 8: 880–890.

Lindberg, D.R., and Guralnick, R.P. 2003. Phyletic patterns of early development in gastropod mollusks. *Evolution and Development* 5: 494–507.

Lindberg, D.R., and Hickman, C.S. 1985. A new anomalous giant limpet from the Oregon Eocene (Mollusca: Patellida). *Journal of Paleontology* 60: 661–668.

Lindberg, D.R., and Marincovich, L., Jr. 1988. New species of limpets from the Neogene of Alaska. *Arctic* 41: 167–172.

Lindberg, D.R., and Pearse, J.S. 1990. Experimental manipulations of shell color and morphology of the limpets *Lottia asmi* and *Lottia digitalis* (Mollusca: Patellogastropoda). *Journal of Experimental Marine Biology and Ecology* 140: 173–185.

Lindberg, D.R., and Ponder W.F. 1996. An evolutionary tree for the Mollusca: branches or roots? In *Origin and Evolutionary Radiation of the Mollusca*. Edited by J. Taylor. Oxford: Oxford University Press, pp. 67–75

Lindberg, D.R., and Squires, R.1990. Patellogastropods (Mollusca) from the Eocene Tejon Formation of southern California. *Journal of Paleontology* 64: 578–587.

Lindberg, D.R., and Vermeij, G.J. 1985. *Patelloida chamorrorum* spec. nov.: a new member of the Tethyan *Patelloida profunda* group (Gastropoda: Acmaeidae). *The Veliger* 27: 411–417.

Little, C. 1990. *The terrestrial invasion. An ecophysiological approach to the origin of land animals*. Cambridge, UK: Cambridge University Press.

MacClintock, C. 1967. Shell structure of patelloid and bellerophontoid gastropods (Mollusca). *Peabody Museum, Natural History Bulletin* 22: 1–140.

Marshall, B. A. 1985. Recent and Tertiary Cocculinidae and Pseudococculinidae (Mollusca: Gastropoda) from New Zealand and New South Wales. *New Zealand Journal of Zoology* 12: 505–546.

McArthur, A. G., and Harasewych, M. G. 2003. Molecular systematics of the major lineages of the Gastropoda. In *Molecular Systematics and Phylogeography of Mollusks*. Edited by C. Lydeard and D. R. Lindberg. Washington, DC: Smithsonian Institution Press, pp. 140–160.

McLean, J. H. 1990. Neolepetopsidae, a new docoglossate limpet family from hydrothermal vents and its relevance to patellogastropod evolution. *Journal of Zoology* 222: 485–528.

———. 1992. Cocculiniform limpets (Cocculinidae and Pyropeltidae) living on whale bone in the deep sea off California. *Journal of Molluscan Studies* 58: 401–414.

Morton, J. E., and Yonge, C. M. 1964. Classification and structure of the Mollusca. In *Physiology of the Mollusca*, Vol. 1. Edited by K. M. Wilbur and C. M. Yonge. New York: Academic Press, pp. 1–58.

Moskalev, L. I. 1971. New data on the systematic position of the gastropod molluscs of the order Cocculinida Thiele 1908. In *Molluscs: Ways, Methods, and Results of their Investigation*. Abstracts of Reports of the 4th Conference on the Investigations of Molluscs. Leningrad: Academy of Sciences of the USSR (Nauka), pp. 59–60

———. 1976. Concerning the generic diagnosis of the Cocculinidae (Gastropoda, Prosobranchia). *Trudy P. P. Shirshov Institute of Okeanology* 99: 59–70.

Nakano, T., and Ozawa, T. 2004. Phylogeny and historical biogeography of limpets of the order Patellogastropoda based on mitochondrial DNA Sequences. *Journal of Molluscan Studies* 70: 31–41.

Navrot, J., Amiel, A. J., and Kronfeld, J. 1974. *Patella vulgata*: A biological monitor of coastal metal pollution—a preliminary study. *Environmental Pollution* 7: 303–308.

Pavlichenko, V. I. 1977. Natural enemies of gnats *Diptera simuliidae*. *Biologicheskie Nauki* (Moscow) 20: 44–46.

Ponder, W. F. 1998. Superorder Neritopsina. In *Mollusca: The Southern Synthesis*. Vol. 5, *Fauna of Australia* (Part B). Edited by P. L Beesley., G. J. B. Ross, and A. Wells. Melbourne: CSIRO Publishing, pp. 693–694.

Ponder, W. F., and. Creese, R. G. 1980. A revision of the Australian species of *Notoacmea*, *Collisella* and *Patelloida* (Mollusca: Gastropoda: Acmaeidae).

Journal of the Malacological Society of Australia 4: 167–208.

Ponder, W. F., and Lindberg, D. R. 1996. Gastropod phylogeny: challenges for the 90's. In *Origin and Evolutionary Radiation of the Mollusca*. Edited by J. Taylor. Oxford, UK: Oxford University Press, pp. 135–154

———. 1997. Towards a phylogeny of gastropod molluscs—a preliminary anaylsis using morphological characters. *Zoological Journal of the Linnean Society* 119: 83–265.

Powell, A. W. B. 1973. The patellid limpets of the world (Patellidae). *Indo-Pacific Mollusca* 35: 75–206.

Praveenkumar, K., and Devi, U. 1998. Temperature related metabolism in two intertidal, snails *Nerita albicilla* and *Nerita chamaeleon* (Gastropoda/Heritaceae) of Visakhapatnam coast. *Indian Journal of Marine Sciences* 27: 197–200.

Purchon, R. D. 1968. *The Biology of the Mollusca*. London: Permagon Press.

Richling, I. 2004. Classification of the Helicinidae: review of morphological characteristics based on a revision of the Costa Rican species and application to the arrangement of the Central American mainland taxa (Mollusca: Gastropoda: Neritopsina). *Malacologia* 45: 195–440.

Ridgway, S. A., Reid, D. G., Taylor, J. D., Branch, G. M., and Hodgson, A. N. 1998. A cladistic phylogeny of the family Patellidae (Mollusca: Gastropoda). *Philosophical Transactions of the Royal Society of London. Series B, Biological Sciences* 353: 1645–1671.

Rosenberg, G., and Muratov, I. V. 2006. Status report on the terrestrial Mollusca of Jamaica. *Proceedings of the Academy of Natural Sciences of Philadelphia* 155: 117–161.

Roy, K., Collins, A. G., Becker, B. J., Begovic, E., and Engle, J. M. 2003. Anthropogenic impacts and historical decline in body size of rocky intertidal gastropods in southern California. *Ecology Letters* 6: 205–211.

Ruwa, R. K., and Jaccarini, V. 1988. Nocturnal feeding migrations of *Nerita plicata*, *Nerita undata* and *Nerita textilis* (Prosobranchia, Neritacea) on the rocky shores at Mkomani Mombasa, Kenya. *Marine Biology* 99: 229–234.

Salvini-Plawen, L. v., and Steiner, G. 1996. Synapomorphies and plesiomorphies in higher classification of Mollusca. In *Origin and Evolutionary Radiation of the Mollusca*. Edited by J. Taylor. Oxford: Oxford University Press, pp. 29–51.

Sars, G. O. 1878. *Mollusca regionis arcticae Norvegiae*. Cristiana. 466 pp.

Sasaki, T. 1998. Comparative anatomy and phylogeny of the Recent Archaeogastropoda. *The University of Tokyo, Bulletin* 38: 1–224.

Sasaki, T., and Okutani, T. 1993. New genus *Nipponacmea* (Gastropoda, Lottiidae): a revision of Japanese limpets hitherto allocated in *Notoacmea*. *Venus* 52: 1–40.

Sasaki, T., Okutani, T., and Fujikura, K. 2003. New taxa and new records of patelliform gastropods associated with chemoautosynthesis-based communities in Japanese waters (Mollusca: Gastropoda). *The Veliger* 46: 189–210.

———. 2005. Molluscs from hydrothermal vents and cold seeps in Japan: a review of taxa recorded in twenty recent years (1984–2004). *Venus* 64: 87–133.

———. 2006a. Anatomy of *Shinkailepas myojinensis* Sasaki, Okutani & Fujikura, 2003 (Gastropoda: Neritopsina). *Malacologia* 48: 1–26.

———. 2006b. Anatomy of *Bathyacmaea secunda* Okutani, Fujikura and Sasaki, 1993 (Patellogastropoda: Acmaeidae). *Journal of Molluscan Studies* 72: 295–309.

Scott, B. J., and Kenny, R. 1998. Superfamily Neritoidea. In *Mollusca: The Southern Synthesis*. Vol. 5, *Fauna of Australia* (Part B). Edited by P. L Beesley, G. J. B. Ross, and A. Wells. Melbourne: CSIRO Publishing, pp. 694–702.

Sepkoski, J. J., Jr., 2002. A compendium of fossil marine animal genera. *Bulletins of American Paleontology* 363: 1–560.

Solem, A., and Yochelson, E. L. 1979. North American Paleozoic land snails, with a summary of other Paleozoic nonmarine snails. *United States Geological Survey Professional Paper* 1072: 1–42.

Sorensen, F., and. Lindberg, D. R. 1991. Preferential predation by American Black Oystercatcher on transitional ecophenotypes of the limpet *Lottia pelta* (Rathke). *Journal of Experimental Marine Biology and Ecology* 154: 123–136.

Speed, E. 1969. Prehistoric shell collectors. *South African Archaeological Bulletin* 24: 193–196.

Steiner, G., and Dreyer, H. 2002. Cephalopoda and Scaphopoda are sister taxa: an evolutionary scenario. *Zoology* 105 Suppl. 5: 95.

Strong, E. E., Harasewych, M. G., and Haszprunar, G. 2003. Phylogeny of the Cocculinoidea (Mollusca, Gastropoda). *Invertebrate Zoology* 122: 114–125.

Sutton, M. D., Briggs, D. E. G., Siveter, D. J., and Siveter, D. J. 2006. Fossilized soft tissues in a Silurian platyceratid gastropod. *Proceedings of the Royal Society of London. Series B, Biological Sciences* 273: 1039–1044.

Thiele, J. 1903. Die Anatomie und systematische Stellung der Gattung *Cocculina*. *Wissenschaftliche Ergebnisse der Deutschen Tiefsee-Expedition auf dem Dampfer "Valdivia"* 7: 147–179.

———. 1925. Gastropoda In *Handbuch der Zoologie* 5. Edited by W. Kükenthal and T. Krumbach. Berlin: de Gruyter, pp. 38–15

———. 1935. Handbuch der systematischen Weichtierkunde. Teil 4. Jena: G. Fischer.

Tracey, S., Todd, J. A., and Erwin, D. H. 1993. Mollusca: Gastropoda. In *The Fossil Record 2*. Edited by M. J. Benton. London: Chapman and Hall, pp. 131–168

Troschel, F. H. 1866–93. *Das Gebiss der Schnecken, zur Begründung einer natürlichen Classification*. Berlin: Nicolaische Verlagsbuchhandlung.

Underwood, A. J. 1979. The ecology of intertidal gastropods. *Advances in Marine Biology* 16: 110–210.

———. 2004. Landing on one's foot: small-scale topographic features of habitat and the dispersion of juvenile intertidal gastropods. *Marine Ecology Progress Series* 268: 173–182.

Van Gaest, A. L. 2006. Ecology and early life history of *Bathynerita naticoidea*: evidence for long-distance larval dispersal of a cold seep gastropod. M. S. Thesis, University of Oregon.

van den Biggelaar, J. A. M., and Haszprunar, G. 1996. Cleavage patterns and mesentoblast formation in the Gastropoda: an evolutionary perspective. *Evolution* 50: 1520–1540.

Wägele, H. 2004. Potential key characters in Opisthobranchia (Gastropoda, Mollusca) enhancing adaptive radiation. *Organisms Diversity and Evolution* 4: 175–188.

Waller, T. R. 1998. Origin of the molluscan class Bivalvia and a phylogeny of the major groups. In *Bivalves: An Eon of Evolution*. Edited by P. A. Johnston and J. W. Haggard. Alberta: University of Calgary Press, pp. 1–45

Wiens, J. J. 2005. Can incomplete taxa rescue phylogenetic analyses from long-branch attraction? *Systematic Biology* 54: 731–742.

———. 2006. Missing data and the design of phylogenetic analyses. *Journal of Biomedical Informatics* 39: 34–42.

Yochelson, E. L. 1988. A new genus of Patellacea (Gastropoda) from the Middle Ordovician of Utah: the oldest known example of the superfamily. *New Mexico Bureau of Mine and Mineral Resources Memior* 44: 195–200.

Yonge, C. M. 1947. The pallial organs in the aspidobranch Gastropoda and their evolution throughout the Mollusca. *Philosophical Transactions of the Royal Society B* 232: 443–518.

12

Vetigastropoda

Daniel L. Geiger, Alexander Nützel, and Takenori Sasaki

Vetigastropoda, introduced by Salvini-Plawen (1980), is a clade of basal marine snails comprising abalone (Haliotidae: Figures 12.1B, S), keyhole limpets (Fissurellidae: Figures 12.1C, D, Q, R), top and turban snails (Trochoidea: Figures 12.1L–P, T), slit shells (Pleurotomariidae: Figure 12.1I), little slit shells (Scissurellidae *sensu lato*: Figure 12.1J), seguenziids (Seguenziidae: Figure 12.1K), and a number of hydrothermal vent taxa (Lepetodrilidae, Figure 12.1E; Neomphalidae, Figure 12.1F; Peltospiridae, Figure 12.1G) and extinct lineages (e.g., Anomphaloidea, Cirroidea, Euomphaloidea). There are approximately 3,700 described living species (Table 12.1), which are all exclusively marine and are found from the intertidal to the deep sea, including hydrothermal vents, cold seeps, and wood falls. It is an old lineage dating back to the Paleozoic, with many basal characters, including paired organs in the mantle cavity. Members of the clade have served as model organisms in intertidal ecology, larval biology, and biomineralization. It also contains economically important species (e.g., edible abalone and the California giant keyhole limpet *Megathura crenulata*, Figure 12.1R, the source of a cancer drug).

Vetigastropoda is part of Thiele's (1931) Archaeogastropoda; the latter grade additionally contained a variety of basal groups, including Patellogastropoda, Neritimorpha, and Cocculiniformia, that are now recognized as distinct clades (see Bieler 1992; Ponder and Lindberg 1997; Aktipis *et al.*, Chapter 9; Lindberg, Chapter 11 for review of historical developments). In current classification schemes, the archaeogastropod grade is subdivided into several clades, the two largest of which are Patellogastropoda (see Lindberg, Chapter 11) and Vetigastropoda. The latter contains rhipidoglossate taxa having an epipodium with epipodial sense organs and gills with sensory structures known as bursicles (Salvini-Plawen and Haszprunar 1987; Ponder 1998; Figures 12.2A–F). Nondiagnostic characters often associated with Vetigastropoda include a slit or hole(s) in the shell (Figures 12.1B–D, I–K, R, S), paired gills, a heart with two auricles, and a nacreous inner shell layer (Figures 12.1I, 12.2J–K, 12.3H, 12.4I). The name Zeugobranchia has been used for those taxa with two gills and Diotocardia for those with two auricles. Vetigastropoda is an enormously diverse group, containing limpet, calyptraeiform, neritiform, biconical, and trochiform species

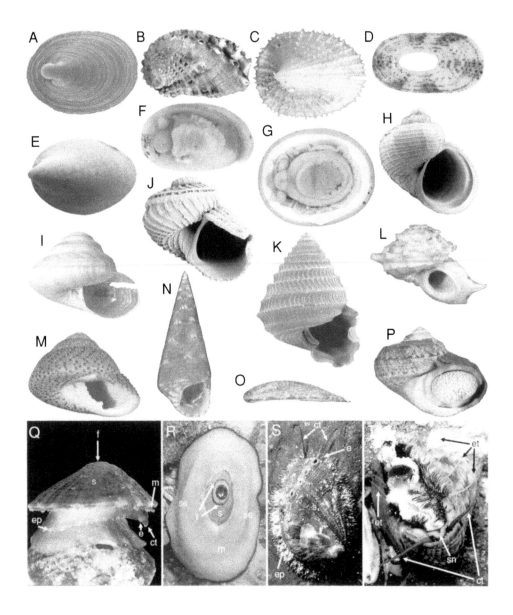

(Figures 12.1, 12.4, 12.5); however, high-spired shells similar to those of many caenogastropods (e.g., *Turritella* or *Terebra*) are rarely found among Recent or fossil (e.g., Figure 12.5T) Vetigastropoda; the most high-spired living taxa include the phasianellid *Phasianotrochus* (Figure 12.1N) and the trochid *Botelloides*, which are approximately twice as high as wide.

A brief summary of the major groups within Vetigastropoda is given in Table 12.1 including their key characters. Taxa at the family or superfamily rank are used to delineate cohesive lineages or groups; membership of some superfamilies (e.g., Pleurotomarioidea, with or without Haliotidae and Scissurellidae) has been controversial. The presence/absence of the bursicles as a key character is considered in detail in the subsequent discussion and in Table 12.2.

When Vetigastropoda first became recognized as a distinct clade of "archaeogastropods," membership was restricted to Fissurelloidea, Pleurotomarioidea (Pleurotomariidae, Haliotidae, Scissurellidae), and Trochoidea (Salvini-Plawen 1980; Salvini-Plawen and Haszprunar 1987).

FIGURE 12.1. (Opposite.) Diversity of vetigastropods. Size is indicated as maximum dimension. (A) *Cocculina* cf. *surugensis* ("cocculiniform"), 1 mm, a limpet-shaped group often found on sunken wood; possibly part of Vetigastropoda. (B) *Haliotis jacnensis* (Haliotidae), 28 mm. The slit of this abalone, commonly found in shallow tropical and temperate waters, has been subdivided into a series of holes. (C) *Emarginula striatula* (Fissurellidae), 15 mm, an exemplar of a fissurellid with anterior slit. (D) *Fissurellidea bimaculata* (Fissurellidae), 13 mm, an example of a fissurellid with an apical foramen. (E) *Lepetodrilus elevatus* (Lepetodrilidae), 3.9 mm, one of the limpet-shaped vetigastropods from a hot vent chemosynthetic habitat. Dorsal view of shell. (F) *Peltospira delicata* (Peltospiridae), 12 mm, a vetigastropod with rapidly expanding coiling from hot vent habitats. Ventral view with animal. Note epipodial tentacles at posterior end towards right side. (G) *Rhynchopelta concentrica* (Neomphalidae), 13 mm, another limpet-shaped vetigastropod from hot vent habitats. Ventral view with animal. (H) *Pendroma* sp. (Pendromidae), 1.8 mm. Photograph courtesy A. Warén. (I) *Perotrochus midas* (Pleurotomariidae), 120 mm. Recent pleurotomariids are found on the continental slope, whereas fossil representatives are known from shallow-water settings. (J) *Scissurella alto* (Scissurellidae), 1.1 mm. (K) *Seguenzia miriabilis* (Seguenzoidea), 4.5 mm, showing the deep sinus in the adsutural region of the aperture. (L–P) Trochoidea. L: *Bathyliotina glassi*, 13 mm, liotiid with strong spiny axial sculpture. M: *Clanculus undatus*, 31 mm, one of the few vetigastropods with apertural teeth. N: *Phasianotrochus apicinus*, 19 mm, one of the few high-spired vetigastropods. O: *Gena planulata*, 23 mm, a limpet-shaped trochoidean (stomatellid) in apertural view. P: *Turbo sarmaticus*, 55 mm; note the strongly crenulated, calcareous operculum. (Q–T) live animals (photographs by DLG). Q: *Fissurella volcano* (Fissurellidae) from Los Angeles, California, United States. Animal size *ca.* 2.5 cm, lateral view. Note the thin skirt forming the epipodium with very short epipodial tentacles. R: *Megathura crenulata* (Fissurellidae) from Los Angeles. Animal size *ca.* 12 cm, dorsal view. The reduced shell can be covered by the mantle. S: *Haliotis tuberculata* (Haliotidae) from Menorca, Balearic Islands, Spain. Animal size *ca.* 5 cm. Note the multiple openings in the shell and the strongly developed epipodium densely covered in epipodial tentacles. T: *Calliostoma zizyphinus* from Erquy, Brittany, France. Animal sized *ca.* 2.5 cm. Note the long epipodial tentacles. *Abbreviations*: ct: cephalic tentacle; e: eye; ep: epipodium; et: epipodial tentacle; f: foramen; m: mantle; s: shell; se: shell edge; sn: snout.

Several basal gastropod groups from hydro-thermal vents (Lepetodrilidae, Peltospiridae, Neomphalidae) were initially of uncertain affinity following their description in the 1980s. Originally, these were placed in their own higher categories, either explicitly or implicitly by placing them outside Vetigastropoda (e.g., McLean 1988, 1989); it has become increasingly clear that they belong within Vetigastropoda. The same also applies to seguenziids. While some authors tried to retain the familiar taxon Archaeogastropoda for Vetigastropoda (e.g., Hickman 1988; Haszprunar 1993; Bandel and Geldmacher 1996), this sugges-tion has not gained general acceptance, and Archaeogastropoda is now usually used as a reference to a historical, non-monophyletic concept. The monophyly and affinity of the "cocculiniform" taxa is not resolved. Although some authors maintain it as a higher taxon (e.g., Haszprunar 1998), some phylogenetic analyses have placed some of the "cocculiniform" taxa within Vetigastropoda (for details see subse-quent discussion), while others have found them more closely related to Patellogastropoda (e.g., Roldán *et al.* 2001). We use the group name in quotation marks to reflect that state of uncertainty and include references to "cocculini-forms" to illustrate similarities and differences to Vetigastropoda.

PHYLOGENY

The first phylogenetic analyses within Vetigas-tropoda produced largely unresolved topologies (e.g., Haszprunar 1988a), while early molecu-lar studies included only a few taxa (e.g., Tillier *et al.* 1994). Indications on sister group rela-tionships among vetigastropod lineages are widely diverging, with little common ground (Figure 12.3) and even a 50% majority rule con-sensus tree of published topologies produces a completely unresolved polytomy because of data conflict. The problem is not due to the use of morphological versus molecular data, because consensus topologies for each of these data types would still show completely unresolved polytomies. If studies with vetigastropod taxa in a larger phylogenetic context are included (e.g., Colgan *et al.* 2000, 2003; Giribet and Wheeler 2002; Schwarzpaul 2002), the dis-crepancies are further exaggerated. Inadequate

TABLE 12.1

Major Vetigastropod Groups, Their Characters and Recent Species-Level Diversity

	SHELL SHAPE	SHELL MARGIN	NACRE	SIZE (MM)	MANTLE ORGANS	RADULA	FOSSIL RECORD	SPECIES	
								RECENT	FOSSIL
Anomphalidae[a]	Trochiform	Straight	+	3–30	0	—	Silurian–Permian	0	(Few)
Cirroidea[a]	Trochiform, planispiral	Straight	+	5–20	0	—	Silurian–Cretaceous	0	(~50–100)
Clypeosectidae	Limpet	Foramen	−	5	2	Three to nine laterals	Recent	3	(0)
"Cocculiniforms"	Limpet	Straight	−	1–5	1\0	Variable	Cenozoic–Recent	130	(Few)
Euomphaloidea[a]	Trochiform, planispiral	Sinus	−	5–20	0	—	Ordovician–Cretaceous	0	(~100)
Fissurellidae	Limpet	Slit, sinus, foramen	−	3–150	2	With latero-marginal plate	Triassic–Recent	606[b]	(~100)
Haliotidae	Limpet	Foramina	+	12–320	2	Central tooth trapezoid	Cretaceous–Recent	55	(40)
Lepetodrilidae	Limpet	Straight	−	3–15	1	Enlarged lateral 1	Cenozoic–Recent	12	(Few)
Peltospiridae	Limpet, trochiform	Straight	−	5–15	1	Central tooth similar to laterals	Recent	12	(0)
Pendromidae	Trochiform	Straight	−	5	1	Absent	Recent	10	(0)
Pleurotomarioidea	Trochiform	Slit, sinus	+	30–200	2	Reduced central tooth, brush-like marginals	Cambrian–Recent	26	(~1,000–2,000)
Neomphaloidea	Limpet, trochiform	Straight	−	1–5	1	Four marginals	Cenozic–Recent	25	(Few)
Scissurellidae sensu lato	Limpet to trochiform	Slit, foramen, straight	−	0.6–11	2	Serrated central tooth	Triassic–Recent	140[c]	(60)
Seguenzioidea	Trochiform	Sinus	+	3–25	1	Few laterals and marginals	Cretaceous–Recent	139	(Few)
Trochoidea	Trochiform	Straight	+	2–100	1	Variable	Ordovician–Recent	2500[d]?	(~1,000–2,000)

NOTE: Based on Hickman 1984a; McLean 1988, 1989, 1992; Anseeuw and Goto 1996; Beesley et al. 1998; Geiger and Poppe 2000; Van Dover et al. 2001; Geiger 2003, 2006a; Geiger and Jansen 2004a, b. Species-level diversity is in many cases based on estimates by the authors, as there are few reliable sources. Species numbers in the fossil record is less certain than that of Recent taxa.
[a]Extinct lineage.
[b]Species count based on Geiger (unpubl. data).
[c]Approximately 90 additional undescribed Recent species (Geiger, unpubl. data).
[d]Diversity of Recent Trochoidea is estimated based on extrapolated ratios of Fissurellidae: Trochoidea from Wilson (1993) and Okutani (2000).

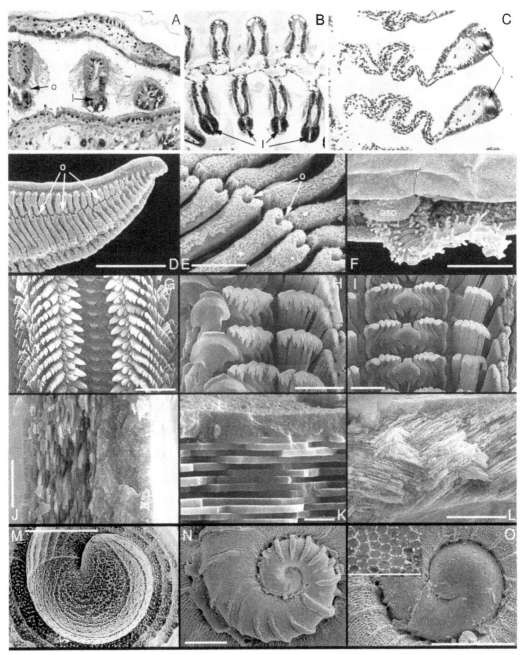

FIGURE 12.2. Vetigastropod characters. (A–C) Bursicles showing opening (o) and lumen (l). A: *Sinezona rimuloides* (Scissurellidae). B: *Puncturella cooperi* (Fissurellidae). C: *Haliotis discus* (Haliotidae). (D, E) Scanning electron micrograph (SEM) images of bursicles in *Turbo stenogyrum*. D: Overview of gill showing gill leaflets and row of bursicles. Scale bar = 500 μm. E: Detail of D; note frontal cilia on gill and distinct opening of bursicles (o). Scale bar = 100 μm. (F) Epipodial sense organ (eso) and associated papillated epipodial tentacle of *Broderipia iridescens* (Trochidae). Scale bar = 200 μm. (G–I) Radulae. G: *Haliotis coccoradiata* (Haliotidae). Scale bar = 500 μm. H: *Leurorhyncha caledonica* (Trochoidea: Skeneidae *sensu lato*). Scale bar = 20 μm. I: Liotiid (Trochoidea: aff. *Dentarene loculosa*). Scale bar = 20 μm. (J–L) Shell structure. J: Nacre and ostracum of *Seguenzia magaloconcha* (Seguenziidae). Scale bar = 10 μm. K: Nacre of *Haliotis cracherodii* (Haliotidae). Scale bar = 2 μm. L: Crossed-lamellar/cross-platy structure without nacre of *Anatoma* sp. (Scissurellidae *sensu lato*: Anatomidae). Scale bar = 10 μm. (M–O) Protoconchs. M: *Tricolia* (Trochoidea: Phasianellidae) with flocculant sculpture and spiral elements. Scale bar = 100 μm. N: *Sinezona beddomei* (Scissurellidae) with strong axials and apertural varix. Scale bar = 100 μm. O: *Satondella tabulata* (Scissurellidae) with microhexagonal sculpture. Scale bar = 100 μm. Inset: Partial enlargment of protoconch surface. Scale bar inset = 10 μm.

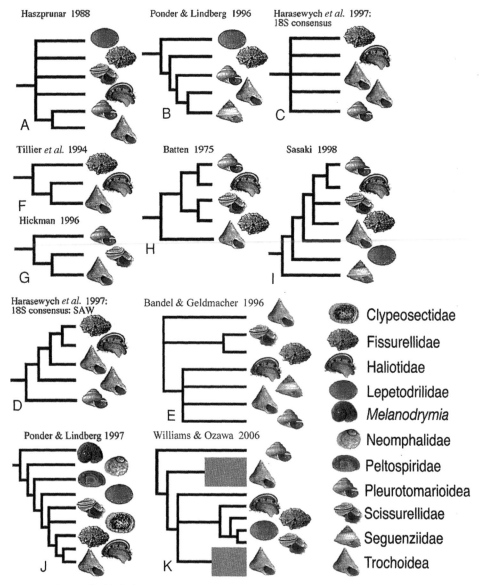

FIGURE 12.3. Selection of published phylogenies: (A) Haszprunar (1988a: morphology). (B) Ponder and Lindberg (1996, 1997: morphology). (C, D) Harasewych et al. (1997: DNA). (E) Bandel and Geldmacher (1996, simplified: fossil). (F) Tillier et al. (1994: DNA). (G) Hickman (1996: morphology). (H) Batten (1975: mineralogy). (I) Sasaki (1998: morphology). (J) Ponder and Lindberg (1997: morphology). (K) Williams and Ozawa (2006: DNA). Note the limited taxon sampling and the lack of agreement among studies. SAW: successive approximations weighting: the procedure transformed the polytomy into a fully resolved topology of Harasewych et al. (1997). The gray boxes in K represent complex relationships within Trochoidea.

taxon sampling and mistakenly assumed monophyly of "families" are the two most likely sources of conflict.

The most comprehensive morphological study of Sasaki (1998) shares only limited elements with Ponder and Lindberg's (1997) topology. These morphological as well as mineralogical studies (Batten 1975; Haszprunar 1988a; Bandel and Geldmacher 1996; Hedegaard 1997) tend to place Pleurotomariidae in a derived position.

Molecular studies have not converged on many common elements either. Harasewych et al. (1997), Yoon and Kim (2005), Geiger and

FIGURE 12.4. Selected Paleozoic (Carboniferous/Permian) vetigastropods. (A, B) Pleurotomarioidea: Phymatopleuridae, Upper Carboniferous (Gzhelian, *ca.* 300 Mya), Finis Shale, north central Texas, United States. A: 1.7 mm wide; B: Protoconch typical of vetigastropods, less than one whorl; 0.55 mm wide. (C) *Glabrocingulum* (Pleurotomarioidea: Eotomariidae), Upper Carboniferous (Gzhelian, *ca.* 300 Mya), Finis Shale, north central Texas, United States, 15 mm wide; one of the most abundant late Paleozoic soft-bottom-dwelling gastropod genera. (D) Juvenile *Glabrocingulum* (Pleurotomarioidea: Eotomariidae) from the Visean (Lower Carboniferous, *ca.* 330 Mya), Imo Formation, Arkansas, United States; 2 mm wide. (E) Juvenile *Trepospira* (Pleurotomarioidea: Raphistomatidae) Visean (Lower Carboniferous, *ca.* 330 Mya), Arkansas, United States; this rotelliform, low-spired genus is widespread in the Late Paleozoic of North America and other regions and lived preferably on soft bottoms; 2 mm wide. (F, G) *Hesperiella* (Cirroidea: Porcelliidae), Tournaisian (Lower Carboniferous, *ca.* 350 Mya), New South Wales, Australia, with heterostrophy from dextral to sinistral in early teleoconch. F: Detail of the base with selenizone; 1 mm across. G: Oblique side view showing teleoconch heterostrophy; 1.9 mm wide. (H, I) *Microdoma* (Trochoidea: Microdomatoidea: Microdomatidae), Upper Carboniferous (Late Moscovian, *ca.* 305 Mya), Buckhorn Asphalt deposit, Oklahoma, United States; microdomatoids are presumably closely related to the Mesozoic Eucyclidae and modern trochids. H: Lateral view of teleoconch, 4.7 mm high. I: Nacreous shell microstructure 100 μm wide (from Bandel *et al.* 2002). (J, K) Juvenile Euomphalidae, Visean (Lower Carboniferous, *ca.* 330 Mya), Arkansas, United States; euomphalids are an important Paleozoic gastropod group; its phylogenetic place is debated because some members have an openly coiled protoconch. However, the protoconch of this euomphalid specimen does not differ from those of modern vetigastropods. J: Width 0.38 mm. K; Width 1 mm (from Nützel 2002). (L) *Anomphalus* (Anomphaloidea: Anomphalidae), Upper Carboniferous (Late Moscovian, *ca.* 305 Mya), Buckhorn Asphalt deposit, Oklahoma, United States; anomphalids are a common group of low-spired Paleozoic to early Mesozoic vetigastropods with nacreous shell and trochoid protoconch; they are probably related to modern trochids; width 1 mm (from Bandel *et al.* 2002).

Thacker (2005), and Williams and Ozawa (2006) show Pleurotomariidae toward the base as well as Haliotidae and Fissurellidae in a derived position. The latter three analyses all show a Scissurellidae *sensu stricto* + Lepetodrilidae clade. Schwarzpaul (2002) focused on hydrothermal vent taxa only, making comparisons with other studies difficult. The most phylogenetically diverse sampling (40 taxa) was that of Geiger and Thacker (2005), while Williams and Ozawa (2006) included 59 taxa with a focus on the Trochoidea.

ORIGIN AND FOSSIL DIVERSITY

It is very likely that vetigastropods appeared together with several other major gastropod clades at the Cambrian/Ordovician boundary and diversified during the Early Ordovician radiation, approximately 490 Mya (Frýda *et al.*, Chapter 10). They probably contain Cambrian members (e.g., Knight *et al.* 1960; Tracey *et al.* 1993; Wagner 2002; Frýda and Rohr 2004), implying that they are as old as (or probably even older than) Neritimorpha, Caenogastropoda, and Heterobranchia (see also Frýda *et al.*, Chapter 10). Slit- or sinus-bearing forms appeared somewhat earlier than slitless forms (Frýda and Rohr 2004). The known fossil record also suggests vetigastropods are older than patellogastropods (Docoglossa), although this group is considered to be more basal (Golikov and Starobogatov 1975; Haszprunar 1988a; Ponder and Lindberg 1997), as were the limpet-shaped "cocculiniform" taxa by Haszprunar (1988a). However, early fossil data are equivocal because they are based on gross shell morphology, whereas important additional shell features (shell microstructure and protoconch morphology) are currently available only from the Middle/Late Paleozoic onward, i.e., long after the major gastropod clades had formed (see also Frýda *et al.*, Chapter 10). The phylogenetic relationship of the first certain vetigastropods to bellerophontiform molluscs and Cambrian mollusc limpets is still uncertain. McLean (1984) suggested a close relationship of fissurel-lids and bellerophontoids, supported by shell structure and the presence of a slit. However, bellerophontoid adult shells are bilaterally symmetrical throughout their ontogeny, whereas fissurellids are anisostrophically coiled in early adult stages, suggesting that their bilateral cap shape comes from a very high expansion rate of an anisostrophically coiled ancestor. However, it is possible that bellerophontoids are polyphyletic (e.g., Harper and Rollins 2000; Frýda *et al.*, Chapter 10), and it might be that some of them are members or close relatives of Fissurellidae.

Vetigastropods have a rich fossil record, with 30 of the 50 recognized families extinct (Bouchet and Rocroi 2005) and many of the extant families having a deep fossil record. The older families are mostly restricted to the Paleozoic, suggesting the Permian/Triassic mass extinction (250 Mya) influenced the evolutionary history of the group. From the Jurassic onward, taxa attributed to most of the extant vetigastropod families begin to appear. Slit-bearing vetigastropods were called Selenimorpha by Bandel and Geldmacher (1996), a name that is probably a synonym of Pleurotomariiformes and largely contains the traditional Pleurotomarioidea (Knight *et al.* 1960). They appear slightly earlier (Late Cambrian or Early Ordovician) than slitless vetigastropods (Trochomorpha). Slit-bearing vetigastropods (e.g., Raphistomatidae, Phymatopleuridae, Eotomariidae and Lophospiridae, all Pleurotomarioidea) are among the most abundant Late Paleozoic invertebrates, especially in shallow-water soft-bottom communities (Figures 12.4A–E), although some families are known by a few fossil forms only, such as the Triassic Laubellidae (Figures 12.5B–C) or Zygitidae. The Late Paleozoic/Triassic *Worthenia*-like gastropods (Figure 12.5A) comprised approximately 200 nominate species, making it one of the most diverse groups in that period. Pleurotomariidae *sensu stricto*, with an Early Jurassic type species, occurred in shallow-water deposits from the Triassic (when they were often abundant) to the Cretaceous (Begg and Grant-Mackie 2003; Kiel and Bandel 2004). In contrast, Recent pleurotomariids are relatively rare and live in deep water

TABLE 12.2

Synapomorphies of Vetigastropoda Revealed by Published Cladistic Analyses. Data from Salvini-Plawen and Steiner (1996); Ponder and Lindberg (1997); and Sasaki (1998).

CHARACTER	VALUE	TYPE OF CHANGE ON CLADOGRAM	REFERENCE	CHARACTER NUMBER
Digestive organ				
Posterior depressor muscles of the odontophore	Present	Ambiguous	Sasaki 1998	31
Jaw, anterior edge	Fimbriate	Unambiguous	Sasaki 1998	37
Radula, number of lateral teeth	5 pairs	Ambiguous	Sasaki 1998	49
Esophageal pouch	Expanded pouches	Unambiguous	Sasaki 1998	62
Esophageal gland of mid-esophagus	Papillate	Unambiguous	Sasaki 1998	63
Sculpturing of fecal material	One or more longitudinal grooves	Ambiguous	Ponder and Lindberg 1997	81
Excretory organ				
Number of kidneys	Single asymmetrical pair	Unambiguous	Sasaki 1998	72
Position of renal organs relative to pericardium in adult	Renal organs on either side of pericardium	Unambiguous	Ponder and Lindberg 1997	30
			Sasaki 1998	73
Nephridial gland	Absent	Unambiguous	Ponder and Lindberg 1997	28
Genital pore	In excretory organ	(?)	Salvini-Plawen and Steiner 1996: fig. 2.6	
Circulatory organ				
Transverse pallial vein	Present	Ambiguous (secondarily lost in dibranchiate taxa)	Sasaki 1998	71
Sense organ				
Ctenidial bursicles	Present	Unambiguous	Ponder and Lindberg 1997 Salvini-Plawen and Steiner 1996: fig. 2.6	107
Sensory papillae	Present	Unambiguous	Ponder and Lindberg 1997	108
(= micropapillae on cephalic and epipodial tentacles)	Present	Ambiguous	Sasaki 1998	14
Epipodial sense organs	Present	Unambiguous	Sasaki 1998 Salvini-Plawen and Steiner 1996: fig. 2.6	13
Eyes	On tentacles	(?)	Salvini-Plawen and Steiner 1996: fig. 2.6	
Statoliths and statoconia	Multiple statoconia	Unambiguous	Ponder and Lindberg 1997	99

FIGURE 12.5. (Opposite.) Selected Mesozoic (Triassic/Jurassic) vetigastropods. (A) *Worthenia* (or *Wortheniella*: Pleurotomarioidea: Lophospiridae), Late Triassic (Carnian, *ca.* 230 Mya), Cassian Formation, northern Italy; this has Paleozoic ancestry and is the most diverse Triassic gastropod genus; there are a few Jurassic representatives that then became extinct; height 8.3 mm. (B, C) *Laubella* (Pleurotomarioidea/Seguenzoidea: Laubellidae), Late Triassic (Carnian, *ca.* 230 Mya), Cassian Formation, northern Italy; Laubellidae is restricted to the Triassic; they are characterized by a slit and selenizone in a subsutural position. B: 1.7 mm high. C: Apex 0.6 mm across. (D, E) *Emarginula* (Fissurellidae), Late Triassic (Carnian, *ca.* 230 Mya), Cassian Formation, northern Italy; one of the earliest known fissurellids; length 11.4 mm. (F–I) *Eunemopsis* (Trochoidea?: Eucyclidae), Late Triassic (Carnian, *ca.* 230 Mya), Cassian Formation, northern Italy; this is one of the most abundant gastropods of the Cassian Formation. F: 4.5 mm high. G: Smooth vetigastropod protoconch and strong axial ribs on early teleoconch, 0.4 mm across. H: thick layer of columellar nacre, 0.11 mm wide. (I) Teleoconch fragment showing pronounced columellar fold, 2.6 mm high. (J) Trochoid-type protoconch with larval projection from a vetigastropod, Late Triassic (Carnian, *ca.* 230 Mya), Cassian Formation, northern Italy. (K) *Amphitrochus* (Trochoidea?: Nododelphinulidae), Early/Middle Jurassic (Toarcian/Aalenian, *ca.* 180 Mya), southern Germany; height 1.5 mm. (L) *Eucyclus elegans* (Trochoidea?: Eucyclidae), Early Jurassic (185 Mya), south Germany. (M) *Brochidium* (Trochoidea?: Brochidiidae), Late Triassic (Carnian, *ca.* 230 Mya), Cassian Formation, northern Italy; 0.9 mm wide. (N, O) *Blodgetella enormecostata* (Trochoidea), Late Triassic, Norian (*ca.* 210 Mya), Idaho, United States, with large ribs; 6.3 mm wide (from Nützel and Erwin 2004). (P, Q) *Spiniomphalus duplicatus* (family uncertain), Late Triassic, Norian (*ca.* 210 Mya), Idaho, United States; planispiral, with double row of prominent spines (from Nützel and Erwin 2004); P: 19.4 mm wide. Q: 15 mm wide. (R, S) *Brochidiella idahoensis* (Trochoidea: Liotiidae), Late Triassic, Norian (*ca.* 210 Mya), Idaho, United States, 16 mm wide (from Nützel and Erwin 2004). (T) *Sororcula gracilis* (Cirroidea: Cirridae), Late Triassic, Norian (*ca.* 210 Mya), Idaho, United States, unusually high-spired member of the sinistral cirrids; 15.6 mm high (from Nützel and Erwin 2004).

(Harasewych 2002). A Carboniferous supposed fissurellid (Mazaev 1998) needs comfirmation, because neither the protoconch, selenizone, or shell structure have been reported. The Late Triassic slit-bearing limpet *Emarginula muensteri* (Cassian Formation) is most likely a member of the Fissurelloidea (Figures 12.5D–E). Other reports from the Triassic (Bandel 1998) need to be confirmed. Recently, a *Temnocinclis*-like fossil was discovered in the Cassian Formation (Nützel and Geiger 2006), showing that the diversity of early slit-bearing vetigastropods was greater than previously assumed.

Cirroidea (= Porcelloidea) is an important extinct group of vetigastropods (Silurian to Cretaceous) that are dextral as early juveniles and become sinistral in mature teleoconch whorls (Figures 12.4F–G, 12.5T) (Bandel 1993b; Frýda 1997; Kiel and Frýda 2004). This coiling mode is shared by the slit-bearing Porcelliidae and slitless Cirridae. Both groups have a trochoid protoconch (Bandel 1993; Yoo 1994), and at least the Cirridae have a nacreous shell, making a placement in Vetigastropoda likely (Kiel and Frýda 2004).

The fossil history of Trochomorpha (vetigastropods without slits, such as Trochidae and Turbinidae) undoubtedly goes back to the

Devonian (e.g., Knight *et al.* 1960; Frýda and Bandel 1997), and it is likely that the group originated in the Ordovician (e.g., Knight *et al.* 1960). Certain representatives with preserved nacreous shell and trochoid protoconch have been reported from the Carboniferous (Bandel *et al.* 2002) and the Late Triassic Cassian Formation (Bandel 1993a). The Microdomatidae and Eucyclidae (treated within the Trochidae by Hickman and McLean 1990) represent examples of fossil taxa close to extant groups of Trochomorpha (Figures 12.4H–I, 12.5F–L). The higher classification of these slitless shells is particularly difficult, because many taxa are diagnosed by few characters. For Skeneidae, Tracey *et al.* (1993) reported a Cenozoic origin, while Kaim (2004) included some Middle Jurassic to Early Cretaceous forms, and Bandel (1993a) traced them back to the Late Triassic Cassian Formation. The Liotiidae and related forms have a rich fossil record that goes back to the Triassic (Nützel and Erwin 2004) (Figures 12.5R–S).

Several problematic fossil gastropod groups are either closely related to vetigastropods or are members of this clade. For instance, Anomphalidae (Figure 12.4L) have a nacreous inner shell layer and a vetigastropod type of protoconch (Bandel *et al.* 2002; Pan and Erwin 2002).

Similarly, protoconch morphology suggests that either euomphalids are vetigastropods or that they are closely related to them (Nützel 2002; Figures 12.4J–K). The Platyceratoidea, treated as "archaeogastropods" by Knight et al. (1960), are probably not vetigastropods, because they differ in protoconch morphology and shell structure, and they are probably close to the Neritimorpha (Bandel and Frýda 1999; Frýda et al., Chapter 10). No attempt has yet been made to undertake a phylogenetic analysis encompassing the many Late Paleozoic and Mesozoic vetigastropod genera, but Wagner (2001, 2002) included putative Early Paleozoic members of the Vetigastropoda in phylogenetic analyses (see also Frýda et al., Chapter 10).

MORPHOLOGICAL DATA SETS
SHELL CHARACTERS

Fossil vetigastropods can be recognized by teleoconch shape, shell structure, and protoconch morphology. None of these character complexes is unequivocally diagnostic (synapomorphic); identification of fossil vetigastropods can only be assessed provisionally. There are examples of convergent teleoconch morphology of Vetigastropoda and other major gastropod clades (although the importance of this phenomenon is generally overemphasized), especially with caenogastropods. Nevertheless, if a fossil gastropod has, for example, a trochi- or turbiniform teleoconch, a nacreous shell, a slit, and a typical vetigastropod protoconch, the likelihood that this is the shell of a vetigastropod is very high. Such vetigastropods are known from exceptionally well-preserved deposits from the Devonian/Carboniferous onward (400/300 Mya). Shell reduction has repeatedly occurred in Vetigastropoda, for example Stomatellidae (Trochoidea), Haliotidae (*Haliotis asinina*) and Fissurellidae (*Macroschisma, Megathura*, Figure 12.1R), and even complete shell loss in *Buchanania* (Fissurellidae).

Limpet shape has evolved several times independently in Gastropoda, including Patellogastropoda, Neritimorpha (*Phenacolepas*), Pulmonata (e.g., Siphonariidae), and within Vetigastropoda. In the latter, true limpet shape is found in Fissurellidae (Figures 12.1C, D, Q, R), Trochoidea (Stomatellidae: *Roya, Broderipia*), Lepetodriloidea (*Lepetodrilus*: Figure 12.1E), Neomphaloidea (*Neomphalus, Rhynchopelta*: Figure 12.1G) and "cocculiniforms" (Figure 12.1A). Functional limpets produced by spirally coiled shells with rapidly expanding whorls are encountered in Haliotidae (Figures 12.1B, S), Peltospiridae (*Peltospira*: Figure 12.1F), Trochoidea (Stomatellidae: Figure 12.1O), and Scissurellidae (*Depressizona*).

PROTOCONCH

Recent vetigastropod larvae do not feed on plankton and do not form multiwhorled larval shells, having a protoconch of nearly one whorl (Bandel 1982; Sasaki 1998; Hickman 1992) or, rarely, slightly more (e.g., *Scissurella cyprina*, Geiger and Jansen 2004b) (Figures 12.2M–O). The typical protoconch was referred to by Haszprunar (1993) as the "trochoid condition." The vetigastropod protoconch is initially organic and egg-shaped, then spirally deformed, and finally mineralized (Bandel 1982; Hickman 1992). Lateral pouches or folds are common and probably the result of mechanical bending. However, Page (1997) reported also a nonmechanical formation of these structures in *Haliotis*. The trochoid-type protoconch is relatively conservative. It can have diagnostic ornamentation (Figures 12.2M–O), as is the case in calliostomatids, but preservation of such ornamentation is extremely rare in fossil material. According to Bandel (1982) and Marshall (1995a), this typical surface sculpture is secreted after protoconch deformation (folding), presumably via solute transmission through the semipermeable outer shell layer at an early stage of mineralization. Generally, protoconch morphology has limited use as a phylogenetic character within Vetigastropoda (e.g., Hayashi 1983; Geiger 2003). However, it has value for higher systematics (e.g., Bandel 1988; Frýda 1999).

There are few reports of Paleozoic slit-bearing gastropods with a typical vetigastropod teleoconch (trochoid shape and selenizone) which possess a protoconch of distinctly more than one whorl (Dzik 1994; Nützel and Mapes 2001; Kaim 2004), suggesting a placement in the Caenogastropoda. However, the question whether such species are vetigastropods with a caenogastropod-like protoconch or caenogastropods with a vetigastropod-like teleoconch cannot be answered conclusively. The fact that planktotrophy is unknown in all Recent vetigastropod clades makes the assumption that planktotrophy existed in the past unparsimonous.

The typical vetigastropod-type (i.e., trochoid type) protoconch is also present in the Patellogastropoda (Docoglossa) and in "cocculiniform" gastropods (Bandel 1982; Warén 1996; Haszprunar 1998; Sasaki 1998; Nützel 2002). Interestingly, similar protoconchs of about one whorl were also present in Carboniferous euomphalids (Nützel 2002; Figures 12.4J–K), and the protoconch can be umbilicate or openly coiled in some (Yoo 1994; Bandel and Frýda 1998). Open protoconch coiling was present in several Paleozoic clades, although this morphology was increasingly abandoned during the Paleozoic. Its loss appears to represent a macroevolutionary trend that was complete by the early Mesozoic (Nützel and Frýda 2003). Apart from euomphalids, openly coiled protoconchs have not been reported from other supposed vetigastropods. The early ontogenetic development was used for a phylogenetic scheme by Bandel and Geldmacher (1996), who identified the Docoglossa (i.e., Patellogastropoda) as sister group of mostly unresolved vetigastropods (as Archaeogastropoda). The trochoid protoconch is seemingly linked to external fertilization, which probably represents the plesiomorphic state in molluscs and gastropods, including Vetigastropoda. Paleozoic vetigastropods with trochoid protoconchs were reported from the Early Devonian (see Frýda et al., Chapter 10; Frýda and Bandel 1997; Frýda and Manda 1997), the Carboniferous (Figure 12.4; Bandel et al. 2002) and the Permian (i.e., Anomphalus and Luogospira, Pan and Erwin 2002). Protoconchs of several Triassic vetigastropods have been reported from the Cassian Formation (Bandel 1991, 1993a; Schwardt 1992; Figures 12.5C, G, J).

TELEOCONCH

Teleoconch shape and geometry, as well as the presence and position of slits and sinuses, provide a rich source of characters, which were recently used by Wagner (2002) for a comprehensive analysis of early Gastropoda. He proposed an early split between bellerophontoideans and Murchisoniina. Among the latter, eotomarioids were hypothesized as being ancestral to the modern vetigastropods. Protoconch and shell structure data were not at hand, yet a link between the Late Paleozoic and the modern vetigastropods is still warranted. The placement of the slit is related to the shell geometry (e.g., Yochelson 1984). Slits or sinuses serve to facilitate water exchange in the mantle cavity. If the slit lies in a median position (e.g., modern pleurotomariids), it generally reflects the presence of a pair of gills. If the slit is situated in a distinctly adapical or abapical position, as is the case in several fossil vetigastropod groups (e.g., the pleurotomarioidean Laubellidae, Figures 12.5B–C), a symmetrical arrangement in the mantle cavity is less likely, and the slit was not necessarily linked to the presence of two gills (Bandel and Geldmacher 1996), despite counter examples such as modern Haliotidae and Scissurella. High-spired, slit-bearing shells with a selenizone were attributed to the Murchisonioidea. They are either interpreted as exceptionally high-spired pleurotomariids (e.g., Gordon and Yochelson 1987; Mazaev 2002) or as ancestral caenogastropods (Cox 1960; Ponder and Warén 1988). Middle Paleozoic Murchisonia-like shells have a vetigastropod protoconch (Frýda and Bandel 1997; Frýda 1999), and Ordovician representatives posses nacre (Carter and Hall 1990). However, Late Paleozoic shells of this type have a caenogastropod larval shell and lack nacre (Nützel and Bandel 2000; Bandel et al. 2002). Therefore, Murchisonioidea, in a traditional sense, is polyphyletic, containing both vetigastropods and caenogastropods (see Ponder et al., Chapter 13). A marked ontogenetic change in teleoconch morphology can be observed in the Jurassic/Cretaceous turbinid *Torallochus*, in which the early whorls are almost planispiral and the later part of the teleoconch is turreted (Kiel and Bandel 2002; Kaim 2004). Heterostrophy occurs within the extinct Cirroidea (= Porcelloidea), which have a dextral protoconch and early teleoconch but become sinistral in mature teleoconch whorls (Figures 12.4F, G, 12.5T). This should not be confused with the heterostrophy of the marine Heterobranchia, which have a sinistral larval shell and a dextral teleoconch.

Teleoconch ornamentation can be a useful character at low taxonomic levels. Many vetigastropod species or genera have highly characteristic sculptures. The onset of the ornament, as well as other ontogenetic changes, are a good source of characters for alpha-taxonomy. Teleoconch ornamentation has rarely been used as a diagnostic character above the genus level, and in most cases this character complex does not help to infer suprageneric sister group relationships. However, in particular groups (e.g., at the family group level using the fine lamellar axial ornament of Liotiinae; see Hickman and McLean 1990), such ornament may be useful for higher systematics. Apertural modifications such as columellar folds and teeth are uncommon (e.g., in the Triassic trochoid *Eunemopsis*, Figure 12.5I; and in living Clanculus, Figure 12.1M) but have been used at the genus level in Trochoidea.

SHELL MICROSTRUCTURE

Shell microstructure has been used in the discussion of vetigastropod relationships. For example, McLean (1984) suggested that Fissurellidae are closely related to Bellerophontida, while Batten (1975) and Kiel (2004) related Fissurellidae to Scissurellidae based on microstructures.

Hedegaard (1997: 375) proposed four diagnostic shell structure characters for Recent vetigastropods: (1) intersected crossed platy structure, (2) absence of crossed-lamellar structures, (3) presence of distinctive columnar nacre, and (4) presence of spherulitic prismatic structure.

The columnar nacreous structure (Figures 12.2J, K), one type of molluscan nacre, has received most attention among microstructural characters. Nacre has generally been considered plesiomorphic, because of its occurrence in primitive representatives of conchiferan groups such as protobranch bivalves, monoplacophorans, and *Nautilus* (e.g., Wingstrand 1985; Salvini-Plawen and Steiner 1996). In vetigastropods, nacre is absent in Lepetodrilidae, Fissurellidae, Scissurellidae *sensu lato* (Figure 12.2L) and in some trochoideans (Phasianellidae, Halistylinae, and Skeneidae; see Carter and Hall 1990).

Studies of crystallographic textures and microstructures of molluscan shells suggest that vetigastropod nacre is homologous with that of cephalopods (Frýda *et al.* 2004), implying that nacre is plesiomorphic in vetigastropods (i.e., stem vetigastropods had a nacreous shell) with the lack of nacre the derived condition. In contrast, Hedegaard (1997) considered crossed-lamellar structures as basal in gastropods and inferred that nacre was acquired subsequently within Vetigastropoda.

Ponder and Lindberg (1997: 216) included bellerophontoideans in vetigastropods. Bellerophontoidea represent an ancient (Cambrian–Early Triassic) group of isostrophic, slit-bearing molluscs that had an aragonitic crossed-lamellar shell structure and no nacre (e.g., Carter and Hall 1990). If they represent the sister group of all other Gastropoda, nacre would be derived in vetigastropods. However, the relationships of bellerophontoideans are not yet resolved satisfactorily (McLean 1984; Harper and Rollins 2000; Frýda *et al.*, Chapter 10). The oldest report of preserved nacreous shell structure is from Carboniferous vetigastropods (e.g., Batten 1972; Carter and Hall 1990; Bandel *et al.* 2002). Some fossil groups have both nacreous and crossed-lamellar structures in the same shell (Carter and Hall 1990; Ponder and Lindberg 1997: 102). These shell microstructure characters may be difficult to assess in the most controversial, Early Paleozoic forms because of diagenetic recrystallization, rendering interpretations questionable.

ANATOMICAL CHARACTERS

The original definition of Vetigastropoda by Salvini-Plawen (1980) based on shell muscle characters was refuted and redefined with 13 characters by Salvini-Plawen and Haszprunar (1987). Since then various characters have been discussed to test and clarify the phylogenetic status of vetigastropods. Sasaki (1998: 211) listed the diagnostic characters of Vetigastropoda. However, it should be noted that such characters are not always synapomorphic but are often plesiomorphic. To avoid this confusion, synapomorphies of vetigastropods revealed by cladistic analyses are listed in Table 12.2. The characters, such as traditionally known common characters, true or possible synapomorphies, and insufficiently explored characters potentially useful in future phylogenetic research, are also briefly discussed in the following paragraphs.

HEAD-FOOT

The most visible external characteristic of the vetigastropod head-foot is the possession of well-developed sensory projections. The head-foot typically has epipodial tentacles (Figures 12.1Q, S, T, 12.2F), epipodial sense organs (Figure 12.2F), cephalic tentacles (Figures 12.1Q, S, T), neck lobes and cephalic lappets (see also subsequent discussion of sense organs).

The mantle slit or hole(s), which corresponds to the shell slit or foramen, is a well-known character of dibranchiate vetigastropods ("zeugobranchs") having paired ctenidia. A bisinuate mantle and a single left ctenidium is seen in *Seguenzia*. Thus, different types of slits have evolved independently within vetigastropods, and some caenogastropods (e.g., turritellids, siliquariids, and turrids) also have slits or sinuses.

MUSCULATURE

Vetigastropoda was originally proposed by Salvini-Plawen (1980) on the basis of the predominance of the supposed post-torsional right shell muscle. However, the unpaired gastropod shell muscle (columellar muscle) was revealed to be post-torsional left in light of its innervation (Haszprunar 1985b). Within vetigastropods, shell muscles are paired in Haliotidae, Lepetodrilidae, and *Tricolia* but, as far as is known, unpaired in the remaining groups.

PALLIAL CAVITY

Vetigastropods represent the only molluscan clade showing both paired and unpaired pallial organs. Ctenidia, hypobranchial glands, osphradi, and kidney openings are paired and/or unpaired in various combinations, depending on the taxon (Sasaki 1998: 173, table 4). Because all major non-gastropod molluscan groups possess paired pallial organs, the unpaired state is regarded as apomorphic (Ponder and Lindberg 1997: 108–114, 192–193; Lindberg and Ponder 2001). It might be more parsimonious to interpret paired pallial organs as secondary multiplication in specific tree topology (e.g., Ponder and Lindberg, 1997: fig. 2; Sasaki 1998: fig. 106; Figure 12.6), but the majority of authors regard their number in gastropods as irreversible from paired to unpaired conditions (e.g., Ponder and Lindberg 1997).

CTENIDIUM   The ctenidium of vetigastropods consists of ctenidial lamellae with exterior ciliation (frontal and lateral cilia); internal skeletal rods, blood vessels, and nerves; and ctenidial axes attached by the afferent and efferent membranes to the mantle skirt (except Pleurotomariidae and Haliotidae, which lack afferent membranes). The lamellae are bipectinate in most vetigastropods but may be modified into a partly or completely monopectinate condition in some taxa. This specialization occurs mostly in small-sized species and is possibly correlated with body size. Examples are known in Seguenziidae, Scissurellidae *sensu lato*, Skeneidae, and part of Fissurellidae (McLean and Geiger 1998; see Ponder and Lindberg 1997: 110 for references before 1997).

The most confusing problem relevant to the vetigastropod ctenidium is the homology with the gills of "cocculiniform limpets." Haszprunar (1988a, b) considered their gills secondary but Ponder and Lindberg (1997: character 14) treated cocculinoidean and pseudococculinid gills as reduced ctenidia mainly on the basis of their lateral ciliated bands, which are comparable to

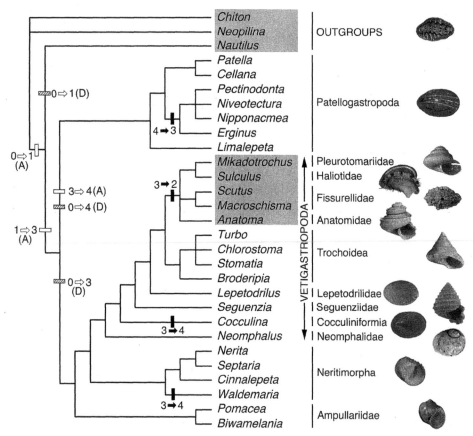

FIGURE 12.6. One example of phylogenetic reconstruction of vetigastropod relationships based on morphological data (from Sasaki 1998, modified). The changes of the organization of the gills are mapped. o: multiple pairs. 1: two pairs. 2: single pair. 3: post-torsionally left gill only. 4: gills absent. Open arrow: ambiguous change. Solid arrow and box: unambiguous change. Open box and A: ACCTRAN optimization. Hatched box and D: DELTRAN optimization. Taxa in shaded boxes have paired gills.

those seen in typical ctenidia. This view was further supported by the presence of skeletal rods in Addisoniidae and Choristellidae and bursicle-like structures in Pyropeltidae, Lepetellidae, and Pseudococculinidae (Haszprunar 1988b). However, the general criteria of ctenidial homology (i.e., position, innervation, blood supply, and skeletal rods) cannot settle this problem at present (see Ponder and Lindberg 1997: 111 for discussion). It should be noted that none of the "cocculiniform" taxa possesses skeletal rods and bursicle-like sensory pockets in the same gill. More thorough investigation of ctenidial morphology is necessary for future research.

HYPOBRANCHIAL GLAND Most vetigastropods have well-developed, paired hypobranchial glands (Sasaki 1998: 173, table 4). Although

some have been reported to have greatly reduced hypobranchial glands or to lack them, earlier descriptions are often ambiguous and need histological confirmation (see also Colgan *et al.* 2003: 139–140).

DIGESTIVE SYSTEM

Vetigastropoda possess a number of characters that differentiate them from other gastropods in the digestive system.

BUCCAL MASS The buccal mass is an insufficiently studied character in the digestive system. Sasaki (1998) compared buccal mass structures of various basal gastropods and identified the homologies of muscles and cartilages (Sasaki 1998: 179, table 5). Vetigastropods generally have primitive structures inherited from the poly- and monoplacophoran plan, but some muscles are newly acquired at the "rhipidoglossate" level of organization. The posterior depressor of the odontophore is one of the apomorphies of vetigastropods, but it is also homoplastic to that of Neritoidea. Some of the other specific muscles are generally shared not only among vetigastropods but also with *Cocculina* and/or Neritoidea.

An elongate anterior and small posterior pair of odontophoral cartilages represents a distinctive character for most vetigastropods ("zeugobranchs" plus Trochoidea), but *Lepetodrilus* and *Seguenzia* have a single pair, which might be regarded as a fused anterior-posterior pair.

RADULA Five pairs of lateral teeth and numerous marginal teeth on either side of the central tooth are a common ground plan for vetigastropod radulae (Figures 12.2G–I). However, some taxa do not conform to this generalization. For example, tooth number is greatly reduced in Seguenziidae, possibly the result of paedomorphic simplification (Ponder and Lindberg 1997: 141) or highly multiplied in Pleurotomariidae (see Harasewych 2002 for review). The traditional way of assessing homology of radular teeth seems effective within vetigastropods, but it is seriously questioned when all gastropods are taken into consideration in a phylogenetic analysis (Ponder and Lindberg 1997: 138–143).

RADULAR SAC Ponder and Lindberg (1997: 144, character 66) proposed that the morphology of the distal end of the radular sac is of phylogenetic value, with all vetigastropods having a deeply bifurcated end. A somewhat similar, although less pronounced, state is also found in *Cocculina* (Sasaki 1998: fig. 70).

JAW The jaws are paired, attached to the inner surface of the anterior snout (free from the buccal mass) in vetigastropods. Their anterior margin is brush-like due to fine scaly elements, an unambiguous synapomorphy of the group in Sasaki's (1998) analysis and in contrast to the jaws of "lower" caenogastropods, where these scales are present on the entire inner surface.

SALIVARY GLANDS Vetigastropod salivary glands are characteristically ramified into numerous microtubules that open directly into the buccal cavity without a salivary duct (Sasaki 1998: character 61). This condition is not synapomorphic for all vetigastropods, because the glands are absent in *Lepetodrilus*, and the condition in *Seguenzia* is uncertain (Sasaki 1998).

ESOPHAGUS The esophageal structure of vetigastropods is markedly different from those of other gastropods. In cross section it is not dorsoventrally depressed as in other basal gastropods but greatly dilated. Esophageal glands lining its inner wall are projected as numerous, distinct papillae.

STOMACH The stomach of vetigastropods is composed of common elements to that of other orthogastropods (e.g., cuticularized gastric shield, ciliated sorting area, paired typhlosoles, and gastric cecum). The presence of a spiral gastric cecum is a conspicuous feature of representative taxa of vetigastropods (Haliotidae, Pleurotomariidae, Trochoidea), but is greatly reduced or absent in others. The reduction of the spiral cecum seems to be linked to the acquision of limpet-like morphology, since it is reduced or lost in limpet-shaped groups (Fissurellidae, Lepetodrilidae, trochid *Broderipia*) (Sasaki 1998: 182). The stomach's internal structure is a potentially useful but unexplored source of characters that may reveal interrelationships of vetigastropods and other basal gastropods. This character suite should be further investigated, as it was in Caenogastropoda by Strong (2003).

FECAL PELLETS AND RECTUM Ponder and Lindberg (1997: character 81) showed that fecal material with one or more longitudinal grooves is a possible (ambiguous) synapomorphy of Vetigastropoda. Fecal morphology presumably reflects the internal structure of the rectum, a system that should be compared in detail in future studies.

CIRCULATORY SYSTEM

The heart of vetigastropods typically consists of paired auricles on the either side of a median ventricle traversed by the rectum. However, in *Seguenzia*, only the left auricle is present, and the pericardium and ventricle are not penetrated by the rectum (Haszprunar 1988a: fig. 2q; Sasaki 1998). The circulation pattern is one of the major characters separating "proso-" and heterobranchs (Ponder and Lindberg 1997: character 23). The blood in the pallial cavity returns to the auricle(s) from the ctenidium(-a) and mantle in "prosobranchs," but from a pallial kidney in heterobranchs. In vetigastropods, vessels passing through the mantle and ctenidium(-a) are separated in dibranchiate families or partially connected in trochoideans. The transverse pallial vessel, which conveys venous blood from the right kidney to the afferent ctenidial vessel (Sasaki 1998: character 71) is well developed in trochoideans and also present in *Lepetodrilus*. The state of *Seguenzia* is unknown.

EXCRETORY SYSTEM

The number and position of excretory organs (variously termed kidneys, nephridia, or renal organs; e.g., Fretter and Graham 1962; Haszprunar 1988a; Ponder and Lindberg 1997) and their inner structures are of obvious phylogenetic value (Ponder and Lindberg 1997: characters 24–30). The kidneys of vetigastropods are paired without exception. The paired condition is also found in patellogastropods and non-gastropod molluscs and is therefore plesiomorphic for gastropods. However, both kidneys are located on the right side of the pericardium in patellogastropods, but in vetigastropods the right and left kidneys are found on the posterior right side of the pericardium and in the pallial cavity on the anterior left side of the pericardium, respectively.

NEPHRIDIAL GLAND The nephridial gland (Ponder and Lindberg 1997: character 28) is absent in non-gastropod molluscs and most basal gastropods but, probably homoplastically, present in caenogastropods and the vetigastropod Trochidae. The reconstruction of character change revealed that the loss of a nephridial gland is apomorphic at the basal node of vetigastropods, and the reacquisition occurred in

Trochidae by reversal (Ponder and Lindberg 1997: appendix 2).

PAPILLARY SAC The inner wall of the left kidney of vetigastropods is typically roughened by elongated porous projections, and it is termed "papillary sac" (Fretter and Graham 1994: 639; Sasaki 1998: character 74). This specialization of the nephridial epithelium is found only within vetigastropods. However, it is not clearly differentiated in Lepetodrilidae and Seguenziidae.

REPRODUCTIVE SYSTEM

The reproductive system of vetigastropods is markedly simple compared to that of "higher" gastropods. There are no glandular tissues along the oviduct and vas deferens and no external copulatory organ (except Skeneidae; see Hickman 1992: 247; Sasaki 1998: 183 for review including vent taxa). The gonoduct opens into the pallial cavity through the right kidney. This direct connection of the gonoduct with the kidney may be regarded as plesiomorphic for gastropods (Ponder and Lindberg 1997: 128).

EGG The egg is coated with an external gelatinous envelope by the ovary in most vetigastropods (except Lepetodriloidea) and Lepetelloidea (Ponder and Lindberg 1997: 134). It is absent in Peltospiridae, *Neomphalus*, Cocculinidae, and other "higher" gastropods as well as non-gastropod molluscs. Thus, this is an undoubted diagnostic character of most vetigastropods.

URINOGENITAL PAPILLA Females of some trochoideans are known to develop an enlarged excretory opening termed "urogenital papilla" (Hickman 1992: 246). It is presumed to provide the jelly coating to eggs. Such a specialization has not been known in other externally fertilizing gastropods. Its phylogenetic significance is still uncertain.

SPERM Sperm ultrastructure has attracted considerable attention in gastropod phylogenetics (e.g., Healy 1988, 1996; Ponder and Lindberg 1997: characters 35–50). Healy (1988) categorized gastropod euspermatozoa into five types, and vetigastropod sperm was grouped into the externally fertilizing type together with patellogastropod ones. Basically, sperm of most vetigastropods is further distinguished from those of patellogastropods by an anterior fossa of the nucleus and a posterior invagination of the acrosome (Hodgson 1995: 174). Nevertheless, the sperm morphology may be highly modified by different modes of reproduction, as was shown in internally or pallially fertilizing species (e.g., Hodgson 1995; Healy 1996).

NERVOUS SYSTEM

Vetigastropods generally have the plesiomorphic condition for gastropods: (1) hypoathroid arrangement of circumesophageal nerve ring, (2) streptoneurous visceral nerve loop with sub-/supraesophageal and visceral ganglia (without parietal ganglia), and (3) scalariform pedal cords with many cross-connections. These states are shared with various basal gastropods and are not specific to vetigastropods. A characteristic branchial ganglion is formed at the base of each ctenidium in some vetigastropods. It originates from the visceral loop and innervates the ctenidium and osphradium. This ganglion is restricted to vetigastropods and is assumed to be homologous to the osphradial ganglion (not subesophageal ganglion) of caenogastropods (Ponder and Lindberg 1997: 162).

Some particular modifications are known within vetigastropod subgroups. In Pleurotomariidae, the visceral loop arises from the cerebropleural connective, not from the pleural ganglia as in other gastropods. Fissurellidae has unique and very specialized pedal cords, which are exposed on the pedal musculature and concentrated in a specific way (Sasaki 1998: characters 91 and 92).

BURSICLES These sense organs are well known as the most obvious hallmark of vetigastropods in a phylogenetic and systematic context (Salvini-Plawen and Haszprunar 1987; Haszprunar 1988a; Ponder and Lindberg 1997; Sasaki 1998; Figures 12.2A–E). They are located on the efferent side of each ctenidial leaflet and consist of a ciliary pocket with a slit-like opening to the exterior. The pocket contains sampling cilia continuous with lateral and frontal cilia (Haszprunar 1987; Sasaki 1998: figs. 43e, 47, and 93). Their position and structure are particularly important as criteria for the identification of this organ.

Although the bursicles have been believed to be a true synapomorphy of vetigastropods, there are some confusing situations (Table 12.3). In Pleurotomariidae, Haszprunar (1987) reported their presence in *Perotrochus caledonicus*, but Sasaki (1998) did not find them in *Mikadotrochus beyrichii*. The state in *Seguenzia* is still insufficiently known. Sasaki (1998) identified knobby parts marked with longitudinal grooves as bursicles in the monopectinate ctenidium of *Seguenzia*, but this identification remains questionable because openings from the lumen of the pocket were not observed (Sasaki 1998: 102, fig. 67d). The protuberance of the tip of ctenidial leaflets in *Lepetodrilus* (Fretter 1988: fig. 7b) has been assumed to represent bursicles (Ponder and Lindberg 1997: 172). However, Sasaki (1998: 138) did not support this view, because of the apparent absence of the ciliary pocket in that part. Bursicle-like structures of *Melanodrymia* (see Haszprunar 1989 for description) are also treated as homologous (Haszprunar 1993; Ponder and Lindberg 1997: 172). Ponder and Lindberg (1997: 172–173) argued that at least some pseudococculinids have ctenidia with bursicles. These examples suggest that further comparative study is necessary for all types of buriscle-like structures in terms of homology.

EPIPODIAL TENTACLES AND SENSE ORGANS The epipodium of vetigastropods is fringed by a row of tentacles (Figures 12.1F, Q, S, T) typically accompanied by the epipodial sense organs (ESO) at their bases (Figure 12.2F). They are universally present in vetigastropods, although in some taxa they are somewhat reduced (Lepetodrilidae: Fretter 1988; Fissurellidae: McLean and Geiger 1998; Pleurotomariidae: Harasewych 2002). The presence of ESOs in *Seguenzia* needs to be confirmed.

TABLE 12.3

Taxonomic Distribution of Bursicles in Putative Vetigastropod Lineages

TAXON	BURSICLES	REFERENCE
Fissurellidae		
Emarginula elongata	Present	Haszprunar 1987
Diodora italica	Present	Haszprunar 1987
Scutus sinensis	Present	Sasaki 1998
Macroschisma dilatatum	Present	Sasaki 1998
Pseudorimula marianae	Present	Haszprunar 1989
Fissurella volcano	Present	Szal 1971[a]
Megathura crenulata	Present	Szal 1971[a]
Puncturella cooperi	Present	Figure 12.2B
Clypeosectidae		
Clypeosectus delectus	Present	Haszprunar 1989
Clypeosectus curvus	Present	Haszprunar 1989
Haliotidae		
Haliotis lamellosa	Present	Haszprunar 1987
Haliotis aquatilis	Present	Sasaki 1998
Haliotis discus	Present	Figure 12.2C
Trochidae		
Gibbula varia	Present	Haszprunar 1987
Tegula eiseni	Present	Szal 1971
Calliostoma sp.	Present	Szal 1971
Umbonium sp.	Present	McLean 1986
Lirularia lirularia	Present	McLean 1986
Turbinidae		
Astraea rugosa	Present	Haszprunar 1987
Turbo stenogyrum	Present	Sasaki 1998
Chlorostoma lischkei	Present	Sasaki 1998
Broderipia iridescens	Present	Sasaki 1998
Homalopoma sp.	Present	Szal 1971
Stomatellidae		
Stomatia phymotis	Present	Sasaki 1998
Phasianellidae		
Tricolia pullus	Present	Haszprunar 1987
Pleurotomariidae		
Perotrochus caledonicus	Present	Haszprunar 1987
Perotrochus sp.	Absent	Geiger pers. obs.
Mikadotrochus beyrichii	Absent	Sasaki 1998
Scissurellidae s.l.		
Incisura lytteltonensis	Present	Bourne 1910
Sinezona rimuloides	Present	Geiger 2004; Figure 12.2A
Anatominae		
Thieleella reticulata[b]	Present	Sasaki 1998
Temnocinclidae		
Temnocinclis euripes	Present	Haszprunar 1989
Temnozaga parilis	Present	Haszprunar 1989

TABLE 12.3

(continued)

TAXON		BURSICLES	REFERENCE
Seguenziidae			
	Seguenzia sp.	Present?	Sasaki 1998, Haszprunar 1993
Peltospiridae			
	Melanodrymia aurantiaca	Present	Haszprunar 1989
Neomphalidae			
	Neomphalus fretterae	Absent	cf. Sasaki 1998, Haszprunar 1988a
Lepetodrilidae			
	Lepetodrilus sp.	Present	Ponder and Lindberg 1997
	Lepetodrilus nux	Absent	Sasaki 1998
Lepetelloidea			
	Bathyphytophilus caribaeus	Present	Haszprunar and McLean 1996
	Bathyphytophilus diegensis	Present	Haszprunar and McLean 1996
Cocculinidae			
	Cocculina nipponica	Inapplicable	Sasaki 1998
Outgroups: Neritimorpha			
Neritidae			
	Nerita albicilla	Absent	Sasaki 1998
Phenacolepadidae			
	Cinnalepeta pulchella	Absent	Sasaki 1998
	Shinkailepas myojinensis	Absent	Sasaki *et al.* 2006

[a]Szal (1971) only identified genera; the species are extrapolated for monotypic genera or genera with single species found in the geographic area studied by Szal (California).
[b]As *Anatoma* sp., see Bandel (1998) for description of species.

SENSORY PAPILLAE The cephalic and epipodial tentacles carry sensory papillae in vetigastropods (Figure 12.2F) except Pleurotomariidae, Fissurellidae, and Lepetodrilidae. Similar sensory papillae are present in the tentacles of certain pseudococculinids (Haszprunar 1988c; Dantart and Luque 1994). Künz and Haszprunar (2001: 161) noted that this state may support the inclusion of lepetelloideans in vetigastropods, as was proposed by Ponder and Lindberg (1997).

OSPHRADIUM The vetigastropod osphradium is different in position and structure from that in other gastropods, with the osphradial epithelium covering the efferent axis of the free tip of the ctenidium and with cilia bottles present in most of them (Haszprunar 1985a: 486, table 4). The absence of the lateral zone with so-called Si cells (see Welsch and Storch 1969; Haszprunar 1985a for terminology) in basal gastropods, including vetigastropods, is a plesiomorphic condition for gastropods (Ponder and Lindberg 1997: 172).

STATOCYSTS The statocyst contains multiple statoconia in all vetigastropods examined so far (Ponder and Lindberg 1997: character 99). By out group comparison, multiple statoconia are regarded as plesiomorphic, but the polarity is opposite according to the ontogenetic criterion (Ponder and Lindberg 1997: 169). Statocysts are located on the antero-dorsal side of the pedal ganglia in vetigastropods, a state shared with "cocculiniform" taxa and Neritimorpha.

MOLECULAR STUDIES

As pointed out previously, molecular phylogenies show limited agreement among each other, largely because of inadequate taxon sampling, too few genes, or too low a number of base pairs. Thus far, traditional family concepts have often been uncritically accepted, particularly with respect to the affinity of the vent slit limpets: Clypeosectidae to Fissurelloidea and Sutilizoninae and Temnocinclinae to Scissurellidae (e.g., Schwarzpaul 2002). Habitat-specific focus (e.g., hot vents) may have obscured the relationships rather than contributed to their resolution. Hence, representatives of as many lineages as possible should be included in future analyses. Most studies have been based on single genes or gene fragments. It has become increasingly clear that Vetigastropoda is too old for resolution from single genes or gene fragments. At present, a convergence on three markers is emerging: histone 3 (H3), cytochrome c oxidase subunit 1 (COI), and 18S rRNA. COI provides resolution between species and genera, while 18S helps with deep divergences. H3 is short, and its specific utility is uncertain, but it is likely to provide characters relevant to deep divergences. Molecular studies have been involved in identifying problem taxa, such as Scissurellidae *sensu lato* (Geiger and Thacker 2005; Figure 12.7), and Trochoidea *sensu lato* (Williams and Ozawa 2006). At present, these studies have pointed out areas of further study,

while still being incomplete. There is ample room for both detailed within-family studies and higher-level investigations.

MONOPHYLY OF MAJOR RECENT VETIGASTROPOD LINEAGES

While some vetigastropod lineages are clearly monophyletic, others remain problematic, particularly hydrothermal vent groups that have been aligned with well-known, shallow-water forms. Clypeosectidae had been classified with Fissurellidae in Fissurelloidea by McLean (1989), a view supported by Schwarzpaul's (2002) analysis of morphological characters, while her molecular analyses resulted in various placements depending on the reconstruction algorithm used. Warén and Bouchet (2001) suggested a closer relationship of Clypeosectidae with Lepetodrilidae, while Geiger and Thacker's (2005) molecular analysis suggests a placement basal to Fissurellidae plus Anatomidae.

The monophyly of the problematic Cocculiniformia ("cocculiniforms") is disputed (Colgan *et al.* 2003; Strong *et al.* 2003). The highly modified, pseudoplicate gill in Cocculinidae makes one of the potential synapomorphies unobservable (Sasaki 1998). Most authors still recognize Cocculinoidea at a higher level outside Vetigastropoda (e.g., Haszprunar 1998; Lindberg *et al* 2004; Lindberg, Chapter 11). Geiger and Thacker's (2005) molecular study placed *Cocculina* between Peltospiridae (with bursicles) and Pleurotomariidae (bursicle condition uncertain), Colgan *et al.* (2003) showed the pseudococculinid *Notocrater* in a derived position within Vetigastropoda, and Giribet *et al.* (2006) placed *Cocculina* within Vetigastropoda. These results suggest that at least some "cocculiniforms" are members of Vetigastropoda. Roldán *et al.* (2001), on the other hand, found a closer affinity of "cocculiniforms" with Patellogastropoda.

Scissurellidae has been a long overlooked family. Six subfamilies (Scissurellinae, Anatominae, Depressizoninae, Larocheinae, Sutilizoninae, and Temnocinclinae) have been

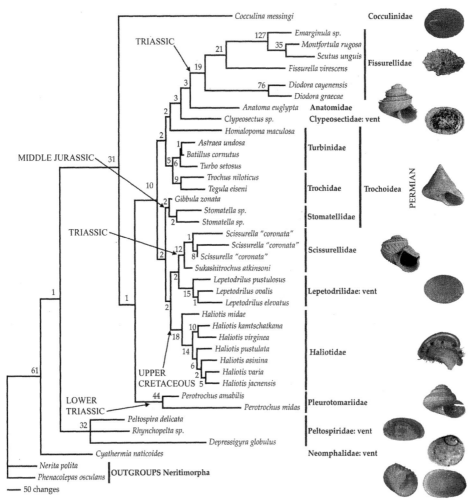

FIGURE 12.7. Vetigastropod phylogeny based on molecular data, from Geiger and Thacker (2005, modified), with superimposed ages of select fossil lineages. Note multiple colonization of chemosynthetic habitats (represented by vent taxa Neomphalidae-Peltospiridae, Lepetodrilidae, Clypeosectidae).

recognized on the basis of overall shell shape and radular patterns (Geiger 2003). The two that are most similar, both in terms of shell as well as radular morphology, are Scissurellinae and Anatominae, but with molecular data they do not form a clade (Geiger and Thacker 2005) and have been elevated to family rank. Members of the other lineages traditionally classified as subfamilies have not been included in broad vetigastropod phylogenies, and their placement is entirely open. Different opinions have been voiced, but without rigorous arguments. The vent lineages have been placed with lepetodrilids

(e.g., Warén and Bouchet 2001; Bouchet and Rocroi 2005), or the four minor lineages have been labeled *incertae sedis* (Geiger 2004).

It is illustrative to reexamine, with the benefit of hindsight, the morphological characters that were used to unite the various scissurellid subfamilies, particularly the radular characteristics. Since Thiele (1931), common radular patterns are viewed as indicative of family level membership in many gastropods. The unique characteristics of the radulae found in the traditional subfamilies of Scissurellidae *sensu lato* have, in the past, been regarded as minor modifications of the ground pattern; *viz.* Scissurellinae, R + 5 + ∞* with fifth lateral tooth broadened; Anatominae, R + 5 + ∞ with fifth lateral tooth elongated; Larocheinae, Sutilizoninae, Temnocinclinae, R + 3–4 + ∞ (Sasaki 1998; Geiger 2003). However, the radula with a serrated central tooth, supposedly diagnostic for Scissurellidae, is of questionable value, because it is likely a common ancestral condition for vetigastropods, based on ontogeny as discussed next.

Radulae of juvenile trochoideans (Warén 1990) and abalone (Dinamani and McRae 1986; Roberts *et al.* 1999; Kawamura *et al.* 2001; Onitsuka *et al.* 2004) are very similar to scissurellid radulae and may reflect a condition similar to that in ancestral vetigastropods. Consequently, the distinct radulae of large-bodied forms (e.g., adult trochids, abalone) are derived by peramorphy. This interpretation is diametrically opposed to Batten's (1975) argument that Scissurellidae are paedomorphic. Recognition of a polyphyletic Scissurellidae *sensu lato* changes the interpretation of radular characteristics from a synapomorphy for the family to an overall constraint imposed by small size and minor, lineage-specific modification from an ancestral vetigastropod ground pattern. The fifth lateral tooth is broadened in Scissurellidae and elongated in Anatomidae, whereas there is poor separation of lateral and marginal teeth in Larocheidae, Temnocinclidae, and Sutilizonidae. In this new light, the modifications of the lateromarginal field become paramount in diagnosing these (and probably other) micro-vetigastropod families if the common serrated central tooth is recognized as a symplesiomorphy retained in juveniles of large-bodied forms by peramorphic heterochrony. This hypothesis is testable by examining the ontogeny of the radulae of early juveniles of other vetigastropod lineages.

Trochoidea is increasingly recognized as a non-monophyletic group (Geiger and Thacker 2005; Williams and Ozawa 2006), although it has been treated until recently as a clade (Hickman and McLean 1990; Hickman 1996). The characters used by Hickman and McLean (1990: 31) to diagnose Trochoidea can be grouped into two categories. General vetigastropod characters include papillate cephalic and epipodial tentacles, neck lobes, lateromarginal radular plate (shared with Fissurellidae), and shafts of the marginal radular teeth with a curved profile. Second, the organs of the pallial cavity complex are affected by heterochrony (paedomorphosis followed by varying degrees of hypermorphosis): lack of slit, emargination, or holes, and the undeveloped right gill (Crofts 1937; "reduced" right gill in Hickman and McLean 1990). In typical vetigastropods the left gill forms before the right one and is always larger (e.g., Crofts 1929). Trochoideans therefore demonstrate decoupling of major organ systems (mantle cavity, radula) with respect to heterochronic processes that affect them.

WITHIN-FAMILY PHYLOGENIES

The two best-studied groups are Haliotidae and Trochoidea. Because of its commercial importance, Haliotidae has received early and repeated attention. Brown (1993) used allozymes, whereas Lee and Vacquier (1995) sequenced the sperm lysin (data of Brown and Lee and Vacquier combined by Geiger 2000), Estes *et al.* (2005) used COI and 16S, and Streit *et al.* (2006) introduced hemocyanin as a novel marker; all show ocean basin–specific clades, indicating limited dispersal. Unfortunately, the early data sets cannot be expanded, because the homology/orthology of the lysin gene of abalone with that of other groups (e.g., Trochidae, Hellberg and Vacquier 1999) is doubtful. More species of abalone have been sampled by Ozawa and Mouri (2004), broadly showing the same topologies as the previous phylogenies.

Trochoideans have been the focus of several studies (Hickman and McLean 1990; Hickman 1996; Hellberg and Vacquier 1999; Williams and Ozawa 2006). Some of the earlier work was based on assumed monophyly that is questionable in the light of more recent evidence.

---

*Radula formula for a half-row of teeth. R = rachidian or central tooth; number = number of lateral teeth on either side of R; ∞ = many marginal teeth on either side of the lateral teeth.

Williams and Ozawa (2006) have shown with dense taxon sampling in Trochoidea that various lineages are not monophyletic and that certain diagnostic characters (calcareous operculum of "Turbinidae") have evolved more than once. Phasianellidae is certainly distinct, as also indicated by the presence of two shell muscles in *Tricolia*, although the condition in *Phasianella* is unknown. Nangammbi *et al.* (2004) are currently studying *Tricolia*. Some other groups have been investigated, particularly deep-sea and hydrothermal vent taxa (Schwarzpaul 2002; Little and Vrijenhoek 2003). McLean and Geiger (1998) produced the only phylogeny of some Fissurellidae. The three traditional genera in Pleurotomariidae (*Mikadotrochus, Entemnotrochus, Perotrochus*) have been confirmed using DNA sequences, and a new lineage (*Bayerotrochus*) was segregated from *Perotrochus* (Harasewych 2002).

ADAPTIVE RADIATION

With an estimated 3,700 Recent species and a temporal record reaching back to the early Paleozoic, Vetigastropoda is a successful lineage. Like many other invertebrates, vetigastropods radiated considerably during the Early Ordovician, and their diversity steadily increased and culminated in the Devonian (Sepkoski and Hulver 1985; Signor 1985) but then suffered a considerable decline during the end-Permian mass extinction (Sepkoski and Hulver 1985; Signor 1985; Erwin 1990). They recovered quickly and radiated in the Late Triassic, forming an important part of the remarkable gastropod fauna from the Cassian Formation (e.g., Zardini 1978; Bandel 1991; 1993a; Schwardt 1992). This Triassic radiation was largely formed by trochoid vetigastropods (Erwin 1990). Vetigastropods suffered another setback at the Triassic-Jurassic boundary, and family diversity seems to have remained relatively stable since then (Sepkoski and Hulver 1985. Their "archaeogastropod" family diversity largely reflects the evolution of diversity in vetigastropods. Sepkoski and Hulver (1985), as well as Signor (1985), used the traditional

paraphyletic taxon Archaeogastropoda; that is, they included neritimorphs and patellids as well as several problematic Paleozoic taxa, but those groups form only a minor componenet of the traditional archaeogastropods. What is striking in this diagram is that in the Paleozoic, "archaeogastropods" outnumbered all other major gastropod clades including caenogastropods. Then, at least from the Jurassic onward, "archaeogastropod" (mainly vetigastropod) diversity remained approximately stable while caenogastropods radiated tremendously (see Ponder *et al.*, Chapter 13). Thus, although vetigastropods may be called successful because they formed an important part of the marine benthos throughout the Phanerozoic, they are certainly not as successful as caenogastropods. This is also reflected in local abundances; for example, the Bouchet *et al.* (2002) survey in a Recent tropical ecosystem (New Caledonia) showed that vetigastropods contribute only 6% of all gastropod species encountered and 16% of all individuals, whereas caenogastropods comprise about 70% of all species and individuals. Jurassic to modern soft-bottom shallow-water communities are normally dominated by caenogastropods and heterobranchs, whereas in the Paleozoic such environments are commonly very rich in slit-bearing vetigastropods. The question why Caenogastropoda and Heterobranchia produced more taxa in modern times is probably related to a wider range of diet in those groups. They also occupy a greater range of habitats, including soft sediments (infaunalization), water column, freshwater, and land. There are no vetigastropods in the water column, freshwater, or land, although a limited number of taxa are found in soft sediments. External fertilization and lack of planktotrophy may be the cause of their absence in these habitats. On the other hand, trochoideans and haliotids are important constituents in intertidal and shallow subtidal communities and are also found in seagrass meadows (e.g., Shepherd 1973; Hickman 2005). It is also possible that mass extinctions selected against vetigastropods, especially the end-Triassic event, but comprehensive studies are lacking.

INNOVATIONS

The earliest vetigastropods were trochiform. Several lineages have modified this basic body plan to a limpet shape (see above). In the case of Fissurellidae and Haliotidae, this body morphology has allowed the animals to survive in environments with high wave action or strong currents. Many lineages have modified the deep slit into a foramen, a series of foramina, or have lost the opening. Geiger (2003) showed for scissurellids that the closure of the foramen evolves by a driven trend: there are significantly more closures of the foramen than secondary reopenings. An open slit makes the anterior shell margin more susceptible to breakage; hence, a closure to a foramen seems advantageous and has evolved in Scissurellidae *sensu stricto* (*Scissurella–Sinezona*), Anatomidae (*Anatoma–Sasakiconcha*: Geiger 2006b), Fissurellidae (*Emarginula–Puncturella*; see also Figures 12.1C, D), and Haliotidae (Figure 12.1B). Geiger (2006b) showed, with an aberrant specimen of a *Scissurella* with repeatedly closed foramen, how the transition from slit to multiple foramina can be accomplished.

The slitless groups (Trochoidea, Lepetodrilidae, Neomphalidae, Peltospiridae) generally have an undeveloped complement of organs on the right side of the mantle cavity, which implies a modified water circulation pattern in the mantle cavity (see Lindberg and Ponder 2001). The majority of all Vetigastropoda are Trochoidea (Table 12.1). Their lack of a slit or foramen is most likely an important part of their success. This lack is combined with a wide variety of feeding strategies and associated modifications of the radula (Hickman and McLean 1990).

BIOLOGY

Vetigastropods utilize a wide variety of foods; however, the diet of relatively few species has been investigated. Bacterial films are used by at least one shallow-water species (*Haliotis cracherodii*, Stein 1984) and possibly by hot-vent as well as by wood and whale fall inhabitants, in which grit and detritus were found as their gut contents (Haszprunar 1989; Warén and Bouchet 1989, 2001). Some trochoideans (*Lirularia*, *Bankivia*) are filter feeders (Fretter 1975; Hickman 1984b, 1996; McLean 1986), and the fissurellid *Puncturella* feeds on Foraminifera (Herbert 1991). Living Pleurotomariidae (Harasewych 2002) and Trochaclididae (Marshall 1995b) are specialized spongivores. Calliostomatidae are obligate carnivores feeding on cnidarians (commonly hydroids) and sponges (e.g., Marshall 1988, 1995a). The vast majority of species are herbivores feeding on incrusting algae to macroalgae and seagrass (Haliotidae, which also often use drift algae, Shepherd 1973; Shepherd and Steinberg 1992; Fissurellidae, Franz 1990; Trochoidea, Hickman and McLean 1990).

Reproductive strategies seem to correlate with size. Many large-bodied species are broadcast spawners (Haliotidae, Shepherd 1986; Pleurotomariidae, Harasewych 2002; Fissurellidae, Soliman 1987), while some minute species with few eggs attach them to the substrate (e.g., Strasoldo 1991). *Melanodrymia brightae* (Neomphalidae) has been shown to produce spermatophores (Warén and Bouchet 2001), and some trochoideans deposit eggs as a gelatinous mass (Soliman 1987; Hickman 1992). However, data on reproduction for most species are wanting.

Vetigastropods have independently colonized chemosynthetic habitats (hot vents, cold seeps, whale and wood falls) at least four times (Figure 12.7), more than any other molluscan group with a comparable number of species. However, they have not invaded land or freshwater, unlike their relatives the Neritimorpha.

GAPS IN KNOWLEDGE

The major obstacle to a better understanding of the evolution of Vetigastropoda is the need for better taxon sampling of microsized groups. Most scissurellids, skeneids, pendromids, and seguenziids are minute (1–5 mm in maximum size) species that require special collecting and processing techniques (see Geiger *et al.* 2007 for discussion). Many groups are found in deep water; although chemosynthetic deep-sea habitats receive much attention, the rest of the deep-sea fauna, and particularly its microfauna, needs much more study.

The characters are known from a relatively small number of rather medium- to large-bodied taxa, and within-group variation and odd taxa have rarely been added. Too often place-holder taxa are relied upon to provide character data that are (often wrongly) presumed to apply to all species included in that group, with the chance of such descrepancies increasing with the size of the group.

Problem lineages include Seguenziidae (including Guttulinae: see Quinn 1991); Eucyclinae, "Skeneidae," Solariellidae, Margaritinae, Halistylinae, and Pendromidae, along with most aberrant taxa from chemosynthetic deep-sea habitats. The extremely diverse trochoideans deserve more attention, because several lineages are strongly modified from the typical vetigastropoda/trochoidean plan (Hickman and McLean 1990). For instance, Tysanodontinae, Ataphridae (= Acremondontidae, Trochaclidinae, Paratrophininae: see Bouchet and Rocroi 2005), and Cataeginae have aberrant radulae; because their anatomy is little known, their placement within Vetigastropoda requires confirmation.

The chief problem facing morphological investigations are the vastly different body sizes and potential resulting functional constraints not related to phylogeny. The smallest species are 0.6 mm, while the largest measure over 300 mm; in linear dimension a difference of a factor of 500, by volume a factor of 125,000,000. Identifying morphological characters unaffected by such enormous size differentials is daunting. Coding of size-specific characters entails its own set of problems in phylogenetic methodologies, particularly how to code the inevitable inapplicables. Sperm ultrastructure may be an interesting candidate (e.g., Healy 1988, 1996; Hodgson 1995), but ideally it requires specialized fixation procedures.

Molecular studies are making progress, though there is a lack of suitably fixed material of some key groups (e.g., seguenziids, larocheines).

Minute species are also difficult to identify; hence, voucher specimens should always be deposited in a public collection. The supposedly universal COI primers (Folmer et al. 1994) do not amplify for several vetigastropods, and the abalone-specific primers of Hamm and Burton (2000) work only for abalone. This difficulty is particularly significant given that COI is the main fragment proposed for the DNA barcoding initiative (Hebert et al. 2003), although modified primers such as those of Williams and Ozawa (2006) may help to overcome these problems. The recently introduced hemocyanin marker (Streit et al. 2006) is promising, although its function as an oxygen carrier may impose functional constraints in sulfide-rich environments; on the other hand, its investigation may well lead to some very interesting studies in molecular evolution, as it may show particular adaptation of the oxygen carrier molecule to cope with the high sulfide concentration, which may be independent of its systematic placement (molecular functional convergence).

OUTLOOK

Vetigastropoda is a fascinating group of marine molluscs deserving more attention. The most significant insights can be gained mainly from groups composed of small-sized taxa that remain poorly defined, in part because of their rarity. A multilevel approach consisting of museum studies combined with new field collecting and live observations to advance morphological (morphometric, anatomical) as well as molecular and mineralogical investigations would probably yield the most useful results.

ACKNOWLEDGMENTS

We thank the organizers of the symposium and editors of this volume, David Lindberg and Winston Ponder, as well as all conference participants for a successful and stimulating event. James McLean and Anders Warén provided various insights; Anders Warén also kindly provided the photograph of Pendroma. Carole Hickman kindly provided information on trochoideans. Christine Thacker read the manuscript and helped improve it. This work was supported in part by NSF grant MRI0402726.

REFERENCES

Anseeuw, P., and Goto, Y. 1996. *The Living Pleurotomariidae*. Osaka, Japan: Elle Scientific Publications.

Bandel, K. 1982. Morphologie und Bildung der frühontogenetischen Schale bei conchiferen Mollusken. *Facies* 7: 1–198.

———. 1988. Early ontogenetic shell and shell structure as aids to unravel gastropod phylogeny and evolution. In *Prosobranch Phylogeny*. Edited by W. F. Ponder. *Malacological Review* Suppl. 4: 273–283.

———. 1991. Schlitzbandschnecken mit perlmutteriger Schale aus den triassischen St. Cassian Schichten der Dolomiten. *Annalen des Naturhistorischen Museums Wien* 92: 1–53.

———. 1993a. Trochomorpha (Archaeogastropoda) aus den St. Cassian-Schichten (Dolomiten, Mittlere Trias). *Annalen des Naturhistorischen Museums Wien* 95: 1–99.

———. 1993b. Evolutionary history of sinistral archaeogastropods with and without slit (Cirroidea, Vetigastropoda). *Freiberger Forschungshefte* C450: 41–81.

———. 1998. Scissurellidae als Modell für die Variationsbreite einer natürlichen Einheit der Schlitzbandschnecken (Mollusca, Archaeogastropoda). *Mitteilungen aus dem Geologisch-Paläontologischen Institut der Universität Hamburg* 81: 1–120.

Bandel, K., and Frýda, J. 1998. The systematic position of the Euomphalidae. *Senckenbergiana lethaea* 78: 103–131.

———. 1999. Notes on the evolution and higher classification of the subclass Neritimorpha (Gastropoda) with the description of some new taxa. *Geologica et Palaeontologica* 33: 219–235.

Bandel, K., and Geldmacher, W. 1996. The structure of the shell of *Patella crenata* connected with suggestions to the classification and evolution of the Archaeogastropoda. *Freiberger Forschungshefte* C 464: 1–71.

Bandel, K., Nützel, A., and Yancey, T. E. 2002. Larval shells and shell microstructures of exceptionally well-preserved Late Carboniferous gastropods from the Buckhorn Asphalt deposit (Oklahoma, USA). *Senckenbergiana lethaea* 82: 639–690.

Batten, R. L. 1972. The ultrastructure of five common Pennsylvanian pleurotomarian gastropod species of eastern United States. *American Museum Novitates* 2501: 1–34.

———. 1975. The Scissurellidae—are they neotenously drived Fissurellidae? (Archaeogastropoda). *American Museum Novitates* 2567: 1–29.

Beesley, P. L., Ross, G. J. B., and Wells, A., eds. *Mollusca: the Southern Synthesis*. Melbourne: CSIRO Publishing.

Begg, J. G., and Grant-Mackie, J. A. 2003. New Zealand and New Caledonian Triassic Pleurotomariidae (Gastropoda, Mollusca). *Journal of the Royal Society of New Zealand* 33: 223–268.

Bieler, R. 1992. Gastropod phylogeny and systematics. *Annual Review of Ecology and Systematics* 23: 311–338.

Bouchet, P., Lozouet, P. Maestrati, P., and Héros, V. 2002. Assessing the magnitude of species richness in tropical marine environments: exceptionally high numbers of molluscs at a New Caledonia site. *Biological Journal of the Linnean Society* 75: 421–436.

Bouchet, P., and Rocroi, J.-P. 2005. A nomenclator and classification of gastropod family-group names. *Malacologia* 47: 1–397.

Bourne, G. C. 1910. On the anatomy and systematic position of *Incisura* (*Scissurella*) *lytteltonensis*. *Quarterly Journal of Microscopical Sciences* 55: 1–45.

Brown, L. D. 1993. Biochemical genetics and species relationships within the genus *Haliotis* (Gastropoda: Haliotidae). *Journal of Molluscan Studies* 59: 429–443.

Carter, J. G., and Hall, R. M. 1990. Polyplacophora, Archaeogastropoda and Paragastropoda (Mollusca). In *Skeletal Biomineralization: Patterns, Processes and Evolutionary Trends*. Vol. 2. *Atlas and Index*. Edited by J. G. Carter. New York: Van Nostrand Reinhold, pp. 29–134.

Colgan, D. J., Ponder, W. F., and Eggler, P. E. 2000. Gastropod evolutionary rates and phylogenetic relationships assessed using partial 28S rDNA and histone H3 sequences. *Zoologica Scripta* 29: 29–63.

Colgan, D. J., Ponder, W. F., Beacham, E., and Macaranas, J. M. 2003. Gastropod phylogeny based on six segments from four genes representing coding or non-coding and mitochondrial or nuclear DNA. *Molluscan Research* 23: 123–148.

Cox, L. R. 1960. Gastropoda. General characteristics of Gastropoda. In *Treatise on Invertebrate Paleontology*, Part I, *Mollusca*. Edited by R. C. Moore. Lawrence, KS: Geological Society of America and University of Kansas Press. Pp. I85–I169.

Crofts, D. R. 1929. *Haliotis. Liverpool Marine Biology Committee Memoirs* 29: 1–174, pls. 1–8.

———. 1937. The development of *Haliotis tuberculata* with special reference to organogenesis during torsion. *Philosophical Transactions of the Royal Society of London Series. B: Biological Sciences* 228: 219–266.

Dantart, L., and Luque, A. 1994. Cocculiniformia and Lepetidae (Gastropoda: Archaeogastropoda) from the Iberian coasts. *Journal of Molluscan Studies* 60: 277–314.

Dinamani, M., and McRae, C. 1986. Paua settlement: the prelude. *Catch Shellfisheries Newsletter* 30: 9.

Dzik, J. 1994. Evolution of "small shelly fossils" assemblages of the Early Paleozoic. *Acta Palaeontologica Polonica* 39: 247–313.

Erwin, D. H. 1990. Carboniferous-Triassic gastropod diversity patterns and the Permo-Triassic mass extinction. *Paleobiology* 16: 187–203.

Estes, J. A., Lindberg, D. R., and Wray, C. 2005. Evolution of large body size in abalone (*Haliotis*): patterns and implications. *Paleobiology* 31: 591–606.

Folmer, O., Black, M., Hoew, W., Lutz, R., and Vrijenhoek, R. 1994. DNA primers for amplification of mitochondrial cytochrome *c* oxidase subunit I from diverse metazoan invertebrates. *Molecular Marine Biology and Biotechnology* 3: 294–299.

Franz, C. J. 1990. Differential algal consumption by three species of *Fissurella* (Mollusca: Gastropoda) at Isla de Margarita, Venezuela. *Bulletin of Marine Science* 46: 735–748.

Fretter, V. 1975. *Umbonium vestiarum*, a filter feeding trochid. *Journal of Zoology* 177: 541–522.

———. 1988. New archaeogastropod limpets from hydrothermal vents; superfamily Lepetodrilacea. II. Anatomy. *Philosophical Transactions of the Royal Society of London Series B: Biological Sciences* 218: 123–169.

Fretter, V., and Graham, A. 1962. *British Prosobranch Molluscs*. London: Ray Society.

———, 1994. *British Prosobranch Molluscs*. Rev. and updated ed. London: Ray Society.

Frýda, J. 1997. Oldest representatives of the superfamily Cirroidea (Vetigastropoda) with notes on early phylogeny. *Journal of Paleontology* 71: 839–847.

———. 1999. Higher classification of Paleozoic gastropods inferred from their early shell ontogeny. *Journal of the Czech Geological Society* 44: 137–154.

Frýda, J., and Bandel, K. 1997: New early Devonian gastropods from the *Plectonotus* (*Boucotonotus*)–Palaeozygopleura community in the Prague Basin (Bohemia). *Mitteilungen des Geologisch-Paläontologischen Instituts der Universität Hamburg* 80: 1–57.

Frýda, J., and Manda, S. 1997. A gastropod faunule from the *Monograptus uniformis* graptolite biozone (Early Lochkovian, Early Devonian) in Bohemia. *Mitteilungen des Geologisch-Paläontologischen Instituts der Universität Hamburg* 80: 59–121.

Frýda, J., Rieder, M. Klementová, M., Weitschat, W., and Bandel, K. 2004. Discovery of gastropod-type nacre in fossil cephalopods: a tale of two crystallographic textures. In *Molluscan Megadiversity: Sea, Land and Freshwater. World Congress of Malacology Perth, Western Australia 11–16 July 2004*. Edited by F. E. Wells. Perth: Western Australian Museum, p. 43.

Frýda, J., and Rohr, D. M. 2004. Gastropods. In *The Great Ordovician Biodiversification Event*. Edited by B. D. Webby, F. Paris, M. L. Droser, and I. G. Percival. New York: Columbia University Press, pp. 184–195.

Geiger, D. L. 2000. Distribution and biogeography of the Haliotidae (Gastropoda: Vetigastropoda) world-wide. *Bollettino Malacologico* 35: 57–120.

———. 2003. Phylogenetic assessment of characters proposed for the generic classification of Recent Scissurellidae (Gastropoda: Vetigastropoda) with a description of one new genus and six new species from Easter Island and Australia. *Molluscan Research* 23: 21–83.

———. 2004. Made for SEM: the fascinating world of scissurellids. *American Conchologist* 32(4): 4–7.

———. 2006a. Eight new species of Scissurellidae and Anatomidae (Mollusca: Gastropoda: Vetigastropoda) from around the world, with discussion of two new senior synonyms. *Zootaxa* 1128: 1–33.

———. 2006b. *Sasakiconcha elegantissima* new genus and species (Gastropoda: Vetigastropoda: Anatomidae?) with disjointly coiled base. *The Nautilus* 120: 45–51.

Geiger, D. L., and Jansen, P. 2004a. Revision of the Australian species of Anatomidae (Gastropoda: Vetigastropoda). *Zootaxa* 435: 1–35.

———. 2004b. New species of Australian Scissurellidae (Mollusca: Gastropoda: Vetigastropoda) with remarks on Australian and Indo-Malayan species. *Zootaxa* 714: 1–72.

Geiger, D. L., Marshall, B. A., Ponder, W. F., Sasaki, T., and Warén, A. 2007. Techniques for collecting, handling, preparing, storing and examining small molluscan specimens. *Molluscan Research* 27: 1–50.

Geiger, D. L., and Poppe, G. T. 2000. Haliotidae. In *Conchological Iconography*. Edited by G. T. Poppe and K. Groh. Hackenheim: Conch Books, pp. 1–135, pls. 1–82.

Geiger, D. L., and Thacker, C. E. 2005. Molecular phylogeny of Vetigastropoda reveals non-monophyletic Scissurellidae, Trochoidea, and Fissurelloidea. *Molluscan Research* 25: 47–55.

Giribet, G., and Wheeler, E. 2002. On bivalve phylogeny: a high-level analysis of the Bivalvia (Mollusca) based on combined morphology and DNA sequence data. *Invertebrate Biology* 121: 271–324.

Giribet, G., Okusu, A., Lindgren, A. R, Huff, S. W., Schrödl, M., and Nishiguchi, M. K. 2006. Evidence for a clade composed of molluscs with serially repeated structures: Monoplacophora are related to chitons. *Proceedings of the National Academy of Sciences of the United States of America* 103: 7723–7728.

Golikov, A. N., and Starobogatov, Y. I. 1975. Systematics of prosobranch gastropods. *Malacologia* 15: 185–232.

Gordon, M., Jr., and Yochelson, E. L. 1987. Late Mississippian gastropods from the Chainman Shale, west-central Utah. *United States Geological Survey Professional Paper* 1368: 1–112.

Hamm, D. E., and Burton, R. S. 2000. Population genetics of black abalone, *Haliotis cracherodii*, along the central California coast. *Journal of Experimental Marine Biology and Ecology* 254: 235–247.

Harasewych, M. G. 2002. Pleurotomarioidean gastropods. *Advances in Marine Biology* 42: 238–294.

Harasewych, M. G., Adamkewicz, S. L., Blake, J. A., Saudek, D., Spriggs, T., and Bult, C. J. 1997. Phylogeny and relationships of pleurotomariid gastropods (Mollusca: Gastropoda) and assessment based on partial 18S rDNA and cytochrome *c* oxidase I sequences. *Molecular Marine Biology and Biotechnology* 6: 1–20.

Harper, J. A., and Rollins, H. B. 2000. The bellerophont controversy revisited. *American Malacological Bulletin* 15: 157–156.

Haszprunar, G. 1985a. The fine morphology of the osphradial sense organs of the Mollusca. Part 1: Gastropoda Prosobranchia. *Philosophical Transactions of the Royal Society of London Series B: Biological Sciences* 307: 457–496.

———. 1985b. On the innervation of gastropod shell muscles. *Journal of Molluscan Studies* 51: 309–314.

———. 1987. The fine structure of the ctenidial sense organs (bursicles) of Vetigastropoda (Zeugobranchia, Trochoidea) and their functional and phylogenetic significance. *Journal of Molluscan Studies* 53: 46–51.

———. 1988a. On the origin and evolution of major gastropod groups, with special reference to the Streptoneura (Mollusca). *Journal of Molluscan Studies* 54: 367–441.

———. 1988b. Comparative anatomy of cocculiniform gastropods and its bearing on archaeogastropod systematics. In *Prosobranch Phylogeny*. Edited by W. F. Ponder. *Malacological Review* Suppl. 4: 64–84.

———. 1988c. Anatomy and affinities of pseudococculinid limpets (Mollusca, Archaeogastropoda). *Zoologica Scripta* 17: 161–179.

———. 1989. The anatomy of *Melanodrymia aurantiaca* Hickman, a coiled archaeogastropod from the East Pacific hydrothermal vents (Mollusca: Gastropoda). *Acta Zoologica* 70: 175–186.

———. 1993. The Archaeogastropoda. A clade, a grade or what else? *American Malacological Bulletin* 10: 165–177.

———. 1998. Superorder Cocculiniformia. In *Mollusca: the Southern Synthesis*. Edited by P. L. Beesley, G. J. B. Ross, and A. Wells. Melbourne: CSIRO Publishing, pp. 653–664.

Haszprunar, G., and McLean, J. H. 1996. Anatomy and systematics of bathyphytophilid limpets (Mollusca, Archaeogastropoda) from the northeastern Pacific. *Zoologica Scripta* 25: 35–49.

Hayashi, I. 1983. Larval shell morphology of some Japanese *Haliotis* for the identification of their veliger larvae and early juveniles. *Venus* 42: 49–58.

Healy, J. M. 1988. Sperm morphology and its systematic importance in the Gastropoda. In *Prosobranch Phylogeny*. Edited by W. F. Ponder. *Malacological Review* Suppl. 4: 251–266.

———. 1996. Molluscan sperm ultrastructure: correlation with taxonomic units within the Gastropoda, Cephalopoda and Bivalvia. In *Origin and Evolutionary Radiation of the Mollusca*. Edited by J. Taylor. Oxford: Oxford University Press, pp. 99–113.

Hebert, P. D. N., Cywinska, A., Ball, S. L., and de Waard, J. R. 2003. Biological identifications through DNA barcodes. *Proceedings of the Royal Society of London Series B: Biological Sciences* 270: 313–322.

Hedegaard, C. 1997. Shell structures of the Recent Vetigastropoda. *Journal of Molluscan Studies* 63: 369–377.

Hellberg, M. E., and Vacquier, V. D. 1999. Rapid evolution of fertilization selectivity and lysin cDNA sequences in teguline gastropods. *Molecular Biology and Evolution* 16: 839–848.

Herbert, D. G. 1991. Foraminiferivory in a *Puncturella* (Gastropoda: Fissurellidae). *Journal of Molluscan Studies* 57: 127–129.

Hickman, C. S. 1984a. A new archaeogastropod (Rhipidoglossa, Trochacea) from hydrothermal vents on the East Pacific Rise. *Zoologica Scripta* 13: 19–25.

———. 1984b. Implications of radular tooth-row functional integration for archaeogastropod systematics. *Malacologia* 25: 143–160.

————. 1988. Archaeogastropod evolution, phylogeny and systematics: a re-evaluation. In *Prosobranch Phylogeny*. Edited by W. F. Ponder. *Malacological Review* Suppl. 4: 17–34.

————. 1992. Reproduction and development of trochacean gastropods. *The Veliger* 35: 245–272.

————. 1996. Phylogeny and patterns of evolutionary radiation in trochoidean gastropods. In *Origin and Evolutionary Radiation of the Mollusca*. Edited by J. Taylor. Oxford: Oxford University Press, pp. 171–176.

————. 2005. Seagrass fauna of the temperate southern coast of Australia I: The cantharidine trochid gastropods. In *The Marine Flora and Fauna of Esperance, Western Australia*. Edited by F. E. Wells, D. I. Walker, and G. A. Kendrick. Perth: Western Australian Museum, pp. 199–220.

Hickman, C. S., and McLean, J. H. 1990. Systematic revision and suprageneric classification of trochacean gastropods. *Science Series, Natural History Museum of Los Angeles County* 35: 1–169.

Hodgson, A. N. 1995. Spermatozoal morphology of Patellogastropoda and Vetigastropoda (Mollusca: Prosobranchia). *Mémoirs du Muséum National d'Histoire Naturelle* 166: 167–177.

Kaim, A. 2004. The evolution of conch ontogeny in Mesozoic open sea gastropods. *Palaeontologia Polonica* 62: 1–183.

Kawamura, T., Takami, H., Roberts, R. D., and Yamashita, Y. 2001. Radula development in abalone *Haliotis discus hannai* from larva to adult in relation to feeding transition. *Fisheries Science* 67: 596–605.

Kiel, S. 2004. Shell structures of selected gastropods from hydrothermal vents and seeps. *Malacologia* 46: 169–183.

Kiel, S., and Bandel, K. 2002. Further Archaeogastropoda from the Campanian of Torallola, northern Spain. *Acta Geologica Polonica* 52: 239–249.

————. 2004. The Cenomanian Gastropoda of the Kassenberg quarry in Mühlheim (Germany, Late Cretaceous). *Paläontologische Zeitschrift* 78: 97–103.

Kiel, S., and Frýda, J. 2004. Nacre in the Late Cretaceous *Sensuitrochus ferreri*—implications for the taxonomic affinities of the Cirridae (Gastropoda). *Journal of Paleontology* 78: 795–797.

Knight, J. B., Cox, L. R., Keen, A. M., Batten, R. L., Yochelson, E. L., and Robertson, R. 1960. Systematic descriptions. In *Treatise on Invertebrate Paleontology*, part I, *Mollusca* 1. Edited by R. C. Moore. Lawrence: University of Kansas Press, pp. I169–I331.

Künz, E., and Haszprunar, G. 2001. Comparative ultrastructure of gastropod cephalic tentacles: Patellogastropoda, Neritaemorphi and Vetigastropoda. *Zoologischer Anzeiger* 240: 137–165.

Lee, Y.-H., and Vacquier, V. D. 1995. Evolution and systematics in Haliotidae (Mollusca, Gastropoda): inference from DNA sequences of sperm lysin. *Marine Biology* 124: 267–278.

Lindberg, D. R., and Ponder, W. F. 2001. The influence of classification on the evolutionary interpretation of structure: a re-evaluation of the evolution of the pallial cavity in gastropod molluscs. *Organisms Diversity and Evolution* 1: 273–299.

Lindberg, D. R., Ponder, W. F., and Haszprunar, G. 2004. The Mollusca: relationships and patterns from their first half-billion years. In *Assembling the Tree of Life*. Edited by J. Cracraft and M. J. Donoghue. New York: Oxford University Press, pp. 252–278.

Little, C. T. S., and Vrijenhoek, R. C. 2003. Are hydrothermal vent animals living fossils? *Trends in Ecology and Evolution* 18: 582–588.

Marshall, B. A. 1988. Thysanodontinae: a new subfamily of the Trochidae (Gastropoda). *Journal of Molluscan Studies* 54: 249–282.

————. 1995a. Calliostomatidae (Gastropoda: Trochoidea) from New Caledonia, the Loyalty Islands, and the northern Lord Howe Rise. *Mémoires du Muséum National d'Histoire Naturelle* 167: 381–458.

————. 1995b. Recent and Tertiary Trochaclididae from the southwest Pacific (Mollusca: Gastropoda: Trochoidea). *The Veliger* 38: 92–115.

Mazaev, A. V. 1998. A new genus of Fissurelloidea (Gastropoda) from the Upper Carboniferous of Moscow Basin: the oldest known example of the suborder. *Ruthenica* 8: 13–15.

————. 2002. Some murchisoniid gastropods from the Middle and Upper Carboniferous part of Russian Plate. *Ruthenica* 12: 89–106.

McLean, J. H. 1984. A case for derivation of the Fissurellidae from the Bellerophontacea. *Malacologia* 25: 3–20.

————. 1986. The trochid genus *Lirularia* Dall, 1909: A filter feeder? *The Western Malacological Society Annual Report 1985* 18: 24–25.

————. 1988. New archaeogastropod limpets from hydrothermal vents; Superfamily Lepetodrilacea I. Systematic descriptions. *Philosophical Transactions of the Royal Society London Series B: Biological Sciences* 319: 1–32.

————. 1989. New slit-limpets (Scissurellacea and Fissurellacea) from hydrothermal vents. Part 1 Systematic description and comparison based on shell and radular characters.. *Contributions in Science of the Los Angeles County Museum of Natural History* 407: 1–29.

———. 1992. A new species of *Pseudorimula* (Fissurellidae: Clypeosectidae) from hydrothermal vents of the mid Atlantic ridge. *The Nautilus* 106: 115–118.

McLean, J. H., and Geiger, D. L. 1998. New genera and species having the *Fissurisepta* shell form, with a generic-level phylogenetic analysis (Gastropoda: Fissurellidae). *Contributions in Science, Natural History Museum of Los Angeles County* 475: 1–32.

Nangammbi, T. C., Herbert, D. G., and Mitchell, A. 2004. Phylogeny of southern African *Tricolia* (Gastropoda: Turbinidae) based on mtDNA sequences. In *Molluscan Megadiversity: Sea, Land and Freshwater. World Congress of Malacology, Perth, Western Australia, 11–16 July 2004.* Edited by F. E. Wells. Perth: Western Australian Museum, pp. 104–105.

Nützel, A. 2002. An evaluation of the recently proposed Palaeozoic gastropod subclass Euomphalomorpha. *Palaeontology* 45: 259–266.

Nützel, A., and Bandel, K. 2000. Goniasmidae and Orthonemidae: two new families of Palaeozoic Caenogastropoda. *Neues Jahrbuch für Geologie und Paläontologie, Monatshefte* 9: 557–569.

Nützel, A., and Erwin, D. 2004. Late Triassic (Late Norian) gastropods from the Wallowa terrane (Idaho, USA). *Paläontologische Zeitschrift* 78: 361–416.

Nützel, A., and Frýda, J. 2003. Paleozoic plankton revolution: Evidence from early gastropod ontogeny. *Geology* 31: 829–831.

Nützel, A., and Geiger, D. L. 2006. A new scissurelloid genus and species (Mollusca, Gastropoda) from the Late Triassic Cassian Formation. *Paläontologische Zeitschrift* 80: 277–283.

Nützel, A., and Mapes, R. H. 2001. Larval and juvenile gastropods from a Mississippian Black Shale: Paleoecology, and implications for the evolution of the Gastropoda. *Lethaia* 34: 143–162.

Okutani, T. 2000. *Marine Mollusks in Japan.* Tokyo: Tokai University Press.

Onitsuka, T., Kawamura, T., Ohashi, S., Horii, T., and Watanabe, Y. 2004. Morphological changes in the radula of abalone *Haliotis diversicolor aquatilis* from post-larva to adult. *Journal of Shellfish Research* 23: 1079–1085.

Ozawa, T., and Mouri, Y. 2004. Molecular phylogeny and historical biogeography of Haliotidae (Gastropoda: Vetigastropoda) based on mitochondrial DNA sequences. In *Molluscan Megadiversity: Sea, Land and Freshwater. World Congress of Malacology, Perth, Western Australia, 11–16 July 2004.* Edited by F. E. Wells. Perth: Western Australian Museum, pp. 112–113.

Page, L. R. 1997. Ontogenetic torsion and protoconch form in the archaeogastropod *Haliotis kamtschatkana*: evolutionary implications. *Acta Zoologica* 78: 227–245.

Pan, H.-Z., and Erwin, D. H. 2002. Gastropods from the Permian of Guanxi and Yunnan Provinces, South China. *Journal of Paleontology* 76 Suppl. 1: 1–49.

Ponder, W. F. 1998. Vetigastropoda. In *Mollusca: the Southern Synthesis.* Edited by P. L. Beesley, G. J. B. Ross, and A. Wells. Melbourne: CSIRO Publishing, p. 664.

Ponder, W. F., and Lindberg, D. R. 1996. Gastropoda phylogeny—challenges for the 90s. In *Origin and Evolutionary Radiation of the Mollusca.* Edited by J. Taylor. Oxford: Oxford University Press, pp. 135–154.

———. 1997. Towards a phylogeny of gastropod molluscs: an analysis using morphological characters. *Zoological Journal of the Linnean Society* 119: 83–265.

Ponder, W. F., and Warén, A. 1988. Appendix, classification of the Caenogastropoda and Heterostropha—a list of the family-group names and higher taxa. *Malacological Review* 4: 288–326.

Quinn, J. F., Jr. 1991. Systematic position of *Basilissopsis* and *Guttula*, and a discussion of the phylogeny of the Seguenzioidea (Gastropoda: Prosobranchia). *Bulletin of Marine Science* 49: 575–598.

Roberts, R. D., Kawamura, T., and Takami, H. 1999. Morphological changes in the radula of abalone (*Haliotis iris*) during post-larval development. *Journal of Shellfish Research* 18: 637–644.

Roldán, E., Backeljau, T., Breugelmans, K., and Luque, Á. 2001. New molecular data on cocculiniform relationships. In *Abstracts, World Congress of Malacology 2001, Vienna, Austria.* Edited by L. v. Salvini-Plawen, J. Voltzow, H. Sattmann, and G. Steiner. Vienna: Unitas Malacologica, p. 295.

Salvini-Plawen, L. v. 1980. A consideration of systematics in the Mollusca (Phylogeny and higher classification. *Malacologia* 19: 249–278.

Salvini-Plawen, L. v., and Haszprunar, G. 1987. The Vetigastropoda and the systematics of streptoneurous Gastropoda (Mollusca). *Journal of Zoology London* 211: 747–770.

Salvini-Plawen, L. v., and Steiner, G. 1996. Synapomorphies and plesiomorphies in higher classification of Mollusca. In *Origin and Evolutionary Radiation of the Mollusca.* Edited by J. Taylor. Oxford: Oxford University Press, pp. 29–52.

Sasaki, T. 1998. Comparative anatomy and phylogeny of the Recent Archaeogastropoda (Mollusca: Gastropoda). *The University of Tokyo Bulletin* 38: 1–223.

Sasaki, T., Okutani, T. and Fujikura, K. 2006. Anatomy of *Shinkailepas myojinensis* Sasaki, Okutani and Fujikura, 2003. *Malacologia* 48: 1–26.

Schwardt, A. 1992. Revision der *Wortheniella*-Gruppe (Archaeogastropoda) der Cassianer Schichten (Trias, Dolomiten). *Annalen des Naturhistorischen Museums, Wien* 94 A: 23–57.

Schwarzpaul, K. 2002. Phylogenie hydrothermaler "Archaeogastropoden" der Tiefsee—morphologische und molekulare Untersuchungen. Doctoral dissertation, Philipps University, Marburg, Germany.

Sepkoski, J. J. Jr., and Hulver, M. L. 1985. An atlas of Phanerozoic clade diversity diagrams. In *Phanerozoic diversity patterns*. Edited by J. W. Valentine. Princeton, NJ: Princeton University Press, pp. 11–39.

Shepherd, S. A. 1973. Studies on southern Australian abalones 1: ecology of five sympatric species. *Australian Journal of Marine and Freshwater Research* 24: 217–257.

———. 1986. Movement of the Southern Australian abalone *Haliotis laevigata* in relation to crevice abundance. *Australian Journal of Ecology* 11: 295–302.

Shepherd, S. A., and Steinberg, P. D. 1992. Food preferences of three Australian abalone species with a review of the algal food of abalone. In *Abalone of the World. Biology, Fisheries and Culture*. Edited by S. A. Shepherd, M. J. Tegner, and S. A. Guzmán del Próo. Oxford: Fishing News Press, pp. 169–181.

Signor, P. W. 1985. Gastropod evolutionary history. In *Mollusks. Notes for a Short Course*. Edited by T. W. Broadhead. University of Tennessee Department of Geological Sciences Studies in Geology 13: 157–173.

Soliman, G. N. 1987. Scheme for classifying gastropod egg masses with special reference to those from the northwestern Red Sea. *Journal of Molluscan Studies* 53: 1–12.

Stein, J. L. 1984. Subtidal gastropods consume sulfur-oxidizing bacteria: evidence from coastal hydrothermal vents. *Science* 223: 696–698.

Strasoldo, M. 1991. Anatomie und Ontogenie von *Scissurella jucunda* (Smith, 1890) und Anatomie von *Anatoma* sp. Doctoral Dissertation. University of Vienna, Austria.

Streit, K., Geiger, D. L., and Lieb, B. 2006. Molecular phylogeny and the geographic origin of Haliotidae traced by hemocyanin sequences. *Journal of Molluscan Studies* 72: 111–116.

Strong, E. 2003. Refining molluscan characters: morphology, character coding and a phylogeny of the Caenogastropoda. *Zoological Journal of the Linnean Society* 137: 447–554.

Strong, E. E., Harasewych, M. G. and Haszprunar, G. 2003. Phylogeny of Cocculinoidea (Mollusca, Gastropoda). *Invertebrate Biology* 122: 114–125.

Szal, R. 1971. "New" sense organ of primitive gastropods. *Nature* 229: 490–492.

Thiele, J. H. 1931. *Handbuch der systematischen Weichtierkunde*. Jena: Gustav Fischer.

Tillier, S., Masselot, M., Guerdoux, J., and Tillier, A. 1994. Monophyly of major gastropod taxa tested from partial 28S rRNA sequences, with emphasis on Ethyneura and hot-vent limpets Peltospiroidea. *The Nautilus* Suppl. 2: 122–140.

Tracey, S., Todd, J. A., and Erwin, D. H. 1993. Mollusca: Gastropoda. In *The Fossil Record 2*. Edited by M. J. Benton. London: Chapman and Hall, pp. 131–167.

Van Dover, C. L., Humpris, S. E., Fornari, D., Cavanaugh, C. M., Collier, R., Goffredi, S. K., Hashimoto, J., Lilley, M. D., Reysenbach, A. L., Shank, T. M., Von Damm, K. L., Banta, A., Gallanz, R. M., Gotz, D., Green, D., Hall, J., Harmer, T. L., Hurtado, L. A., Johnson, P., McKiness, Z. P., Meredith, C., Olson, E., Pan, I. L., Turnipsees, M., Won, Y., Young, C. R., and Vrijenhoek, R. C. 2001. Biogeography and ecological setting of Indian Ocean hydrothermal vents. *Science* 294: 818–823.

Wagner, P. J. 2001. Rate heterogeneity in shell character evolution among lophospiroid gastropods. *Paleobiology* 27: 290–310.

———. 2002. Phylogenetic relationships of the earliest anisostrophically coiled gastropods. *Smithsonian Contributions to Paleobiology* 88: 1–152.

Warén, A. 1990. Ontogenetic changes in the trochoidean (Archaeogastropoda) radula, with some phylogenetic interpretations. *Zoologica Scripta* 19: 179–187.

———. 1996. Description of *Bathysciadium xylophagum* Warén and Carozza, sp. n. and comments on *Addisonia excentrica* (Tiberi), two Mediterranean cocculiniform gastropods. *Bollettino Malacologico* 31: 231–266.

Warén, A., and Bouchet, P. 1989. New gastropods from East Pacific hydrothermal vents. *Zoologica Scripta* 18: 67–102.

———. 2001. Gastropoda and Monoplacophora from hydrothermal vents and seeps; new taxa and records. *The Veliger* 44: 116–231.

Welsch, U., and Storch, V. 1969. Über das Osphradium der prosobranchen Schnecken *Buccinum undatum* and *Neptunea antiqua*. *Zeitschrift für Zellforschung und Mikroskopische Anatomie* 95: 317–330.

Williams, S. T., and Ozawa, T. 2006. Molecular phylogeny suggest polyphyly of both the turban shells (family Turbinidae) and the superfamily Trochoidea (Mollusca: Vetigastropoda) *Molecular Phylogenetics and Evolution* 39: 33–51.

Wilson, B. 1993. *Australian Marine Shells, Vol. 1*. Kallaroo, Western Australia: Odyssey.

Wingstrand, K. G. 1985. On the anatomy and relationships of Recent Monoplacophora. *Galathea Report* 16: 7–94, 12 pls.

Yochelson, E. L. 1984. Historic and current considerations for revision of Paleozoic gastropod classification. *Journal of Paleontology* 58: 259–269.

Yoo, E. K. 1994. Early Carboniferous Gastropoda from the Tamworth Belt, New South Wales, Australia. *Records of the Australian Museum* 46: 63–110.

Yoon, S. H., and Kim, W. 2005. Phylogenetic relationships among six vetigastropoda subgroups (Mollusca, Gastropoda) based on 18S rDNA sequences. *Molecules and Cells* 19: 283–288.

Zardini, R. 1978. *Fossili Cassiani (Trias Medio-Superiore). Atlante dei Gasteropodi della formazione di S. Cassiano raccolti nella regione Dolomitica attorno a Cortina D'Ampezzo.* Cortina d'Ampezzo: Edizione Ghedina.

13

Caenogastropoda

Winston F. Ponder, Donald J. Colgan,
John M. Healy, Alexander Nützel,
Luiz R. L. Simone, and Ellen E. Strong

Caenogastropods comprise about 60% of living gastropod species and include a large number of ecologically and commercially important marine families. They have undergone an extraordinary adaptive radiation, resulting in considerable morphological, ecological, physiological, and behavioral diversity. There is a wide array of often convergent shell morphologies (Figure 13.1), with the typically coiled shell being tall-spired to globose or flattened, with some uncoiled or limpet-like and others with the shells reduced or, rarely, lost. There are also considerable modifications to the head-foot and mantle through the group (Figure 13.2) and major dietary specializations. It is our aim in this chapter to review the phylogeny of this group, with emphasis on the areas of expertise of the authors.

The first records of undisputed caenogastropods are from the middle and upper Paleozoic, and there were significant radiations during the Jurassic, Cretaceous, and Paleogene (see subsequent section on the fossil record). They have diversified into a wide range of habitats and have successfully invaded freshwater and terrestrial ecosystems multiple times.

Many caenogastropods are well-known marine snails and include the Littorinidae (periwinkles), Cypraeidae (cowries), Cerithiidae (creepers), Calyptraeidae (slipper limpets), Tonnidae (tuns), Cassidae (helmet shells), Ranellidae (tritons), Strombidae (strombs), Naticidae (moon snails), Muricidae (rock shells, oyster drills, etc.), Volutidae (balers, etc.), Mitridae (miters), Buccinidae (whelks), Terebridae (augers), and Conidae (cones). There are also well-known freshwater families such as the Viviparidae, Thiaridae, and Hydrobiidae and a few terrestrial groups, notably the Cyclophoroidea.

Although there are no reliable estimates of named species, living caenogastropods are one of the most diverse metazoan clades. Most families are marine, and many (e.g., Strombidae, Cypraeidae, Ovulidae, Cerithiopsidae, Triphoridae, Olividae, Mitridae, Costellariidae, Terebridae, Turridae, Conidae) have large numbers of tropical taxa. A few families have diversified more in cooler waters (e.g., Buccinidae, Eatoniellidae, Struthiolariidae), and many others are diverse in both temperate and tropical seas. Caenogastropod diversity has increased, especially since the Mesozoic (Sepkoski and Hulver

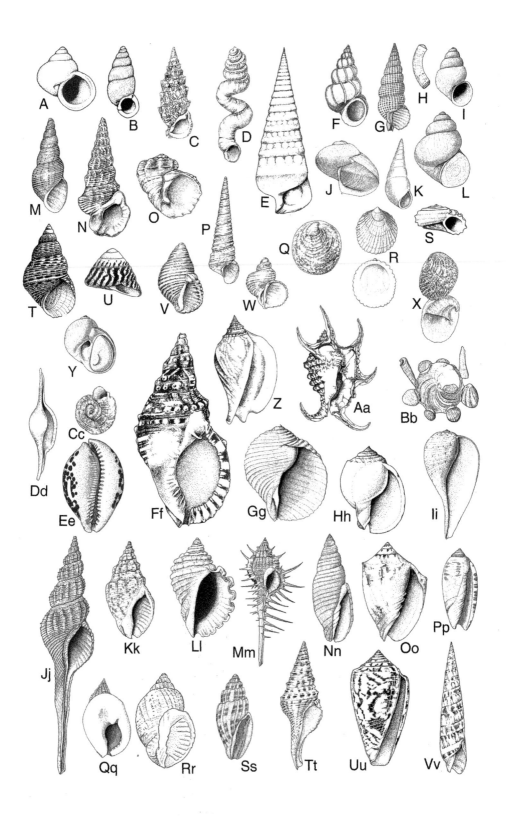

FIGURE 13.1. (Opposite.) Shells of some Recent caenogastropods showing the range of morphology. (A) *Leptopoma* (Cyclophoridae); (B) *Pupinella* (Pupinidae); (C) *Pseudovertagus* (Cerithiidae); (D) *Tenagodus* (Siliquariidae); (E) *Campanile* (Campanilidae); (F) *Epitonium* (Epitoniidae); (G) *Ataxocerithium* (Newtoniellidae); (H) *Caecum* (Caecidae); (I) *Austropyrgus* (Hydrobiidae *sensu lato*); (J) *Janthina* (Janthinidae); (K) *Monogamus* (Eulimidae); (L) *Gabbia* (Bithyniidae); (M) *Melanoides* (Thiaridae); (N) *Pyrazus* (Batillariidae); (O) *Modulus* (Modulidae); (P) *Colpospira* (Turritellidae); (Q) *Capulus* (Capulidae); (R) *Sabia* (Hipponicidae); (S) *Circulus* (Vitrinellidae); (T) *Littoraria* (Littorinidae); (U) *Bembicium* (Littorinidae); (V) *Planaxis* (Planaxidae); (W) *Sirius* (Capulidae); (X) *Crepidula* (Calyptraeidae); (Y) *Notocochlis* (Naticidae); (Z) *Strombus* (Strombidae); (Aa) *Lambis* (Strombidae); (Bb) *Xenophora* (Xenophoridae); (Cc) *Serpulorbis* (Vermetidae); (Dd) *Volva* (Ovulidae); (Ee) *Cypraea* (Cypraeidae); (Ff) *Charonia* (Ranellidae); (Gg) *Tonna* (Tonnidae); (Hh) *Semicassis* (Cassidae); (Ii) *Ficus* (Ficidae); (Jj) *Fusinus* (Fasciolariidae); (Kk) *Cominella* (Buccinidae); (Ll) *Dicathais* (Muricidae); (Mm) *Murex* (Muricidae); (Nn) *Cancilla* (Mitridae); (Oo) *Cymbiola* (Volutidae); (Pp) *Oliva* (Olividae); (Qq) *Nassarius* (Nassariidae); (Rr) *Cancellaria* (Cancellariidae); (Ss) *Eucithara* (Turridae *sensu lato*); (Tt) *Lophiotoma* (Turridae); (Uu) *Conus* (Conidae); (Vv) *Terebra* (Terebridae). All figures reproduced with permission from Beesley *et al.* (1998), some slightly modified. Not to scale.

1985; Signor 1985), accelerating during the Cretaceous with the radiation of neogastropods and other predatory gastropods (Sohl 1964; Taylor *et al.* 1980), suggesting that diet and competition (e.g., Vermeij 1978, 1987) played a significant role in their adaptive radiation (discussed later in this chapter). The pattern of steady diversification in caenogastropods differs from that of most other major groups of marine metazoans, which are characterized by marked waxing and waning or even complete extinction (e.g., Ammonoidea).

Since the late Paleozoic, caenogastropods form large portions of the richest known gastropod faunas, both in numbers of species and in abundance of individuals (Table 13.1).

PHYLOGENY AND CLASSIFICATION

SISTER GROUP RELATIONSHIPS

Heterobranchia are usually shown as the sister group to caenogastropods in recent analyses involving extant taxa using morphological (e.g., Ponder and Lindberg 1997) or molecular data (e.g., Tillier *et al.* 1992; McArthur and Harasewych 2003). However, this is not always the case, with neritimorphs being the sister taxon in some molecular analyses such as those of Colgan *et al.* (2000, 2003, 2007) and Aktipis *et al.* (Chapter 9). Whereas modern neritimorphs have highly convolute larval shells with resorbed inner walls, the assumed early members of this clade had planktotrophic larval shells, which are not fundamentally different from those of

caenogastropods (as discussed subsequently). A sister group relationship of Caenogastropoda with Neritimorpha may be just as feasible as one with the heterobranchs, especially as all three groups share aragonitic crossed-lamellar shell structure and lack of nacre. While heterobranchs have a heterostrophic larval shell, in neritimorphs and caenogastropods the larval shell is orthostrophic.

Neritimorphs are sister to the apogastropods (Caenogastropoda + Heterobranchia) in the supertree analysis of published molecular and morphological trees in McArthur and Harasewych (2003, fig. 6.2) and Aktipis *et al.* (Chapter 9), sharing (perhaps convergently) the development of complex genital ducts, the reduction of the pallial organs on the right side, and larval planktotrophy. Some other analyses (Ponder and Lindberg 1997; Colgan *et al.* 2003) show vetigastropods as the sister to the apogastropods.

Heterobranchs have been regarded to be the sister group to caenogastropods by many paleontologists (e.g., Bandel and Geldmacher 1996; Kaim 2004; Figure 13.5A), but some evidence suggests they may not be (Nützel, unpublished data; Figure 13.5B). Putative early heterobranchs have subulitid-like or turreted teleoconchs (Frýda and Bandel 1997), perhaps suggesting a caenogastropod relationship. However, they coexisted with several caenogastropod clades, suggesting that the stem group evolved earlier than indicated by the fossil record. An alternative explanation might be that Heterobranchia

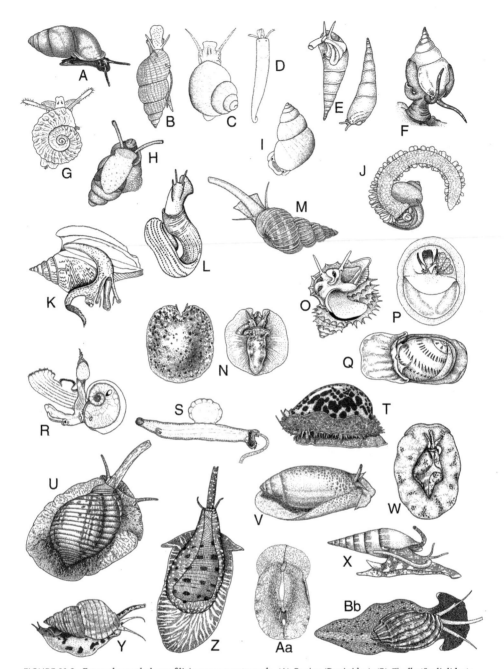

FIGURE 13.2. External morphology of living caenogastropods. (A) *Pupina* (Pupinidae); (B) *Finella* (Scaliolidae); (C) *Gabbia* (Bithyniidae); (D) *Parastrophia* (Caecidae); (E) *Vitreolina* (Eulimidae); (F) *Echineulima* (Eulimidae); (G) *Pseudoliotia* (Vitrinellidae); (H) *Ascorhis* (Hydrobiidae *sensu lato*); (I) *Cryptassiminea* (Assimineidae); (J) *Janthina* (Janthinidae); (K) *Strombus* (Strombidae); (L) *Aletes* (Vermetidae); (M) *Epitonium* (Epitoniidae); (N) *Lamellaria* (Velutinidae); (O) *Trichotropis* (Capulidae); (P) *Sabia* (Hipponicidae); (Q) *Euspira* (Naticidae); (R) *Atlanta* (Atlantidae); (S) *Firoloida* (Pterotracheidae); (T) *Cypraea* (Cypraeidae); (U) *Tonna* (Tonnidae); (V) *Oliva* (Olividae); (W) *Austroginella* (Marginellidae); (X) *Mitrella* (Columbellidae); (Y) *Nassarius* (Nassariidae); (Z) *Ficus* (Ficidae); (Aa) *Ancillista* (Olividae); (Bb) *Harpa* (Harpidae). All figures reproduced with permission from Beesley *et al.* (1998), some slightly modified. Not to scale.

TABLE 13.1

Proportions of Caenogastropod Species and Individuals

FORMATION	REFERENCE	MYA	% SPECIES	% INDIVIDUALS
Recent, New Caledonia[a]	Bouchet *et al.* 2002	0	72	63
Late Triassic, Mission Creek, United States	Nützel and Erwin 2004	205	52	88
Late Triassic, Pucara Formation, Peru	Haas 1953	210	41	29
Late Carboniferous, United States	Kues and Batten 2001	310	56	32

NOTE: Proportions are relative to the total numbers of all gastropods of some rich gastropod collections from the late Paleozoic, early Mesozoic and from a Recent Indo-West Pacific site. Caenogastropods form a major or even dominant part of these faunas.
[a]This site also includes shell-less taxa, so it is not strictly comparable with the fossil faunas; with the shell-less taxa excluded, the proportion of caenogastropods would be higher.

is nested within Caenogastropoda (e.g., Nützel unpublished, fig. 13.5B); although this possibility is concordant with views expressed by some (e.g., Gosliner 1981), we consider it highly unlikely on the basis of current molecular, morphological, and ultrastructural evidence (e.g., Ponder and Lindberg 1997).

In summary, although the sister group of caenogastropods is often the heterobranchs in analyses involving extant taxa, it is also possible that an extinct, non-heterobranch taxon (e.g., a neritimorph) is the actual sister group. Although fossil taxa shed some light on sister taxon relationships, the data are by no means clear (see also Frýda *et al.*, Chapter 10).

OUTLINE OF CLASSIFICATION/PHYLOGENY

RECOGNITION OF THE CLADE CAENOGASTROPODA

Although the name Caenogastropoda was used by Cox (1960a, b) nearly half a century ago to encompass Thiele's (1925–1926) Mesogastropoda and Stenoglossa (= Neogastropoda Wenz, 1938–1944), the general recognition of this group has been relatively recent. Caenogastropods were incorporated in a paraphyletic subclass, Prosobranchia, in synoptic works (e.g., Thiele 1929–1931; Wenz 1938–1944; Cox 1960b) (Figure 13.3A) and the taxonomic overviews of Taylor and Sohl (1962), Ponder and Warén (1988) and Vaught (1989), and they often continue to be treated as such in many textbooks (e.g., Brusca and Brusca 2002) and other literature, keys, indexing systems, and checklists.

Golikov and Starobogatov's (1975) revolutionary classification of gastropods included what we now know as caenogastropods within a subclass Pectinibranchia, which also included neritimorphs and some vetigastropods, notably Trochoidea. Extant caenogastropods, in this scheme, were diphyletic, with two superorders, Cerithiimorpha and Littorinimorpha, both of which included non-caenogastropod taxa, derived independently from "Anisobranchia," a group containing several vetigastropod families. Graham (1985: 174) also found that "the boundary between animals that are clearly archaeogastropod or caenogastropod is extremely blurred," and, in particular, he referred to the architaenioglossan groups and the vent-living neomphaloideans (see Geiger *et al.*, Chapter 12) as being problematic.

Haszprunar's (1988) ground-breaking analysis of gastropod relationships using morphology, including osphradial ultrastructure, showed an unresolved Caenogastropoda (Figure 13.3B) included with Architaenioglossa in a paraphyletic "Archaeogastropoda." This arrangement was modified by Ponder and Warén (1988) (Figure 13.3C), who, as in the morphological cladistic analyses of Ponder and Lindberg (1996, 1997) (Figure 13.3D), had the architaenioglossans as the sister group to the rest of the caenogastropods (the Sorbeoconcha), with the "mesogastropod" groups Cerithioidea and Campaniloidea as sister taxa to the remaining caenogastropods (Hypsogastropoda). The great majority

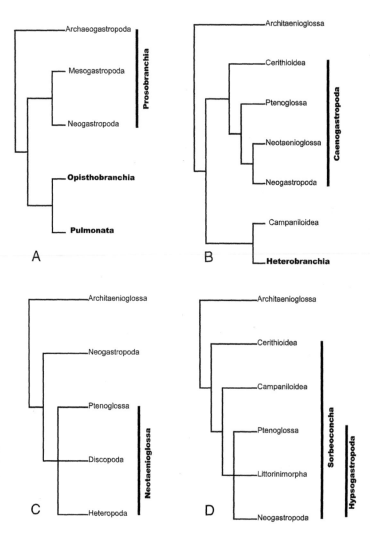

FIGURE 13.3. Some alternative hypotheses from morphology prior to 1999. (A) Thiele (1929–1931). (B) Haszprunar (1988). (C) Ponder and Warén (1988). (D) Ponder and Lindberg (1997).

of caenogastropods are contained within Hypsogastropoda, where there is little resolution to date. Strong's (2003) morphological analysis maintained a monophyletic Architaenioglossa and Neogastropoda, but the sorbeoconchan taxa were contained in two separate clades, one of which also included the only cerithioidean in her analysis (Figure 13.4A).

The major burst of interest in gastropod phylogeny in the last three decades (see Bieler 1992; Ponder and Lindberg 1997; Aktipis *et al.*, Chapter 9) has identified and delineated the major monophyletic groups and most analyses have recognized Caenogastropoda as a clade (Salvini-Plawen and Haszprunar 1987; Ponder

and Warén 1988; Bieler 1992; Tillier *et al.* 1992, 1994; Rosenberg *et al.* 1994; Ponder and Lindberg 1996, 1997; Taylor 1996; McArthur and Koop 1999; Colgan *et al.* 2000, 2003, 2007; Strong 2003; McArthur and Harasewych 2003).

In terms of rank, Cox (1960b) treated Caenogastropoda as an order, but Bandel (1991b, 1993; Bandel and Riedel 1994) used Caenogastropoda as a subclass. It was treated as a superorder by Ponder and Warén (1988; within Prosobranchia) and Beesley *et al.* (1998; within Orthogastropoda), while Ponder and Lindberg (1997), Strong (2003), and Bouchet and Rocroi (2005) treated caenogastropods as an unranked major clade.

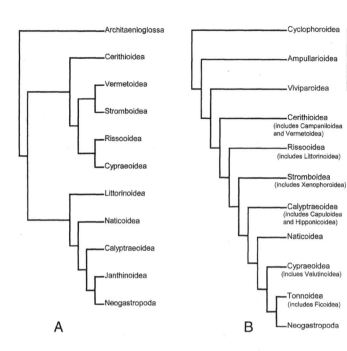

A   B

FIGURE 13.4. Phylogenies from the most recent analyses of caenogastropod phylogeny based on morphology. (A) Strong (2003). (B) Simone (2000a).

In summary, caenogastropods are currently thought to comprise the majority of the Mesogastropoda of Thiele (1929–1931) and all of the Neogastropoda. Several groups (Architectonicoidea, Rissoellidae, Omalogyridae, Pyramidellidae, Valvatidae) previously included in Mesogastropoda are now included in Heterobranchia (Haszprunar 1985b, 1988; Ponder and Warén 1988; Healy 1990d, 1993b; Bouchet and Rocroi 2005). Although the monophyly of Caenogastropoda is well supported in recent morphological analyses (Ponder and Lindberg 1997; Strong 2003), it is often not strongly supported in molecular analyses (see discussion below).

MAIN DISTINGUISHING FEATURES

Caenogastropods are defined by a number of significant characters, including a shell that is typically coiled, with a multispiral, orthostrophic protoconch and crossed-lamellar shell structure. The foot is typically simple and usually bears an operculum. The mantle cavity organs are reduced, including a single (left) monopectinate ctenidium with skeletal rods; a single left osphradium, which is typically hypertrophied and has unique histology; and a single (left) hypobranchial gland. The heart has a single auricle, and the rectum never passes through the ventricle. Only the left kidney remains, although elements of the right kidney are incorporated in the oviduct. There is a single pair of buccal cartilages, and the radula is plesiomorphically taenioglossate. The esophagus lacks conspicuous ventral folds, the intestine is not markedly looped, and fecal pellets are produced. Pallial genital ducts enable internal fertilization and, consequently, the production of encapsulated eggs and nonplanktonic early development (i.e., lacking a trochophore stage). Planktonic larvae are often planktotrophic. The nervous system is concentrated with well-defined cerebral and pedal ganglia. Most caenogastropods are epiathroid (the pleural ganglia lie close to, or are fused with, the cerebral ganglia), in contrast to the condition in vetigastropods, where the pleural ganglia are close to the pedal ganglia (the hypoathroid condition).

PHYLOGENY OF CAENOGASTROPODA

Despite the great diversity and extensive fossil record of caenogastropods, detailed relationships within the group have remained largely unresolved, although a few broad groups have usually been recognized (Table 13.2). Release

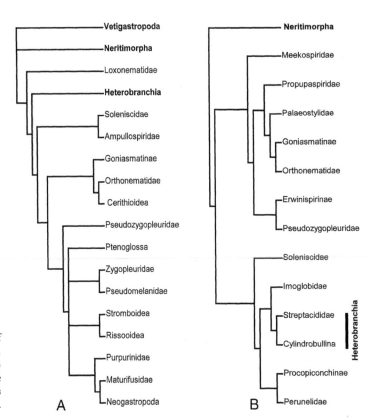

FIGURE 13.5. Examples of phylogenies based on fossil taxa. (A) Tree based on Kaim's (2004) hypothesis; (B) majority-rule tree based on 25 shell characters (Nützel, unpublished).

A

Vetigastropoda
Neritimorpha
Loxonematidae
Heterobranchia
Soleniscidae
Ampullospiridae
Goniasmatinae
Orthonematidae
Cerithioidea
Pseudozygopleuridae
Ptenoglossa
Zygopleuridae
Pseudomelanidae
Stromboidea
Rissooidea
Purpurinidae
Maturifusidae
Neogastropoda

B

Neritimorpha
Meekospiridae
Propupaspiridae
Palaeostylidae
Goniasmatinae
Orthonematidae
Erwinispirinae
Pseudozygopleuridae
Soleniscidae
Imoglobidae
Streptacididae
Cylindrobullina
Procopiconchinae
Perunelidae

Heterobranchia

from the constraints of ancestral adult morphologies (Ponder and Lindberg 1997) appears to have played a major part in the evolution of the group, although the differentiation of innovation and ancestral conditions in various traits has often been difficult or impossible to determine in the absence of a robust phylogeny. Compounding this lack of resolution are the significant and rapid radiations that occurred during caenogastropod history and, in particular, during the latter part of the Mesozoic.

MAIN GROUPS RECOGNIZED WITHIN CAENOGASTROPODS

ARCHITAENIOGLOSSA This exclusively nonmarine grouping, which may represent a grade, not a clade, is usually regarded as the sister to all other living caenogastropods. It comprises the terrestrial cyclophoroideans (the major group of operculate land snails) and two freshwater families, previously in Ampullarioidea but now included in separate superfamilies,

Ampullarioidea (Ampullariidae) and Viviparoidea (Viviparidae). While having some shared plesiomorphic characters, including a partially or fully hypoathroid nervous system and subradular organ, the included taxa do not share any obvious synapomorphies. Considerable modification in some features has occurred as a result of the nonmarine habitat of all living architaenioglossans. These include the protoconch, which is typical of many direct-developing caenogastropods; in cyclophorids, modifications due to terrestriality, notably loss of the pallial organs; in ampullariids, development of a separate lung in the mantle cavity as a consequence of their amphibious habits; and, in viviparids, modifications due to filter feeding.

Architaenioglossa was included as part of the Mesogastropoda by Thiele (1929–1931) and Wenz (1938–1944) and included in the caenogastropods by Cox (1960b) and Ponder and Warén (1988) but excluded by Haszprunar (1988, 1993). In Simone's (2004b) analysis the

TABLE 13.2

Some Recent Classifications of Living Caenogastropoda

HASZPRUNAR 1985a	PONDER AND WARÉN 1988	HASZPRUNAR 1988[a]	PONDER AND LINDBERG 1997	BOUCHET AND ROCROI 2005
Caenogastropoda	Caenogastropoda	Archaeogastropoda	Caenogastropoda	Caenogastropoda
Architaenioglossa (including Valvatoidea)	Architaenioglossa	Architaenioglossa	Architaenioglossa	Architaenioglossa (informal)
Neotaenioglossa	Neotaenioglossa	Apogastropoda[b]	Sorbeoconcha (Cerithioidea and Campaniloidea)	Sorbeoconcha (Cerithioidea and Campaniloidea)
Heteroglossa	Discopoda (including Campanilidae)	Caenogastropoda	Hypsogastropoda	Hypsogastropoda
Stenoglossa	Ptenoglossa	Cerithiimorpha	Ptenoglossa	Littorinimorpha
	Heteropoda	Ctenoglossa	Littorinimorpha	Ptenoglossa (informal)
	Neogastropoda	Neotaenioglossa	Neogastropoda	Neogastropoda
		Stenoglossa		
		Campanilimorpha		
		(Campanilidae)		

NOTE: See text for explanation of names.

[a]Haszprunar (1988) provided several alternative classifications; the one summarized in the table is based on his figure 5, "the most probable phylogenetic reconstruction."

[b]Haszprunar also included in Apogastropoda Ectobranchia (Valvatidae), Allogastropoda (Architectonicidae, Rissoellidae, Omalogyridae), while the remaining heterobranchs (*sensu* Haszprunar 1985b; Ponder and Warén 1988; Ponder and Lindberg 1997) were included in the "subclass" Euthyneura. Ponder and Lindberg (1997) used Apogastropoda to encompass both Caenogastropoda and Heterobranchia.

architaenioglossan taxa were paraphyletic, forming three branches of basal caenogastropods.

SORBEOCONCHA   This term was introduced by Ponder and Lindberg (1997) to include all caenogastropods other than the architaenioglossan clade or grade. Basal members are Cerithioidea and Campaniloidea, the former including numerous freshwater taxa, notably the Thiaridae and several related families (e.g., Lydeard *et al.* 2002). This grouping differs primarily from the architaenioglossan grade in having a primary emphasis on control of the inhalant water flow rather than the exhalant flow (Ponder and Lindberg 1997; Lindberg and Ponder 2001), with corresponding emphasis on the chemosensory role of the osphradium. This is correlated with three synapomorphic osphradial characters (Haszprunar 1985a, 1988; Ponder and Lindberg 1997): an increase in size, the presence of ciliated lateral fields, and Si4 cells (Haszprunar 1985a). Other synapomorphies identified by Ponder and Lindberg (1997) include an epiathroid nervous system, the formation of a seminal vesicle, a coiled radular sac, and the formation of a polar lobe in early development (Freeman and Lundelius 1992).

HYPSOGASTROPODA   This term was introduced by Ponder and Lindberg (1997) to include the great majority of extant caenogastropods (most "mesogastropods" and all neogastropods)—that is, all caenogastropods other than architaenioglossans, Cerithioidea, and Campaniloidea. It equates with the "higher caenogastropods" of Healy (1988a) and is defined by Ponder and Lindberg (1997) by some sperm characters (Healy 1988a); a single statolith in each statocyst, rather than several statoconia; exophallic penis (Simone 2000a); osphradial Si1 and Si2 cells (Haszprunar 1985a); and absorptive cells in the larvae (Ruthensteiner and Schaefer 1991). Members are largely marine and include families of medium to large size such as the Littorinidae, Cypraeidae, Calyptraeidae, Tonnidae, Cassidae, Ranellidae, Strombidae, and Naticidae; small-sized families such as the very diverse Rissoidae, Triphoridae, and Eulimidae; and the pelagic heteropods. Freshwater families include the rissooidean Hydrobiidae and some related families, some of these groups having undergone large radiations. A few terrestrial groups are found in the Littorinoidea (Pomatiidae) and Rissooidea (some Truncatellidae, Pomatiopsidae, and Assimineidae). A major subgroup of Hypsogastropoda is the Neogastropoda (= Stenoglossa), members of which are almost exclusively marine and virtually all are carnivorous. It contains well-known, diverse, and ecologically significant families such as Muricidae, Volutidae, Mitridae, Buccinidae, and conoideans (Turridae *sensu lato*, Terebridae and Conidae). Members of this large clade share several apomorphies (Ponder and Lindberg 1997; Strong 2003) related to the digestive system, including unique structures such as a rectal (=anal) gland, tubular accessory salivary glands, and the possession of either a stenoglossan or a toxoglossan radula. Additional significant characters include the salivary gland ducts not passing through the nerve ring, the esophageal gland separated from the esophagus (as the gland of Leiblein or poison gland), and the enlargement of the ventral tensor muscle of the radula (m11 of Simone 2003), working to enable the sliding movement of the radula (this muscle does not function in this way in other caenogastropods; Simone 2000a).

OTHER GROUPINGS   Neotaenioglossa is a paraphyletic (e.g., Ponder and Lindberg 1997) grouping used by Haszprunar (1988) and Ponder and Warén (1988), but with different concepts. The latter encompassed all the non-architaenioglossan "mesogastropods" (other than those now treated as basal heterobranchs), whereas the former excluded the Campaniloidea, Cerithioidea, and Ptenoglossa.

The higher category names Cerithimorpha and Littorinimorpha were used by Golikov and Starobogatov (1975), the former being used for the basal group of caenogastropods (from which architaenioglossans were excluded) by Haszprunar (1988). Most recently Bouchet and Rocroi (2005) have used Littorinimorpha to encompass the taenioglossate Hypsogastropoda.

Heteropoda comprises only the pelagic Ptero-tracheoidea (= Carinarioidea) and was used as a high-rank taxon until recently. It is included in the Littorinimorpha by Bouchet and Rocroi (2005).

Ptenoglossa (= Ctenoglossa, e.g., Bandel 1993) is a probably polyphyletic grouping (see following paragraphs) of Eulimoidea, Janthinoidea, and Triphoroidea. This assemblage was based on the presence of an acrembolic[1] proboscis and, in some members, a ptenoglossate radula and two pairs of salivary glands. The group, with Eulimoidea excluded, is treated as monophyletic by Nützel (1998) in a study using fossil and Recent taxa. Ponder and Lindberg (1997) argue that the broader concept of Ptenoglossa (including Eulimoidea) is polyphyletic, and this view is supported in molecular analyses (Colgan et al. 2000, 2003, 2007).

Suggestions of heterobranch affinities of Epitoniidae have been made by Robertson (1985) and Collin (1997) on the basis of the supposed homology of pigmented mantle organs and shared hydrophobic larval shells, respectively, but other characters and molecular data do not support such a relationship.

Heterogastropoda was erected by Kosuge (1966) to encompass Ptenoglossa and Architectonicoidea,[2] which he considered to lie between the meso- and neogastropods.

For further discussion of the status and composition of some of these groups, see the later section "Summary of Major Groups."

Present caenogastropod classifications are essentially based on a few key shell and anatomical (including radular) details, although Healy (e.g., 1988a, 1996b; see also below) used sperm ultrastructure to determine the relationships of several groups. Available data suggests that some of the currently recognized higher taxa (orders, suborders) are probably paraphyletic or even polyphyletic, and the

relationships of intermediate groups (superfamilies, families) are unresolved. Only one phylogenetic hypothesis has been previously published for caenogastropods as a whole[3] based on morphological data (Strong 2003), but phylogenies of some family group (or higher) taxa within caenogastropods have been proposed, examples being Rissooidea (Ponder 1988; Wilke et al. 2001); Cerithioidea (Houbrick 1988; Ponder 1991; Lydeard et al. 2002; Simone 2001); Neogastropoda (Taylor and Morris 1988; Kantor 1996); Littorinidae (Reid 1989; Williams et al. 2003); Ampullariidae (Berthold 1991; Bieler 1993); Olivioidea (Kantor 1991; Kantor and Pavlinov 1991); Muricidae (Rapaninae) (Kool 1993); toxoglossans (Conoidea) (Taylor et al. 1993; Kantor 1996; Rosenberg 1998; Simone 2000b; Kantor and Taylor 2002); "neomesogastropods" (Bandel and Riedel 1994); Tonnoidea (as Cassoidea) (Riedel 1995, 2000); Ptenoglossa (excluding Eulimoidea) (Nützel 1998); Columbellidae (deMaintenon 1999); Nassariidae (Haasl 2000); Calyptraeoidea (Simone 2002, 2006; Collin 2003); Muricidae (Oliverio et al. 2002); Cypraeidae (Meyer 2003, 2004); Stromboidea (Simone 2005), Cypraeoidea (Simone 2004a); Architaenioglossa (Simone 2004b); and Buccinidae (Hayashi 2005).

THE PALEONTOLOGICAL PERSPECTIVE

Although several higher-level caenogastropod taxa have been proposed in the paleontological literature, there are few attempts to frame explicit hypotheses using cladistic methodology and even fewer involving fossil caenogastropods using maximum-parsimony methods. Frýda (1999) introduced Perunelomorpha, a group with open-coiled protoconchs, initially as a sister taxon to Caenogastropoda, but they were later (Bouchet and Rocroi 2005) incorporated within it. Bandel (1991b, 1993, 2002) proposed higher taxa, some based primarily on the time of their appearance in the fossil record

---

1. In an acrembolic proboscis, the proboscis retractor muscles are attached to the distal end of the proboscis so that the buccal mass lies behind the retracted proboscis (Fretter and Graham 1962).

2. Now in Heterobranchia.

3. Simone (2000a) presented a phylogeny (see Figure 13.4B), but as of this writing the full details have not yet been published.

and not on explicit phylogenetic hypotheses. These include the Palaeo-Caenogastropoda for those with Paleozoic origins, the Meta-Mesogastropoda for those first appearing in mid Mesozoic times, and the Neo-Mesogastropoda, which appear in the late Mesozoic and are united by an "expanded ontogeny," but does not include the Neogastropoda. Bandel (1993) also erected the Scaphoconchoidea for taxa with larvae that have their true larval shell surrounded by a pseudoshell—the echinospira and limacosphaera types of larvae. Riedel (2000) proposed two additional high-level taxa: the Latrogastropoda and Vermivora. Latrogastropoda included the Pleurembolica [encompassing Troschelina (Troschelina Bandel and Riedel, 1994 composed of Laubierinioidea and Calyptraeoidea) + Vermivora]. Vermivora included Ficoidea, Tonnoidea, and Neogastropoda.

Although comprehensive analyses of fossil caenogastropods are difficult because there are comparatively few shell characters, there are cladistic hypotheses using parsimony methods for a few groups (Roy 1994 for Aporrhaidae; Nützel *et al.* 2000 for Subulitoidea; Wagner 2002 for early Paleozoic gastropods including caenogastropod ancestors). Other phylogenetic studies involving fossil taxa but not using parsimony methods include studies on "higher" caenogastropods (Latrogastropoda: Neomesogastropoda + Neogastropoda) by Bandel and Riedel (1994) and Riedel (2000). According to Riedel (2000), the origin of Latrogastropoda is obscure, but he suggested a relationship with rissooideans and proposed Ficoidea as the sister taxon of neogastropods. Kowalke (1998: figs. 12, 13) presented phylogenetic hypotheses for Cerithimorpha (*sensu* Golikov and Starobogatov 1975 (= Cerithioidea *sensu lato*)) and vermetoideans that are almost exclusively based on larval shell morphology (especially ornament) of a few Cretaceous, Cenozoic, and Recent representatives of these groups. Nützel (1998) investigated ptenoglossans (excluding Eulimoidea) and their possible stem groups. Modern Triphoroidea formed a clade with the extinct Protorculidae as the sister group. Janthinoidea and

the fossil Zygopleuridae were the sister groups to that clade. The Paleozoic Pseudozygopleuridae were identified as the extinct stem group to Triphoroidea + Janthinoidea, while the Paleozoic precursors of the Cerithioidea were shown to be the possible sister group to the combined grouping. Although the monophyly of Recent ptenoglossans is widely seen as unlikely (see subsequent discussion), the long separation of the triphorid/cerithiopsid line from that of the Janthinoidea, as suggested by the fossil record, could explain the marked disparity of their living representatives. Nützel *et al.* (2000) analyzed the Late Paleozoic Subulitoidea, and although several genera could be arranged in family level groups (Soleniscidae, Meekospiridae, Imoglobidae), monophyly of the ingroup could not be established with various outgroups, suggesting probable nonmonophyly of the traditional Subulitoidea. Polyphyly has also been hypothesized for the Early Paleozoic Subulitoidea (Wagner 2001, 2002).

MORPHOLOGICAL DATA

Traditional taxonomic work on caenogastropods has mainly focused on shell and radular characters, and this is one of the reasons there is a paucity of anatomical information for many groups.

SHELL/PROTOCONCH

The shell (Figures 13.1, 13.2, 13.6, 13.13) is typically coiled, very elongate to flattened, loosely coiled to uncoiled, as in Vermetidae (Figure. 13.1Cc) and Caecidae (Figure 13.1H), or openly coiled, as in Siliquariidae (Figure 13.1D) and a few members of other families (e.g., Cyclophoridae, Epitoniidae, Hydrobiidae *sensu lato*; Rex and Boss 1976). Others have secondarily become limpet-like (Capulidae, Figure 13.1Q; Calyptraeidae, Figure 13.1X; and Hipponicidae), while a few families produce one or two limpet-like taxa (e.g., *Thyca* in Eulimidae; *Concholepas* in Muricidae; *Quoyula* in Coralliophilidae). Remarkably, only two small groups of caenogastropods have lost their postlarval shell: a few endoparasitic, worm-like Eulimidae and the pelagic Pterotracheidae.

FIGURE 13.6. Selected Paleozoic (Carboniferous/Permian) caenogastropods and putative outgroup taxa. (A, B) Protoconch of a naticopsid (Neritimorpha) from the Mississippian (Lower Carboniferous, ca. 330 Mya; Ruddle Shale, Arkansas, United States). Protoconchs of Recent neritimorphs are highly convolute with resorbed inner whorls; however, protoconchs of Naticopsidae show no major differences from those of caenogastropods, except for a relatively high whorl expansion rate. Width 0.8 mm (from Nützel and Mapes 2001). (C, D) Cerithimorph caenogastropod from the Upper Carboniferous (Late Moscovian, c. 305 Mya; Buckhorn Asphalt deposit, Oklahoma, United States); this small heliciform, planktotrophic larval shell resembles the protoconch of some modern cerithioids. C, height 2.0 mm; D, height 0.4 mm. (E–G) *Stegocoelia* (Goniasmatidae, Palaeostyloidea), a slit-bearing caenogastropod from the Upper Carboniferous (Late Moscovian, c. 305 Mya; Buckhorn Asphalt deposit, Oklahoma, United States), representative of a rich late Paleozoic group of *Murchisonia*-resembling caenogastropods; E, teleoconch detail showing slitlike structure (selenizone) slightly above mid-whorl; width 0.5 mm; F, height 2.0 mm; G, protoconch in side view, a lecithotrophic larval shell with a distinct sinusigera. Protoconchs of planktotrophic species of this group resemble Figure 13.6D; height 0.36 mm (from Bandel *et al.* 2002). (H) *Soleniscus*, a widespread subulitoid (Soleniscidae) from the Upper Carboniferous (Gzhelian, c. 300 Mya; Finis Shale, Texas, United States), showing a distinct, twisted siphonal canal, a columellar fold, and a smooth larval shell; height 3.2 mm (from Nützel *et al.* 2000). (I–K) *Imogloba* (Imoglobidae) from the Mississippian (Lower Carboniferous, c. 330 Mya of Arkansas, United States); these globular, subulitoid gastropods have an open coiled initial whorl followed by early whorls (probably larval shell) with a very characteristic ornament of noncollabral threads; I, height 2.5 mm; J, probably isolated larval shell, height 0.85 mm; K, width 4.4 mm. (L, M) *Pseudozygopleura* (Pseudozygopleuridae) from the Late Carboniferous (ca. 300 Mya, Gzhelian, Ames Shale, West Virginia, United States); pseudozygopleurids were abundant and diverse for about 100 million years (during the late Paleozoic) and became extinct at the end-Permian mass extinction event; they have highly characteristic larval shells with an ornament of curving, collabral ribs that form a spiral thread (from Nützel 1998); L, larval shell; height 0.86 mm; M, height 3.0 mm.

The sluglike Velutinidae (= Lamellariidae) (Fig. 13.2N) have a reduced internal shell. Most caenogastropods have dextral shells, with members of only one family (Triphoridae) being almost entirely sinistral, although sinistral taxa occur sporadically in some families (Robertson 1993).

Preliminary phylogenetic analyses (Nützel, unpublished) suggest that high-spired shells could be diagnostic for some Paleozoic and early

Mesozoic clades. Although there are many cases of convergent teleoconch morphology with representatives of each of the other major gastropod clades (vetigastropods, neritimorphs, heterobranchs), the protoconch morphology and shell microstructure can be used to determine the group. Specialists are usually able to recognize members of a particular group reliably by the teleoconch morphology alone, suggesting that this is often diagnostic, although differences can be subtle and difficult to quantify, resulting in a low number of scorable characters. It is also frequently difficult to establish homology and consistent coding of shell characters because many are not sufficiently complex to reject convergence convincingly. Nevertheless, Wagner (2002) comprehensively coded teleoconch characters of Early Paleozoic gastropods, and Schander and Sundberg (2001) have shown that shell characters can provide a similar level of resolution to other data sets in some analyses.

Growth lines and the shape of the outer lip can reflect the organization of the mantle cavity (especially inhalant and exhalant flows) and the orientation of the shell. For paleontologists, these not only provide clues about the way the organism functioned but can have phylogenetic significance. Many modern caenogastropods have straight or slightly opisthocline growth lines, with inhalant and exhalant flows restricted to the anterior and posterior corners of the aperture, which are often modified with siphonal notches or canals. This configuration is seen in the Late Paleozoic Pseudozygopleuridae and Subulitoidea, but many other fossil caenogastropods have strongly parasigmoidal (loxonematoid) growth lines or possess slits, sinuses, and selenizones. A slit occurs in Late Paleozoic probable caenogastropods[4] of the family Goniasmatidae (Nützel and Bandel 2000; Bandel et al. 2002; Nützel and Pan 2005). Although it has been suggested that the slit or deep sinus in this group may indicate the presence of a pair

of ctenidia (Cox 1960b: 143), slits (e.g., Siliquariidae) or deep notches (e.g., some Turritellidae) are known in a few modern caenogastropods with a single monopectinate gill.

When shell growth ceases and the lips of the aperture thicken, they are sometimes modified in shape or have special ornament. This determinate growth is found in many caenogastropods but is typically clade specific (Vermeij and Signor 1992). Apertural thickening may also occur intermittently during growth, and the thickened part of the outer lip may be retained as a distinct varix.

The shell microstructure of caenogastropods is aragonitic crossed-lamellar, and although comprehensive comparative studies are lacking, there are some indications of significant variation (e.g., Falniowski 1989), although these differences can sometimes be correlated with environment (e.g., Taylor and Reid 1990). Nacre is absent in all Caenogastropoda, and calcite is rare (e.g., Epitoniidae) (see Bandel 1990 for a review of caenogastropod shell microstructure).

Protoconchs, as extensively shown by the work of Bandel (1982 and subsequently), are a rich source of characters and reflect life history. The following parameters have proved useful for defining at least species and sometimes genera: size and ornament of embryonic shell; size and shape of entire protoconch; number of larval whorls; transition from protoconch to teleoconch (abrupt, presence or absence of sinusigera, fluent) and whorl shape. Protoconch morphology can be very useful for taxonomy at generic and species levels (e.g., Triphoridae, Marshall 1983), but they are also assumed to be diagnostic for caenogastropods because larvae of modern representatives of basal clades (vetigastropods, patellogastropods) are never planktotrophic. Heterobranchia have sinistral protoconchs, and modern neritimorphs have highly characteristic convolute larval shells (e.g., Bandel 1982). Thus, multi-whorled, orthostrophic larval shells characteristic of planktotrophic larvae are present only in Caenogastropoda. As indicated previously, the situation appears to

---

4. They have crossed-lamellar shell structure, have multiwhorled planktotrophic larval shells, and are high spired.

be more complicated in the Paleozoic because presumed early neritimorphs (Naticopsidae and Trachyspiridae) can have planktotrophic larval shells (Figure 13.6A, B) that show no signs of resorbed inner whorls (Nützel and Mapes 2001; Nützel *et al.* 2007) and thus are not fundamentally different from those of caenogastropods. Moreover, some Paleozoic gastropods with typical pleurotomarioid teleoconch morphology (and thus assumed to be vetigastropods) have simple, smooth larval shells of about two whorls (Nützel and Mapes 2001; Kaim 2004).

In caenogastropods, it is especially easy to infer larval feeding strategies (i.e., planktotrophic vs. non-planktotrophic) from the larval shell with well-preserved material (e.g., Bandel 1982). Jablonski (1986) tested possible selectivity of larval strategies at the end-Cretaceous extinction event, and Nützel (1998) separated larval strategies of Late Paleozoic and Triassic zygopleuroid gastropods based on protoconch measurements.

Although the phylogenetic utility of shell characters is considerably increased with protoconch data, the latter have greater value if coded from species with the same or a similar ontogenetic strategy (e.g., only from planktotrophic species). Often the protoconch shows less evolutionary change than the teleoconch (e.g., in the families Pseudozygopleuridae, Epitoniidae, and Cerithiidae) and, in such cases, can provide apomorphies for families or even higher taxa. However, there are other cases in which protoconch morphology is highly variable within families, especially in groups with diverse life history traits. Also, protoconchs, like any other character complex, are subject to convergence and other homoplastic phenomena.

HEAD-FOOT, OPERCULUM, AND MANTLE EDGE

Head-foot characters (Figure 13.2) have not been greatly used in morphological analyses to date. The foot is plesiomorphically elongate-oval but has been extensively modified in many groups. Lateral expansion of the foot has occurred in several groups (e.g., many neogastropods, tonnoideans) or it has become disk-like for clamping in

limpet-like taxa. The foot is laterally compressed in stromboideans and xenophorids, where it can be used as a lever or for leaping, and in the actively swimming heteropods.

In some, lateral flaps emerge from the sides of the neck or foot (e.g., Viviparidae, Vanikoridae), while in others the shell can be covered by lateral or anterior extensions of the foot and/or mantle in Naticidae (Figure 13.2Q), Triviidae, Olividae (Figure 13.2V, Aa), some Volutidae, Marginellidae (Figure 13.2W), and Cypraeoidea (Cypraeidae, Figure 13.2T; Ovulidae), with the shell becoming internal and reduced in Velutinidae (Figure 13.2N).

Some amphibious or terrestrial rissooidean taxa (e.g., Assimineidae, Pomatiopsidae) have a deep omniphoric groove running down each side of the neck, which carries mucus and waste to the sides of the foot. A few taxa possess tentacles that emerge from the sides of the foot (e.g., some Cerithiidae [Bittiinae] and Litiopidae) or posteriorly (e.g., the neogastropod Nassariidae, Figure 13.2Y, and some rissooideans: Stenothyridae, some Rissoidae, and Vitrinellidae, Figure 13.2G). Tentacles emerge laterally from the opercular lobe in some Eatoniellidae (Ponder 1965). The foot is very reduced and lacks a sole in vermetids.

The cephalic tentacles are typically long and narrow, with the eyes on swellings at their outer bases (Figure 13.2A–H, L–P), although there is considerable variation in the length of the tentacles, and they are lost in some Assimineidae (Figure 13.2I). The tentacles, especially of small-sized taxa, can have complex patterns of ciliation, with some developing long compound cilia distally. Although the eyes may be on short stalks or situated along the tentacle (Figure 13.2K, Bb), they are always located on the outer side of the tentacles (plesiomorphic in gastropods), in contrast to most basal Heterobranchia. A cephalic penis arises from the right to center of the head behind the base of the tentacles in many hypsogastropods, where it appears to have been derived independently in several groups. In contrast, cerithioideans, cingulopsoideans, vermetoideans, triphoroideans, and janthinoideans lack a penis, and penial structures are differently derived in the three architaenioglossan clades: two from noncephalic structures (Viviparoidea from right tentacle, Ampullarioidea from mantle) and one cephalic (Cyclophoroidea).

An operculum is usually present but is lost in the adults of a few groups (e.g., Cypraeidae, Triviidae, Ovulidae, Velutinidae, Carinariidae, Pterotracheidae, Calyptraeoidea *sensu stricto*, Mitridae, Marginellidae, and many Volutidae). Checa and Jiménez-Jiménez (1998) distinguished three main types of operculum: (i) flexiclaudent spiral (mostly multispiral) operculum, the shape of which does not fit the aperture, (ii) rigiclaudent spiral, fitting the aperture and usually paucispiral, and (iii) rigiclaudent concentric, also fitting the aperture. Their study showed that the rigiclaudent spiral type predominates in caenogastropods, with flexiclaudent spiral opercula found in some cerithioideans. Concentric opercula are predominant in higher caenogastropods.

A well-developed, narrow snout, typical of basal caenogastropods, is sometimes very extensile (e.g., some rissooideans, stromboideans, and cerithioideans) and is used to assist in locomotion in some Truncatellidae and Pomatiopsidae. The snout has become infolded to form a proboscis (introvert) convergently in several groups (e.g., Ptenoglossa, Neogastropoda, Tonnoidea), and in many such taxa the proboscis is capable of considerable extension (e.g., Figure 13.2M, X). Unique structures can be associated with the snout/proboscis; a pseudoproboscis is formed in the Capuloidea (e.g., Pernet and Kohn 1998) and an epiproboscis in the Mitridae (Ponder 1972; West 1990). Two main types of proboscis are usually recognized in caenogastropods: the acrembolic and pleurembolic (Fretter and Graham 1962), which differ in the way in which they lie in the body in their retracted state. In the acrembolic type of proboscis, the tip of the proboscis is fully inverted by retractor muscles that attach mostly to the anterior buccal region. With the pleurembolic type, the retractor muscles are inserted on the sides of the proboscis, so that the anterior part is not

inverted on retraction. The acrembolic type has appeared independently several times in gastropods, including some heterobranchs, while the pleurembolic proboscis may have appeared only once. This latter type is a synapomorphy of the node that precedes the calyptraeoideans in Simone's (2000a) cladogram (Figure 13.4B) and includes all the higher hypsogastropods, while in Strong's (2003) analysis (Figure 13.4A) the proboscidate taxa also form a separate clade, although it comprises both pleurembolic and acrembolic taxa.

The mantle edge can have one or more short to moderately long papillae or one or two ciliated anterior or posterior tentacles, rarely very long, as in *Finella* (Figure 13.2B). The anterior mantle edge is extended as a siphon in many caenogastropods, although this may have arisen independently in Cerithioidea, Stromboidea, and the higher Hypsogastropoda (Simone 2005) and probably also Triphoroidea. A siphonal notch or canal in the shell is found in many cerithioideans and stromboideans, but they do not have any clear modification of the mantle edge (Simone 2001, 2005). On the other hand, a clearly defined siphon at the mantle border (i.e., a long fold in the inhalant canal separated from the mantle edge) occurs in most higher hypsogastropods and is elongated, mobile, and exploratory in many (Figures 2U–Bb). The specialized inhalant siphon is a synapomorphy of the node that precedes the calyptraeoideans in Simone's (2000a) cladogram (Figure 13.4B), with a reversal in the naticoideans, while in Strong's (2003) analysis (Figure 13.4A) the siphonate hypsogastropods are found in two separate clades.

RADULA

The radula has been used as an important character set at both the species level and higher levels within caenogastropods, representing the basis of such names as Stenoglossa, Ptenoglossa, and Taenioglossa. By far the most significant historical review is the work of Troschel and Thiele (1856–1893), synthesized and added to by Thiele (1929–1931). These early workers relied on light microscopy, but the advent of scanning electron microscopy provided an excellent tool for more detailed examination and illustration of radulae. Most subsequent studies have focused on lower-rank taxa; a notable exception being Bandel's (1984) study of the radulae of Caribbean caenogastropods.

The taenioglossate condition, with seven teeth in each row (two marginals and a lateral on each side and a central tooth) is found in all architaenioglossans and many sorbeoconch caenogastropods. Remarkably, this type of radula is retained by the majority of the group with very little modification to reflect the enormous diversity of feeding strategies (see subsequent section on Adaptive Radiation). Exceptions are the neogastropods and some of the ptenoglossate groups. Within neogastropods the toxoglossan type (5–1 teeth per row) is found in the Conoidea, with the most extreme modification being the toxoglossan "harpoon" tooth (Shimek and Kohn 1981; Kantor 1990). An unusual, very elongate tooth type (nematoglossan) is found in Cancellarioidea, while the remaining neogastropods have a rachiglossan (3–1 teeth per row) radula.

The typical ptenoglossate radula occurs in Epitoniidae and Janthinidae, exhibiting numerous similar simple teeth in each radular row. Multiple similar teeth are also known from a few eulimids (Warén 1984), and tooth multiplication is seen in some cerithiopsids (Marshall 1978) and triphorids (Marshall 1983), although some other triphoroideans have taenioglossate radulae. The spongivorous triphorids possess a modified radula with 5-63 teeth per row (the rhinioglossate condition). Although cerithiopsids are also spongivorous, they usually retain a taenioglossate radula but the tooth morphology is extremely variable, with some genera having very elongated teeth (Marshall 1978; Nützel 1998).

The radula has been lost in some conoidean taxa (e.g., Ponder 1974; Kantor and Sysoev 1989), some other neogastropods (Ponder 1974), and many eulimids (Warén 1984).

ANATOMY

Major synoptic anatomical studies on caenogastropods were undertaken in the late nineteenth century, including those on the nervous system (Bouvier 1887), kidney (Perrier 1889), pallial cavity (Bernard 1890) and anterior gut (Amaudrut 1898). Many anatomical accounts have been published since. Notably, these include a large body of work by Fretter and Graham and their students on mainly European taxa (summarized in Fretter and Graham 1962, 1994) while the Marcuses provided detailed accounts of mainly Brazilian taxa (e.g., Marcus 1956; Marcus and Marcus 1963, 1965). Many aspects of caenogastropod anatomy and histology were reviewed by Fretter and Graham (1962, 1994), Hyman (1967), and Voltzow (1994). Reviews of the evolution of organ systems for gastropods in general, including the gut (Salvini-Plawen 1988), kidney (Andrews 1988), and mantle cavity (Lindberg and Ponder 2001) are also relevant.

Recent morphological studies include Strong's and Simone's detailed investigations. Strong's (2003) study, based on the examination of 16 caenogastropods (Figure 13.4A), resulted in the reformulation of homologies for several key characters (jaw, subradular organ, buccal pouches) and the identification of characters new to caenogastropod systematics, including those of the kidney (blood circulation patterns), nervous system (tentacular nerve, siphonal ganglion), foregut (esophageal ventral folding) and stomach (= midgut; details of the gastric shield and style sac ciliary tracts). As an illustration of how reinterpretations of some of these structures change our ideas of homology, we detail the following examples from the anterior gut. Based on criteria of position (Figure 13.5A–C) and histological detail (Figure 13.7D–F), Strong (2003) proposed that the glandular structures below the radula in many sorbeoconchan taxa are homologous to the subradular organ (Figure 13.7A–F, sro) of architaenioglossans. Also, in contrast to long-held views (Graham 1939; Salvini-Plawen and Haszprunar 1987; Haszprunar 1988; Ponder and Lindberg 1997), the buccal cavity and anterior esophagus of some basal caenogastropods were shown to possess the ventral folding characteristic of vetigastropods (Figure 13.7G–I), including a mid-ventral fold (Figure 13.7G–I, vf)

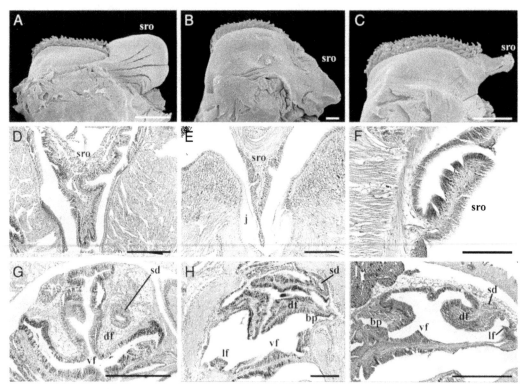

FIGURE 13.7. Foregut characters. (A–C) Scanning electron micrographs of subradular organ. Right lateral view of radular apparatus, ventral is to the right. (D–F) Histology of subradular organ. Transverse section through subradular organ, ventral is below. (G–I) Histology of foregut showing ventral folding. Transverse section through anterior esophagus, ventral is below. A. *Neocyclotus dysoni*. Scale bar, 0.5 mm. B. *Lampanella minima*. Scale bar, 100 µm. C. *Littorina littorea*. Scale bar, 0.5 mm. D. *Neocyclotus dysoni*. Scale bar, 0.25 mm. E. *Lampanella minima*. Scale bar, 1 mm. F. *Littorina littorea*. Scale bar, 1 mm. G. *Neocyclotus dysoni*. Scale bar, 0.5 mm. H. *Lampanella minima*. Scale bar, 1 mm. I. *Littorina littorea*. Scale bar, 0.5 mm. *Abbreviations:* bp, buccal pouch; df, dorsal fold; j, jaw; lf, ventro-lateral fold; sd, salivary gland duct; sro, subradular organ; vf, mid-ventral fold. Reproduced from Strong (2003).

and two ventro-lateral folds (Figure 13.7H, I, lf). The ventro-lateral folds, commonly associated with the inner margins of the buccal pouches, are retained in many caenogastropods, even in those lacking buccal pouches. In contrast to vetigastropods, the mid-ventral fold is not associated with an underlying gland.

Strong's (2003) analysis also included detailed stomach morphology that built on earlier functional studies, particularly that of Graham (1949). She identified several stomach characters found in a broad spectrum of taxa, regardless of feeding mode, which demonstrated an underlying phylogenetic, rather than functional, signal. This was also shown by Kantor (2003) in buccinoidean neogastropods, and it is likely that this set of characters can be

extended given the diversity of neogastropod stomach morphology (e.g., Smith 1967). Some important caenogastropod stomach characters are shown in Figure 13.8. They include a ventral gastric shield (gs) and a ciliary tract along the major (and sometimes minor) typhlosole of the style sac (ct). Complex typhlosolar folding and the presence of a discrete ciliary rejection current in the proximal stomach are plesiomorphic for caenogastropods (Strong 2003). The simplification of these and other features, such as development of the sorting area (sa) and complexity of the style sac (ss), are not necessarily related to the innovation of carnivory. The two architaenioglossans included in Strong's (2003) analysis were characterized by the presence of mucus-secreting glandular pouches (gap), a

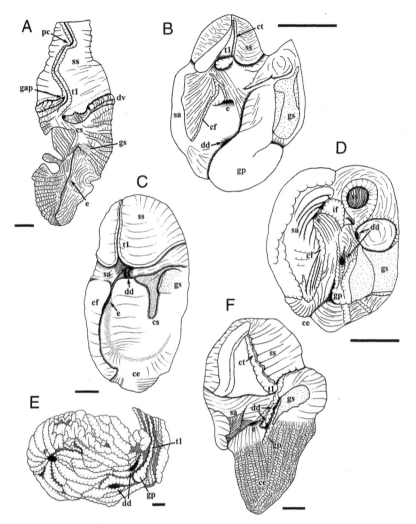

FIGURE 13.8. Stomach characters. (A) *Neocyclotus dysoni* (macrophagous grazer). (B) *Lampanella minima* (microphagous grazer). (C) *Littorina littorea* (microphagous grazer). (D) *Strombus mutabilis* (microphagous grazer). (E) *Conus jaspideus* (carnivorous predator). Scale bar, 100 μm. (F) *Ilyanassa obsoleta* (opportunistic scavenger). Scale bar, 100 μm except E, 1 mm. *Abbreviations:* ce, cecal extension; cf, ciliated fold; cs, ciliated strip; ct, ciliary tract; dd, duct of digestive gland; dv, digestive diverticulum; e, esophagus; gap, gastric pouch; gp, glandular pad; gs, gastric shield; if, intestinal flap; pc, pyloric cecum; sa, sorting area; ss, style sac; t, major typhlosole. Reproduced from Strong (2003).

digestive gland vestibule bearing the apertures of the digestive gland ducts (dv) and pyloric cecae at the distal end of the style sac (pc). The latter were also found to be present in cypraeids, where they are probably convergent, as are various cecal extensions (ce), which originate from different portions of the stomach.

Simone has studied the anatomy of over 250 species in most extant families and carried out cladistic analyses at the level of superfamily or groups of closely related superfamilies (1999, 2000b, 2001, 2002, 2004a, b, 2005, 2006). These and ongoing studies of some superfamilies, including representatives for most caenogastropod families, have generated much new comparative anatomical data. The overall intention of this work was to obtain a better definition of each superfamily and their relationships. Some details resulting from his unpublished analyses are outlined later in the Summary of Major Groups section.

Simone's work has paid particular attention to the odontophore and buccal muscles, clarifying their homology and standardizing their terminology. These muscles are reduced in number in comparison with those of vetigastropods and neritimorphs, in part related to the possession of a single pair of odontophoral cartilages. Additionally, the muscles between the odontophoral cartilages and the radula stretch over the odontophoral cartilages (Figure 13.9). The main muscles of the caenogastropod odontophore (Figure 13.10) are the pair of lateral-dorsal tensor muscles of the radula (called "m4" in Simone's papers). These large muscles are located mostly between the cartilages and subradular membrane, an arrangement that prevents the two structures from sliding across each other, except in neogastropods, in which sliding movements between the radula and cartilages occur. In that group, the pair of ventral tensor muscles of the radula, that is responsible for the sliding movement, appears to be derived from median fibers of the m4 muscles, not the ventral tensor muscles as in neritimorphs and vetigastropods.

Ontogenetic data for some apomorphic structures in the anterior gut have proved useful in testing putative homologies. Such studies include the development of the anterior gut (e.g., Page and Pedersen 1998; Page 2000, 2002, 2005), the neogastropod accessory salivary glands (Ball *et al.* 1997b), and the proboscis (Ball *et al.* 1997a; Ball 2002). Similarly, recent detailed studies of the anatomy, histology, and ultrastructure of gut structures, such as those by E. Andrews (e.g., anal gland, Andrews 1992; digestive gland, Andrews 2000; salivary glands, Andrews 1991; Andrews *et al.* 1999) have provided data to test phylogenetic relationships.

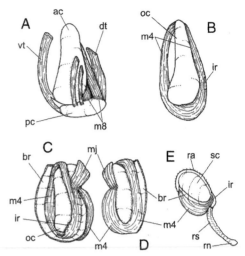

FIGURE 13.9. Modification of the main tensor muscle of the radula (m4) as an example of odontophoral muscle modifications in caenogastropods. Most other muscles and structures are not shown. (A) Typical vetigastropod (e.g., *Haliotis* [Haliotidae], *Calliostoma* [Trochidae]) with two pairs of cartilages and several pairs of muscles; showing left side of odontophore (internal view), with most structures except cartilages and adjacent muscles removed. (B–E) Typical basal caenogastropod (e.g., *Cerithium* [Cerithiidae]; B, corresponding view to A, with a single muscle (m4) present. C, ventral view, right half (left in figure) removed, some adjacent structures also shown; connection between subradular membrane (br) and tensor muscle (m4) suggesting its function as a tensor. D, dorsal view; E, schematic representation of position of left m4 in odontophore, most other muscles not shown. Not to scale. *Abbreviations:* ac, anterior cartilage; br, subradular membrane; dt, dorsal tensor muscle of radula; ir, insertion of m4 in radular sac; m4, main tensor muscle of radula; m8, approximator muscle of cartilages; mj, peribuccal muscles; oc, odontophore cartilage; pc, posterior cartilage; ra, radula; rn, radular nucleus; rs, radular sac; sc, subradular cartilage; vt, ventral tensor muscle of radula. Illustrations from Simone (2000a).

ULTRASTRUCTURAL DATA

Ultrastructural information on sperm and sensory organs has contributed significantly to our current understanding of gastropod phylogeny, but other tissues may prove equally useful in the future if investigated in comparable detail. The ultrastructural findings of Haszprunar (1985a)

on the osphradium have been valuable in gastropod phylogenetic studies but do not provide resolution within Sorbeoconcha. Scanning electron micrographic (SEM) studies of osphradial surface morphology (Taylor and Miller 1989) showed many features with potential phylogenetic utility (e.g., leaflet structure, ciliary patterns/type) in the families they examined, but most of these results have not yet been sufficiently developed or scored in enough taxa to be incorporated in analyses.

Spermatozoa of caenogastropods are structurally complex, are usually strongly dimorphic

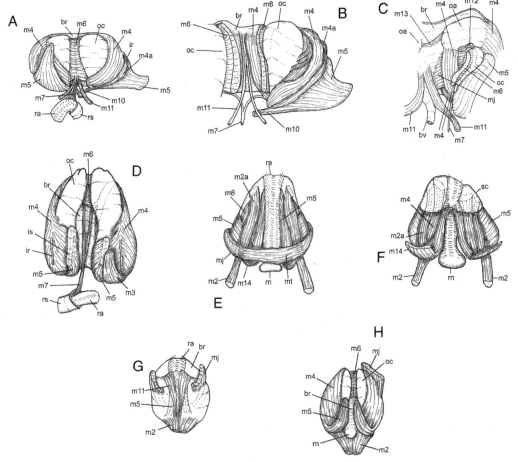

FIGURE 13.10. Main odontophore intrinsic muscles showing modifications in different clades. (A, B) Cyclophoridae (*Neocyclotus*), A, dorsal view of whole odontophore with superficial layers of muscles and structures removed, both cartilages deflected from each other, middle region of radular ribbon also shown, left m5 (right in figure) deflected; B, left half mainly shown, horizontal muscle (m6) sectioned longitudinally, most muscles deflected showing multiple components of m4 (m4 + m4a and branches) suggesting multiple origins. (C) left half of an ampullariid (*Pomacea*) odontophore in dorsal view with most muscles and cartilage deflected; m4 still multiple (bottom in figure) but simpler than in cyclophorids. (D) Dorsal view of the odontophore of an annulariid (*Annularia*),with superficial structures and muscles removed. This represents the basic type of odontophore of Viviparoidea + Sorbeoconcha (except neogastropods) with each m4 a simple, strong muscular mass. Both cartilages are deflected and the middle part of the radular sac also shown. (E–F) odontophore of a calyptraeid (*Crepidula*), with most muscles seen as if the superficial structures were transparent. This represents further modification in Calyptraeoidea, where most of the intrinsic muscles have become directly attached to the subradular membrane; E, ventral view, F, dorsal view. (G–H) Odontophore of a pseudolivid (*Benthobia*), representing a basal neogastropod, where the pair of ventral tensor muscle of the radula (m11) become stronger, indicating that the radula has reverted to undertaking sliding movements; G, ventral view, some structures seen by transparency; H, ventral view, superficial structures and dorsal portion of radular ribbon removed, both cartilages slightly deflected. *Abbreviations:* br, subradular membrane; bv, blood vessel; ir, insertion of m4 in radular sac; is, insertion of m5 in radular sac; m2, retractor muscle of odontophore; m2a, continuation of m2 connected also in cartilages and subradular membrane (only in some calyptraeoideans); m3, superficial circular muscle; m4, main dorsal tensor muscle of radula; m5, accessory dorsal tensor muscle of radula; m6, horizontal muscle; m7, muscle running inside radular sac; m8, bending muscle of cartilage (only in some calyptraeoideans); m10, protractor muscle of odontophore; m11, ventral tensor muscle of radula; m12–m13, accessories of m6; m14, ventral protractor muscle of odontophore; mj, jaw or peribuccal muscle; mt, transversal superficial muscle; oa, auxiliary cartilage; oc, odontophore cartilage; ra, radula; rn, radular nucleus; rs, radular sac; sc, subradular cartilage. A–D from Simone (2004b); E–F from Simone (2002); G–H from Simone (2003).

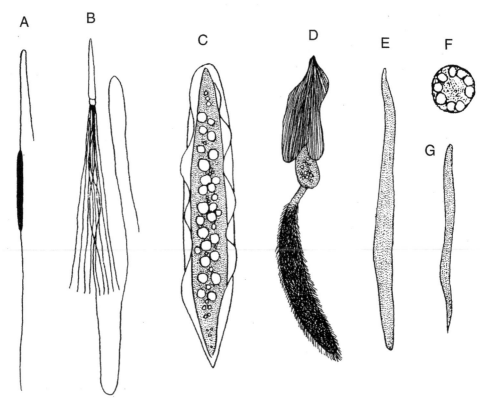

FIGURE 13.11. Types of gastropod paraspermatozoa: (A) with anterior and posterior tail (Neritidae, Neritimorpha); (B) with head and posterior tail tuft (e.g., Cerithiidae, Cerithioidea); (C) with undulating lateral wings (Strombidae, Stromboidea); (D) with numerous attached eusperm (e.g., Epitoniidae, Janthinoidea); (E) vermiform (Cypraeidae, Cypraeoidea); (F) round (*Drupa*, Muricidae, Muricoidea); (G) vermiform (Terebridae, Conoidea). Source of figures: Nishiwaki (1964).

(for reviews see Healy 1988a, 1996a; Buckland-Nicks 1998), and have provided many important characters. For this reason, a detailed summary follows.

Much of the work on sperm morphology has been directed towards the phenomenon of sperm dimorphism (sometimes polymorphism) seen in many caenogastropods[5] (for reviews or comparative accounts see Melone *et al.* 1980; Healy and Jamieson 1981; Giusti and Selmi 1982; Healy 1988a; Hodgson 1997; Buckland-Nicks 1998; Bulnheim 2000). In such instances, fertile sperm (euspermatozoa) are accompanied by co-occurring nonfertile sperm (paraspermatozoa),

the latter type often showing distinctive shapes that are taxon specific even at the light-microscopic level (see Figure 13.11; Nishiwaki 1964; Tochimoto 1967). For example, the parasperm in Strombidae (Figure 13.11C) have lateral, undulating "wings." and the ptenoglossan families (except Eulimidae) (Figure 13.11D) have large paraspermatozoa bearing hundreds of attached euspermatozoa—an association known as a "spermatozeugma." Paraspermatozoa are apparently absent in some taxa (e.g., Eulimoidea, Naticoidea, Rissooidea) (Kohnert and Storch 1984; Koike 1985; Healy 1988a, 1996a).

Transmission electron microscopy (TEM) has revealed taxa-specific features within euspermatozoa that are not visible at the light-microscopic level or with SEM, namely, internal details of the acrosome, nucleus, midpiece, and

5. Paraspermatozoa are also known in Neritimorpha and some Vetigastropoda (see Nishiwaki 1964; Healy 1988a, 1990e).

glycogen deposits (Figure 13.12).[6] Although not all groups of caenogastropods show wide divergences in eusperm or parasperm ultrastructure (e.g., some families of Neogastropoda), the internal structure of the eusperm midpiece in particular can be very informative, with caenogastropod higher taxa differing in the number of periaxonemal mitochondria, their arrangement relative to the axoneme, and the structure of the mitochondrial cristae (plates) or their derivatives (Figure 13.12H–L; contrast with Figure 13.12D, F showing midpiece of gastropod aquasperm of a vetigastropod). Thus, for example, Campaniloidea are characterized by seven to eight straight mitochondria with unmodified cristae, partly enclosed by a sheath of dense (probably non-mitochondrial), segmented structures (Figure 13.12L); Cerithioidea have four straight mitochondria with complex, parallel cristae and lack any segmented sheath (see Figure 13.12K); and Cyclophoroidea are intermediate between these two, though, like the Cerithioidea, lack the campaniloidean dense sheath (Figure 13.12I). Families such as Campanilidae, Cerithiopsidae, Provannidae, and Vermetidae have been excluded from Cerithioidea largely based on the results of comparative eusperm ultrastructure (Healy 1983a, 1988b, 1990a, b; for discussion and literature see Healy 1996a, 2000). Some families whose affinities were uncertain, even after anatomical study, could be placed within the systematic framework of the Caenogastropoda on examination of their sperm. For example, *Plesiotrochus*, long considered a cerithiid and later separated as a distinct cerithioidean family (Plesiotrochidae, Houbrick 1990), was shown to have *Campanile*-like eusperm and parasperm (Healy 1993a; see Figure 13.12L). Moreover, *Plesiotrochus* and *Campanile* share a simple larval shell morphology (Kiel *et al.* 2000), which is also present in other Recent and fossil basal caenogastropods (Nützel and Pan 2005).

---

6. For broad, comparative studies see Giusti 1971; Healy 1983a, 1988a, 1996a; Kohnert and Storch 1984; Koike 1985 and literature therein.

While a few vetigastropods (e.g., Healy 1990e) have parasperm, heterobranch euspermatozoa are not accompanied by paraspermatozoa. Heterobranch eusperm are characterized by, among other things, a rounded acrosomal vesicle and a very complex, continuous mitochondrial sheath (Figure 13.12C, E, M, N) (often with paracrystalline layers and an enclosed glycogen helix) (e.g., Thompson 1973; Anderson and Personne 1976; Healy 1983b, 1988a, 1990d, 1993b, 1996a; Healy and Willan 1984, 1991; Giusti *et al.* 1991; Hodgson and Healy 1998; Wilson and Healy 2002; Fahey and Healy 2003).

DEVELOPMENTAL DATA

While there is a great deal of data on caenogastropod larval and intracapsular development (e.g., Thorson 1946; Fretter and Graham 1962, 1994; Fioroni 1966, 1982; Bandel 1975), good data is lacking for some groups.

Detailed studies on early cleavage and embryology have been undertaken on relatively few taxa, these including two viviparids (Johansson 1951; Tanaka *et al.* 1987), an ampullariid (Demian and Yousif 1973a), a turritellid (Kennedy and Keegan 1992), a strombid (D'Asaro 1965), a calyptraeid (Conklin 1897), a naticid (Bondar and Page 2003), a bursid and personid (D'Asaro 1969), two muricids (D'Asaro 1966; Stockmann-Bosbach 1988), two nassariids (Tomlinson 1987), a melongenid (Conklin 1907), and a columbellid (Bondar and Page 2003).

Different modes and timings of D quadrant formation during early embryo development are characteristic of different gastropod clades (Freeman and Lundelius 1992; van den Biggelaar, 1996; van den Biggelaar and Haszprunar 1996; Guralnick and Lindberg 2001; Lindberg and Guralnick 2003). These include the presence of either unequal cell cleavage or polar lobes and the cell stage (i.e., number of cells) at which the 4d cell forms and timing of the formation of the 2a–2d and 3a–3d lineages. Although known only from a relatively small number of taxa, these characters show that caenogastropods have a unique cleavage pattern, with the first cleavages associated with

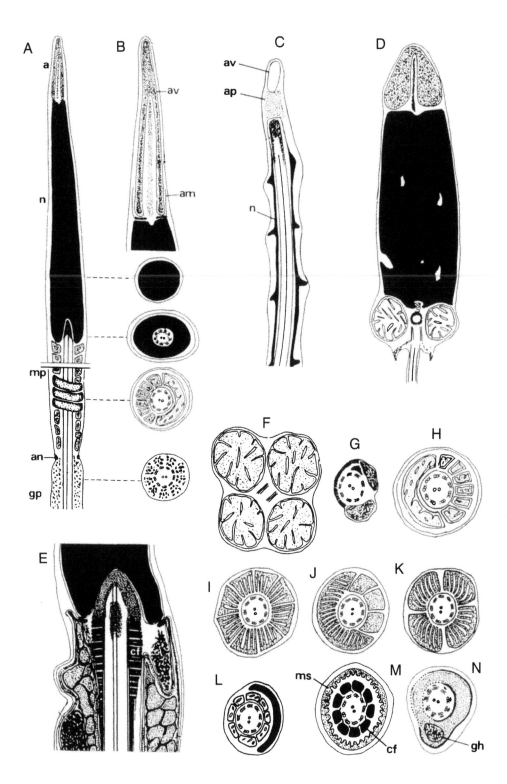

FIGURE 13.12. (Opposite.) Gastropod euspermatozoan features. (A) Basic features of caenogastropod euspermatozoon in littorinid *Bembicium auratum* (a, acrosomal complex, an, annulus, gp, glycogen piece, mp, midpiece, n, nucleus). (B) Acrosomal vesicle (av) (conical) of *Bembicium auratum*: note the accessory membrane (am). (C) Acrosomal vesicle (rounded) associated with acrosomal pedestal (ap) and anterior portion of nucleus and axoneme in the heterobranch *Rissoella micra*. (D) Eusperm of pleurotomariid, *Perotrochus westralis*, showing irregular spaces in nucleus (nuclear lacunae) and short midpiece with relatively unmodified mitochondria. (E) Nucleus-midpiece junction of nudibranch euspermatozoon, showing coarse fibers (cf) and complex mitochondrial sheath (with internal glycogen helix). (F–N) eusperm midpiece in transverse section (F, Pleurotomariidae, 4 round, "unmodified" mitochondrial cristae; G, Neritidae, 2 straight mitochondria; H, Littorinidae, 6–10 helical mitochondria with "unmodified" cristae, the commonest pattern in Caenogastropoda; I–K, with parallel cristal plates, Cyclophoroidea, 7–8 straight mitochondria; J, most Ampullarioidea, 4 helical mitochondria; K, Cerithioidea, 4 straight mitochondria; L, Campaniloidea, 7–8 straight mitochondria, accompanied by a dense segmented sheath; M, Architectonicoidea, Heterobranchia, continuous sheath [ms] with helical grooves, and thick coarse fibers; N, most other Heterobranchia, continuous sheath with paracrystalline layers and glycogen helix [gh]). Sources of figures: A, B, H: Healy (1996b); C, M, N: Healy (1993b); D: Healy (1990c); E: Healy and Willan (1991), F: Healy (1988c); G: Healy (1988a, 1993b); I–L: Healy (1993a).

polar lobe formation (van den Biggelaar and Haszprunar 1996).

Studies on organogenesis are relatively few, with the most comprehensive being those of Demian and Yousif (1973a–d, 1975) on *Marisa* (Ampullariidae). Recent detailed studies by Ball *et al.* (1997a, b), Ball (2002), Page (2000, 2002, 2005), Page and Pederson (1998), Pederson and Page (2000), and Parries and Page (2003) on the ontogenetic development of the anterior gut have substantially added to the available data, although some key taxa (e.g., ptenoglossan groups) have not yet been studied. In some proboscis-bearing gastropods—*Marsenina* (Velutinidae) (Page 2002), the naticid *Euspira* (Page and Pedersen 1998, Pedersen and Page 2000), and the direct-developing neogastropods *Nucella* (Ball *et al.* 1997a) and *Conus* (Ball 2002)—the new anterior gut develops independently to the larval gut and opens at the larval mouth at metamorphosis. However, during the development of the anterior gut in the planktonic larva of *Nassarius* (Fretter 1969; Page 2000, 2005), the larval mouth is sealed off and a new postmetamorphic mouth, develops into which the new anterior gut structures (including the proboscis) open. This separated development has been postulated to facilitate a rapid switch from larval microherbivory to postlarval carnivory. It will be of great interest to see whether this arrangement is common in other carnivorous proboscidate caenogastropods with planktotrophic development, and, if not, how this transition is achieved.

There is considerable morphological variation in caenogastropod spawn, ranging from gelatinous masses to benthic or pelagic capsules. The capsules may be complex, with delicate to tough walls, and their shape and size can vary within families and genera (e.g., Thorson 1946; Fretter and Graham 1962). This variation may be in response to selective pressures, because the encapsulating structures reduce embryo mortality through protection from predation, salinity stress, desiccation, bacterial attack, and, possibly, ultraviolet light (Rawlings 1994; Przeslawski 2004). Some adults actively defend their spawn from predators, whereas others brood capsules in the mantle cavity, oviduct, foot, shell umbilicus, or even a special chamber in the head.

FOSSIL RECORD

The origin and the early evolution of Caenogastropoda are summarized by Frýda *et al.* (Chapter 10). Putative caenogastropods first appeared during the Early Ordovician gastropod radiation (c. 490 Mya) (e.g., Loxonematidae, Subulitidae), although their identity as caenogastropods is not confirmed because information on shell microstructure and protoconchs is not available. It is likely that the last common ancestor of all extant caenogastropods lived in the Paleozoic prior to the Carboniferous and that some stem groups of the various crown group clades were present in the Early Paleozoic.

Recently suggested hypotheses regarding caenogastropod ancestory include Wagner's (2002) analysis of early gastropods based on teleoconch characters. He suggested that in the Early Paleozoic the Murchisoniinae gave rise to four groups, among them the hormotomoids and eotomarioids, the latter being possible precursors of modern vetigastropods. Hormotomoids split into loxonematids, subulitids, and, according to his hypothesis, apogastropods (i.e., Caenogastropoda + Heterobranchia) arose from the hormotomoid-loxonematid lineage (i.e., subulitids are the sister group of loxonematids and apogastropods).

Frýda (1999) assumed that Caenogastropoda and Heterobranchia arose from the Perunelomorpha, a Paleozoic (Ordovician to Devonian) group characterized by an open-coiled initial protoconch whorl and, commonly, fusiform teleoconchs (Frýda 2001; Frýda et al., Chapter 10, Figures 10.4, 10.9) (Figure 13.6I–K). Perunelomorpha was left without assignment to a higher category by Frýda (1999). Bandel (2002) included them in the Procaenogastropoda, a poorly characterized and heterogeneous assemblage of Paleozoic caenogastropods (Nützel and Pan 2005) (discussed subsequently). To date, no explicit phylogenetic hypothesis has been presented clarifying the systematic placement of Perunelomorpha. Late Paleozoic perunelomorphs (Family Imoglobidae) seem to be caenogastropods, as suggested by their teleoconch morphology and dextral planktotrophic larval shell (Nützel et al. 2000; Nützel and Pan 2005). Frýda (1999: fig. 7) suggested that Perunelomorpha split into Caenogastropoda and Heterobranchia, but Frýda was not explicit about possible sister group relationships, and no apomorphies were given.

Kaim's (2004: fig. 140) phylogenetic scheme (Figure 13.5A) assumes a sister group relationship of Heterobranchia and Caenogastropoda. As in several previous scenarios, Kaim (2004) placed Loxonematidae as the stem group of apogastropods, while the apogastropod sister group was an unresolved cluster of taxa previously included in "Archaeogastropoda." Kaim (2004: fig. 140) noted loss of nacre, high-spired

shell, and closure of the protoconch umbilicus as apogastropod apomorphies.

Caenogastropods have a rich fossil record from the Devonian, with about one-third of the approximately 190 families extinct (Bouchet and Rocroi 2005). Many superfamilies and orders with living representatives (e.g., Rissooidea, Cerithioidea, Stromboidea, some "Ptenoglossa"[7]) have a fossil record from the early Mesozoic or even late Paleozoic (e.g., Bandel 1993; Nützel 1998; Nützel and Erwin 2004; Kaim 2004). Supposed basal caenogastropods, such as Cerithioidea, Rissooidea, and Littorinoidea, as well as some fossil groups were assigned to a paraphyletic Palaeo-Caenogastropoda (Bandel 1993, 2002). Protoconch morphology and shell microstructure is currently available from the Devonian/Carboniferous onward (ca. 350–400 Mya).

The oldest known gastropod with a preserved caenogastropod-type larval shell, the Devonian *Pragoscutula*, has a limpet-shaped teleoconch (Frýda 2001; see Frýda et al., Chapter 10, Figure 10.10E) and was included in Neritimorpha by Bouchet and Rocroi (2005) because the larval shell of the Devonian *Pragoscutula* is similar to that of some fossil neritimorphs (Figure 13.6A, B; see subsequent discussion). However, this conclusion was reassessed following examination of recently discovered Early Carboniferous pragoscutulid limpets from Australia that have slender, caenogastropod-like larval shells. Pragoscutulids are now interpreted as early (but derived) caenogastropods (Cook et al., in press). It is unlikely that the first caenogastropods were limpets, because if there had been a reversal from limpet to coiled shells in ancestral caenogastropods, there would be significant anatomical implications, of which there is no hint in modern taxa.

Open-coiled protoconchs are present in several Paleozoic gastropod clades (notably the Perunelomorpha), but this character was lost by the Mesozoic (Nützel and Frýda 2003).

---

7. Triphoroidea, Janthinoidea, and Nystiellidae.

Open-coiled protoconchs may represent the plesiomorphic state in caenogastropods. Based on exceptionally well-preserved material, several Late Paleozoic families have been reported that possess multiwhorled larval shells (indicating planktotrophy) with well-separated, orthostrophic whorls, aragonitic crossed-lamellar shell structure, a high-spired or fusiform shape, and, in some, anterior siphonal structures (e.g., Yoo 1994; Nützel and Cook 2002; Bandel et al. 2002; Bandel 2002; Pan and Erwin 2002; Nützel and Pan 2005) (Figures 13.6C–M), characters strongly suggestive of caenogastropod affinities (see also Frýda et al., Chapter 10). However, their relationships to Mesozoic and extant caenogastropod clades are unclear.

Late Paleozoic Orthonematidae and the slit-bearing Goniasmatidae (Figure 13.6E–G) are probably stem group members of the Cerithioidea; they share a simple caenogastropod-type larval shell similar to that seen in extant basal caenogastropods such as Cerithioidea (Nützel 1998; Nützel and Bandel 2000; Bandel et al. 2002; Nützel 2002; Nützel and Pan 2005). The same type of larval shell was also reported in Cretaceous Campaniloidea, corroborating the basal position of that group (Kiel et al. 2000). The Devonian genus *Murchisonia* (Murchisoniidae) superficially resembles these Late Paleozoic high-spired caenogastropods, but *Murchisonia* is a high-spired vetigastropod (Frýda et al., Chapter 10).

A scenario that sees the nonmarine Architaenioglossa as the sister group of all other caenogastropods (Sorbeoconcha) is, not surprisingly, unsupported by the fossil record. Because all extant architaenioglossans are nonmarine and direct developers, they do not build larval shells. Thus, their protoconchs cannot be meaningfully compared with those of marine planktotrophic caenogastropods. Similarly, the recognition of potential ancestral teleoconchs may be complicated by the assumed Paleozoic marine representives having shell morphologies distinct from those of modern architaenioglossans. The earliest nonmarine gastropods that may be related to architaenioglossans are of Carboniferous age (Wenz 1938–1944; Knight et al. 1960)

and were united in the order Procyclophorida by Bandel (2002). Others, however (e.g., Solem and Yochelson 1979), interpreted most of these Paleozoic nonmarine snails as pulmonates. Even if the oldest (Carboniferous) nonmarine snails were architaenioglossans, it still remains an open question as to which marine group (or groups) gave rise to them. Caenogastropoda are almost certainly much older than Carboniferous, but the assumption of a basal split between architaenioglossan groups and other caenogastropods (Sorbeoconcha) is neither supported or rejected by the fossil record. Because there are several synapomorphies for the sorbeoconchs that suggest that the architaenioglossan grade/clade lies outside them, it remains a plausible hypothesis that the architaenioglossan lineage (or lineages) is older than the known fossil record suggests.

The Paleozoic/Mesozoic Zygopleuroidea (Figures 13.6L, M, 13.13C–E) contains some groups of the traditional polyphyletic Loxonematoidea (Bandel 1991a; Nützel 1998). The zygopleuroid group represents a grade (Nützel 1998), or, in its present composition, even a polyphyletic assemblage, which encompasses close relatives of extant Janthinoidea and Triphoroidea (Bandel 1991a; Nützel 1998). It is also possible that the zygopleuroid group (grade) contains the ancestors of the Rissooidea and Stromboidea, as indicated by a zygopleuroid teleoconch morphology (high spire with axial ribs) present in several early members of these superfamilies. The oldest certain Rissooidea were reported from the earliest Middle Jurassic (Gründel 1999b; Kaim 2004) (Figure 13.13F). The earliest stromboideans are Aporrhaidae from the Early Jurassic (Figure 13.13K, L). Houbrick (1979) suggested that the non-planktotrophic Abyssochrysidae were modern representatives of the Loxonematoidea (i.e., zygopleuroid group).

The traditional Subulitoidea from the middle and late Paleozoic form a polyphyletic assemblage of caenogastropods (Nützel et al. 2000; Frýda 2001) (Figures 13.6H–K). Their relationship to the early Paleozoic subulitids is unclear, as are their links to modern groups. It is very likely

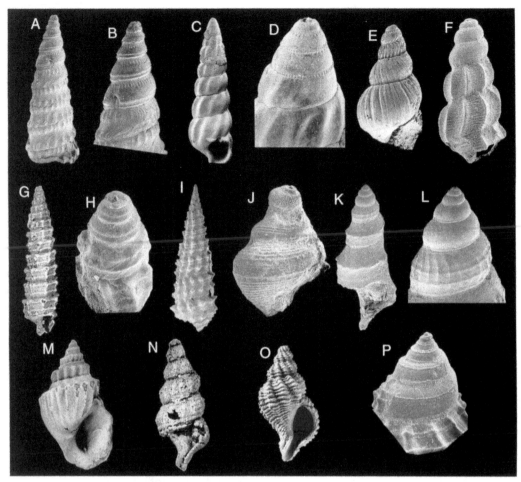

FIGURE 13.13. Selected Mesozoic (Triassic/Jurassic) caenogastropods. (A, B) *Protorcula* (Protorculidae) from the Late Triassic (Carnian, c. 230 Mya; Cassian Formation, northern Italy), a close relative of the modern cerithiopsids (from Nützel 1998); A, height 3.8 mm; B, larval shell, height 2.2 mm. (C, D) *Zygopleura* (Zygopleuridae) from the Late Triassic (Carnian, c. 230 Mya; Cassian Formation, northern Italy); the Zygopleuridae may be related to either modern Janthinoidea or Rissooidea (from Nützel 1998); C, height 5.9 mm; D, larval shell, height 1.0 mm. (E) *Ampezzopleura* (Zygopleuridae) from the Early Triassic Monekopi Formation (c. 247 Mya; Utah, United States); this genus appeared shortly after the end-Permian mass extinction and may be related to the late Paleozoic Pseudozygopleuridae (see Figure 13.6L, M); height 1.4 mm. (F) *Bralitzia* (= *Palaeorissoina*) (Rissoidae) from the Early/Middle Jurassic (Toarcian/Aalenian, ca. 180 Mya; southern Germany); *Bralitzia* is an early rissoid; the teleoconch morphology indicates a possible close relationship with the Zygopleuroidea; height 1.5 mm. (G–I) Jurassic/Triassic species of *Cryptaulax*, a typical, widespread genus of early Mesozoic Cerithioidea (Procerithiidae or Cryptaulacidae); like many modern Cerithioidea, they commonly posses bicarinate larval shells; G, Late Triassic (Norian, c. 210 Mya, western United States), height 10.5 mm; H, Middle Jurassic (southern Germany), showing typical bicarinate larval shell; height 0.8 mm. I, Early/Middle Jurassic (c. 180 Mya, S Germany); height 15.4 mm. (J) Cerithimorph caenogastropod from the Late Triassic (Carnian, c. 230 Mya, Cassian Formation, northern Italy); perhaps an offshoot of the Paleozoic Orthonematidae; height 1.6 mm. (K, L) *Dicroloma* (Aporrhaidae) from the Early/Middle Jurassic (c. 180 Mya; southern Germany); this early member of the Stromboidea has a relatively large, smooth larval shell; the typical apertural spines are broken away; K, height 8.7 mm; L, height 2.5 mm. (M) *Angularia* (Purpurinidae) from the Norian (c. 210 Mya; Idaho, United States); members of this family were sometimes thought to be ancestral or closely related to neogastropods; height 11.8 mm (from Nützel and Erwin 2004). (N) *Astandes* (?) (= *Maturifusus*) (Maturifusidae) from the Norian (ca. 210 Mya; Idaho, United States), one of the earliest distinctly siphonostomatous gastropods with cancellate teleoconch ornament. It could be ancestral or closely related to neogastropods; height 11.8 mm (from Nützel and Erwin 2004). (O, P) *Astandes* (= *Maturifusus*) from the early Late Jurassic (Oxfordian) (c. 154 Mya; Russia) (from Guzhov 2004, by courtesy of A. Guzhov); O, teleoconch showing distinct siphonal canal, height 13.3 mm; P, relatively large planktotrophic larval shell; height 1.2 mm.

that the Early Paleozoic (Ordovician/Silurian) Subulitidae ("true subulitids") are caenogastropods, as indicated by their fusiform shape and the presence of anterior siphonal canals. However, no data about their protoconch morphology and shell microstructure are available. The same is true for the Permian Ischnoptygmatidae, which are probably closely related to the Soleniscidae. Some subulitoid families (Soleniscidae, Meekospiridae, Imoglobidae, Sphaerodomidae) were placed in a subclass Procaenogastropoda by Bandel (2002), a taxon largely based on a single character: a seemingly fluent protoconch/teleoconch transition. However, the material on which this observation was based was worn, with better-preserved material indicating that members of this group have a normal caenogastropod protoconch (Nützel and Pan 2005). These families are now regarded as early Caenogastropoda (see Bouchet and Rocroi 2005 for current classification).

Subulitoid gastropods, especially Soleniscidae, survived the end-Permian mass extinction event (Nützel 2005) and are possible ancestors of Mesozoic to modern caenogastropod clades. Mesozoic descendants are possibly included in the families Coelostylinidae and Pseudomelaniidae, which represented extremely diverse, largely Mesozoic caenogastropod groups. Generally they possessed high-spired to conical, smooth teleoconchs and probably contained descendants of late Paleozoic subulitoids as well as other caenogastropods of yet unknown affinity. In their present composition, both families are almost certainly polyphyletic.

The Late Triassic to Jurassic/Cretaceous Procerithiidae (Figure 13.13G–I) were globally distributed and diverse (e.g., Gründel 1999a; Nützel and Erwin 2004, Kaim 2004; Guzhov 2004). Many of them have bicarinate larval shells, as in some Recent Cerithioidea (e.g., Kowalke 1998, pl. 2). Procerithiidae were included in a separate superfamily (Procerithioidea) and distinguished from modern cerithioidean families largely on the lack of a pronounced anterior siphonal notch or canal. In all other respects the shells of both groups are very similar, and,

moreover, a number of modern cerithioideans lack an anterior canal whereas some putative procerithiids have a well-developed canal. Thus, the separation of these two superfamilies is not justified. Several families (Ladinulidae, Lanascalidae, Popenellidae, and Prostyliferidae) from the Late Triassic Cassian Formation (Bandel 1992), based on their larval shell morphology, can be included within Cerithioidea.

One of the reasons why gastropods (and particularly caenogastropods) are so diverse is their supposed "extinction resistance" (Erwin and Signor 1990). However, it is obvious that the end-Permian mass extinction event about 250 Mya ago had a major impact on the evolution of caenogastropods (Nützel and Erwin 2002; Nützel 2005). Important late Paleozoic families (e.g., the Pseudozygopleuridae; Figure 13.6L, M) became extinct or were marginalized. During the recovery period there was a high degree of turnover, and many new genera and several families appeared (e.g., Ampezzopleura, Figure 13.13E; Nützel and Erwin 2002; Nützel 2005). The rise of the modern, strongly ornamented cerithioideans (Figure 13.13G–I) started in the Triassic and could represent a recovery phenomenon. The impact of the end-Triassic mass extinction event on gastropods has not been well studied. Some widespread genera, such as Protorcula (Protorculidae; Figure 13.13A, B), seemingly became extinct in the latest Triassic. As far as we know, the end-Cretaceous mass extinction event did not have a major impact on caenogastropod evolution, as no major group became extinct. Highly diverse extant groups (such as neogastropods, tonnoideans, Turritellidae, cerithiopsoideans, eulimoideans, and rissooideans) started to radiate in the Cretaceous and continued to undergo major radiations in the Cenozoic. Conversely, a few previously diverse families in the late Mesozoic–early Cenozoic (e.g., Campanilidae; Aporrhaidae) have few living species.

MOLECULAR STUDIES

Some of the more significant molecular analyses, including smaller groups of caenogastropods, are listed above. To date, DNA sequence

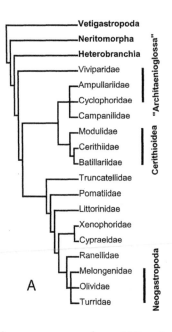

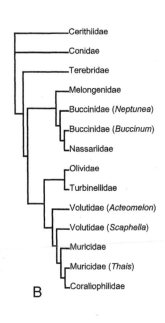

FIGURE 13.14. Some previous caenogastropod phylogenies using molecular data. (A) Strict consensus tree from a maximum-likelihood (ML) analysis using partial 18S rDNA of mainly lower caenogastropod taxa (Harasewych *et al.* 1998). (B) Single ML tree obtained from partial 18S rDNA of neogastropods using *Cerithium* as the outgroup (Harasewych *et al.* 1997).

data has also been used to address more general questions regarding caenogastropod evolution. These include the monophyly of Caenogastropoda itself, Architaenioglossa, Sorbeoconcha, Hypsogastropoda, Ptenoglossa, Neogastropoda, and several superfamilies for which multiple exemplars are available. The published molecular investigations that included a wide range of caenogastropod taxa are summarized below. Harasewych *et al.* (1997) sequenced parts of the 18S rDNA gene in 21 caenogastropods (17 neogastropods) and part of cytochrome *c* oxidase 1 for 17 of these (16 Neogastropoda) (Figure 13.14B). Harasewych *et al.* (1998) sequenced parts of the 18S rDNA gene in 19 caenogastropods, including five Architaenioglossa (two species of Cyclophoroidea, two Ampullariidae, and one Viviparidae), Campaniloidea, and Cerithioidea (three species) (Figure 13.14A). McArthur and Harasewych (2003) included 18S rDNA data from 23 caenogastropods in their Bayesian analysis of overall gastropod phylogeny but obtained very little resolution within a monophyletic Caenogastropoda with Heterobranchia as the monophyletic sister taxon. Included in Colgan *et al.*'s (2000) study of overall gastropod phylogeny were 17 caenogastropod taxa scored for two segments of 28S rDNA and histone H3. Colgan *et al.* (2003), with 16 caenogastropod taxa, added three extra genes, an

additional segment of 28S ribosomal DNA, small nuclear RNA U2, and part of cytochrome *c* oxidase subunit 1. A larger survey of caenogastropod molecular phylogeny (Colgan *et al.* 2007) added additional data (two segments of 28S ribosomal DNA, histone H3, and cytochrome *c* oxidase subunit 1) and new data from additional genes (part of 12S rDNA domain III, another region of the 28S rDNA) and part of the 18S ribosomal DNA and elongation factor 1 alpha) were also added. The data set comprised more than four thousand aligned bases for 29 caenogastropods (six non-hypsogastropods and 23 Hypsogastropoda) and six outgroup taxa. One of the trees resulting from this analysis is shown in Figure 13.15. Of particular interest is the division into "asiphonate" and "siphonate" clades, support for the Hypsogastropoda, the non-monophyly of the neogastropods, and the long branch length exhibited by the eatoniellid (see Colgan *et al.* 2007, for detailed discussion).

Caenogastropod monophyly has generally been supported in synoptic molecular studies of Gastropoda, although these included limited representation. This was observed for the six caenogastropods included in Tillier *et al.*'s (1992) study of D1 28S rDNA. The five representatives included in McArthur and Koop (1999) were monophyletic in parsimony analyses (but

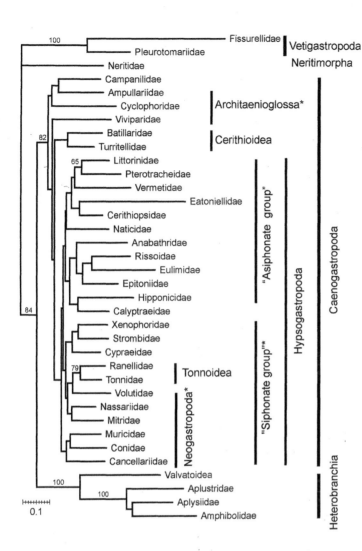

FIGURE 13.15. Maximum likelihood tree obtained by Colgan *et al.* (2007: fig. 2). Numbers on the branches are the maximum-likelihood bootstrap support percentages over 50. Higher-level taxa are indicated by bars to the right of the topology. * indicates a nonmonophyletic group. The scale bar is graduated in units of 0.01 substitutions per site.

not maximum likelihood) when data from both D1 and D6 28S rDNA expansion regions was included. In the larger studies, monophyly was observed in Harasewych *et al.* (1998) and McArthur and Harasewych (2003), in all analyses in Colgan *et al.* (2007), and in some but not all analyses in Colgan *et al.* (2003), with the exceptions mostly due to the inclusion in Caenogastropoda of the sequenced neritimorph or the exclusion of the cyclophorid. Monophyly has never been strongly contradicted.

MORPHOLOGICAL ANALYSIS

Ponder and Lindberg's (1997) data set of 117 morphological characters was used as a starting point to develop characters appropriate to the finer levels of resolution required in this

analysis. In addition to the more traditional characters associated with the shell, radula, head-foot, nervous system, alimentary canal, kidney, and reproductive system, recent studies have targeted the complex musculature of the buccal mass/ proboscis (Simone 2001, 2002, 2004a, b, 2005; Kantor 1988, 1990, 1991; Kantor and Taylor 1991), the stomach (Strong 2003; Kantor 2003) and radular development (Guralnick and Smith 1999). The full character list and data set are available at http://www.ucmp.berkeley.edu/science/archived_data.php.

The morphological terminals coded in the analysis that generated the tree in Figure 13.16 are composite family-level terminals in order to minimize ambiguity and decrease the amount of missing data. Although not ideal, this approach

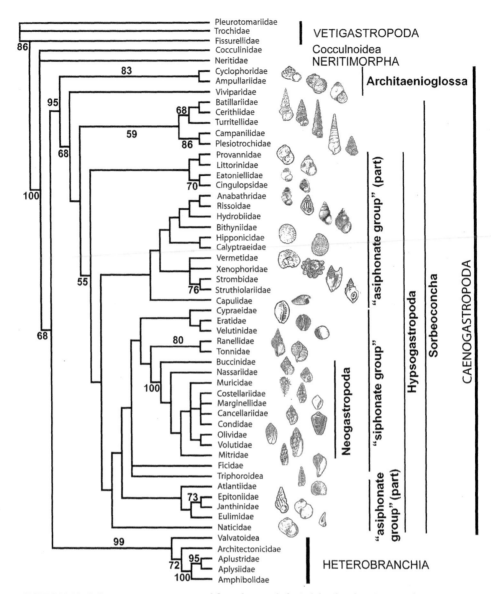

FIGURE 13.16. Strict-consensus tree generated from the morphological data by a heuristic search using PAUP* with 1,000 replications using random addition sequence, tree bisection-recombination (TBR) and with steepest descent not invoked, resulted in 72 trees (length 668, CI 0.436, RI 0.713, RC 0.311, HI 0.564). Bootstrap values > 50 resulting from 100 replications are shown.

has been necessary as a consequence of the often-fragmented anatomical data, with studies focusing on a single organ system or on a subset of characters. Thus, there are few caenogastropod species for which large data sources are available, including (but not universally) comprehensive anatomy for each major organ system, ultrastructure, histochemistry, as well as developmental, ecological, karyological, and cytological information. These few "model" species include *Marisa cornuarietis, Crepidula fornicata, Littorina littorea, Nucella lapillus, Ilyanassa obsoleta,* and *Viviparus viviparus.* Some large families have remained virtually untouched by comparative approaches, with only incomplete information available in a few scattered publications (e.g., cerithiopsids, triphorids), while with others only anatomical information may be available.

Characters and character states in the Ponder and Lindberg (1997) data set that are rendered inapplicable by the restricted taxonomic scope of this analysis have been omitted (for details see data set). Additional characters comprise features of external anatomy and shell morphology that are informative within caenogastropods. The most significant additions are in the areas of buccal musculature, stomach, and nervous system. In the last decade, as discussed previously, the morphological data set for caenogastropods has benefited from broad comparative surveys that have significantly added to our knowledge of caenogastropod anatomy for poorly understood groups (e.g., epitoniids, cyclophorids, and hipponicids) and provided characters new to caenogastropod systematics (Simone 1999, 2000a, 2001, 2002, 2004a, b, 2005; Strong 2003).

The eusperm characters included in the present study are those we consider the most robust and for which substantial amounts of data are available (and for which comparable data is available for the out groups). The parasperm are divided into different types, which are treated as separate characters, as they are likely not homologous.

The dataset of 55 taxa and 164 characters was used for a maximum-parsimony analysis using PAUP* ver. 4.0b10 (Swofford 2001); for details see caption of Figure 13.16. The strict consensus tree is shown in Figure 13.16. Caenogastropoda is strongly supported, with a bootstrap value of 95. The architaenioglossan taxa are basal but paraphyletic, and Cerithioidea and Neogastropoda are monophyletic. Sorbeoconcha is weakly supported, and, while Hypsogastropoda is not supported by bootstrap values greater than 50%, the Campaniloidea + Cerithioidea are well supported as a clade separate from the rest of the sorbeoconchs. The "asiphonate" clade seen in the molecular analysis, is paraphyletic, but the siphonate clade is supported, with Naticidae being the sister of that group.

COMBINED ANALYSIS

A combined analysis, the first carried out for caenogastropods, was conducted using the molecular data set of Colgan *et al.* (2007)[8] with areas of uncertain alignment removed and a pruned (taxa only) version of the morphological data set referred to above. As the molecular dataset is based on species, these were aligned with the taxa in the morphological data set at the family level. The full data set is available at http://www.ucmp.berkeley.edu/science/archived_data.php. A Bayesian analysis was performed using MrBayes version 3.1.2 (Huelsenbeck and Ronquist 2001) with two million iterations and the following parameters: DNA data nst = 6 rates = invgamma; unlink shape = (all); pinvar = (all); statefreq = (all); revmat = (all); prset ratepr = variable; out group Fissurellidae; mcmcp ngen = 2,000,000; nruns = 4; printfreq = 100; samplefreq = 100 nchains = 4 savebrlens = yes startingtree = random. 5,000 trees were discarded to allow for convergence.

A strict consensus of the sampled trees is shown in Figure 13.17. The resulting tree contains a monophyletic Caenogastropoda, with Heterobranchia the sister taxon. Within Caenogastropoda, Architaenioglossa is paraphyletic, the Sorbeoconcha is well supported, with Campanilidae sister to the remaining sorbeoconchs and the Cerithioidea are sister to the rest (Hypsogastropoda). The Hypsogastropoda is composed of asiphonate and siphonate clades, similar to those found in the molecular analysis (only the position of Calyptraeidae has changed), although neither is strongly supported.

The neogastropods are monophyletic, with the tonnoideans as their sister group, but Latrogastropoda is not recovered because of the inclusion of Cypraeidae as the sister to the tonnoideans, even though this has poor support. The ptenoglossans are polyphyletic, with the eulimids sister to the rissoids; the triphoroidean is in a poorly supported clade including the littorinid and the heteropod; and the position of the epitoniid is unresolved within the "asiphonate group."

---

8. The data set used included the architectonicid *Philippia lutea*, which was not in the Colgan *et al.* (2007) dataset and 18S data for the pleurotomariid was not included. This latter data did not influence the topology.

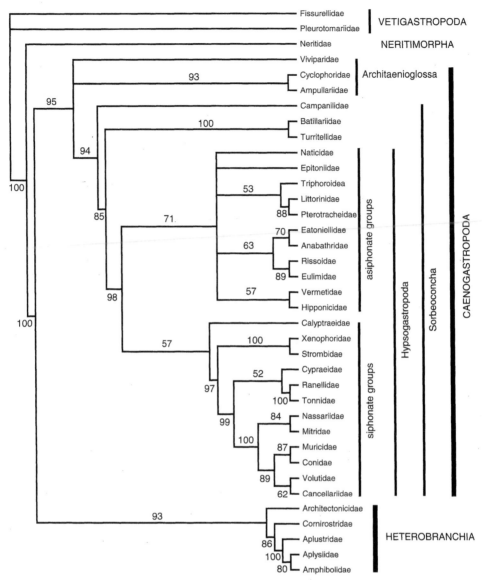

FIGURE 13.17. Strict-consensus tree obtained from a Bayesian analysis using the combined morphological dataset and the molecular data from Colgan *et al.* (2007), with the regions of questionable alignment excluded and an architectonicid (*Philippea lutea*) included. Named higher-level taxa are indicated by bars to the right of the topology. Clade credibility values (>50) are indicated.

SUMMARY OF MAJOR GROUPS

The following discussion summarizes the major groups of caenogastropods as determined by our analyses and other recent work.

ARCHITAENIOGLOSSA

Ponder and Lindberg's (1996, 1997) morphological analyses placed Architaenioglossa as the sister to the rest of the caenogastropods (Sorbeoconcha). Haszprunar (1988) could not find support for the monophyly of Architaenioglossa, but this grouping, using morphological characters, is weakly supported in Ponder and Lindberg's (1997), Strong's (2003),[9] and in our morphological analyses (Figure 13.16).

9. Only two (Cyclophoroidea, Ampullarioidea) of the three architaenioglossan superfamilies were represented in this analysis.

However, the contained superfamilies are paraphyletic in Simone's (2000a, 2004b) analyses (Figure 13.4B) and in our molecular and combined analyses (Figures 13.15, 13.17). Similarly, in most molecular analyses to date, the architaenioglossan groups are basal caenogastropods but are not monophyletic (e.g., Figure 13.14A). Harasewych *et al.* (1998), the first molecular study to include a substantial number of non-hypsogastropod caenogastropods, found that non-hypsogastropod taxa comprised a monophyletic group in maximum-likelihood analyses and in parsimony analyses considering only transversions. Monophyly was not found in other analyses by Harasewych *et al.* (1998) or in later analyses of 18S data (McArthur and Harasewych 2003) or other genes (Colgan *et al.* 2007), but it remains an hypothesis worthy of consideration.

Although monophyly of Architaenioglossa is rare in analysis of DNA sequences, Colgan *et al.* (2000) showed monophyly of the two included Architaenioglossa, but as sister to Caenogastropoda plus Heterobranchia and *Nerita* and *Nautilus*. In Colgan *et al.* (2003), the same two Architaenioglossa formed a monophyletic group only in maximum-likelihood analysis of a reduced data set and in Bayesian analyses. In Harasewych *et al.* (1998), Cerithioidea, Ampullariidae, and Cyclophorioidea were monophyletic. However, Ampullarioidea and "Architaenioglossa" were never monophyletic in these analyses because of the (variable) position of Viviparidae.

While molecular studies have consistently included Campaniloidea in Caenogastropoda (Harasewych *et al.* 1998; McArthur and Koop 1999; McArthur and Harasewych 2003; Colgan *et al.* 2000, 2003, 2007), this taxon is grouped with Cyclophoridae (Colgan *et al.* 2003, some analyses; McArthur and Harasewych 2003) or Ampullariidae (McArthur and Koop 1999; Colgan *et al.* 2007, maximum likelihood and some maximum parsimony analyses) rather than as a basal sorbeoconchan as suggested by morphology (Ponder and Lindberg 1997; Simone 2001) and our morphological and combined analyses (Figures 13.16, 13.17).

With the three superfamilies (Cyclophoroidea, Ampullarioidea, and Viviparoidea) comprising the architaenioglossans forming two or three branches of basal caenogastropods, there remains the puzzling question as to why nonmarine taxa are the likely sisters to the largely marine caenogastropods, with the most likely assumption being that their marine ancestors are extinct. However, some molecular analyses (as discussed previously) suggest that the marine Campaniloidea may be associated with one of these branches. Such a relationship is intriguing given the report of a hypoathroid nervous system in the possible campaniloidean *Cernina* (discussed later).

SORBEOCONCHA

This group contains caenogastropods other than the architaenioglossan clade or grade. Within this grouping, the relative position/relationships of the Cerithioidea and Campaniloidea (see also previous discussions), which probably represent the most primitive extant marine caenogastropods, are unclear because morphological (Ponder and Lindberg 1996, 1997; Simone 2000a, Figure 13.4B; Strong 2003, Figure 13.4A) and molecular data (Harasewych *et al.* 1998, Colgan *et al.* 2000, McArthur and Harasewych 2003), give different results (see previous discussion). The enigmatic Campanilidae is represented by a single extant species endemic to southwestern Australia. Plesiotrochidae (Houbrick 1990) is included in Campaniloidea based on sperm morphology (Healy 1993a). *Cernina fluctuata* (previously included in Naticidae) has a hypoathroid nervous system (Kase 1990), suggesting architaenioglossan affinities, but its sperm morphology is like that of campaniloideans (Healy and Kase, unpublished data; J. M. H., personal observation), other members of which are epiathroid. Simone's (2001) morphological analysis of Cerithioidea (which lacks sperm data) has *Campanile* closely related to turritellids and vermetids, with the distinctive characters seen in *Campanile* and Vermetidae interpreted as autapomorphies.

Despite some seemingly good morphological support for Sorbeoconcha (see earlier discussion of main groups recognized within caenogastropods), this grouping is only weakly supported in the strict-consensus tree in our new morphological analysis (Figure 13.16). Although the Sorbeoconcha has not been supported by any large molecular data set, generally because of the inclusion of a cerithioidean or campaniloidean in the group containing the architaenioglossans (discussed earlier), it is recovered in our combined analysis (Figure 13.17).

CERITHIOIDEA

In the morphological analysis of Simone (2001) the Cerithioidea also included the Campaniloidea and Vermetoidea. Molecular (Lydeard et al. 2002) and ultrastructural sperm data (Healy 1983a, 1988a, 1996a), however, support the restricted interpretation of Cerithioidea. In our new morphological analysis the strict-consenus tree (Figure 13.16) has the Cerithioidea and Campaniloidea as sister taxa and the Viviparidae as sister to this clade + Hypsogastropoda. In the majority-rule tree of the combined analysis the Campaniloidea are sister to Cerithioidea + Hypsogastropoda, with Vermetidae included among the partially resolved "lower" hypsogastropods (Figure 13.17), forming an "asiphonate" group.

HYPSOGASTROPODA

Hypsogastropoda (Ponder and Lindberg 1997) includes all the extant caenogastropods other than architaenioglossans, Cerithioidea, and Campaniloidea. According to the latest classification (Bouchet and Rocroi 2005), this group contains two orders, three suborders, and 30 superfamilies. It is not known whether the "mesogastropod" component of the hypsogastropods (the "higher mesogastropods" *sensu* Healy 1988a) is monophyletic, but the late appearance of clearly identifiable neogastropods relative to "higher mesogastropods" (discussed subsequently) strongly suggests that the latter group is paraphyletic.

Hypsogastropoda is often observed in molecular analyses (all analyses in Harasewych et al. 1998; McArthur and Harasewych 2003; most analyses in Colgan et al. 2007, Figure 13.15) but is weakly contradicted in Colgan et al. (2003). In the analyses in Colgan et al. (2003, 2007) that

do contradict monophyly, a clade containing all studied hypsogastropods is still seen, but it also includes cerithioideans (either Turritellidae or Batillariidae, but not both) and, often, Viviparidae. The hypsogastropods are recovered in both our new morphological (Figure 13.16) and combined (Figure 13.17) analyses.

Relationships within Hypsogastropoda are poorly resolved in all molecular and morphological analyses to date. Within Hypsogastropoda, the molecular analysis of Colgan et al. (2007) (Figure 13.15) and our combined analysis (Figure 13.17) revealed two main clades: a siphonate paraphyletic clade and one in which the anterior siphon is lacking, apart from Cerithiopsidae, suggesting that the siphon in that group (Triphoroidea) may not be homologous with those in other caenogastropods. The only exception was that the Calyptraeidae changed from the asiphonate clade in the molecular analysis to the siphonate clade in the combined analysis. The siphonate clade was also recovered in the morphological analysis, but the asiphonate clade was paraphyletic.

Apart from the asiphonate and siphonate clades, little consistent structure above the superfamily level has been observed within Hypsogastropoda in published molecular analyses. In Harasewych et al. (1998), two of the three Littorinoidea were monophyletic in most analyses. In Colgan et al. (2003), Vermetidae plus Epitoniidae was supported in all analyses, with the Cerithiopsidae being the sister group to this pair in a number of analyses, notably when third base positions are excluded.

In Simone's (2000a) analysis, a sister group relationship between Tonnoidea (including Ficidae) + Neogastropoda (= Latrogastropoda) is supported by several characters including a pleurembolic proboscis.[10] Latrogastropoda is

---

10. If Tonnoidea + Neogastropoda is a monophyletic group, there remains the question as to whether their proboscis is homologous. Kantor (2002) argued that a short proboscis is plesiomorphic for neogastropods, and the probably basal tonnoidean families Pisanianuridae and Laubierinidae also have a short proboscis, but detailed comparative studies are lacking.

also recovered in part in our morphological analysis (Figure 13.16), with the Ficidae (not included in the molecular analyses), considered to be the sister to the Neogastropoda by Riedel (2000), well outside Tonnoidea + Neogastropoda. In the combined analysis, a similar grouping is achieved, although the cypraeid is in the same monophyletic group as the Tonnoidea (Figure 13.17).

Kosuge (1966) suggested that his Heterogastropoda lay between the meso- and neogastropods, but Graham (1985: 177) argued that "this suggestion is not acceptable." Interestingly, Strong's (2003) analysis has an epitoniid (the only ptenoglossan in her analysis) as the sister to the neogastropods. However, in other analyses involving ptenoglossans and neogastropods (Ponder and Lindberg 1997; Colgan et al. 2000, 2003, 2007) there is little or no support for this relationship. This is also the case in our morphological analysis, where the triphorids and Ficidae are unresolved sisters to the "higher" hypsogastropods while the remaining ptenoglossans + Atlantidae are sister to that clade (Figure 13.16). In the combined analysis the ptenoglossan taxa are scattered among the other "lower" hypsogastropod taxa (Figure 13.17).

The placement of the heteropod Pterotrachoidea in our morphological analysis (Figure 13.16) as sister to the eulimids, epitoniids, and janthinids is at variance to a littorinoidean relationship suggested by morphological similarities, primarily in reproductive anatomy (Gabe 1965; Martoja and Thiriot-Quiévreux 1975) and supported with molecular data (Strong and Harasewych 2004; Colgan et al. 2007; Harasewych and Strong, unpublished data; our combined analysis, Figure 13.17). Bandel and Hemleben (1987) and Bandel (1993) suggested that the sister taxon to the heteropods was Vanikoridae or Pickworthiidae based on the similarity of their larval shells with those of atlantids; these relationships remain untested, because no molecular or morphological analyses have included representatives of these suggested sister taxa to date. Vanikoridae and Hipponicidae are usually included in

separate superfamilies but are both regarded as Calyptraeoidea in Simone's (2002) morphological analysis. In contrast, Collin's (2003) molecular analysis involving three genes indicated that vanikorids are well removed from both hipponicids and calyptraeids and that both the latter groups are separate clades. In that analysis vanikorids were sister to the two tonnoideans, although with a very long branch, while *Littorina* and naticids were sisters (Collin 2003: fig. 6).

NEOGASTROPODA

The Neogastropoda are usually considered to be monophyletic, this view being supported by several morphological synapomorphies (Ponder 1974; Taylor and Morris 1988; Kantor 1996; Ponder and Lindberg 1996, 1997; Strong 2003). They tend to have higher chromosome numbers and larger cellular DNA content than other gastropods have (see Thiriot-Quiévreux 2003 and Gregory 2005 for references). In our morphological analysis the neogastropods are recovered as a group, with the Buccinidae as the sister to the remainder, in the consensus tree (Figure 13.16). They were also strongly supported in the combined analysis (Figure 13.17). However, results from molecular data are ambivalent (discussed subsequently).

The origin and subsequent radiation of the Neogastropoda represent a major event in caenogastropod evolutionary history, but the identity of the sister taxon to this group remains unresolved. Different hypotheses have been presented by neontologists: that the neogastropods arose from an "archaeogastropod" or primitive "mesogastropod" (Ponder 1974) or that they arose from a "higher mesogastropod," usually considered to be a tonnoidean (Graham 1941; Ponder 1974 [gives summary of early literature]; Taylor and Morris 1988; Kantor 2002), or Ficoidea (Riedel 2000), or noncommittal (e.g., Kantor 1996). The "higher mesogastropod" hypothesis has support from morphological (Ponder and Lindberg 1996, 1997; Strong 2003), molecular (Tillier et al. 1992, 1994; but not Rosenberg et al. 1994, in which the neogastropods form a clade with the only vetigastropod in their analysis) and ultrastructural data (sperm and osphradium) (e.g., Haszprunar 1985a; Healy 1988a, 1996a). Buckland-Nicks and Tompkins (2005) have recently suggested the Ranellidae (and presumably therefore the Tonnoidea) are the sister taxon to the Neogastropoda, based on shared occurrence of a particular type of vermiform parasperm ('lancet parasperm'), and this is also consistent with eusperm morphology (see Kohnert and Storch 1984; Koike 1985; and Healy 1996a for comparative figures of eusperm and literature). In Colgan et al. (2007), Tonnoidea are the sister group to Volutidae in all likelihood-based analyses, sometimes with substantial support. This set of three taxa was generally included in a monophyletic clade including two other neogastropods (Nassariidae and Mitridae).

The identity of fossil sister taxa has been equally contentious. Suggestions that Paleozoic siphonate subulitoid gastropods (Figure 13.6H) or other basal caenogastropods are precursors or close relatives of neogastropods (Cox 1960a; Ponder 1974) have not attracted recent support (e.g., Riedel 2000). The Triassic/Jurassic Purpurinidae (Figure 13.13M) have been hypothesized as close relatives, or even members of, Neogastropoda (Taylor et al. 1980; Kaim 2004). These assumptions are based on the more or less fusiform shell shape, teleoconch ornament, and presence of an anterior siphonal notch/canal. Kaim (2004) placed the Purpurinidae in the Neogastropoda, implying that this lineage was present as early as Middle to late Triassic. However, Bandel (1993) and Riedel (2000) doubted that Purpurinoidea are neogastropods, because they have a small larval shell as exemplified in a Triassic member (*Angularia*) (Figure 13.13M; Bandel 1993: pl. 14 fig. 4). The Early Jurassic to Cretaceous genus *Astandes* (= *Maturifusus*, family Maturifusidae) has been repeatedly suggested as a possible early member or close relative of the Neogastropoda (Szabó 1983; Schröder 1995; Bandel 1993; Riedel 2000; Kaim 2004) (Figure 13.13N–P). The cancellate teleoconch ornament, distinctly siphonostomatous

aperture, and the relatively large larval shell suggest that *Astandes* is a neogastropod or closely related. A possible Late Triassic maturifusid was reported by Nützel and Erwin (2004) as the earliest member of this group (Figure 13.13N). However, even if the fossil ancestor is convincingly identified, the identity of the living sister taxon remains necessary for the interpretation of the homology of anatomical characters.

Neogastropoda has usually been contradicted, albeit weakly, in molecular analyses, with monophyly supported by Tillier et al. (1992) and questionably by Rosenberg et al. (1994) or weakly contradicted by Tiller et al. (1994) and Colgan et al. (2000, 2003, 2007). Harasewych et al. (1997) failed to resolve neogastropod monophyly in their 18S rDNA analyses. Monophyly of Neogastropoda could not be tested with the CO1 data of Harasewych et al. (1997), but three of the four superfamilies (Conoidea, Buccinoidea, and Muricoidea) within the group were supported in all analyses. Harasewych et al. (1998) recovered Neogastropoda (three taxa) in a few of their analyses (MP ML), but not all. Riedel (2000) was also unable to retrieve a monophyletic Neogastropoda using 16S and 18S sequence data. In Colgan et al. (2003), at most two of the five studied neogastropods were included in a monophyletic clade exclusive of other taxa. In one analysis, three Neogastropoda (Conidae, Mitridae, and Cancellariidae) were included in a monophyletic clade with Cerithiopsidae. At most, five of the six studied Neogastropoda were in the same monophyletic clade in parsimony analyses conducted by Colgan et al. (2007). Given that the position of any recognizable neogastropod group was predominantly basal within Hypsogastropoda, Colgan et al. (2007) argued that the group has a deep phylogenetic history in Hypsogastropoda, implying that the stem group is not yet recognized or extinct.

This large and important group is still in a state of flux, with even some supposedly well-known taxa recently reinterpreted. For example, *Morum* was considered to be a cassid (Tonnoidea) until shown to be a harpid (Neogastropoda) (Hughes 1986b), and the commercially important genus

Babylonia, the subject of many studies and long considered to be a buccinid, was recently included in a separate family related to Olividae and Volutidae (Harasewych and Kantor 2002).

SUMMARY OF WHAT WE KNOW ABOUT THE MAJOR CAENOGASTROPOD LINEAGES

It is apparent that the Caenogastropoda is a monophyletic group and, as far as living taxa are concerned, is sister to the heterobranchs. The basal taxa within the caenogastropods are the architaenioglossan groups, with some molecular results suggesting also the possible inclusion of the Campaniloidea and thus violating the monophyly of the original concept of Sorbeoconcha (all caenogastropods other than architaenioglossans). Cerithioidea is sister to the remaining caenogastropods (Hypsogastropoda) in most molecular and morphological analyses. Hypsogastropods are poorly resolved in morphological analyses. In the latest molecular analysis (Colgan *et al.* 2007) (Figure 13.15) and in our combined analysis (Figure 13.17), two groups of hypsogastropods are identified: asiphonate and siphonate clades. The asiphonate clade is not retrieved in the morphological analysis (Figure 13.16), but the siphonate clade is. In morphological analyses and our combined analysis the neogastropods are monophyletic, but rarely so in molecular analyses. The general lack of resolution in the hypsogastropods may be due to their very rapid radiation, particularly in the early Cenozoic.

ADAPTIVE RADIATIONS

Given our reassessed phylogeny, we now discuss how this assists in interpreting the extraordinary adaptive radiations undergone by caenogastropods and expressed in the great morphological, ecological, physiological, and behavioral diversity in the group.

While most are marine, benthic, and epifaunal, including many of the minute cerithioidean and hypsogastropod taxa that live on algae beneath stones and rocks, some burrow in sediment or even live interstitially. A few hypsogastropods are pelagic (Janthinidae), or active swimmers (Pterotracheoidea, the heteropods). From our phylogeny, it is clear that these invasions of the water column have occurred independently from different benthic lineages. Other habits, such as burrowing (e.g., Naticidae, Terebridae, Olividae, Struthiolariidae) have also occurred independently. Vermetids are one of only two caenogastropod lineages that can directly cement their shells to hard substrates, sometimes even forming conspicuous intertidal zones, while hipponicids secrete a shelly plate with their foot, which is fused to the substrate independently of the shell. The other direct-cementing caenogastropod is an enigmatic freshwater taxon, *Helicostoa* (Lamy 1926), of probable rissooidean affinities. Some other taxa have independently adopted stationary lifestyles, embedded in corals or sponges (e.g., some Coralliophilidae, Siliquariidae), or their echinoderm host (some parasitic Eulimidae).

Several caenogastropod lineages have members that have independently undergone extensive freshwater radiations (Strong *et al.,* in press). It is uncertain as to whether the architaenioglossan Viviparidae and Ampullariidae shared a marine or a freshwater ancestor. The five families of cerithioideans that live in freshwater entered that habitat independently at least twice (Lydeard *et al.* 2002). The hypsogastropod rissooidean families Hydrobiidae *sensu lato,* Pomatiopsidae, and Bithyniidae probably entered freshwater independently, but there is insufficient evidence to demonstrate this in published analyses. However, some members of the predominantly brackish-water rissooidean families Assimineidae and Stenothyridae have also entered freshwater, as has one genus of Littorinidae. Interestingly, the movement into freshwater habitats is rare in higher hypsogastropods, with only a couple of Recent genera of neogastropods (Buccinidae and Marginellidae) having managed this transition, presumably via estuaries (Strong *et al.,* in press). Similarly, terrestriality has evolved several times. All members of the architaenioglossan Cyclophoroidea and the hypsogastropod littorinoidean Pomatiidae are terrestrial, as are many species in the rissooidean Assimineidae and Truncatellidae, with members of the latter family having been shown to independently become fully terrestrial a number of times (Rosenberg 1996). No higher hypsogastropods occupy this habitat.

Most caenogastropods are small (< 10 mm in maximum dimension). For example, Bouchet *et al.* (2002) found that small taxa are among the most abundant and diverse caenogastropods in a tropical reef environment. The members of some families are mostly smaller than 5 mm, for example the hypsogastropod rissooideans, cingulopsoideans, most Triphoroidea, Eulimoidea, and Marginellidae, many Columbellidae and Turridae *sensu lato,* with some a little less than a millimeter in maximum shell dimension. At the other end of the scale, members of some of the higher hypsogastropod carnivorous groups achieve a size of 30 cm or more (e.g., Ranellidae, Bursidae, Cassidae, Volutidae, Fasciolariidae, Turbinellidae, and Melongenidae), while members of other hypsogastropod families attain more than 20 cm (e.g., Strombidae, Tonnidae, Buccinidae, and Muricidae). Large size is also attained by some non-hypsogastropods, including Ampullariidae, Campanilidae, Potamididae, and Cerithiidae.

One of the hallmarks of caenogastropod evolution is modification of the mantle cavity organs, with only the left (monopectinate) gill, osphradium, and hypobranchial gland being retained. The move from exhalant to inhalant control of the water currents through the mantle cavity was seen by Ponder and Lindberg (1997) as a critical innovation of the Sorbeoconcha that enabled better utilization of the chemosensory function of the osphradium and is correlated with its enlargement and the development (probably independently in several lineages) of an anterior siphon. Adoption of a semiterrestrial or terrestrial lifestyle resulted in the atrophy or loss of the ctenidium in those lineages and with the development of a lung in the amphibious Ampullariidae.

Feeding strategies range from the supposedly ancestral deposit feeding and surface grazing in most architaenioglossans, cerithioideans, and lower hypsogastropods to herbivory, grazing carnivory, or active predation and, in Eulimidae and a few neogastropods, parasitism. Some hypsogastropod families show a range of feeding habits, with cypraeids grazing on algae or sponges and triviids on algae, sponges, and tunicates. By way of contrast, the related ovulids feed exclusively on soft corals and the velutinids on tunicates. Such specialization in carnivorous feeding appears to be the norm in many groups; for example, within the ptenoglossan taxa the Triphoroidea (Cerithiopsidae and Triphoridae) feed exclusively on sponges, the epitonoideans on cnidarians, and the eulimoideans on echinoderms. In the neogastropods mitrids feed exclusively on sipunculids, coralliophillines on soft and hard corals, and turrids (*sensu lato*) on polychaetes. Some neogastropod families appear to be more generalist; in marginellids (*sensu lato*), for example, some species graze on bryozoans or tunicates, a few are mollusc shell drillers, and some are suctorial fish feeders (discussed later). Some columbellids are carnivores, and others have become herbivores (deMaintenon 1999). Nassariidae are primarily scavengers (e.g., Morton 2003), but at least one has reverted to selective deposit feeding (Connor and Edgar 1998). Tonnoideans feed on a range of prey including echinoderms, molluscs, and tunicates, although some are more specialized at the family level.

Most carnivorous hypsogastropods feed by biting or rasping at their prey, but the ability to swallow large prey intact has evolved in at least two lineages: the tonnids (Tonnoidea), which engulf entire holothurians, and the fish-eating conids (Conoidea). Other conids and other conoideans (Turridae *sensu lato* and Terebridae) engulf polychaete worms. A few groups feed suctorially, with some cancellariids (O'Sullivan *et al.* 1987) feeding on resting rays (Elasmobranchia), while some colubrariids and marginellids suctorially feed on sleeping parrot fishes (Bouchet and Perrine 1996). Eulimidae feed suctorially on their echinoderm hosts. The hypsogastropod families Naticidae and Muricidae have long been known to drill holes in the shells of their prey and have independently evolved accessory boring organs (Carriker and Gruber 1999). A few other groups also drill their prey (marginellids, Ponder and Taylor 1992; buccinids, Morton 2006; cassids, Hughes and Hughes 1971; Hughes 1986a), but the mechanisms involved have not yet been investigated in detail. Other ways of entering shelled prey have been evolved, including forced entry to bivalves using a spine on the aperture of the predator, a mechanism that has independently evolved in several neogastropod families (Vermeij and Kool 1994).

Macroherbivory is uncommon, with the Strombidae and Ampullariidae being the main groups engaging in this type of feeding, and it is seen in some cypraeids and columbellids (Neogastropoda). This feeding mode is derived within the columbellids (deMaintenon 1999) and Stromboidea because, in the latter case, the ancestral (and paraphyletic; Roy 1994) Aporrhaidae are deposit feeders (as are the stromboideans' sister group, Xenophoroidea). Another stromboidean family, Struthiolariidae, filter feeds (Morton 1951). Other filter (suspension)-feeding taxa have also evolved independently: architaenioglossan Viviparidae (Viviparoidea), the cerithioidean Turritellidae and Siliquariidae, and, within the hypsogastropods, the Vermetidae (Vermetoidea),[11] Bithyniidae (Rissooidea), Calyptraeidae (Calyptraeoidea), and Capulidae (*Trichotropis*) (Capuloidea) as well as the Struthiolariidae previously noted. Similar structures (elongated gill filaments, endostyle, food groove) associated with this feeding mode have convergently

developed in these taxa. Many vermetids use a mucus net secreted by pedal glands to ensnare floating food.

Apart from overall shell thickening and the adoption of cryptic habits, some other adaptive changes to caenogastropod shells appear to be responses to predation. These include the development of terminal growth, allowing thickening of the aperture, including the formation of varices and/or spines, as a defense against crab predation in particular (e.g., Vermeij and Signor 1992). Some velutinids employ chemical defense (e.g., Andersen *et al.* 1985), and dramatic escape responses have also evolved in some taxa, as has the convergent autotomy of the posterior end of the foot (e.g., some cypraeids, Burgess 1970; *Harpa*, e.g., Liu and Wang 2002) or mantle (e.g., *Ficus*, Liu and Wang 2002), a mechanism that has also independently evolved in some other groups of gastropods (notably stomatelline trochids).

One of the major innovations in caenogastropod ancestors was internal fertilization. The advent of planktotrophy was also a significant hallmark of the group and may have been associated with internal fertilization, enabling the production of encapsulated eggs and thus allowing larvae to undergo their early development in a protected environment to be released as veligers. A consequence of this was the ability to forgo a planktonic larval stage—so-called direct development, either within an external capsule or in capsules or eggs retained in a brood pouch within the animal. Various mechanisms have been evolved within caenogastropods to provide nutrients to such embryos. These include yolk, albumen, infertile eggs ("nurse eggs"), and cannibalism (adelphophagy). Although there appears to be considerable variation in some groups, particularly the neogastropods, mapping the distribution of these strategies phylogenetically within families and superfamilies will undoubtedly be informative. Egg encapsulation and intracapsular development also facilitated invasion of marginal marine and nonmarine habitats.

---

11. For an alternative view see Simone (2001), who treats vermetids and turritellids as closely related taxa within Cerithioidea, as was the case in earlier literature (e.g., Thiele 1929).

GAPS IN KNOWLEDGE AND FUTURE STUDIES

Following are some of the significant phylogenetic questions relating to caenogastropods that remain to be resolved using a variety of approaches:

Which extinct lineage is the sister taxon of caenogastropods?
What are the monophyletic groups in the "architaenioglossan" grade?
Is Sorbeoconcha a monophyletic group?
What are relationships within the Hypsogastropoda?
What are the main monophyletic groupings?
What are the relationships of "ptenoglossan" groups?
Establish the composition of the smallest monophyletic group including neogastropods and, if Neogastropoda are monophyletic, identify their sister taxon.
Are some of the large, diverse groups currently recognized as families or superfamilies (e.g., Rissooidea) monophyletic?

To achieve answers to these questions, the numerous gaps in our knowledge need to be addressed. Many characters and character complexes need to be reexamined in a phylogenetic context, and more taxa need to be examined in detail, using histology and ultrastructure as well as examining physiological and functional aspects.

With morphological data, there are still few published studies on the anatomy of many family-level taxa, and in some cases we are still relying on accounts over 100 years old. While there are good data sets available for some families, for others typically only one or a few species have been anatomically described, and often these are known somewhat superficially. Histological data is often not provided, and details are sometimes superficial or lacking for organ systems such as the renopericardial and nervous systems. A good sampling of ultrastructural data is lacking for all systems other than sperm and osphradia. Given that these two systems have contributed so much to our understanding of gastropod phylogenetics, the potential for substantial new contributions via ultrastructure is considerable.

DEVELOPMENTAL AND GENOMIC DATA

Although the few available items of developmental data appear to be informative with very promising potential, many additional studies are required. Even basic information on spawn and egg capsules has phylogenetic potential but has been little utilized because of many gaps in the available data. Similarly, organogenesis is a key to understanding many of the homology issues, but, again, very few studies have been undertaken.

DNA SEQUENCE DATA

DNA sequence data has only recently begun to add significant insights into caenogastropod phylogeny, but interesting hypotheses are now emerging. These include the close relationships of Littorinidae with heteropods and the association of Campaniloidea with Ampullariidae and Cyclophoridae. A stable understanding of the main lineages within Caenogastropoda will, however, likely require at least double or triple the numbers of aligned base positions and three times the taxa that have already been sequenced. Mitochondrial gene order has hardly been looked at despite its significance in some other gastropods, with changes recently demonstrated even within single families (Vermetidae, Rawlings *et al.* 2001; Ampullariidae, Rawlings *et al.* 2003).

ACKNOWLEDGMENTS

W.F.P. acknowledges discussion and collaboration with many colleagues other than those involved in this chapter, including David Lindberg, Gerhard Haszprunar, Jeffrey Stilwell, John Taylor, Jerry Harasewych, Yuri Kantor, Philippe Bouchet, Rosemary Golding, and Louise Page. The Australian Research Council provided some financial support to W.F.P. and D.J.C. for caenogastropod studies. A.N. acknowledges collaboration with R.B. Blodgett, Alex Cook, D.H. Erwin, J. Frýda, A.

Guzhov (for the reproduction of *Astandes*), A. Kaim, and Peter Wagner. Financial support of the Deutsche Forschungsgemeinschaft (NU 96/3-1, 3-2; 96/6-1, 6-2) is also acknowledged. D. J. C. thanks Australian Museum colleagues Peter Eggler, Julie Macaranas, Emma Beacham, Edwina Rickard, and Andrea Feilen for their help with the project. L. R. L. S.'s work on caenogastropods is supported by Fundação de Amparo à Pesquisa do Estado de São Paulo (FAPESP). J. M. H. thanks many colleagues including Barrie Jamieson, Lina Daddow, Alan Hodgson, Rüdiger Bieler, Paula Mikkelsen, Folco Giusti, Richard Willan, Don Anderson, John Buckland-Nicks, John Taylor, David Reid, Emily Glover, Liz Platts, Gerhard Haszprunar, Jerry Harasewych, Philippe Bouchet, and Australian Research Council, Queensland Museum and University of Queensland for past funding. Oxford University Press, Blackwell Publishing, and The California Malacozoological Society are thanked for allowing the use of previously published figures. E. S. thanks, for discussions and guidance, Philippe Bouchet, Jerry Harasewych, Gerhard Haszprunar, Yuri Kantor, David Lindberg, and Winston Ponder. Useful comments by Rosemary Golding, Yuri Kantor, Steffen Kiel, and two anonymous reviewers improved the manuscript.

REFERENCES

Amaudrut, M. A. 1898. La partie antérieure du tube digestif de la torsion chez les mollusques gastéropodes. *Annales des Sciences Naturelles. Zoologie* 8: 1–291.

Andersen, R. J., Faulkner, D. J., He, C. H., and Clardy, J. 1985. Metabolites of the marine prosobranch mollusc *Lamellaria* sp. *Journal of the American Chemical Society* 107: 5492–5495.

Anderson, W., and Personne, P. 1976. The molluscan spermatozoon: dynamic aspects of its structure and function. *American Zoologist* 16: 293–313.

Andrews, E. B. 1988. Excretory systems of molluscs. In *The Mollusca. Form and Function* 11. Edited by E. R. Trueman and M. R. Clarke. San Diego: Academic Press, 381–448.

———. 1991. The fine structure and function of the salivary glands of *Nucella lapillus* (Gastropoda:

Muricidae). *Journal of Molluscan Studies* 57: 111–126.

———. 1992. The fine structure and function of the anal gland of the muricid *Nucella lapillus* (Neogastropoda) and a comparison with that of the trochid *Gibbula cineraria*. *Journal of Molluscan Studies* 58: 297–313.

———. 2000. Ultrastructural and cytochemical study of the digestive gland cells of the marine prosobranch mollusc *Nucella lapillus* (L.) in relation to function. *Malacologia* 42: 103–112.

Andrews, E. B., Page, A. M., and Taylor, J. D. 1999. The fine structure and function of the anterior foregut glands of *Cymatium intermedius* (Cassoidea: Ranellidae). *Journal of Molluscan Studies* 65: 1–19.

Ball, A. D. 2002. Foregut ontogeny of the Neogastropoda: comparison of development in *Nucella lapillus* and *Conus anemone*. *Bollettino Malacologico* 38 Suppl. 4: 51–78.

Ball, A. D., Andrews, E. B., and Taylor, J. D. 1997a. The ontogeny of the pleurembolic proboscis in *Nucella lapillus* (Gastropoda: Muricidae). *Journal of Molluscan Studies* 63: 87–99.

Ball, A. D., Taylor, J. D., and Andrews, E. B. 1997b. Development of the acinous and accessory salivary glands in *Nucella lapillus* (Neogastropoda: Muricoidea). *Journal of Molluscan Studies* 63: 245–260.

Bandel, K. 1975. Embryonalgehäuse karibischer Meso- und Neogastropoden (Mollusca). *Akademie der Wissenschaften und der Literatur Mainz, Abhandlungen der Mathematisch-Naturwissenschaftlichen Klasse* 1: 1–133, 21 pls.

———. 1982. Morphologie und Bildung der frühontogenetischen Gehäuse bei conchiferen Mollusken. *Facies* 7: 1–198.

———. 1984. The radulae of Caribbean and other Mesogastropoda and Neogastropoda. *Zoologische Verhandelingen* 214: 1–188.

———. 1990. Shell structure of the Gastropoda excluding Archaeogastropoda. In: ed. *Skeletal Biomineralization: Patterns, Processes and Evolutionary trends*. Vol. I. Edited by J. G. Carter. New York: Van Nostrand-Reinhold, pp. 117–134.

———. 1991a. Über triassische "Loxonematoidea" und ihre Beziehungen zu rezenten und paläozoischen Schnecken. *Paläontologische Zeitschrift* 65: 239–268.

———. 1991b. Character of a microgastropod fauna from a carbonate sand of Cebu (Philippines). *Mitteilungen aus dem Geologisch-Paläontologischen Institut der Universität Hamburg* 71:441–485.

———. 1992. Über Caenogastropoden der Cassianer Schichten (Obertrias) der Dolomiten (Italien) und ihre taxonomische Bewertung. *Mitteilungen aus dem Geologisch-Paläontologischen Institut der Universität Hamburg* 73: 37–97.

———. 1993. Caenogastropoda during Mesozoic times. *Scripta Geologica* Special Issue 2: 7–56.

———. 2002. Reevaluation and classification of Carboniferous and Permian Gastropoda belonging to the Caenogastropoda and their relation. *Mitteilungen aus dem Geologisch-Paläontologischen Institut der Universität Hamburg* 86: 81–188.

Bandel, K., and Geldmacher, W. 1996. The structure of the shell of *Patella crenata* connected with suggestions to the classification and evolution of the Archaeogastropoda. *Freiberger Forschungshefte* C 464: 1–71.

Bandel, K., and Hemleben, C. 1987. Jurassic heteropods and their modern counterparts (planktonic Gastropoda: Mollusca). *Neues Jahrbuch für Geologie und Paläontologie, Abhandlungen* 174: 1–22.

Bandel, K., Nützel, A., and Yancey, T. E. 2002. Larval shells and shell microstructures of exceptionally well-preserved Late Carboniferous gastropods from the Buckhorn Asphalt deposit (Oklahoma, USA). *Senckenbergiana Lethaea* 82: 639–690.

Bandel, K., and Riedel, F. 1994. Classification of fossil and Recent Calyptraeoidea (Caenogastropoda) with a discussion on neomesogastropod phylogeny. *Berliner Geowissenschaftliche Abhandlungen* 13: 329–367.

Beesley, P. L., Ross, G. J. B., and Wells, A., eds. 1998. *Fauna of Australia,* Vol. 5 Part B, *Mollusca: The Southern Synthesis.* Melbourne: CSIRO Publishing.

Bernard, F. 1890. Recherches sur les organes palléaux des Gastéropodes Prosobranches. *Annales des Sciences Naturelles. Zoologie* (7) 9: 89–404.

Berthold, T. 1991. *Vergleichende Anatomie, Phylogenie und historische Biogeographie der Ampullariidae* (Mollusca, Gastropoda). Abhandlungen des naturwissenschaftlichen Vereins in Hamburg (NF) 29. Hamburg: Verlag Paul Parey.

Bieler, R. 1992. Gastropod phylogeny and systematics. *Annual Review of Ecology and Systematics* 23: 311–338.

———. 1993. Ampullariid phylogeny—book review and cladistic re-analysis. *The Veliger* 36: 291–299.

Bondar, C. A., and Page, L. R. 2003. Development of asymmetry in the caenogastropods *Amphissa columbiana* and *Euspira lewisii*. *Invertebrate Biology* 122: 28–41.

Bouchet, P., Lozouet, P., Maestrati, P., and Heros, V. 2002. Assessing the magnitude of species richness in tropical marine environments: exceptionally high numbers of molluscs at a New Caledonia site. *Biological Journal of the Linnean Society* 75: 421–436.

Bouchet, P., and Perrine, D. 1996. More gastropods feeding at night on parrotfishes. *Bulletin of Marine Science* 59: 224–228.

Bouchet, P., and Rocroi, J.-P., eds. 2005. A nomenclator and classification of gastropod family-group names. With classification by J. Frýda, B. Hausdorf, W. Ponder, A. Valdes, and A. Warén. *Malacologia* 47:1–397.

Bouvier, E. L. 1887. Système nerveux, morphologie générale et classification des Gastéropodes Prosobranches. *Annales des Sciences Naturelles. Zoologie* 7: 1–510.

Brusca, R. C., and Brusca, G. J. 2002. *Invertebrates,* 2nd ed. Sunderland, MA: Sinauer Associates.

Buckland-Nicks, J. 1998. Prosobranch parasperm: sterile germ cells that promote paternity? *Micron* 29: 267–280.

Buckland-Nicks, J., and Tompkins, G. 2005. Paraspermatogenesis in *Ceratostoma foliatum* (Neogastropoda): confirmation of programmed nuclear death. *Journal of Experimental Zoology* 303A: 723–741.

Bulnheim, H.-P. 2000. Warum produzieren zahlreiche marine Schnecken zwei verschiedene Typen von Spermien? *Biologie in unserer Zeit* 30: 105–110.

Burgess, C. M. 1970. *The Living Cowries.* Cranbury, NJ: A. S. Barnes and Company.

Carriker, M. R., and Gruber, G. L. 1999. Uniqueness of the gastropod accessory boring organ (ABO): Comparative biology, an update. *Journal of Shellfish Research* 18: 579–595.

Checa, A. G., and Jiménez-Jiménez, A. P. 1998. Constructional morphology, origin, and evolution of the gastropod operculum. *Paleobiology* 24: 109–132.

Colgan, D., Ponder, W. F., and Eggler, P. E. 2000. Gastropod evolutionary rates and phylogenetic relationships assessed using partial 28s rDNA and histone H3 sequences. *Zoologica Scripta* 29: 29–63.

Colgan, D. J., Ponder, W. F., Beacham, E., and Macaranas, J. M. 2003. Gastropod phylogeny based on six segments from four genes representing coding or non-coding and mitochondrial or nuclear DNA. *Molluscan Research* 23: 101–148.

Colgan, D. J., Ponder, W. F., Beacham, E., and Macaranas, J. 2007. Molecular phylogenetics of Caenogastropoda (Gastropoda: Mollusca). *Molecular Phylogenetics and Evolution* 42: 717–737.

Collin, R. 1997. Hydrophobic larval shells: Another character for higher level systematics of gastropods. *Journal of Molluscan Studies* 63: 425–430.

———. 2003. Phylogenetic relationships among calyptraeid gastropods and their implications for the biogeography of marine speciation. *Systematic Biology* 52: 618–640.

Conklin, E. G. 1897. The embryology of *Crepidula*, a contribution to the cell lineage and early development of some marine gasteropods. *Journal of Morphology* 13: 1–226.

———. 1907. The embryology of *Fulgar*: a study of the influence of yolk on development. *Proceedings of the Academy of Natural Sciences of Philadelphia* 1907: 320–359, pls. XXIII–XXVIII.

Connor, M. S., and Edgar, R. K. 1998. Selective grazing by the mud snail *Ilyanassa obsoleta*. *Oecologia* 53: 271–275.

Cook, A., Nützel, A., and Frýda, J. In press. Two Carboniferous caenogastropod limpets from Australia and their meaning for the ancestry of the Caenogastropoda. *Journal of Paleontology*.

Cox, L. R. 1960a. Thoughts on the classification of the Gastropoda. *Proceedings of the Malacological Society of London* 33: 239–261.

———. 1960b. General characteristics of the Gastropoda. In *Treatise on Invertebrate Paleontology, Part I, Mollusca 1*. Edited by R. C. Moore. Lawrence, KS: Geological Society of America and University of Kansas Press, pp. I84–I169.

D'Asaro, C. N. 1965. Organogensis, development, and metamorphosis in the Queen Conch, *Strombus gigas*, with notes on breeding habits. *Bulletin of Marine Science* 15: 359–416.

———. 1966. The egg capsules, embryogenesis, and early organogenesis of a common oyster predator, *Thais haemastoma floridana* (Gastropoda: Prosobranchia). *Bulletin of Marine Science* 16:884–914.

———. 1969. The comparative embryogensis, and early organogenesis of *Bursa corrugata* Pery and *Distorsio clathrata* Lamarck (Gastropoda: Prosobranchia). *Malacologia* 9: 349–389.

deMaintenon, M. J. 1999. Phylogenetic analysis of the Columbellidae (Mollusca: Neogastropoda) and the evolution of herbivory from carnivory. *Invertebrate Biology* 118: 253–258.

Demian, E. S., and Yousif, F. 1973a. Embryonic development and organogenesis in the snail *Marisa cornuarietis* (Mesogastropoda: Ampullariidae). I. General outlines of development. *Malacologia* 12: 123–150.

———. 1973b. Embryonic development and organogenesis in the snail *Marisa cornuarietis* (Mesogastropoda:Ampullariidae).II.Development of the alimentary system. *Malacologia* 12: 1 51–174.

———. 1973c. Embryonic development and organogenesis in the snail *Marisa cornuarietis* (Mesogastropoda: Ampullariidae). III. Development of the circulatory and renal systems. *Malacologia* 12: 175–194.

———. 1973d. Embryonic development and organogenesis in the snail *Marisa cornuarietis* (Mesogastropoda: Ampullariidae). IV. Development of the shell gland, mantle and respiratory organs. *Malacologia* 12: 195–211.

———. 1975. Embryonic development and organogenesis in the snail *Marisa cornuarietis* (Mesogastropoda: Ampullariidae). V. Development of the nervous system. *Malacologia* 15: 29–42.

Erwin, D. H., and Signor, P. W. 1990. Extinction in an extinction-resistant clade: the evolutionary history of the Gastropoda. In *The Unity of Evolutionary Biology* 1. *Proceedings of the Fourth International Congress of Systematics and Evolutionary Biology*. Edited by E. C. Dudley. Portland: Dioscorides Press, pp. 152–160.

Fahey, S. J., and Healy, J. M. 2003. Sperm ultrastructure in the nudibranch genus *Halgerda* with reference to other Discodorididae and the Chromodorididae (Mollusca: Opisthobranchia). *Journal of Morphology* 257: 9–21.

Falniowski, A. 1989. Przodoskrzelne (Prosobranchia, Gastropoda, Mollusca) Polski. 1. Neritidae, Viviparidae, Valvatidae, Bithyniidae, Rissoidae, Aciculidae. *Zeszyty Naukowe Uniwersytetu Jagiellońskiego, Prace Zoologiczne* 35: 1–148.

Fioroni, P. 1966. Zur Morphologie und Embryogenese des Darmtraktes und der Transitorischen Organe bei Prosobranchiern (Mollusca, Gastropoda). *Revue Suisse de Zoologie* 73: 621–876.

———. 1982. Larval organs, larvae, metamorphosis and types of development of Mollusca, a comprehensive review. *Zoologische Jahrbücher. Abteilung für Anatomie und Ontogenie der Tiere* 108: 375–420.

Freeman, G., and Lundelius, J. W. 1992. Evolutionary implications of the mode of D quadrant specification in coelomates with spiral cleavage. *Journal of Evolutionary Biology* 5: 205–248.

Fretter, V. 1969. Aspects of metamorphosis in prosobranch gastropods. *Proceedings of the Malacological Society of London* 38: 375–386.

Fretter, V., and Graham, A. 1962. *British Prosobranch Molluscs. Their Functional Anatomy and Ecology*. London: Ray Society.

———. 1994. *British prosobranch molluscs. Their Functional Anatomy and Ecology*. Revised ed. London: Ray Society.

Frýda, J. 1999. Higher classification of Paleozoic gastropods inferred from their early shell ontogeny. *Journal of the Czech Geological Society* 44:137–154.

———. 2001. Discovery of a larval shell in Middle Paleozoic subulitoidean gastropods with description of two new species from the early Devonian of Bohemia. *Bulletin of the Czech Geological Survey* 76: 29–38.

Frýda, J., and Bandel, K. 1997. New Early Devonian gastropods from the *Plectonotus (Boucotonotus)– Palaeozygopleura* community in the Prague Basin

(Bohemia). *Mitteilungen aus dem Geologisch-Paläontologischen Institut der Universität Hamburg* 80: 1–57.

Gabe, M. 1965. Données morphologiques et histologiques sur l'appareil génital male des Hétéropodes. *Zeitschrift für Morphologie und Ökologie der Tiere* 55: 1024–1079.

Giusti, F. 1971. L'ultrastruttura dello spermatozoo nella filogenesis e nelle sistematica dei molluschi gasteropodi. *Atti della Societa Italiana di Scienze Naturali e Museuo Civico di Storia Naturale Milano* 112: 381–402.

Giusti, F., Manganelli, G., and Selmi, G. 1991. Spermatozoon fine structure in the phylogenetic study of the Helicoidea (Gastropoda, Pulmonata). In *Proceedings of the Tenth International Malacological Congress, Tübingen, 1989*. Edited by C. Meier-Brook. Tübingen: Unitas Malacologica, pp. 611–616.

Giusti, F., and Selmi, M. G. 1982. The atypical sperm in the prosobranch molluscs. *Malacologia* 22: 171–181.

Golikov, A. N., and Starobogatov, Y. 1975. Systematics of prosobranch gastropods. *Malacologia* 15: 185–232.

Gosliner, T. 1981. Origins and relationships of primitive members of the Opisthobranchia (Mollusca: Gastropoda). *Biological Journal of the Linnean Society* 16: 197–225.

Graham, A. 1939. On the structure of the alimentary canal of style-bearing prosobranchs. *Proceedings of the Zoological Society of London Series B: Biological Sciences* 109: 75–112.

———. 1941. The oesophagus of the stenoglossan prosobranchs. *Proceedings of the Royal Society of Edinburgh B* 61: 1–23.

———. 1949. The molluscan stomach. *Transactions of the Royal Society of Edinburgh* 61: 737–778.

———. 1985. Evolution within the Gastropoda: Prosobranchia. In *The Mollusca*, Vol. 10, *Evolution*. Edited by E. R. Trueman and M. R. Clark. New York: Academic Press, 151–186.

Gregory, T. R. 2005. Animal genome size database. http://www.genomesize.com.

Gründel, J. 1999a. Procerithiidae (Gastropoda) aus dem Lias und Dogger Deutschlands und Polens. *Freiberger Forschungshefte C* 481: 1–37.

———. 1999b. Truncatelloidea (Littorinimorpha, Gastropoda) aus dem Lias und Dogger Deutschlands und Nordpolens. *Berliner Geowissenschaftliche Abhandlungen Reihe E, Paläobiologie* 30: 89–119.

Guralnick, R. P., and Lindberg, D. R. 2001. Reconnecting cell and animal lineages: What do cell lineages tell us about the evolution and development of Spiralia? *Evolution* 55:1501–1519.

Guralnick, R., and Smith, K. 1999. Historical and biomechanical analysis of integration and dissociation in molluscan feeding, with special emphasis on the true limpets (Patellogastropoda: Gastropoda). *Journal of Morphology* 241: 175–195.

Guzhov, A. 2004. Jurassic gastropods of European Russia (orders Cerithiiformes, Bucciniformes, and Epitoniiformes). *Paleontological Journal* 38 Suppl. 5: 457–562.

Haas, O. 1953. Mesozoic invertebrate faunas of Peru. *Bulletin of the American Museum of Natural History* 101: 1–328.

Haasl, D. M. 2000. Phylogenetic relationships among nassariid gastropods. *Journal of Paleontology* 74: 839–852.

Harasewych, M. G., Adamkewicz, S. L., Blake, J. A., Saudek, D., Spriggs, T., and Bult, C. J. 1997. Neogastropod phylogeny: a molecular perspective. *Journal of Molluscan Studies* 63: 327–351.

Harasewych, M. G., Adamkewicz, S. L., Plassmeyer, M., and Gillevet, P. M. 1998. Phylogenetic relationships of the lower Caenogastropoda (Mollusca, Gastropoda, Architaenioglossa, Campaniloidea, Cerithioidea) as determined by partial 18S rDNA sequences. *Zoologica Scripta* 27: 361–372.

Harasewych, M. G., and Kantor, Y. I. 2002. On the morphology and taxonomic position of *Babylonia* (Neogastropoda: Babyloniidae). *Bollettino Malacologico*, 38 Suppl.: 19–36.

Haszprunar, G. 1985a. The fine morphology of the osphradial sense organs of the Mollusca. Part 1: Gastropoda—Prosobranchia. *Philosophical Transactions of the Royal Society of London Series B: Biological Sciences* 307: 457–496.

———. 1985b. The Heterobranchia—a new concept of the phylogeny and evolution of the higher Gastropoda. *Zeitschrift für Zoologische Systematik und Evolutionsforschung* 23: 15–37.

———. 1988. On the origin and evolution of major gastropod groups, with special reference to the Streptoneura. *Journal of Molluscan Studies* 54: 367–441.

———. 1993. Sententia. The Archaeogastropoda: a clade, a grade or what else? *American Malacological Union, Bulletin* 10: 165–177.

Hayashi, S. 2005. The molecular phylogeny of the Buccinidae (Caenogastropoda: Neogastropoda) as inferred from the complete mitochondrial 16S rRNA gene sequences of selected representatives. *Molluscan Research* 25: 85–98.

Healy, J. M. 1983a. Ultrastructure of euspermatozoa of cerithiacean gastropods (Prosobranchia: Mesogastropoda). *Journal of Morphology* 178: 57–75.

———. 1983b. An ultrastructural study of basommatophoran spermatozoa. *Zoologica Scripta* 12: 57–66.

———. 1988a. Sperm morphology and its systematic importance in the Gastropoda. In *Prosobranch Phylogeny*. Edited by W. F. Ponder. *Malacological Review* Suppl. 4: 251–266.

———. 1988b. Sperm morphology in *Serpulorbis* and *Dendropoma* and its relevance to the systematic position of the Vermetidae (Gastropoda). *Journal of Molluscan Studies* 54: 295–308.

———. 1988c. Ultrastructural observations on the spermatozoa of *Pleurotomaria africana* Tomlin (Gastropoda). *Journal of Molluscan Studies* 54: 309–316.

———. 1990a. Systematic importance of spermatozeugmata in triphorid and cerithiopsid gastropods (Caenogastropoda: Triphoroidea). *Journal of Molluscan Studies* 56: 115–118.

———. 1990b. Taxonomic affinities of the deep-sea genus *Provanna* (Caenogastropoda): new evidence from sperm ultrastructure. *Journal of Molluscan Studies* 56: 119–122.

———. 1990c. Sperm structure in the scissurellid gastropod *Sinezona* sp. (Prosobranchia, Pleurotomarioidea). *Zoologica Scripta* 19: 189–193.

———. 1990d. Spermatozoa and spermiogenesis of *Cornirostra, Valvata* and *Orbitestella* (Gastropoda: Heterobranchia) with a discussion of valvatoidean sperm morphology. *Journal of Molluscan Studies* 56: 557–566.

———. 1990e. Euspermatozoa and paraspermatozoa in the trochoid gastropod *Zalipais laseroni* (Trochoidea: Skeneidae), *Marine Biology* 105: 497–507.

———. 1993a. Transfer of the gastropod family Plesiotrochidae to the Campaniloidea based on sperm ultrastructural evidence. *Journal of Molluscan Studies* 59: 135–146.

———. 1993b. Comparative sperm ultrastructure and spermiogenesis in basal heterobranch gastropods (Valvatoidea, Architectonicoidea, Rissoelloidea, Omalogyroidea and Pyramidelloidea) (Mollusca). *Zoologica Scripta* 22: 263–276.

———. 1996a. Molluscan sperm ultrastructure: correlation with taxonomic units within the Gastropoda, Cephalopoda and Bivalvia. In *Origin and Evolutionary Radiation of the Mollusca*. Edited by J. Taylor. Oxford: Oxford University Press, pp. 99–113.

———. 1996b. Euspermatozoan ultrastructure in *Bembicium auratum* (Gastropoda): comparison with other caenogastropods especially other Littorinidae. *Journal of Molluscan Studies* 62: 57–63.

———. 2000. Mollusca—relict taxa. In *Reproductive Biology of Invertebrates*, Vol. 9. Part B, *Progress in Male Gamete Biology*. Edited by B. G. M. Jamieson. New Delhi: Oxford & IBH Publishing, pp. 21–79.

Healy, J. M., and Jamieson, B. G. M. 1981. An ultrastructural examination of developing and mature paraspermatozoa in *Pyrazus ebeninus* (Mollusca, Gastropoda, Potamididae). *Zoomorphology* 98: 101–119.

Healy, J. M., and Willan, R. C. 1984. Ultrastructure and phylogenetic significance of notaspidean spermatozoa (Mollusca, Gastropoda, Opisthobranchia). *Zoologica Scripta* 13: 107–120.

———. 1991. Nudibranch spermatozoa: comparative ultrastructure and systematic importance. *The Veliger* 34: 134–165.

Hodgson, A. N. 1997. Paraspermatogenesis in gastropod molluscs. *Invertebrate Reproduction and Development* 31: 31–38.

Hodgson, A. N., and Healy, J. M. 1998. Comparative sperm morphology of the pulmonate limpets *Trimusculus costatus, T. reticulatus* (Trimusculidae) and *Burnupia stenochorias* and *Ancylus fluviatilis* (Ancylidae). *Journal of Molluscan Studies* 64: 447–460.

Houbrick, R. S. 1979. Classification and systematic relationships of the Abyssochrysidae, a relict family of bathyal snails (Prosobranchia: Gastropoda. *Smithsonian Contributions to Zoology* 290: 1–21.

———. 1988. Cerithioidean phylogeny. In *Prosobranch Phylogeny*. Edited by W. F. Ponder. *Malacological Review*, Suppl. 4: 88–128.

———. 1990. Aspects of the anatomy of *Plesiotrochus* (Plesiotrochidae, fam. n.) and its systematic position in Cerithioidea (Prosobranchia, Caenogastropoda). In *Proceedings of the Third International Marine Biological Workshop: The Marine Flora and Fauna of Albany, Western Australia*, Western Australian Museum, Perth. Edited by F. E. Wells, D. I. Walker, H. Kirkman, and R. Lethbridge. 1: 237–249. Perth: Western Australian Museum.

Huelsenbeck, J. P., and Ronquist, F. 2001. MRBAYES: Bayesian inference of phylogeny. *Bioinformatics* 17: 754–755.

Hughes, R. N. 1986a. Laboratory observations on the feeding behaviour, reproduction and morphology of *Galeodea echinophora* (Gastropoda: Cassidae). *Zoological Journal of the Linnean Society* 86: 355–365.

———. 1986b. Anatomy of the foregut of *Morum* Röding, 1798 (Gastropoda, Tonnoidea) and the taxonomic misplacement of the genus. *The Veliger* 29: 91–100.

Hughes, R. N., and Hughes, H. P. I. 1971. A study of the gastropod *Cassis tuberosa* (L.) preying upon sea urchins. *Journal of Experimental Marine Biology and Ecology* 7: 305–314.

Hyman, L. H. 1967. *The Invertebrates,* Vol. VI. *Mollusca I.* New York: McGraw-Hill.

Jablonski, D. 1986. Larval ecology and macroevolution in marine invertebrates. *Bulletin of Marine Science* 39: 565–587.

Johansson, J. 1951. The embryology of *Viviparus* and its significance for the phylogeny of the the Gastropoda. *Arkiv för Zoology* 1: 173–177.

Kaim, A. 2004. The evolution of conch ontogeny in Mesozoic open sea gastropods. *Palaeontologia Polonica* 62: 1–183.

Kantor, Y. I. 1988. On the anatomy of Pseudomelatominae (Gastropoda, Toxoglossa, Turridae) with notes on functional morphology and phylogeny of the subfamily. *Apex* 3: 1–19.

———. 1990. Anatomical basis for the origin and evolution of the toxoglossan mode of feeding. *Malacologia* 32: 3–18.

———. 1991. On the morphology and relationships of some oliviform gastropods. *Ruthenica* 1: 17–52.

———. 1996. Phylogeny and relationships of Neogastropoda. In *Origin and Evolutionary Radiation of the Mollusca.* Edited by J. D. Taylor. Oxford: Oxford University Press, pp. 221–230.

———. 2002. Morphological prerequisites for understanding neogastropod phylogeny. *Bollettino Malacologico* 38 Suppl. 4: 161–174.

———. 2003. Comparative anatomy of the stomach of Buccinoidea (Neogastropoda). *Journal of Molluscan Studies* 69: 203–220.

Kantor, Y. I., and Pavlinov, I. Y. 1991. Cladistic analysis of oliviform gastropods (Gastropoda, Pectinibranchia, Olividae *s. lato*). *Zhurnal Obshchei Biologii* 52: 356–371.

Kantor, Y. I., and Sysoev, A. V. 1989. The morphology of toxoglossan gastropods lacking a radula with a description of new species and genus of Turridae. *Journal of Molluscan Studies* 55: 537–550.

Kantor, Y. I., and Taylor, J. D. 1991. Evolution of the toxoglossan feeding mechanism: new information on the use of the radula. *Journal of Molluscan Studies* 57: 129–134.

———. 2002. Foregut anatomy and relationships of raphitomine gastropods (Gastropoda: Conoidea: Raphitominae). *Bollettino Malacologico* 38 Suppl. 4: 83–110.

Kase, T. 1990. Research report on ecology of a living fossil of extinct naticids, *Globularia fluctuata* (Sowerby) (Gastropoda, Mollusca) in Palawan, The Philippines—II. *Journal of Geography, Tokyo* 99: 92–95 (in Japanese).

Kennedy, J. J., and Keegan, B. F. 1992. The encapsular developmental sequence of the mesogastropod *Turritella communis* (Gastropoda: Turritellidae).

Journal of Marine Biological Association of the United Kingdom 72: 783–805.

Kiel, S., Bandel, K., Banjac, N., and Perrilliat, M. C. 2000. On Cretaceous Campanilidae. *Freiberger Forschungshefte* C490: 15–26.

Knight, J. B., Batten, R. L., Yochelson, E. L., and Cox, L. R. 1960. Paleozoic and some Mesozoic Caenogastropoda and Opisthobranchia. Suppl. In *Treatise on Invertebrate Paleontology.* Part I, *Mollusca* 1. Edited by R. C. Moore. Lawrence, KS: University of Kansas Press, pp. I310–I324.

Kohnert, R., and Storch, V. 1984. Vergleichend-ultrastrukturelle Untersuchungen zur Morphologie eupyrener Spermien der Monotocardia (Prosobranchia). *Zoologischer Jahrbucher* 111: 51–93.

Koike, K. 1985. Comparative ultrastructural studies on the spermatozoa of the Prosobranchia (Mollusca: Gastropoda). *Science Report of the Faculty of Education, Gunma University* 34: 33–153.

Kool, S. P. 1993. Phylogenetic analysis of the Rapaninae (Neogastropoda: Muricinae). *Malacologia* 35: 155–259.

Kosuge, S. 1966. The family Triphoridae and its systematic position. *Malacologia* 4: 297–324.

Kowalke, T. 1998. Bewertung protoconchmorphologischer Daten basaler Caenogastropoda (Cerithiimorpha und Littorinimorpha) hinsichtlich ihrer Systematik und Evolution von der Kreide bis Rezent. *Berliner Geowissenschaftliche Abhandlungen, Reihe E, Paläobiologie* 27: 1–121.

Kues, B. S., and Batten, R. L. 2001. Middle Pennsylvanian gastropods from the Flechado Formation, north-central New Mexico. *Journal of Paleontology* 75, Suppl. 1: 1–95.

Lamy, E. 1926. Sur une coquille enigmatique. *Journal de Conchyliologie* 70: 51–56.

Lindberg, D. R., and Guralnick, R. P. 2003. Phyletic patterns of early development in gastropod mollusks. *Evolution and Development* 5:494–507.

Lindberg, D. R., and Ponder, W. F. 2001. The influence of classification on the evolutionary interpretation of structure—a re-evaluation of the evolution of the pallial cavity. *Organisms, Diversity and Evolution* 1: 273–299.

Liu, L. L., and Wang, S. P. 2002. Histology and biochemical composition of the autotomy mantle of *Ficus ficus* (Mesogastropoda: Ficidae). *Acta Zoologica* 83: 111–116.

Lydeard, C., Holznagel, W. E., Glaubrecht, M., and Ponder, W. F. 2002. Molecular phylogeny of a circum-global, diverse gastropod superfamily (Cerithioidea: Mollusca: Caenogastropoda): pushing the deepest phylogenetic limits of mitochondrial LSU rDNA sequences. *Molecular Phylogenetics and Evolution* 22: 399–406.

Marcus, E. B.-R. 1956. On some Prosobranchia from the coast of São Paulo. *Boletim do Instituto Oceanográfico* 7: 3–29.

Marcus, E., and Marcus, E. 1963. Mesogastropoden von der Küste Sao Paulos. *Abhandlungen der Mathematisch-Naturwissenschaftlichen Klasse. Akademie der Wissenschaften und der Literatur, Mainz* 1963 (1): 1–105.

———. 1965. On Brazilian supratidal and estuarine snails. *Boletim da Faculdade de Filosofia, Ciencias e Letras. Universidade de Sao Paulo, Zoologica* 25: 19–82.

Marshall, B.A. 1978. Cerithiopsidae (Mollusca: Gastropoda) of New Zealand and a provisional classification of the family. *New Zealand Journal of Zoology* 5: 47–120.

———. 1983. A revision of Recent Triphoridae of Southern Australia. *Records of the Australian Museum* Suppl. 12: 1–119.

Martoja, C., and M. Thiriot-Quiévreux. 1975. Convergence morphologique entre l'appareil copulateur des Hétéropoda et des Littorinidae (Gastropoda, Prosobranchia). *Netherlands Journal of Zoology* 25: 243–246.

McArthur, A.G., and Harasewych, M.G. 2003. Molecular systematics of the major lineages of the Gastropoda. In *Molecular Systematics and Phylogeography of Mollusks*. Edited by C. Lydeard and D. R. Lindberg. Washington, DC: Smithsonian Books, pp. 140–160.

McArthur, A.G., and Koop, B.F. 1999. Partial 28S rDNA sequences and the antiquity of the hydrothermal vent endemic gastropods. *Molecular Phylogenetics and Evolution* 13: 255–274.

Melone, G., Donin, L.L.D., and Cotelli, F. 1980. The paraspermatic cell (atypical spermatozoon) of Prosobranchia: a comparative ultrastructural study. *Acta Zoologica* 61: 191–201.

Meyer, C.P. 2003. Molecular systematics of cowries (Gastropoda: Cypraeidae) and diversification patterns in the tropics. *Biological Journal of the Linnean Society* 79: 401–459.

———. 2004. Toward comprehensiveness: Increased molecular sampling within Cypraeidae and its phylogenetic implications. *Malacologia* 46: 127–156.

Morton, B. 2003. Observations on the feeding behaviour of *Nassarius clarus* (Gastropoda: Nassariidae) in Shark Bay, Western Australia. *Molluscan Research* 23: 239–249.

———. 2006. Diet and predation behaviour exhibited by *Cominella eburnea* (Gastropoda: Caenogastropoda: Neogastropoda) in Princess Royal Harbour, Albany, Western Australia, with a review of attack strategies in the Buccinidae. *Molluscan Research* 26: 39–50.

Morton, J. E. 1951. The ecology and digestive system of the Struthiolariidae (Gastropoda). *The Quarterly Journal of Microscopical Science* 92: 1–25.

Nishiwaki, S. 1964. Phylogenetical study on the type of the dimorphic spermatozoa in Prosobranchia. *Science Report of the Tokyo Kyoiku Daigaku B* 11: 237–275.

Nützel, A. 1998. Über die Stammesgeschichte der Ptenoglossa (Gastropoda). *Berliner Geowissenschaftliche Abhandlungen, Reihe E, Paläobiologie* 26: 1–229.

———. 2002. The Late Triassic species *Cryptaulax bittneri* (Mollusca: Gastropoda: Procerithiidae) and the Mesozoic marine revolution. *Paläontologische Zeitschrift* 76: 57–63.

———. 2005. Recovery of gastropods in the Early Triassic. *Comptes Rendus Palevol* 4: 1–17.

Nützel, A., and Bandel, K. 2000. Goniasmidae and Orthonemidae: two new families of the Palaeozoic Caenogastropoda (Mollusca, Gastropoda). *Neues Jahrbuch für Geologie und Paläontologie, Abhandlungen* 9: 557–569.

Nützel, A., and Cook, A.G. 2002. *Chlorozyga*, a new caenogastropod genus from the Early Carboniferous of Australia. *Alcheringa* 26: 151–157.

Nützel, A., and Erwin, D.H. 2002. *Battenizyga*, a new early Triassic gastropod genus with a discussion on the gastropod evolution at the Permian/Triassic boundary. *Paläontologische Zeitschrift* 76: 21–26.

———. 2004. Late Triassic (Late Norian) gastropods from the Wallowa terrane (Idaho, USA). *Paläontologische Zeitschrift* 78: 361–416.

Nützel, A., Erwin, D.H., and Mapes, R.H. 2000. Identity and phylogeny of the late Paleozoic Subulitoidea (Gastropoda). *Journal of Paleontology* 74: 575–598.

Nützel, A., and Frýda, J. 2003. Paleozoic plankton revolution: evidence from early gastropod ontogeny. *Geology* 31: 829–831.

Nützel, A., Frýda, J., Yancey, T.E., and Anderson, J.R. 2007. Larval shells of Late Palaeozoic naticopsid gastropods (Neritopsoidea: Neritimorpha) with a discussion of the early neritimorph evolution. *Paläontologische Zeitschrift* 81: 213–228.

Nützel, A., and Mapes, R.H. 2001. Larval and juvenile gastropods from a Mississippian black shale: paleoecology, and implications for the evolution of the Gastropoda. *Lethaia* 34: 143–162.

Nützel, A., and Pan, H.-Z. 2005. Late Paleozoic evolution of the Caenogastropoda: larval shell morphology and implications for the Permian/Triassic mass extinction event. *Journal of Paleontology* 79: 1175–1188.

Oliverio, M., Cervelli, M., and Mariottini, P. 2002. ITS2 rRNA evolution and its congruence with the phylogeny of muricid neogastropods (Caenogastropoda, Muricoidea). *Molecular Phylogenetics and Evolution* 25: 63–69.

O'Sullivan, J. B., McConnaughey, R. R., and Huber, M. E. 1987. A bloodsucking snail—the Cooper Nutmeg, *Cancellaria cooperi* Gabb, parasitizes the California electric ray, *Torpedo californica* Ayres. *Biological Bulletin* 172: 362–366.

Page, L. R. 2000. Development and evolution of adult feeding structures in caenogastropods: overcoming larval functional constraints. *Evolution and Development* 2: 25–34.

———. 2002. Larval and metamorphic development of the foregut and proboscis in the caenogastropod *Marsenina* (*Lamellaria*) *stearnsii*. *Journal of Morphology* 252: 202–217.

———. 2005. Development of foregut and proboscis in the buccinid neogastropod *Nassarius mendicus*: evolutionary opportunity exploited by a developmental module. *Journal of Morphology* 264: 327–338.

Page, L. R., and Pedersen, R. V. K. 1998. Transformation of phytoplanktivorous larvae into predatory carnivores during the development of *Polinices lewisii* (Mollusca, Caenogastropoda). *Invertebrate Biology* 117: 208–220.

Pan, H.-Z., and Erwin, D. H. 2002. Gastropods from the Permian of Guanxi and Yunnan Provinces, South China. *Journal of Paleontology* 76 Suppl. 1: 1–49.

Parries, S. C., and Page, L. R. 2003. Larval development and metamorphic transformation of the feeding system in the kleptoparasitic snail *Trichotropis cancellata* (Mollusca, Caenogastropoda). *Canadian Journal of Zoology* 81: 1650–1661.

Pedersen, R. V. K., and Page, L. R. 2000. Development and metamorphosis of the planktotrophic larvae of the moon snail, *Polinices lewisii* (Gould, 1847) (Caenogastropoda: Naticoidea). *Veliger* 43: 58–63.

Pernet, B., and Kohn, A. J. 1998. Size-related obligate and facultative parasitism in the marine gastropod *Trichotropis cancellata*. *The Biological Bulletin* 195: 349–356.

Perrier, R. 1889. Recherches sur l'anatomie et l'histologie du rein des gastéropodes. Prosobranchiata. *Annales des Sciences Naturelles. Zoologie et Biologie Animale* 8: 61–192.

Ponder, W. F. 1965. The family Eatoniellidae in New Zealand. *Records of the Auckland Institute and Museum* 6: 47–99.

———. 1972. The morphology of some mitriform gastropods with special reference to their alimentary canal and productive systems (Mollusca: Neogastropoda). *Malacologia* 11: 295–342.

———. 1974. The origin and evolution of the Neogastropoda. *Malacologia* 12: 295–338.

———. 1988. The truncatelloidean (= Rissoacean) radiation—a preliminary phylogeny. In *Prosobranch Phylogeny*. Edited by W. F. Ponder. *Malacological Review* Suppl. 4: 129–166.

———. 1991. The anatomy of *Diala*, with an assessment of its taxonomic position (Mollusca: Cerithioidea). In *Proceedings of the Third International Marine Biological Workshop: the Marine Flora and Fauna of Albany, Western Australia*, Vol. 2. Edited by F. E. Wells, D. I. Walker, H. Kirkman, and R. Lethbridge. Perth: Western Australia Museum, pp. 499–519.

Ponder, W. F., and Lindberg, D. R. 1996. Gastropod phylogeny—challenges for the 90's. In *Origin and Evolutionary Radiation of the Mollusca*. Edited by J. D. Taylor. Oxford: Oxford University Press, pp. 135–154.

———. 1997. Towards a phylogeny of gastropod molluscs—an analysis using morphological characters. *Zoological Journal of the Linnean Society* 119: 83–265.

Ponder, W. F., and Taylor, J. D. 1992. Predatory shell drilling by two species of *Austroginella* (Gastropoda: Marginellidae). *Journal of Zoology* 228: 317–328.

Ponder, W. F., and Warén, A. 1988. Classification of the Caenogastropoda and Heterostropha—a list of the family-group names and higher taxa. In *Prosobranch Phylogeny*. Edited by W. F. Ponder. *Malacological Review* Suppl. 4: 288–328.

Przeslawski, R. 2004. A review of the effects of environmental stress on embryonic development within intertidal egg masses. *Molluscan Research* 24: 43–63.

Rawlings, T. A. 1994. Encapsulation of eggs by marine gastropods: effect of variation in capsule form on the vulnerability of embryos to predation. *Evolution* 48: 1301–1313.

Rawlings, T. A., Collins, T. M., and Bieler, R. 2001. A major mitochondrial gene rearrangement among closely related species. *Molecular Biology and Evolution* 18: 1604–1609.

———. 2003. Changing identities: tRNA duplication and remolding within animal mitochondrial genomes. *Proceedings of the National Academy of Sciences of the United States of America* 100: 15700–15705.

Reid, D. G. 1989. The comparative morphology, phylogeny and evolution of the gastropod family Littorinidae. *Philosophical Transactions of the Royal Society of London Series B: Biological Sciences* 324: 1–110.

Rex, M. A., and Boss, K. J. 1976. Open coiling in Recent gastropods. *Malacologia* 15: 289–297.

Riedel, F. 1995. An outline of cassoidean phylogeny (Mollusca, Gastropoda). *Contributions to Tertiary and Quaternary Geology* 32: 97–132.

——. 2000. Ursprung und Evolution der "höheren" Caenogastropoda. Eine paläobiologische Konzeption. *Berliner Geowissenschaftliche Abhandlungen, Reihe E, Paläobiologie* 32: 1–240.

Robertson, R. 1985. Four characters and the higher category systematics of gastropods. *American Malacological Bulletin* Special Edition 1: 1–22.

——. 1993. Snail handedness. *National Geographic Research and Exploration* 9: 104–119.

Rosenberg, G. 1996. Parallel evolution of terrestriality in Atlantic truncatellid gastropods. *Evolution* 50: 682–693.

——. 1998. Reproducibility of results in phylogenetic analysis of mollusks: a reanalysis of the Taylor, Kantor and Sysoev (1993) data set for conoidean gastropods. *American Malacological Bulletin* 14: 219–228.

Rosenberg, G., Kuncio, G. S., Davis, G. M., and Harasewych, M. G. 1994. Preliminary ribosomal RNA phylogeny of gastropod and unionoidean bivalve mollusks. *Nautilus* Suppl. 2: 111–121.

Roy, K. 1994. Effects of the Mesozoic marine revolution on the taxonomic, morphologic, and biogeographic evolution of a group: aporrhaid gastropods during the Mesozoic. *Paleobiology* 20: 274–296.

Ruthensteiner, B., and Schaefer, K. 1991. On the protonephridia and "larval kidneys" of *Nassarius (Hina) reticulatus* (Linnaeus) (Caenogastropoda). *Journal of Molluscan Studies* 57: 323–329.

Salvini-Plawen, L. v. 1988. The structure and function of molluscan digestive systems. In *The Mollusca. Form and Function*, 11. Edited by E. R. Trueman and M. R. Clarke. Orlando: Academic Press, 301–379

Salvini-Plawen, L. v., and Haszprunar, G. 1987. The Vetigastropoda and the systematics of streptoneurous Gastropoda (Mollusca). *Journal of Zoology, London* 211: 747–770.

Schander, C., and Sundberg, P. 2001. Useful characters in gastropod phylogeny: Soft information or hard facts? *Systematic Biology* 50: 136–141.

Schröder, M. 1995. Frühontogenetische Schalen jurassischer und unterkretazischer Gastropoden aus Norddeutschland und Polen. *Palaeontographica, Abteilung A* 283: 1–95.

Sepkoski, J. J. Jr., and Hulver, M. L. 1985. An atlas of Phanerozoic clade diversity diagrams. In (ed.) *Phanerozoic Diversity Patterns*. Edited by J. W. Valentine. Princeton, NJ: Princeton University Press, pp. 11–39.

Shimek, R. L., and Kohn, A. J. 1981. Functional morphology and evolution of the toxoglossan radula. *Malacologia* 20: 423–438.

Signor, P. W. 1985. Gastropod evolutionary history. In *Mollusks. Notes for a Short Course*. Edited by T. W.

Broadhead. University of Tennessee Department of Geological Sciences Sudies in Geology 13. Knoxville, TN: University of Tennessee, pp. 157–173.

Simone, L. R. L. 1999. Comparative morphology study and systematics of Brazilian Terebridae (Mollusca, Gastropoda, Conoidea), with descriptions of three new species. *Zoosystema* 21: 199–248.

——. 2000a. Filogenia das superfamílias de Caenogastropoda (Mollusca) com base em morfologia comparativa. PhD. Thesis, Instituto de Biociências, Universidade de São Paulo, 164 pp., 45 text figs.

——. 2000b. A phylogenetic study of the Terebrinae (Mollusca, Caenogastropoda, Terebridae) based on species from the Western Atlantic. *Journal of Comparative Biology* 3: 137–150.

——. 2001. Phylogenetic analyses of Cerithioidea (Mollusca, Caenogastropoda) based on comparative morphology. *Arquivos de Zoologia, São Paulo* 36: 147–263.

——. 2002. Comparative morphological study and phylogeny of representatives of the Superfamily Calyptraeoidea (including Hipponicoidea) (Mollusca, Caenogastropoda). *Biota Neotropica* 2(2): 1–137.

——. 2003. Revision of the genus *Benthobia* (Caenogastropoda, Pseudolividae). *Journal of Molluscan Studies* 69: 245–262.

——. 2004a. *Morphology and Phylogeny of the Cypraeoidea (Mollusca, Caenogastropoda)*. Rio de Janeiro: Papel & Virtual Editora.

——. 2004b. Comparative morphology and phylogeny of representatives of the superfamilies of architaenioglossans and the Annulariidae (Mollusca, Caenogastropoda). *Arquivos do Museu Nacional (Rio de Janeiro)* 64: 387–504.

——. 2005. Comparative morphological study of representatives of the three families of Stromboidea and the Xenophoroidea (Mollusca, Caenogastropoda), with an assessment of their phylogeny. *Arquivos de Zoologia, São Paulo* 37: 141–267.

——. 2006. Morphological and phylogenetic study of the Western Atlantic *Crepidula plana* complex (Caenogastropoda, Calyptraeidae), with description of three new species from Brazil. *Zootaxa* 1112: 1–64.

Smith, E. H. 1967. The neogastropod midgut, with notes on the digestive diverticula and intestine. *Transactions of the Royal Society of Edinburgh* 67: 23–42.

Sohl, N. D. 1964. Neogastropoda, Opisthobranchia, and Basommatophora of the Ripley, Owl Creek and Prairie Bluff Formations. *United States Geological Survey Professional Paper* 331-B: 153–344.

Solem, A., and Yochelson, E. L. 1979. North American Palaeozoic land snails, with a summary of other

Palaeozoic non-marine snails. *United States Geological Survey Professional Paper* 1072: 1–42.

Stockmann-Bosbach, R. 1988. Early stages of the encapsulated development of *Nucella lapillus* (Linnaeus) (Gastropoda, Muricidae). *Journal of Molluscan Studies* 54: 181–196.

Strong, E. E. 2003. Refining molluscan characters: morphology, character coding and a phylogeny of the Caenogastropoda. *Zoological Journal of the Linnean Society* 137: 447–554.

Strong, E. E., Gargominy, O., Ponder, W. F., and Bouchet, P. In press. Global diversity of gastropods (Gastropoda; Mollusca) in freshwater. *Hydrobiologia.*

Strong, E. E., and Harasewych, M. G. 2004. On the origin of heteropods. In *Molluscan Megadiversity: Sea, Land and Freshwater. World Congress of Malacology, Perth, Western Australia, 11–16 July 2004.* Edited by F. E. Wells. Perth: Western Australian Museum, p. 141.

Swofford, D. L. 2001. *PAUP*. Phylogenetic Analysis Using Parsimony (*and Other Methods).* Version 4. Sunderland, MA: Sinauer Associates.

Szabó, J. 1983. Lower and Middle Jurassic gastropods from the Bakony Mountains (Hungary), Part 5. Supplement to Archaeogastropoda; Caenogastropoda. *Annales Historico-Naturales Musei Nationalis Hungarici* 75: 27–46.

Tanaka, M., Asahina, H., Yamada, N., Osumi, M., Wada, A., and Ishihara, K. 1987. Pattern and time of course of cleavages in early development of the ovoviviparous pond snail, *Sinotaia quadratus historica. Development, Growth and Differentiation* 29: 469–478.

Taylor, D. W., and Sohl, N. F. 1962. An outline of gastropod classification. *Malacologia* 1: 7–32.

Taylor, J. D., ed. 1996. *Origin and evolutionary radiation of the Mollusca.* Oxford: Oxford University Press.

Taylor, J. D., Kantor, Y. I., and Sysoev, A. V. 1993. Foregut anatomy, feeding mechanisms, relationships and classification of the Conoidea (= Toxoglossa) (Gastropoda). *Bulletin of the Natural History Museum, London (Zoology)* 59: 125–170.

Taylor, J. D., and Miller, J. A. 1989. The morphology of the osphradium in relation to feeding habits in meso- and neogastropods. *Journal of Molluscan Studies* 55: 227–237.

Taylor, J. D., and Morris, N. J. 1988. Relationships of neogastropods. In *Prosobranch Phylogeny.* Edited by W. F. Ponder. *Malacological Review* Suppl. 4: 167–179.

Taylor, J. D., Morris, N. J., and Taylor, C. N. 1980. Food specialization and the evolution of predatory prosobranch gastropods. *Palaeontology* 23: 375–409.

Taylor, J. D., and Reid, D. G. 1990. Shell microstructure and mineralogy of the Littorinidae: ecological and evolutionary significance. *Hydrobiologia* 193: 199–215.

Thiele, J. 1925–1926. Mollusca = Weichtiere. In: *Handbuch der Zoologie,* Vol. 5. Edited by W. Kükenthal and T. Krumbach. Berlin and Leipzig: De Gruyter.

———. 1929–1931. *Handbuch der systematischen Weichtierkunde.* Vol. 1. Jena: Gustav Fischer Verlag.

Thiriot-Quiévreux, C. 2003. Advances in chromosome studies of gastropod molluscs. *Journal of Molluscan Studies* 69: 197–201.

Thompson, T. E. 1973. Euthyneuran and other molluscan spermatozoa. *Malacologia* 14: 167–206, plus addendum pp 443–444.

Thorson, G. 1946. Reproduction and larval development of Danish marine bottom invertebrates, with special reference to the planktonic larvae in the Sound (Øresund). *Meddelelser Kommn Havunders, Series Plankton* 4: 1–523.

Tillier, S., Masselot, M., Guerdoux, J., and Tillier, A. 1994. Monophyly of major gastropod taxa tested from partial 28S rRNA sequences, with emphasis on Euthyneura and hot vent limpets Peltospiroidea. *Nautilus* Suppl. 2: 122–140.

Tillier, S., Masselot, M., Phillippe, H., and Tillier, A. 1992. Phylogénie moléculaire des Gastropoda (Mollusca) fondée sur le séquençage partiel de l'ARN ribosomique 28S. *Comptes Rendus de l'Académie des Sciences Paris* 314: 79–85.

Tochimoto, T. 1967. Comparative histochemical study on the dimorphic spermatozoa of the Prosobranchia with special reference to polysaccharides. *Science Report of the Tokyo Kyoiku Daigaku B* 13: 75–109.

Tomlinson, S. G. 1987. Intermediate stages in the embryonic development of the gastropod *Ilyanassa obsoleta:* a scanning electron microscope study. *Invertebrate Reproduction and Development* 12: 253–280.

Troschel, F. H. (and Thiele, J.). 1856–1893. *Das Gebiss der Schnecken, zur Begründung einer natürlichen Classification.* Berlin.

van den Biggelaar, J. A. M. 1996. The significance of the early cleavage pattern for the reconstruction of gastropod phylogeny. In *Origin and evolutionary radiation of the Mollusca.* Edited by J. D. Taylor. Oxford: Oxford University Press, pp. 155–160.

van den Biggelaar, J. A. M., and Haszprunar, G. 1996. Cleavage patterns in the Gastropoda: an evolutionary approach. *Evolution* 50: 1520–1540.

Vaught, K. C. 1989. *A Classification of the Living Mollusca.* Melbourne, FL: American Malacologists.

Vermeij, G. J. 1978. *Biogeography and Adaptation: Patterns of Marine Life.* Cambridge, MA: Harvard University Press.

———. 1987. *Evolution and Escalation: An Ecological History of Life.* Princeton, NJ: Princeton University Press.

Vermeij, G. J., and Kool, S. P. 1994. Evolution of labral spines in *Acanthais*, new genus, and other rapanine muricid gastropods. *Veliger* 37: 414–424.

Vermeij, G. J., and Signor, P. W. 1992. The geographic, taxonomic and temporal distribution of determinate growth in marine gastropods. *Biological Journal of the Linnean Society* 47: 233–247.

Voltzow, J. 1994. Gastropoda: Prosobranchia. In *Microscopic Anatomy of Invertebrates, 5, Mollusca I.* Edited by F. W. Harrison and A. J. Kohn. New York: Wiley-Liss, pp. 111–252.

Wagner, P. J. 2001. Gastropod phylogenetics: progress, problems, and implications. *Journal of Paleontology* 75: 1128–1140.

———. 2002. Phylogenetic relationships of the earliest anisostrophically coiled gastropods. *Smithsonian Contributions to Paleobiology* 88: 1–152.

Warén, A. 1984. A generic revision of the family Eulimidae (Gastropoda, Prosobranchia). *Journal of Molluscan Studies* Suppl. 13.

Wenz, W. 1938–1944. *Gastropoda. Teil 1, Allgemeiner Teil und Prosobranchia.* Handbuch der Paläozoologie 6. Edited by O. H. Schindewolf. Berlin: Gebrüder Bornträger.

West, T. L. 1990. Feeding behavior and functional morphology of the epiproboscis of *Mitra idae* Melvill (Mollusca: Gastropoda: Mitridae). *Bulletin of Marine Science* 46: 761–799.

Wilke, T., Davis, G. M., Falniowski, A., Giusti, F., Bodon, M., and Szarowska, M. 2001. Molecular systematics of Hydrobiidae (Mollusca: Gastropoda: Rissooidea): testing monophyly and phylogenetic relationships. *Proceedings of the Academy of Natural Sciences of Philadelphia* 151: 1–21.

Williams, S. T., Reid, D. G., and Littlewood, D. T. J. 2003. A molecular phylogeny of the Littorininae (Gastropoda: Littorinidae): Unequal evolutionary rates, morphological parallelism, and biogeography of the Southern Ocean. *Molecular Phylogenetics and Evolution* 28: 60–86.

Wilson, N. G., and Healy, J. M. 2002. Comparative sperm ultrastructure in five genera of the nudibranch family Chromodorididae (Gastropoda: Opisthobranchia). *Journal of Molluscan Studies* 68: 133–145.

Yoo, E. K. 1994. Early Carboniferous Gastropoda from the Tamworth Belt, New South Wales, Australia. *Records of the Australian Museum* 46: 63–110.

14

Heterobranchia I

THE OPISTHOBRANCHIA

Heike Wägele, Annette Klussmann-Kolb,
Verena Vonnemann, and Monica Medina

The opisthobranchs comprise the bubble shells, sea hares, and sea slugs, almost all of which are marine. This major group of heterobranch gastropods is composed of approximately 6,000 species, some of which have attracted the interest of naturalists for more than 2,000 years (Carefoot 1987). The Greeks used sea hare extracts (Anaspidea; Figure 14.1D) for medical treatment (Nicandros CE 150, quoted in Caprotti 1977 and Eales 1921), whereas the Romans considered them to be highly toxic. Numerous chemical substances, sequestered from food or formed *de novo* and used as repellents, make opisthobranchs an important target group for natural products research and bioprospecting for pharmaceutical purposes (see Fontana *et al.* 2000; Faulkner 2001; Avila 1995). However, it is the beauty of the colorful animals and the many unusual natural history traits that attract large numbers of scientists and amateur naturalists alike.

SYSTEMATICS

Nearly every major opisthobranch subgroup exhibits unique specialized traits that, to a certain extent, characterized the groups recognized in classifications. For example: (1) the Cephalaspidea, including the Acteonoidea (Figure 14.1A, B) by the cephalic shield and Hancock's organ; (2) the Anaspidea (Figure 14.1D) by two pairs of tentacles; (3) the Sacoglossa (Figure 14.1C) by only one radular tooth per row and their suctorial feeding on green algal cells; (4) the Tylodinoidea by a limpet-shaped shell (Figure 14.1E); (5) the Pleurobranchoidea (Figure 14.1H) by a large acid gland; (6) the Nudibranchia (Figure 14.1G, I) by a special vacuolated epithelium (Figure 14.2D); (7) the Thecosomata (Figure 14.1F) and Gymnosomata (Pteropoda) by their pelagic lifestyle and their modified foot; (8) the Acochlidoidea by their reduction of size and the migration of some of them into freshwater; and (9) the enigmatic Rhodopidae by their minute size and turbellarian-like body. Assignment of the Rhodopidae in the Opisthobranchia or Pulmonata was long debated, but recent studies favor affiliation with opisthobranchs (Haszprunar and Huber 1990; Burn 1998). Early delineations of these major groups (e.g., Thiele 1931) have changed only in minor details and are still used in modern synopses (e.g., Rudman and Willan 1998).

FIGURE 14.1. Representatives of the major opisthobranch groups to show the range of body form within the group.
(A) *Bulla quoyii* (Cephalaspidea; Albany, Western Australia). (B) *Hydatina physis* (Acteonoidea; Hastings Point, N.S.W., Australia). (C) *Elysia ornata* (Sacoglossa; Magnetic Island, Queensland, Australia). (D) Aplysia californica (Anaspidea; Florida, USA). (E) *Umbraculum umbraculum* (Tylodinoidea; Wollongong, N.S.W., Australia). (F) *Creseis* sp. (Thecosomata; Florida, United States). (G) *Notaeolidia gigas* (Nudibranchia, Cladobranchia; King George Island, Antarctica). (H) *Pleurobranchaea meckelii* (Pleurobranchoidea; Banyuls-sur-Mer, France). (I) *Chromodoris westraliensis* (Nudibranchia, Anthobranchia; Rottnest Island, Western Australia).

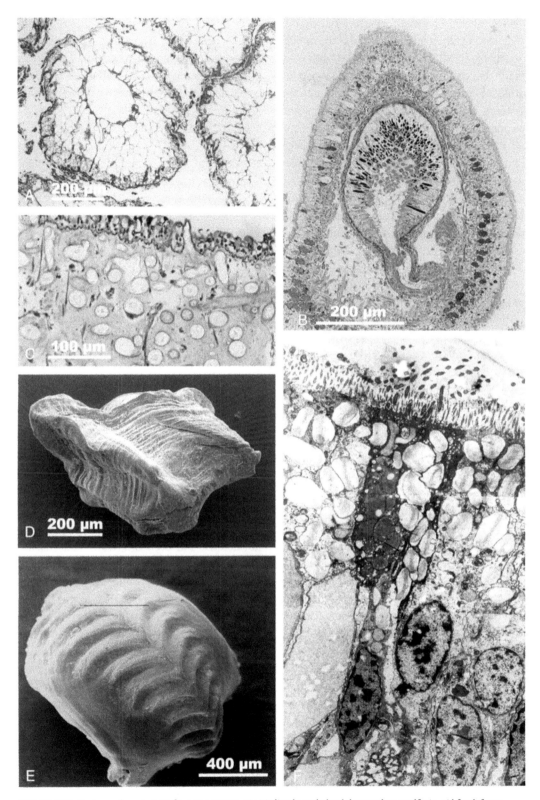

FIGURE 14.2. Features characteristic for some groups. (A) Median buccal gland that produces sulfuric acid for defense (*Bathyberthella antarctica*, Pleurobranchoidea). (B) Longitudinal section of dorsal appendage with cnidosac. Incorporated cnidocysts are cut longitudinally as well as in cross section (*Aeolidia papillosa*, Nudibranchia, Cladobranchia). (C) Cross section of mantle with spicules (*Notodoris citrina*, Nudibranchia, Anthobranchia). (D) Gizzard plate (*Dolabrifera dolabrifera*, Anaspidea). (E) Gizzard plate (*Haminoea callidegenita*, Cephalaspidea). (F) Ultrastructure of outer epithelium with specialized vacuolated cells (*Dermatobranchus semistriatus*, Nudibranchia, Cladobranchia).

The shell is reduced, internalized, or lost (Figure 14.1A, B) in more than half of the known opisthobranchs. This affected the evolution of many biological features, leading to parallel evolution in (1) shape (slug form); (2) defensive systems (e.g., many glandular structures and coloration); (3) feeding (grazing or hunters); and (4) mating strategies (Harris 1973; Thompson 1976; Schmekel and Portmann 1982; Gosliner 1994; Wägele and Klussmann-Kolb 2005). Gosliner and Ghiselin (1984) and Gosliner (1991) noted that parallel evolution in many characters obscured phylogenetic reconstruction. The trend toward shell loss has also impeded the use of fossils in evolutionary studies of opisthobranchs. No fossil record is available for several groups, including the Nudibranchia, the largest living opisthobranch taxon (*ca.* 3,000 species; Wägele 2004).

DEFENSE

A correlation has been observed in many opisthobranch lineages between shell reduction and loss with both chemical defense systems and warning coloration patterns (Thompson 1976; Edmunds 1987; Cimino and Ghiselin 1998, 1999; Gosliner 2001; Wägele and Klussmann-Kolb 2005). Several studies include the investigation of toxic compounds (reviews in Avila 1995; Cimino and Ghiselin 1998, 1999; Cimino *et al.* 1999; Faulkner 2001). Acidic substances (sulfuric acid with pH 1–2) are known from members of the Cephalaspidea (Figure 14.1A), Pleurobranchoidea (Figure 14.1H), and Doridoidea (Figure 14.1G). Thompson investigated the glandular structures involved in the production of acidic products, such as the median buccal gland (Figure 14.2A) and the mantle glands in several pleurobranch species (1960, 1969, 1983, 1986, 1988; Thompson and Colman 1984). Wägele *et al.* (2006a) gave a thorough description of many glandular structures in opisthobranchs, including mantle dermal formations (MDFs; Rudman 1984; García-Gómez *et al.* 1990, 1991; Avila and Paul 1997; Cimino *et al.* 1999) within the context of current knowledge of natural products'

chemistry and phylogenetic hypotheses. Within the Aeolidioidea (Figure 14.1G), cnido-cysts taken from their cnidarian prey and stored in the cnidosacs (Figure 14.2B) can be used against potential predators (Edmunds 1966; Streble 1968; Thompson 1976; Kälker and Schmekel 1976; Schmekel and Portmann 1982; Greenwood and Mariscal 1984a, b). Other defense devices include spicules in Pleurobranchoidea, Doridoidea (Figure 14.2C), and Acochlidoidea (Willan 1987; Cattaneo-Vietti *et al.* 1993, 1995; Challis 1970) or cryptic shape and coloration (Burghardt and Wägele 2004).

DIET

Opisthobranchs, like other gastropod groups, have a diverse diet. While some feed on algae, others are specialized feeders on poriferans, cnidarians, and tunicates, these invertebrate groups being rarely preyed upon by other animal phyla. Some opisthobranchs are specialized carnivores, feeding on polychaetes (e.g., Acteonoidea; Figure 14.1B), bivalves, crustaceans, other gastropods, or even congeners (Aglajidae). Several groups have a gizzard, which might be composed of several plates in Anaspidea (Figure 14.2D), or of only three in cephalaspideans (Figure 14.2E), which allow the trituration of hard food such as algae, foraminiferans, and even bivalves or polychaetes. Several nudibranchs live in a mutualistic symbiosis with the unicellular, photosynthetic dinoflagellate *Symbiodinium* and share metabolites with them (Kempf 1984; Hoegh-Guldberg *et al.* 1986a, b; Rudman 1987, 1991; Wägele and Johnson 2001; Wägele 2004; Burghardt and Wägele 2004; Burghardt *et al.* 2005; Burghardt and Wägele 2006). Some members of the Sacoglossa (Figure 14.1C) incorporate the chloroplasts of their algal food in their digestive gland and make use of their metabolites (Clark and Busacca 1978; Clark *et al.* 1990; Marin and Ros 1992; Mujer *et al.* 1996; Rumpho *et al.* 2000, 2001; Wägele and Johnsen 2001). A comprehensive list of nudibranch food items has been compiled by McDonald and Nybakken (1991).

FOSSIL RECORD

Fossils are available for many shell-bearing taxa, but their simple shells, with few diagnostic characters, and are reduced in most opisthobranch subgroups, are an impediment to interpreting stem lines. The earliest suspected heterobranch taxa, *Kuskokwimia* (Frýda and Blodgett 2001; Frýda *et al.*, Chapter 10) and *Palaeocarboninia* (Bandel and Heidelberger 2002), lived about 370 Mya and may be related to the Recent Valvatoidea (Bandel 2002). According to Nützel (personal communication, 2004; Frýda *et al.*, Chapter 10), the earliest unequivocal heterobranch fossil is *Cylindrobullina* (240 Mya), which is also considered a basal opisthobranch lineage. Better known lower heterobranchs, such as Mathilididae or Architectonicidae, probably appeared about 230 and 210 Mya, respectively (Tracey *et al.* 1993; Kiel *et al.* 2002), and are documented from St. Cassian Formations (about 200 Mya; Bandel 1995). First undoubted pyramidellids appear only 70 Mya ago (Kiel *et al.* 2002), and it seems likely that they are an offshoot of a yet undetermined stem line.

The earliest opisthobranchs (Hydatinidae) appeared about 190 Mya, but assignment of these fossils to extant taxa is problematic (K. Bandel, personal communication). Bandel (1994, 2002) considered that the oldest opisthobranchs appeared in the Triassic (about 220 Mya) and the pulmonates in the Jurassic (about 190 Mya). Major diversification of opisthobranch lineages may have been as late as the beginning of the Cenozoic (about 60 Mya), with the first fossil record of the Bulloidea, Sacoglossa, Anaspidea, and Thecosomata from that time (Bandel 1994). The Pleurobranchoidea are the last shelled group to appear, about 30 Mya (Valdés and Lozouet 2000). Table 14.1 gives an overview of known ages of heterobranch groups.

PHYLOGENETICS

It was Gray (1840, according to Haszprunar 1985) who recognized the close relationship of Opisthobranchia and Pulmonata, which he combined in the Heterobranchia (this grouping was usually called Euthyneura by later authors). Haszprunar (1985, 1988) revived the name Heterobranchia and widened the definition by adding several other groups (Valvatoidea, Architectonicoidea, Rissoelloidea, Glacidorbdoidea, Pyramidelloidea).

Since Haszprunar's (1985, 1988) analyses of Heterobranchia, the old concept of the sister taxon relationship of "monophyletic" Opisthobranchia and "monophyletic" Pulmo-nata has been debated, as well as the relationship of these groups with basal heterobranchs.

Whereas Salvini-Plawen (1990) based his cladogram of euthyneurans on "sequences of organ systems" (1990: 7), Salvini-Plawen and Steiner (1996; see Figure 14.3A) used parsimony methods. In the latter analysis, the monophyly of the Opisthobranchia and Pulmonata within the Euthyneura, and the sister taxon relationship of Nudibranchia and Pleurobranchoidea, were recovered. Ponder and Lindberg's (1997) analysis of gastropods included several heterobranchs, including two opisthobranch and two pulmonate taxa. Although their analyses recovered the monophyly of heterobranchs, the taxon sampling of euthyneurans was too limited to address the monophyly of either opisthobranchs or pulmonates. The most recent morphological analysis with new characters that included basal heterobranchs, opisthobranchs, and pulmonates is that of Dayrat and Tillier (2002). With no assumptions applied, the Pulmonata were monophyletic in that analysis, but Opisthobranchia were paraphyletic (Figure 14.3C). The authors reanalyzed their data under several assumptions and finally favored a tree with very low resolution (Figure 14.3D; see also Dayrat and Tillier 2003).

Several phylogenies of Pulmonata include a few opisthobranchs (as outgroups as well as ingroups; Tillier *et al.* 1996; Wade and Mordan 2000; Yoon and Kim 2000; Barker 2001), but these are biased toward pulmonates and do not reliably test the monophyly of either pulmonates or opisthobranchs.

TABLE 14.1
Earliest Fossil Occurrences for Heterobranch Subgroups Taken from the Literature

	TAXON	AGE IN MYA (APPROXIMATION)	AUTHOR
Basal Heterobranchia and *incertae sedis*	*Kuskokwimia*	370	Fryda and Blodgett 2001
	Palaeocarboninia	380	Bandel and Heidelberger 2002
	Valvatoidea	150	Tracey *et al.* 1993
		136	Bandel (personal communication)
	Pyramidellidae	70	Kiel *et al.* 2002
	Mathilididae	200	Gründel 1997a
		195–225	Bandel 1995 and personal communication
	Architectonicidae	210	Tracey *et al.* 1993
	Modern Architectonicidae	70	Kiel *et al.* 2002
Modern Opisthobranchia		200	Bandel 1995 Kiel *et al.* 2002
Opisthobranchia	Acteonidae	160	Wenz and Zilch 1959
		190–160	Bandel 1994; Bandel *et al.* 2000
	Hydatinidae	190*	Tracey *et al.* 1993
	Ringiculidae	160	Gründel 1997b
		190–160	Bandel (personal communication)
	Bullidae	180*	Tracey *et al.* 1993
	Scaphandridae	150	Wenz and Zilch 1959
	Retusidae	160	Wenz and Zilch 1959
		80	Bandel 1994
	Cylichnidae	150	Wenz and Zilch 1959
	Philinidae	90	Wenz and Zilch 1959
	Haminoeidae	70	Tracey *et al.* 1993
	Diaphanidae	160*	Wenz and Zilch 1959
	Thecosomata	60	Tracey *et al.* 1993
	Akera	190	Tracey *et al.* 1993
	Anaspidea	54	Valdés and Lozouet 2000
	Sacoglossa	50	Tracey *et al.* 1993
	Tylodinoidea	50	Wenz and Zilch 1959
	Pleurobranchoidea	30	Valdés and Lozouet 2000
Pulmonata	Ellobiidae	140	Bandel 1994
	Siphonariidae	150*	Tracey *et al.* 1993
	Physidae	345*	Tracey *et al.* 1993
	Carychidae	330*	Tracey *et al.* 1993
		300	Bandel 1994
	Lymnaeidae	140	Tracey *et al.* 1993
	Planorbidae	150	Tracey *et al.* 1993
	Acroloxidae	80	Tracey *et al.* 1993
	Buliminidae	290	Tracey *et al.* 1993
	Amphibolidae	50	Bandel 1994
	Clausiliidae	80	Tracey *et al.* 1993
	Stylommatophora	100	Bandel 1994

NOTE: The age is only an approximation and in some cases extrapolated from the stratigraphic formation given in the literature. *Indicate problematic groups, where assignment or postulated relationship to recent taxa is difficult (K. Bandel, personal communication).

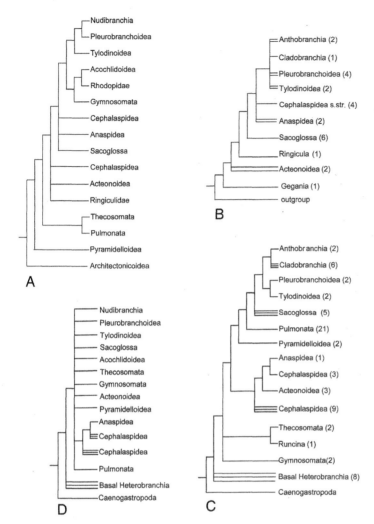

FIGURE 14.3. Published trees based on morphology. Individually mentioned genera indicate that these do not group with the taxon they are usually assigned to, or that the genus is the sole member of a larger taxon included in the original analysis. Otherwise only higher taxa are noted. Numbers in parentheses following the taxon names are the number of species included in the original analyses. (A) Salvini-Plawen and Steiner 1996: Maximum Parsimony tree. (B) Mikkelsen 2002: strict consensus tree, all characters unordered and unweighted. (C) Dayrat and Tillier 2002: strict consensus tree with all characters unordered, undirected, and unweighted. (D) Dayrat and Tillier 2002: proposed phylogeny of Heterobranchia with results of different analyses combined by hand.

Prior to the general acceptance of cladistic methodology, several evolutionary scenarios of opisthobranch taxa were suggested, some dating back to the nineteenth century (e.g., Bergh 1891; Pelseneer 1893; Pilsbry 1894–95; Guiart 1901). These were often based on one or few organ systems, for example, Boettger (1954) on the nervous system, Ghiselin (1965) on the genital system, Rudman (1978) on the digestive system, and Edlinger (1980) on Hancock's organ. These evolutionary schemes were developed without invoking explicit phylogenetic criteria (i.e., synapomorphies to define monophyletic clades), and we refer the reader to extensive recent reviews for a more in-depth account of the early literature (e.g., Haszprunar 1988; Ponder and Lindberg 1997; Mikkelsen 1996, 2002; Wägele and Willan 2000).

Schmekel (1985) presented phylogenetic hypotheses of Opisthobranchia and its subgroups and discussed apomorphic characters, although she did not employ parsimony criteria. Within Opisthobranchia, the first morphological investigation using parsimony-based cladistics was an analysis on the Pleurobranchoidea and Tylodinoidea by Willan (1987). Based on the knowledge at the time, he assumed both groups were sister taxa (Notaspidea) without delineating synapomorphies. Later, analyses of the Sacoglossa were undertaken (Jensen 1996) and used (Jensen 1997) for an *a posteriori* evaluation of their assumed coevolution with their food, the green algae Chlorophyta. Mikkelsen's (1996) investigation of the phylogeny of several cephalaspidean groups, with some Anaspidea and Sacoglossa as additional groups, used many histological characters. Her results excluded *Acteon* from the Opisthobranchia, but her taxon sampling was limited to shelled opisthobranchs. Wägele and Willan's (2000) phylogeny of the Nudibranchia was based on anatomical and histological data. They showed that the Anthobranchia (comprising the widely distributed doridoideans with a posterior gill circle, Figure 14.1I, and the bathydoridoideans from the polar regions and the deep sea) and Cladobranchia (comprising taxa with dorsal papillae used as secondary gills, Figure 14.1G) are well-defined groups and established that Pleurobranchoidea and Nudibranchia were sister taxa. Their findings confirmed former hypotheses of relationship (Schmekel 1985; Salvini-Plawen and Steiner 1996), and they named this combined clade Nudipleura. Monophyly of the Cladobranchia (Wägele and Willan 2000) is disputed in some analyses (e.g., Dayrat and Tillier 2002). The most recent morphological investigation on a major subgroup of the Opisthobranchia is Klussmann-Kolb's (2004) study of Anaspidea, which included histological data. A detailed analysis of the enigmatic Acochlidioidea is currently in progress (Schrödl and Neusser, personal communication).

Mikkelsen (2002) reanalyzed and reevaluated several of these earlier data sets of opisthobranch subgroups to elucidate opisthobranch phylogeny. She used taxa and data from Mikkelsen (1996), Jensen (1996), Willan (1987), and Wägele and Willan (2000) to produce a matrix of 40 characters and 26 taxa that resulted in a cladogram with low resolution (Figure 14.3B). Wägele and Klussmann-Kolb (2005) undertook an analysis of 79 taxa covering all major opisthobranch groups, as well as basal heterobranchs and pulmonates, using 111 morphological and histological characters.[1] Most of the characters were reinvestigated, but in a few cases information is taken

from the literature. The phylogeny obtained from this study (Figure 14.4) showed that many opisthobranch subgroups are monophyletic, a result congruent with the other most recent morphological analyses (summarized in Figure 14.3; Salvini-Plawen and Steiner 1996; Mikkelsen 1996, 2002; Dayrat and Tillier 2002). Resolution within each group differs to some extent, and in a few cases there are apparent anomalies (Figure 14.4). For instance, the basal position of *Elysia* within the Sacoglossa seems unlikely and contradicts former analyses of Jensen (1996, 1997) and Mikkelsen (1996, 2002). According to their results, taxa with a shell are more basal and shell reduction occurred at least twice within the Sacoglossa. This latter scenario is more likely than assuming the reappearance of a coiled shell from a "naked" sacoglossan.

Molecular phylogenies including more than 15 opisthobranchs became available in the late 1990s (Thollesson 1999; Wollscheid and Wägele 1999), with the number of published analyses that included a variety of euthyneuran taxa having increased considerably since then (Table 14.2). Analyzed markers comprise nuclear genes, especially the small and large ribosomal subunit (18S and 28S; Figure 14.5A, C, Figure 14.6; Wollscheid and Wägele 1999; Dayrat et al. 2001; Wollscheid-Lengeling et al. 2001; Wägele et al. 2003; Vonnemann et al. 2005), and mitochondrial genes, such as the large ribosomal subunit (16S), cytochrome oxidase subunit 1 (CO1), and several others (Figure 14.5B, D, E; Thollesson 1999; Wollscheid-Lengeling et al. 2001; Dayrat et al. 2001; Wägele et al. 2003; Grande et al. 2004a, b). Figure 14.6 shows the most recently published gene tree (Vonnemann et al. 2005), a combined analysis of complete 18S rDNA and partial 28S rDNA. This tree and a few other gene trees (Wollscheid-Lengeling et al. 2001; Dayrat et al. 2001; Wägele et al. 2003; Grande et al. 2004a, b; Figure 14.5) are congruent in supporting

---

1. A table of the characters and the data matrix is available in Wägele and Klussmann-Kolb (2005) and histological methods are outlined in Wägele (1997).

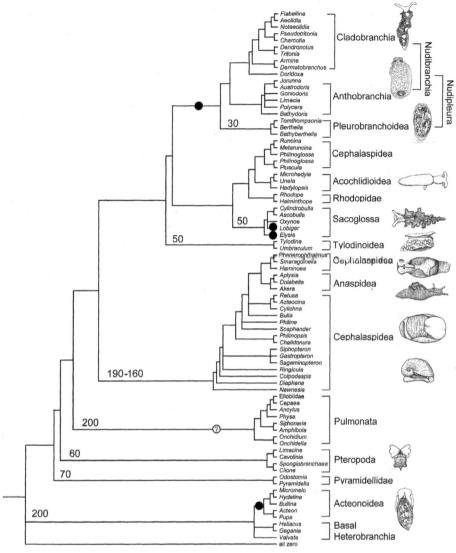

FIGURE 14.4. Strict consensus tree based on 111 morphological and histological characters (see Wägele and Klussmann-Kolb 2005). All characters treated as unordered, with equal weight. Evolution of diaulic genital system (male duct separate from female duct) indicated by a black circle (genital system in pulmonates treated as not comparable with opisthobranchs). The oldest fossil records (in Mya) of members of the stem lines are given.

the monophyly of the major groups. At present, however, there is either incongruence between phylogenies based on different genes and different subsets of taxa or absence of resolution (see following).

Irrespective of whether the data matrix is based on morphological or molecular data, only few studies support monophyly of Opisthobranchia (Salvini-Plawen and Steiner 1996; Vonne-

mann et al. 2005). Many are inconclusive regarding the monophyly of Opisthobranchia but support the monophyly of Pulmonata (Salvini-Plawen and Steiner 1996; Thollesson 1999; Dayrat et al. 2001; Dayrat and Tillier 2002, 2003; Wägele et al. 2003; Wägele and Klussmann-Kolb 2005), while a few analyses lack support for both groups (e.g., Grande et al. 2004a). The position of Pulmonata within Opisthobranchia is found

TABLE 14.2
Published Trees on Opisthobranch Groups Based on Different Genes

	18S	28S	16S	CO1	OTHERS
Heterobranchia	A	A	—	—	—
Pyramidellidae	A	A	—	F	F
Acochlidoidea	A	A	—	—	—
Pteropoda	—	D	C	C	—
Acteonoidea	A	A, D	C	C, F	F
Cephalaspidea *sensu stricto*	A, B, C	A, D	B, C	B, C, F	F
Anaspidea	A, B, C	A, D	B, C, E	B, C, F	F
Sacoglossa	A, B, C	A, D	B, C	C, F	—
Tylodinoidea	A, C	A, D	—	B, C, F	F
Pleurobranchoidea	A, B, C	A, D	C	B, C, F	F
Nudibranchia	A, B, C	A, D	B, C	B, C, F	F
Pulmonata	A, C	A, D	C	C, F	F

NOTE: A, Vonnemann *et al.* 2005; B, Wollscheid *et al.* 2001; C, Wägele *et al.* 2003; D, Dayrat and Tillier 2001; E, Medina and Walsh 2000; F, Grande *et al.* 2004a, b.

in morphological (Dayrat and Tillier 2002) and molecular (16S; Wägele *et al.* 2003) analyses. Several analyses suggest that Pyramidellidae is basal and sister to the opisthobranchs (Salvini-Plawen and Steiner 1996; Wägele and Klussmann-Kolb 2005).

Within the Opisthobranchia, the sister taxon relationships of Anthobranchia and Clado-branchia, supporting the monophyly of Nudibranchia, are found in many studies (Wollscheid-Lengeling *et al.* 2001; Dayrat *et al.* 2001; Dayrat and Tillier 2002; Mikkelsen 2002; Wägele *et al.* 2003; Grande *et al.* 2004a, b; Figure 14.5), although there are exceptions in Thollesson (1999) and Grande *et al.* (2004a, b). Inconsistency in the phylogenetic placement of these groups is more common in molecular than in morphological analyses, even when using the same genes but different representative taxa (e.g., with 28S in Dayrat and Tillier 2001; Vonnemann *et al.* 2005; or 18S in Wägele *et al.* 2003; Vonnemann *et al.* 2005). Other sister group relationships that generally emerge are Nudibranchia and Pleurobranchoidea, rendering the Nudipleura monophyletic (but not in Thollesson 1999 or Grande *et al.* 2004a, b).

Several analyses indicate that the Sacoglossa might have a closer relationship to the Nudipleura (Dayrat and Tillier 2002; Vonnemann *et al.* 2005; Wägele and Klussmann-Kolb 2005) than to the former Cephalaspidea, as was previously favored (e.g., Salvini-Plawen and Steiner 1996; Mikkelsen 1996; Wägele *et al.* 2003). But results are too controversial and preliminary and warrant further investigation.

Cephalaspidea has previously been an amalgamation of many different groups, some of them characterized by a muscular gizzard with typical gizzard plates (Figure 14.2E). Gizzard-bearing cephalaspidean families (Cylichnidae, Retusidae, Philinidae, Scaphandridae, Bullidae, Haminoeidae) and the Anaspidea (also characterized by esophageal plates; Figure 14.2D) form a monophyletic clade, with the Anaspidea grouping within the gizzard-bearing cephalaspids (Mikkelsen 1996; Dayrat and Tillier 2002; Wägele and Klussmann-Kolb 2005). The cephalaspidean families Gastropteridae and Aglajidae, which both lack gizzards, are separate monophyletic groups with a more basal position than the gizzard-bearing ones. This contradicts currently used classifications (e.g., Rudman 1978; Burn and Thompson 1998)

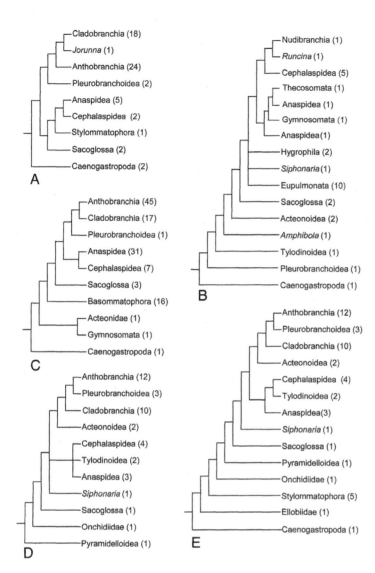

A

B

C

D

E

FIGURE 14.5. Published molecular analyses. Numbers in parentheses following taxon names are the number of species included in the original analyses. (A) Wollscheid-Lengeling *et al.* 2001: Maximum Parsimony tree based on 18S rDNA. (B) Dayrat and Tillier 2001: Maximum Parsimony tree based on partial 28S rDNA. (C) Wägele *et al.* 2003: Maximum Parsimony tree based on mitochondrial 16S rDNA. (D) Grande *et al.* 2004b: Maximum parsimony tree based on mitochondrial COI, trnV, and rrnL genes. (E) Grande *et al.* 2004a: same data set as in D, but more taxa included.

that imply the independent loss of a gizzard in both groups. The family Diaphanidae (*Newnesia, Diaphana,* and *Colpodaspis*) is not monophyletic in the Wägele and Klussmann-Kolb (2005) analysis (Figure 14.4), but unfortunately no published molecular analyses exist that include these problematic diaphanid taxa.

Mikkelsen (1996) defined Cephalaspidea *sensu stricto* after excluding Acteonidae and Hydatinidae. She considered *Ringicula* to be more derived than Acteonidae and Hydatinidae, a result confirmed by Wägele and Klussmann-Kolb (2005; Figure 14.4). Dayrat and Tillier (2002) also falsified monophyly of the Acteonoidea

and excluded *Ringicula*. Although sister group status of Acteonidae and Hydatinidae was not supported in the analyses of Mikkelsen (1996, 2002), it is in other morphology- and gene-based studies (Dayrat and Tillier 2002; Vonnemann *et al.* 2005; Wägele and Klussmann-Kolb 2005). Thus we regard Acteonoidea as comprising Acteonidae and Hydatinidae. Placement of Acteonoidea within Cephalaspidea, as suggested by Dayrat and Tillier (2002), is unlikely because it postulates the acquisition of a complicated gizzard and then its complete loss, as well as the reoccurrence of an operculum—a scenario not found in their favored hypothesis (Figure 14.3D).

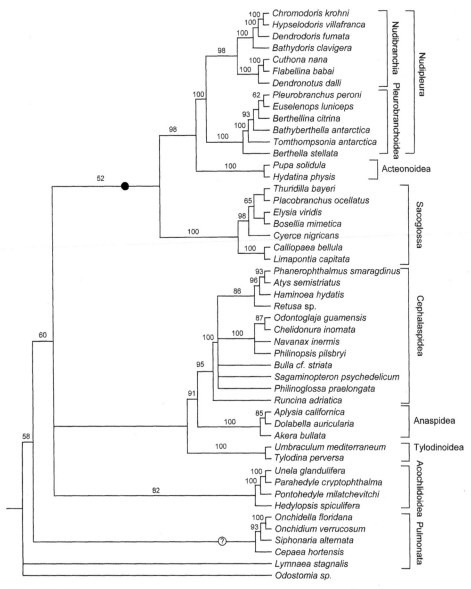

FIGURE 14.6. Strict consensus tree (maximum parsimony, eight shortest trees) based on a combined analysis of 18S and 28S gene (after Vonnemann *et al.* 2005). Numbers above stem lines indicate bootstrap indices. Evolution of diaulic genital system (male duct separate from female duct) indicated by a black circle, while the open circle with a question mark indicates that the situation in pulmonates is unclear.

Contradictory to these morphological studies, several molecular studies (Grande *et al.* 2004a, b; Vonnemann *et al.* 2005), as well as mitochondrial protein-encoding gene phylogenies (Medina, personal communication), place the Acteonidae and Hydatinidae as the likely sister group to the Nudipleura. Although the morphological analysis summarized in Figure 14.4 indicates a basal position for Acteonoidea, some morphological characters suggest an affinity with Nudipleura. These include a diaulic genital system with a separate vas deferens and joined oviduct and vaginal duct, but similar systems with separate ducts also evolved within the Sacoglossa and Pulmonata and the morphological details of the ducts differ in these groups suggesting convergence. Possible evolution of a diaulic genital system as a single event within Opisthobranchia is indicated in the stem line in Figure 14.6 and as several independent events in Figure 14.4.

The grouping of the cephalaspidean taxa Runcinidae and Philinoglossidae with Acochlidoidea and Rhodopidae in Wägele and Klussmann-Kolb's (2005) cladogram (Figure 14.4) is intriguing, although evaluation of these findings must await further study as these groups are poorly represented in other analyses. The morphological characters coded for these small animals, some of which are interstitial, support their monophyly; but parallel evolution of some features is also likely, in particular the reduction of the mantle cavity and associated organs. Molecular phylogenies (28S, 18S; Vonnemann *et al.* 2005) indicate that *Runcina* and *Philinoglossa* belong within the Cephalaspidea, as has been generally assumed (e.g., Burn and Thompson 1998). Runcinidae are known to have a gizzard, although composed of four plates (Schmekel and Cappellato 2001). Wägele and Klussmann-Kolb (2005) included in their analysis an undescribed *Philinoglossa*, which has a gizzard with three large plates, the usual number for cephalaspideans, supporting the hypothesis of a close relationship to the gizzard-bearing cephalaspideans.

Acochlidoideans are monophyletic, but their position varies from being members of a Cephalaspidea/Anaspidea clade (18S), to being a basal euthyneuran group (28S; Vonnemann *et al.* 2005). In the combined analysis (Figure 14.6), Acochlidoidea belongs to the opisthobranch clade, but no resolution was obtained (Vonnemann *et al.* 2005).

Rhodopidae has been included only twice in phylogenetic reconstructions (Salvini-Plawen and Steiner 1996; Wägele and Klussmann-Kolb 2005), and both analyses support a close relationship with Acochlidoidea.

Because of the difficulties in collecting, identifying, and preserving Pteropoda, overall information on this group is poor (Lalli and Gilmer 1989). Monophyly of Pteropoda is supported by the recent morphological studies of Wägele and Klussmann-Kolb (2005) but was earlier refuted by Salvini-Plawen and Steiner (1996). They excluded the Thecosomata from the Opisthobranchia, considering them to be closely related to pulmonates, while the Gymnosomata were considered opisthobranchs. Dayrat *et al.* (2001) showed a well-supported relationship of both taxa with Anaspidea in their analyses. Although they had no resolution in their morphology-based tree (Dayrat and Tillier 2002), they later presented an evolutionary hypothesis for the esophageal gizzard that could represent a synapomorphy for the Anaspidea/Pteropoda clade, with secondary loss in the Gymnosomata (Dayrat and Tillier 2003).

Within opisthobranch taxa, the fossil record agrees with proposed phylogenies in several cases (see Figure 14.4). Within the earliest "true" opisthobranchs (about 200 Mya), the cephalaspideans *sensu stricto* evolved at least 180 Mya ago (Bullidae: 180 Mya, Tracey *et al.* 1993). The Anaspidea probably split off from this group around the same time (*Akera*: 190 Mya, Tracey *et al.* 1993; Anaspidea: 54 Mya, Valdés and Lozouet 2000; see Figures 14.3B, C, D, 14.4, 14.5A, D, E). The ages of Tylodinoidea (50 Mya, Wenz and Zilch 1959), Sacoglossa (50 Mya, Tracey *et al.* 1993), and Pleurobranchoidea (30 Mya, Valdés and Lozouet 2000) also match branching patterns in most analyses (see Figures 14.3, 14.4, 14.5). Some scenarios seem very unlikely according to our knowledge of the fossil record. In Figure 14.6, the Sacoglossa (50 Mya, Tracey *et al.* 1993) are the sister group of a monophyletic clade Acteonoidea + Nudipleura with the Acteonoidea having an assumed age of 160 Mya (Wenz and Zilch 1959). Also in Figure 14.6, the Tylodinoidea (50 Mya) are the sister taxon to the Anaspidea/ Cephalaspidea clade (190 Mya). A basal position of the Pteropoda (fossil record only available for Thecosomata: 60 Mya, Tracey *et al.* 1993; Figure 14.4) is questionable, their age indicating that they are a much younger clade, for example, nested within the gizzard-bearing cephalaspideans and anaspideans (as suggested by Dayrat and Tillier 2001, 2003).

A BIOGEOGRAPHIC HYPOTHESIS

When analyzing opisthobranch distributional data, it has been observed that basal members of each major lineage, especially of the subgroups of the Nudipleura, are known only from Antarctic waters, such as *Notaeolidia*, *Pseudotritonia*, *Charcotia*, *Bathydoris*, *Tomthompsonia*, and *Bathyberthella*. Other widespread genera, such as *Tritonia* and *Flabellina*, also have some Antarctic species. *Doridoxa*, the sister taxon of the Cladobranchia (Figure 14.4; see Schrödl *et al.* 2001), as well as some members of the genus *Bathydoris*, are only known from deep-sea areas, which are mainly under the influence of Antarctic bottom water (Dietrich *et al.* 1975). Schrödl (2003: 139) assumed, that "the splitting of the *Tylodina/Umbraculum* and Nudipleura clades was due to vicariance with the first ones remaining related to shallow Tethyan warm waters while early nudipleurans drifted to the south with Antarctica, and later on adapted to cooler conditions." He came to these conclusions based on the assumed sister taxa relationship of Tylodinoidea and Nudipleura and the minimum age of the Tylodinoidea (47 Mya), which are not represented in Antarctic waters (Wägele *et al.* 2006b). The phylogenetic position of the Tylodinoidea is, however, still not clear. Wägele and Willan's (2000) analysis of Nudipleura, with a sister taxa relationship of Nudibranchia and Pleurobranchoidea, and with the most basal species from Antarctic waters, suggests an alternative scenario. The earliest fossils of Pleurobranchoidea are 30 Mya (no fossils are known for the shell-less Nudibranchia; Valdés and Lozouet 2000). Recent ideas on the origin of Antarctic marine benthic communities (Thatje *et al.* 2005) suggest that the ancestor of the Nudipleura could have survived the cooling of Antarctica 30 Mya ago in the deep sea around the Antarctic shelf, or even in sheltered areas of the shelf. Under this scenario, radiation followed with invasions to temperate and tropic seas, as well as other parts of the deep sea, by following

the northward currents of the Antarctic bottom water. A recolonization of the Antarctic shelf is also likely. This does not contradict the vicariance hypothesis of Schrödl (2003), which explained the absence of Tylodinoidea in Southern Polar regions, but gives an alternative hypothesis for the origin of the Nudipleura. More thorough phylogenetic analyses, especially including more taxa from Antarctica and the deep sea, are needed to clarify the origin of the Nudipleura. The phylogenetic analysis of one of the key genera, *Bathydoris*, by Valdés (2002) does not contradict this scenario, since the Antarctic species *B. clavigera* is sister taxon to other recent doridoideans, and *B. hodgsoni* from Antarctic waters forms a non-resolved monophyletic clade with all deep-sea bathydorids. At present we lack sufficient data for other opisthobranch lineages to suggest other biogeographical and evolutionary scenarios.

ADAPTIVE RADIATIONS

The loss of the shell was certainly a prerequisite for the evolution of many outstanding biological features in gastropod evolution. Cimino and Ghiselin (1999) discussed incorporation of natural products from the food into the slug's body for defensive devices as a major driving force of opisthobranch evolution. Wägele (2004) reviewed and summarized several characters that might have led to an adaptive radiation within opisthobranch subgroups. These comprise gizzard plates in the Cephalaspidea, incorporation of chloroplasts in Sacoglossa, incorporation of cnidocysts in Aeolidioidea, incorporation of zooxanthellae in Aeolidioidea, and mantle dermal formations (MDFs) in Opisthobranchia. While studying MDFs, the difficulties in identifying key characters become evident, especially when functional constraints are not well understood. MDFs are a characteristic feature of the Chromodorididae (Figure 14.1I), the largest family within the Anthobranchia, and are interpreted as defensive organs, in which secondary metabolites from the sponge prey are stored (e.g., Cimino and Ghiselin 1999). This renders the chromodoridids unpalatable to putative predators. Wägele *et al.* (2006a) described analogous structures in other opisthobranch taxa, but because all those groups have similar numbers of species, the MDFs are not considered to be key characters as they have not led to a higher speciation rate.

Jensen (1997) convincingly showed that the evolution of a single tooth per row was a major step in the radiation of sacoglossans, and the change of the shape from triangular to blade-shaped allowed their expansion into other algal food niches. Similar investigations on coevolutionary processes are warranted for all the other possible key characters discussed previously.

Unfortunately, current phylogenetic analyses are too unreliable to pinpoint certain characters as key characters that have enhanced radiations. Based on the phylogenetic congruence for monophyly of certain groups, some characters seem to have evolved in stem lines and might be interpreted as key characters (e.g., the single tooth in Sacoglossa, and the incorporation of cnidocysts in Aeolidioidea). However, the acquisition of certain characters is usually combined with other traits. For example, although we know that most members of the Cladobranchia feed on cnidarians, the key character for radiation of Aeolidioidea might have been the switch to a particular cnidarian taxon, which secondarily enabled the incorporation of cnidocysts.

SUMMARY OF WHAT WE KNOW

The subgroups of Opisthobranchia have remained largely unchanged since the 1960s (e.g., Odhner 1968), with some exceptions, the most notable being the cephalaspideans. Nevertheless, despite recent phylogenetic analyses based on different techniques and data, we still lack a reliable phylogeny of Opisthobranchia (reviewed in Dayrat and Tillier 2003). Recent morphological analyses have verified the assumption of high parallelism in many traits, especially loss of shell, reproductive anatomy, nervous system, and probably also body shape (reviewed in Wägele and Klussmann-Kolb 2005). Not only are analyses hampered by this

morphological parallelism, but also molecular studies to date lack informative positions (apomorphies at base pair levels) that convincingly support basal lineages.

The Nudipleura (Wägele and Willan 2000) combines Nudibranchia and Pleurobranchoidea. Another well-defined group is the gizzard-bearing cephalaspidean and anaspidean clade, the latter probably comprising also the pteropods (Dayrat *et al.* 2001; Dayrat and Tillier 2003). According to Wägele and Klussmann-Kolb (2005), the gizzard evolved in taxa feeding on vagile fauna, such as polychaetes, and, later, a switch to algal food occurred in some lineages. All algal-feeding opisthobranchs, including the Thecosomata, have a gizzard, which probably enables the destruction of tough plant cell walls. The Sacoglossa, another well-defined group with no muscular stomach or gizzard, uses plant food differently: by piercing the algal cells with the radular teeth and sucking out their contents.

GAPS IN KNOWLEDGE

A great deal of new data is required to resolve the current incongruencies in both molecular and morphological data. The latest analyses (Mikkelsen 1996; Wägele and Willan 2000; Klussmann-Kolb 2004; Wägele and Klussmann-Kolb 2005) show the importance of histological data. Some characters were identified as synapomorphies for specific groups, such as the specialized vacuolated cells for Nudibranchia, while others helped to clarify possible homology (or vice versa) of morphological characters by analyzing their fine structure (e.g., the gizzard is considered to be a homologous structure, whereas the MDFs are not homologous throughout the opisthobranchs). Therefore, histological investigations remain a very important technique for analyzing characters (Wägele 1997). In a few cases, ultrastructural investigation is needed to clarify further substructures, for instance (1) the specialized vacuolated epithelium, which is considered to be an apomorphic character for Nudibranchia, (2) the blood gland, which occurs in several opisthobranch and probably pulmonate groups, and (3) the medio-dorsal bodies*, which are typical of pulmonates, have been detected in at least one member of the Acochlidoidea (Sommerfeldt and Schrödl 2005). Taxon-specific analyses of evolutionary rates, improved methods of aligning sequence data, and the application of programs analyzing phylogenetic information in these alignments (e.g., Splitstree by Huson and Bryoant 2006) might provide information on the reliability of reconstructed cladograms and retesting of earlier results using alternative alignments.

TAXON SAMPLING

Mikkelsen (2002) noted that many analyses are biased toward Pulmonata or Opisthobranchia and do not include enough groups of lower heterobranchs. Only by inclusion of more basal heterobranchs from the Valvatoidea, Architectonicoidea, Rissoelloidea, and Omalogyroidea, as well as an inclusion of a representative selection of opisthobranchs and pulmonates, can we generate reliable phylogenetic hypotheses for the main lineages within the Heterobranchia.

A more complete analysis of opisthobranch subgroups is still necessary. Identifying and understanding the more basal taxa within each group will help establish the ground pattern of the groups. Analysis of these ground patterns will reduce data sets to those characters that are apomorphies for the stem lines of the subgroups and might therefore help in analyzing relationships among opisthobranch clades otherwise characterized by a high degree of parallelism.

NEW APPROACHES NEEDED

New approaches are required to deal with incongruence between morphological and gene-based trees, as well as between different gene trees. Some promising genes that could be investigated in the future are new and longer regions

*Structures associated with the cerebral ganglia: see Chapter 15.

of the mitochondrial genome, for example, concatenated proteins and tRNAs. Gene expression patterns of several developmental genes such as the *Hox* genes (e.g., Degnan *et al.* 1995) seem to have potential for phylogenetic analysis. Further possible data sources for phylogenetic investigation might be other nuclear genes such as shell protein genes (e.g., Bowen and Tang 1996) and hemocyanins (e.g., Lieb *et al.* 2000; see also Medina and Collins 2003 for a review of these potential nuclear markers). Sequences of histones, particularly of the histone gene cluster H3/H4, have been used in phylogenetic analyses of molluscan taxa at different taxonomic levels (Gastropoda, e.g., Colgan *et al.* 2000, 2003; Stylommatophora, Armbruster *et al.*, unpublished data) as well as in non-gastropod studies (e.g., Huff *et al.* 2004; Okusu *et al.* 2003).

With two molluscan genomes being sequenced, the limpet *Lottia gigantea* by the U.S. Department of Energy (http://genome.jgi-psf.org/Lotgi1/Lotgi1.home.html) and the sea hare *Aplysia californica* by the U.S. National Institutes of Health (Nahir *et al.*, unpublished: http://www.broad.mit.edu/tools/data/data-vert.html), as well as a broad range of molluscan EST (expressed sequence tags) projects, a large suite of nuclear genes will be available for phylogenetic purposes.

In recent years, as a result of higher sequencing capacity, mitochondrial genomics has become an alternative method for analyzing molluscan phylogenetics (e.g. Hatzoglou *et al.* 1995, Yamazaki *et al.* 1997, Ueshima 2004; Medina *et al.* 2004). Gene content in metazoan mitochondrial genomes is highly conserved (usually 37 genes), whereas they can be highly rearranged in gene order, notably in opisthobranchs (see Simison and Boore, Chapter 17). Protein-encoding genes tend to rearrange more slowly than tRNAs; thus, gene order can provide insight at multiple levels of divergence (Boore and Brown 1998). The mitochondrial genome of opisthobranchs is characterized by its small size (around 14 Kb) compared to the majority of other mollusc mitochondrial genomes sequenced at present (16 Kb or higher). The noncoding regions are few and short in

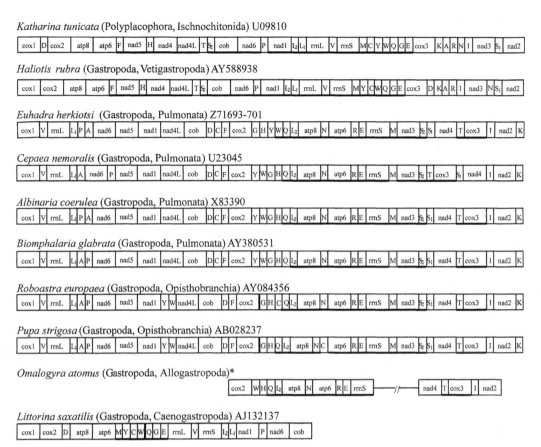

FIGURE 14.7. Mitochondrial gene arrangements of complete and partial published euthyneuran taxa and two outgroups, the chiton *Katharina tunicata* and the vetigastropod *Haliotis rubra*. Genomes are graphically linearized at *cox1*. All genes are transcribed from left to right except those underlined to indicate opposite orientation. NCBI accession numbers depicted next to each genome. *In Kurabayashi and Ueshima (2000b). Gene designations: Cytochrome *c* oxidase subunit I, II, III (*cox1, cox2, cox3*), cytochrome *b* (*cob*), NADH dehydrogenase subunits 1–6, 4L (*nad1–6, 4L*), ATP synthase subunits 6, 8 (*atp6, atp8*), large ribosomal subunit (*rrnL*), small ribosomal subunit (*rrnS*), 18 transfer RNAs specifying a single amino acid (*trnX*), two transfer RNAs specifying leucine L_1 = L(CUN) and L_2 = L(UUR), two transfer RNAs specifying serine S_1 = S(AGN) and S_2 = S(UCN).

length, and genes are highly rearranged compared to other published gastropod genomes such as *Haliotis rubra* (Maynard *et al.*, 2005) and *Littorina saxatilis* (Wilding *et al.* 1999), although they share similar gene junctions with several pulmonate taxa such as *Biomphalaria glabrata* (DeJong *et al.* 2004; Figure 14.7). *Haliotis* and *Littorina* share more gene junctions with *Katharina tunicata* (a chiton) than with the euthyneuran genomes, indicating high rates of rearrangement before that group radiated. The cephalaspidean *Pupa strigosa* (Kurabayashi and Ueshima 2000a) has the same gene order as the anaspidean *Aplysia californica*

(Knudsen *et al.* 2006), both differing from the nudibranch *Roboastra europaea* (Grande *et al.* 2002) by one tRNA rearrangement. There are, however, several segmental rearrangements in other opisthobranch taxa that have potential as informative phylogenetic characters (Medina, personal observation).

Further research should be encouraged involving *a posteriori* analyses on published phylogenies. Sperm ultrastructure (see, e.g., Healy 1988, 1996; Healy and Willan 1991; Fahey and Healy 2004; Wilson *et al.* 2004) has been applied in a few analyses, as have cell lineage data and other features linked to development (Page 1998, 2003).

A good example involving novel informative characters is the investigation of homologies within the nervous system and sensory organs by Klussmann-Kolb *et al.* (2004), using neurobiological techniques such as axonal tracing. Characters such as these can then be mapped on phylogenetic reconstructions to support or falsify certain clades. Preliminary results already show that the rhinophores in Anaspidea probably did not evolve from Hancock's organ in Cephalaspidea (Annette Klussmann-Kolb, unpublished data), as had been proposed previously (e.g., Hoffmann 1939; Huber 1993). Other prospects lie in the investigation of biological compounds used for defense and their location in the body (Avila 1995; Cimino and Ghiselin 1998, 1999; Cimino *et al.* 1999; Fahey 2004). Finally, investigation of biogeographical data (Kolb and Wägele 1998; Schrödl 2003) can potentially contribute to understanding certain nodes on a phylogenetic tree.

One important taxonomic requirement is, however, that new descriptions of new and previously named taxa be more detailed than in the past. They should involve many more characters than just those necessary for reidentification. It is only with such new data and a multidisciplinary approach that we can understand the complexity of this extraordinary taxon, the Opisthobranchia.

ACKNOWLEDGMENTS

Many colleagues helped in collecting the material for the phylogenetic studies, particularly Gilianne Brodie (Townsville, Australia) and Michael Schrödl (Munich, Germany).

We thank Alexander Nützel (Erlangen, Germany) and Klaus Bandel (Hamburg, Germany) for discussion on opisthobranch fossils. We are also grateful to Benoît Dayrat and Yvonne Vallès for helpful discussions on opisthobranch evolution. Petra Wahl and Claudia Brefeld helped in histological preparations. Anja Schulze (Fort Pierce, Florida, United States) kindly gave us the picture of the thecosome. This project was supported by the German Science Foundation in several projects to H. W. (Wa 618/5, Wa 618/6, Wa 618/7) and to A. K.-K. (KL 1313/1). Part of this research was supported by NSF Grant OCE 0313708 to M. M. The first author is especially grateful to Peter Hoffmann (Bochum, Germany) for his continuous support.

REFERENCES

Avila, C. 1995. Natural products of opisthobranch molluscs: a biological review. *Oceanography and Marine Biology, an Annual Review* 33: 487–559.

Avila, C., and Paul, V. J. 1997. Chemical ecology of the nudibranch *Glossodoris pallida*: is the location of diet-derived metabolites important for defense? *Marine Ecology Progress Series* 150: 171–180.

Bandel, K. 1994. Triassic Euthyneura (Gastropoda) from St. Cassian Formation (Italian Alps) with a discussion on the evolution of the Heterostropha. *Freiberger Forschungshefte* C452: 79–100.

———. 1995. Mathildoidea (Gastropoda, Heterostropha) from the Late Triassic St Casian Formation. *Scripta Geologica* 111: 1–83.

———. 2002. About the Heterostropha (Gastropoda) from the Carboniferous and Permian. *Mitteilungen des Geologisch-Paläontologischen Institutes der Universität Hamburg* 86: 45–80.

Bandel, K., Gründel, J., and Maxwell, P. 2000. Gastropods from the upper Early Jurassic/early Middle Jurassic of Kaiwara Valley, North Canterbury, New Zealand. *Freiberger Forschungshefte* C490: 67–132.

Bandel, K., and Heidelberger, D. 2002. A Devonian member of the subclass Heterostropha (Gastropoda) with valvatoid shell shape. *Neues Jahrbuch für Geologie und Paläontologie Monatshefte* 9: 533–550.

Barker, G. M. 2001. Gastropods on land: phylogeny, diversity and adaptive morphology. In *The Biology of Terrestrial Mollusks*. Edited by G. M. Barker. London: CAB International, pp. 1–130.

Bergh, L. S. R. 1891. Die cladohepatischen Nudibranchien. *Zoologische Jahrbücher, Systematik* 5: 1–75.

Boettger, C. R. 1954. Die Systematik der euthyneuren Schnecken. *Zoologischer Anzeiger* 18 Suppl.: 253–280.

Boore, J, and Brown, W. M. 1998. Big trees from little genomes: mitochondrial gene order as a phylogenetic tool. *Current Opinion in Genetics and Development* 8: 668–674.

Bowen, C. E., and Tang, H. 1996. Conchiolin-protein in aragonite shells of mollusks. *Comparative Biochemistry and Physiology Part A* 115: 269–275.

Burghardt, I., Evertsen, J., Johnsen, G., and Wägele, H. 2005. Solar powered seaslugs—mutualistic symbiosis of aeolid Nudibranchia (Mollusca, Gastropoda, Opisthobranchia) with *Symbiodinium*. *Symbiosis* 38: 227–250.

Burghardt, I., and Wägele, H. 2004. A new solar powered species of the genus *Phyllodesmium* Ehrenberg, 1831 (Mollusca: Nudibranchia: Aeolidoidea) from Indonesia with analysis of its photosynthetic activity and notes on biology. *Zootaxa* 596: 1–18.

———. 2006. Interspecific differences in the efficiency and photosynthetic characteristics of the symbiosis of "solarpowered" Nudibranchia (Mollusca: Gastropoda) with zooxanthellae. *Records of the Western Australian Museum Suppl.* 69: 1–9.

Burn, R. 1998. Order Rhodopemorpha. In *Mollusca: The Southern Synthesis*. Vol. 5, *Fauna of Australia*. Edited by P. L. Beesley, G. J. B. Ross, and A. Wells. Melbourne: CSIRO Publishing, pp. 960–961.

Burn, R., and Thompon, T. E. 1998. Order Cephalaspidea. Vol. 5, *Fauna of Australia*. Edited by P. L. Beesley, G. J. B. Ross, and A. Wells. Melbourne: CSIRO Publishing, pp. 943–959.

Caprotti, E. 1977. Molluschi e medicina nel Io secolo dC. *Conchiglie* 13: 137–144.

Carefoot, T. H. 1987. *Aplysia*: its biology and ecology. *Oceanography and Marine Biology, Annual Review* 25: 167–284.

Cattaneo-Vietti, R., Angelini, S., and Bavestrello, G. 1993. Skin and gut spicules in *Discodoris atromaculata* (Bergh, 1880) (Mollusca: Nudibranchia). *Bolletim Malacologico* 29: 173–180.

Cattaneo-Vietti, R, Angelini, S., Gaggero, L., and Lucchetti, G. 1995. Mineral composition of nudibranch spicules. *Journal of Molluscan Studies* 61: 331–337.

Challis, D. A. 1970. *Hedylopsis cornuta* and *Microhedyle verrucosa*, two new Achlidiacea (Mollusca: Opisthobranchia) from the Solomon Islands Protectorate. *Transactions of the Royal Society of New Zealand, Biological Sciences* 12: 29–38.

Cimino, G., Fontana, A., and Gavagnin, M. 1999. Marine opisthobranch molluscs: chemistry and ecology in sacoglossans and dorids. *Current Organic Chemistry* 3: 327–372.

Cimino, G., and Ghiselin, M. T. 1998. Chemical defense and evolution in Sacoglossa (Mollusca: Gastropoda: Opisthobranchia). *Chemoecology* 8: 51–60.

———. 1999. Chemical defense and evolutionary trends in biosynthetic capacity among dorid nudibranchs (Mollusca: Gastropoda: Opisthobranchia). *Chemoecology* 9: 187–207.

Clark, K. B., and Bussacca, M. 1978. Feeding specifity and chloroplast retention in four tropical Ascoglossa, with a discussion of the extent of chloroplast symbiosis and the evolution of the order. *Journal of Molluscan Studies* 44: 272–282.

Clark, K. B., Jensen, K. R., and Stirts, H. M. 1990. Survey for functional kleptoplasty among West Atlantic Ascoglossa (= Sacoglossa) (Mollusca: Opisthobranchia). *Veliger* 33: 339–345.

Colgan, D. J., Ponder, W. F., and Eggler, P. E. 2000. Gastropod evolutionary rates and phylogenetic relationships assessed using partial 28S rDNA and histone H3 sequences. *Zoologica Scripta* 29: 29–65.

Colgan, D. J., Ponder, W. F., Beacham, E., and Macaranas, J. M. 2003. Molecular phylogenetic studies of Gastropoda based on six gene segments representing coding or non-coding and mitochondrial or nuclear DNA. *Molluscan Research* 23: 123–148.

Dayrat, B., and Tillier, S. 2002. Evolutionary relationships of euthyneuran gastropods (Mollusca): a cladistic re-evaluation of morphological characters. *Zoological Journal of the Linnean Society* 135: 403–470.

———. 2003. Goals and limits of phylogenetics. The euthyneuran gastropods. In *Molecular systematics and phylogeography of Mollusks*. Edited by C. Lydeard and D. Lindberg. Washington, London: Smithsonia Books, pp. 161–184.

Dayrat, B., Tillier, A., Lecointre, G., and Tillier, S. 2001. New clades of euthyneuran gastropods (Mollusca) from 28S rRNA sequences. *Molecular Phylogenetics and Evolution* 19: 225–235.

Degnan, B. M., Degnan, S. M., Giusti, A., and Morse, D. E. 1995. A *hox/hom* homeobox gene in sponges. *Gene* 155: 175–177.

DeJong, R. J., Emery, A. M., and Adema, C. M. 2004. The mitochondrial genome of *Biomphalaria glabrata* (Gastropoda, Basommatophora), intermediate host of *Schistosoma mansoni*. *Journal of Parasitology* 90: 991–997.

Dietrich, G., Kalle, K., Krauss, W., and Siedler, G., eds. 1975. *Allgemeine Meereskunde*. Berlin: Gebrüder Bornträger, pp. 1–593.

Eales, N. B. 1921. *Aplysia*. *Liverpool Marine Biological Comments and Memoirs* 24: 183–266.

Edlinger, K. 1980. Zur Phylogenie der chemischen Sinnesorgane einiger Cephalaspidea (Mollusca–Opisthobranchia). *Zeitschrift für zoologische Systematik und Evolutionsforschung* 18: 241–256.

Edmunds, M. 1966. Protective mechanisms in the Eolidacea (Mollusca Nudibranchia). *Journal of the Linnean Society, Zoology* 46: 27–71.

———. 1987. Color in opisthobranchs. *American Malacological Bulletin* 5: 185–196.

Fahay, S. J. 2004. Molluscs as models for examining evolutionary complexities. In *Molluscan Megadiversity: Sea, Land and Freshwater. World Congress of Malacology, Perth, Western Australia.* Edited by F. E. Wells. Perth: Western Australian Museum, p. 39.

Fahay, S. J., and Healy, J. 2004. Using sperm ultrastructure to examine a species-level phylogeny. In *Molluscan Megadiversity: Sea, Land and Freshwater. World Congress of Malacology, Perth, Western Australia.* Edited by F. E. Wells. Perth: Western Australian Museum, p. 40.

Faulkner, J. 2001. Marine natural products. *Natural Products Reports* 18: 1–49.

Fontana, A., Cavaliere, P., Wahidulla, S., Naik, C. G., and Cimino, G. 2000. A new antitumor isoquinoline alkaloid from the marine nudibranch *Jorunna festiva. Tetrahedron* 56: 7305–7308.

Frýda, J., and Blodgett, R. B. 2001. The oldest known heterobranch gastropod, *Kuskokwimia* gen. nov., from the early Devonian of west-central Alaska, with notes on the early phylogeny of higher gastropods. *Vestnik Ceskeho geologickeho ustavu* 76: 39–53.

García-Gómez, J. C., Cimino, G., and Medina, A. 1990. Studies on the defensive behaviour of *Hypselodoris* species (Gastropoda: Nudibranchia): ultrastructure and chemical analysis of mantle dermal formations (MDFs). *Marine Biology* 106: 245–250.

García-Gómez, J. C., Medina, A., and Coveñas, R. 1991. Study of the anatomy and histology of the mantle dermal formations (MDFs) of *Chromodoris* and *Hypselodoris* (Opisthobranchia: Chromodorididae). *Malacologia* 32: 233–240.

Ghiselin, M. T. 1965. Reproductive function and the phylogeny of opisthobranch gastropods. *Malacologia* 3: 327–378.

Gosliner, T. M. 1991. Morphological parallelism in opisthobranch gastropods. *Malacologia* 32: 313–327.

———. 1994. Gastropoda. Opisthobranchia. In *Microscopic Anatomy of Invertebrates.* Vol. 5, *Mollusca I.* Edited by F. W. Harrison and A. J. Kohn. New York: Wiley-Liss, pp. 253–355.

———. 2001. Aposematic coloration and mimicry in opisthobranch mollusks: new phylogenetic and experimental data. *Bollettino Malacologico, Roma* 37: 163–170.

Gosliner, T. M., and Ghiselin, M. T. 1984. Parallel evolution in opisthobranch gastropods and its implications for phylogenetic methodology. *Systematic Zoology* 33: 255–274.

Grande, C., Templado, J., Cervera, J. L., and Zardoya, R. 2002. The complete mitochondrial genome of the nudibranch *Roboastra europaea* (Mollusca: Gastropoda) supports the monophyly of opisthobranchs. *Molecular Biology and Evolution* 19: 1672–1685.

———. 2004a. Molecular phylogeny of Euthyneura (Mollusca: Gastropoda). *Molecular Biology and Evolution* 21: 303–313.

———. 2004b. Phylogenetic relationships among Opisthobranchia (Mollusca: Gastropoda) based on mitochondrial cox 1, trnV, and rrnL genes. *Molecular Phylogenetics and Evolution* 33: 378–388.

Greenwood, P. G., and Mariscal, R. N. 1984a. The utilization of cnidarian nematocysts by aeolid nudibranchs: nematocyst maintenance and release in *Spurilla. Tissue Cell* 16: 719–730.

———. 1984b. Immature nematocyst incorporation by the aeolid nudibranch *Spurilla neapolitana. Marine Biology* 80: 35–38.

Gründel, J. 1997a. Heterostropha (Gastropoda) aus dem Dogger Norddeutschland und Nordpolens. I. Mathildoidea (Mathildidae). *Berliner geowissenschaftliche Abhandlungen* E 25: 131–175.

———. Heterostropha (Gastropoda) aus dem Dogger Norddeutschland und Nordpolens. III. Opisthobranchia. *Berliner geowissenschaftliche Abhandlungen* E 25: 177–223.

Guiart, J. 1901. Contributions à l'étude des Gastéropodes Opisthobranches. *Mémoires de la Societé Zoologique de France* 14: 5–219.

Harris, L. G. 1973. Nudibranch association. *Current Topics of Comparative Pathobiology* 2: 213–315.

Haszprunar, G. 1985. The Heterobranchia—a new concept of the phylogeny of the higher Gastropoda. *Zeitschrift für zoologische Systematik und Evolutionsforschung* 23: 15–37.

———. 1988. On the origin and evolution of major gastropod groups, with special reference to the Streptoneura. *Journal of Molluscan Studies* 54: 367–441.

Haszprunar, G., and Huber, G. 1990. On the central nervous system of Smeagolidae and Rhodopidae, two families questionably allied with the Gymnomorpha (Gastropoda: Euthyneura). *Journal of Zoology, London* 220: 185–199.

Hatzoglou, E., Rodakis, G. C., and Lecanidou, R. 1995. Complete sequence and gene organization of the mitochondrial genome of the land snail *Albinaria coerulea. Genetics* 140: 1353–1366.

Healy, J. M. 1988. Sperm morphology and its systematic importance in the Gastropoda. In *Prosobranch Phylogeny.* Edited by W. F. Ponder. *Malacological Review Suppl.* 4: 251–266.

———. 1996. Molluscan sperm ultrastructure: correlation with taxonomic units within the Gastropoda, Cephalopoda and Bivalvia. In *Origin and Evolutionary Radiation of the Mollusca.* Edited by J. Taylor. London: Oxford University Press, pp. 99–113

Healy, J. M., and Willan, R. C. 1991. Nudibranch spermatozoa: comparative ultrastructure and systematic importance. *Veliger* 34: 134–165.

Hoegh-Guldberg, O., Hinde, R., and Muscatine, L. 1986a. Studies on a nudibranch that contains zooxanthellae. I. Photosynthesis respiration and the translocation of newly fixed carbon by zooxanthellae in *Pteraeolidia ianthina*. *Proceedings of the Royal Society of London. Series B, Biological Sciences* 228: 493–509.

———. 1986b. Studies on a nudibranch that contains zooxanthellae. II. Contribution of zooxanthellae to animal respiration (CZAR) in *Pteraeolidia ianthina* with high and low densities of zooxanthellae. *Proceedings of the Royal Society of London. Series B, Biological Sciences* 228: 511–521.

Hoffmann, H. 1939. Opisthobranchia. In *Klassen und Ordnungen des Tierreichs. 3. Band: Mollusca, II. Abteilung: Gastropoda. 3. Buch: Opisthobranchia.* Edited by H. G. Bronns. Leipzig: Akademische Verlagsgesellschaft, pp. 1–1247.

Huber, G. 1993. On the cerebral nervous system of marine Heterobranchia (Gastropoda). *Journal of Molluscan Studies* 59: 381–420.

Huff, S. W., Campbell, D., Gustafson, D. L., Lydeard, C., Altaba, C. R., and Giribet, G. 2004. Investigations into the phylogenetic relationships of the threatened freshwater pearl-mussels (Bivalvia, Unionoidea, Margaritiferidae) based on molecular data: implications for their taxonomy and biogeography. *Journal of Molluscan Studies* 70: 379–388.

Huson, D. H., and Bryant, D. 2006. Application of phylogenetic networks in evolutionary studies. *Molecular Biology and Evolution* 23: 254–267.

Jensen, K. R. 1996. Phylogenetic systematics and classification of the Sacoglossa (Mollusca, Gastropoda, Opisthobranchia). *Philosophical Transactions of the Royal Society of London. Series B, Biological Sciences* 351: 91–122.

———. 1997. Evolution of the Sacoglossa (Mollusca, Opisthobranchia) and the ecological associations with their food plants. *Evolutionary Ecology* 11: 301–335.

Kälker, H., and Schmekel, L. 1976. Bau und Funktion des Cnidosacks der Aeolidoidea (Gastropoda Nudibranchia). *Zoomorphology* 86: 41–60.

Kempf, S. C. 1984. Symbiosis between the zooxanthella *Symbiodinium* (*Gymnodinium*) *microadriaticum* (Freudenthal) and four species of nudibranchs. *Biological Bulletin* 166: 110–126.

Kiel, S., Bandel, K., and Perrilliat, M. D. C. 2002. New gastropods from the Maastrichtian of the Mexcala Formation in Guerrero, southern Mexico, part II: Archaeogastropoda, Neritimorpha and Heterostropha. *Naturwissenschaftliche Jahrbücher der Geolo-gischen und Paläontologischen Abhandlungen* 226: 319–342.

Klussmann-Kolb, A. 2004. Phylogeny of the Aplysiidae (Gastropoda, Opisthobranchia) with new aspects of the evolution of seahares. *Zoologica Scripta* 33: 439–462.

Klussmann-Kolb, A., Staubach, S., and Croll, R. P. 2004. Comparative study of the cephalic sensory organs in the Opisthobranchia. In *Molluscan Megadiversity: Sea, Land and Freshwater. World Congress of Malacology, Perth, Western Australia.* Edited by F. E. Wells. Perth: Western Australian Museum, p. 80.

Knudsen, B., Kohn, A. B., Nahir, B., McFadden, C. S., and Moroz, L. L. 2006. Complete DNA sequence of the mitochondrial genome of the sea-slug, *Aplysia californica*: conservation of the gene order in Euthyneura. *Molecular Phylogenetics and Evolution* 38(2): 459–469.

Kolb, A., and Wägele, H. 1998. On the phylogeny of the Arminidae (Gastropoda, Opisthobranchia, Nudibranchia) with considerations of biogeography. *Journal of Zoological Systematics and Evolutionary Research* 36: 53–64.

Kurabayashi, A., and Ueshima, R. 2000a. Complete sequence of the mitochondrial DNA of the primitive opisthobranch gastropod *Pupa strigosa*: systematic implication of the genome organization. *Molecular Biology and Evolution* 17: 266–277.

———. 2000b. Partial mitochondrial genome organization of the heterostrophan gastropod *Omalogyra atomus* and its systematic significance. *Venus* 59: 7–18.

Lalli, C. M., and Gilmer, R. W. 1989. *Pelagic Snails. The Biology of Holoplanktonic Gastropod Mollusks.* Palo Alto, CA: Stanford University Press.

Lieb, B., Altenhein, B., and Markl, J. 2000. The sequence of a gastropod hemocyanin (HtH1) from *Haliotis tuberculata*. *Journal of Biological Chemistry* 275: 5675–5681.

Marin, A., and Ros, J. D. 1992. Dynamics of a peculiar plant-herbivore relationship: the photosynthetic ascoglossan *Elysia timida* and the chlorophycean *Acetabularia acetabulum*. *Marine Biology* 112: 677–682.

Maynard, B. T., Kerr, L. J., McKiernan, J. M., Jansen, E. S., and Hanna, P. J. 2005. Mitochondrial DNA sequence and gene organization of the Australian blacklip abalone, *Haliotis rubra* (leach). *Marine Biotechnology* 7(6): 645–658.

McDonald, G. R., and Nybakken, J. W. 1991. *A List of the Worldwide Food Habits of Nudibranchs.* Long Marine Laboratory. http://people.ucsc.edu/~mcduck/nudifood.htm. Accessed March 15, 2006.

Medina, M., and A. Collins. 2003. The role of molecules in understanding molluscan evolution. In *Molecular Systematics and Phylogeography of Mollusks*. Edited by C. Lydeard and D. Lindberg. Washington, DC: Smithsonian Institution Press, pp. 14–44.

Medina, M., Takaoka, T., Valles, Y., Gosliner, T., and Boore, J. 2004. Opisthobranch genomics: mitochondria and beyond. In *Molluscan Megadiversity: Sea, Land and Freshwater. World Congress of Malacology, Perth, Western Australia*. Edited by F. E. Wells. Perth: Western Australian Museum, 99.

Medina, M., and Walsh, P. J. 2000. Systematics of the order Anaspidea based on mitochondrial DNA sequence (12S, 16S and COI). *Molecular Phylogenetics and Evolution* 15: 41–58.

Mikkelsen, P. M. 1996. The evolutionary relationships of Cephalaspidea s. l. (Gastropoda; Opisthobranchia): a phylogenetic analysis. *Malacologia* 37: 375–442.

———. 2002. Shelled opisthobranchs. *Advances in Marine Biology* 42: 67–136.

Mujer, C. V., Andrews, D. L., Manhart, J. R., Pierce, S. K., and Rumpho, M. E. 1996. Chloroplast genes are expressed during intracellular symbiotic association of *Vaucheria litorea* plastids with the sea slug *Elysia chlorotica*. *Proceedings of the National Academy of Sciences of the United States of America* 93: 12333–12338.

Odhner, N. H. 1968. Sous-classe des opisthobranches. In *Traité de Zoologie 5(3)*. Edited by P.-P. Grassé. Paris: Masson, pp. 608–893.

Okusu, A., Schwabe, E., Eernisse, D. J., and Giribet, G. 2003. Towards a phylogeny of chitons (Mollusca, Polyplacophora) based on combined analysis of five molecular loci. *Organism, Diversity and Evolution* 3: 281–302.

Page, L. R. 1998. Sequential developmental programs for retractor muscles of a caenogastropod: reappraisal of evolutionary homologues. *Proceedings of the Royal Society of London. Series B, Biological Sciences* 265: 2243–2250.

———. 2003. Gastropod ontogenetic torsion: Developmental remnants of an ancient evolutionary change in body plan. *Journal of Experimental Zoology (Molecular and Developmental Evolution)* 297B: 11–26.

Pelseneer, P. 1893. Sur quelques points d'organisation des Nudibranches et sur leur phylogénie. *Bulletin de la Societé Malacologique de Belgique* 26: 68–71.

Pilsbry, H. A. 1894–95. Order Opisthobranchia. *Manual of Conchology* 15: 134–436.

Ponder, W. F., and Lindberg, D. R. 1997. Towards a phylogeny of gastropod molluscs: an analysis using morphological characters. *Zoological Journal of the Linnean Society* 119: 83–265.

Rudman, W. B. 1978. A new species and genus of the Aglajidae and the evolution of the philinacean opisthobranch molluscs. *Zoological Journal of the Linnean Society of London* 62: 89–107.

———. 1984. The Chromodorididae (Opisthobranchia: Mollusca) of the Indo-West Pacific: a review of the genera. *Zoological Journal of the Linnean Society of London* 81: 115–273.

———. 1987. Solar-powered animals. *Natural History* 10: 50–52.

———. 1991. Further studies on the taxonomy and biology of the octocoral-feeding genus *Phyllodesmium* Ehrenberg, 1831 (Nudibranchia: Aeolidoidea). *Journal of Molluscan Studies* 57: 167–203.

Rudman, W. B., and Willan, R. C. 1998. Opisthobranchia, Introduction. In *Mollusca: The Southern Synthesis*. Vol. 5, *Fauna of Australia*. Edited by P. L. Beesley, G. J. B. Ross, and A. Wells. Melbourne: CSIRO Publishing, pp. 915–942.

Rumpho, M. E., Summer, E. J., Green B. J., Fox, T. C., and Manhart, J. R. 2001. Mollusc/algal chloroplast symbiosis: how can isolated chloroplasts continue to function for months in the cytosol of a sea slug in the absence of an algal nucleus? *Zoology* 104: 303–312.

Rumpho, M. E., Summer, E. J., and Manhart, J. R. 2000. Solar-powered seaslugs. Mollusc/Algal chloroplast symbiosis. *Plant Physiology* 123: 29–38.

Salvini-Plawen, L. v. 1990. Origin, phylogeny and classification of the phylum Mollusca. *Iberus* 9: 1–33.

Salvini-Plawen, L. v., and Steiner, G. 1996. Synapomorphies and plesiomorphies in higher classification of Mollusca. In *Origin and Evolutionary Radiation of the Mollusca*. Edited by J. Taylor. London: Oxford University Press, pp. 29–51.

Schmekel, L. 1985. 5. Aspects of evolution within the opisthobranchs. In *The Mollusca*. Vol. 10, *Evolution*. Edited by R. Trueman and M. R. Clarke. New York: Academic Press, pp. 221–267.

Schmekel, L., and Cappellato, D. 2001. Contributions to the Runcinidae: six new species of the genus *Runcina* (Opisthobranchia Cephalaspidea) in the Mediterranean. *Vie et Milieu* 51: 141–160.

Schmekel, L., and Portmann, A. 1982. *Opisthobranchia des Mittelmeeres. Nudibranchia und Saccoglossa*. Berlin, Heidelberg, New York: Springer Verlag.

Schrödl, M. 2003. *Sea slugs of southern South America*. Meckenheim: Conch Books.

Schrödl, M., Wägele, H., and Willan, R. C. 2001. Taxonomic redescription of the Doridoxidae (Gastropoda: Opisthobranchia), an enigmatic family of deep water nudibranchs, with discussion of basal nudibranch phylogeny. *Zoologischer Anzeiger* 240: 83–97.

Sommerfeldt, N., and Schrödl, M. 2005. Microanatomy of *Hedylopsis ballantinei*, a new interstitial acochlidian gastropod from the Red Sea, and its sitgnificance for phylogeny. *Journal of Molluscan Studies* 71: 153–165.

Streble, H. 1968. Bau und Bedeutung der Nesselsäcke von *Aeolidia papillosa* L., der Breitwarzigen Fadenschnecke (Gastropoda, Opisthobranchia). *Zoologischer Anzeiger* 180: 356–372.

Thatje, S., Hillenbrand, C.-D., and Larter, R. 2005. On the origin of Antarctic marine benthic community structure. *Trends in Ecology and Evolution* 20: 534–540.

Thiele, J. 1931. *Handbuch der systematischen Weichtierkunde. II. Subclassis Opisthobranchia.* Bd. 1, Teil 2. Jena: Gustav Fischer Verlag, pp. 377–461.

Thollesson, M. 1999. Phylogenetic analysis of Euthyneura (Gastropoda) by means of the 16S rRNA gene: use of a "fast" gene for "higher-level" phylogenies. *Proceedings of the Royal Society of London. Series B: Biological Sciences* 266: 75–83.

Thompson, T. E. 1960. Defensive adaptations in opisthobranchs. *Journal of the Marine Biological Association U.K.* 39: 123–134.

———. 1969. Acid-secretion in Pacific Ocean gastropods. *Australian Journal of Zoology* 17: 755–764.

———. 1976. *Biology of Opisthobranch Mollusks.* Vol. I. London: The Ray Society.

———. 1983. Detection of epithelial acid secretions in marine molluscs: review of techniques and new analytical methods. *Comparative Biochemistry and Physiology* 74A: 615–621.

———. 1986. Investigation of the acidic allomone of the gastropod mollusc *Philine aperta* by means of ion chromatography and histochemical localisation of sulphate and chloride ions. *Journal of Molluscan Studies* 52: 38–44.

———. 1988. Acidic allomones in marine organisms. *Journal of the Marine Biological Association, U.K.* 68: 499–517.

Thompson, T. E., and Colman, J. G. 1984. Histology of acid glands in Pleurobranchomorpha. *Journal of Molluscan Studies* 50: 66–67.

Tillier, S., Masselot, M., and Tillier, A. 1996. Phylogenetic relationships of the pulmonate gastropods from rRNA sequences, and tempo and age of the stylommatophoran radiation. In *Origin and Evolutionary Radiation of the Mollusca.* Edited by J. Taylor. London: Oxford University Press, pp. 267–284.

Tracey, S., Todd, J. A., and Erwin, D. H. 1993. Mollusca: Gastropoda. In *The Fossil Record.* Edited by M. J. Benton. London: Chapman and Hall, pp. 1–845.

Ueshima, R. 2004. Comparative genomics of molluscan mitochondrial DNA and the systematic significance. In *Molluscan Megadiversity: Sea, Land and Freshwater. World Congress of Malacology, Perth, Western Australia.* Edited by F. E. Wells. Perth: Western Australian Museum, p. 151.

Valdés, A. 2002. Phylogenetic systematics of "Bathydoris" s.l. Bergh, 1884 (Mollusca, Nudibranchia), with the description of a new species from New Caledonian deep waters. *Canadian Journal of Zoology* 80: 1084–1099.

Valdés, A., and Lozouet, P. 2000. Opisthobranch molluscs from the Tertiary of the Aquitaine Basin (South-Western France), with descriptions of seven new species and a new genus. *Palaeontology* 43: 457–479.

Vonnemann, V., Schrödl, M., Klussmann-Kolb, A., and Wägele, H. 2005. Reconstruction of the phylogeny of the Opisthobranchia (Mollusca, Gastropoda) by means of 18S and 28S rRNA gene sequences. *Journal of Molluscan Studies* 71: 113–125.

Wade, C. M., and Mordan, P. B. 2000: Evolution within the gastropod molluscs; using the ribosomal RNA gene-cluster as an indicator of phylogenetic relationships. *Journal of Molluscan Studies* 66: 565–570.

Wägele, H. 1997. Histological investigation of some organs and specialized cellular structures in Opisthobranchia (Gastropoda) with the potential to yield phylogenetically significant characters. *Zoologischer Anzeiger* 236: 119–131.

———. 2004. Potential key characters in Opisthobranchia (Gastropoda, Mollusca) enhancing adaptive radiation. *Organisms, Diversity and Evolution* 4: 175–188.

Wägele, H., Ballesteros, M., and Avila, C. 2006a. Defensive glandular structures in opisthobranch molluscs—from histology to ecology. *Oceanography and Marine Biology: An Annual Review* 44: 197–276.

Wägele, H., and Johnsen, G. 2001. Observations on the histology and photosynthetic performance of "solar-powered" opisthobranchs (Mollusca, Gastropoda, Opisthobranchia) containing symbiotic chloroplasts or zooxanthellae. *Organisms, Diversity and Evolution* 1: 193–210.

Wägele, H., and Klussmann-Kolb, A. 2005. Opisthobranchia (Mollusca, Gastropoda)—more than just slimy slugs. Shell reduction and its implications on defence and foraging. *Frontiers in Zoology* 2: 1–18.

Wägele, H., Vonnemann, V., and Rudman, W. B. 2006b. *Umbraculum umbraculum* (Lightfoot, 1786) (Gastropoda, Opisthobranchia, Tylodinoidea) and the synonymy of *U. mediterraneum*

(Lamarck, 1812). *Records of the Western Australian Museum Suppl.* 69: 69– 82.

Wägele, H, Vonnemann, V., and Wägele, J. W. 2003. Towards a phylogeny of the Opisthobranchia. In *Molecular Systematics and Phylogeography of Mollusks.* Edited by C. Lydeard and D. Lindberg. Washington, DC: Smithsonian Institution Press, pp. 185–228.

Wägele, H., and Willan, R. C. 2000. On the phylogeny of the Nudibranchia. *Zoological Journal of the Linnean Society* 130: 83–181.

Wenz, W., and Zilch, A. 1959. Gastropoda. II. Euthyneura. In *Handbuch der Paläozoologie.* Edited by O. H. Schindewolf. Berlin: Verlag Gebrüder Bornträger.

Wilding, C. S., Mill, P. J., and Graham, J. 1999. Partial sequence of the mitochondrial genome of *Littorina saxatilis*: relevance to gastropod phylogenetics. *Journal of Molecular Evolution* 48: 348–359.

Willan, R. C. 1987. Phylogenetic systematics of the Notaspidea (Opisthobranchia) with reappraisal of families and genera. *American Malacological Bulletin* 5: 215–241.

Wilson, N. G., Healy, J. M., and Lee, M. S. Y. 2004. Molecular phylogeny and sperm morphology of *Chromodoris* (Gastropoda: Nudibranchia). In *Molluscan Megadiversity: Sea, Land and Freshwater.* World Congress of Malacology, Perth, Western Australia. Edited by F. E. Wells. Perth: Western Australian Museum, p. 164.

Wollscheid, E., and Wägele, H. 1999. Initial results on the molecular phylogeny of the Nudibranchia (Gastropoda, Opisthobranchia) based on 18S rDNA data. *Molecular Phylogenetics and Evolution* 13: 215–226.

Wollscheid-Lengeling, E., Boore, J., Brown, W., and Wägele, H. 2001. The phylogeny of Nudibranchia (Opisthobranchia, Gastropoda, Mollusca) reconstructed by three molecular markers. *Organism, Diversity and Evolution* 1: 241–256.

Yamazaki, N., Ueshima, R., Terrett, J. A., Yokobori, S., Kaifu, M., Segawa, R., Kobayashi, T., Numachi, K., Ueda, T., Nishikawa, K., Watanabe, K., and Thomas, R. H. 1997. Evolution of pulmonate gastropod mitochondrial genomes: comparisons of gene organizations of *Euhadra, Cepaea* and *Albinaria* and implications of unusual tRNA secondary structures. *Genetics* 145: 749–758.

Yoon, S. H., and Kim, W. 2000. Phylogeny of some gastropod mollusks derived from 18S rDNA sequences with emphasis on the Euthyneura. *Nautilus* 114: 85–92.

15

Heterobranchia II

THE PULMONATA

Peter Mordan and Christopher Wade

The pulmonates include the large majority of land snails and slugs, as well as many freshwater snails and a few small families of estuarine and shallow-water marine slugs and snails. In total there are thought to be some 25,000–30,000 living pulmonate species worldwide. As such they represent by far the most significant invasion of nonmarine environments by the Mollusca. This habitat shift has necessitated numerous adaptive changes in their respiratory, nervous, excretory, and reproductive systems, as well as to their behavior and physiology. Although these changes might reasonably be expected to have yielded a series of phylogenetically informative morphological characters, so far, this hope has not been realized, and our understanding of pulmonate relationships remains sketchy. This is especially regrettable because an accurate knowledge of their phylogeny is essential to any real understanding of the sequence of evolutionary events leading to their occupation of land and freshwater habitats.

The pulmonates were first defined by Cuvier (1817) as one of his seven orders of gastropods. In early editions of his *Règne Animal*, the *Pulmonés* included many of the familiar land and freshwater groups, but also two coastal marine families: the onchidiid slugs and ellobiid snails. He defined pulmonates as hermaphrodite slugs and snails that breathed air through a contractile pneumostome and lacked an operculum. Although our modern concept of the group includes many more families and is based on additional morphological characters (discussed subsequently), Cuvier's definition was nevertheless remarkably perceptive. Today three anatomical characters are generally accepted as true apomorphies of the Pulmonata: a pallial cavity opening by means of a pneumostome and the presence of a procerebrum and medio-dorsal (cerebral) bodies in the nervous system.

Many pulmonate land slugs such as *Arion* and *Deroceras* are agricultural pests of major importance in temperate regions, as is the giant African land snail *Achatina*, which has been spread by humans throughout the tropics (Mead 1961; Barker 2002). Conversely, it has recently been shown that grazing by the slugs that are pests in agricultural situations is beneficial in promoting plant diversity in grassland habitats (Bachmann *et al.* 2005). Land snails, most notably the European banded snail *Cepaea*

TABLE 15.1

Major Taxonomic Groupings Included Within the Pulmonata

TAXON	HABITAT[a]	NUMBER OF SPECIES	EARLIEST RECORDED FOSSIL
Otinidae	M	1	?Upper Jurassic Kimmeridgian/Tithonian
Amphibolidae	M	10	Eocene (Harbeck 1996)
Carychiidae	T	10	?Middle Carboniferous, Bashkirian
Chilinidae	F	10	Upper Pliocene
Glacidorbidae	F	10	—
Smeagolidae	M	10	—
Trimusculidae	M	10	Oligocene
Rathousiidae	T	10	—
Vaginulidae	T	100	—
Ellobiidae	M(T)	100	Upper Jurassic, Kimmeridgian/Tithonian
Siphonariidae	M	100	Upper Jurassic, Oxfordian
Onchidiidae	M(T)	100	—
Hygrophila	F	1,000	Middle Jurassic, Doggerian (Planorbidae)
Stylommatophora	T	10,000 +	?Carboniferous, Moscovian

NOTE: Numbers of species are given to the nearest order of magnitude. Minimum ages of fossils are based largely on Tracey *et al.* (1993). [a]Habitats: M, marine; F freshwater; T terrestrial.

(e.g., Clarke *et al.* 1978), the West Indian *Cerion* (Woodruff 1978) and the Pacific-island tree snails of the family Partulidae (Cowie 1992), have proved to be important models for studies on the mechanisms of evolution. However, the recent introduction of the predatory snail *Euglandina* into Pacific islands in an attempt to control *Achatina* has led to the extinction of many of the native tree snail species (Cowie 1992). Several freshwater pulmonate snails such as *Bulinus* act as intermediate hosts for major parasites of humans and farm animals, such as *Schistosoma* (which causes schistosomiasis) and *Fasciola* (the sheep liver fluke) (see review by Brown 1978), while some land snails harbor parasites such as the nematode *Angiostrongylus*, which can cause meningitis in humans (Kliks and Palumbo 1992).

MAJOR GROUPS

There are several valuable modern reviews of pulmonate anatomy (Nordsieck 1992; Smith and Stanisic 1998; Barker 2001). Nordsieck also includes a classification that is broadly representative of several of the more important recent pulmonate classifications based on anatomical characters (Mol 1967; Tillier 1984b; Haszprunar and Huber 1990; Salvini-Plawen and Steiner 1996). Although these all differ in detail, there are nevertheless important areas of broad agreement between them. Table 15.1 lists the main groups currently included in the Pulmonata.

GEOPHILA

The Geophila, introduced by Férussac (1819), comprises the main terrestrial mollusc group Stylommatophora, normally treated as an order or suborder in the literature, as well as three families of slugs, all lacking any form of shell, which appear to be closely related to it: the Vaginulidae (= Veronicellidae), Rathouisiidae, and Onchidiidae.

About 95% of all pulmonate species are land slugs and snails (Figures 15.1D, F) belonging to the Stylommatophora (Table 15.1), which are grouped into about 90 families. Although commonest in moist environments, they can occupy habitats ranging from within the Arctic Circle to all but the most extreme hot deserts. Most are ground

FIGURE 15.1. Pulmonate diversity. (A) *Bulinus* (Bulinidae), Africa. Hygrophilan snail showing eyes positioned at base of a single pair of nonretractable tentacles. (B) *Siphonaria* (Siphonariidae), Andaman Islands. Dorsal and ventral views, the latter showing the pale siphonal groove on the right-hand side. (C) *Onchidium* (Onchidiidae), Australia. Photograph by W. Rudman. (D) *Ratnadvipia* (Ariophantidae). Sri Lanka. Stylommatophoran land slug showing the pneumostome. (E) *Myosotella* (Ellobiidae), Azores. Photograph by A. M. F. Martins. (F) *Beddomea* (Camaenidae), Sri Lanka. Stylommatophoran land snail showing the two pairs of retractile tentacles, with terminal eyes on the cephalic pair. (G) *Salinator* (Amphibolidae), Australia. Photograph by R. Golding. (H) *Angustipes* (Vaginulidae), Jamaica. Frontal view of crawling slug showing the contractile cephalic tentacles with terminal eyes. Photographs not otherwise credited and image editing by H. Taylor, NHM.

dwellers, but many are arboreal. Anatomically the Stylommatophora form a highly cohesive group and, despite the large number of taxa, are almost invariably treated as a single taxonomic unit. They have two pairs of retractile tentacles, with eyes on the tips of the posterior, cephalic pair. The hermaphrodite reproductive system is monotrematous, opening by means of a single genital pore.

The terrestrial vaginulid and rathouisiid slugs live in the tropics and were grouped as the

Soleolifera by Thiele (1931). Both have separate male and female genital openings and, like the Stylommatophora, two pairs of tentacles with eyes at the tips of the posterior pair (Figure 15.1H), but these are contractile rather than retractile. Vaginulids have no true lung, but breathe through the skin and a system of subcutaneous tubules. Rathouisiids lack a jaw, and their radula has long, sharp teeth, because they are at least partly predatory. These two families were later combined with the littoral marine onchidiid slugs as the Systellommatophora (Pilsbry 1948). The Onchidiidae (Figure 15.1C) are almost all coastal-marine species with a near-worldwide distribution, inhabiting sheltered intertidal or supratidal sites where they live in holes or under boulders or wood, though a few fully terrestrial species are known living up to 600 meters above sea level (Tillier 1984a). They have only one pair of cephalic tentacles, with eyes at the tips. Male and female openings are separate, and eggs are laid in ribbon-shaped masses.

HYGROPHILA

The principal freshwater pulmonate group is the Hygrophila (or Lymnaeoidea). Introduced by Férussac (1822), it comprises most of the well-known limnic families such as the Planorbidae, Lymnaeidae, and Physidae, as well as several families of limpets such as the New Zealand Latiidae, the north American Lancidae, and the predominantly Eurasian Acroloxidae. It includes no slugs. A useful account of their comparative morphology is given by Hubendick (1978). The single pair of cephalic tentacles, with the eyes at their inner bases (Figure 15.1A), cannot be retracted. Several species have developed external respiratory gills in addition to the enclosed, air-breathing lung. The family Chilididae from South America has several seemingly primitive features of the pallial cavity, such as retaining an osphradium (Dayrat and Tillier 2003), but is also normally included in the Hygrophila.

OTHER GROUPS

The Glacidorbidae are a family of tiny orthostrophic to hyperstrophic freshwater operculate snails restricted to Australia and South America. The pallial cavity is open, lacking a pneumostome, and the genital openings are separate. Their nervous system also lacks many important pulmonate characters. As a result some workers (Haszprunar and Huber 1990; Dayrat and Tillier 2002) exclude them, but others (Ponder 1986; Smith and Stanisic 1998; Ponder and Avern 2000) believe them to be a highly paedomorphic basal group of pulmonates. In contrast, it has now been clearly established that the Rhodopidae, a family of minute, interstitial, subtidal, wormlike molluscs first described as pulmonates, are opisthobranchs (Tillier 1984b; Salvini-Plawen 1991).

All the other pulmonates are heterostrophic; that is, the embryonic (or larval) shell and the adult shell coil in different directions, as opposed to orthostrophic (Dayrat and Tillier 2003). The ellobioids are the largest of these groups, comprising five subfamilies (Martins 1996) and, like the marine limpet family Trimusculidae (discussed subsequently), have variously been included in the Basommatophora or Eupulmonata; as a consequence, these two higher groupings have little real value in our current state of knowledge. Most ellobioids are tropical and inhabit mangrove areas, but temperate species live in salt marsh or under rocks in the upper littoral zone. One family, the Northern Hemisphere Carychiidae, is wholly terrestrial. Ellobioids have a single pair of cephalic tentacles with the eyes at their base (Figure 15.1E); genital systems are very variable, but the male and female pores are always separate.

The remaining families live in the marine intertidal. Smeagolids are known from Australasia and Japan (H. Fukuda, personal communication) and are minute, interstitial slugs, whereas *Otina*, the sole genus in the family Otinidae, has a small, cap-like shell and lives on rocks in the northwestern Atlantic. Both have a gastric cecum and a contractile pneumostome, show secondary loss of the pallial ciliary tracts, and lack a trochophore (Tillier 1984b); the two families are currently united in the Otinoidea (Tillier and Ponder 1992).

Two littoral marine groups, the Siphonariidae and Amphibolidae, are usually paired together as either the Amphiboloidea (Tillier 1984b) or the Thalassophila (Nordsieck 1992) and are often placed with the Hygrophila in the Basommatophora (Mol 1967; Haszprunar and Huber 1990; Nordsieck 1992; Salvini-Plawen and Steiner 1996). Amphibolids are small to relatively large, globose, operculate snails that live mainly in estuarine conditions in the Indo-West Pacific and Australasia (Golding et al. 2007). They are deposit feeders that lack a jaw. The eyes are situated at the base of the very short tentacles (Figure 15.1G), and the pallial cavity contains a small osphradium. The Siphonariidae and Trimusculidae are both widely distributed families of intertidal limpets. Siphonariids lack tentacles, as well as several key pulmonate characters such as the contractile pneumostome and pulmonary blood vessels, but have an osphradium and pallial raphes. The family includes the genus *Williamia,* most species of which are entirely subtidal. Siphonariids derive their name from the prominent siphonal groove on the right side of the shell (Figure 15.1B). The shell of the Trimusculidae also has a siphonal grove, but this is situated more anteriorly than in the siphonariids; their cephalic tentacles are reduced, with the eyes at the base, and the genital pores are separate.

FOSSIL HISTORY

Pulmonate fossil history is patchy. Their shells are formed of aragonite, which does not preserve well, and many are slugs, in which the shell is either lacking or extremely reduced. Furthermore, the systematic interpretation of those fossils that are known is often extremely difficult, because there is much convergence in shell form in unrelated taxa. Consequently much reliance is placed on anatomical characters in modern pulmonate classification. Clearly the fossil data that are available must be interpreted with extreme caution, and ideas can change rapidly. Tracey et al. (1993) give a systematic review of the earliest fossil occurrence of the various

pulmonate families (Table 15.1), and Bandel (1997) presented a narrative account of their history. Solem and Yochelson (1979) undertook a major review of Paleozoic land snails, but their attribution of various Carboniferous fossils to several Recent stylommatophoran families is highly controversial. According to Bandel (1997), the earliest pulmonates may have been present during the Late Carboniferous but were possibly Carychiidae, with the other ellobioids, as well as the siphonariids, otinids, and freshwater Hygrophila, not appearing until the Middle or Upper Jurassic. Undoubted stylommatophoran land snails are present around the Jurassic/Cretaceous boundary, but it is not until well into the Cenozoic that certain marine groups such as the Trimusculidae first appear in the fossil record. Tillier et al. (1996) proposed two possible schemes for early pulmonate evolution: either emergence of both the pulmonates and opisthobranchs during the mid-Paleozoic, or alternatively a Triassic origin for the opisthobranchs and a later origin for the pulmonates. In the first case the Carboniferous fossil land snails were interpreted as terrestrial pulmonates, including possibly carychiids (as suggested by Bandel 1997) but not stylommatophorans; in the second, none of these Paleozoic fossils were considered pulmonates. In either event, Tillier et al. (1996) favored a Jurassic origin for both the Stylommatophora and Hygrophila, with most diversification around 60 Mya in the Cenozoic.

Hrubesch (1965) described a diverse terrestrial fauna from the Santonian (Late Cretaceous 85 Mya) of Austria, including representatives of several Recent stylommatophoran families, but perhaps the best-preserved fossil land snail assemblages are those reported by Pickford (1995) from the early Miocene of East Africa. The age of these deposits has been accurately determined using K-Ar dating, and the land snail fossils show a remarkably close resemblance to the present-day fauna in the area, with over half of the Recent taxa represented. Indeed, all the fossil genera, and many of the species, appear to be the same as those living today, suggesting that there has been little evolution above the species level in that area in the intervening 20 million years.

PHYLOGENY

SISTER GROUP RELATIONSHIPS

Classically the pulmonates have been thought to have a sister group relationship with the almost exclusively marine opisthobranchs, the two forming the gastropod crown group Euthyneura. However, several molecular studies in particular have questioned the monophyly of both the Opisthobranchia (e.g., Thollesson 1999; Grande *et al.* 2004b) and the Pulmonata (e.g., Grande *et al.* 2004a). Consequently the precise nature of the relationship is currently unclear. There is rather stronger support for the monophyly of the Euthyneura as a whole, both from molecules and from morphology (discussed subsequently).

MORPHOLOGY

The early classification of pulmonates was based on gross morphological structures, but recently the central nervous system (CNS) has played an increasingly important role in our understanding of pulmonate relationships. Bargmann (1930) was the first to use variations in the pulmonate visceral nerve chain as a basis for phylogenetic understanding, but concentration of the ganglia has probably occurred independently several times and consequently is a poor phylogenetic character, especially at older, deeper levels in the evolutionary tree (Bishop 1978; Haszprunar and Huber 1990). An important breakthrough came with the work of Mol (1967), who concentrated on the pulmonate cerebral ganglion and, in particular, a unique pulmonate structure, the procerebrum. His detailed study produced an explicit phylogenetic hypothesis, though his methodology and conclusions have been criticized (Bishop 1978). Haszprunar and Huber (1990) reviewed a range of pulmonate nervous systems. They argued that there was a correlation between neural and respiratory morphology and, on the basis of characters such as the size of the cells in the procerebrum, the degree of proximity of the pleural, pedal and cerebral ganglia, and the presence or absence of an osphradium and pallial ciliary tracts (or raphes), recognized three higher groupings of pulmonates, including a new order Eupulmonata for the Trimusculidae, Ellobiidae *sensu lato,* and Stylommatophora.

Two phylogenetic trees have recently been published that have a reasonably comprehensive coverage of pulmonate families, are based on a broad range of morphological characters, and use modern cladistic methodology. Dayrat and Tillier (2002) analyzed a suite of 77 morphological characters for a large group of Euthyneura, including 20 pulmonate families. Several analyses were undertaken, the results of which differed widely, suggesting that the data were inconsistent and included a high proportion of homoplasic character states. They presented a tree (Figure 15.2A) showing the relationships that could be inferred from these analyses, together with those synapomorphies that were considered to be unambiguous. This supports a monophyletic Pulmonata including all traditional taxa except the Glacidorbidae (which appears outside the Pulmonata with the Valvatidae and *Rissoella*). The Stylommatophora are monophyletic and form a clade with the vaginulid and onchidiid slugs (the Geophila). Surprisingly, their tree offers no support for the freshwater Hygrophila. Monophyly of the Pulmonata is supported by several characters: acquisition of a pneumostome and well-developed pulmonary blood vessels and the presence of a procerebrum and medio-dorsal bodies in the CNS. These characters can be associated with a life at least partly outside the sea. The Geophila were characterized by eyes located on the tips of the posterior (cephalic) tentacles (Figure 15.2), acquisition of a long pedal gland located on the floor of the visceral cavity, and an unpaired jaw. Again, adaptations of the eyes and of the pedal gland, which is associated with mucus production during locomotion (Luchtel and Deyrup-Olsen 2001), as well as of the respiratory system, can be linked to a terrestrial mode of life.

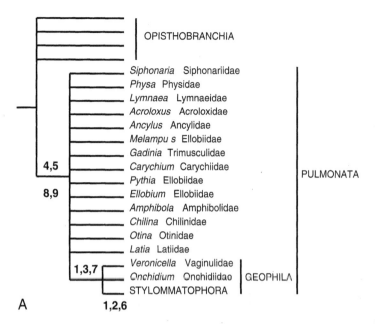

A

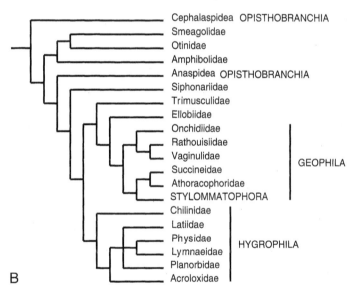

B

FIGURE 15.2. Morphological phylogenies. (A) Phylogeny proposed by Dayrat and Tillier (2002), based on a series of parsimony analyses of 77 characters. Characters mapped on tree as follows: 1. Long anterior pedal gland lying free on the floor of visceral cavity (Geophila) or under a membrane (Stylommatophora); 2. retractile cephalic tentacles; 3. eyes at tip of cephalic tentacles; 4. pallial cavity with pneumostome; 5. pulmonary vessels on suprapallium; 6. secondary ureter; 7. unpaired dorsal jaw; 8. procerebrum; 9. medio-dorsal bodies. (B) Phylogeny of Barker (2001) based on a parsimony analysis of 72 characters.

Barker's (2001) parsimony tree is based on a matrix of 72 characters. It does not support pulmonate monophyly, as the opisthobranch Anaspidea fall between a basal clade, including the Smeagolidae, Otinidae, and Amphibolidae, and a second clade comprising the remaining pulmonates; the sister group of this entire clade is the Cephalaspidea (Figure 15.2B). A "Geophila" clade comprising the onchidiid, rathouisiid, and vaginulid slugs and the Stylommatophora is resolved, and there is also a freshwater "Hygrophila" clade encompassing the Physidae, Lymnaeidae, Planorbidae, and Acroloxidae and the aberrant Chilinidae and Latiidae. Synapomorphies of the Hygrophila are the suprapedal gland and a eusperm glycogen piece with tracts of glycogen granules around the axonome. The Geophila are defined by the loss of gizzard plates, eyes situated at the tips of retractile cephalic tentacles, and concentration of the CNS.

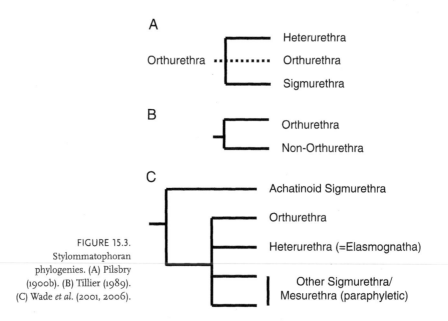

FIGURE 15.3.
Stylommatophoran
phylogenies. (A) Pilsbry
(1900b). (B) Tillier (1989).
(C) Wade *et al.* (2001, 2006).

The Stylommatophora are the by far the most speciose pulmonate group and can be defined, according to Dayrat and Tillier (2002), by three synapomorphies: the ability to retract (as opposed to contract) the cephalic tentacles, the presence of a membrane covering the pedal gland, and the acquisition of a secondary ureter. Their broad classification was first laid out by Pilsbry (1900a, b), who even illustrated a basic phylogeny (Figure 15.3A). He distinguished three primary divisions based on the morphology of the pallial system: the Orthurethra, with a straight ureter and ancestral to the other two groups; the Heterurethra, comprising the single family Succineidae with a so-called "heterurethrous" ureter, which runs along the kidney and then at right-angles along the rectum; and the Sigmurethra, with an S-shaped ureter and including all the remaining families. Later a fourth group of equal status, the Mesurethra, was added for four families lacking a ureter (Baker 1955). This classic Pilsbry-Baker system has only recently begun to be questioned. A useful comparison of classifications of stylommatophoran families is given by Emberton *et al.* (1990).

Despite their diversity, there have been few serious attempts at a morphology-based cladistic analysis of the Stylommatophora. Tillier (1989) examined the nervous, pallial, and alimentary systems of a wide range of families but undertook his analysis using an algorithm (Delattre 1988) that is phenetic rather than cladistic (Emberton and Tillier 1995). His principal finding was a basal split between the Orthurethra and all other stylommatophorans (Figure 15.3B). The large "nonorthurethran" clade was in turn divided into two new suborders, the Dolichonephra and the Brachynephra, based on the relative chronology of kidney shortening and ureteric-tube formation. However, only the basal Orthurethra/non-Orthurethra dichotomy survived a proper cladistic re-evaluation (Emberton and Tillier 1995). Barker (2001) undertook a parsimony analysis based on 57 characters, but the results of his analysis are often bizarre, showing, for example, the Orthurethra split into two well-separated groups: one the sister group of the Acavidae and the other in a clade that includes the Sphincterochilidae and Urocoptidae. A second tree was produced with the *a priori* constraint that it should be compatible with the molecular phylogeny of Wade *et al.* (2001) at superfamily/infraorder level. The elasmognaths, comprising the families Succineidae and Athoracophoridae, fell at the base of both Barker's trees.

THE IMPACT OF MOLECULES

Until very recently, only ribosomal genes had been used to examine euthyneuran relationships in any detail, and no study has been directed specifically at the Pulmonata as a whole. As a consequence of the latter, several key pulmonate taxa have been omitted in every case. Wade and Mordan (2000) tested the monophyly of the Stylommatophora using part of the ribosomal gene cluster (5.8S, ITS2, 28S). Their analysis also included ellobiids, carychiids, siphonariids, vaginulids, and rathouisiids, as well as opisthobranchs and a caenogastropod outgroup (Figure 15.4B). It confirmed monophyly of the relatively few included basal pulmonate families, as well as of the Stylommatophora. However, the only other robust structure within the pulmonate clade was the positioning of the Siphonariidae, which fell at its base. Dayrat *et al.* (2001) used partial 28S sequences of 32 species of Euthyneura, including 14 pulmonates, and presented the results of a parsimony analysis (Figure 15.4A). Few clades were well supported (bootstrap values >61%), but of particular interest was a Hygrophila clade comprising *Lymnaea* and *Chilina*. Neither the Pulmonata nor the Geophila were supported, but nor were they refuted as the nodes that caused them to be paraphyletic had low bootstrap values. Yoon and Kim (2000) used nearly complete 18S data to look at euthyneuran relationships. Their analyses supported monophyly of the Stylommatophora and Systellommatophora (in this case the families Onchidiidae and Vaginulidae), but not of the Pulmonata or Geophila (Figure 15.4D). Similar results were obtained by Dutra-Clarke *et al.* (2001) using complete 18S sequences from many of the same euthyneuran taxa, but with a different outgroup (Figure 15.4E).

Recently the use of mitochondrial genes alone by Grande *et al.* (2004a) has yielded some challenging results that question many of our preconceptions. Trees incorporating an ellobiid,

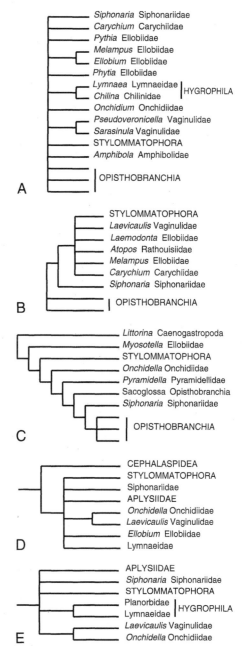

FIGURE 15.4. Pulmonate phylogenies. (A) Maximum-parsimony analysis using partial 28S rRNA sequences, from Dayrat *et al.* (2001). (B) Tree based on partial 28S, ITS2, and 5.8S rRNA sequences, using neighbor-joining, Fitch-Margoliash, maximum likelihood, and maximum parsimony, from Wade and Mordan (2000). (C) Bayesian tree inferred from nucleotide sequences of the mitochondrial *rrnL* gene and deduced amino acid sequences of the mitochondrial *cox 1* gene, from Grande *et al.* (2004a). (D) Maximum-parsimony tree based on nearly complete 18S rDNA sequences, from Yoon and Kim (2000). (E) Tree from parsimony analysis of 18S rRNA sequences, from Dutra-Clarke *et al.* (2001).

a siphonariid, an onchidiid, and several sty-lommatophorans as well as a large number of opisthobranchs, suggest that the pulmonates are not monophyletic because of the occurrence of *Siphonaria* within an otherwise monophy-letic Opisthobranchia and because the remain-ing pulmonates form a paraphyletic stem group from which the opisthobranchs evolved (Figure 15.4C). Their results do, however, sup-port monophyly of the Euthyneura. The use of mitochondrial genes for resolving such deep-level evolutionary relationships is, however, questionable, and it is important that their find-ings are confirmed using conventional deep-level nuclear markers.

CONSENSUS

Rather than adopt a total-evidence approach, Dayrat *et al.* (2001) attempted to produce a con-sensus tree for the Euthyneura by combining their own molecular and morphological trees (Dayrat and Tillier 2002) using a technique called combinable component consensus (Bremer 1990). Their resulting consensus tree sup-ported monophyly of the Pulmonata, Geophila, Stylommatophora, and Hygrophila (including *Chilina*), but morphological synapomorphies could be found only for the first three of these (Figure 15.5A). Though Nordsieck (1992) had proposed the "nephridial ureter" and "anterior gizzard" as synapomorphies of the Hygrophila, Dayrat and Tillier (2002) showed that they were not, suggesting instead that the "early devel-opment of the kidney" might be (Dayrat *et al.* 2001). The relationships of the various genera of Ellobiidae *sensu lato* were unresolved, as were the affinities of the Amphibolidae, Trimusculidae, Otinidae, and Smeagolidae. Subsequently Dayrat and Tillier (2003) utilized trees from a further five molecular studies (Thollesson 1999; Wollsheid and Wägele 1999; Wade and Mordan 2000; Wade *et al.* 2001; Dutra-Clarke *et al.* 2001) to produce a revised consensus tree (Figure 15.5B). In this the Pulmonata, Hygrophila, Geophila, and Stylommatophora were again resolved, but additionally the Eupulmonata (Geophila + Ellobioidea) emerged as a clade but unsupported by morphological characters. Ellobioids were not shown to be monophyletic, and there were again a number of basal pulmonate groups (Siphonariidae; Amphibolidae; Hygrophila) whose relationships remained uncertain.

THE STYLOMMATOPHORA

The most comprehensive molecular phylog-enies of the Stylommatophora are those of Wade *et al.* (2001, 2006), the second of which used partial 5.8S, ITS-2, and 28S sequences from over 144 genera in 61 families. It resulted in a fully supported stylommatophoran clade, with an achatinoid group, comprising the families Achatinidae, Subulinidae, Coeliaxidae, Ferussaciidae, and Streptaxidae, falling at its base (Figure 15.3C). This achatinoid clade had very high support (98% bootstrap) and closely resembles the Achatinoidea *sensu* Tillier (1989), which he defined mainly on the basis of several characters of the CNS. This basal dichotomy between achatinoids and all other stylomma-tophoran taxa was an unexpected finding and is controversial for two reasons: the level of bootstrap support for the clade containing the remaining families, the so-called "nonach-atinoid" clade, was rather low (63%), and its unity was difficult to justify on morphological grounds. However, this nonachatinoid clade was consistently recovered by a variety of tree-building techniques, and both this and the ach-atinoid clade showed posterior probabilities of 1.0 in a Bayesian analysis. Deep-level relation-ships within the "nonachatinoid" clade are, however, very unclear, and this was attributed to either a phase of explosive radiation or pos-sibly a failure of resolution by the genes. Tillier *et al.* (1996) favored the former explanation for the Stylommatophora, arguing that the lack of resolution was real, as both fossil (Zilch 1959–1960; Bandel 1991, 1997) and molecular stud-ies represented independent data sets that led to the same conclusion. The Succineidae have previously been thought by some to be opistho-branchs (e.g., Rigby 1965), but were fully sup-ported as the sister group of the athoracophorid slugs, as earlier shown by Tillier *et al.* (1996).

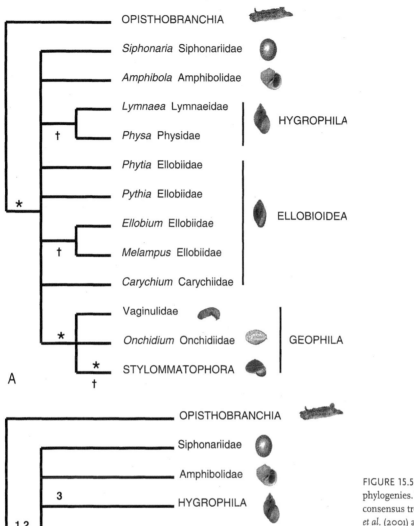

A

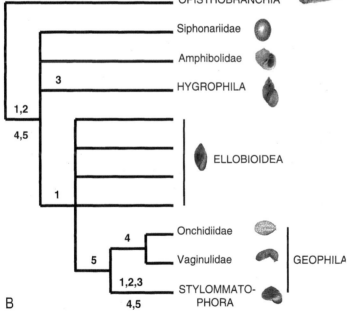

B

FIGURE 15.5. Consensus phylogenies. (A) Bremer consensus tree based on Dayrat et al. (2001) and Dayrat and Tillier (2002), from Dayrat et al. (2001); † signifies support from molecular data; * signifies support from morphology. (B) Bremer consensus tree from Dayrat and Tillier (2003), based on support at individual nodes from (1) Wade and Mordan (2000), (2) Wade et al. (2001), (3) Dayrat et al. (2001), (4) Dutra-Clarke et al. (2001), and (5) Dayrat and Tillier (2002); support for individual nodes is expressed by the respective numbers.

However, although this elasmognath clade appeared within the main nonachatinoid assemblage, it lay on a relatively long branch, perhaps rendering its placement problematic. Importantly, there was no suggestion that the supposedly primitive Orthurethra were a stem group—rather, they occupied a position equivalent to that of some of the larger superfamilies, such as the Limacoidea or Helicoidea. This is of particular interest in view of the traditional orthodoxy that the Orthurethra form the paraphyletic stem group of the Stylommatophora (e.g., Pilsbry 1900b; Nordsieck 1985) and suggests that the orthurethrous pallial system is relatively advanced. Indeed, if the tree topology of Wade et al. (2001, 2006) is broadly correct, it questions much of the Pilsbry-Baker system.

Dutra-Clarke et al. (2001) set out specifically to test the phylogenetic relationships of the Succineidae and Athoracophoridae using complete sequences of the 18S rRNA gene. They concluded that the athoracophorid slugs were an ingroup of a nonmonophyletic Succineidae and that the resulting elasmognath clade "was positioned among the derived stylommatophorans." Yoon and Kim (2000), using the same gene, similarly found the Elasmognatha fell within the Stylommatophora. Grande et al. (2004a) used mitochondrial genes but only incorporated taxa from three stylommatophoran superfamilies; they found the Achatinoidea to be basal to the Helicoidea and Clausilioidea, in agreement with the analyses of Wade et al. (2001, 2006).

At shallower levels in the tree the situation is somewhat clearer. Wade et al. (2001, 2006) found that many of the conventional stylommatophoran superfamilial and even infraordinal groups are monophyletic. The trees provided strong support not only for the Achatinoidea and Elasmognatha (see previous paragraphs) but also for the Helicoidea, Limacoidea, Clausilioidea, and Orthurethra. The trees also suggested that several important families such as the Camaenidae, Subulinidae, Helicarionidae, and Achatinidae are not monophyletic, but in general it supported, or at least did not contradict, the monophyly of many other recognized families.

ADAPTIVE RADIATIONS

The key to successful invasion of nonmarine habitats is regulation of the salt and water composition of body fluids. Pulmonates have achieved this to such a degree that they have now radiated to all areas except the most extreme polar regions and deserts. A major adaptive pulmonate feature, from which they derive their name, is the enclosed, heavily vascularized, air-breathing lung found in most families, which is homologous with the pallial cavity (Ruthensteiner 1997). The contractile pneumostome through which air is exchanged can be considered an apomorphy of the group. In addition to its respiratory function, the lung acts as a water reservoir. However, many of the adaptations that enable land snails to survive even severe desert conditions are behavioral or physiological; in both land and freshwater pulmonates, the latter relate largely to the excretory system. The move to brackish water by ancestral marine groups presumably demanded increased osmoregulatory ability; marine otinids and ellobiids, for example, lack a ureter. Freshwater pulmonates produce copious hyposmotic urine, with the heart being the main site of filtration, while the kidney and mesodermal ureter are specialized for the resorption of ions. The capacity of hygrophilids to excrete uric acid perhaps suggests a terrestrial or amphibious phase during their evolution (Andrews 1988). Most Stylommatophora additionally develop an ectodermal "secondary" ureter, which is important in water resorption. It is uncertain whether or not veronicellid and rathouisiid slugs also have a true secondary ureter. Many pulmonates also show a greatly increased tolerance of desiccation: the ellobiid Melampus, for example, can survive a loss of almost 80% of its water (Price 1980), and the eggs of some stylommatophoran slugs are even more tolerant of desiccation than this. In the only groups retaining the operculum in adults (the amphibolids and glacidorbids), that structure can greatly reduce water loss from the aperture during periods of drought. The calcareous or mucous epiphragmata secreted by many stylommatophoran snails to cover the aperture do not significantly prevent water loss, but seem to provide a physical barrier against predators. It appears to be the exposed mantle skin of the estivating snail that prevents desiccation by becoming almost as impervious as the shell itself (Newell and Appleton 1979). Many estivating land snails seal the shell aperture to flat surfaces such as rocks or trees, further reducing the potential for water loss. Some desert snails have additionally become extremely heat tolerant, surviving body temperatures of up to 50 °C.

Among aquatic groups, several have responded to intertidal or fast-flowing freshwater environments by adopting a limpet shape: the marine trimusculids and siphonariids and freshwater ancylids and acroloxids have apparently each evolved their limpet-like shells quite independently. Numerous unrelated pulmonate clades show reduction or complete loss of the shell, becoming semislugs or slugs. This makes them more vulnerable to predation and desiccation but does allow increased mobility and the ability to burrow and reduces dependence on environmental calcium. The minute smeagolids live interstitially in littoral gravel or coarse sand. Most onchidiid slugs are marine, but the few terrestrial forms show no obvious morphological adaptations to life on land (Tillier 1984a). A secondary ureter is highly developed in the sigmurethran and heterurethran Stylommatophora, and it is probably this factor that has allowed only these groups of stylommatophorans to evolve into slugs. There are no known freshwater pulmonate slugs, perhaps because of the high osmotic burden they would have to endure, although this problem appears to have been overcome by the freshwater acochlidioidian opisthobranchs.

Some of the best apomorphies defining the Pulmonata are found in the nervous system and can be related to life outside the sea. The procerebrum has direct nervous connections with the cephalic tentacles and, according to Cooke and Gelperin (2001), is the major central site of olfactory information processing in terrestrial forms. The medio-dorsal bodies appear to act on the development of both male and female cells in the gonad (Gomot de Vaufleury 2001), probably under the control of photoperiod. In onchidiids and the Geophila, the eyes are situated on the tips of the cephalic tentacles, where presumably they can be used to greater effect. The principal olfactory organ is situated just below the eye. Additionally, Geophila develop a second, shorter pair of tentacles with both a tactile and olfactory capability (Chase 2001), which probe the ground in front of the crawling animal and have been implicated in trail-following (Cook 2001). It is uncertain whether the paired papillae found on the head of ellobiids are homologous. Behavioral adaptations are exceedingly important in reducing water loss in terrestrial forms, and most inhabit sheltered environments. Desert snails estivate either buried, in shade or in crevices, or well above ground level, where the air is cooler; they may remain thus for most of the year, or even for several years, becoming active only to feed and reproduce when it rains (Yom-Tov 1971; Schmidt-Nielsen et al. 1971).

Both the opisthobranchs and pulmonates are hermaphrodites with internal fertilization, an evident preadaptation for life on the land. Amphibolids, siphonariids, many onchidiids, and a few ellobiids such as Melampus are wedded to the sea because they retain a free-living aquatic larval stage, but in most opisthobranchs and pulmonates indirect development takes place within the egg. Cross-fertilization is the norm in pulmonates, though many groups are capable of selfing. This ability, along with ovoviviparity, characteristic of several stylommatophoran families and the freshwater glacidorbids, is advantageous in the successful invasion of isolated environments. Courtship can be highly elaborate, and in a number of stylommatophoran families "love darts" are exchanged, which carry a chemical that stimulates the partner's reproductive system to become more receptive to sperm (Koene and Chase 1998).

Locomotion on land presents its own particular problems. Only one ellobiid, Carychium, is fully terrestrial and has a pedal gland to aid locomotion, though Dayrat and Tillier (2002) regarded it as plesiomorphic. Pedal glands appear to have arisen independently in the Geophila, where they represent an apomorphy of the group.

Unlike the opisthobranchs, most pulmonates are predominantly detritivores, fungivores or herbivores, though some families, particularly in the Stylommatophora, have become predatory, with accompanying modifications to the alimentary system—the jaw is often lost, the radular teeth become sharp and elongated, and the gastric crop becomes enlarged.

GAPS IN KNOWLEDGE

The application of modern cladistic techniques, whether based on morphology or on molecules, has so far failed to provide the hoped-for resolution of pulmonate phylogeny. It is still uncertain whether Pulmonata as currently conceived is monophyletic, or what its sister group is. Nor do we understand the deep-level relationships of major component groups of Pulmonata or of its two principal nonmarine clades: the Stylommatophora and Hygrophila. This lack may be due to the use of inappropriate genes or to periods of explosive cladogenesis occurring as new environments or novel morphological breakthroughs appeared. Additionally, the various minor groups of pulmonates, although relatively insignificant in terms of species diversity, have largely been neglected, even though they are of considerable potential in trying to establish the history of pulmonate evolution and, in particular, details of the major invasions of land and freshwater environments. A further limitation of many morphology-based studies is that we still have a relatively poor understanding of the true homologies of many key anatomical characters within the pulmonates.

In their review of euthyneuran relationships, Dayrat and Tillier (2003) made the important point that we should not fear lack of resolution as a result of tree-building methodology, as it may indeed accurately reflect the true pattern of evolution, rather than simply represent a gap in our understanding. For example, a rapid diversification of the Stylommatophora in the very early Cenozoic, as evidenced by the first appearance at that time of the majority of Recent families in the fossil record (Tillier et al. 1996) and possibly also apparent in areas of their molecular phylogeny (Wade et al. 2001, 2006), might well be expected in view of the major extinction of terrestrial biota which occurred at the Cretaceous/Cenozoic Tertiary (K/T) boundary some 65 Mya (Benton 1995; Vajda et al. 2001).

FUTURE STUDIES NEEDED

A persistent theme throughout this review has been the inadequacy of taxon sampling. To date, studies have been directed toward either the Euthyneura as a whole (Yoon and Kim 2000; Dayrat et al. 2001; Grande et al. 2004a) or to a subgroup such as the Stylommatophora (Tillier et al. 1996; Wade and Mordan 2000; Wade et al. 2001, 2006; Dutra-Clarke et al. 2001). What is now required is a program of sampling targeted specifically at resolution of deep-level pulmonate relationships, encompassing representatives of all the principal putative pulmonate groups. It is especially important that any such survey include enigmatic taxa such as the Glacidorbidae, and it is self-evident that a range of key opisthobranch and lower heterobranch groups should also be represented.

Although most of the main "basal" family-level pulmonate groups appear to be monophyletic, the ellobioids probably are not, and, as they include a variety of coastal marine and even fully terrestrial taxa, it is particularly important that they be comprehensively sampled. At higher levels of the tree, basal relationships within the main terrestrial (Stylommatophora) and freshwater (Hygrophila) lineages urgently need to be resolved, and, to have real value, coverage must be comprehensive at the family level. As in the case the Pulmonata as a whole, it is likely that some of the seemingly minor families, such as the terrestrial Aillyidae or the freshwater Lancidae or Latiidae, might hold important keys to the interrelationships of the more speciose groups.

There already exists an extensive literature on pulmonate anatomy, but it is vital that homologies are correctly interpreted, where possible based on developmental and ultrastructural studies. The latter have proved especially fruitful in the field of comparative sperm morphology (e.g., Healy 1996), but there is considerable potential in other areas such as the heart-kidney complex (Delhaye and Bouillon 1972) and various components of the reproductive system (e.g., Hodgson 1996). Computer-based 3D reconstruction techniques also need to be more fully exploited to unravel complex anatomical regions such as the fertilization pouch–spermatheca complex (Tompa 1984). Although it seems improbable that many major breakthroughs remain to be made from an anatomical approach alone, it is likely that the results of molecular studies will prompt a reassessment of the comparative morphology of various organ systems in an attempt to corroborate new molecular hypotheses of relationship. The mapping of anatomical or behavioral characters onto the stylommatophoran molecular trees has already proved highly instructive (Roth 2001; Davison *et al.* 2005).

Whereas analyses based on the same genes have tended to yield broadly similar results, important differences have shown up in the trees produced using different genes, and it is unlikely that better taxon sampling will, of itself, reduce the level of conflict. It is therefore highly desirable that future molecular analyses utilize several different genes in building phylogenetic trees. The recent study by Grande *et al.* (2004a), using various mitochondrial genes, presents a radically different view of pulmonate evolution from that indicated by the rRNA gene cluster. Instead of supporting a broadly monophyletic Pulmonata as the sister taxon of a paraphyletic Opisthobranchia (e.g., Thollesson 1999), it suggests that much of the radiation of the pulmonates, including the origin of the Stylommatophora, predates that of an essentially monophyletic Opisthobranchia. This has profound implications for our, admittedly sketchy, understanding of the invasion of nonmarine habitats by the early pulmonate stock. Until we can base an interpretation on sound phylogenetic information, we must continue to rely heavily on other evidence, such as that provided by comparative ecophysiological studies (Little 1983).

Rare genomic changes (RGCs) offer a possible solution to the problem of conflicting trees derived from sequence data (Rokas and Holland 2000). Mitochondrial gene order is unusually variable in heterobranch gastropods (Kurabayashi and Ueshima 2000), and there is already some published data demonstrating the phylogenetic potential of RGCs in the Euthyneura (Grande *et al.* 2004a; see also Chapters 14 and 17). Differences in the secondary structure of genes such as mitochondrial large-subunit rRNA might also provide useful molecular markers (Lydeard *et al.* 2002).

The fossil record of pulmonates adds a further level of confusion. Some seemingly primitive groups appear late in the fossil record (Table 15.1), and the interpretation of certain fossils is highly questionable. What is needed in this case is a thorough reinterpretation of all the available pulmonate fossil data: in conjunction with a robust phylogenetic hypothesis for the pulmonates as a whole, the results might finally lead to a real understanding of the "mode and tempo" of pulmonate evolution.

REFERENCES

Andrews, E. B. 1988. Excretory systems of molluscs. In *The Mollusca*. Edited by K. M. Wilbur. Vol. 11. San Diego: Academic Press, pp. 381–448.

Bachmann, H, Keller, M., Porret, N., Dietz, H. and Edwards, P. J. 2005. The effect of slug grazing on vegetation development and plant species diversity in an experimental grassland. *Functional Ecology* 19: 291–298.

Baker, H. B. 1955. Heterurethrous and aulacopod. *The Nautilus* 68: 109–112.

Bandel, K. 1991. Gastropods from brackish and freshwater of the Jurassic–Cretaceous transition (a systematic reevaluation). *Berliner Geowissenschaftliche Abhandlungen* 134: 9–55.

———. 1997. Higher classification and pattern or evolution of the Gastropoda. *Courier Forschungsinstitut Senckenberg* 201: 57–81.

Bargmann, H. E. 1930. The morphology of the central nervous system in the Gastropoda Pulmonata. *Journal of the Zoological Society of London* 37: 1–59.

Barker, G. M. 2001. Gastropods on land: phylogeny, diversity and adaptive morphology. In *The Biology of Terrestrial Molluscs*. Edited by G. M. Barker. Wallingford, UK: CABI Publishing, pp 1–146.

———. 2002. *Molluscs as Crop Pests*. Wallingford, UK: CABI Publishing.

Benton, M. J. 1995. Diversification and extinction in the history of life. *Science* 268: 52–58.

Bishop, M. 1978. The value of the pulmonate central nervous system in phylogenetic hypothesis. *Journal of Molluscan Studies* 44: 116–119.

Bremer, K. 1990. Combinable component consensus. *Cladistics* 6:369–372.

Brown, D. S. 1978. Pulmonate molluscs as intermediate hosts for digenetic trematodes. In *Pulmonates*, 2A. Edited by V. Fretter and J. Peake. London, Academic Press: pp. 287–333.

Chase, R. 2001. Sensory organs and the nervous system. In *The Biology of Terrestrial Molluscs*. Edited by G. M. Barker. Wallingford, UK: CABI Publishing, pp. 179–211.

Clarke, B, Arthur, W., Horsley, D. T., and Parkin, D. T. 1978. Genetic variation and natural selection in pulmonate molluscs. In *Pulmonates*, 2A. Edited by V. Fretter and J. Peake. London: Academic Press, pp. 219–270.

Cook, A. 2001. Behavioural ecology: on doing the right thing, in the right place at the right time. In *The Biology of Terrestrial Molluscs*. Edited by G. M. Barker. Wallingford, UK: CABI Publishing, pp. 447–487.

Cooke, I. R., and Gelperin, A. 2001. *In vivo* recordings of spontaneous and odor-modulated dynamics in the olfactory lobe. *Journal of Neurobiology* 46: 126–141.

Cowie, R. H. 1992. Evolution and extinction of Partulidae, endemic Pacific land snails. *Philosophical Transactions of the Royal Society of London Series B: Biological Sciences* 335: 167–191.

Cuvier, G. 1817. *Le Régne Animal*. Paris: Libraire Déterville.

Davison, A., Wade, C. M., Mordan, P. B., and Chiba, S. 2005. Sex and darts in slugs and snails (Mollusca: Gastropoda: Stylommatophora). *Journal of Zoology, London* 267: 329–338.

Dayrat, B., Tillier, A., Lecointre, G., and Tillier, S. 2001. New clades of euthyneuran gastropods (Mollusca) from 28S rRNA sequences. *Molecular Phylogenetics and Evolution* 19: 225–235.

Dayrat, B., and Tillier, S. 2002. Evolutionary relationships of the euthyneuran gastropods (Mollusca): a cladistic re-evaluation of morphological characters. *Zoological Journal of the Linnean Society* 135: 403–470.

———. 2003. Goals and limits of phylogenetics. The euthyneuran gastropods. In *Molecular Systematics and Phylogeography of Mollusks*. Edited by C. Lydeard and D. R. Lindberg. Washington, DC: Smithsonan Books, pp. 161–183.

Delattre, P. 1988. Sur la recherché des filations en phylogénèse. *Revue International de Systématique* 2: 479–504.

Delhaye, W., and Bouillon, J. 1972. L'évolution et adaptation de l'organe excréteur chez les gastéropodes pulmonées. III. Histophysiologie comparée du rene chez les soléolifères et conclusions générales pour tous les stylommatophores. *Bulletin Biologique* 106: 295–314.

Dutra-Clarke, A. V. C., Williams, C., Dickstein, R., Kaufer, R., and Spotila, J. R. 2001. Inferences on the phylogenetic relationships of Succineidae (Mollusca, Pulmonata) based on 18S rRNA gene. *Malacologia* 43: 223–236.

Emberton, K. C., Kuncio, G. S., Davis, G. M., Phillips, S. M., Monderewicz, K. M., and Guo, Y. H. 1990. Comparison of recent classifications of stylommatophoran land-snail families, and evaluation of large-ribosomal-RNA sequencing for their phylogenetics. *Malacologia* 31: 327–352.

Emberton, K. C., and Tillier, S. 1995.Clarification and evaluation of Tillier's (1989) stylommatophoran monograph. *Malacologia* 36: 203–208.

Férussac, A. E. J. de. 1819. *Histoire naturelle des pulmonés sans opercules*. Paris: Bertrand.

———. 1822. *Tableaux systématiques des animeaux mollusques*. Paris: Bertrand.

Golding, R. E., Ponder, W. F. and Byrne, M. 2007. Taxonomy and anatomy of Amphiboloidea (Gastropoda: Heterobranchia: Archaeopulmonata). *Zootaxa* 1476: 1–50.

Gomot de Vaufleury, A. 2001. Regulation of growth and reproduction. In *The Biology of Terrestrial Molluscs*. Edited by G. M.Barker. Wallingford, UK: CABI Publishing, pp. 331–355.

Grande, C., Templado, J., Cervera, J. L., and Zardoya, R. 2004a. Molecular phylogeny of Euthyneura (Mollusca:Gastropoda). *Molecular Biology and Evolution* 21: 303–313.

———. 2004b. Phylogenetic relationships among Opisthobranchia (Mollusca: Gastropoda) based on mitochondrial *cox1*, *trnV*, and *rrnL* genes. *Molecular Phylogenetics and Evolution* 33: 378–388.

Harbeck, K. 1996. Die evolution der Archaeopulmonata. *Zoologische Verhandelingen* 305: 1–133.

Haszprunar, G., and Huber, G. 1990. On the central nervous system of Smeagolidae and Rhodipidae, two families questionably allied with the

Gymnomorpha (Gastropoda:Euthyneura). *Journal of Zoology, London* 220: 185–199.

Healy, J. M. 1996. Molluscan sperm ultrastructure: correlation with taxonomic units within the Gastropoda, Cephalopoda and Bivalvia. In *Origin and Evolutionary Radiation of the Mollusca*. Edited by J. D. Taylor. Oxford: Oxford University Press, pp. 99–113.

Hodgson, A. 1996. The structure of the seminal vesicle region of the hermaphrodite duct of some pulmonate snails. In *Molluscan Reproduction*. Edited by N. Runham and W. H. Heard. *Malacological Review* Suppl. 6: 89–99.

Hrubesch, K. 1965. Die santone Gosau-Landschneckenfauna von Glanegg bei Salzburg. *Mitteilungen der Bayerischen Staatssammlung für Paläontologie und Historische Geologie* 5: 83–120.

Hubendick, B. 1978. Systematics and comparative morphology of the Basommatophora. In *Pulmonates*. 2A. Edited by V. Fretter and J. Peake. London: Academic Press, pp. 1–47.

Kliks, M. M., and Palumbo, N. E. 1992. Eosinophilic meningitis beyond the Pacific basin: the global dispersal of a peridomestic zoonosis caused by *Angiostrongylus cantonensis*, the nematode lungworm of rats. *Social Science Medicine* 34: 199–212.

Koene, J. M., and Chase, R. 1998. Changes in the reproductive system of the snail *Helix aspersa* caused by mucus from the love dart. *The Journal of Experimental Biology* 201: 2313–2319.

Kurabayashi, A., and Ueshima, R. 2000. Complete sequence of the mitochondrial DNA of the primitive opisthobranch gastropod *Pupa strigosa*: systematic implication of the genome organization. *Molecular Biology and Evolution* 17: 266–277.

Little, C. 1983. *The Colonisation of Land. Origins and Adaptations of Terrestrial Animals*. Cambridge, UK: Cambridge University Press.

Luchtel, D. L., and Deyrup-Olsen, I. 2001. Body wall: form and function. In *The Biology of Terrestrial Molluscs*. Edited by G. M. Barker. Wallingford, UK: CABI Publishing, pp. 147–178.

Lydeard, C., Holznagel, W. E., Ueshima, R., and Kurabayashi, A. 2002. Systematic implications of extreme loss or reduction of mitochondrial LSU rRNA helical-loop structures in gastropods. *Malacologia* 44: 349–352.

Martins, A. M. F. 1996. Anatomy and systematics of the Western Atlantic Ellobiidae (Gastropoda: Pulmonata). *Malacologia* 37: 163–332.

Mead, A. R. 1961. *The Giant African Snail: A Problem in Economic Malacology*. Chicago: University of Chicago Press.

Mol, J.-J. v. 1967. Étude morphologique et phylogénétique du ganglion cérébroïde des gastéropodes pulmonés (Mollusques). *Mémoires Académie Royale de Belgique, Classe des Sciences* 37: 1–168.

Newell, P. F., and Appleton, T. C. 1979. Aestivating snails—the physiology of water regulation in the mantle of the terrestrial pulmonate *Otala lactea*. *Malacologia* 18: 575–581.

Nordsieck, H. 1985. The system of the Stylommatophora (Gastropoda), with special regard to the systematic position of the Clausiliidae, I. Importance of the excretory and genital systems. *Archiv für Molluskenkunde* 116: 1–24.

———. 1992. Phylogeny and system of the Pulmonata (Gastropoda). *Archiv für Molluskenkunde* 121: 31–52.

Pickford, M. 1995. Fossil land snails of East Africa and their palaeoecological significance. *Journal of African Earth Sciences* 20: 167–226.

Pilsbry, H. A. 1900a. On the zoological position of *Achatinella* and *Partula*. *Proceedings of the Academy of Natural Sciences, Philadelphia* 52: 561–567.

———. 1900b. The genesis of mid-Pacific faunas. *Proceedings of the Academy of Natural Sciences, Philadelphia* 52: 568–575.

———. 1948. Land Mollusca of North America. *Academy of Natural Sciences Philadelphia Monographs* 3 II (2): 521–1113.

Ponder, W. F. 1986. Glacidorbidae (Glacidorbacea: Basommatophora), a new family and superfamily of operculate freshwater gastropods. *Zoological Journal of the Linnean Society* 87: 53–83.

Ponder, W. F., and Avern, G. J. 2000. The Glacidorbidae (Mollusca: Gastropoda: Heterobranchia) of Australia. *Records of the Australian Museum* 52: 307–353.

Price, C. H. 1980. Water relations and physiological ecology of the salt marsh snail, *Melampus bidentatus* Say. *Journal of Experimental Marine Biology and Ecology* 45: 51–67.

Rigby, J. E. 1965. *Succinea putris*: a terrestrial opisthobranch mollusc. *Proceedings of the Zoological Society of London* 144: 445–486.

Rokas, A., and Holland, P. W. H. 2000. Rare genomic changes as a tool for phylogenetics. *Trends in Ecology and Evolution* 15: 454–459.

Roth, B. 2001. Phylogeny of pneumostomal area morphology in terrestrial Pulmonata (Gastropoda). *The Nautilus* 115: 140–146.

Ruthensteiner, B. 1997. Homology of the pallial and pulmonary cavity of gastropods. *Journal of Molluscan Studies* 63: 353–367.

Salvini-Plawen, L. v. 1991. The status of the Rhodopidae (Gastropoda: Euthyneura). *Malacologia* 32: 301–311.

Salvini-Plawen, L. v., and Steiner, G. 1996. Synapomorphies and plesiomorphies in higher classification of the Mollusca. In *Origin and Evolutionary Radiation of the Mollusca*. Edited by J. D. Taylor. Oxford: Oxford University Press, pp. 29–51.

Schmidt-Nielsen, K., Taylor, C. R., and Shkolnik, A. 1971. Desert snails: problems of heat, water and food. *Journal of Experimental Biology* 55: 385–398.

Smith, B. J., and Stanisic, J. 1998. Pulmonata. In *Mollusca: The Southern Synthesis*. Fauna of Australia 5. Edited by P. L. Beesley, G. J. B. Ross, and A. Wells. Melbourne: CSIRO Publishing, pp. 1037–1125.

Solem, A., and Yochelson, E. L. 1979. North American Paleozoic land snails, with a summary of other Paleozoic non-marine snails. *U.S. Geological Survey Professional Paper* 1072: 1–42.

Thiele, J. 1931. Pulmonata. In *Handbuch der systematischen Weichtierkunde*. 2. Edited by J. Theile. Jena: Verlag von Gustav Fischer, pp. 461–734.

Thollesson, M. 1999. Phylogenetic analysis of Euthyneura (Gastropoda) by means of the 16S rRNA gene: use of a "fast" gene for "higher-level" phylogenies. *Proceedings of the Royal Society of London Series B: Biological Sciences* 266: 75–83.

Tillier, S. 1984a. A new mountain *Platevindex* from Philippine islands (Pulmonata: Onchidiidae). *Journal of Molluscan Studies* Suppl. 12A: 198–202.

———. 1984b. Relationships of the gymnomorph gastropods (Mollusca:Gastropoda). *Zoological Journal of the Linnean Society* 82: 345–362.

———. 1989. Comparative morphology, phylogeny and classification of land slugs and snails (Gastropoda: Pulmonata: Stylommatophora). *Malacologia* 30: 1–303.

Tillier, S., Masselot M., and Tillier, A. 1996. Phylogenetic relationships of the pulmonate gastropods from rRNA sequences, and tempo and age of the stylommatophoran radiation. In *Origin and Evolutionary Radiation of the Mollusca*. Edited by J. D. Taylor. Oxford: Oxford University Press, pp. 267–734.

Tillier, S., and Ponder, W. F. 1992. New species of *Smeagol* from Australia and New Zealand, with a discussion of the affinities of the genus. *Journal of Molluscan Studies* 58: 135–155.

Tompa, A. S. 1984. Land snails (Stylommatophora). In *The Mollusca*. Edited by K. M. Wilbur. Vol. 7. San Diego: Academic Press, pp. 47–140.

Tracey, S., Todd, J. A., and Erwin, D. H. 1993. Mollusca; Gastropoda. In *The Fossil Record*. Edited by M. J. Benton. London: Chapman and Hall, pp. 131–167.

Vajda, V., Raine, J. I., and Hollis, C. J. 2001. Indication of global deforestation at the Cretaceous-Tertiary boundary by New Zealand fern spike. *Science* 294: 1700–1702.

Wade, C. M., and Mordan, P. B. 2000. Evolution within the gastropod molluscs; using the ribosomal RNA gene-cluster as an indicator of phylogenetic relationships. *Journal of Molluscan Studies* 66: 565–570.

Wade, C. M., Mordan, P. B., and Clarke, B. 2001. A phylogeny of the land snails (Gastropoda: Pulmonata). *Proceedings of the Royal Society of London Series B: Biological Sciences* 268: 413–422.

Wade, C. M., Mordan P. B., and Naggs, F. 2006. Evolutionary relationships among the pulmonate land snails and slugs. *Biological Journal of the Linnean Society* 87: 593–610.

Wollsheild, E., and Wägele, H. 1999. Initial results on the molecular phylogeny of the Nudibranchia (Gastropoda, Opisthobranchia) based on 18S rDNA data. *Molecular Phylogenetics and Evolution* 13: 215–226.

Woodruff, D. S. 1978. Evolution and adaptive radiation of *Cerion*: a remarkably diverse group of West Indian land snails. *Malacologia* 17: 223–239.

Yom-Tov., Y. 1971. The biology of two desert snails *Trochoidea* (*Xerocrassa*) *seetzeni* and *Sphincterochila boissieri*. *Israel Journal of Zoology* 20:231–248.

Yoon, S. H., and Kim, W. 2000. Phylogeny of some gastropod mollusks derived from 18S rDNA sequences with emphasis on the Euthyneura. *The Nautilus* 114: 84–92.

Zilch, A. 1959–1960. Euthyneura. In *Handbuch der Paläozoologie*, Vol. 6 (2). Edited by O. Schindewolf. Berlin: Gebrüder Bornträger.

16

Molluscan Evolutionary Development

Andreas Wanninger, Demian Koop, Sharon Moshel-Lynch,
and Bernard M. Degnan

Recent phylogenetic analyses have led to a re-evaluation of metazoan relationships, resulting in an assemblage of a number of invertebrate phyla, such as annelids, sipunculans, and molluscs, known as Spiralia (Aguinaldo *et al.* 1997; de Rosa *et al.*, 1999; Halanych *et al.* 1995; Mallatt and Winchell 2002; Philippe *et al.* 2005). The taxa of this "superclade" share highly conserved cleavage patterns, cell lineages, and embryogenesis. The stereotypic cleavage pattern of spiralians enables the identification of homologous blastomeres and the elucidation of their cell fate (Verdonk and van den Biggelaar 1983). This has provided a useful tool for analyzing the relationships among and within spiralian taxa as well as revealing the ontogeny and cell lineage of some morphological structures such as the larval prototroch (Damen and Dictus 1994a, b, 1996; Dictus and Damen 1997; Henry *et al.* 2004; Maslakova *et al.* 2004). However, despite the highly conserved spiralian cleavage pattern and cell lineages, molluscs have one of the highest diversities of body plans of any metazoan group, ranging from vermiform taxa such as Solenogastres (= Neomeniomorpha) and Caudofoveata (= Chaetodermorpha) to highly modified conchiferans such as gastropods and cephalopods. In addition, there are cases of secondary reduction or loss of features such as the shell or radula, or even simplification of the whole body, resulting in a secondary worm-shaped appearance (e.g., some interstitial gastropods and shipworms (Teredinidae; Bivalvia)). The remarkable plasticity of the molluscan body plan renders this group unique for the study of genetic and morphological mechanisms underlying metazoan body plan diversification. Considering this, it is not surprising that new comparative data on gene expression profiles or morphogenetic events, for example, have recently become available for Mollusca. At the same time, the increasing amount of information for other metazoan phyla, especially those from other trochozoans such as polychaetes (e.g., Irvine and Martindale 2000; Peterson *et al.* 2000; Kulakova *et al.* 2002; Raible *et al.* 2005; Fröbius and Seaver 2006; Seaver and Kaneshige 2006), nemerteans (Maslakova *et al.* 2004), or sipunculans (Wanninger *et al.* 2005) provide excellent comparative data on gene functions, gene expression patterns, and modes of organogenesis. In addition, the application of new micro-morphological research tools such as fluorescence-coupled antibody staining and confocal laser scanning microscopy has allowed detailed reconstruction of key events in molluscan ontogeny such as the formation of larval and post-metamorphic nervous and muscle systems (e.g., Dickinson *et al.* 1999; Wanninger *et al.* 1999a, 2005; Friedrich *et al.* 2002; Page 2002; Voronezhskaya *et al.* 2002; Wanninger and Haszprunar 2002a, b, 2003; Dickinson and Croll 2003; Wanninger 2004, 2005; Kempf and Page 2005). Such comparative analyses across several molluscan classes have led to a recent revival in interest in higher-level phylogeny. Similarly, phylogenetic studies spanning the entire Lophotrochozoa have contributed to insights concerning the phylogenetic relationships of Mollusca to other phyla. In this review we mainly focus on the recent contributions of developmental studies to our understanding of the evolution of the diversity of molluscan body plans as well as the bearing of these findings on molluscan phylogenetics. In particular, we concentrate on data acquired by gene expression analyses and on the morphogenesis of the molluscan nervous system, musculature, and shells.

CLEAVAGE AND CELL LINEAGES

Molluscan (except cephalopod) cleavage results in the formation of four vegetal macromeres and four animal micromeres at the eight-cell stage. From the third cleavage onward the mitotic spindles form obliquely, and during successive cleavages the chirality alternates clockwise and counter-clockwise, generating the typical spiral pattern of cleavage. In addition, the vegetal macromeres give rise to successive tiers of animal micromeres. Cleavage results in an embryo divided into quadrants of blastomeres designated A, B, C, and D, which correspond to the left, ventral, right, and dorsal regions of the embryo, respectively. Cell lineages from these quadrants give rise to a highly conserved set of larval and adult morphological structures (van Dongen and Geilenkirchen 1974; Verdonk and van den Biggelaar 1983; Damen and Dictus 1994a, b, 1996; Dictus and Damen 1997).

The cell lineages of molluscan embryos have been studied in some detail, in particular as part of early embryological and developmental studies in the late nineteenth and early twentieth century (reviews in van den Biggelaar and Haszprunar 1996; Guralnick 2002; Nielsen 2004). Cell ablation experiments in the caenogastropod whelk *Ilyanassa obsoleta* (Clement 1962, 1967, 1976) and cell lineage studies in the scaphopod *Antalis* (as *Dentalium;* van Dongen and Geilenkirchen 1974; van Dongen 1976) revealed the contribution of micromere lineages to the normal development of larval structures. Recently there have been a number of studies that have utilized cell-labeling techniques to develop fate maps for several molluscs, including *Ilyanassa obsoleta* (Render 1997) and the patellogastropod limpet *Patella vulgata* (Dictus and Damen 1997) as well as the polyplacophoran *Chaetopleura apiculata* (Henry *et al.* 2004). These studies demonstrated that the first-quartet micromeres (1a–1d) contribute to the pretrochal ectoderm, which corresponds to the future "head" domain, as well as the prototroch, that is, the ciliated locomotory organ of the trochophore larva. Photoreceptors ("eyes") are formed from the descendants of 1a and 1c in virtually all spiralians studied, except the polyplacophoran, where they are derived from second-quartet micromeres (Henry *et al.* 2004). The second-quartet micromeres (2a–2d) also give rise to secondary trochoblasts as well as the majority of the posttrochal ectoderm (the shell field, mantle, foot, and stomodeum). The third-quartet micromeres (3a–3d) contribute to the ectomesoderm and the stomodeum. The macromeres (3A–D) form the endodermal structures as well as cells containing the yolk supply (Render 1997). In addition, the 3D cell gives rise to the mesentoblast, which forms a significant proportion of the mesodermal structures such as the heart and muscles and plays an important role in gastrulation.

THE ROLE OF THE D QUADRANT DURING EARLY DEVELOPMENT

The ability of isolated blastomeres to form tissues led to the hypothesis of an archetypal "mosaic" development in molluscs (Wilson 1904).

However, blastomere deletion experiments made it clear that inductive interactions are also involved in this process (Clement 1976; Cather and Verdonk 1979; van den Biggelaar and Guerrier 1979, 1983). In *Ilyanassa obsoleta*, while deletion of micromeres produced specific abnormalities in the larva, there was also strong evidence for regulation in molluscan development (Clement 1962, 1967, 1976). In some molluscs (e.g., *Ilyanassa obsoleta*), the first two cleavages are unequal, producing a cytoplasmic protrusion—the polar lobe—inherited by one of the blastomeres. This results in this blastomere being larger than the others, and it is specified to form the dorsal (D) quadrant. Deletions of the D macromere were performed in the gastropod *Ilyanassa* and the scaphopod *Antalis* (as *Dentalium*) with similar results (Clement 1962; Cather and Verdonk 1979). In these experiments, deletion of the polar lobe or the D macromere during the first two cleavages resulted in larvae that had everted stomodea and a disorganized velum and lacked eyes, foot, heart, intestine, and shell. Furthermore, experiments in which 3D induction had been prevented in the abalone *Haliotis* (Vetigastropoda) resulted in radialization of gene expression patterns in these embryos (Koop *et al.*, unpublished data). In the post-trochal ectoderm, preventing 3D induction does not prevent dorso-ventral patterning in at least parts of the vegetal ectoderm. This suggests that there is either 3D-independent induction or a regulatory process involved in the axial patterning of molluscs (Koop *et al.*, unpublished data; see also Freeman 2006).

These experiments demonstrated that the D quadrant acts as an inductive center and organizer of molluscan development. In equally cleaving gastropods such as *Patella* and *Haliotis*, one macromere at the 32-cell stage is induced by the overlying micromeres to become the D quadrant macromere, 3D (van den Biggelaar and Guerrier 1979; Arnolds *et al.* 1983; Boring 1989). While the process is not well studied in other molluscan classes, it appears that a similar process takes place in polyplacophorans (van den Biggelaar 1996). Inhibiting this interaction

results in larvae similar to those produced from polar lobe deletions (van den Biggelaar 1977; van den Biggelaar and Guerrier 1979; Arnolds *et al.* 1983; Martindale *et al.* 1985; Martindale 1986; Kuhtreiber *et al.* 1988; Boring 1989; Damen and Dictus 1996; Dictus and Damen 1997; Lambert and Nagy 2003). While the mechanism of D quadrant specification depends on whether a certain molluscan species is an equal or an unequal cleaver, the timing of the D macromere induction and its subsequent role in establishing the organization of the molluscan embryo appears to be the same (Clement 1962, 1976; Cather and Verdonk 1979; Martindale *et al.* 1985; Martindale 1986; Damen and Dictus 1996; Lambert and Nagy 2001, 2003). These studies showed that the D quadrant is required for the establishing of a dorso-ventral cleavage pattern, the organization of the molluscan embryo, and the formation of the endomesoderm.

MESODERM FORMATION

In molluscs, the mesoderm is derived from two embryonic sources. The endomesoderm is derived from the daughter cell of the 3D macromere, the mesentoblast 4d, which gives rise to much of the mesoderm including most of the musculature. The other type of mesoderm is the ectomesoderm, derived from ectodermal micromeres, usually 2b, 3a, and 3b derivatives. The ectomesoderm forms in the anterior portion of the embryo and trochophore larva, while the mesodermal bands are situated just underneath the foot anlage and the neuroectoderm (Hinman and Degnan 2002; Hinman *et al.* 2003; Le Gouar *et al.* 2004).

A number of genes commonly associated with mesodermal specification in vertebrates and in the endoderm of cnidarians (Martindale *et al.* 2004), including *forkhead*, *goosecoid*, and *twist*, have been characterized in the patellogastropod *Patella vulgata*, where they are expressed in the anterior ectomesoderm (Lartillot *et al.* 2002a; Nederbragt *et al.* 2002a). In the caenogastropod *Ilyanassa obsoleta*, however, the expression pattern of *twist* differs considerably from that in *Patella*, as it is expressed in the former species as a maternal gene in the germinal vesicle prior to first cleavage, with expression continuing throughout early cleavage showing no specificity (Moshel-Lynch and Collier, unpublished data). Later in development of *Ilyanassa*, *twist* becomes localized in derivatives of the ectomesoderm such as head, foot, and shell gland mesenchyme (Moshel-Lynch and Collier, unpublished data). At the veliger stage, *twist* is localized in both ecto-mesoderm and endomesoderm derivatives in *Ilyanassa*. It is seen in the heart and kidney of the larva, both derivatives of the D quadrant (Moshel-Lynch and Collier, unpublished data). Since *twist* expression differs in the two gastropods studied so far, its ancestral role in gastropod and molluscan development remains unclear.

Forkhead is first expressed following fifth cleavage and 3D specification, in the 3A, 3B, and 3C macromeres (Lartillot *et al.* 2002a). During early embryogenesis it is expressed in all endodermal lineage cells as well as in some of the ectodermal cell lineages that give rise to the ectomesoderm. In the trochophore of *Patella vulgata*, *forkhead* expression is maintained in the endoderm and ectomesoderm behind the stomodeum. Similarly, *goosecoid* is expressed in the 3a and 3b lineages during embryogenesis, and this expression is maintained in the cells anterior to the vegetal plate and in cells that form the stomodeum (Lartillot *et al.* 2002a). In the *P. vulgata* trochophore, *goosecoid* is expressed in the stomodeum and in the mesoderm posterior to it. The *Patella* homolog of *twist* is first expressed in the early trochophore and has an expression domain that corresponds to the 3a and 3b lineages in the anterior mesoderm (Nederbragt *et al.* 2002a).

Mox is a homeobox gene that was initially isolated from vertebrates, where it was found to be involved in mesoderm formation (Candia *et al.* 1992; Candia and Wright 1995). In the amphibian *Xenopus*, *Mox* is expressed in the mesoderm spatially and temporally immediately downstream of the site of mesoderm induction, namely, in the dorsal lip of the blastopore. *Mox*

expression has also been characterized in the abalone *Haliotis asinina* (Hinman and Degnan 2002), where it was found to play a conserved role in mesoderm and muscle differentiation. *Mox* is first expressed in the putative mesodermal bands of the trochophore larva of *H. asinina* and then in the mesoderm of the developing foot. In the *H. asinina* veliger, *Mox* expression is localized to the foot muscle, which develops into the pedal muscle of the adult following metamorphosis. To date this is the only gene expressed specifically in the mesodermal bands of the molluscan trochophore.

BRACHYURY EXPRESSION AND THE MAPK SIGNALING PATHWAY

Over the past several years various evolutionarily conserved regulatory genes have been isolated from molluscs (e.g., O'Brien and Degnan 2002a, b, 2003; Hinman *et al.* 2003; Lee *et al.* 2003). Lambert and Nagy (2001) were the first to demonstrate the localization of a regulatory pathway in 3D and its subsequent derivatives in *Ilyanassa obsoleta*. Subsequently, it has been demonstrated in both equally and unequally cleaving gastropods, polyplacophorans, and in the 4d cell in annelids (Lambert and Nagy 2003). The mitogen-activated protein kinase (MAPK) signaling cascade, known to regulate a number of cellular activities such as mitosis, transcription, gene expression, and cell differentiation in both vertebrates and invertebrates including molluscs (Iakovleva *et al.* 2006), is activated in the D quadrant, and inhibition of this pathway leads to a defective larva very similar to those that have had the polar lobe removed (see previous section). While the gene battery regulated by the MAPK pathway has yet to be determined, a *brachyury* ortholog in *Patella* was shown to be expressed in 3D and repressed in embryos where MAPK activation was inhibited (Lartillot *et al.* 2002b). *Brachyury* is an evolutionarily conserved gene that has a fundamental role in gastrulation in vertebrates (Schulte-Merker *et al.* 1994; Wilson *et al.* 1995; Melby *et al.* 1996; Wilson and Beddington 1997). In invertebrates, brachyury orthologs have been shown to have a similar conserved role (Arendt *et al.* 2001). Although *brachyury* expression has been implied in various organisms for mesoderm formation (Smith *et al.* 1991; Isaacs *et al.* 1994), in other organisms the role of *brachyury* seems to be important for morphogenic movements essential for normal development (Wilson *et al.* 1995; Melby *et al.* 1996; Wilson and Beddington 1997). Microarray analysis of MAPK-treated embryos of *Haliotis* showed a knockdown of genes known to be involved in cell conformation and movement, suggesting that the MAPK cascade is responsible for the transition from a radial to bilateral cleavage pattern in Mollusca (Koop *et al.*, unpublished). Since cell fate specification is dependent upon both maternal determinants and cell-cell interactions, establishing the correct cell orientation early in development is essential. It appears that MAPK and *brachyury* are essential for this task.

THE ROLE OF *ENGRAILED* DURING ORGANOGENESIS

The expression of the homeobox gene *engrailed*, known from a number of studies to be involved in invertebrate neurogenesis and in establishing cell-cell boundaries (e.g., Weisblat *et al.* 1980; Patel *et al.* 1989; Abzhanov and Kaufman 2000), has been studied in a variety of molluscs, including the gastropods *Ilyanassa* and *Patella* (Moshel *et al.* 1998; Nederbragt *et al.* 2002b), the polyplacophoran *Lepidochitona* (Jacobs *et al.* 2000), the bivalve *Transennella* (Jacobs *et al.* 2000), and the scaphopod *Antalis* (Wanninger and Haszprunar 2001). In molluscs, *engrailed* is consistently expressed in the leading edge of the shell field(s) during shell formation (Moshel *et al.* 1998; Jacobs *et al.* 2000; Wanninger and Haszprunar 2001; Nederbragt *et al.* 2002b; see subsequent discussion). The conserved role for *engrailed* in Bilateria is probably compartmentalization, enabling differentiation to occur, which may result in body plan patterning processes as diverse as segmentation, central nervous system (CNS) differentiation, or shell development. In annelids and molluscs *engrailed* expression has been conserved in the primary somatoblast, the 2d lineage. Rapid proliferation of 2d gives rise to the posterior growing point of numerous ectodermal cells (Nielsen 2004). In annelids, these cells form the ventral plate from which the ventral nerve cord will be formed (Ackermann *et al.* 2005). In molluscs, these cells contribute to the shell gland and produce parts of the shell. What appears to be a divergent relationship in *engrailed* expression between annelids and molluscs can be explained by the homologous lineage of the primary somatoblast. Using this as an example when studying regulatory gene expression in early development of organisms as diverse as, for example, molluscs and annelids, regulatory gene pathways should be determined by both conserved gene expression patterns and cell lineage conservation.

CONSERVED AND CO-OPTED DEVELOPMENTAL GENE EXPRESSION

Apart from *engrailed* expression studies, comparative gene expression analyses in Mollusca are still scarce. Some progress, however, has been made regarding the expression of a number of developmental genes in gastropods—chiefly the patellogastropod *Patella vulgata* (Lartillot *et al.* 2002a; Lespinet *et al.* 2002; Nederbragt *et al.* 2002a, b, c, d; Le Gouar *et al.* 2003, 2004) and the vetigastropod *Haliotis asinina* (Degnan and Morse 1993; Giusti *et al.* 2000; Hinman and Degnan 2002; O'Brien and Degnan 2002a, b, 2003; Hinman *et al.* 2003), but also in cephalopods (Lee *et al.* 2003). These studies have sought to document the expression of highly conserved bilaterian developmental genes in molluscs and to interpret the data with respect to the striking number of body plan novelties that evolved in Mollusca. In general, sophisticated methodologies such as gene knockout experiments, used to understand gene function in ecdysozoans (primarily the insect *Drosophila* and the nematode *Caenorhabditis elegans;* see, e.g., Venken and Bellen 2005; Labbé and Roy 2006) and deuterostomes (a range of vertebrates,

TABLE 16.1

*Examples for Developmental Genes That Have Acquired Additional Functions in
Certain Molluscs with Respect to their Conserved Roles in Bilaterian Ontogeny*

GENE	CONSERVED FUNCTIONS IN BILATERIA	CO-OPTION IN MOLLUSCA
engrailed	Neurogenesis, compartmentalization	Shell formation (*Lepidochitona, Transennella, Antalis, Ilyanassa*)
Hox1	Antero-posterior axis patterning, neurogenesis	Shell formation, loss of function in neurogenesis (*Haliotis*)
Hox4	Antero-posterior axis patterning, neurogenesis	Shell formation (*Haliotis*)
POUIII	Neurogenesis, (geo)sensory, secretory	Foot mucus cells, radular sac (*Haliotis*)
POUIV, Pax258	Neurogenesis, (geo)sensory	Sensory cells in foot, eyes, tentacles, ctenidia (*Haliotis*)

NOTE: Data taken from Jacobs *et al.* (2000), Moshel *et al.* (1998), Wanninger and Haszprunar (2001), O'Brien and Degnan (2002a, b, 2003), and Hinman *et al.* (2003).

the ascidian *Ciona*, and sea urchins; see, e.g., Meinertzhagen and Okamura 2001; Heasman 2002), has yet to be developed in molluscs. To date, genes known to have conserved roles in bilaterian body axis patterning and organogenesis have been characterized as summarized in the following paragraphs.

The function in anterior-posterior axis patterning of developmental genes belonging to the *Hox* cluster has been shown for many bilaterian phyla (see review by Garcia-Fernàndez 2005) and recently also in the anthozoan *Nematostella* (Finnerty *et al.* 2004). Of the highly conserved developmental genes, *Hox* genes have been the most intensely investigated. These are clustered in the genomes of bilaterians, where they are expressed in nested patterns along the anterior-posterior axis (reviewed in McGinnis and Krumlauf 1992; Manak and Scott 1994; Carroll *et al.* 2004; Garcia-Fernàndez 2005) and lead to cellular identity along this axis, thus contributing to body plan patterning. *Hox* genes also function in the patterning of limbs in vertebrates and in the development of mesodermal structures in penta-radially symmetrical echinoderms, suggesting that they have been co-opted during bilaterian evolution into also patterning other body axes than the anterior-posterior one

(Arenas-Mena *et al.* 2000). In *Haliotis* (Giusti *et al.* 2000; Hinman *et al.* 2003), as well as in the bobtail squid *Euprymna* (Lee *et al.* 2003), several *Hox* genes have been shown to be expressed in body axis patterning and nervous system formation in a manner that appears homologous to their roles in other bilaterians. In *Haliotis*, anterior *Hox* genes (*Has-Hox2*, *Has-Hox3*, and *Has-Hox4*) are expressed in overlapping patterns in the trochophore on the post-trochal ventral ectoderm prior to the formation of pleural, pedal, and esophageal ganglia (Tables 16.1, 16.2). *Has-Hox5* is expressed shortly after torsion in mantle cells that give rise to the branchial ganglia in *Haliotis*. Expression is maintained in these fields of neuroectoderm during neuroblast ingression and gangliogenesis. These observations are consistent with the role of these *Hox* genes in patterning the gastropod CNS—and probably the molluscan nervous system in general—in a manner similar to that observed in other bilaterians. However, some of these genes have co-opted additional functions in patterning the molluscan body plan (Table 16.1). In the cephalopod *Euprymna*, *Hox* expression also appears to be colinear, although the nested patterns observed in *Haliotis* and other bilaterians appear to be absent. As is the case in other bilaterians,

TABLE 16.2

A Comparison of Gastropod and Cephalopod Gene Expression Patterns

GENE	FUNCTIONS IN THE GASTROPOD *HALIOTIS*	FUNCTIONS IN THE SQUID *EUPRYMNA*
Has-Hox1, Esc-Lab	Shell formation	Neurogenesis, stellate ganglia, arms
Has-Hox3, Esc-Hox3	Neurogenesis	Neurogenesis, stellate ganglia, photosensory vesicles, funnel tube
Has-Hox5, Esc-Scr	Neurogenesis	Neurogenesis, arms, funnel tube

NOTE: This comparison shows that recruitment of novel gene functions may be a driving force for the establishment of new body plan phenotypes within Mollusca. *Has-* marks genes isolated from the gastropod *Haliotis asinina*, *Esc-* are the respective gene orthologs from the bobtail squid *Euprymna scolopes*. Data taken from Hinman *et al.* (2003) and Lee *et al.* (2003), respectively.

Hox gene expression in *Haliotis* and *Euprymna* was not detected in the most anterior portion of the CNS (i.e., the cerebral ganglia). Other transcription factors, such as *Otx* and *Emx*, may be involved in the formation of that portion of the nervous system. *Cdx* appears to play a conserved role in specifying the posterior neuroectoderm of gastropods (Le Gouar *et al.* 2003).

In both *Haliotis* and *Euprymna*, a number of *Hox* genes have lost their role in patterning the CNS. *Has-Hox1* is never detected in the forming *Haliotis* CNS (Hinman *et al.* 2003), and *Esc-Lox5* and *Esc-Post1* are not expressed in the developing CNS of the squid (Lee *et al.* 2003). Instead, these genes, along with some of the other *Hox* genes, have been co-opted into taxon-specific roles. Specifically, *Has-Hox1* is expressed in the shell gland and larval mantle and as such probably plays a role in shell production in *Haliotis* (Tables 16.1, 16.2). In the mature veliger larva of *Haliotis*, *Has-Hox4* is expressed in the mantle together with *Has-Hox1* (Hinman *et al.* 2003), and later, in the juvenile of *Haliotis*, all five anterior *Hox* genes are expressed in the mantle (McDougall and Degnan, unpublished). Comparison of *Hox* expression patterns in *Haliotis* and in *Euprymna* suggests that innovations that occurred in cephalopods, specifically ectodermally derived components of the brachial crown, funnel tube, stellate ganglia, metabrachial vesicles, buccal crown, and light organ, were due to recruitment of *Hox* genes into new roles (Lee *et al.* 2003; Table 16.2).

Other regulators of cell fate and differentiation have been shown to have conserved roles in gastropods. Nerve and secretory cells in *Haliotis* express members of *Sox*, *POU*, and *Pax* gene families, as has been observed in a wide range of bilaterian taxa (O'Brien and Degnan 2002a, b, 2003; Le Gouar *et al.* 2004). A *snail* gene is expressed in *Patella* in an epithelial-mesenchymal transition, as occurs in vertebrates (Lespinet *et al.* 2002).

ORGANOGENESIS AND MOLLUSCAN PHYLOGENY: NERVES, MUSCLES, SHELLS, AND BEYOND

MOLLUSCAN SISTER TAXON RELATIONSHIPS

One of the key issues in molluscan evolution is the question regarding its direct sister taxon. Traditionally, molluscs have been aligned with either Annelida or Sipuncula (e.g., Scheltema 1996; Nielsen 2001; but see Haszprunar *et al.*, Chapter 2). The former hypothesis received particular attention because it was used to argue that annelids and molluscs both stem from a segmented last common ancestor. The eight-"metameric" arrangement of polyplacophoran shell plates and dorso-ventral muscles were regarded as evidence for a segmented polyplacophoran body plan, despite the clades often considered to be earlier extant molluscan offshoots, Solenogastres and Caudofoveata (e.g., Haszprunar 2000; Haszprunar and Wanninger

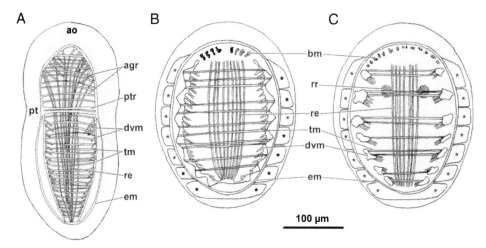

FIGURE 16.1. Semischematic line drawings illustrating the muscle anatomy in the metamorphic competent larva and early postmetamorphic stages of the polyplacophoran *Mopalia muscosa* and showing the "eight-seriality" of the polyplacophoran dorso-ventral shell musculature as a secondary condition. All images are in dorsal view with apical end uppermost. The cilia of the apical organ (ao) and the prototroch (pt) are omitted in A. Drawn after data from Wanninger and Haszprunar (2002a). (A) Metamorphic competent larva with an apical muscle grid (agr) comprising outer ring and inner diagonal muscles interspersed with longitudinal fibers belonging to the dorsal rectus muscle (re). A ring muscle underlying the prototroch (ptr) marks the transition between the pre- and the post-trochal larval body region. The ventral enrolling muscle (em) and the dorsal transverse muscles (tm) have already been formed. Note the homogeneous distribution of the dorso-ventral muscles (dvm) along the anterior-posterior axis. (B) Early juvenile at one day after metamorphosis showing the beginning of the concentration of the (initially seven) dorso-ventral muscle attachment sites under the shell plates (asterisks). The larval apical muscle grid and the prototroch muscle ring have been lost during metamorphosis while the anlagen of the buccal musculature (bm) are already visible. (C) Juvenile at 10 days after metamorphosis. The seven dorso-ventral muscle units are now distinct and correspond to the first seven shell plates. The eighth shell plate with its associated musculature forms during subsequent development. Note the paired radular retractor muscle (rr).

2000) not showing any signs of "segmented" organization. Moreover, both myogenesis and neurogenesis have shown that the ontogenetic patterns of these organ systems differ significantly in polyplacophorans from those found in polychaete annelids. First, the dorso-ventral muscle units in the polyplacophoran larva arise simultaneously as a meshwork of multiple muscle fibers, while the segmental ring muscles in polychaete larvae are formed subsequently one after another, as expected for organs derived from a preanal growth zone (Hill and Boyer 2001; Wanninger and Haszprunar 2002a). In chiton larvae, the concentration into the eight functional muscle units occurs after metamorphosis, thus representing a secondary condition (Figure 16.1). Second, the ventral commissures of the nervous system in polyplacophoran trochophores are not formed in an anterior-posterior direction as would be expected for

annelid-like segmented taxa, but instead the most anterior commissure arises considerably later than more posterior ones (Friedrich *et al.* 2002; Voronezhskaya *et al.* 2002).

These findings substantiate reports on shell plate development in the polyplacophoran *Lepidochitona caverna*, where the homeobox gene *engrailed* is expressed synchronously in the first seven shell plate rudiments and not sequentially (Jacobs *et al.* 2000; see Figure 16.2). Ontogenetic data on a solenogaster (= neomeniomorph) and a caudofoveate (= chaetodermomorph) corroborate these findings: during ontogeny of *Epimenia babai* (Okusu 2002) and *Chaetoderma* sp. (Nielsen *et al.* 2007) no "segmented" mode of spicule formation was observed, and ontogeny of the body wall ring musculature in *Chaetoderma* did not proceed from anterior to posterior (Nielsen *et al.* 2007). The combination of

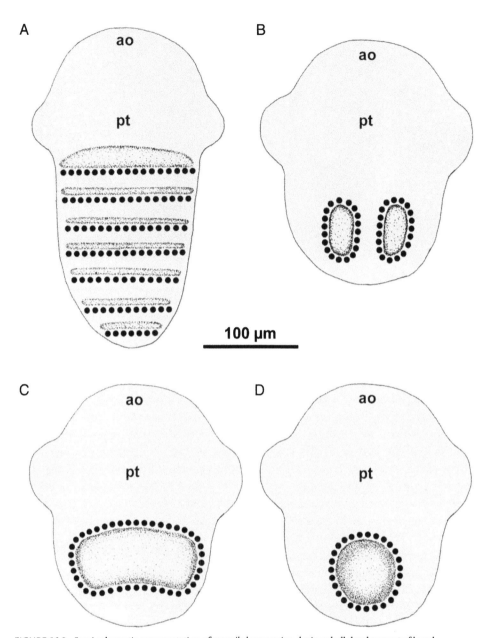

FIGURE 16.2. Semi-schematic representation of *engrailed* expression during shell development of larval Polyplacophora, Bivalvia, Scaphopoda, and Gastropoda. All images are in dorsal view with anterior facing upward. Cilia of the apical organ (ao) and the prototroch (pt) are omitted. Stippled areas show the anlagen of the shell fields; black dots mark cells expressing the homeobox gene *engrailed*. Drawn after data from Moshel *et al.* (1998), Jacobs *et al.* (2000), Wanninger and Haszprunar (2001), and Nederbragt *et al.* (2002b). (A) The polyplacophoran *Lepidochitona caverna*. The *en* transcript is found at the posterior margin of each of the forming shell plates. *Engrailed* is also expressed in spicule-forming cells of the mantle epithelium that surrounds the entire post-trochal area of the larva (omitted herein for reasons of clarity). (B) The bivalve *Transennella tantilla*, showing *en* expression in cells surrounding the two distinct shell fields. (C) The scaphopod *Antalis entalis*. In contrast to bivalves, *en*-expressing cells surround one single shell field indicating the univalved character of the scaphopod shell. (D) The gastropods *Patella vulgata* and *Ilyanassa obsoleta* show a similar condition to the scaphopod *Antalis*, with *en*-positive cells encircling a single shell field.

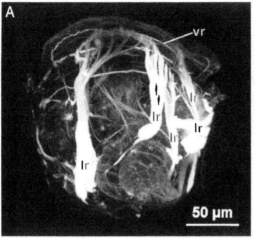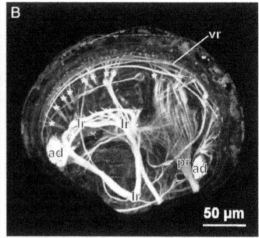

FIGURE 16.3. Confocal laser scanning micrographs of unidentified bivalve veligers stained with FITC-coupled phalloidin to label the larval and juvenile or adult retractor muscle systems. Lateral views with anterior facing upward. (A) Larva with velum expanded showing a complex arrangement of larval retractor muscles (lr) inserting at the velum muscle ring (vr). (B) Veliger retracted into the larval shell. Note the already prominent adductor muscles (ad) as well as the pedal retractor (pr). Asterisks mark muscle fibers of the larval retractors attaching at the velum muscle ring.

these data suggests a nonsegmented ancestor at the base of Mollusca and argues against a close annelid-mollusc relationship (see also Haszprunar and Wanninger 2000). Instead, recent data on the anatomy of the serotonergic nervous system of the creeping-type larva of the entoproct *Loxosomella murmanica* revealed a number of shared neural features with both adult solenogasters and larval polyplacophorans. Among these are a complex apical organ consisting of several central flask-shaped cells and some more basally situated peripheral cells, two pairs of longitudinal nerve cords interpreted as mollusc-like tetraneury, commissures that interconnect the ventral nerve cords, and four pairs of serotonergic perikarya that are in contact with the pedal nerve cords by short cell protrusions (Wanninger *et al.* 2007). The combination and the degree of complexity of these similarities in the neural body plan of the *Loxosomella* larva and in basal adult and larval molluscs provide strong support for a mollusc-entoproct clade (Wanninger *et al.* 2007), whereas neuromuscular development in the sipunculan *Phascolion strombus* favors an annelid-sipunculan assemblage (Wanninger *et al.* 2005), a hypothesis that has recently gained additional support from molecular phylogenetic analyses

(Boore and Staton 2002; Staton 2003, Jennings and Halanych 2005).

In summary, instead of a previous alignment of Mollusca and Sipuncula (Scheltema 1996), recent results on molluscan organogenesis are congruent with molecular phylogenies and suggest two groups of trochozoan phyla, one a taxon comprising Annelida and Sipuncula, the other Mollusca and Entoprocta (see also Haszprunar *et al.*, Chapter 2 and Simison and Boore, Chapter 17).

SCAPHOPOD RELATIONSHIPS

For a long time it had been argued that scaphopods may develop through a larval stage during which the embryonic shell (protoconch) passes through a bilobed stage, thus phylogenetically linking Scaphopoda to Bivalvia ("Diasoma concept," see Runnegar and Pojeta 1974). Lindberg and Ponder (1996) argued against this relationship and proposed a Gastropoda-Cephalopoda-Scaphopoda clade based on the presence of ano-pedal flexure in these taxa. Wanninger and Haszprunar (2001) examined engrailed protein expression in the scaphopod *Antalis entalis* and demonstrated that *engrailed* is expressed in cells surrounding a single shell field, similar to the condition found in gastropods (Moshel *et al.* 1998; Nederbragt *et al.* 2002b;

see Figure 16.3). Furthermore, development of the musculature in *Antalis entalis* revealed the existence of a distinct cephalic retractor, which otherwise is found only in gastropods and cephalopods (Wanninger and Haszprunar 2002b). This finding was unexpected because, unlike gastropods and cephalopods, scaphopods do not have a free, movable head but instead have a rather reduced "buccal cone." However, the possession of a distinct head retractor in combination with the singularity of the scaphopod shell field represents a strong argument for a clade comprising Scaphopoda + (Gastropoda + Cephalopoda). Such a phylogenetic scenario is in agreement with recent molecular and morphology-based cladistic analyses (e.g., Waller 1998; Haszprunar 2000; Haszprunar *et al.*, Chapter 2; Reynolds and Steiner, Chapter 7).

LARVAL NERVOUS SYSTEMS

Traditionally, the development and anatomy of the gastropod larval nervous system has received much attention (e.g., Bonar 1978; Chia and Koss 1984; Goldberg *et al.* 1994; see also review by Croll and Dickinson 2004). As a result, taxa that acted as model system animals for neurofunctional and behavioral studies, such as the heterobranch gastropods *Aplysia, Lymnaea, or Helix,* are known in considerable detail (e.g., Goldberg *et al.* 1994; Kandel 1979; Kuang *et al.* 2002; see also review by Chase 2002). With the recently increasing focus on evolutionary questions, neurogenesis has gained a wider interest, and new data on polyplacophorans (Friedrich *et al.* 2002; Voronezhskaya *et al.* 2002), scaphopods (Wanninger and Haszprunar 2003), bivalves (Croll *et al.* 1997; Plummer 2002), and, in particular, gastropods (e.g., Dickinson *et al.* 1999; Dickinson and Croll 2003) have become available in recent years. A detailed review of the anatomy of the molluscan larval nervous system is provided by Croll and Dickinson (2004). Herein, we concentrate on evolutionary and phylogenetically relevant conclusions based on a summary of the comparative neurodevelopmental data available to date.

As in other Trochozoa, molluscan larvae typically form an apical organ that comprises cells bearing the apical ciliary tuft and a more or less complex array of serotonergic, FMRFamidergic[1], and catecholaminergic cells. Other neurotransmitters may also be involved, but only a few neuronal reagents such as leu-enkephalin (Dickinson and Croll 2003) or small cardiac peptide-B (Barlow and Truman 1992) have been explored to date. Because of its proposed sensory function, the apical complex has recently been termed "apical sensory organ" (ASO). To date, it is generally assumed that bilaterian larval apical organs play an important role in metamorphic processes, although loss of the ASO in the scaphopod *Antalis* before metamorphic competence challenges this general assumption (Wanninger and Haszprunar 2003).

The nerve cells of the ASO form the major component of the molluscan larval nervous system and are typically reduced during or before metamorphosis (Wanninger and Haszprunar 2003; Gifondorwa and Leise 2006). In most gastropods and bivalves, the ASO comprises several central, flask-shaped cells that bear the cilia of the apical ciliary tuft (Croll *et al.* 1997; Dickinson *et al.* 1999; Plummer 2002; Dickinson and Croll 2003). These cells are usually surrounded by a number of equally flask-like so-called ampullary cells and may also include more rounded parampullary cells (Bonar 1978; Chia and Koss 1984; Dickinson *et al.* 1999; Croll and Dickinson 2004). Most of these cells have been shown to be serotonin positive, and some also contain the neuropeptide FMRFamide (Dickinson *et al.* 1999). A similar scenario is found in polyplacophoran trochophores, although FMRFamide activity has been reported only in the apical organ of one of the two species investigated so far, *Ischnochiton hakodadensis* (Voronezhskaya *et al.* 2002), but not in *Mopalia muscosa* (Friedrich *et al.* 2002). In

---

1. FMRFamide = L-phenylalanyl-L-methionyl-L-arginyl-L-phenylalaninamide

addition, Polyplacophora express a distinct larval accessory pretrochal organ—possibly of chemosensory function—that contains FMRFamidergic and serotonergic ampullary cells (Haszprunar et al. 2002; Voronezhskaya et al. 2002). Similar to the ASO, this system is also lost during metamorphosis. In larvae of the sole scaphopod species investigated so far, *Antalis entalis*, the ASO comprises only four central and two lateral serotonergic apical cells, while no larval FMRFamidergic components have been found (Wanninger and Haszprunar 2003).

An additional feature of the molluscan larval nervous system is a serotonergic nerve ring underlying (and probably innervating) the prototroch or velum. Such a structure has been found in many lophotrochozoan larvae and thus most likely represents a plesiomorphic condition for Mollusca (see Hay-Schmidt 2000). Although present in all other molluscs investigated, it is absent in the larva of the scaphopod *Antalis entalis* (Wanninger and Haszprunar 2003).

In most taxa, the formation of major parts of the adult nervous system (cerebral or anterior commissure or ganglion, ventral and lateral nerve cords) starts with the "brain" (i.e., the polyplacophoran anterior commissure or the conchiferan cerebral ganglion), which develops at the base of the ASO (Kempf et al. 1997; Friedrich et al. 2002; Voronezhskaya et al. 2002; Wanninger and Haszprunar 2003). The onset of development of these adult neural components is usually heterochronically shifted into the larval stages (Dickinson et al. 1999; Friedrich et al. 2002; Voronezhskaya et al. 2002; Wanninger and Haszprunar 2003), and the formation of the cerebral ganglion seems to be induced by the ASO, with which it forms physical contact (e.g., Friedrich et al. 2002; Voronezhskaya et al. 2002; Wanninger and Haszprunar 2003). The larval ASO, the serotonergic prototroch nerve ring, and the polyplacophoran pretrochal sensory organ are lost before, during, or shortly after metamorphosis (Friedrich et al. 2002; Voronezhskaya et al. 2002; Wanninger and Haszprunar 2003),

although parts of the ASO of some gastropods have been shown to persist for considerable time after metamorphosis (Bonar 1978; Page 2002). Thus, many taxa show a striking distinction between larval and adult nervous systems that, in later larval stages, may partly coexist beside each other but largely originate independently.

LARVAL MUSCLES AND THE ONTOGENY OF GASTROPOD TORSION

The process of torsion, that is, the 180° twist of the cephalopedal body region relative to the visceral part, marks a key event in gastropod evolution. The wide variety of data on the timing and the proposed ontogenetic mechanisms involved in this process indicates that torsion itself has undergone mechanistic evolution within the several gastropod taxa (for recent critical discussions on the subject, see Wanninger et al. 2000; Page 2006). However, it remains clear that torsion is a primarily larval process that in the most basal gastropods investigated so far (the patellogastropods *Patella vulgata* and *P. caerulea*) is caused by activity of the asymmetrically positioned main and accessory larval retractor muscles and is completed prior to the formation of the adult shell muscles (Wanninger et al. 1999b, 2000). The independence of the torsion process from the adult shell musculature has significant implications for molluscan evolution and especially for the interpretation of fossil data based on muscle scar imprints on internal molds of univalved Paleozoic molluscs, because it may render the arrangement of these fossilized adult muscle scars inappropriate for deductions as to whether such an univalved animal was torted (and thus a gastropod) or not (Wanninger et al. 2000; see also Yochelson 1978 and Chapters 3 and 10).

Gastropods (Degnan et al. 1997; Wanninger et al. 1999a; Page 1995, 1997), have one or multiple larval retractor muscles that enable retraction of the larval prototroch or velum into the mantle cavity (Degnan et al. 1997; Wanninger et al. 1999a; Page 1995, 1997) and these are also found in bivalves (Figure 16.3). Most of these retractors insert at a muscle ring that underlies the prototroch or velum. The existence of such a prototroch ring muscle, however, is not necessarily correlated with the expression of a larval retractor system, because the former has also been reported for polyplacophoran larvae, which lack larval retractor muscles (Wanninger and Haszprunar 2002a). Interestingly, polyplacophoran larvae exhibit a pretrochal three-dimensional muscle grid that resembles the architecture of the body wall musculature of adult Solenogastres, Caudofoveata, and other vermiform trochozoans and can thus be interpreted as an ontogenetic repetition of parts of the body wall muscles of a proposed worm-shaped molluscan ancestor (Figure 16.1; see also Wanninger and Haszprunar 2002a). The existence of a prototroch muscle ring in other trochozoan phyla such as Polychaeta (Wanninger, personal observation) indicates that this structure may represent a plesiomorphic character for Mollusca. Scaphopoda have a lecithotrophic larva and a short planktonic phase, and the scaphopod larva lacks a number of features typically attributed to molluscan trochophores or veligers, such as a prototroch muscle ring and distinct larval retractors (Wanninger and Haszprunar 2003). Instead, the adult cephalic and pedal retractors, already present in the late larva of *Antalis entalis,* form some additional fibers that project into the prototroch and are lost during metamorphosis (Wanninger and Haszprunar 2002b).

In general, formation of the adult molluscan shell musculature usually starts in late larval stages (Wanninger *et al.* 1999a, 2002a, b). Most of the larval muscle systems, such as the prototroch or velum muscle ring or the polyplacophoran pretrochal muscle grid, are lost during metamorphosis, although the larval retractor muscles may persist for some time in juveniles of basal gastropod taxa such as *Patella* (Wanninger *et al.* 1999a).

In *Patella caerulea* alone, the existence of a larval "extensor muscle" has recently been proposed, and its function was interpreted as being involved in extending the retracted larval body out of the shell (Damen and Dictus

2002). However, comparison with the work by Wanninger *et al.* (1999a) on the ontogeny of the larval and adult musculature of the same species suggests that Damen and Dictus (2002) probably misinterpreted this structure, which more likely constitutes the paired *anlage* of the future adult shell muscles, instead of being a distinct, independent muscle system (cf. Wanninger *et al.* 1999a).

CONCLUSIONS, FUTURE PERSPECTIVES, AND GAPS IN KNOWLEDGE

The ontogenetic data sets at the molecular, cellular, and organ system levels that have recently become available illustrate the importance of such data sets for intraphyletic deductions as well as for assessing molluscan sister relationships and plesiomorphic molluscan body plan features. Accordingly, the data currently available and reviewed herein suggests that the larva of the hypothetical ancestral mollusc (HAM) possessed a prototroch muscle ring; a three-dimensional body wall musculature comprising outer ring, intermediate oblique, and inner longitudinal muscles; a serotonergic prototroch nerve ring; and an apical organ containing centrally positioned flask-shaped cells as well as more laterally positioned peripheral cells. In addition, the HAM was unsegmented and exhibited a tetraneurous nervous system with ventral nerve commissures. The dorso-ventral musculature comprised numerous serially arranged muscle fibers. Shared ontogenetic features with the entoproct creeping-type larva suggest a mollusc–entoproct clade. Furthermore, the demonstration of the singularity of the scaphopod shell throughout ontogeny and the presence of a paired head retractor system argues strongly against the "Diasoma concept" and instead favors a scaphopod-gastropod-cephalopod assemblage. Many of the conserved developmental genes involved in cell specification, determination, and differentiation, as well as in pattern formation in bilaterians, appear to be operational in extant molluscs, suggesting that they were also present in the HAM. In some cases (e.g., *engrailed,* several genes of the *Hox* cluster, etc.) it appears that developmental genes have acquired new functions early in molluscan evolution that may explain the wide range of morphological novelties that evolved within Mollusca.

New developmental data sets employing molecular, cell lineage, and immunocytochemical techniques, especially on the two aplacophoran groups, putatively the most basal molluscs, will be particularly promising to substantiate the current hypotheses regarding molluscan phylogeny and evolution. For example, comparative cell lineage studies may help to clarify questions concerning the homology of certain organ systems such as the ciliated test cells of protobranch bivalve larvae and the prototroch or velum of other molluscan larvae. Further analysis of conserved regulator genes in a diversity of molluscs may be the key to understanding the high body plan diversity in Mollusca. Especially the combination of different methodologies, such as comparative developmental gene expression studies, functional gene analyses, and micromorphological studies, promise to be highly rewarding for future malacological research.

ACKNOWLEDGMENTS

We thank Winston Ponder and David R. Lindberg for initiating this project and for inviting us to participate in it. We extend our thanks to Henrike Semmler (Copenhagen) for help with the line drawings and to Roger P. Croll (Halifax) for providing a copy of a manuscript on molluscan larval nervous systems that was in press when we started working on this chapter. Research from the Degnan laboratory has been supported by grants from the Australian Research Council. Andreas Wanninger's research was funded by grants from the German Science Foundation and the Danish Research Agency.

REFERENCES

Abzhanov, A., and Kaufman, T.C. 2000. Evolution of distinct expression patterns for *engrailed* paralogues in higher crustaceans (Malacostraca). *Development, Genes and Evolution* 210: 493–506.

Ackermann, C., Dorresteijn, A., and Fischer, A. 2005. Clonal domains in postlarval *Platynereis dumerilii* (Annelida: Polychaeta). *Journal of Morphology* 266: 258–280.

Aguinaldo, A.M.A., Turbeville, J.M., Linford, L.S., Rivera, M.C., Garey, J.R., Raff, R.A., and Lake, J.A. 1997. Evidence for a clade of nematodes, arthropods and other moulting animals. *Nature* 387: 489–493.

Arenas-Mena, C., Cameron, A.R., and Davidson, E.H. 2000. Spatial expression of Hox cluster genes in the ontogeny of a sea urchin. *Development* 127: 4631–4643.

Arendt, D., Technau, U., and Wittbrodt, J. 2001. Evolution of the bilaterian larval foregut. *Nature* 409: 81–85.

Arnolds, J.A., van den Biggelaar, J.A.M., and Verdonk, N.H. 1983. Spatial aspects of cell interactions involved in the determination of dorsoventral polarity in equally cleaving gastropods and regulative abilities in their embryos, as studied by micromere deletions in *Lymnaea* and *Patella*. *Roux's Archives of Developmental Biology* 192: 75–85.

Barlow, L.A., and Truman, J.W. 1992. Patterns of serotonin and SCP immunoreactivity during metamorphosis of the nervous system of the red abalone, *Haliotis rufescens*. *Journal of Neurobiology* 23: 829–844.

Bonar, D.B. 1978. Ultrastructure of a cephalic sensory organ in the larvae of the gastropod *Phestilla sibogae* (Aeolidiacea, Nudibranchia). *Tissue Cell* 10: 153–165.

Boore, J.L., and Staton, J.L. 2002. The mitochondrial genome of the sipunculid *Phascolopsis gouldii* supports its association with Annelida rather than Mollusca. *Molecular Biology and Evolution* 19: 127–137.

Boring, L. 1989. Cell-cell interactions determine the dorsoventral axis in embryos of an equally cleaving opisthobranch mollusc. *Developmental Biology* 136: 239–253.

Candia, A.F., Hu, J., Crosby, J., Lalley, P.A., Noden, D., Nadeau, J.H., and Wright, C.V.E. 1992. *Mox-1* and *Mox-2* define a novel homeobox subfamily and are differentially expressed during early mesodermal patterning in mouse embryos. *Development* 116: 1123–1136.

Candia, A.F., and Wright, C.V.E. 1995. The expression of *Xenopus Mox-2* implies a role in initial mesoderm differentiation. *Mechanisms of Development* 52: 27–36.

Carroll, S.B., Grenier, J.K. and Weatherbee, S.D. 2004. *From DNA to Diversity: Molecular Genetics and the Evolution of Animal Design*. Malden: Blackwell Science.

Cather, J. N., and Verdonk, N. H. 1979. Development of *Dentalium* following removal of D-quadrant blastomeres at successive cleavage stages. *Roux's Archives of Developmental Biology* 187: 355–366.

Chase, R. 2002. *Behavior and its neural control in gastropod molluscs*. New York: Oxford University Press.

Chia, F., and Koss, R. 1984. Fine structure of the cephalic sensory organ in the larva of the nudibranch *Rostanga pulchra* (Mollusca, Opisthobranchia, Nudibranchia). *Zoomorphology* 104: 131–139.

Clement, A. C. 1962. Experimental studies on germinal localization in *Ilyanassa*. I. The role of the polar lobe in determination of the cleavage pattern and its influence in later development. *Journal of Experimental Zoology* 121: 593–625.

———. 1967. The embryonic value of the micromeres in *Ilyanassa*, as determined by deletion experiments. I. The first quartet cells. *Journal of Experimental Zoology* 166: 77–88.

———. 1976. Cell determination and organogenesis in molluscan development: a reappraisal based on deletion experiments in *Ilyanassa*. *American Zoologist* 16: 447–453.

Croll, R. P., and Dickinson, A. J. G. 2004. Form and function of the larval nervous system in molluscs. *Invertebrate Reproduction and Development* 46: 173–187.

Croll, R. P., Jackson, D. L., and Voronezhskaya, E. E. 1997. Catecholamine-containing cells in larval and post-larval bivalve molluscs. *Biological Bulletin* 193: 116–124.

Damen, P., and Dictus, W. J. A. G. 1994a. Cell lineage of the prototroch of *Patella vulgata* (Gastropoda, Mollusca). *Developmental Biology* 162: 364–383.

———. 1994b. Cell-lineage analysis of the prototroch of the gastropod mollusk *Patella vulgata* shows conditional specification of some trochoblasts. *Roux's Archives of Developmental Biology* 203: 187–198.

———. 1996. Organiser role of the stem cell of the mesoderm in prototroch patterning in *Patella vulgata* (Mollusca, Gastropoda). *Mechanisms of Development* 56: 41–60.

———. 2002. Newly discovered muscle in the larva of *Patella coerulea* (Mollusca, Gastropoda) suggests the presence of a larval extensor. *Contributions to Zoology* 71: 37–45.

Degnan, B. M., Degnan, S. M., and Morse, D. E. 1997. Muscle-specific regulation of tropomyosin gene expression and myofibrillogenesis differs among muscle systems examined at metamorphosis

of the gastropod *Haliotis rufescens*. *Development, Genes and Evolution* 206: 464–471.

Degnan, B. M., and Morse, D. E. 1993. Programmed cell death at metamorphosis—induction of muscle-specific protease gene expression in the mollusk *Haliotis rufescens*. *Journal of Cellular Biochemistry* Suppl. 17D: 148.

de Rosa, R., Grenier, J. K., Andreeva, T., Cook, C., Adoutte, A., Akam, M., Carroll, S. B., and Balavoine, G. 1999. *Hox* genes in brachiopods and priapulids and protostome evolution. *Nature* 399: 772–776.

Dickinson, A. J. G., and Croll, R. P. 2003. Development of the larval nervous system of the gastropod *Ilyanassa obsoleta*. *The Journal of Comparative Neurology* 466: 197–218.

Dickinson, A. J. G., Nason, J., and Croll, R. P. 1999. Histochemical localization of FMRF-amide, serotonin and catecholamines in embryonic *Crepidula fornicata* (Gastropoda, Prosobranchia). *Zoomorphology* 119: 49–62.

Dictus, W. J. A. G., and Damen, P. 1997. Cell-lineage and clonal-distribution map of the trochophore larva of *Patella vulgata* (Mollusca). *Mechanisms of Development* 62: 213–226.

Finnerty, J. R., Pang, K., Burton, P., Paulson, D., and Martindale, M. Q. 2004. Origins of bilateral symmetry: *Hox* and *dpp* expression in a sea anemone. *Science* 304: 1335–1337.

Freeman, G. 2006. Oocyte and egg organization in the patellogastropod *Lottia* and its bearing on axial specification during early embryogenesis. *Developmental Biology* 295: 141–155.

Friedrich, S., Wanninger, A., Brückner, M., and Haszprunar, G. 2002. Neurogenesis in the mossy chiton, *Mopalia muscosa* (Gould; Polyplacophora): evidence against molluscan metamerism. *Journal of Morphology* 253: 109–117.

Fröbius, A. C., and Seaver, E. C. 2006. *ParaHox* gene expression in the polychaete annelid *Capitella* sp. I. *Development, Genes and Evolution* 216: 81–88.

García-Fernández, J. 2005. Hox, ParaHox, ProtoHox: facts and guesses. *Heredity* 94: 145–152.

Gifondorwa, D. J., and Leise, E. M. 2006. Programmed cell death in the apical ganglion during larval metamorphosis of the marine mollusc *Ilyanassa obsoleta*. *Biological Bulletin* 210: 109–120.

Giusti, A. F., Hinman, V. F., Degnan, S. M., Degnan, B. M., and Morse, D. E. 2000. Expression of a *Scr/Hox5* gene in the larval central nervous system of the gastropod *Haliotis*, a non-segmented spiralian lophotrochozoan. *Evolution and Developement* 2: 294–302.

Goldberg, J. I., Koehncke, N. K., Christopher, K. J., Neumann, C., and Diefenbach, T. J. 1994. Pharmacological characterization of a serotonin receptor involved in an early embryonic behavior of *Helisoma trivolvis*. *Journal of Neurobiology* 25: 1545–1557.

Guralnick, R. P. 2002. A recapitulation of the rise and fall of the cell lineage research program: the evolutionary-developmental relationship of cleavage to homology, body plans and life history. *Journal of the History of Biology* 35: 537–567.

Halanych, K. M., Bacheller, J. D., Aguinaldo, A. M. A., Liva, S. M., Hillis, D. M., and Lake, J. A. 1995. Evidence from 18S ribosomal DNA that the lophophorates are protostome animals. *Science* 267: 1641–1643.

Haszprunar, G. 2000. Is the Aplacophora monophyletic? A cladistic point of view. *American Malacological Bulletin* 15: 115–130.

Haszprunar, G., Friedrich, S., Wanninger, A., and Ruthensteiner, B. 2002. Fine structure and immunocytochemistry of a new chemosensory system in the chiton larva (Mollusca: Polyplacophora). *Journal of Morphology* 252: 210–218.

Haszprunar, G., and Wanninger, A. 2000. Molluscan muscle systems in development and evolution. *Journal of Zoological Systematics and Evolutionary Research* 38: 157–163.

Hay-Schmidt, A. 2000. The evolution of the serotonergic nervous system. *Proceedings of the Royal Society of London. Series B, Biological Sciences* 267: 1071–1079.

Heasman, J. 2002. Morpholino oligos: making sense of antisense? *Developmental Biology* 243: 209–214.

Henry, J. Q., Okusu, A., and Martindale, M. Q. M. 2004. The cell lineage of the polyplacophoran, *Chaetopleura apiculata*: variation in the spiralian program and implications for molluscan evolution. *Developmental Biology* 272: 145–160.

Hill, S. D., and Boyer, B. C. 2001. Phalloidin labeling of developing muscle in embryos of the polychaete *Capitella* I. *Biological Bulletin* 201: 257–258.

Hinman, V. F., and Degnan, B. M. 2002. Mox homeobox expression in muscle lineage of the gastropod *Haliotis asinina*: evidence for a conserved role in bilaterian myogenesis. *Development, Genes and Evolution* 212: 141–144.

Hinman, V. F, O'Brien, E. K, Richards, G. S., and Degnan, B. M. 2003. Expression of anterior *Hox* genes during larval development of the gastropod *Haliotis asinina*. *Evolution and Development* 5: 508–521.

Iakovleva, N. V., Gorbushin, A. M., and Zelck, U. E. 2006. Partial characterization of mitogen-activated protein kinases (MAPK) from haemocytes of the common periwinkle, *Littorina littorea*

(Gastropoda: Prosobranchia). *Fish and Shellfish Immunology* 20: 665–668.

Irvine, S. Q., and Martindale, M. Q. 2000. Expression patterns of anterior Hox genes in the polychaete *Chaetopteros*: correlation with morphological boundaries. *Developmental Biology* 217: 333–351.

Isaacs, H. V., Pownall, M. E., and Slack, J. M. W. 1994. eFGF regulates *Xbra* expression during *Xenopus* gastrulation. *EMBO Journal* 13: 4469–4481.

Jacobs, D. K., Wray, C. G., Wedeen, C. J., Kostriken, R., DeSalle, R., Staton, J. L., Gates, R. D., and Lindberg, D. R. 2000. Molluscan *engrailed* expression, serial organization, and shell evolution. *Evolution and Development* 2: 340–347.

Jennings, R. M., and Halanych, K. M. 2005. Mitochondrial genomes of *Clymenella torquata* (Maldanidae) and *Riftia pachyptila* (Siboglinidae): evidence for conserved gene order in Annelida. *Molecular Biology and Evolution* 22: 210–222.

Kandel, E. R. 1979. *Behavioral Biology of Aplysia*. San Francisco: WH. Freeman and Company.

Kempf, S. C., and Page, L. R. 2005. Anti-tubulin labeling reveals ampullary neuron ciliary bundles in opisthobranch larvae and a new putative neural structure associated with the apical ganglion. *Biological Bulletin* 208: 169–182.

Kempf, S. C., Page, L. R., and Pires, A. 1997. Development of serotonin-like immunoreactivity in the embryos and larvae of nudibranch mollusks with emphasis on the structure and possible function of the apical sensory organ. *The Journal of Comparative Neurology* 386: 507–528.

Kuang, S., Doran, S. A., Wilson, R. J., Goss, G. G., and Goldberg, J. I. 2002. Serotonergic sensory-motor neurons mediate a behavioral response to hypoxia in pond snail embryos. *Journal of Neurobiology* 52: 73–83.

Kuhtreiber, W. M., Vantil, E. H., and van Dongen, C. A. M. 1988. Monensin interferes with the determination of the mesodermal cell-line in embryos of *Patella vulgata*. *Roux's Archives of Developmental Biology* 197: 10–18.

Kulakova, M. A., Kostyuchenko, R. P., Andreeva, T. F., and Dondua, A. K. 2002. The *Abdominal-B*-like gene expression during larval development of *Nereis virens* (Polychaeta). *Mechanisms of Development* 115: 177–179.

Labbé, J. C., and Roy, R. 2006. New developmental insights from high-throughput biological analysis in *Caenorhabditis elegans*. *Clinical Genetics* 69: 306–314.

Lambert, J. D., and Nagy, L. M. 2001. MAPK signaling by the D quadrant embryonic organizer of the mollusc *Ilyanassa obsoleta*. *Development* 128: 45–56.

———. 2003. The MAPK cascade in equally cleaving spiralian embryos. *Developmental Biology* 263: 231–241.

Lartillot, N., Le Gouar, M., and Adoutte, A. 2002a. Expression patterns of *fork head* and *goosecoid* homologues in the mollusc *Patella vulgata* supports the ancestry of the anterior mesendoderm across Bilateria. *Development, Genes and Evolution* 212: 551–561.

Lartillot, N., Lespinet, O., Vervoort, M., and Adoutte, A. 2002b. Expression pattern of *brachyury* in the mollusc *Patella vulgata* suggests a conserved role in the establishment of the AP axis in Bilateria. *Development* 129: 1411–1421.

Lee, P.N., Callaerts, P., de Couet, H.G., and Martindale, M.Q.M. 2003. Cephalopod *Hox* genes and the origin of morphological novelties. *Nature* 424: 1061–1065.

Le Gouar, M., Guillou, A., and Vervoort, M. 2004. Expression of a *SoxB* and a *Wnt2/13* gene during the development of the mollusc *Patella vulgata*. *Development, Genes and Evolution* 214: 250–256.

Le Gouar, M., Lartillot, N., Adoutte, A., and Vervoort, M. 2003. The expression of a *caudal* homologue in a mollusc, *Patella vulgata*. *Gene Expression Patterns* 3: 35–37.

Lespinet, O., Nederbragt, A.J., Cassan, M., Dictus, W.J.A.G., van Loon, A.E., and Adoutte, A. 2002. Characterisation of two *snail* genes in the gastropod mollusc *Patella vulgata*. Implications for understanding the ancestral function of the *snail*-related genes in Bilateria. *Development, Genes and Evolution* 212: 186–195.

Lindberg, D.R., and Ponder, W.F. 1996. An evolutionary tree for the Mollusca: Branches or roots? Pp. 67–75 *In* J. Taylor, ed. *Origin and Evolutionary Radiation of the Mollusca*, Oxford University Press, Oxford.

Mallat, J., and Winchell, C.J. 2002. Testing the new animal phylogeny: first use of combined large-subunit and small-subunit rRNA gene sequences to classify the protostomes. *Molecular Biology and Evolution* 19: 289–301.

Manak, J.R., and Scott, M.P. 1994. A class act: conservation of homeodomain protein functions. *Development* Suppl.: 61–77.

Martindale, M.Q. 1986. The organizing role of the D quadrant in an equal-cleaving spiralian, *Lymnaea stagnalis* as studied by UV laser deletion of macromeres at intervals between 3rd and 4th quartet formation. *International Journal of Invertebrate Reproduction and Development* 9: 229–242.

Martindale, M.Q., Doe, C.Q., and Morrill, J.B. 1985. The role of animal-vegetal interaction with respect to the determination of dorsoventral polarity in the equal-cleaving spiralian, *Lymnaea palustris*. *Roux's Archives of Developmental Biology* 194: 281–295.

Martindale, M.Q., Pang, K., and Finnerty, J.R. 2004. Investigating the origins of triploblasty: "mesodermal" gene expression in a diploblastic animal, the sea anemone *Nematostella vectensis* (phylum, Cnidaria; class, Anthozoa). *Development* 131: 2463–2474.

Maslakova, S.A, Martindale, M.Q., and Norenburg, J.L. 2004. Vestigial prototroch in a basal nemertean *Carinoma tremaphorus* (Nemertea, Palaeonemertea). *Evolution and Development* 6: 219–226.

McGinnis, W., and Krumlauf, R. 1992. Homeobox genes and axial patterning. *Cell* 68: 283–302.

Meinertzhagen, I.A., and Okamura, Y. 2001. The larval ascidian nervous system: the chordate brain from its small beginnings. *Trends in Neurosciences* 24: 401–410.

Melby, A.E., Warga, R.M., and Kimmel, C.B. 1996. Specification of cell fates at the dorsal margin of the zebrafish gastrula. *Development* 122: 2225–2237.

Moshel, S.M., Levine, M., and Collier, J.R. 1998. Shell differentiation and *engrailed* expression in the *Ilyanassa* embryo. *Development, Genes and Evolution* 208: 135–141.

Nederbragt, A.J., Lespinet, O., van Wageningen, S., van Loon, A.E., Adoutte, A., and Dictus, W.J.A.G. 2002a. A lophotrochozoan *twist* gene is expressed in the ectomesoderm of the gastropod mollusk *Patella vulgata*. *Evolution and Development* 4: 334–343.

Nederbragt, A.J., van Loon, A.E., and Dictus, W.J.A. G. 2002b. Expression of *Patella vulgata* orthologs of *engrailed* and *dpp-BMP2/4* in adjacent domains during molluscan shell development suggests a conserved compartment boundary mechanism. *Developmental Biology* 246: 341–355.

Nederbragt, A.J., van Loon, A.E., and Dictus, W.J.A.G. 2002c. Hedgehog crosses the snail's midline. *Nature* 417: 811–812.

Nederbragt, A.J., Welscher, P.T., van den Driesche, S., van Loon, A.E., and Dictus, W.J.A. G. 2002d. Novel and conserved roles for *orthodenticle/otx* and *orthopedia/otp* orthologs in the gastropod mollusc *Patella vulgata*. *Development, Genes and Evolution* 212: 330–337.

Nielsen, C. 2001. *Animal Evolution*. Oxford: Oxford University Press.

———. 2004. Trochophore larvae: cell-lineages, ciliary bands, and body regions. 1. Annelida and Mollusca. *Journal of Experimental Zoology (Molecular and Developmental Evolution)* 302B: 35–68.

Nielsen, C., Haszprunar, G., Ruthensteiner, B., and Wanninger, A. 2007. Early development of the aplacophoran mollusc *Chaetoderma*. *Acta Zoologica* (Stockholm) 88: 231–247.

O'Brien, E. K., and Degnan, B. M. 2002a. Pleiotropic developmental expression of *HasPOU-III*, a class III *POU* gene, in the gastropod *Haliotis asinina*. *Mechanisms of Development* 114: 129–132.

———. 2002b. Developmental expression of a class IV *POU* gene in the gastropod *Haliotis asinina* supports a conserved role in sensory cell development in bilaterians. *Development, Genes and Evolution* 212: 394–398.

———. 2003. Expression of *Pax258* in the gastropod statocyst: insights into the antiquity of metazoan geosensory organs. *Evolution and Development* 5: 572–578.

Okusu, A. 2002. Embryogenesis and development of *Epimenia babai* (Mollusca Neomeniomorpha). *Biological Bulletin* 203: 87–103.

Page, L. R. 1995. Similarities in form and development sequence for three larval shell muscles in nudibranch gastropods. *Acta Zoologica* 76: 177–191.

———. 1997. Larval shell muscles in the abalone *Haliotis kamtschatkana*: evolutionary implications. *Acta Zoologica* 78: 227–245.

———. 2002. Apical sensory organ in larvae of the patellogastropod *Tectura scutum*. *Biological Bulletin* 202: 6–22.

———. 2006. Modern insights on gastropod development: reevaluation of the evolution of a novel body plan. *Integrative and Comparative Biology* 46: 134–143.

Patel, N. H., Martin-Blanco, E., Coleman, K. G., Poole, S. J., Ellis, M. C., Kornberg, T. B., and Goodman, C. S. 1989. Expression of *engrailed* proteins in arthropods, annelids, and chordates. *Cell* 58: 955–968.

Peterson, K. J., Irvine, S. Q., Cameron, R. A., and Davidson, E. H. 2000. Quantitative assessment of *Hox* complex expression in the indirect development of the polychaete annelid *Chaetopteros* sp. *Proceedings of the National Academy of Sciences of the United States of America* 97: 4487–4482.

Philippe, H., Lartillot, N., and Brinkmann, H. 2005. Multigene analyses of bilaterian animals corroborate the monophyly of Ecdysozoa, Lophotrochozoa, and Protostomia. *Molecular Biology and Evolution* 22: 1246–1253.

Plummer, J. T. 2002. The bivalve larval nervous system. M.Sc. thesis, Dalhousie University, Halifax, Nova Scotia.

Raible, F., Tessmar-Raible, K., Osoegawa, K., Wincker, P., Jubin, C., Balavoine, G., Ferrier, D., Benes, V., de Jong, P., Weissenbach, J., Bork, P., and Arendt, D. 2005. Vertebrate-type intron-rich genes in the marine annelid *Platynereis dumerilii*. *Science* 310: 1325–1326.

Render, J. 1997. Cell fate maps in the *Ilyanassa obsoleta* embryo beyond the third division. *Developmental Biology* 189: 301–310.

Runnegar, B., and Pojeta, J., Jr. 1974. Molluscan phylogeny: the paleontological viewpoint. *Science* 186: 311–317.

Scheltema, A. H. 1996. Aplacophora as progenetic aculiferans and the coelomate origin of mollusks as the sister taxon of Sipuncula. *Biological Bulletin* 184: 57–78.

Schulte-Merker, S., van Eeden, F. J., Halpern, M. E., Kimmel, C. B., and Nüsslein-Volhard, C. 1994. *No tail* (*ntl*) is the zebrafish homologue of the mouse *T* (*brachyury*) gene. *Development* 120: 1009–1015.

Seaver, E. C., and Kaneshige, L. M. 2006. Expression of "segmentation" genes during larval and juvenile development in the polychaetes *Capitella* sp. I and *H. elegans*. *Developmental Biology* 289: 179–194.

Smith, J. C., Armes, N. A., Conlon, F. L., Tada, M., Umbhauer, M., and Weston, K. M. 1991. Upstream and downstream from *brachyury*, a gene required for vertebrate mesoderm formation. *Cold Spring Harbor Symposium on Quantitative Biology* 62: 337–346.

Staton, J. L. 2003. Phylogenetic analysis of the mitochondrial cytochrome c oxidase subunit 1 gene from 13 sipunculan genera: intra- and interphylum relationships. *Invertebrate Biology* 122: 252–264.

van den Biggelaar, J. A. M. 1977. Development of dorsoventral polarity and mesentoblast determination in *Patella vulgata*. *Journal of Morphology* 154: 157–186.

———. 1996. Cleavage pattern and mesentoblast formation in *Acanthochiton crinitus* (Polyplacophora, Mollusca). *Developmental Biology* 174: 423–430.

van den Biggelaar, J. A. M., and Guerrier, P. 1979. Dorsoventral polarity and mesentoblast determination as concomitant results of cellular interactions in the mollusk *Patella vulgata*. *Developmental Biology* 68: 462–471.

———. 1983. Origin of spatial organization. In *The Mollusca*. Vol. 3. *Development*. Edited by N. H. Verdonk, J. A. M. van den Biggelaar, and A. S. Tompa. New York: Academic Press.

van den Biggelaar, J. A. M., and Haszprunar, G. 1996. Cleavage patterns and mesentoblast formation in the Gastropoda: an evolutionary perspective. *Evolution* 50: 1520–1540.

van Dongen, C. A. M. 1976. The development of *Dentalium* with special reference to the significance of the polar lobe. V. and VI. Differentiation of the cell pattern in lobeless embryos of *Dentalium vulgare* (da Costa) during late larval development. *Verhandelingen der Koninklijke Nederlandse Akademie van Wetenschappen* C 79: 245–266.

van Dongen, C.A.M., and Geilenkirchen, W.C.M. 1974. The development of *Dentalium* with special reference to the significance of the polar lobe. I, II and III. Division chronology and and development of the cell pattern in *Dentalium dentale* (Scaphopoda). *Verhandelingen der Koninklijke Nederlandse Akademie van Wetenschappen C* 77: 57–100.

Venken, K.J.T., and Bellen, H.J. 2005. Emerging technologies for gene manipulation in *Drosophila melanogaster*. *Nature Reviews Genetics* 6: 167–178.

Verdonk, N.H., and van den Biggelaar, J.A.M. 1983. Early development and the formation of germ layers. In *The Mollusca*. Vol. 3. *Development*. Edited by N.H. Verdonk, J.A.M. van den Biggelaar, and A.S. Tompa. New York: Academic Press.

Voronezhskaya, E.E., Tyurin, S.A., and Nezlin, L.P. 2002. Neuronal development in larval chiton *Ischnochiton hokadodensis* (Mollusca, Polyplacophora). *The Journal of Comparative Neurology* 444: 25–38.

Waller, T.R. 1998. Origin of the molluscan class Bivalvia and a phylogeny of major groups. In *Bivalves: An Eon of Evolution*. Edited by P.A. Johnston and J.W. Haggart. Calgary: University of Calgary Press.

Wanninger, A. 2004. The myo-anatomy of juvenile and adult loxosomatid Entoprocta and the use of muscular body plans for phylogenetic inferences. *Journal of Morphology* 261: 249–257.

———. 2005. Immunocytochemistry of the nervous system and the musculature of the chordoid larva of *Symbion pandora* (Cycliophora). *Journal of Morphology* 265: 237–243.

Wanninger, A., Fuchs, J., and Haszprunar, G. 2007. The anatomy of the serotonergic nervous system of an entoproct creeping-type larva and its phylogenetic implications. *Invertebrate Biology* (in press).

Wanninger, A., and Haszprunar, G. 2001. The expression of an engrailed protein during embryonic shell formation of the tusk-shell, *Antalis entalis* (Mollusca, Scaphopoda). *Evolution and Development* 3: 312–321.

———. 2002a. Chiton myogenesis: perspectives for the development and evolution of larval and adult muscle systems in molluscs. *Journal of Morphology* 251: 103–113.

———. 2002b. Muscle development in *Antalis entalis* (Mollusca, Scaphopoda) and its significance for scaphopod relationships. *Journal of Morphology* 254: 53–64.

———. 2003. The development of the serotonergic and FMRF-amidergic nervous system in *Antalis entalis* (Mollusca, Scaphopoda). *Zoomorphology* 122: 77–85.

Wanninger, A., Koop, D., Bromham, L., Noonan, E., and Degnan, B.M. 2005. Nervous and muscle system development in *Phascolion strombus* (Sipuncula). *Development, Genes and Evolution* 215: 509–518.

Wanninger, A., Ruthensteiner, B., Dictus, W.J.A.G., and Haszprunar, G. 1999b. The development of the musculature in the limpet *Patella* with implications on its role in the process of ontogenetic torsion. *Invertebrate Reproduction and Development* 36: 211–215.

Wanninger A., Ruthensteiner B., and Haszprunar, G. 2000. Torsion in *Patella caerulea* (Mollusca, Patellogastropoda): ontogenetic process, timing, and mechanisms. *Invertebrate Biology* 119: 177–187.

Wanninger, A., Ruthensteiner, B., Lobenwein, S., Salvenmoser, W., Dictus, W.J.A.G., and Haszprunar, G. 1999a. Development of the musculature in the limpet *Patella* (Mollusca, Patellogastropoda). *Development, Genes and Evolution* 209: 226–238.

Weisblat, D.A., Harper, G., Stent, G.S., and Sawyer, R.T. 1980. Embryonic cell lineage in the nervous system of the glossiphoniid leech *Helobdella triserialis*. *Developmental Biology* 76: 58–78.

Wilson, E.B. 1904. On germinal localization in the egg. II. Experiments on the cleavage mosaic in *Patella*. *Journal of Experimental Zoology* 1: 197–268.

Wilson, V., and Beddington, R. 1997. Expression of T protein in the primitive streak is necessary and sufficient for posterior mesoderm movement and somite differentiation. *Developmental Biology* 192: 45–58.

Wilson, V., Manson, L., Skarnes, W.C., and Beddington, R.S. 1995. The *T* gene is necessary for normal mesodermal morphogenetic cell movements during gastrulation. *Development* 121: 877–886.

Yochelson, E.L. 1978. An alternative approach to the interpretation of the phylogeny of ancient mollusks. *Malacologia* 17: 165–191.

17

Molluscan Evolutionary Genomics

W. Brian Simison and Jeffrey L. Boore

In the last twenty years there have been dramatic advances in techniques of high-throughput DNA sequencing, most recently accelerated by the Human Genome Project, which determined our (approximately) three-billion-base-pair code (Lander *et al.* 2001; Venter *et al.* 2001). Now this capability is being directed at other genomes across the entire range of life (see <http://www.genomesonline.org/>). This opens up opportunities as never before for evolutionary and organismal biologists to address questions of both processes and patterns of organismal change. We are at the dawn of a new "modern synthesis," paralleling that of the early twentieth century, when the fledgling field of genetics first identified the underlying basis for Darwin's theory of natural selection. We must now unite the efforts of systematists, paleontologists, computational biologists, mathematicians, computer programmers, molecular biologists, developmental biologists, and others in the pursuit of discovering what genomics can teach us about the diversity of life.

Genome-level sampling for molluscs to date has mostly been limited to mitochondrial genomes, and it is likely that these will continue to provide the best targets for broad phylogenetic sampling in the near future. However, we are just beginning to see the start of complete nuclear genome sequencing, with several molluscs and other eutrochozoans having been selected for sequencing. Here, we provide an overview of the state of molluscan mitochondrial genomics, outline the promise of broadening this dataset, show how comparisons of mitochondrial genomes have reconstructed the relationships among Annelida, Sipuncula, and Mollusca, describe upcoming projects to sequence whole mollusc nuclear genomes, and challenge the community to prepare for making the best use of these data.

MOLLUSC MITOCHONDRIAL GENOMES

The typical animal mitochondrial genome is about 15,000 base pairs (bp) in size and contains 37 genes: 13 coding protein genes, 2 ribosomal RNA genes, and 22 tRNAs (reviewed in Boore 1999). In addition to the genes, there is often one large noncoding region that is thought to contain the signals for controlling transcription and replication, although these signals, if they exist, are not well conserved and, except for a few model organisms, remain unidentified. For some animals all genes are on one strand, but for others they are divided between both. Most animal mtDNAs are A+T-rich, and there is often an asymmetry in nucleotide composition between the two strands, with one being more rich in G and T at the expense of C and A (reviewed in Asakawa et al. 1991; Perna and Kocher 1995). In the few cases where it has been studied, all genes on each strand are expressed as a single polycistronic mRNA (mRNA strand containing multiple genes), which is then enzymatically cleaved to yield gene-specific messages (Ojala et al. 1980, 1981; Rossmanith et al. 1995; Dubrovsky et al. 2004; Levinger et al. 2004). Mitochondrial DNA (mtDNA), in general, is maternally inherited and therefore, does not experience allelic segregation (but see subsequent discussion).

There are several features that make it feasible and important to sample mitochondrial genomes across a broad phylogenetic range. Since this DNA is structurally a closed circle, supercoiled and extrachromosomal, they can be physically isolated from the nuclear genome (Boore et al. 2005). They are known to play important roles in cellular metabolism, including apoptosis (Nieminen 2003), genetic disease (Wallace 1999), and embryological development (Yost et al. 1995; Krakauer and Mira 1999), and their biochemistry is relatively well understood. Their diminutive size facilitates sampling, yet they contain many elements of a complete genomic system, including genes for all three primary transcript types (protein, tRNA, rRNA; see Boore 1999). They have been used extensively in studies of population structure (e.g., Nyakaana et al. 2002) and forensics (Budowle et al. 2003). Many aspects of genome evolution are being modeled in these systems, including the role of biased nucleotide mutations on amino acid substitution patterns (e.g., Helfenbein et al. 2001), codon usage changes (e.g., Boore and Brown 2000) and mechanisms of gene order rearrangement (e.g., Boore 2000;

Rawlings et al. 2001, 2003; Lavrov et al. 2002; Mueller and Boore 2005).

Even the limited sampling of complete mollusc mtDNA sequences has revealed that they have many unusual characteristics. Some molluscs have very large mtDNAs, up to 42 Kb (La Roche et al. 1990). In some lineages there have been duplications of genes, the establishing of a secondary copy of *trnM*, splitting of an rRNA gene, and the loss of the gene for the subunit 8 protein of ATP synthase (Hoffmann et al. 1992; Boore et al. 2004; Yokobori et al. 2004; Mizi et al. 2005; Milbury and Gaffney 2005; Akasaki et al. 2006; GenBank accessions NC_003354, NC_001276, AB055624, and AB055625). Although some mitochondrial lineages have been stable in gene arrangement over long periods of evolutionary time (e.g., Boore and Brown 1994), others have rearranged nearly every gene (e.g., Hoffmann et al. 1992). At least some bivalves have a bizarre exception to maternal inheritance that has been dubbed "doubly-uniparental inheritance" (Zouros 2000; Passamonti et al. 2003), whereby females have one type of mitochondrial genome and males have, in their gonads, another haplotype that differs by as much as 20% in sequence (Mizi et al. 2005); male somatic tissues have a mixture of the two types, but only the male-specific haplotype is passed on to male offspring. This rich variation in mitochondrial features makes molluscs a great model system for understanding general principles of genome evolution and mitochondrial molecular biology in particular.

CHARACTERS FOUND FROM MOLLUSC MITOCHONDRIAL GENOMES

Most molecular phylogenetic studies of molluscs to date have focused on the comparisons of a handful of partial nuclear and mitochondrial gene sequences, typically using from less than 1,000 to a few thousand bp per organism, and most commonly have used the sequences of small-subunit (i.e., 18S) rRNA encoded by the nucleus. In general, these studies have failed to demonstrate robust monophyly for any but a few of the major classes and have yielded poorly resolved or contradictory phylogenetic results for molluscs and for the broader Eutrochozoa. For most lineages, mtDNA has a higher substitution rate than nuclear DNA and thus is well suited to resolving relationships among closely related taxa, although it has also provided strong signal for deep divergences (e.g., Nardi *et al.* 2003; Helfenbein *et al.* 2001; Ruiz-Trillo *et al.* 2004). Studies using complete mitochondrial genomes have over 15 Kb of sequence per organism, and it has been shown that complete mtDNA sequences give much stronger trees than comparisons of individual mitochondrial genes, as has been done in several studies, most commonly using portions of *cox1*, *cob*, or *rrnL* (Ingman *et al.* 2001; Boore *et al.* 2004; Mueller *et al.* 2004; Parham *et al.* 2006).

In addition to these sequences, there are other features that can be compared among mitochondrial genomes for phylogenetic reconstruction. These are so-called "genome-level" features (Boore 2006), the most useful of which to date has been comparisons of gene arrangements (Boore *et al.* 1995; Yamazaki *et al.* 1997; Boore and Brown 1998; Boore *et al.* 1998; Nickisch-Rosenegk *et al.* 2001; Boore and Staton 2002; Helfenbein and Boore 2004; Lavrov *et al.* 2004). These genes can potentially rearrange into a great number of states, and gene order does not seem to be commonly under selection, so the likelihood of two lineages independently adopting identical states or of a lineage reverting to an earlier state is low. In fact, comparisons of hundreds of mtDNAs have revealed only a very few cases of convergent evolution (Boore and Brown 1998; Boore 1999; Macey *et al.* 2004). Compared to most phyla, molluscs have a higher rate of gene rearrangement (Table 17.1), extending the range of phylogenetic levels where these might be informative. Other characters might include modes of replication, transcription, translation, or protein movement (Clayton 2000, 2003; Beddoe and Lithgow 2002), genetic code changes, such as those characterizing echinoderms and flatworms (Osawa *et al.* 1992) and, within this latter group, the Rhabditophora (Telford *et al.* 2000), patterns of RNA editing (e.g., Lavrov *et al.* 2000; Masta and Boore 2004), or the natural secondary structures of tRNAs or rRNAs coded as a morphological entity (Lydeard *et al.* 2000).

Unfortunately, to date, comparisons of even whole mtDNA sequences have not given strong resolution of deep-level relationships within molluscs, perhaps because of the small and taxonomically biased sample of taxa included, even in analyses that provided strong support for relationships among major groups of arthropods and of vertebrates (e.g., see Boore *et al.* 2004). Two studies using partial and complete sequences of mitochondrial genomes, however, have been very important to our understanding of the evolution of the molluscan body plan, the first showing the close relationship of Annelida to Mollusca (rather than to Arthropoda; Boore and Brown 2000) and the second showing that Sipuncula is not the sister phylum to Mollusca (but rather more closely related to Annelida; Boore and Staton 2002).

The evolutionary association between Annelida and Arthropoda has been long held, based in large part on narratives (e.g., Snodgrass 1938; Brusca and Brusca 1990) that describe the plausible scenario of the progressive body segment specialization of an annelidan ancestor through to onychophorans, myriapods, then insects, in some cases even complete with speculation on the underlying genetic changes (e.g., Raff and Kaufman 1983). This view has been challenged by studies using morphological characters (e.g., Eernisse *et al.* 1992), molecular sequences (e.g., Ghiselin 1988; Lake 1990; Garcia-Machado *et al.* 1999), and fossil evidence (i.e., the halkieriids; see Conway Morris and Peel 1995), all of which argued for an Annelida-Mollusca association instead. This case was greatly bolstered by the comparison of mitochondrial genomes, which provided not only very strongly supported trees based on larger-scale sequence comparisons, but also trees based on the relative arrangements of mitochondrial genes (Boore and Brown 2000), making clear that segmentation is not a reliable character for inferring phylogeny, but perhaps that sharing a trochophore larva is (see Eernisse *et al.* 1992).

Early in embryological development, sipunculids and molluscs share a particular pattern

TABLE 17.1

All Available Complete Gene Arrangements for Mollusc mtDNAs

NAME AND CLASSIFICATION	GENE ARRANGEMENT	REFERENCE
Polyplacophora		
Katharina tunicata (Chitonida; Mopaliidae)	cox1, D, cox2, atp8, atp6, -F, -nad5, -H, -nad4, -nad4L, T, -S(nga), -cob, -nad6, P, -nad1, -L(yaa), -L(nag), -rrnL, -V, -rrnS, -M, -C, -Y, -W, -Q, -G, -E, cox3, K, A, R, N, I, nad3, S(nct), nad2	Boore and Brown 1994, NC_001636
Bivalvia		
Mytilus edulis and *M. galloprovincialis* (Pteriomorphia; Mytilidae)	cox1, atp6, T, nad4L, nad5, nad6, F, rrnS, G, N, E, C, I, Q, D, rrnL, Y, cob, cox2, K, M, L1, L2, nad1, V, nad4, cox3, S, M, nad2, R, W, A, S, H, P, nad3	Boore et al. 2004; Hoffmann et al. 1992; Mizi et al. 2005, NC_006161, NC_006886
Mytilus trossulus (Pteriomorphia Mytilidae)	cox1, atp6, T, nad4L, nad5, nad6, F, rrnS, G, N, E, C, I, Q, D, rrnL, Q, Y, cob, cox2, K, M, L1, L2, nad1, V, nad4, cox3, S, M, nad2, R, W, A, S, H, P, nad3	Breton et al. 2006, NC_007687
Crassostrea virginica (Pteriomorphia; Ostreidae)	cox1, rrnL (split), cox3, I, T, E, cob, cox2, S1, L1, P, G, rrnS, M, K, C, V, D, rrnL (split), M, S2, Y, atp6, nad2, R, H, nad4, N, nad5, nad6, Q, nad3, L2, F, A, nad1, nad4L, W	Milbury and Gaffney 2005, NC_007175
Crassostrea gigas (Pteriomorphia; Ostreidae)	cox1, rrnL, cox3, I, T, E, cob, D, cox2, M, L1, P, rrnS, K, C, N, rrnS, Y, atp6, G, V, nad2, R, H, nad4, nad5, nad6, Q, nad3, L2, nad1, nad4L, W	Kim et al. unpublished, NC_001276
Placopecten magellanicus (Pteriomorphia; Pectinidae)	cox1, nad5, nad3, M, nad6, M, F, M, F, M, atp6, L1, cob, cox2, Q, E, K, F, A, cox3, F, M, M, M, N, G, M, M, N, E, D, K, S1, rrnL, rrnS, S2, nad4L, nad2, nad4, H, W, Y, T, I, M, C, L2, nad1	Milbury and Gaffney 2005, NC_007175
Argopecten irradians (Pteriomorphia; Pectinoida)	cox1, F, nad6, L2, cob, cox2, nad4L, cox3, nad3, S1, N, nad4, G, S2, nad2, D, I, H, W, P, nad5, atp6, C, Y, T, A, L1, M, E, K, rrnS, Q, V, nad1, R, rrnL	Petten and Snyder, unpublished, NC_009687
Mizuhopecten yessoensis (Pteriomorphia; Pectinidae)	cox1, D, C, nad5, nad4L, nad6, L2, cob, cox3, K, E, atp6, cox2, nad2, T, P, I, L1, nad3, nad4, V, N, nad1, R, rrnL, M, rrnS	*Sato and Nagashima 2001, NC_009081

Species	Gene order	Reference
Lampsilis ornate (Palaeoheterodonta; Unionidae)	cox1, -cox2, -nad3, -H, A, S(nga), E, nad2, M, W, R, rrnS, K, T, Y, rrnL, L1, N, P, cob, F, -nad5, Q, C, I, V, L2, nad1, G, nad6, -nad4, -nad4L, -atp8, -D, -atp6, -cox3	Serb and Lydeard 2003, NC_005335
Inversidens japanensis (female type) (Palaeoheterodonta; Unionidae)	cox1, cox3, atp6, D, nad4L, nad4, -nad6, -G, -nad1, -L1, V, -I, -C, -Q, nad5, -P, -F, -cob, -N, -L2, -rrnL, -Y, -T, -K, -rrnS, -R, -W, -E, -S(nga), -A, nad3, -M, -nad2, -S(rct), H, cox2	Okazaki and Ueshima, unpublished, AB055625
Inversidens japanensis (male type) (Palaeoheterodonta; Unionidae)	cox1, cox3, atp6, -D, nad4L, nad4, -nad6, -G, -nad1, -L1, V, -I, -C, -Q, nad5, -F, -cob, -P, -N, -L2, -rrnL, -Y, -T, -K, -rrnS, R, -W, -M, -nad2, -E, -S, -S, -A, nad3, cox2, H	Okazaki and Ueshima unpublished, AB055624
Hiatella arctica (Heteroconchia; Myoida; Hiatellidae)	cox1, atp6, F, K, M, cox2, rrnS, H, nad5, nad1, G, D, Y, nad4, P, T, A, atp8, I, L2, L1, V, R, C, N, nad4L, nad3, S1, E, rrnL, Q, cox3, nad6, M, W, S2, cob, nad2	Dreyer and Steiner 2006, NC_008452
Acanthocardia tuberculata (Heteroconchia; Veneroida)	cox1, P, nad4L, L2, nad6, cox2, D, M, H, nad3, M, W, K, L1, nad1, F, rrnS, Q, R, I, cox3, S1, T, nad5, S2, cob, rrnL, N, nad4, Y, atp6, E, G, V, nad2, A	Dreyer and Steiner 2006, NC_008452
Venerupis (Ruditapes) philippinarum (Heteroconchia; Veneridae)	cox1, L1, nad1, nad2, nad4L, I, cox2, P, cob, rrnL, nad4, H, E, S(nga), atp6, nad3, nad5, Y, M, M, D, V, nad6, K, V, F, W, R, L2, G, Q, N, T, C, A, cox3, rrnS	Okazaki and Ueshima, unpublished, NC_003354

Cephalopoda

Species	Gene order	Reference
Nautilus macromphalus (Nautilidae)	cox1, cox2, D, atp8, -F, -L1, -L2, -rrnL, -V, -rrnS, -M, -C, -Y, -W, -Q, T, -G, atp6, -nad5, -H, -nad4, -nad4L, -S2, -cob, -nad6, -P, -nad1, -E, cox3, A, R, N, I, nad3, S1, n2	Boore 2006, NC_007980
Vampyroteuthis infernalis (Coleoidea; Vampyroteuthidae)	cox1, cox2, D, atp8, atp6-F, -nad5, -H, -nad4, -nad4L, T, -S2, -cob, -nad6, -P-nad1, -L2, -L1, -rrnL, -V, -rrnS, -M, -C, -Y, -W, -Q, -G, E, cox3, K, A, R, N, I, nad3, S1, nad2	Yokobori et al. 2007, NC_009690
Octopus ocellatus and O. vulgaris (Coleoidea; Octopodidae)	cox1, cox2, D, atp8, atp6, -F, -nad5, -H, -nad4, -nad4L, T, -S1, -cob, -nad6, -P, -nad1, -L1, -L2, -rrnL, -V, -rrnS, -M, -C, -Y, -W, -Q, -G, -E, cox3, K, A, R, N, I, nad3, S2, nad2	Akasaki et al. 2006, AB240156; Yokobori et al. 2004, NC_006353

TABLE 17.1
(continued)

NAME AND CLASSIFICATION	GENE ARRANGEMENT	REFERENCE
Watasenia scintillans (Coleoidea; Enoploteuthidae)	cox1, cox2, D, atp8, atp6, -F, -V, -rrnS, -M, -C, -Q, cox3, K, R, S1, nad2, cox1, cox2, D, atp8, atp6, -nad5, -H, -nad4, -nad4L, T, S2, -cob, -nad6, -P, -nad1, -L1, -L2, -rrnL, -Y, -W, -G, -E, cox3, A, N, I, nad3	Akasaki *et al.* 2006, NC_007893
Dosidicus gigas (Coleoidea; Ommastrephidae)	cox1, cox2, D, atp8, atp6, F, V, rrnS, C, Q, cox3, K, R, S1, nad2, cox1, cox2, D, atp8, atp6, -nad5, -H, -nad4, -nad4L, T, -S2, -cob, -nad6, -P, -nad1, -L2, -L1, - rrnL, -M, -Y, -W, -G, -E, cox3, A, N, I, nad3	Elliger *et al.* unpublished, NC_009734
Loligo bleekeri (Coleoidea; Loliginidae)	cox1, -C, -Y, E, N, cox2, -M, R, -F, -nad5, -nad4, -nad4L, T, -L1, -G, A, D, atp8, atp6, -H, -L2, cox3, nad3, -S1, -cob, -nad6, -P, -nad1, -Q, I, -rrnL, -V, -rrnS, -W, K, S2, nad2	Tomita *et al.* 2002; Sasuga *et al.* 1999; Tomita *et al.* 1998, NC_002507
Sepioteuthis lessoniana (Coleoidea; Loliginidae)	cox1, -C, -Y, -E, N, cox2, -M, R, -F, -nad5, -nad4, -nad4L, T, -L1, -G, I, -rrnL, -V, -rrnS, -W, A, D, atp8, atp6, -H, -L2, cox3, nad3, -S1, -cob, -nad6, -P, -nad1, -Q, K, S2, nad2	Akasaki *et al.* 2006, NC_007894
Sepia esculenta (Coleoidea; Sepiidae)	cox1, cox2, atp8, atp6, -F, -nad1, -L2, -L1, -rrnL, -V, -rrnS, -C, -Y, -Q, -G, N, I, nad3, D, -nad5, -H, -nad4, -nad4L, T, -S2, -cob, -nad6, -P, -M, -W, -E, cox3, K, A, R, S1, nad2	Yokobori *et al.* 2007, NC_009690
Sepia officinalis (Coleoidea; Sepiidae)	cox1, cox2, atp8, atp6, -F, -nad1, -L1, -L2, -rrnL, -V, -rrnS, -C, -Y, -Q, -G, N, I, nad3, D, -nad5, -H, -nad4, -nad4L, T, -S1, -cob, -nad6, -P, -M, -W, -E, cox3, K, A, R, S2, nad2	Akasaki *et al.* 2006, NC_007895
Scaphopoda *Graptacme eborea* (Dentaliidae)	cox1, S, N, nad2, cob, H, -cox2, -Q, G, -cox3, -Y, R, S, -nad6, -P, -nad1, -atp8, -I, -T, rrnS, -M, -rrnL, V, A, nad3, L1, L2, E, W, -F, -K, -nad5, -D, -nad4, -nad4L, atp6, C	Boore *et al.* 2004, NC_006162

Taxon	Gene order	Reference
Siphonodentalium lobatum (Siphonodentaliidae)	cox1, L1, G, -T, R, nad2, nad4, I, nad1, nad5, -Y, -nad4L, -atp8, -H, -A, -W, -M, -V, -nad6, -Q, -K, rrnS, -P, -N, -S1, -cob, -cox2, -cox3, -C, -atp6, -S2, -nad3, -E, -D, -F, -rrnL, -L2	Dreyer and Steiner 2004, NC_005840
Gastropoda		
Lottia digitalis (Patellogastropoda; Lottiidae)	cox1, N, cox3, F, V, cox2, E, T, P, rrnL, M, rrnS, Y, -cob, atp8, Q, nad4L, nad4, L1, I, -R, nad2, W, -nad3, -atp6, -A, -G, D, nad5, S, L2, -H, -nad1, C, nad6	Simison *et al.* 2006, NC_007792
Haliotis rubra (Vetigastropoda; Haliotidae)	cox1, cox2, atp8, atp6, -F, -nad5, -H, -nad4, -nad4L, T, -S1, -cob, -nad6, -P, -nad1, -L1, -L2, -rrnL, -V, -rrnS, -M, -Y, -C, -W, -Q, -G, -E, cox3, D, K, A, R, I, nad3, N, S2, nad2	Maynard *et al.* 2005, NC_005940
Ilyanassa obsolete (Caenogastropoda; Buccinoidea; Nassariidae)	cox1, cox2, D, atp8, atp6, -M, -Y, -C, -W, -Q, -G, -E, rrnS, V, rrnL, L1, L2, nad1, P, nad6, cob, S2, -T, nad4L, nad4, H, nad5, F, cox3, K, A, R, N, I, nad3, S1, nad2	Simison *et al.* 2006, NC_007781
Lophiotoma cerithformis (Caenogastropoda; Neogastropoda; Conidae)	cox1, cox2, D, atp8, atp6, -M, -Y, -C, -W, -Q, -G, -E, rrnS, V, rrnL, L1, -L2, nad1, P, nad6, cob, S, T, nad4L, nad4, H, nad5, F, cox3, K, A, R, N, I, nad3, S, nad2	Bandyopadhyay *et al.* unpublished
Conus textile (Caenogastropoda; Neogastropoda; Conoidea)	cox1, cox2, D, atp8, atp6, M, Y, C, W, L2, Q, G, E, rrnS, V, rrnL, L1, nad1, P, nad6, cob, S2, T, nad4L, nad4, H, nad5, F, cox3, K, A, R, N, I, nad3, S1, nad2	Bandyopadhyay *et al.* unpublished, NC_00879
Aplysia californica (Heterobranchia; Opisthobranchia; Aplysiidae)	cox1, V, rrnL, L1, A, P, nad6, nad5, nad1, Y, W, nad4L, cob, -D, F, cox2, G, H, -Q, L2, -atp8, -N, C, -atp6, -R, -E, -rrnS, -M, -nad3, -S1, S2, nad4, -T, -cox3, I, nad2, K	Knudsen *et al.* 2005, NC_005827
Pupa strigosa (Heterobranchia; Opisthobranchia; Acteonidae)	cox1, V, rrnL, L1, A, P, nad6, nad5, nad1, Y, W, nad4L, cob, D, F, cox2, G, H, -Q, -L2, -atp8, -N, C, -atp6, -R, -E, -rrnS, -M, -nad3, -S1, S2, nad4, T, -cox3, I, nad2, K	Kurabayashi and Ueshima 2000, NC_002176
Roboastra europaea (Heterobranchia; Opisthobranchia; Polyceridae)	cox1, V, rrnL, L1, A, P, nad6, nad5, nad1, Y, W, nad4L, cob, D, F, cox2, G, H, C, -Q, -L2, -atp8, -N, -atp6, -R, -E, -rrnS, -M, -nad3, -S1, S2, nad4, -T, -cox3, I, nad2, K	Grande *et al.* 2002, NC_004321

TABLE 17.1
(continued)

NAME AND CLASSIFICATION	GENE ARRANGEMENT	REFERENCE
Biomphalaria glabrata (Heterobranchia; Pulmonata; Planorbidae)	cox1, V, rrnL, L1, A, P, nad6, nad5, nad1, nad4L, cob, D, C, F, cox2, Y, W, G, H, Q, -L2, -atp8, -N, -atp6, -R, -E, -rrnS, M, -nad3, -S1, S2, nad4, -T, -cox3, I, nad2, K	DeJong *et al.* 2004, NC_005439
Albinaria coerulea (Heterobranchia; Pulmonata; Clausiliidae)	cox1, V, rrnL, L1, P, A, nad6, nad5, nad1, nad4L, cob, D, C, F, cox2, Y, W, G, H, -Q, -L2, -atp8, -N, -atp6, -R, -E, -rrnS, -M, -nad3, -S1, S2, nad4, -T, -cox3, I, nad2, K	Hatzoglou *et al.* 1995, NC_001761
Cepaea nemoralis (Heterobranchia; Pulmonata; Helicidae)	cox1, V, rrnL, L1, A, nad6, P, nad5, nad1, nad4L, cob, D, C, F, cox2, Y, W, G, H, -Q, -L2, -atp8, -N, -atp6, -R, -E, -rrnS, -M, -nad3, -S1, -T, -cox3, S2, nad4, I, nad2, K	Yamazaki *et al.* 1997; Terrett *et al.* 1996, NC_001816
Euhadra herklotsi (Sequence not complete)[1] (Heterobranchia; Pulmonata; Bradybaenidae)	cox1, V, rrnL, L1, P, A, nad6, nad5, nad1, nad4L, cob, D, C, F, cox2, G, H, Y, -W, -Q, -L2, -atp8, -N, -atp6, -R, -E, -rrnS, -M, -nad3, -S1, S2, nad4, -T, -cox3, I, nad2, K	Yamazaki *et al.* 1997, Z71693 701

NOTE: All that are cited as "unpublished" are available in GenBank. tRNA genes are abbreviated by the one-letter code for the corresponding amino acid and anticodons are shown in parentheses to differentiate the two leucine and two serine tRNAs. A minus symbol indicates reverse transcriptional orientation.
[1]This annotation omits 9 of the tRNA genes (A, F, G, H, Q, S1, S2, W, and Y) ordinarily found in animal mtDNAs, which may be due to a failure by the authors to recognize them in the sequence.

of blastomere cleavage termed the "molluscan cross," long thought to be a synapomorphy uniting these groups (Scheltema 1993, 1996), in contrast to the "annelid cross" pattern (Gerould 1907) shared between annelids and echiurans. Based on the hypothesis of a close relationship between Sipuncula and Mollusca, Scheltema (1996) hypothesized homology between several sipunculan larval features to those of larval and adult molluscs, including the ventral buccal organ of sipunculan larvae with the odontophore of molluscs, the ciliated lip below the mouth of sipunculan pelagosphera-stage larvae with the foot of a molluscan pediveliger larva, and the lip glands of sipunculan larvae with pedal glands of larval chitons and adult neomenioid aplacophorans. However, this phylogenetic association is refuted not only by mtDNA sequence comparisons but also by gene arrangement characters (Boore and Staton 2002).

TECHNIQUES FOR INVESTIGATING MOLLUSC MITOCHONDRIAL GENOMES

The most commonly used process of producing a complete mitochondrial genome sequence uses "shotgun sequencing" and involves six steps: (1) extraction of DNA from tissue; (2) polymerase chain reaction (PCR) of long, overlapping fragments summing to the complete mtDNA; (3) creation of genomic libraries by cloning overlapping portions of these fragments into plasmid vectors; (4) sequencing numerous, random clones from the libraries; (5) assembling the sequencing reads into a complete mtDNA sequence; (6) gene annotation (see Wyman *et al.* 2004) and analysis. In some cases, mtDNA may be physically isolated in sufficient quantity that PCR is not necessary, or rolling circle amplification (RCA) (Simison *et al.* 2006) may be substituted for PCR. Details of preparing, sequencing, and interpreting mtDNAs can be found in Boore *et al.* (2005).

MOLLUSC NUCLEAR GENOMES

Whereas mitochondrial genomes of molluscs are known to range in size from ~10 to ~42 kb, nuclear genomes are very much larger, ranging

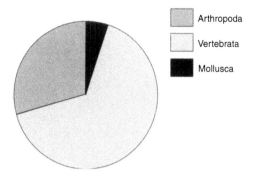

FIGURE 17.1. Pie diagram illustrating the relative numbers of publications found in the *Zoological Record* database for "Mollusca", "Vertebrata," and "Arthropoda."

in size from ~400,000 kb for *Lottia gigantea* to ~5,800,000 kb for *Neobuccinum eatoni* (see http://www.genomesize.com/molluscs.htm). Although cytogenetic studies of chromosome number and structure have been used since the late nineteenth century for inferring molluscan phylogeny, the results have generally proven to be unreliable (Patterson 1969; Thiriot-Quiévreux 2003). However, this literature records data for over 1,000 molluscan species.

Unfortunately, the study of molluscs lags behind that of vertebrates or arthropods, especially in the fields of genomics and molecular evolution. A search of the *Zoological Record*[1] bibliographic database with the keywords "Mollusca," "Vertebrata," and "Arthropoda" returned 120,230; 1,450,563; and 629,499 publications, respectively, making mollusc publications amount to only 8.3% of vertebrate and 19.1% of arthropod publications (Figure 17.1). Combining the keyword "genome" with each of "Mollusca," "Vertebrata," and "Arthropoda" returned 234; 2,804; and 1,833 publications, respectively (Figure 17.2), reinforcing the view that not only is malacology under-represented in the biological sciences compared with the study of vertebrates and arthropods, it is even farther behind in the field of genomics specifically.

---

1. *Zoological Record* does not include all biological publication records, and searches of its database do not reflect actual numbers of publications. Therefore all comparisons made are intended only to illustrate relative trends in the fields of study.

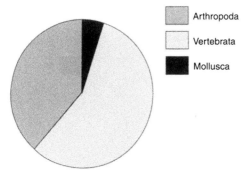

	Arthropoda
	Vertebrata
	Mollusca

FIGURE 17.2. Pie diagram illustrating the relative numbers of publications found in the *Zoological Record* database for "Mollusca," "Vertebrata," and "Arthropoda" when each is combined with the term "Genome."

This disparity is somewhat surprising. Molluscs are the second largest phylum (after arthropods); have an enormous diversity of body plans, and so are of interest to those who study evolution; have an excellent fossil record, enabling correlations with patterns of genomic change; and play a very wide range of ecological roles. Many molluscs are important as food sources (e.g., mussels, clams, oysters, squid), as agricultural pests (e.g., snails), or as secondary hosts for human diseases (e.g., *Biomphalaria* and *Bulimus*). Several molluscs are model organisms for molecular, cell, or developmental biology, including the gastropods *Aplysia* and *Lymnaea* and cephalopods *Loligo* and *Octopus*. Thus, it is imperative that this deficiency be addressed. Fortunately, complete nuclear genome sequences will soon be available for a few molluscs (Table 17.2).

Malacologists need to be actively engaged in determining which genomes are chosen for future sequencing. The two major funding sources in the United States—the National Human Genome Research Institute (NHGRI; part of the National Institutes of Health) and the Department of Energy—each have programs for input from the scientific community in guiding genome sequencing efforts. More information can be found at the following web sites:

http://www.genome.gov/Grants/

http://www.sc.doe.gov/production/ober/LSD/ DNASeq.html

http://www.jgi.doe.gov/CSP/

Evolutionary genomics is a rapidly developing field providing exciting new opportunities to gain previously unimagined insights into molluscan evolution and the processes of natural selection. To take advantage of these opportunities workers need to build on the long history of molluscan evolutionary research and integrate evolutionary genomics into research programs wherever appropriate. Funding agencies are looking for cooperative programs within and between research fields and are actively encouraging collaboration and broader participation by the research community. Because of the high cost of high-throughput sequencing, it will remain limited to a few well-funded projects for at least the next few years. Therefore, it is imperative that malacologists join or form collaborate initiatives involving genomic research so that molluscan studies become a critical component of this new field and thereby increase their presence in the life sciences.

FUTURE STUDIES

Determining complete mtDNA sequences has proceeded more slowly for molluscs than for some other phyla, notably vertebrates and arthropods. However, there are other big and important groups that remain underrepresented, such as annelids, and numerous phyla remain unsampled. Even with this limited sampling, the field of molluscan mitogenomics is particularly promising. Gene rearrangements are common for some groups, bolstering confidence that this will provide a rich phylogenetic signal and offering a data set that may illuminate models of gene order change. Molluscs themselves are highly diverse, and the molecular mechanisms apparent in their mitochondrial systems seem to follow suit. Recent successes of several research groups suggest that this data set is growing exponentially. The field of mitogenomics is at a formative stage, and now is the time to establish molluscs as a model system for addressing many of the processes and patterns of genomic change.

TABLE 17.2
Molluscs Currently Selected for Whole Nuclear Genome Sequencing

SPECIES	GROUP	GENOME SIZE	PRIMARY INSTITUTE	FUNDER	STATUS
Aplysia californica	Gastropoda; Heterobranchia	2.0 Gb	Broad Institute	NHGRI[a]	Pending
Biomphalaria glabrata	Gastropoda; Heterobranchia	0.93 Gb	Washington University	NHGRI	Pending
Mytilus californianus	Bivalvia; Pteriomorpha	1.8 Gb	JGI[b]	DOE[c]	Pending
Lottia gigantea	Gastropoda; Patellogastropoda	0.42 Gb	JGI	DOE	In assembly

[a]National Human Genome Research Institute, one of the National Institutes of Health (NIH).
[b]U.S. Department of Energy (DOE) Joint Genome Institute.
[c]U.S. Department of Energy.

Further, the imminent determination of the complete nuclear genome sequences of several molluscs and other eutrochozoans holds extraordinary promise. In addition to the enormous amount of phylogenetic information that will be made available, this achievement will also reveal the genomic underpinnings of the developmental and morphological characters that comprise the bulk of our traditional systematic studies.

Today's high-throughput genome sequencing centers lack the resources to comprehensively analyze anything but a small portion of these issues. It will take the devoted effort of a broad community with a wide spectrum of expertise to produce the most relevant and exciting biological insights. We urge those who study molluscs to be ready.

ACKNOWLEDGMENTS

Thanks to David Lindberg for critical reading of this manuscript. This work was supported by NSF grants DEB-0089624, EAR-0342392, and EF-0328516 and was performed partly under the auspices of the U.S. Department of Energy's Office of Science, Biological and Environmental Research Program and by the University of California, Lawrence Berkeley National Laboratory under Contract No. DE-AC02-05CH11231.

REFERENCES

Akasaki, T., Nikaido, M., Tsuchiya, K., Segawa, S., Hasegawa, M., and Okada, N. 2006. Extensive mitochondrial gene arrangements in coleoid Cephalopoda and their phylogenetic implications. *Molecular Phylogenetics and Evolution* 38: 648–658.

Asakawa, S., Kumazawa, Y., Araki, T., Himeno, H., Miura, K., and Watanabe, K. 1991. Strand-specific nucleotide composition bias in echinoderm and vertebrate mitochondrial genomes. *Journal of Molecular Evolution* 32: 511–520.

Beddoe, T., and Lithgow, T. 2002. Delivery of nascent polypeptides to the mitochondrial surface. *Biochimica et Biophysica Acta* 1592: 35–39.

Boore, J. L. 1999. Animal mitochondrial genomes. *Nucleic Acids Research* 27: 1767–1780.

———. 2000. The duplication/random loss model for gene rearrangement exemplified by mitochondrial genomes of deuterostome animals. In *Comparative Genomics*. Edited by D. Sankoff and J. Nadeau. Computational Biology Series 1. Dordrecht, Netherlands: Kluwer Academic Publishers, pp. 133–147.

———. 2006. The use of genome-level characters for phylogenetic reconstruction. *Trends in Ecology and Evolution* 21:439–446.

Boore, J. L., and Brown, W. M. 1994. Complete DNA sequence of the mitochondrial genome of the black chiton, *Katharina tunicata*. *Genetics* 138: 423–443.

———. 1998. Big trees from little genomes: Mitochondrial gene order as a phylogenetic tool. *Current Opinion in Genetics and Development* 8(6): 668–674.

──────. 2000. Mitochondrial genomes of *Galathea-linum*, *Helobdella*, and *Platynereis*: sequence and gene arrangement comparisons indicate that Pogonophora is not a phylum and Annelida and Arthropoda are not sister taxa. *Molecular Biology and Evolution* 17: 87–106.

Boore, J. L., Collins, T. M., Stanton, D., Daehler, L. L., and Brown, W. M. 1995. Deducing arthropod phylogeny from mitochondrial DNA rearrangements. *Nature* 376: 163–165.

Boore, J. L., Lavrov, D., and Brown, W. M. 1998. Gene translocation links insects and crustaceans. *Nature* 392: 667–668.

Boore, J. L., Macey J. R., and Medina, M. 2005. Sequencing and comparing whole mitochondrial genomes of animals. In *Molecular Evolution: Producing the Biochemical Data, Part B*. Methods in Enzymology 395. Edited by E. A. Zimmer and E. Roalson. Burlington, MA: Elsevier, pp. 311–348.

Boore, J. L., Medina, M., and Rosenberg, L. A. 2004. Complete sequences of the highly rearranged molluscan mitochondrial genomes of the scaphopod *Graptacme eborea* and the bivalve *Mytilus edulis*. *Molecular Biology and Evolution* 21: 1492–1503.

Boore, J. L., and Staton, J. 2002. The mitochondrial genome of the sipunculid *Phascolopsis gouldii* supports its association with Annelida rather than Mollusca. *Molecular Biology and Evolution* 19: 127–137.

Brusca, R. C., and Brusca, G. J. *Invertebrates*. 1990. Sunderland, MA: Sinauer.

Budowle, B., Allard, M. W., Wilson, M. R., and Chakraborty, R. 2003. Forensics and mitochondrial DNA: Applications, debates, and foundations. *Annual Review of Genomics and Human Genetics* 4: 119–141.

Clayton, D. A. 2000. Transcription and replication of mitochondrial DNA. *Human Reproduction* 15 Suppl. 2: 11–17.

──────. 2003. Mitochondrial DNA replication: what we know. *International Union of Biochemistry and Molecular Biology Life* 55: 213–217.

Conway Morris, S., and Peel, J. S. 1995. Articulated halkieriids from the Lower Cambrian of North Greenland and their role in early protostome evolution. *Philosophical Transactions of the Royal Society of London Series B: Biological Sciences* 347: 305–358.

DeJong, R. J., Emery, A. M., and Adema, C. M. 2004. The mitochondrial genome of *Biomphalaria glabrata* (Gastropoda: Basommatophora), intermediate host of *Schistosoma mansoni*. *The Journal of Parasitology* 90: 991–997.

Dreyer, H., and Steiner, G. 2004. The complete sequence and gene organization of the mitochondrial genome of the gadilid scaphopod *Siphonondentalium lobatum* (Mollusca). *Molecular Phylogenetics and Evolution* 31: 605–617.

Dreyer, H., and Steiner, G. 2006. The complete sequences and gene organisation of the mitochondrial genomes of the heterodont bivalves *Acanthocardia tuberculata* and *Hiatella arctica*—and the first record for a putative Atpase subunit 8 gene in marine bivalves. *Frontiers in Zoology* 3: 1–14.

Dubrovsky, E. B., Dubrovskaya, V. A., Levinger, L., Schiffer, S., and Marchfelder, A. 2004. Drosophila RNase Z processes mitochondrial and nuclear pre-tRNA 3′ ends *in vivo*. *Nucleic Acids Research* 32: 255–262.

Eernisse, D. J., Albert J. S., and Anderson, F. E. 1992. Annelida and Arthropoda are not sister taxa: a phylogenetic analysis of spiralian metazoan morphology. *Systematic Biology* 41: 305–330.

Garcia-Machado, E., Pempera, M., Dennebouy, N. Oliva-Suarez, M., Mounolou, J. C., and Monnerot, M. 1999. Mitochondrial genes collectively suggest the paraphyly of Crustacea with respect to Insecta. *Journal of Molecular Evolution* 49: 142–149.

Gerould, J. H. 1907. The development of *Phascolosoma*. Studies on the embryology of the Sipunculidae. *Zoologische Jahrbücher. Abteilung für Anatomie und Ontogenie der Tiere* 23: 77–162.

Ghiselin, M. T. 1988. The origin of molluscs in the light of molecular evidence. *Oxford Surveys in Evolutionary Biology* 5: 66–95.

Grande, C., Templado, J., Cervera, J. L., and Zardoya, R. 2002. The complete mitochondrial genome of the nudibranch *Roboastra europaea* (Mollusca: Gastropoda) supports the monophyly of opisthobranchs. *Molecular Biology and Evolution* 19: 1672–1685.

Hatzoglou, E., Rodakis, G. C., and Lecanidou, R. 1995. Complete sequence and gene organization of the mitochondrial genome of the land snail *Albinaria coerulea*. *Genetics* 140: 1353–1366.

Helfenbein, K. G., and Boore, J. L. 2004. The mitochondrial genome of *Phoronis architecta*—comparisons demonstrate that phoronids are lophotrochozoan protostomes. *Molecular Biology and Evolution* 21: 153–157.

Helfenbein, K. G., Brown, W. M., and Boore, J. L. 2001. The complete mitochondrial genome of a lophophorate, the brachiopod *Terebratalia transversa*. *Molecular Biology and Evolution* 18: 1734–1744.

Hoffmann, R. J., Boore, J. L., and Brown, W. M. 1992. A novel mitochondrial genome organization for the blue mussel, *Mytilus edulis. Genetics* 131: 397–412.

Ingman, M., Kaessmann, H., Pääbo, S., and Gyllensten, U. 2001. Mitochondrial genome variation and the origin of modern humans. *Nature* 408: 708–713.

Knudsen, B., Kohn, A. B., Nahir, B., McFadden, C. S., and Moroz, L. L. 2005. Complete DNA sequence of the mitochondrial genome of the sea-slug, *Aplysia californica*: Conservation of the gene order in Euthyneura. *Molecular Phylogenetics and Evolution* 38: 459–469.

Krakauer, D. C., and Mira, A. 1999. Mitochondria and germ-cell death. *Nature* 400: 125–126.

Kurabayashi, A., and Ueshima, R. 2000. Complete sequence of the mitochondrial DNA of the primitive opisthobranch gastropod *Pupa strigosa*: systematic implication of the genome organization. *Molecular Biology and Evolution* 17: 266–277.

Lake, J. A. 1990. Origin of the Metazoa. *Proceedings of the National Academy of Sciences of the United States of America* 87: 763–766.

Lander, E. S., Linton, L. M., Birren, B., Nusbaum, C., Zody, M. C., and 250 others; International Human Genome Sequencing Consortium. 2001. Initial sequencing and analysis of the human genome. *Nature* 409: 860–921.

La Roche, J., Snyder, M., Cook, D. I., Fuller, K., and Zouros, E.1990. Molecular characterization of a repeat element causing large-scale size variation in the mitochondrial DNA of the sea scallop *Placopecten magellanicus. Molecular Biology and Evolution* 7: 45–64.

Lavrov, D. V., Boore, J. L., and Brown, W. M. 2000. The complete mitochondrial DNA sequence of the horseshoe crab *Limulus polyphemus. Molecular Biology and Evolution* 17: 813–824.

———. 2002. Complete mtDNA sequences of two millipedes suggest a new model for mitochondrial gene rearrangements: Duplication and non-random loss. *Molecular Biology and Evolution* 19: 163–169.

Lavrov, D., Brown, W. M., and Boore, J. L. 2000. A novel type of RNA editing occurs in the mitochondrial tRNAs of the centipede *Lithobius forficatus. Proceedings of the National Academy of Sciences of the United States of America* 97: 13738–13742.

———. 2004. Phylogenetic position of the Pentastomida and (pan)crustacean relationships. *Proceedings of the Royal Society of London Series B: Biological Sciences* 271(1538): 537–544.

Levinger, L., Morl, M., and Florentz, C. 2004. Mitochondrial tRNA 3´ end metabolism and human disease. *Nucleic Acids Research* 32: 5430–5441.

Lydeard, C., Holznagel, W. E., Schnare, M. N., and Gutell, R. R. 2000. Phylogenetic analysis of molluscan mitochondrial LSU rDNA sequences and secondary structures. *Molecular Phylogenetics and Evolution* 15: 83–102.

Macey, J. R., Papenfuss, T. J., Kuehl, J. V., Fourcade, H. M., and Boore, J. L. 2004. Phylogenetic relationships among amphisbaenian reptiles based on complete mitochondrial genome sequences. *Molecular Phylogenetics and Evolution* 33: 22–31.

Masta, S. E., and Boore, J. L. 2004. The complete mitochondrial genome sequence of the spider *Habronattus oregonensis* reveals rearranged and extremely truncated tRNAs. *Molecular Biology and Evolution* 21: 893–902.

Maynard, B. T., Kerr, L. J., McKiernan, J. M., Jansen, E. S., and Hanna, P. J. 2005. Mitochondrial DNA sequence and gene organization of the Australian blacklip abalone, *Haliotis rubra* (Leach). *Marine Biotechnology* 7: 645–658.

Milbury, C. A., and Gaffney, P. M. 2005. Complete mitochondrial DNA sequence of the Eastern oyster, *Crassostrea virginica. Marine Biotechnology* 7: 697–712.

Mizi, A., Zouros, E., Moschonas, N., and Rodakis, G. C. 2005. The complete maternal and paternal mitochondrial genomes of the Mediterranean mussel *Mytilus galloprovincialis*: implications for the doubly uniparental inheritance mode of mtDNA. *Molecular Biology and Evolution* 22: 952–967.

Mueller, R. L., and Boore, J. L. 2005. Molecular mechanisms of extensive mitochondrial gene rearrangement in plethodontid salamanders. *Molecular Biology and Evolution* 22: 2104–2112.

Mueller, R. L., Macey, J. R., Jaekel, M., Wake, D. B., and Boore, J. L. 2004. Morphological homoplasy, life history evolution, and historical biogeography of plethodontid salamanders: Novel insights from complete mitochondrial genome sequences. *Proceedings of the National Academy of Sciences of the United States of America* 101: 13820–13825.

Nardi, F., Spinsanti, G., Boore, J. L., Carapelli, A., Dallai, R., and Frati, F. 2003. Hexapod origins, monophyletic or paraphyletic? *Science* 299: 1887–1889.

Nickisch-Rosenegk, M. v., Brown, W. M., and Boore, J. L. 2001. Sequence and structure of the mitochondrial genome of the tapeworm *Hymenolepis diminuta*: gene arrangement indicates that platyhelminths are derived eutrochozoans. *Molecular Biology and Evolution* 18: 721–730.

Nieminen, A. L. 2003. Apoptosis and necrosis in health and disease: role of mitochondria. *International Review of Cytology* 224: 29–55.

Nyakaana, S., Arctander, P., and Siegismund, H. 2002. Population structure of the African savannah elephant inferred from mitochondrial control region sequences and nuclear microsatellite loci. *Heredity* 89: 90-98.

Ojala, D., Merkel, C., Gelfand R., and Attardi, G. 1980. The tRNA genes punctuate the reading of genetic information in human mitochondrial DNA. *Cell* 22: 393–403.

Ojala, D., Montoya, J., and Attardi, G. 1981. tRNA punctuation model of RNA processing in human mitochondria. *Nature* 290: 470–474.

Osawa, S., Jukes, T. H., Watanabe, K., and Muto, A. 1992. Recent evidence for evolution of the genetic code. *Microbiology and Molecular Biology Reviews* 56(1): 229–264.

Parham, J. F., Macey, J. R., Papenfuss, T. J., Feldmam, C. R., Türkozan, O., Polymeni, R., and Boore, J. L. 2006. The phylogeny of Mediterranean tortoises and their close relatives based on complete mitochondrial genome sequences from museum specimens. *Molecular Phylogenetics and Evolution* 38: 50–64.

Passamonti, M., Boore, J.L., and Scali, V. 2003. Molecular evolution and recombination in gender-associated mitochondrial DNAs of the Manila clam *Tapes philippinarum*. *Genetics* 164: 603–611.

Patterson, C. M. 1969. Chromosomes of molluscs. *Proceedings of the 2nd Symposium of Mollusca*. Ernakulam, Cochin, India: Marine Biological Association of India, pp. 635–686.

Perna, N.T., and Kocher, T.D. 1995. Patterns of nucleotide composition at fourfold degenerate sites of animal mitochondrial genomes. *Journal of Molecular Evolution* 41: 353–358.

Raff, R.A., and Kaufman, T.C. 1983. *Embryos, Genes and Evolution*. New York: Macmillan, pp. 251–261.

Rawlings, T.A., Collins, T.M., and Bieler, R. 2001. A major mitochondrial gene rearrangement among closely related species. *Molecular Biology and Evolution* 18(8): 1604–1609.

———. 2003. Changing identities: tRNA duplication and remolding within animal mitochondrial genomes. *Proceedings of the National Academy of Sciences of the United States of America* 100(26): 15700–15705.

Rossmanith, W., Tullo, A., Potuschak, T., Karwan, R., and Sbisa, E. 1995. Human mitochondrial tRNA processing. *The Journal of Biological Chemistry* 270: 12885–12891.

Ruiz-Trillo, I., Riutort, M., Fourcade, H.M., Baguñà, J., and Boore, J. L. 2004. Mitochondrial genome data support the basal position of Acoelomorpha and the polyphyly of the Platyhelminthes. *Molecular Phylogenetics and Evolution* 33: 321–332.

Sasuga, J., Yokobori, S.-I., Kaifu, M., Ueda, T., Nishikawa, K., and Watanabe, K. 1999. Gene contents and organization of a mitochondrial DNA segment of the squid *Loligo bleekeri*. *Journal of Molecular Evolution* 48: 692–702.

Sato, M., and Nagashima, K. 2001. Molecular characterization of a mitochondrial DNA segment from the Japanese scallop (Patinopecten yessoensis): demonstration of a region showing sequence polymorphism in the population. *Marine Biotechnology* 3: 370–379.

Scheltema, A. H. 1993. Aplacophora as progenetic aculiferans and the coelomate origin of mollusks as the sister taxon of Sipuncula. *The Biological Bulletin* 184: 57–78.

———. 1996. Phylogenetic position of the Sipuncula, Mollusca and the progenetic Aplacophora. In *Origin and Evolutionary Radiation of the Mollusca*. Edited by J. D. Taylor. Oxford: Oxford University Press, pp. 53–58.

Serb, J.M., and Lydeard, C. 2003. Complete mtDNA sequence of the North American freshwater mussel, *Lampsilis ornata* (Unionidae): An examination of the evolution and phylogenetic utility of mitochondrial genome organization in Bivalvia (Mollusca). *Molecular Biology and Evolution* 20: 1854–1866.

Simison, W. B., Lindberg, D. R., and Boore, J. L. 2006. Rolling circle amplification of metazoan mitochondrial genomes. *Molecular Phylogenetics and Evolution* 39: 562–567.

Snodgrass, R. E. 1938. Evolution of Annelida, Onychophora and Arthropoda. *Smithsonian Miscellaneous Collections* 97: 1–77.

Telford, M. J., Herniou, E. A., Russell, R. B., and Littlewood, D. T. 2000. Changes in mitochondrial genetic codes as phylogenetic characters: two examples from the flatworms. *Proceedings of the National Academy of Sciences of the United States of America* 97: 11359–11364.

Terrett, J.A., Miles, S., and Thomas, R.H. 1996. Complete DNA sequence of the mitochondrial genome of *Cepaea nemoralis* (Gastropoda: Pulmonata). *Journal of Molecular Evolution* 42: 160–168.

Thiriot-Quiévreux, C. 2003. Review of the literature on bivalve cytogenetics in the last ten years. *Cahiers de Biologie Marine* 43: 17–26.

Tomita, K., Ueda, T., and Watanabe, K. 1998. 7-Methylguanosine at the anticodon wobble position of squid mitochondrial tRNA(Ser)GCU: molecular basis for assignment of AGA/AGG codons as serine in invertebrate mitochondria. *Biochimica et Biophysica Acta* 1399: 78–82.

Tomita, K., Yokobori, S.-I., Oshima, T., Ueda, T., and Watanabe, K. 2002. The cephalopod *Loligo bleekeri* mitochondrial genome: multiplied noncoding regions and transposition of tRNA genes. *Journal of Molecular Evolution* 54: 486–500.

Venter, J. C., Adams, M. D., Myers, E. W., Li, P. W., Mural, R. J., and 269 others. 2001. The sequence of the human genome. *Science* 291: 1304–1351.

Wallace, D. C. 1999. Mitochondrial diseases in man and mouse. *Science* 283: 1482–1488.

Wyman, S., Jansen, R. K., and Boore, J. L. 2004. Automatic annotation of organellar genomes with DOGMA. *Bioinformatics* 20: 3252–3255.

Yamazaki, N., Ueshima, R., Terrett, J. A., Yokobori, S.-I., Kaifu, M., Segawa, R., Kobayashi, T., Numachi, K., Ueda, T., Nishikawa, K., Watanabe, K., and Thomas, R. H. 1997. Evolution of pulmonate gastropod mitochondrial genomes: comparisons of complete gene organization of *Euhadra, Cepaea* and *Albinaria* and implications of unusual tRNA secondary structure. *Genetics* 145: 749–758.

Yokobori, S.-I., Fukuda, N., Nakamura, M., Aoyama, T., and Oshima, T. 2004. Long-term conservation of six duplicated structural genes in cephalopod mitochondrial genomes. *Molecular Biology and Evolution* 21: 2034–2046.

Yokobori, S., Lindsay, D.J., Yoshida, M., Tsuchiya, K., Yamagishi, A., Maruyama, T., and Oshima, T. 2007. Mitochondrial genome structure and evolution in the living fossil vampire squid, *Vampyroteuthis infernalis,* and extant cephalopods. *Molecular Phylogenetics and Evolution* 44: 898–910.

Yost, H. J., Phillips, C. R., Boore, J. L., Bertman, J., Whalen, B., and Danilchik, M. V. 1995. Relocation of mitochondrial RNA to the prospective dorsal midline during *Xenopus* embryogenesis. *Developmental Biology* 170: 83–90.

Zouros, E. 2000. The exceptional mitochondrial DNA system of the mussel family Mytilidae. *Genes and Genetic Systems* 75: 313–318.

INDEX

Page numbers in bold font refer to figures and tables.

Milton Keynes UK
Ingram Content Group UK Ltd.
UKHW052120060824
446496UK00001B/8